AF775543

MICROWAVE AND MILLIMETRE-WAVE DESIGN FOR WIRELESS COMMUNICATIONS

MICROWAVE AND MILLIMETRE-WAVE DESIGN FOR WIRELESS COMMUNICATIONS

Ian Robertson
University of Leeds, UK

Nutapong Somjit
University of Leeds, UK

Mitchai Chongcheawchamnan
Prince of Songkla University, Thailand

WILEY

This edition first published 2016
© 2016 John Wiley & Sons, Ltd

First Edition published in 2016

Registered office
John Wiley & Sons, Ltd, The Atrium, Southern Gate, Chichester, West Sussex, PO19 8SQ, United Kingdom

For details of our global editorial offices, for customer services and for information about how to apply for permission to reuse the copyright material in this book please see our website at www.wiley.com.

The right of the author to be identified as the author of this work has been asserted in accordance with the Copyright, Designs and Patents Act 1988.

All rights reserved. No part of this publication may be reproduced, stored in a retrieval system, or transmitted, in any form or by any means, electronic, mechanical, photocopying, recording or otherwise, except as permitted by the UK Copyright, Designs and Patents Act 1988, without the prior permission of the publisher.

Wiley also publishes its books in a variety of electronic formats. Some content that appears in print may not be available in electronic books.

Designations used by companies to distinguish their products are often claimed as trademarks. All brand names and product names used in this book are trade names, service marks, trademarks or registered trademarks of their respective owners. The publisher is not associated with any product or vendor mentioned in this book.

Limit of Liability/Disclaimer of Warranty: While the publisher and author have used their best efforts in preparing this book, they make no representations or warranties with respect to the accuracy or completeness of the contents of this book and specifically disclaim any implied warranties of merchantability or fitness for a particular purpose. It is sold on the understanding that the publisher is not engaged in rendering professional services and neither the publisher nor the author shall be liable for damages arising herefrom. If professional advice or other expert assistance is required, the services of a competent professional should be sought.

Library of Congress Cataloging-in-Publication Data applied for.
ISBN: 9781118917213

A catalogue record for this book is available from the British Library.

Set in 10/12pt Times by Aptara Inc., New Delhi, India

1 2016

Contents

About the Authors **xvii**

Acknowledgements **xix**

Preface **xxi**

1 Introduction **1**
1.1 A Brief Timeline of Consumer Electronics 2
1.2 The Electromagnetic Spectrum 3
1.2.1 Spectrum Licensing and Standards 6
1.3 Industry Trends 7
1.3.1 The Multidisciplinary Nature of Modern RF Engineering 8
1.3.2 Business Matters 10
1.4 Forms of Wireless Communication 12
1.4.1 Terrestrial Television 12
1.4.2 Satellite Communications and Broadcasting 15
1.4.3 Terrestrial Line-of-Sight Links 18
1.4.4 Cellular Mobile Communications 20
1.4.5 Wireless Indoor Networks 22
1.4.6 Hybrid Fibre-Radio Systems 24
1.4.7 RFID, NFC and ICT 25
1.4.8 Wireless Sensor Networks 28
1.4.9 Wearables and Body Area Networks 29
1.4.10 Machine-to-Machine and Robotic Communications 29
1.5 Conclusion 30
References 31

2 Transmitters and Receivers **32**
2.1 Introduction 32
2.2 Transmitter and Receiver Components 33
2.2.1 Amplifier 33
2.2.2 Oscillator 34
2.2.3 Mixer 34
2.2.4 Band-Pass Filter 35
2.2.5 Modulator and Demodulator 36

	2.2.6	*Antenna*	37
	2.2.7	*The Superheterodyne Receiver*	37
2.3	Noise and Interference		38
	2.3.1	*Thermal Noise*	38
	2.3.2	*Shot Noise*	40
	2.3.3	*Flicker Noise*	41
	2.3.4	*The Image Band and Noise*	42
	2.3.5	*Interference*	44
	2.3.6	*Atmospheric Attenuation*	46
	2.3.7	*Path Loss*	46
2.4	Introduction to Modulation		48
	2.4.1	*Amplitude Modulation (AM)*	49
	2.4.2	*Frequency Modulation (FM)*	50
2.5	Digital Modulation		50
	2.5.1	*Channel Capacity*	51
	2.5.2	*Amplitude Shift Keying (ASK)*	52
	2.5.3	*Frequency Shift Keying (FSK)*	54
	2.5.4	*Phase Shift Keying (PSK)*	55
	2.5.5	*Constellation Diagram*	56
	2.5.6	*Quadrature Amplitude Modulation (QAM)*	57
	2.5.7	*Orthogonal Frequency Division Multiplexing (OFDM) Modulation*	58
	2.5.8	*Comparison of E_b/N_0 versus BER for Various Modulation Schemes*	59
	2.5.9	*Bandwidth Utilisation Efficiency and CNR*	60
2.6	Noise Analysis and Link Budget Calculation		61
	2.6.1	*Noise Equivalent Bandwidth*	61
	2.6.2	*Noise Analysis in a Circuit*	61
	2.6.3	*Figures of Merit*	62
	2.6.4	*Noise Performance of Cascaded Networks*	65
	2.6.5	*Noise Calculation Examples*	68
2.7	Some Wireless Transceiver Architectures		71
	2.7.1	*Superheterodyne*	71
	2.7.2	*Double Conversion*	73
	2.7.3	*Direct Conversion*	74
	2.7.4	*Low-IF Architecture*	77
	2.7.5	*Band-Pass Sampling Receiver*	78
2.8	Conclusion		79
	References		80
3	**Scattering Parameters**		**81**
3.1	Introduction		81
3.2	*Z*-Parameters (Open-Circuit Impedance Parameters)		81
3.3	*Y*-Parameters (Short-Circuit Admittance Parameters)		82
3.4	*H*-Parameters (Hybrid Parameters)		83
3.5	*ABCD*-Parameters (Transmission or Chain Parameters)		84
3.6	Summary of Two-Port Parameter Operations		85

3.7 Scattering Parameters 87
3.7.1 Flowgraph Representation of Two-Port S-Parameters 89
3.7.2 Properties of S-Parameters 90
3.7.3 S-Parameters and Decibels 90
3.7.4 N-Port S-Parameters 91
3.7.5 Measuring S-Parameters 94
3.8 Transmission Parameters 95
3.8.1 Definition 95
3.8.2 Relationship Between T- and ABCD-Parameters 96
3.8.3 Application to Cascaded Networks 97
References 98

4 Lumped-Element Filters 99
4.1 Introduction 99
4.2 Filter Theory 100
4.2.1 Transfer Function Analysis 100
4.2.2 The s-Domain 101
4.2.3 Bode Plots 102
4.2.4 Impulse Response and Causality 103
4.2.5 The Need for a System Impedance 105
4.3 Butterworth, Chebyshev and Elliptic Low-Pass Prototypes 107
4.3.1 Butterworth Low-Pass Filter 107
4.3.2 Chebyshev Filter 109
4.3.3 Elliptic Filter 110
4.4 Filter Design Method 111
4.4.1 High-Pass Transformation 113
4.4.2 Band-Pass Transformation 115
4.4.3 Band-Stop Transformation 116
4.4.4 Band-Pass Filter Design Example 117
4.4.5 Group Delay 118
4.5 Practical Lumped Elements 120
4.6 Capacitively-Coupled Resonator Filter 121
References 124

5 Transmission Line Theory 125
5.1 Introduction 125
5.2 Reflections on Transmission Lines 126
5.2.1 Open-Circuited Line 127
5.2.2 Short-Circuited Line 127
5.3 Transmission Line Theory 129
5.3.1 Telegrapher's Equation 129
5.3.2 Lossless Transmission Line 134
5.4 Standing Waves on a Lossless Transmission Line with Mismatched Load 135
5.4.1 Impedance Variation along a Transmission Line 140

5.5 The Smith Chart 142
5.5.1 Standing Wave Ratio Circles 147
5.5.2 Smith Chart Computer Tools 149
5.6 The Signal Flow Graph 150
5.6.1 SFG Diagram 150
5.6.2 SFG Simplification 152
5.7 Conclusion 154
References 154

6 Transmission Line Components 155
6.1 Introduction 155
6.2 Coaxial Components 155
6.3 Twisted Pairs and Twin-Lead 157
6.4 Rectangular Waveguide 158
6.5 Microstrip 161
6.5.1 Analysis and Synthesis 163
6.5.2 Microstrip Frequency Limitations 163
6.5.3 Microstrip Discontinuities and EDA Software 164
6.5.4 Test Fixtures and Housings 164
6.6 Common Microstrip Components 166
6.6.1 Quasi-Lumped Elements 166
6.6.2 Distributed Elements 168
6.6.3 Wilkinson Splitter/Combiner 170
6.6.4 Parallel-Coupled Lines 171
6.6.5 Lange Coupler 173
6.6.6 Branch-Line Coupler 173
6.6.7 Rat-Race Hybrid 174
6.6.8 Planar Marchand Balun 175
6.6.9 Microstrip Resonators 176
6.7 Uniplanar Transmission Lines 177
6.7.1 Coplanar Waveguide 177
6.7.2 Coplanar Strips 178
6.7.3 Slotline 178
6.8 Other Transmission Line Types 179
6.8.1 Stripline 179
6.8.2 Suspended Substrate Techniques 179
6.8.3 Thin-Film Microstrip 180
6.8.4 Lumped and Lumped-Distributed Components 180
6.8.5 Substrate Integrated Waveguide 181
6.8.6 Ridge Waveguide and Finline 182
6.8.7 Dielectric Waveguides 183
6.8.8 Micromachined Lines 184
6.9 Conclusion 184
References 185

7 Transmission Line Filters 187
7.1 Introduction 187
7.2 Unloaded Q of a Transmission Line Resonator 188
7.3 Lumped-to-Distributed Conversion 189
7.3.1 Shunt Stub 189
7.3.2 Series Line 190
7.4 Impedance and Admittance Inverters 191
7.4.1 Definition 191
7.4.2 Network Transformation with Impedance and Admittance Inverters 194
7.4.3 Inverter Realisation 198
7.5 Richards Transformation 202
7.6 Unit Element, Kuroda's Identity and Coupled-Lines Section 203
7.6.1 Unit Element 203
7.6.2 Kuroda's Identity 204
7.6.3 Coupled-Lines Section 207
7.7 Stepped-Impedance Low-Pass Filter 208
7.8 Parallel-Coupled Line Filter 211
7.9 Interdigital Filter 217
7.10 Combline Filter 223
7.11 Hairpin Filter 233
7.12 Cross-Coupled Filters 237
7.13 Conclusion 246
References 247

8 Semiconductor Devices 248
8.1 Introduction 248
8.2 Fabrication Technology 249
8.2.1 Processing Steps 250
8.2.2 Extending Moore's Law 253
8.2.3 More than Moore 255
8.2.4 Other Nanofabrication Approaches 256
8.3 Field-Effect Transistors 256
8.3.1 GaAs and InP HEMTs 257
8.3.2 GaN HEMTs 259
8.3.3 HEMT Small-Signal Equivalent Circuit 259
8.3.4 Dual-Gate FETs 262
8.3.5 CMOS 262
8.3.6 LDMOS 263
8.4 Bipolar Transistors 263
8.4.1 SiGe Heterojunction Bipolar Transistors (HBTs) 265
8.4.2 GaAs and InP HBTs 266
8.5 Package Styles 267
8.6 High-Power Transistors 269
8.6.1 High-Power Device Geometries 270

8.7	RFICs and MMICs	271
	8.7.1 On-Chip Lumped Passive Components	272
8.8	Two-Terminal Devices	275
	8.8.1 Schottky Diodes	275
	8.8.2 PIN Diodes	276
	8.8.3 Sources	276
	References	277
9	**Impedance Matching**	**279**
9.1	Introduction	279
9.2	The Purpose of Impedance Matching	279
9.3	Lumped-Element Matching Networks	282
	9.3.1 Lumped-Element Matching with L-Networks	284
	9.3.2 Matching with π- and T-Networks	287
9.4	Distributed Matching Networks	289
	9.4.1 Stub Matching	290
	9.4.2 Single-Section Quarter-Wave Transformer	293
	9.4.3 Multisection Quarter-Wave Transformer	294
	9.4.4 Other Types of Transmission Line Transformer	294
9.5	The Cyclic Nature of Distributed Circuits	295
9.6	Conclusion	296
	References	296
10	**Amplifiers**	**298**
10.1	Introduction	298
	10.1.1 Gain Definitions	298
10.2	Transistor Configurations	300
	10.2.1 FET Configurations	300
	10.2.2 Bipolar Transistor Configurations	301
10.3	Classical Analysis of Gain and Stability	302
	10.3.1 Stability Circles	304
	10.3.2 Stability Factor	305
	10.3.3 Maximum Gain	306
	10.3.4 Stabilisation Techniques	308
10.4	DC Biasing	309
	10.4.1 Inductors as Bias Chokes	309
	10.4.2 Bias Stubs	309
	10.4.3 Active Loads	310
10.5	Common Amplifier Topologies	310
	10.5.1 Reactively Matched Amplifiers	311
	10.5.2 FET Feedback Amplifier	312
	10.5.3 Darlington Pair	312
	10.5.4 Long-Tailed Pair	313
	10.5.5 Distributed Amplifier	313
	10.5.6 Balanced Amplifier	314

10.6 Low-Noise Amplifiers 315
10.6.1 Cascaded Noise Figures 316
10.6.2 Noise Circles 317
10.7 Nonlinearity and Intermodulation 318
10.7.1 Nonlinear Characteristics 318
10.7.2 Second-Order Intermodulation 319
10.7.3 Third-Order Intermodulation 321
10.7.4 Higher-Order Terms 323
10.7.5 Intermodulation in Cascaded Circuits 323
10.7.6 Other Measures of Distortion in RF Amplifiers 325
10.7.7 Intermodulation in Other Components 325
10.8 Power Amplifier Classes of Operation 326
10.8.1 Class A 327
10.8.2 Class B 329
10.8.3 Class AB 331
10.8.4 Class C 331
10.8.5 Class D 332
10.8.6 Class E 333
10.8.7 Class F 334
10.8.8 Other Classes 334
10.9 Power-Combining Techniques 334
10.9.1 Planar Combining Techniques 335
10.9.2 Coaxial and Waveguide Combiners 337
10.9.3 Spatial and Quasi-Optical Power Combining 338
10.10 Power Amplifier Linearisation 339
10.10.1 Feedback Techniques 339
10.10.2 Predistortion Linearisation 340
10.10.3 Feedforward Linearisation 340
10.11 Conclusion 341
References 342

11 Oscillators 344
11.1 Introduction 344
11.2 Basic Concepts 344
11.2.1 One-Port Model: Negative Resistance 344
11.2.2 Two-Port Model: Positive Feedback 349
11.2.3 Nyquist Theory Applied to the Oscillation Condition 354
11.2.4 Timing Jitter 360
11.2.5 Phase Noise 361
11.3 Resonators 363
11.3.1 Dielectric Resonator 363
11.3.2 Transmission Line Resonator 363
11.3.3 Quartz Crystal Resonator 365
11.4 Some Oscillator Circuits 367
11.4.1 Colpitts 367
11.4.2 Cross-Coupled Oscillator 372

	11.4.3 Voltage-Controlled Oscillator	375
	11.4.4 Phase-Locked Loops	377
11.5	Oscillator Design Procedure	382
	11.5.1 Design Flow	382
	11.5.2 Negative Resistance Technique Applied to Microwave Oscillator Design	384
11.6	Conclusion	389
	References	390
12	**Mixers and Modulators**	**391**
12.1	Introduction	391
12.2	Single-Ended Mixers	392
	12.2.1 Diode Mixers	396
	12.2.2 Active FET Mixers	396
	12.2.3 Resistive FET Mixers	397
12.3	Balanced and Image-Rejection Mixers	398
	12.3.1 Single-Balanced Mixers	399
	12.3.2 Double-Balanced Mixers	399
	12.3.3 Image-Rejection Mixer and SSB Upconverter	401
12.4	Baluns and Couplers	402
	12.4.1 Transmission Line Baluns	402
	12.4.2 Lumped-Element Splitters and Combiners	403
	12.4.3 Active Splitters and Combiners	403
12.5	Common Mixer Circuits	404
	12.5.1 Passive Double-Balanced Diode Mixer	404
	12.5.2 Gilbert Cell Mixer	405
	12.5.3 Double-Balanced Resistive FET Mixer	405
	12.5.4 Distributed FET Mixer	407
	12.5.5 Millimetre-Wave Single-Balanced Diode Mixer	408
12.6	Modulators	409
	12.6.1 Vector Modulators using Mixers	409
	12.6.2 Other Forms of Vector Modulator	410
12.7	Mixer Linearisation and Adaptive Signal Cancellation	412
12.8	Conclusion	412
	References	413
13	**RF MEMS**	**415**
13.1	Introduction	415
13.2	Novel Transceiver Architectures using RF MEMS	416
13.3	Micromachined Transmission Lines and Passive Elements	416
	13.3.1 Rectangular-Coaxial Transmission Lines	417
	13.3.2 Micromachined Coplanar Waveguide	418
	13.3.3 MEMS Inductors	419
	13.3.4 MEMS Tuneable Capacitor	420

13.4	RF MEMS Switches	420
13.5	Reconfigurable Impedance-Matching Networks	424
13.6	MEMS Phase Shifters	425
13.7	Tuneable Filters	426
13.8	MEMS Antennas	427
13.9	RF MEMS Fabrication and Packaging	428
13.10	Reliability and Design Consideration of RF MEMS Devices	430
	References	432
14	**Antennas and Propagation**	**434**
14.1	Introduction	434
14.2	Antenna Systems	437
	14.2.1 Wire Antennas	437
	14.2.2 Travelling-Wave Antennas	442
	14.2.3 Reflector Antennas	443
	14.2.4 Horn Antennas	444
	14.2.5 Microstrip Patch Antennas	445
	14.2.6 NFC Antennas	447
14.3	Transmission Equations, Free-Space Path Loss and Link Budget Calculation	447
	14.3.1 Transmission Equation	447
	14.3.2 Free-Space Path Loss	449
	14.3.3 Link Budget	450
14.4	Other Propagation Effects	451
	14.4.1 Reflection and Refraction	451
	14.4.2 Diffraction	451
	14.4.3 Scattering	452
14.5	Millimetre-Wave and THz Propagation	452
14.6	Indoor Propagation	453
	14.6.1 Long-Distance Models	453
	14.6.2 Partition Losses on the Same Floor	453
	14.6.3 Partition Losses Between Floors	454
	14.6.4 Motley–Keenan Model	454
	14.6.5 Ericsson Multiple Breaking Model	454
14.7	Outdoor Propagation	455
	14.7.1 Theoretical Propagation Prediction Models	455
	14.7.2 Empirical or Statistical Models	455
	14.7.3 Semi-Empirical Models	455
	14.7.4 Okumura Model	455
	14.7.5 Hata Model	456
	14.7.6 Other Models	456
14.8	Multipath Propagation	457
	14.8.1 Multipath Fading	457
14.9	Antenna Arrays	458
14.10	Multiple-Input and Multiple-Output Systems	460
	References	462

15 Digital Signal Processing for Transceivers **464**
15.1 Introduction 464
15.2 RF Performance Challenges 464
15.2.1 Imperfections in the RF Front End 465
15.2.2 Cost and Low Power Consumption 466
15.2.3 Linearity 467
15.3 DSP in Modern Wireless Communications Systems 467
15.4 Signal Conversion and Processing 468
15.4.1 Analogue-to-Digital Conversion 468
15.4.2 Digital-to-Analogue Conversion 473
15.4.3 Discrete Fourier Transform 474
15.4.4 Fast Fourier Transform 475
15.4.5 CORDIC Algorithm 477
15.5 Digital Calibration for *I–Q* Imbalance 482
15.6 Digital Predistortion Techniques 486
15.7 DSP Techniques for OFDM 489
15.8 MIMO 491
15.9 Conclusion 493
References 494

16 Packaging and Assembly **495**
16.1 Introduction 495
16.2 Technology Options and System Partitioning 496
16.3 PCB/Laminate Technology 498
16.3.1 Rapid Prototyping of PCBs 500
16.4 Thin-Film Fabrication 500
16.5 Thick-Film Fabrication 500
16.5.1 Photoimageable Thick-Film Process 502
16.6 LTCC Technology 503
16.7 Chip Packaging 504
16.7.1 Chip-on-Board 504
16.7.2 Lead Frame 505
16.7.3 Pin-Grid Array (PGA) and Ball-Grid Array (BGA) 505
16.7.4 Flip-Chip 505
16.7.5 3-D Packaging 507
16.8 Manufacturing using Surface Mount Technology 508
16.8.1 Solder Deposition 508
16.8.2 Solder Reflow 509
16.9 System-in-Package and System-on-Substrate Technology 509
16.10 Transitions and Antenna-in-Package Techniques 512
16.11 Conclusion 512
References 513

17 Electronic Design Automation **515**
17.1 Introduction 515
17.2 Linear Frequency-Domain Analysis 518

17.3 Time-Domain Simulation 518
17.4 Harmonic Balance 522
17.4.1 Concept 523
17.4.2 Application to Microwave and Millimetre-Wave Circuits 524
17.5 Large-Signal/Small-Signal Simulation 525
17.6 Planar Electromagnetic Simulation 529
17.7 3-D Electromagnetic Simulation 533
17.8 Integrated Circuit Simulation and Layout 534
17.8.1 Process Design Kits for RFIC and MMIC Design 534
17.9 Conclusion 538
References 538

18 Measurement Techniques 539
18.1 Introduction 539
18.2 The Oscilloscope 540
18.3 Function Generator and Arbitrary Waveform Generator 542
18.4 LCR Meters and Component Analysers 542
18.5 Signal Generators 542
18.6 Spectrum and Signal Analysers 544
18.6.1 Vector Signal Analysers 546
18.6.2 Harmonic Mixers 547
18.7 Vector Network Analysers 547
18.7.1 VNA Calibration 549
18.7.2 Calibration Standards 552
18.7.3 Electronic Calibration 554
18.7.4 Calibration Example 554
18.7.5 Connector Care 556
18.7.6 Time-Domain Measurements 557
18.8 Microstrip Test Fixture Measurements 558
18.8.1 In-Fixture TRL Calibration 559
18.9 Probe Station Measurements 560
18.9.1 Probe Calibration Techniques 561
18.9.2 On-Wafer Measurements of Microstrip DUTs 562
18.10 Mixed-Mode *S*-Parameters 563
18.11 Source- and Load-Pull Measurements 565
18.12 *X*-Parameter Measurements 565
References 567

Glossary 569

Index 575

About the Authors

Ian Robertson received his BSc (Eng.) and PhD degrees from King's College London in 1984 and 1990, respectively. From 1984 to 1986 he worked in the GaAs MMIC Research Group at Plessey Research, Caswell. After that he returned to King's College, initially as a Research Assistant working on the T-SAT project and then as a Lecturer, leading the MMIC Research Team and becoming Reader in 1994. In 1998 he was appointed Professor of Microwave Subsystems Engineering at the University of Surrey, where he established the Microwave Systems Research Group and was a founder member of the Advanced Technology Institute. In June 2004 he was appointed to the Centenary Chair in Microwave and Millimetre-Wave Circuits at the University of Leeds. He was Director of Learning and Teaching from 2006 to 2011 and Head of School from 2011 to 2016. He has 30 years of teaching experience in RF and microwave engineering and has published over 400 peer-reviewed research papers. He edited the book *MMIC Design* published by the IEE in 1995 and co-edited the book *RFIC & MMIC Design and Technology*, published in English in 2001 and in Chinese in 2007. He was elected Fellow of the IEEE in 2012 in recognition of his contributions to MMIC design techniques and millimetre-wave system-in-package technology. He was General Technical Programme Committee Chair for the European Microwave Week in 2011 and 2016.

Nutapong Somjit received the Dipl.-Ing. (MSc) degree from Dresden University of Technology in 2005 and the PhD degree from the KTH Royal Institute of Technology in 2012. Then, he returned to Dresden to lead a research team in micro-sensors and MEMS ICs for the Chair for Circuit Design and Network Theory. In 2013, he was appointed Lecturer (Assistant Professor) in the School of Electronic and Electrical Engineering, University of Leeds. His main research focuses on RFICs, RF MEMS, tuneable antennas, and RFIC-MEMS integration. Dr Somjit was the recipient of the Best Paper Award (EuMIC prize) at the European Microwave Week in 2009. He was awarded a Graduate Fellowship from the IEEE Microwave Theory and Techniques Society (MTT-S) in 2010 and 2011, and the IEEE Doctoral Research Award from the IEEE Antennas and Propagation Society in 2012. He was appointed a member of the Engineering, Physical and Space Science Research Panel of the British Council in 2014. He serves as a reviewer for various international journals, including *IEEE Transactions on Microwave Theory and Techniques*. Since 2013, he has been a member of the International Editorial Board of the *International Journal of Applied Science and Technology*. In 2016, he was the Chair of the Student Design Competition for the European Microwave Week.

Mitchai Chongcheawchamnan was born in Bangkok, Thailand. He received the BEng degree in Telecommunication from King Mongkut's Institute of Technology Ladkrabang, Bangkok in 1992, the MSc degree in Communication and Signal Processing from Imperial College London in 1995 and the PhD degree in Electrical Engineering from the University of Surrey, Guildford, UK, in 2001. He joined the Faculty of Engineering, Prince of Songkla University, Thailand, in 2008 as an Associate Professor. He has published more than 100 papers in international journals and conferences in RF, microwave and millimetre-wave engineering and agricultural applications. He has one international patent (pending), five national patents (pending) and three international innovation awards. Currently, he is a Director of Academic Outreach at the Prince of Songkla University, Hat Yai campus, Thailand.

Acknowledgements

First and foremost we would like to thank all those students (undergraduate, masters and PhD) and colleagues from industry and academia that have shared their experience with us over many years. Students, in particular, often asked fundamental questions that challenged our understanding of perceived wisdom in RF engineering – it is widely recognised these days that it is only when you teach a subject that you fully understand it, and we think that this has benefitted this book enormously. We would, in turn, like to thank all our own teachers, from school through to PhD study, who taught, guided and supported us, inspiring us to develop careers in engineering, with special thanks to Charles Turner, E.M. Deeley, P.R. Adby, Hamid Aghvami and Joachim Oberhammer.

We are very grateful to the people at Keysight Technologies, Plextek RFI, Phase 2 Devices, Rako Controls and Instrumentel for providing photographs to help put the text and diagrams into proper context. Some of the content has been informed by our own cutting-edge research and we are grateful to all those funding agencies that have supported us, especially the Engineering and Physical Sciences Research Council (EPSRC, UK), the European Commission, Vinnova (Sweden) and the Thailand Research Fund.

We are grateful for the assistance with diagrams and teaching materials provided by Umer Shah, Nemr Al Thalathiney, Hassan Al Nasser, Ali Alameer, Matthew Gillick, Ayodeji Sunday, Burawich Pamornnak, Sahapong Somwong, Kritsada Poungsuwan, Huda Kosumpan and Silachai Moongdee.

Mitchai Chongcheawchamnan would like to thank his colleagues at the Electrical Engineering (EE) Department, Faculty of Engineering, Prince of Songkla University (PSU) for providing this opportunity and supporting this book. Special thanks are due to Dr Kiattisak Wongsopanakul, Head of the EE department and colleagues from the Faculty of Engineering, who permitted sabbatical leave to allow time to write this book.

Finally, we would all like to thank our families for their love and support and for allowing us to spend so much time on this book instead of being with them.

Ian Robertson
Nutapong Somjit
Mitchai Chongcheawchamnan
April 2016

Acknowledgments

Preface

This book is aimed primarily at final year undergraduates and Masters students. It will also be invaluable as background reading for PhD students and as a primer for practising engineers in industry. Its primary purpose is to explain concepts in clear terms, with an emphasis on modern industry practice, with references for further reading given to provide more detailed information and background theory.

The book describes a full range of contemporary techniques for the design of transmitters and receivers for communications systems operating in the range from 1 through to 300 GHz. Within this frequency range there are many technologies that need to be employed, with silicon integrated circuits at the core but, compared to other electronic systems, with a much greater need for specialist devices and components for high performance – for example, low-loss passive components and low-noise and high-power compound semiconductor devices. Millimetre-waves (frequencies from 30 to 300 GHz) have rapidly been adopted for a wide range of consumer applications such as wireless high-definition video for virtual and augmented reality, 5G mobile systems, and automotive radars. It has taken many years to develop low-cost technologies for suitable transmitters and receivers, so previously these frequencies have been employed only in expensive military and space applications. This really does represent a new era in the use of the millimetre-wave part of the electromagnetic spectrum. The next 20 years will see explosive growth in the use of the frequency range from 30 to 300 GHz, and this requires engineers to be able to understand and apply the techniques of traditional microwave engineering, as well as the newer technologies such as silicon germanium and gallium nitride devices and integrated circuits. These are then assembled into sophisticated multichip modules and system-in-package products with integrated antenna arrays.

The book starts with an emphasis on the applications of microwave and millimetre-wave systems, in Chapter 1, followed by a description of the wide range of modern transmitter-receiver architectures in Chapter 2. The use of quadrature modulation schemes and modern direct-conversion transceivers, that are at the heart of nearly all wireless communication systems, are covered. Chapter 3 introduces *S*-parameters so that they can be used in Chapter 4 on Lumped Element Filters. This introduces Bode plots, transfer functions, poles and zeros and the *s*-domain, followed by design techniques for low-pass, high-pass, band-pass and bandstop filters. However, a key conclusion of this chapter is that lumped elements often suffer from serious limitations in high-frequency design, which is why transmission lines are needed for many circuits.

Chapter 5 introduces the main principles of transmission lines, including the telegrapher's equation, signal flowgraphs, standing waves, impedance transformation along a line and the

Smith chart. This is followed by Chapter 6 on transmission line components, which is focused on the practical aspects of the main transmission line types – coaxial lines, rectangular waveguide, microstrip and various others, including coplanar waveguide (CPW) and substrate integrated waveguide (SIW). The emphasis is on those transmission line media that are used for circuit design, looking at their propagation characteristics, technologies and materials, discontinuities and modelling. A number of key standard components are introduced, such as the Wilkinson power divider, Lange coupler, branch-line coupler and rat-race hybrid.

Having covered the key transmission line media, Chapter 7 returns to the subject of filters. A succinct coverage of distributed filters is given, covering those topologies most commonly encountered in modern microwave and millimetre-wave design. The mathematical conversion from lumped element filters to distributed elements, followed by network transformation to produce realisable practical filters, is described.

Chapter 8 introduces the key semiconductor devices – both discrete low-noise and high-power transistors and those used for integrated circuits. The chapter summarises the key features of silicon, silicon–germanium, GaAs, GaN and InP transistors. Chapter 9 explains the importance of impedance matching, with particular reference to amplifier design. The use of the Smith chart is introduced and a number of common matching techniques is described. Chapter 10 builds on this by describing amplifier design techniques, starting with the classical gain and stability analysis using S-parameter and Mason's rule analysis. The key topologies used in GaAs MMIC and silicon RFIC design are introduced. Distortion and intermodulation in power amplifiers is explained, followed by a summary of key classes of operation (A, AB, C, D, E, F, etc.) and a brief treatment of the main techniques employed for power-combining and linearising power amplifiers.

Chapter 11 introduces the fundamentals of oscillators, followed by coverage of the main oscillator circuit topologies. The importance of resonator Q-factor and phase noise are explained. Phase-locked loop techniques are covered, giving an introduction to integer-N and fractional-N frequency synthesis. Mixers are the core component for up- and down-conversion of signals, including use in 'IQ' direct conversion transceivers. Chapter 12 focuses on the most common mixer design techniques, including the Gilbert cell mixer technique, now adopted even beyond 100 GHz.

Chapter 13 describes radiofrequency microelectromechanical systems (MEMS), which are an important niche technology that can substantially change the architecture of modern transceivers. RF MEMS are of increasing interest as adaptive antennas, and multiband transceivers become key to future communications systems. Switches, phase shifters and reconfigurable circuits are also covered.

Chapter 14 gives an overview of the basics of antennas and propagation since, in order to understand the challenge of exploiting the millimetre-wave part of the electromagnetic spectrum, it is necessary to study the propagation characteristics in the range 1 to 1000 GHz. Millimetre-waves give the very significant advantage of small, high-gain antennas, and modern communications systems are relying more than ever on adaptive and multiple-beam antennas in order to combat propagation effects and increase the capacity of systems. The chapter therefore introduces the key concepts such as multiple input/multiple output (MIMO) systems.

Modern communications systems nearly always use direct conversion architectures, and it is important that RF engineers have some understanding of the digital part of the system. Consequently, Chapter 15 provides an introduction to some of the digital signal processing (DSP) techniques used for communications transceivers. The chapter also provides a tutorial on the key principles of analogue-to-digital and digital-to-analogue converters and DSP techniques. This is important so that RF engineers understand how DSP is used to combat signal impairments and realise functions such as beam forming.

The final three chapters describe how all the devices, components and circuits covered in the previous chapters are assembled, simulated and measured. Chapter 16 describes the main technologies and design techniques for microwave circuit assemblies and multichip modules, as well as emerging 3-D system-in-package and ‘system-on-substrate’ technologies. Chapter 17 covers electronic design automation (EDA), describing the most common design tools used by designers. The aim here is that the reader will fully understand the role and limitations of different simulation techniques, including linear, time-domain nonlinear, harmonic balance, modulation domain, planar electromagnetic, 3-D electromagnetic and system-level simulation. Finally, Chapter 18 covers measurement techniques, including vector network analyser calibration methods. A wide range of test equipment for measuring individual devices and components through to complete transmitters and receivers is described, using Keysight Technologies equipment as examples.

We sincerely hope that you find this book useful. Microwave and millimetre-wave design is a fascinating subject, and is possibly unique in the way it blends electronic circuits into works of art – such is the importance of their physical form in determining function. That, combined with the amazing potential for wireless communications to transform every aspect of human endeavour, makes RF engineering one of the most rewarding careers of all.

Ian Robertson
Nutapong Somjit
Mitchai Chongcheawchamnan
April 2016

1

Introduction

The rapid convergence of wireless communications systems and digital media has been one of the most exciting developments in electronics during recent years. Increasingly, consumer electronics is the driving force behind technological advance, whereas in the past it was military and space projects. Consumer electronics has become the leading edge of technology because of the potentially huge sales volumes. As the market is extremely competitive, major global companies invest large sums in research and development in order to introduce new products with more features and better performance than their rivals. One of the biggest growth areas has been in mobile internet connectivity through smartphones and tablets, leading to a paradigm shift towards cloud computing, shown by the dramatic growth in mobile subscriptions and internet traffic. 5G heralds a new era in which wireless connectivity reaches everything, with the Internet of Things (IoT) creating a revolution in almost every facet of modern life, including healthcare, smart cities, energy, robotics, manufacturing, retail and the creative industries.

In the business of consumer electronics, it is well known that customers generally get (and, indeed, expect) more features for less money as technology advances. The key to this has been the continual and rapid development of silicon technology, allied with the globalisation of the electronics manufacturing business. By putting more transistors onto a single integrated circuit (IC), clearly a given product will require less ICs and less associated components on a circuit board. Gordon Moore, a cofounder of Intel, predicted in 1965 that the complexity of a chip could be expected to increase exponentially with time for a number of years [1]. This famous prediction has developed into Moore's Law, which is an empirical law that accurately charts technological progress in the microelectronics industry, with the number of transistors on a chip doubling every year. Figure 1.1 shows a graph charting this evolution of Intel microprocessor technology, re-plotted on a linear scale to show the dramatic development of computer technology, and illustrates that we have only recently entered the information age. The impressive milestone of two billion transistors on a chip was reached in 2008 [2].

Microwave and Millimetre-Wave Design for Wireless Communications, First Edition.
Ian Robertson, Nutapong Somjit and Mitchai Chongcheawchamnan.
© 2016 John Wiley & Sons, Ltd. Published 2016 by John Wiley & Sons, Ltd.

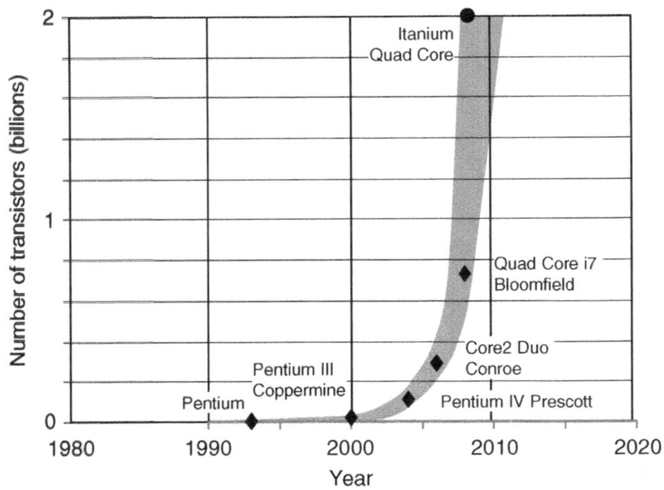

Figure 1.1 Microprocessor transistor count versus year on a linear scale.

1.1 A Brief Timeline of Consumer Electronics

"We only recently entered the information age" is a bold claim that requires substantiation. A worthy method of investigating the claim is to briefly study the history of consumer electronics and develop a timeline of how the main audio, video and communications products have developed. Figure 1.2 shows a simplified timeline of some of the most important developments

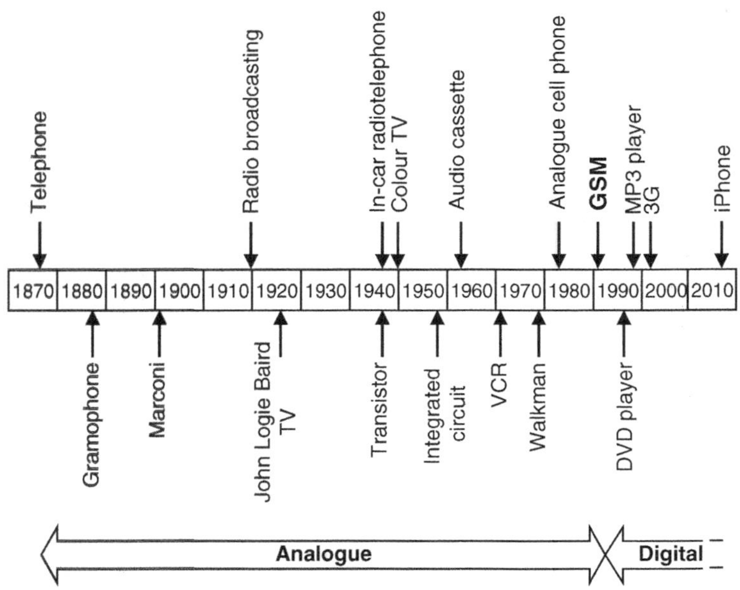

Figure 1.2 A simplified timeline of developments in consumer electronics.

in communications, radio and television. The telegraph came first, with the Cooke and Wheatstone six-wire, five-needle system and the Morse code system being independently introduced in the 1830s in the UK and US, respectively. A number of pioneers had experimented with the speaking telegraph, and it was on 10th March 1876 that Alexander Graham Bell spoke the words "Mr. Watson, come here, I want to see you" to his assistant.

After various pioneering efforts to record audio, including Thomas Edison's cylinder-based Phonograph, in 1887 Emile Berliner patented the disc-based machine (the Gramophone) that that is similar to what we know today. Broadcasting has its roots in Marconi's successful demonstration of radio communications across the Atlantic in 1901. Radio broadcasting became established during the 1920s, and TV started appearing shortly after John Logie Baird's demonstration of an electromechanical television in 1926. In 1928, American inventor Philo Taylor Farnsworth gave the first public demonstration of an all-electronic scanning television system [3]. He was awarded a string of patents, which he later licensed to RCA, which became a giant in the world of TV and radio. The 1950s, 1960s and 1970s witnessed an extensive development of colour television and recording, and the first portable audio products.

The mobile telephone has its origins in Motorola's radiotelephone of 1946. The first automated cellular systems were NTT's in Tokyo in 1979 and the NMT network introduced in Sweden and Norway in 1981. The AMPS system was introduced in the USA in 1983 and the DynaTAC handheld cellular phone was introduced by Motorola, considered to be the first ever handheld phone and now recognisable instantly as the "brick phone". GSM phones were introduced in 1991 – a digital system with greatly enhanced speech quality. Although personal computers were available about a decade before, GSM phones were one of the first consumer products that used digital electronics to perform real-time processing of signals. The 1990s can therefore be considered (debatably) the start of the digital consumer product revolution, with the DVD player (1996) and portable MP3 player (1998) appearing in the same decade. The iPhone was launched in 2007 and, in the short time since, smartphones and tablets have become an almost indispensable part of modern living, being most people's main portal to the internet, with awesome functionality that includes multi-band cellular and internet connectivity, voice-directed navigation, multimedia playback, games, augmented reality, and so on.

1.2 The Electromagnetic Spectrum

Electromagnetic waves are vital to almost all communications systems and modern consumer products. Radio waves, infrared, visible light, ultraviolet (UV) radiation and X-rays are all forms of the same phenomenon of electromagnetic wave propagation. Some of the most important frequency bands and applications of the electromagnetic spectrum are shown in Figure 1.3. This is, once again, on a logarithmic scale: X-rays have a frequency around a billion times the frequency of a mobile phone radio signal. Of course, not all these frequencies can be generated easily (if at all) by conventional transistor-based electronics. High-power microwaves, for example, are generated in a magnetron for some applications. Light, for optical storage and communications, is generated by a laser.

Some of the most common frequencies and applications are listed in Table 1.1. Wavelength (λ) is inversely proportional to frequency and, assuming free space propagation, the relationship $c = f\lambda$ can be used, where c is the speed of light *in vacuo* (3×10^8 m s^{-1}). For radio

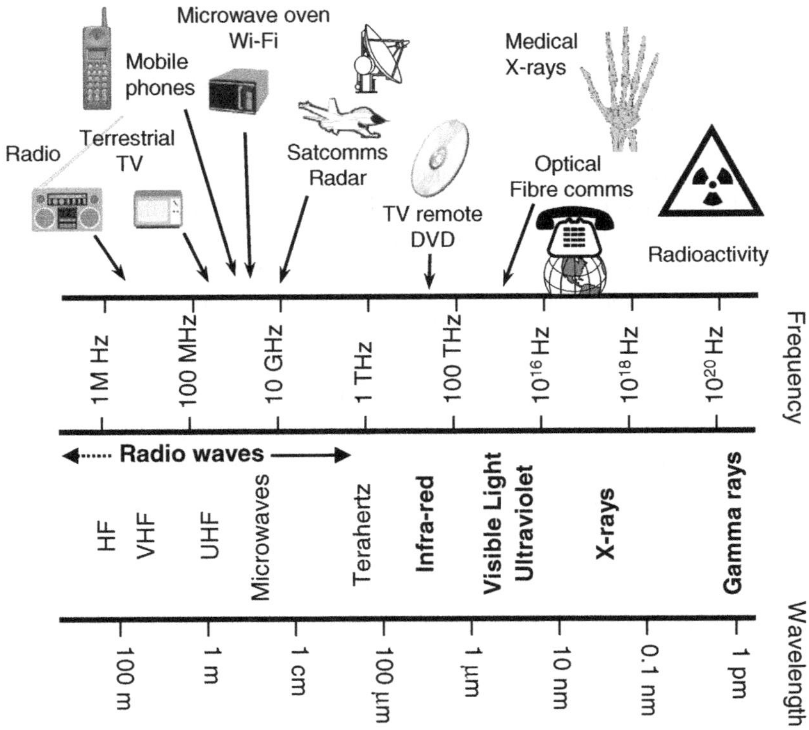

Figure 1.3 Frequency, wavelength and terminology for various applications of electromagnetic waves.

Table 1.1 Example frequencies of some common systems.

Frequency	Application
0.5 to 2.6 MHz	Medium wave AM radio band
220 MHz	Digital audio broadcasting
500–850 MHz	UHF TV band
850, 900 MHz and 1.8, 1.9 GHz	GSM mobile (cell) phones
2.2 GHz	3G mobile phones
2.45 GHz	Microwave ovens, Bluetooth, 802.11b/g/n wireless LANs
10 GHz	Military radar
11–12 GHz	European satellite TV
60 GHz	802.11ad gigabit wireless LANs
77 GHz	Cruise control in some cars
198 THz	Long-distance fibre communications
333 THz	TV infra-red remote control
430 THz	Red light
1 million THz	X-rays in hospital

waves, the wavelength is large and the macro effects of electromagnetism and wave propagation readily apply. At the other extreme, light has a typical wavelength of <1 μm and quantum effects start to become important: In other words, the packet nature of light – the existence of photons – must be taken into account.

Radio transmission is used for hundreds of applications, and modern society could not exist without these radiofrequency (RF) systems. The frequency bands ranging from 100 kHz up to 100 GHz (100 000 MHz) are already extensively used for radio, TV, mobile telephones and the cellular network infrastructure, Wi-Fi, private mobile radio, satellite TV and data links, along with a whole host of other communications systems. Frequencies from 100 GHz to 400 GHz are already starting to be allocated to new systems, as the lower frequencies become congested. In addition to communications, the radio spectrum has slots allocated for other uses, such as radar, navigation, radio astronomy and remote-sensing. Remote-sensing is very important for the environmental monitoring of problems such as the hole in the ozone layer and the destruction of tropical rainforests, while microwave systems are very important because of the need to penetrate cloud cover.

A key principle in radio communications is that antennas, which convert electrical currents into radio waves (or vice versa), need to be of a size comparable to a wavelength. This, essentially, is why high-frequency signals are used for wireless communications – the antennas must be of a practical size. In all cases, the choice of operating frequency is governed by a range of factors, including bandwidth and range required, transmitter power available, propagation loss, antenna gain, cost of the equipment and licensing. The gain of an antenna is a figure of merit that describes how effectively it concentrates power into a desired direction, that is, towards the receiver(s). A perfect omnidirectional antenna spreads power out equally in all directions, as illustrated in Figure 1.4(a), and has a gain of 0 dBi (dBi = decibels relative to the *isotropic* case). For an idealised point source transmitter, the power would be equally distributed over the area of a sphere ($4\pi r^2$, where r is the distance from the transmitter) giving rise to what is known as spreading loss, with signal strength falling as $1/r^2$. Even in a broadcasting application, such as terrestrial TV, if the transmitter really had an omnidirectional antenna like this, a great deal of its precious RF power would be radiated upwards into space and wasted.

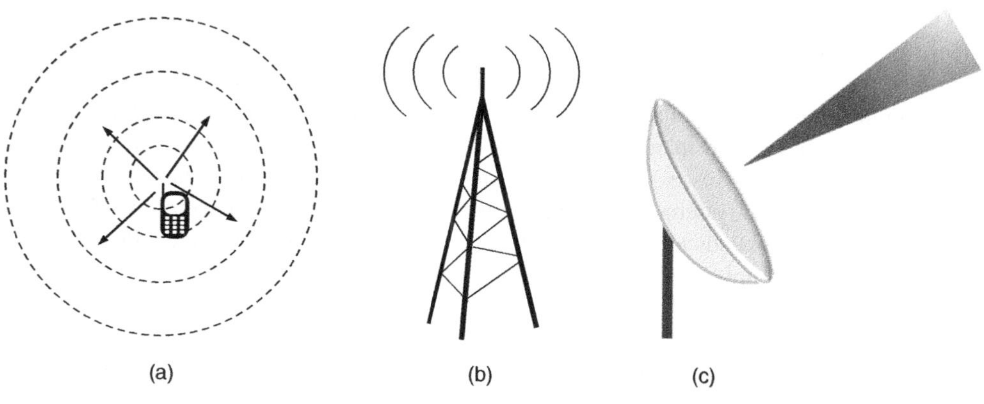

Figure 1.4 (a) Idealised point source transmitter giving omnidirectional coverage. (b) Doughnut-shaped radiation pattern from a typical terrestrial transmitter. (c) Pencil beam from a satellite dish.

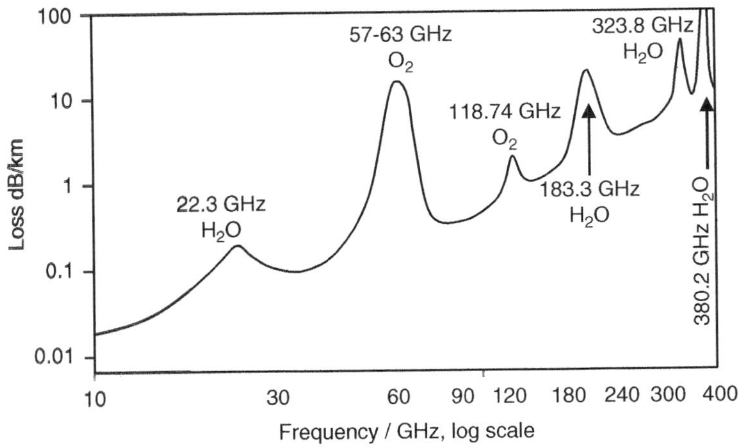

Figure 1.5 Trend of atmospheric attenuation at sea level; see Ref. [4] for data.

Hence, the antenna is designed to concentrate its power in the directions where it is needed – typically the radiation pattern would be doughnut-shaped for a terrestrial TV transmitter, as illustrated in Figure 1.4(b). For direct line-of-sight communications, a parabolic dish has a 'pencil beam' radiation pattern approximately 2–5 degrees wide, as illustrated in Figure 1.4(c), and has high gain of the order 35–40 dBi. Antennas are passive, reciprocal, devices and if an omnidirectional antenna is used for a receiver it will pick up signals equally from all directions, including noise and interference, and so the receiver's antenna gain is equally important.

High-gain antennas are many wavelengths in size, so at low frequencies a high-gain antenna may be unfeasibly large. On the other hand, whilst high-gain antennas are physically small at high frequencies, the propagation loss starts to become more of a problem. Figure 1.5 shows a plot of the atmospheric attenuation (dB km^{-1}) at sea level of a signal [4]. The loss is dominated by the effects of water vapour and oxygen: water vapour causes the steady increase in loss as frequency increases, as well as peaks at 22.3, 183.3 and 323.8 and 380.2 GHz due to rotational modes of the asymmetric H_2O molecule; oxygen absorption causes the peak at 60 GHz (which actually is formed from a number of lines between 57 and 63 GHz) and a clear line at 118.74 GHz. A further complication is the need to consider the intermittent attenuation caused by rain, fog, clouds and dust, particularly at frequencies beyond 10 GHz. For communication that is not line-of-sight, it is also necessary to account for attenuation and blocking due to natural and man-made obstacles – in those situations, low frequencies give a diffraction advantage. This is sea-level data and the atmosphere actually has a number of distinct layers. In mobile systems, the propagation characteristics can change rapidly with time, and the Doppler effect needs to be accounted for in the case of moving vehicles. Propagation effects are a highly complex issue that is covered in more depth in Chapter 14.

1.2.1 Spectrum Licensing and Standards

In order to prevent different users interfering with one another, the use of the radio spectrum is carefully regulated by government agencies (such as the FCC in the USA and OFCOM in

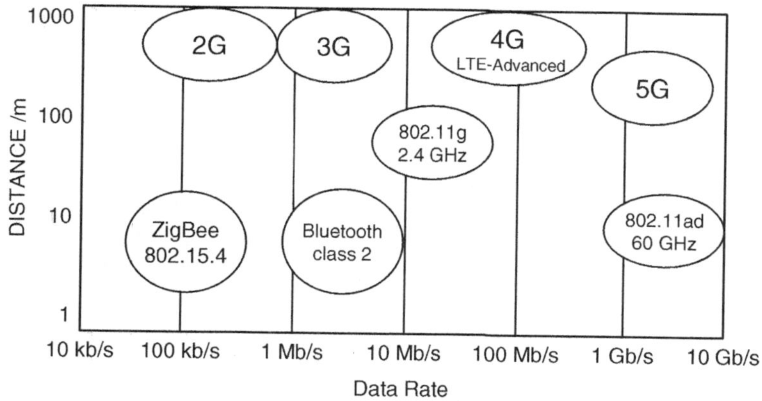

Figure 1.6 Various wireless communications standards compared.

the UK) and international bodies such as the International Telecommunications Union (ITU). These bodies issue licences to bona fide users, seek out unauthorised users, and conduct extensive consultations with stakeholders before allocating new frequency bands. There are many different types of licence, depending on the transmission frequency and application. Frequency allocation charts are available to download, and the number of different bands that have been allocated for diverse purposes is amazing. Wallcharts of ITU allocations are available (e.g., www.cedmagazine.com has a free pdf file) but there are so many allocations that the chart would be unreadable if reproduced here.

To ensure that radio equipment and consumer products are compatible, there is a tireless effort by standards committees to define the hardware and software specifications for different applications. Modern communications systems are so sophisticated that these documents run to hundreds and hundreds of pages. The Institute of Electrical and Electronics Engineers (IEEE) standards are particularly important; for example, the IEEE 802.11 wireless LAN standards were key to the rapid adoption of Wi-Fi around the world. Figure 1.6 shows various IEEE and other wireless communications standards compared on a range versus data rate graph.

1.3 Industry Trends

A general background to the electronics industry as a whole is important in the study of modern wireless communications systems. At the heart of the industry's success is nanometre-scale silicon technology, particularly CMOS technology, capable of realising sophisticated system-on-chip products. In addition to this, many microwave and millimetre-wave systems, and the photonic systems that make up the core network, require specialist compound semiconductor devices such as GaAs, GaN and InP devices. This has led to a great deal of research and development in the area of miniature systems integration, including multi-chip modules, system-in-package and antenna-in-package techniques (these are described in Chapter 16). Silicon and SiGe technology has started to challenge many of the traditional III–V markets and is rapidly opening-up the millimetre-wave part of the spectrum for consumer applications such as 60 GHz wireless networks.

The microwave industry has changed dramatically as consumer-oriented products such as cellular mobile, Wi-Fi, Bluetooth and GPS systems, amongst others, have become mainstream. The volume of this side of the industry is absolutely vast; for example, one billion Android smartphones and nearly 200 million iPhones were sold in 2014. This immense scale has changed the way the industry operates. To be a world leader, a company must have the resources to meet this huge demand. Typical silicon fabs with state-of-art capability cost more than US$3 Bn, while a large display factory, required to process 30 000 glass panels per month, each as large as 2 × 3 metres, typically requires an investment of US$5 Bn. Such a factory will require exceptionally good supply chain management, and often suppliers will be housed on-site in order to keep things running smoothly. The scale of mass production of consumer electronics in China is well known, and there are already signs of the industry entering a new phase of globalisation. For example, at the time of writing Foxconn was planning to invest in 12 new assembly factories in India, potentially creating one million jobs.

This immense scale of the consumer electronics industry, which drives a great deal of microwave and millimetre-wave development, has led to new ways of the industry building critical mass, including IP licensing, joint ventures, alliances, mergers and acquisitions. Fortunately, the industry has largely learned the importance of cooperation on wireless standards, with bodies such as the 3GPP, IEEE and ITU playing a vital role. Electronics is already found almost everywhere and has transformed our lives in many ways, and yet there is a further revolution underway in 5G+ communications, the internet of things, machine-to-machine communications, augmented reality and telepresence, to name a few. Without standards, the vision of everything being connected seamlessly to the Cloud will not be achievable.

The scale of the industry and the spectacular pace of change have given tremendous importance to intellectual property and branding. The value of a brand has become more and more about the user experience and even emotional attachment and loyalty. This has dramatically increased the importance of industrial product design and consideration to the human–computer interface. The style and ergonomics of the product are often a key differentiator in the market place, since the electronics inside is comparable. The skill of the product designers lies in exploiting new materials and manufacturing processes to create a stylish design that is economical to produce. Design-for-manufacture (DFM) and design for the environment (DFE) are important themes of this activity. As consumer products are replaced so quickly, designers need to consider how the whole ensemble of electronic and mechanical components can be assembled quickly, for low production costs, and later disassembled for the efficient recycling of key parts and materials. Electronic waste and protecting the environment are areas where government legislation is crucial to the long-term sustainability of the industry.

1.3.1 The Multidisciplinary Nature of Modern RF Engineering

A state-of-the-art integrated circuit is more complex than the whole US road network. It is not difficult to justify the claim, therefore, that electronic engineering is one of the most challenging careers of all, and engineers have to cope with exceptionally rapid technological change. But at the same time, it is also one of the most exciting and fulfilling careers – one only needs to read *IEEE Spectrum* regularly to know that electronic engineers are capable of amazing feats of technology that can dramatically improve our lives through impact in the fields of healthcare, energy, communications, transport, science and space exploration.

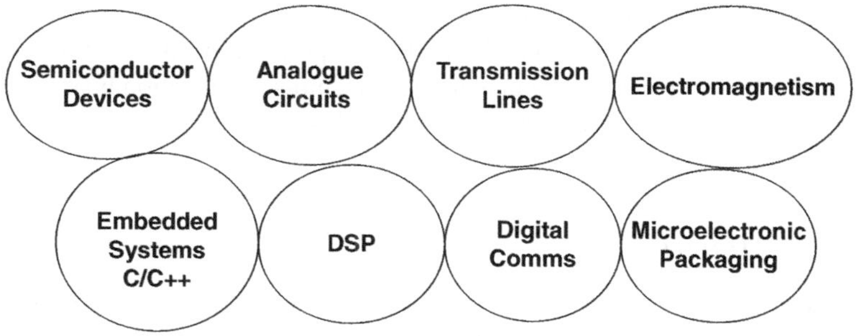

Figure 1.7 Aspects of modern RF engineering.

Microwave and millimetre-wave engineering was, for many years, a niche area in the realm of electromagnetics, transmission lines and the Smith chart, sometimes unfairly referred to as a 'black art'. As digital techniques have advanced, the RF engineer has needed to develop a much wider range of skills, as illustrated in Figure 1.7, with at least a basic understanding of topics from semiconductor physics through to digital electronics:

Semiconductor Physics: RF engineering uses some of the most advanced semiconductor and quantum devices, because achieving low noise and high power at these frequencies is highly challenging and often beyond the capability of regular devices. RF engineers need, therefore, to be conversant with the underlying physics in the device to make informed design choices.

Analogue circuit design: Twenty years ago, a monolithic microwave integrated circuit (MMIC) amplifier could be designed with two transistors and a few inductors and capacitors to match the transistors to 50 Ω. Now, the radiofrequency integrated circuit (RFIC) designer has to make use of the full range of analogue circuit design skills to reduce current consumption and chip area, and realise complex circuits and subsystems with adaptive control loops to maximise performance.

Microelectronics and packaging: The integration of high-frequency analogue and digital circuits has pushed integration levels and packaging technology to new extremes. Twenty years ago, RF designers might be using single transistor die-mounted onto specialist substrates. Nowadays, they are more likely to encounter 3600-pin ball-grid array (BGA) packages that have pushed printed circuit board and assembly technologies to the limit.

Digital communications: The close integration of digital signal processing (DSP) with RF transceivers makes it impossible to work effectively in a team without an understanding of IQ modulation, orthogonal frequency division multiplexing (OFDM), sigma-delta modulators, and so forth. For example, power amplifiers need to be designed to cope with complex varying-envelope signals that maximise spectral efficiency; they cannot be designed in isolation from the digital communications engineers.

Embedded systems design: Modern RF transceivers use extensive digital control for a range of functions. Without the ability to programme in C or a similar language, the RF engineer will be unable to get a chip-set evaluation board or (for example) a ready-made ZigBee or Bluetooth module to work.

1.3.2 Business Matters

The electronics business is extremely competitive; customers demand better and better products every year and expect the price to fall. The need for standardisation sometimes makes it difficult to introduce a new product without extensive industry-wide consultations; format wars such as VHS versus Betamax, Blu-Ray versus HD-DVD, and 8 mm versus VHS-C can be extremely costly and damaging. Competing companies might lower their prices below an economically viable level to encourage consumers to adopt their standard. But sooner or later there have to be winners and losers, and sometimes it is the technically superior technology that loses. Fortunately, with wireless communications standards there is some government control because of spectrum regulation: Thanks to the tireless efforts of engineers on committees, there is a high degree of standardisation. However, this is never a straightforward process as licensing fees for crucial patented technologies might require extensive negotiation.

Whilst consumer electronics is a very competitive business, it can also be extremely profitable. In 2014, the top ten consumer electronics companies were, in alphabetical order; Apple, Canon, Hitachi, HP, Nokia, Panasonic, Philips, Samsung, Sony and Western Digital. The top five smartphone brands were Samsung, Apple, Xiaomi, Lenovo and LG. Apple, the most valuable brand in the world at the time of writing, had sales of US$ 183 Bn and a net income of almost US$ 40 Bn in 2014 [5]. There are other familiar brands, such as Motorola, Dell, LG, Pioneer, Sanyo, Sharp and Toshiba. Some companies are well-known as IC suppliers, including Intel, AMD, Qualcomm, TI, MediaTek, Hynix, Micron, ST, Broadcom, Renesas, Infineon, NXP, Nvidia, Fujitsu and Qorvo. Some of these companies do not actually fabricate any ICs, but there are a number of other companies, including TSMC, UMC, GlobalFoundries and SMIC, which specialise only in fabricating ICs for customers. There are many big companies (e.g., Foxconn, Pegatron and Flextronics) that focus on carrying out electronics manufacturing for the aforementioned global-brand companies. There are many other companies, including Inventec, Quanta Computer, Compal, Elite, WKK and Wistron, that design and manufacture products which are then badged by other companies with their own branding for retail. There is an increasing trend for companies based in China and South East Asia to develop their own brand in the retail market place, with companies including Huawei, ZTE, Lenovo, HTC, Haier, Changhong, TCL, Asus, Panda Electronics, Ningbo Bird, Skyworth, Amoi, SVA, Xiaomi and BenQ amongst those leading the way. What all this highlights is the fact that the global electronics industry is vast (it is estimated to be worth US$ 600 Bn per annum) and relies on a number of different business models, forming a sophisticated network:

Vertically integrated company: The model where a large company makes everything from the semiconductor devices to the branded consumer products.

Contract manufacturer (EMS): A company that concentrates solely on providing an electronics manufacturing service for its customers.

OEM: There are three meanings in use, but in the context used here an original equipment manufacturer produces its own branded products which include components and sub-assemblies from other companies.

ODM: An original design manufacturer designs and manufactures a product that is then sold under another brand name; this is often referred to as 'badge engineering'.

Chip manufacturer: A company that specialises in designing sophisticated high-value ICs and manufactures them for sale as components.

Foundry: A company that carries out semiconductor manufacturing for customers but does not design products of its own.

Fabless design house: A company that specialises in designing ICs and supplies them as components, but subcontracts out the chip manufacture to a foundry.

Chipless design house: A company that has design techniques that are sufficiently novel for other companies to want to pay to use them in their own chip designs. The UK-based company ARM^{TM} is a good example of this; ARM's processors are used in a vast number of consumer products but they supply the design IP, which is then integrated within a larger system-on-chip by the customer.

Material, component and sub-assembly suppliers: Companies that specialise in supplying into the businesses that manufacture electronic products. Sometimes these are household names due to the breadth of their activities (e.g., BASF, 3M, Corning and DuPont), but usually these suppliers operate entirely on a business-to-business (b2b) basis and are not well known outside the industry.

Distributor: With a vast range of products available, distributors provide a key intermediary service between suppliers and customers, giving a broad product portfolio, independent advice and technical support.

Value-added reseller: In the context used here, this model refers to a systems integrator that takes wireless equipment and combines it with other hardware and software to provide a high-level product solution. With mobile and wireless systems having a profound impact in new areas such as retail, commerce and the creative industries, this is a valuable service for companies that do not wish to become involved in technical details but instead want a turn-key solution for their business needs.

These categories are well known for the general electronics industry, whilst in the specific case of the telecommunications industry there are additional company categories relating to the manufacture, installation, operation and marketing of the internet and telephone networks. The internet itself, comprised of a huge conglomeration of networks, is operated by a variety of telecommunications companies at the core and access network levels. Mobile network operators (or carriers) own specific cellular network infrastructure and offer services to customers. There are also virtual network operators that buy air time in bulk and sell packages with a distinct retail brand. At another level there are internet-focused companies such as internet service providers and major web site companies, such as social media sites, that operate numerous large data centres in order to meet the demands of more than a billion customers in some cases.

1.4 Forms of Wireless Communication

A wide variety of forms of wireless communication has evolved since the first demonstration of wireless telegraphy by Guglielmo Marconi [6]. In terms of purpose and physical location of the terminals, wireless links can be classified as point-to-point, point-to-multipoint, broadcasting, or mesh networks. In the following sections, some of the key distinct types of communications systems are described, chosen as examples of the main approaches for wireless communication.

1.4.1 Terrestrial Television

All around the world, the analogue TV system has had to make way for digital systems. The analogue TV standards, NTSC (National Television System Committee), PAL (Phase Alternating Line) and SECAM (Séquentiel couleur à mémoire) are rapidly being confined to history. Digital TV makes use of powerful video compression as well as sophisticated modulation and coding schemes to reduce the bandwidth of the transmitted signal. Standard-definition digital TV signals actually occupy a similar bandwidth to the previous analogue systems (e.g., around 8 MHz bandwidth for PAL). A fair question is, then: why go to all this trouble, when the analogue system had been working very well for around 70 years?

The need to move to digital signals to transmit more TV channels was dictated by analogue signals being sensitive to propagation effects and interference. With the power density of the radiated wave falling off as $1/r^2$, a simple calculation shows that in theory a single transmitter antenna could comfortably cover the whole of the UK or a typical State of the US. However, the curvature of the earth means that there is a *radio horizon* which is determined by the height of the transmitter (TX) and receiver (RX) antennas and the radius of the earth. This is given by $\sqrt{17h}$ km [7], where h is the effective antenna height in metres. This is why the transmitters are mounted on tall towers and situated on hills where possible. A 300-metre transmitter height gives around 70 km radio horizon, so the Crystal Palace transmitter, for example, serves all of London. The Crystal Palace transmitter's effective isotropic radiated power (EIRP) is of the order of one megaWatt. The actual power output is less than this, but it is the effective value, which takes into account the transmitter antenna's gain, that determines the signal strength at the TV receiver antenna.

Figure 1.8 shows a stylised map of the main transmitter sites in the UK for UHF TV. There are many other, smaller transmitters or repeaters required to fill in gaps in the coverage that are caused by blocking of hills and buildings. For example, in the London area alone there are about 50 low-power transmitters in addition to the main Crystal Palace one. With analogue signals, receivers are very sensitive to interference; 'ghosting' is a clear visual indication in analogue TV of how a receiver is affected when it receives two components of the same signal from different paths. The ghosting is caused by the small but noticeable difference in the time that it takes each signal component to reach the receiver. To avoid this interference, with analogue TV it is necessary to ensure that adjacent transmitters use different radio carrier frequencies. The frequency allocations and the coverage must be carefully planned. In the UHF band, some diffraction occurs around buildings and even around the tops of hills because

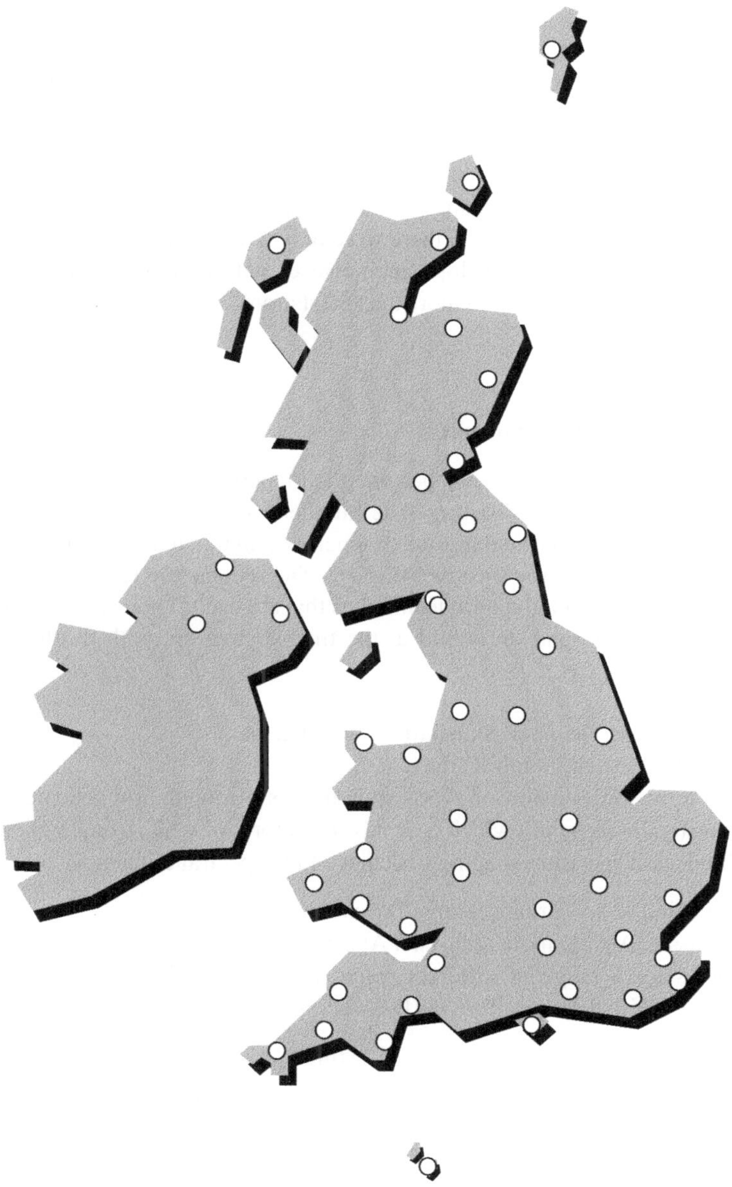

Figure 1.8 The UK's main terrestrial TV transmitter sites.

of the fairly long wavelength of the signal (500 MHz in free space gives 0.6 metre wavelength) but, even so, repeaters are used to fill in poor coverage areas such as valleys where the main transmitter signals cannot reach. The signals from a non-adjacent transmitter with different coverage area can be on the same frequency because of the physical separation; normally, the signals simply do not reach far enough to interfere. However, sometimes the atmospheric conditions can make a signal travel further than normal by refracting (bending) it past the normal horizon. Then, non-adjacent transmitters can interfere; for example, in Kent (in the South East of England) it was sometimes possible to receive TV signals from Belgium in unusual weather conditions.

The need to use many different frequencies in the analogue system to avoid interference, unfortunately, leads to a very inefficient use of the precious UHF spectrum. This is the main reason why digital TV is now dominant. As an indication of how much more efficient the digital TV system is, note that in the UK there were only five terrestrial analogue TV stations, even after many years of technical development: however, within a few years of its introduction to the UK, digital terrestrial TV offered more than 50 stations, even before the analogue system was switched off.

1.4.1.1 Digital Terrestrial TV Formats

On December 11, 2006, the Netherlands became the first country to switch off analogue and move entirely to digital TV broadcasting. The number-one advantage of digital TV is that more channels are possible, but making this happen is a major feat. The first step is to use video data compression standards such as MPEG-2. The second key technology involves the highly advanced coding and modulation techniques that are used. There are four major digital terrestrial TV standards around the world at the time of writing, with the following main deployments:

ATSC: Advanced Television Standards Committee
USA, Canada, Mexico, South Korea
The main terrestrial version of this standard uses a modulation scheme called 8-VSB (vestigial sideband), which is an advanced form of ASK, using eight levels of amplitude and spectrum shaping to achieve high spectral efficiency.

DVB-T: Digital Video Broadcasting (Terrestrial)
Europe, Russia, Africa, India and Asia, Australia
DVB allows for a range of different signal formats for different broadcasting systems: e.g. DVB-S for satellite, DVB-C for cable, DVB-H for handhelds.

ISDB: Integrated Services Digital Broadcasting
Japan, South America, Botswana, Philippines

DTMB: Digital Terrestrial Multimedia Broadcast
China, Hong Kong, Macau

DVB-T, like many other wireless communication standards, uses orthogonal frequency division multiplexing (OFDM). The purpose of OFDM is to combat the effects of multipath propagation. Figure 1.9 illustrates, simplistically, how multipath propagation leads to spreading of the data signal, placing a limit on the maximum data rate because of inter-symbol interference (ISI). In practice, there will rarely be just a few distinct signal components, and it is necessary to analyse the channel statistically and refer to a more continuous effect of *delay spread*. In OFDM, the digital data is split up and sent in parallel using a closely packed block of thousands of RF carriers, each with a much lower individual data rate, as shown in Figure 1.10. This type

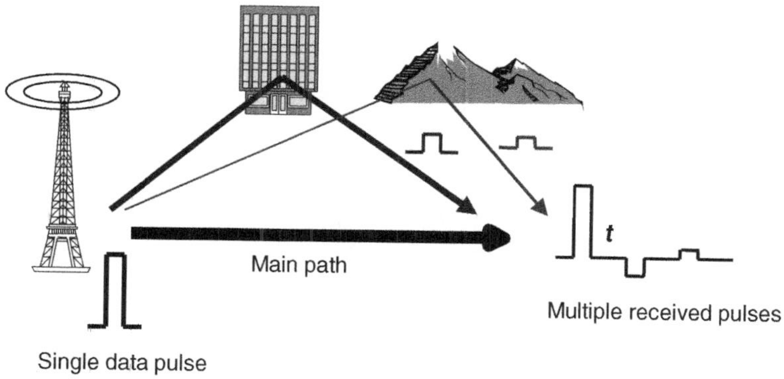

Figure 1.9 Illustration of multipath propagation and its limiting effect on symbol rate.

of signal is more resilient to the multipath fading than a single carrier carrying a high data rate. The multiple carriers are generated entirely within the realm of digital signal processing (DSP) using a fast Fourier transform (FFT), as discussed in Chapter 15. Coded OFDM (COFDM) employs a coding scheme to recover lost data when any carriers are lost due to the multipath propagation. Because these digital signals are much more resilient to interference than the old analogue signal formats, much more efficient use of the spectrum is achievable.

1.4.2 Satellite Communications and Broadcasting

Arthur C. Clarke famously proposed the use of geostationary satellites for communications in 1945 [8, 9]. For a geostationary satellite, by setting the altitude to 35 786 km, the spacecraft naturally orbits the Earth at exactly the same angular velocity as the rotation of the Earth on its axis, appearing therefore to be stationary in the sky. To achieve this, the satellite must be orbiting in a plane perpendicular to the Earth's axis. Hence, in the UK a geostationary Earth orbit (GEO) satellite will be at an elevation of 25.4° (in London) to the South, as shown in Figure 1.11(a). Around the equator, geostationary satellites are directly overhead, but near the

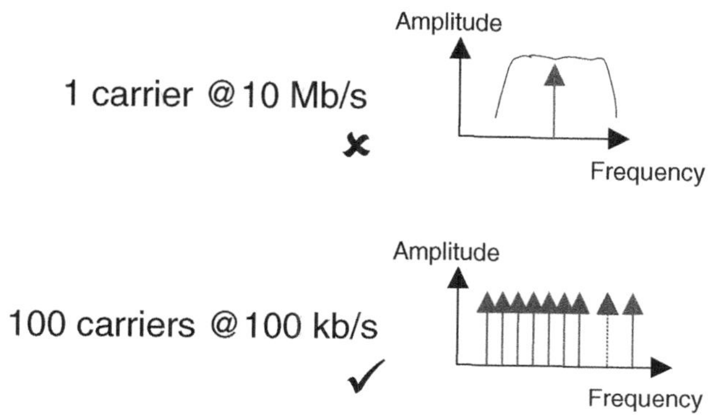

Figure 1.10 The OFDM principle; the data is spread out over many sub-carriers to lower the symbol rate in each.

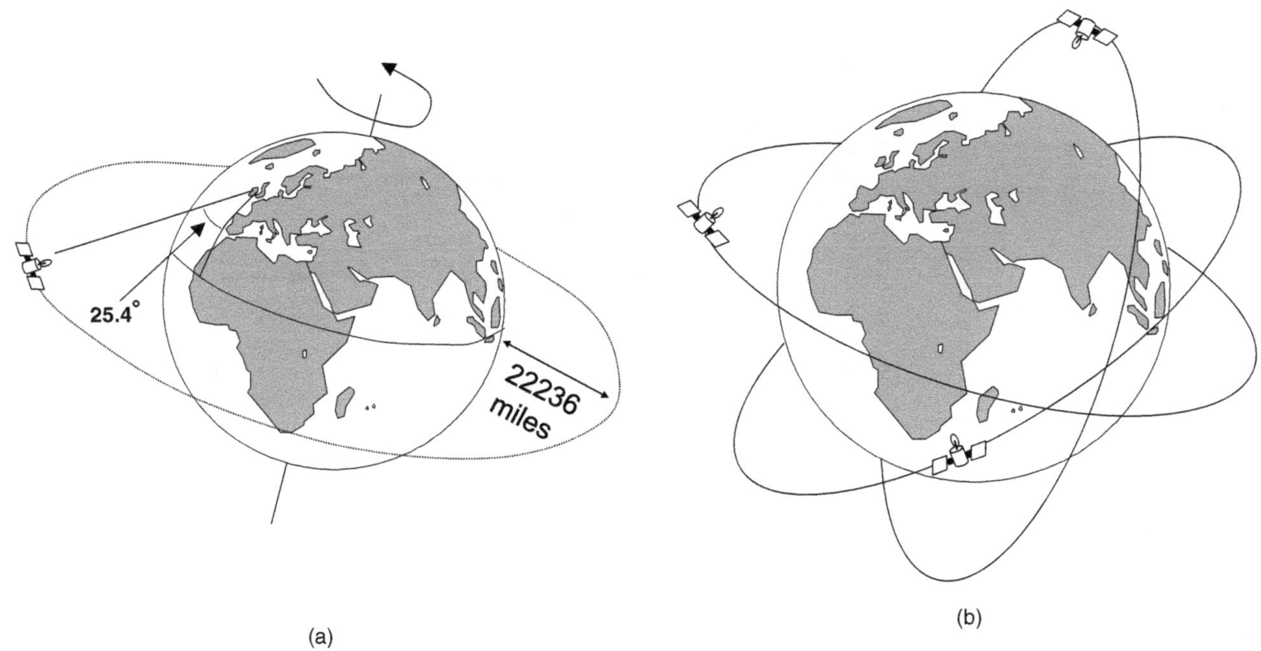

Figure 1.11 (a) TV broadcast satellite in geostationary orbit. The elevation from London is shown. (b) An example LEO satellite constellation.

poles geostationary satellites start to become useless as they are at a very low elevation or below the horizon. A low Earth orbit (LEO), like those shown in Figure 1.11(b), can be useful for these areas and for other applications such as mobile communications as the time delay is reduced, launch costs are reduced, the transmitter power is reduced and the elevation from the horizon is increased. However, as the satellite is moving relative to the Earth and will eventually disappear from view, multiple satellites have to be used to provide a continuous service – the 'constellation' used by communications systems such as Iridium™ and Globalstar™ are examples of this. Global navigation satellite systems such as GPS, Galileo, BeiDou and GLONASS use a constellation like this, and the receiver calculates the time difference between the signals received from different satellites in order to determine its position.

Satellite TV broadcasting uses microwave frequencies, much higher than terrestrial TV, because the antennas must have high gain. There is also plenty of spectrum available, allowing many channels. A high-gain antenna is needed for the transmitter on the spacecraft because only solar power is available, so the transmitter power is severely limited. The narrow beam concentrates the precious transmitter power on the Earth, to the appropriate countries. At the receiver, the signal is extremely weak and a large antenna aperture is necessary. The high gain of the receiver antenna rejects signals from other satellites and background noise from the sky, as illustrated in Figure 1.12. In this way, satellites at different locations in the sky can re-use the same frequencies.

Since, generally, a high-gain antenna must be many wavelengths in size, using microwave frequencies enables a high gain to be achieved with reasonably small dishes (e.g., ca, 60 cm for a domestic satellite TV dish in the UK). This is also critical for the satellite: the dish needs to fit inside the nose of the launch rocket (although, in exceptional cases, folded antennas have been used, deployable once the spacecraft is in position). In Europe, frequencies ranging from 10.7 to 12.75 GHz are used for satellite TV. To receive signals on the ground, a dish with a feed is first required to collect the signal. The dish part is used to reflect the electromagnetic power and focus it onto the feed. The signals are so weak, because of the long transmission distance and low satellite transmitter power, that there must be a direct line-of-sight between the receiver dish and satellite. The feed collects the signal and sends it to a low-noise amplifier (LNA)

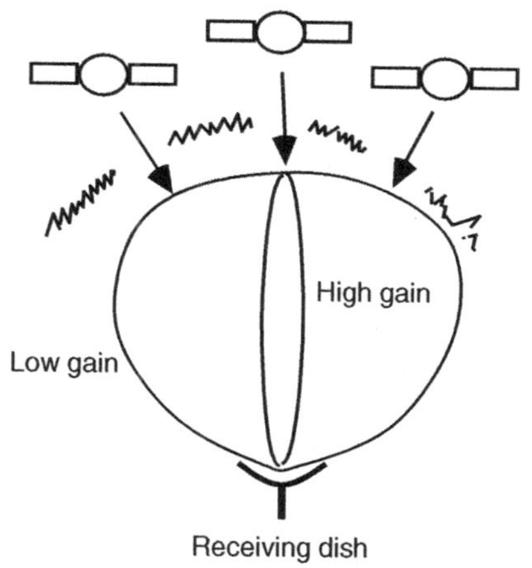

Figure 1.12 Satellite dish beam-width; low- and high-gain antennas compared.

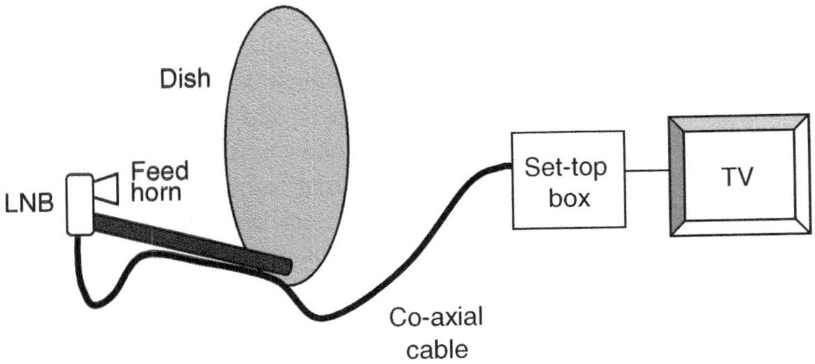

Figure 1.13 Components of a satellite TV installation.

that must be mounted on the dish in order to amplify the weak received signal immediately. After amplification, the signal is frequency-converted because microwave cable is expensive and has a high loss. For example, if the microwave signal were sent directly into the lounge of the house, the 30 feet (approximately) of cable required would be expensive and the loss of the cable would be so high that the signal would disappear into the noise. By first amplifying and then converting the microwave signals down to approximately 1–2 GHz, a much cheaper coaxial cable can be used and the loss is lower. The 1–2 GHz signal is then fed down a cable into the 'set-top box' (STB) where all the complicated signal processing takes place. The unit that amplifies and downconverts the signal is mounted on the dish close to the feed to avoid losses, and is known as the 'low-noise block' (LNB). The components of a typical satellite TV installation are shown in Figure 1.13.

1.4.2.1 Set-Top Box Example

A satellite TV set-top box (STB) serves as a good example of the hardware required to receive and decode digital wireless communications signals. The signal frequencies, modulation schemes and coding vary between terrestrial, cable and satellite systems, and between countries, requiring a sophisticated application-specific system-on-chip (SoC) at the heart of the receiver. The RF engineer's primary contribution is focused on frequency-converting the incoming signal down to an intermediate frequency or baseband signal that this SoC can process. Figure 1.14 shows a simplified block diagram of a typical STB. The heart of the design is a powerful system-on-chip, which is sufficiently power-hungry to require a substantial heatsink. The first functional block in the STB chain is an RF front end that downconverts the incoming signal to analogue baseband (IQ) form. This is followed by a sophisticated demodulator chip that recovers the data stream and performs error correction. This feeds the MPEG decoder, which incorporates MPEG-2 video and audio decoders and a video graphics processor. This chip also includes circuitry to handle the user interface, a wireless network connection, and a wide range of interfaces, such as USB.

1.4.3 Terrestrial Line-of-Sight Links

Often, a private point-to-point system is required and satellite systems have some drawbacks such as the size of the dish required, cost and propagation delay. Line-of-sight (LOS)

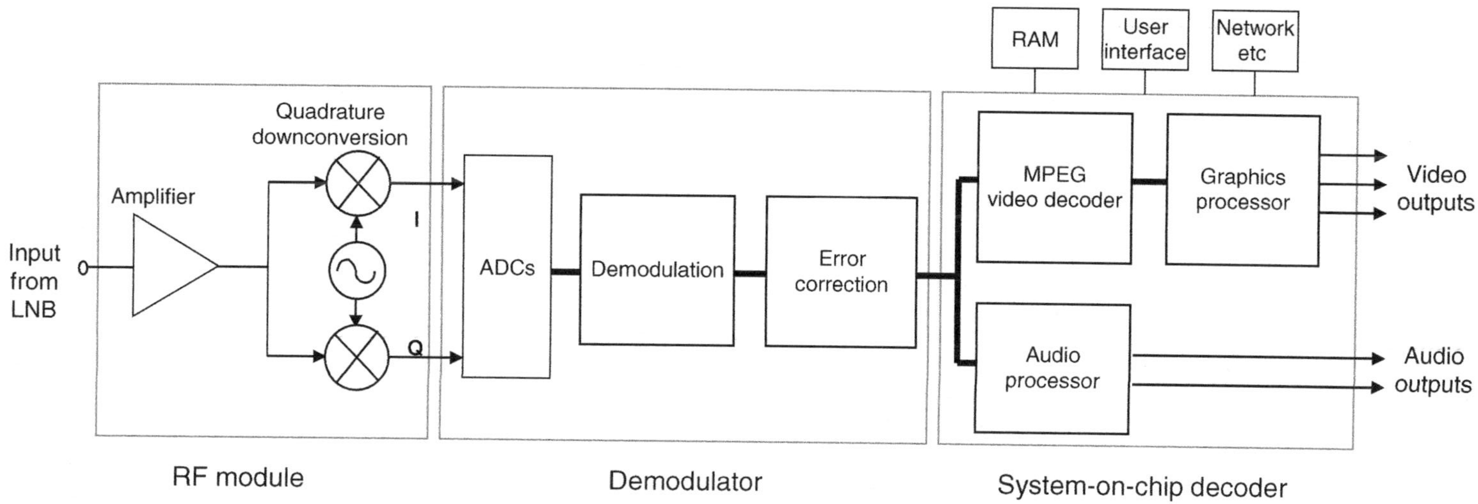

Figure 1.14 Satellite HD TV STB block diagram.

microwave and millimetre-wave fixed links are increasing in popularity as a low-cost alternative to leasing a dedicated wired line for high data rate connections between buildings. The hardware cost has reduced significantly and large bandwidth is available in Ku/K/Ka-band (13, 18, 23, 26 and 38 GHz are common), V-band (the 57–66 GHz band is unlicensed) and E-band (in bands such as 71–76 GHz, 81–86 GHz and 92–95 GHz). Dish sizes are small, especially at V-band and above, and gigabit-per-second equipment can be mounted unobtrusively on rooftops and towers – indeed, the cost of renting suitable space on antenna masts depends on the size. Rain fading can be a major factor constraining the link budget (this is discussed in Chapter 14).

1.4.4 Cellular Mobile Communications

The radio spectrum is an extremely limited resource and, because of the propagation characteristics, only a small part of the spectrum is ideally suited for mobile communications. Low frequencies (up to a few GHz) will readily diffract around buildings and other obstacles and can be received indoors. Higher frequencies, especially above 10 GHz, will travel only along a LOS path. Furthermore, higher frequencies can suffer very high atmospheric loss, especially when it is raining, limiting their use to short ranges. This leaves only frequencies up to a few GHz being suitable for large cells in mobile communications, although higher frequencies can be advantageous when the objective is to have very small, dense cells in urban areas for greater capacity. The bands up to a few GHz are also needed for other applications including TV, radio, GPS, Wi-Fi, Bluetooth and NFC, as well as radar systems, radio telescopes, car alarms, police and emergency services radios, and so forth. As a result, only a limited amount of spectrum can be allocated to mobile services, and network operators have often been required to pay licence fees of the order of billions of pounds just for a few tens of MHz of spectrum. The cellular approach to mobile communications, where many basestations are installed, each serving quite a small area (as shown in Figure 1.15) is essential because:

1. The mobile terminal is battery-powered, so its transmitter power is restricted.
2. The large number of users requires exceptionally efficient use of the spectrum.

In the cellular system, the radio spectrum available for mobile communications can be used over and over again in different cells, as long as they are not adjacent; this is referred to as *frequency re-use*. The radio link is kept to a fairly short range, reducing the handset transmitter power required, and a complex network of interlinked basestations is used to cover a large area. The basestations are linked with optical fibres under the ground, where possible, or with very high-frequency LOS radio links (referred to as wireless backhaul, using frequencies such as 23, 38 and 71 to 76 GHz). The area spectral efficiency (units b/s/Hz/m^2) can be increased by having more, smaller cells: for example, in urban areas where the density of users is higher and propagation more challenging, or when much higher data rates are required, as in 5G.

The first mobile systems were analogue, and the pioneers were in Japan (NTT), Scandinavia (which is why Nokia and Ericsson became leaders in the industry) and the US (Bell Labs and Motorola). In the UK, Racal-Vodafone and Cellnet developed the TACS system which was introduced in 1984. These analogue systems are referred to as the first-generation systems. In mobile systems, the propagation characteristics are challenging; there is rarely a LOS

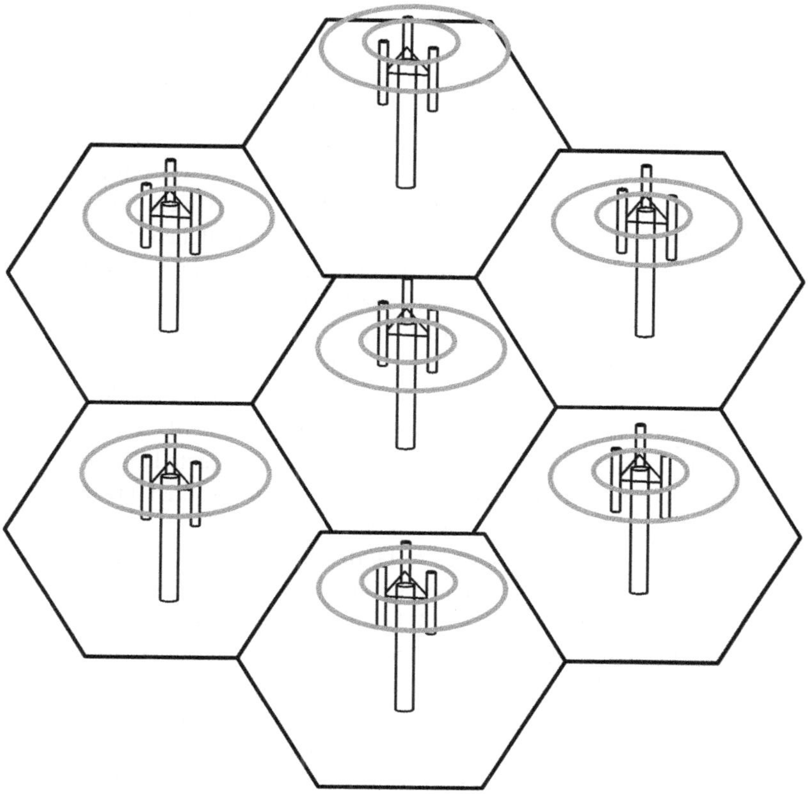

Figure 1.15 A cellular system.

to a basestation and the handset is often indoors, with many obstacles between it and the basestation. With analogue systems it is difficult to counteract all the multipath propagation effects and the voice quality was not good. The early systems used FDMA (frequency division multiple access, where users transmit on different frequencies) which tended to lead to a bulky mobile phone.

Digital systems were introduced in the 1990s. The GSM system introduced in Europe was a result of Groupe Spéciale Mobile, a pan-European standards initiative. This became so successful it was adopted in many countries worldwide, becoming known as the Global System for Mobile communications, which covered more than 80% of subscribers across the world at its peak, before the switch-off started in some countries, from 2015/2016 onwards. These first digital systems, which are referred to as second-generation (2G) systems, gave excellent voice quality, while GSM gave roaming capability in many countries. Different systems and countries use different frequencies, however, which is why there are dual-band/tri-band/etc. phones. The GSM system uses time division multiple access (TDMA) in which different users within one cell can use the same frequency but transmit in different time slots. This enabled the phones to be made smaller and lighter. In the US, the huge size of the country was a barrier to realising a country-wide cellular mobile communications network. The US had different companies and standards operating in different cities and the IS-95 system using CDMA became the most common 2G digital system.

For internet use and e-mailing, the 2G systems were unattractive; the data rate was too low and the system acted like a dial-up service. For this reason, enhanced systems were developed which gave 'always on' connection and higher data rate. In the UK, these included GPRS (general packet radio system) and GSM-EDGE (enhanced data rate system), which are often referred to 2.5G systems. The true third-generation (3G) systems were conceived from the outset as being capable of high data rates and being packet-switched, rather than circuit-switched, for high-speed internet access. The conventional TDMA and FDMA systems were unable to provide this for millions of users. Wideband CDMA (code division multiple access) was the solution. In this approach, many users transmit at the same time and in the same frequency range. However, their signals are all spread out thinly over that frequency range, and the spreading is performed by multiplication with a unique pseudorandom code for each user. In a receiver, only the signal component with the correct code will be correlated with the code in the receiver. By this mechanism, the other signals are all rejected as noise-like. CDMA has proved able to accommodate many more users in a given cell than TDMA and FDMA, although huge technical hurdles had to be overcome.

LTE (long-term evolution) systems started to be deployed in 2009, marketed as 4G or 4G LTE, although it was, strictly speaking, one of a number of pre-4G candidate standards. LTE requires multiple-input-multiple-output (MIMO) antennas in the handset. LTE-Advanced is 4G as defined by the 3GPP (3rd Generation Partnership Project) and was optimised as a purely IP packet-based system for the very high data rates (ca. 300 Mb s^{-1} downlink, up to 8×8 MIMO, OFDM signals) required for routine web browsing on smartphones and tablets. 5G systems aim to significantly reduce latency, so that new applications such as telepresence, augmented reality and cyber-physical systems are feasible, whilst also offering more than Gb per second data rates for very high-speed connections. Another key aim of 5G systems is to achieve new levels of integration of diverse wireless systems, so that the full range of applications – from smart cities and smart grid systems down to small internet of things (IoT) sensor nodes –could be interconnected seamlessly.

Figure 1.16 shows the simplified block diagram of a basic mobile phone. The components in the top half will be immediately familiar; the communications hardware is in the lower half. It might seem illogical that the signal goes from analogue to digital, and then back to analogue, before being modulated onto the RF carrier. The reason is that modulating a pure digital signal onto the RF carrier would lead to a spectrum with many sidelobes. In order to make more efficient use of spectrum the signal waveform envelope has to be carefully shaped, and so a bandlimited analogue version of the digital data is fed into the modulator. The signal is referred to as analogue baseband, usually in Cartesian I-Q form (this is discussed further in Chapter 2).

1.4.5 Wireless Indoor Networks

Indoor wireless networks for internet connectivity are typically based on the IEEE 802.11 Wi-Fi standards for the 2.4 GHz and 5 GHz bands, whilst IEEE802.11ad (60 GHz) is gaining in popularity, with gigabit data rates. It might be imagined that this is the ideal communications medium for all networking indoors, but there are other options. IEEE802.11 transceivers are

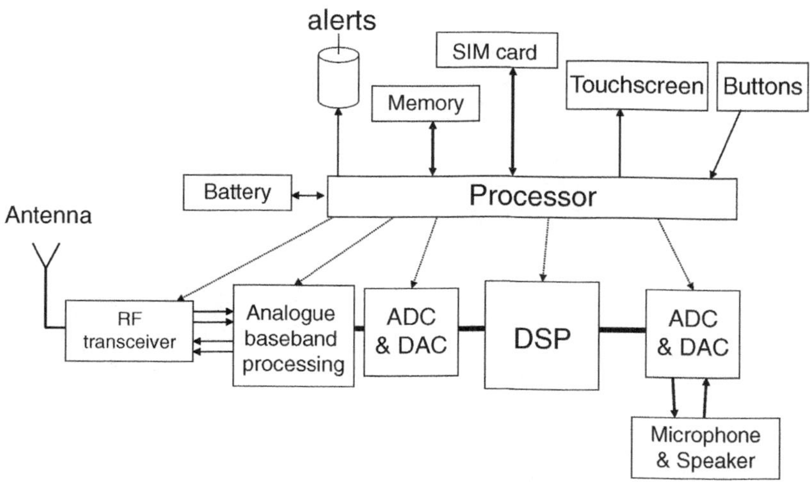

Figure 1.16 A simplified block diagram of a mobile phone.

quite sophisticated and transmit rather high RF power because they are designed to operate at ~10 Mb s^{-1} and above. Devices such as smoke alarms and dimmer switches simply do not require a 10 Mb s^{-1} data rate, and using 802.11 transceivers in them yields a solution that is too expensive and uses more DC power than is necessary. Since many of these devices are battery-powered, this is a major drawback. Hence, there are other radio standards, as well as Li-Fi visible light communication with LED lighting. In many countries, frequencies in the UHF band are allocated for short-range telemetry links; for example, the Rako® wireless control system (as illustrated in Figure 1.17) is available for the 433 MHz band that is allocated in the UK.

The ZigBee® specification is strongly supported by an alliance of companies and has been widely adopted. The ZigBee specification adds additional network, security and application layers on top of the IEEE 802.15.4 radio standard. IEEE 802.15.4 allows operation in unlicensed bands worldwide at 2.4 GHz, at 915 MHz in America and 868 MHz in Europe. It is a low data rate, low-power standard giving a data rate of up to 250 kb s^{-1} in the 2.4 GHz band. A significant advantage of ZigBee is that networks can have very large numbers of nodes; it uses a mesh network topology and nodes may remain dormant for long times using very low DC power. Since a completely wireless, battery-powered product can be installed extremely easily and at low or zero cost, this type of networking is highly attractive as most consumer products retail through the DIY market for retro-fitting (e.g., Wi-Fi itself, which would not have succeeded so well if home users had to pay for a professional installation service). Some of the products that use ZigBee or similar systems include:

- Security systems; motion detectors, pressure pads, cameras, etc.
- HVAC (Heating, Ventilation, Air Conditioning) controls.
- Lighting controls and room-occupancy sensors.
- Curtain and blind motorised controls.
- Utility meter readers.
- Garden watering.

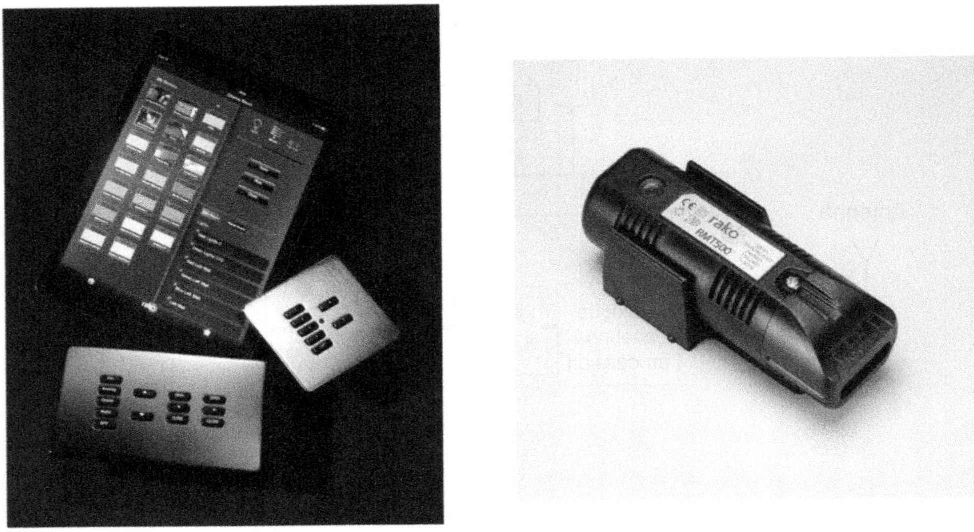

Figure 1.17 Rako® wireless control products using the 433 MHz band. Left: Controls and app. Right: A motorised curtain/blind unit. Images © 2016 Rako Controls Ltd., used with permission.

- Pet feeding systems.
- Entertainment systems.
- Internet-enabled products; e.g., a smart refrigerator.
- Health-monitoring devices.

1.4.6 Hybrid Fibre-Radio Systems

As wireless systems grow in capacity due to increased data rates and higher density of users, it can become very expensive to realise a network that interconnects cells using the raw digital data streams across an optical network. The remote units require Gb per second optical-electrical conversion, followed by DACs, ADCs and up/down frequency translation to convert the digital data to/from modulated microwave or millimetre-wave signals. The result can be expensive radio equipment with significant complexity and DC power. As cell sizes are reduced to increase capacity, this becomes less and less acceptable. A hybrid fibre-radio (or radio-over-fibre) system addresses this problem by transmitting radio signals over the fibre rather than digital data, either at an intermediate frequency or directly at the transmission frequency. This relies on the fibre being a very low-loss conduit for analogue signals modulated onto the optical carrier(s). The remote station can be simplified considerably; for example, the main functional component of the transmitter can simply be an optical-electrical converter (e.g., a broadband photodetector) to extract the desired RF carrier from the optical signal and connect it to the antenna. This type of antenna remoting is already a feature of some sophisticated electronically steered antennas for radar, where there is an advantage in minimising the mass of the antenna array and reducing the spacing between antenna elements by placing the beamforming systems remotely. It has been proposed that this type of remote beamforming be used in basestations for 5G+ millimetre-wave communications. It should be remembered, however, that while the fibre itself is extremely low-loss, the electrical-optical and optical-electrical converters have significant conversion loss. Fibre-radio systems have been very successfully applied already to a wide range of wireless systems, such as in-building distributed antenna systems. The fibre-radio approach is especially relevant where the aim is to improve the indoor reception of cellular, GPS and other signals from outdoors; the fibre can be used as a passive low-loss conduit to indoor repeaters.

1.4.7 RFID, NFC and ICT

Radiofrequency identification (RFID), near-field communications (NFC) and inductively coupled telemetry (ICT) are a related set of contactless systems where the primary aim is to achieve short-range, low-data rate wireless communication for the transfer of basic information. In RFID, a reader communicates with a very simple transponder that can be embedded into a wide variety of objects and relays a unique ID code. NFC can be similar, with a tag inside a poster giving a web link to a phone, for example, but also allows two-way communication, so that two phones could perhaps exchange contact details. Inductively coupled telemetry refers to the situation where the 'tag' includes one or more sensors, such as temperature or pressure, and measured parameters can be relayed back to the reader. These technologies, covered by a wide range of standards, are rapidly being applied to more and more applications and are an important facet of the internet of things (IoT). As a result of the development of miniature integrated circuit transponders, everything can become 'smart'. For example, in the home the washing machine could read the wash labels on clothes, while the refrigerator could monitor food stocks and 'use by' dates and place orders on the internet. In this way, a huge variety of businesses is being transformed, from retail management through to medicine.

Radiofrequency identification is by no means a new idea – transponders have been used on aircraft since World War II as a means of identification for radar systems. However, the development of miniature integrated circuit transponders has advanced so much that it is now technically feasible (if not always economically so) to replace the ubiquitous bar code with an RFID transponder – making it possible to scan a supermarket trolley full of goods automatically as one passes the checkout [10]. The basic idea of RFID (as illustrated in Figure 1.18) is that a tag holds a unique code, and a reader device can determine the code by electromagnetic coupling or transmission of a radio signal to and from the tag. This technology is rapidly combining with smart card and smart label technology to provide a whole new range of possibilities for automation in manufacturing, retailing, transport and security. These new applications are having a profound effect on our everyday lives. Some of the key applications of this technology are listed in Table 1.2. The organisation GS1 [11] sets the standards for identification codes, including optical barcodes, so that they can be understood globally. A number of important RFID standards have been agreed, such as the Electronic Product Code (EPC) Gen 2 standard in the UHF band (860–960 MHz), which is in widespread use.

The simplest and most widespread form of tagging is electronic article surveillance (EAS). In this application, basic security tags are used to set off an alarm when theft is attempted. The tag can be a sticker, embedded metal strip, or a more bulky removable plastic-encapsulated device. These might be thought of as 'one bit' tags. The introduction of 'intelligent tags' which can store information about an item would dramatically increase the number of applications possible, however.

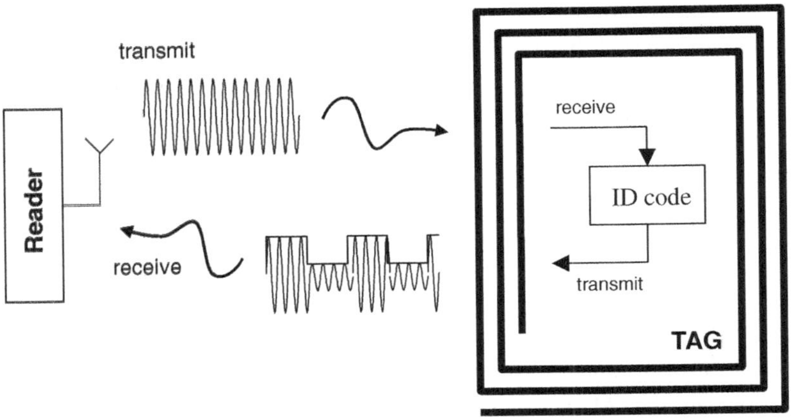

Figure 1.18 A simplified representation of RFID using backscatter modulation.

1.4.7.1 Active and Passive Transponder Tags

There are many types of tag, though not everyone agrees on how they should be classified. The most important types encountered are:

1. Radio tags with a battery or other on-board power source.
2. Radio tags that employ transistor circuits but have no battery or on-board power source, and draw their power from the electromagnetic field of the reader.
3. Completely passive tags; the tag contains no active circuits, and the identity is defined by some geometrical or material structure.

The most widely accepted terminology in RFID is to refer to type 1 as active tags, and type 2 as passive tags since they have no on-board power. Various forms of type 3 tags have been reported, often referred to as chipless RFID [12]. Active tags have the longest read range, but require batteries, solar power or some form of local energy harvesting. The additional DC

Table 1.2 Some applications of radiofrequency identification.

Entry cards	Travel passes
Ski passes	Document identification
Electronic keys	Animal identification
Sports timing	Automotive tolling
Contactless payment cards	Electronic article surveillance (EAS)
Vehicle security	Postal systems – smart stamps
Baggage handling	Laundry
Libraries	Stock control
Offender tagging	Smart coins and tokens
Intelligent appliances	Factory automation
Freight control	Medical records
Post-natal ward security	Asset tagging
Wildlife conservation	Golf ball identification

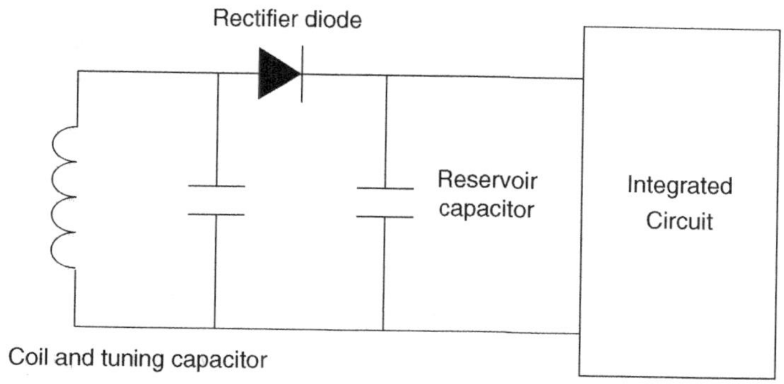

Figure 1.19 A passive RFID tag block diagram.

power makes it possible to transmit much further – examples include the keyfobs of many car alarms and other security and access control systems. Battery life can be as long as 10 years but there is obviously a range/battery life trade-off.

When a passive tag is in the field of the reader antenna, the tag draws power from the field and powers up the chip, as shown in Figure 1.19. If the tag is in the near field of the reader antenna (e.g., inductive coupling in the 125 kHz band) a process of load modulation is generally used for data transfer: the reader can sense the changing load on its antenna. When the tag is further away than approximately one wavelength, it is in the far-field and the data transfer is achieved by means of backscatter modulation, as illustrated in Figure 1.18. With suitable protocols, the data can be transmitted back in such a way as to avoid collision between multiple tags within the reading zone. Since the integrated circuit in a passive RFID tag is powered by the RF field itself, this inevitably limits the read range to 1 m or so in most cases. The biggest application of passive RFID tags is in 'smart cards', and as a result the term 'vicinity card' or 'contactless ID' is often used. This represents an important convergence of technologies and terminologies between tagging and smart cards.

A number of truly passive tags have been developed including electronically readable inks, RF barcodes and SAW devices. The RF barcode, for example, relies on a number of RF resonator strips that have separate resonant frequencies [13]. The SAW devices rely on the much lower speed of propagation of acoustic waves to create a simple geometrically defined transponder that can send back a series of pulses when interrogated by the reader. The pulses are entirely defined by the layout of electrodes on the SAW device; no transistors are required at all.

1.4.7.2 Tags with Position Information

The usefulness of the tag is dramatically increased when the tag is integrated with a satellite positioning receiver, or it can use local Bluetooth/Wi-Fi/cellular signals to deduce its position. The tag is then able to determine its own position and transmit it back to a central location. If, in addition, the tag can connect wirelessly to the internet then there is almost no restriction on the tag location. This combined technology enables freight, vehicles and personnel to be tracked almost without limit. One of the fantasy proposed applications is that if every document in an office had a suitable tag, you would be able to find every document easily with

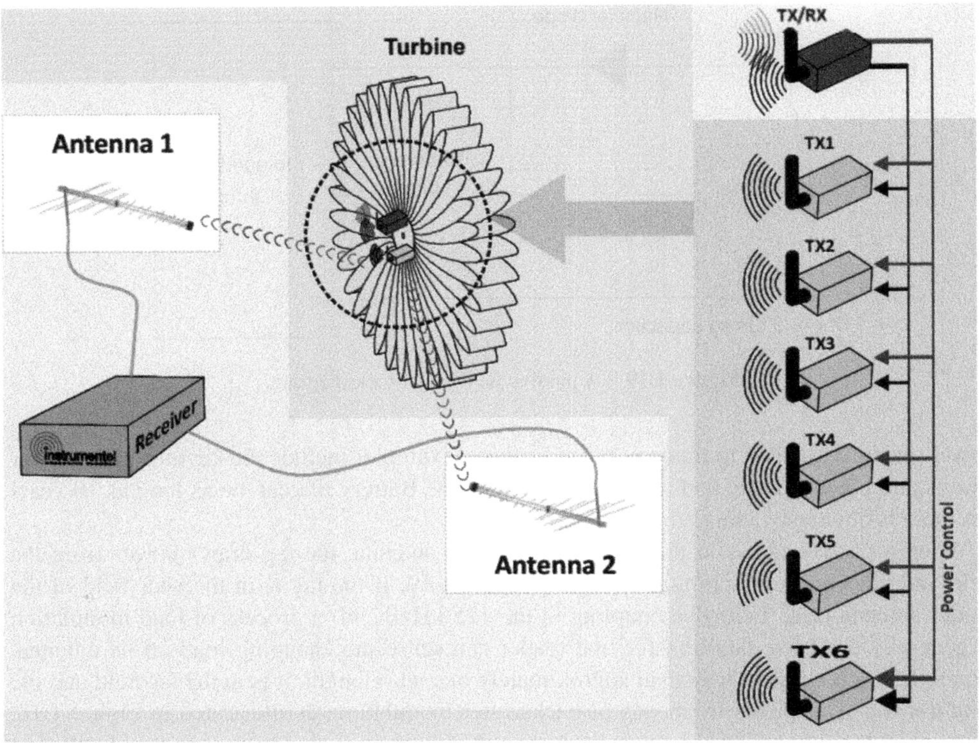

Figure 1.20 Instrumentel™ wireless telemetry system for gas turbine diagnostics. Illustration © Instrumentel Ltd. Used with permission; all rights reserved.

a reader, searching on the reader for the title, keywords, date, and so on, and the document then identifying itself and its location.

1.4.7.3 Inductively Coupled Telemetry

With inductively coupled telemetry, the key additional capability of the tag is the inclusion of sensors, so that parameters such as temperature, pressure or strain can be relayed wirelessly. A battery-less transponder can be made very small and robust, which means it can be embedded in a harsh environment and powered up and interrogated wirelessly through inductive coupling. Instrumentel has been a pioneer of this field [14], and has successfully demonstrated sensors embedded into the pistons and valves of a racing car engine. Figure 1.20 shows a diagram of Instrumentel's gas turbine wireless telemetry system, with sensors embedded in the blades for health monitoring and diagnostics during operation at 100 000 rpm.

1.4.8 Wireless Sensor Networks

A wireless sensor network (WSN) comprises an array of sensing nodes, each of which measures a range of parameter(s) – such as temperature, pressure, sound, vibration, chemical composition, or movement – at its individual location. The node then transmits the data back wirelessly to a central processor at regular intervals. The central processor then collates all the data from the network of nodes and builds up a full picture of the area being monitored. WSNs are a particularly powerful technology when the individual nodes can determine their own position and can form the wireless network autonomously, with no expert set-up. WSNs have many applications, including security and defence, process engineering, environmental monitoring, seismic detection, space exploration, inventory tracking, medicine and smart buildings. The sensor nodes range in size from the smart dust concept to substantial pieces of hardware. The fundamental advantage is the ease of deployment, with particularly important advantages in harsh environments. In process engineering, the low cost of installation means that nodes can be liberally spread around the chemical plant to measure temperature, pressure and vibration. The WSN can then monitor the health of the plant much more effectively than any routine inspection programme, leading to increased safety. In smart buildings, WSNs can monitor the movement of people and reduce energy usage by optimising the HVAC (heating, ventilation and air conditioning) control. Public spaces can be embedded with nodes that sense weapons and explosives, for combating crime and terrorism unobtrusively. In environmental monitoring, WSNs can be rapidly deployed in remote or harsh environments – such as volcanoes, forest fires and inaccessible areas – and used to monitor the situation remotely. In the home, a WSN could monitor the movement and well-being of elderly or other occupants that might occasionally need assistance.

1.4.9 Wearables and Body Area Networks

Wearable electronics has been studied by many research teams and design centres. Wearable electronic products can vary from a smart bracelet to a full body suit. Mobile phones, MP3 players and cameras have been integrated into jackets and Bluetooth used to create a personal area network. There is some fascinating research on the technology of embedding electronics into clothing, including conducting threads, body area communications, and printed or plastic electronics. In the home environment, an important application of wearables is in the unobtrusive monitoring of the wearer's well-being: parameters such as blood pressure, heart rate and temperature can all be measured and the information relayed back wirelessly to the Cloud. A further step is to incorporate therapeutic functions into the clothing, such as massage and vibration functions to relieve pain.

1.4.10 Machine-to-Machine and Robotic Communications

This is a much less-developed application area for microwave and millimetre-wave systems, and yet one that offers many new opportunities. Machine-to-machine (m2m) communications includes both wired and wireless connectivity. One of the key aims is to have equipment, machines of all types, robots and autonomous systems connected to the Cloud. This represents a huge step-change in the number of internet-connected devices, leading the industry to move to IPv6 addressing as the current IPv4 reaches maximum capacity. The range of sectors that will benefit from this machine connectivity is so extensive that it is not possible to do the subject justice in this introduction. One exemplar is the factory of the future, sometimes referred to as Industry 4.0, in which manufacturing systems, RFID-tagged parts and sub-assemblies and robotic assistants all work seamlessly together with human operators. Transport will surely be transformed by connected vehicles, communicating with each other and with smart infrastructure in cities to dramatically reduce congestion, leading to a better quality of life and reduced environmental impact. In healthcare, there is a growing revolution in rehabilitation and surgical robotics, and already studies on networked medical devices and surgical tools have been reported. In security and defence, intelligence is always key to national security, and robotic systems have a tremendous capability to gather data while keeping people out of harm's way. There are, of course, many ethical issues to be debated in that space: Is a robotic mosquito that can sample DNA and administer poison a good idea? And what about a network of autonomous drones with weapons capability? Science fiction has provided many stories about how such technologies can go wrong, but on the other hand we have a very real need to defend ourselves and not put lives at risk.

Connecting machines and robots wirelessly to the Cloud has tremendous potential for providing new capability. For example, currently, a mobile robot needs to carry around its own processing power, and with batteries the amount of computing power that the robot can have is limited. If, instead, the robot is connected to the Cloud wirelessly, then it can use remote processing power to greatly increase its apparent intelligence and give it access to vast databases of information. Clearly, very high data rate communications with low latency are required to make this vision of Cloud Robotics [15, 16] become a reality. From the wireless communications perspective, some of the other challenges for engineers include developing improved techniques for wireless sensor networks, precision localisation techniques, system-in-package technology, microelectromechanical systems (MEMS), high-frequency sensing and imaging, wireless communications networks, MIMO systems, wireless power transfer and energy-harvesting systems. This application sector is therefore predicted to be one that offers RF engineers excellent new career opportunities.

1.5 Conclusion

This introduction has sought to provide a background to microwave and millimetre-wave wireless communications systems by describing some of the main applications and giving some insight into the nature of the industry. Actually, this has included only a very small subset of modern wireless communications systems, focusing on the best-known examples. Today, there is a vast range of other radio communications in use, often for more specialist applications. One of the most recent and impressive demonstrations of the importance of microwave communications was the successful transmission of pictures of Pluto by the New Horizons spacecraft, which used a modest 20 W X-band transmitter to relay data three billion miles back to Earth at a data rate of approximately 1 kb s^{-1}. The best way to fully appreciate the huge variety of applications of the radio spectrum is to study the frequency allocations via the sources already given. Even when focusing only on 5G+ mobile networks, it can be seen that engineers have the ambition to truly change the world by enabling very significant advances in application areas such as mobile health, intelligent transport, future cities and smart grid systems. The prospects for improving lives and creating a more sustainable planet are tremendously exciting, and this is just one of the industry sectors underpinning the IEEE tagline: *Advancing Technology for Humanity*.

References

1. Moore, G.E. (1965) Cramming more components onto integrated circuits. *Electronics*, **38** (8), 114–117.
2. Stackhouse, B., Cherkauer, B., Gowan, M., Gronowski, P., and Lyles, C. (2008) A 65nm 2-Billion-Transistor Quad-Core Itanium® Processor. IEEE International Solid-State Circuits Conference, pp. 92–98. doi: 10.1109/ISSCC.2008.4523072.
3. Burns, R.W. (1998) *Television: An International History of the Formative Years*. Institution of Engineering and Technology. ISBN: 978-0-85296-9-14-4.
4. ITU, Recommendation ITU-R P.676-10 (09/2013). Attenuation by atmospheric gases.
5. Apple Inc., Annual Report, 2014.
6. Belrose, J.S. (1995) Fessenden and Marconi: their differing technologies and transatlantic experiments during the first decade of this century; 100 Years of Radio. Proceedings of the 1995 International Conference, 5–7 September 1995, pp. 32–43. doi: 10.1049/cp:19950787.
7. Hutson, G.H., Shepherd, P., Brice, J. (1990) *Colour Television*. McGraw-Hill Publishing Co., 2nd edition. ISBN-10: 0070841993.
8. Clarke, A.C. (1945) Peacetime Uses for V2. *Wireless World*, February.
9. Clarke, A.C. (1945) Extra-Terrestrial Relays: Can Rocket Stations Give World-wide Radio Coverage? *Wireless World*, October.
10. http://trolleyscan.com/.
11. http://www.gs1.org/.
12. Preradovic, S., Balbin, I., Karmakar, N.C., and Swiegers, G. (2008) A Novel Chipless RFID System Based on Planar Multiresonators for Barcode Replacement. RFID, 2008 IEEE International Conference, 16–17 April, pp. 289–296. doi: 10.1109/RFID.2008.4519383.
13. Jalaly, I. and Robertson, I.D. (2005) RF barcodes using multiple frequency bands. Microwave Symposium Digest, 2005 IEEE MTT-S International, 12–17 June, p. 4. doi: 10.1109/MWSYM.2005.1516542.
14. www.instrumentel.com.
15. Guizzo, E. (2011) Robots with their heads in the clouds. *Spectrum, IEEE*, **48** (3), 16–18. doi: 10.1109/MSPEC.2011.5719709.
16. Kehoe, B., Patil, S., Abbeel, P., and Goldberg, K. (2015) A Survey of Research on Cloud Robotics and Automation. *Automation Science and Engineering, IEEE Transactions*, **12**(2), 398–409. doi: 10.1109/TASE.2014.2376492.

2

Transmitters and Receivers

2.1 Introduction

In a communication system, the transmitter sends information to a receiver some distance away. There are three different communication modes; simplex, half-duplex and full-duplex. In simplex communication, information transfer is entirely in one direction, from transmitter to receiver. Some of the examples given in Chapter 1 are simplex, including TV broadcasting and GPS, but a significant disadvantage of simplex communication for other applications is that there is no way the receiver can even acknowledge successful receipt of the information. In the majority of commercial wireless communications, therefore, half- and full-duplex communications are used. In half-duplex communication, the communication is bi-directional, one direction at a time, while in full-duplex mode the information can be sent in both directions simultaneously. In both half- and full-duplex communication, a transmitter and receiver are required at both ends and are integrated as a 'transceiver'.

In many ways, receiver design is more challenging than transmitter design since the receiver must detect an extremely weak signal in the presence of noise and interference, including other signals that may be much stronger. Receivers can be classified into a number of distinct groups. Some of the earliest receivers used the tuned radiofrequency approach (TRF) where RF amplifiers (tube/valve based) were tuned with an inductor-capacitor resonator directly to the signal frequency. With multiple gain stages it becomes difficult to keep the tuning in track, and fundamentally the selectivity is poor; as the receiver is tuned to a higher frequency the absolute bandwidth increases, meaning that more noise and interference is picked up. A downconversion receiver was proposed by F.M. Colebrook in 1924, called a homodyne receiver (from the Greek, *homo*, meaning the 'same' and *dyne*, meaning 'force' or 'power'). Strictly speaking, homodyne means that the RF is mixed with itself, and the local oscillator (LO) must be derived from the received signal in some way. Various means of extracting an LO from the incoming RF have been devised, including using amplifiers on the verge of instability that lock to the incoming RF and methods to strip the incoming RF signal of its modulation. In fact, earlier – in 1922 – Robinson had patented a carrier reinforcement technique that was effectively a homodyne system. The overlap of terminology was not uncommon in those pioneering years of radio, and other terms such as autodyne receiver and regenerative

Microwave and Millimetre-Wave Design for Wireless Communications, First Edition.
Ian Robertson, Nutapong Somjit and Mitchai Chongcheawchamnan.
© 2016 John Wiley & Sons, Ltd. Published 2016 by John Wiley & Sons, Ltd.

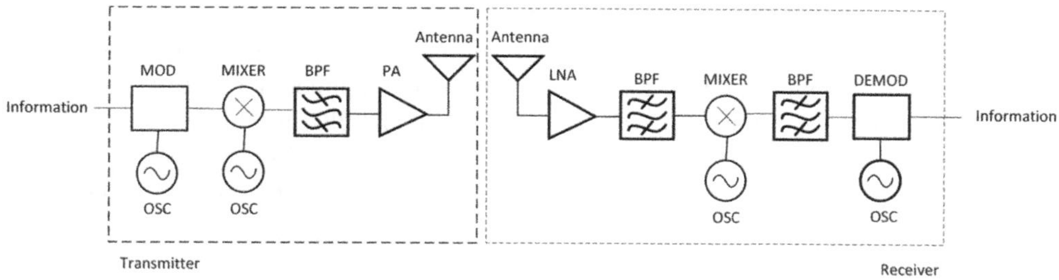

Figure 2.1 Simplified block diagram of a transmitter and receiver for wireless communication.

receiver were in use and not fundamentally different. Today, they could perhaps be categorised alongside self-oscillating mixers. The synchrodyne receiver was developed later than the homodyne, and uses a local oscillator circuit that is synchronised to the incoming RF carrier, in a form of injection locking. It could be argued that many of the techniques to create an LO in a homodyne receiver were already forms of the synchrodyne [1,2].

In the heterodyne receiver (from the Greek, *hetero*, meaning 'other' or 'different' and *dyne*) the RF is mixed with a different signal, an independent local oscillator. Heterodyning was introduced in 1901 by the Canadian radio pioneer Reginald Fessenden, and this work is widely recognised as a very fundamental step in radio engineering [3,4]. The superheterodyne receiver was introduced by Edwin Armstrong in 1918. The superheterodyne receiver is a heterodyne receiver with supersonic IF (intermediate frequency): the LO is deliberately offset from the RF signal so that an IF is produced at the output, which can then be conveniently filtered and amplified before being demodulated. This leads us finally to the most important principle of modern transmitter–receiver architectures, in which frequency conversion is used so that the majority of signal processing can be carried out at low frequency. Figure 2.1 shows the simplified block diagram of a generic wireless transmitter and receiver. The transmitter consists of a modulator (MOD), an upconverter mixer, a band-pass filter (BPF), a power amplifier (PA) and an antenna. The transmitted signal is received by the antenna at the receiver. The electrical signal at the receiver antenna output passes through the low-noise amplifier (LNA), another filter (BPF), a downconversion mixer and then through the IF filter and amplifier into the demodulator (DEMOD). The information is retrieved at the output of the demodulator, the last stage of the receiver. In the following sections, the basic functions of each component in the transmitter and receiver are described.

2.2 Transmitter and Receiver Components

2.2.1 Amplifier

The amplifier is a very crucial two-port active circuit in transmitters and receivers. The purpose of an amplifier is to increase signal power levels to overcome noise, loss and interference in signal transmission from the transmitter to the receiver. The ratio of the output and input signal power is called the power gain. In a wireless transmitter, the amplifier which is adjacent to the antenna is normally required to deliver significant transmitter output power and is called the power amplifier (PA). The power amplifier must amplify the transmitter signal power up to the target power level which ensures that the transmitted information arrives at the receiver.

In the receiver, the received signal is very small and a LNA is commonly required. Practical amplifiers are nonlinear, adding distortion to the amplified signal, and this will affect the performance of a communication system. In the transmitter amplifier, nonlinearity can cause unwanted out-of-band transmitted signals, requiring additional filtering, and distortion of the signal, potentially leading to data errors. Although the desired received signal is weak, linearity of the LNA may also be an important factor if there are strong interfering signals that cannot be filtered out.

2.2.2 *Oscillator*

The oscillator is an electrical circuit with a DC power input that generates a sinusoidal output signal that can be used as the local oscillator signal for a mixer, modulator or demodulator. A pure sinusoidal signal, in the frequency domain, has a spectrum that is a single line at f_c, as shown in Figure 2.2(a). In practice, noise generated by the electronic device(s) in the oscillator circuit is converted and amplified by nonlinear or linear time-varying effects to become amplitude and phase noise in the oscillator output signal. The spectrum of the signal from a practical oscillator exhibits a two-sideband spectrum around the oscillator frequency, f_c, as shown in Figure 2.2(b).

2.2.3 *Mixer*

A mixer (strictly speaking, a frequency mixer) is an electrical circuit that frequency-translates the signal spectrum by multiplication with a second local oscillator signal. The mixer is a three-port device, and mixers can be used for both upconversion and downconversion. The three ports of a mixer are named RF, IF and LO. The signal applied to the LO port is a sinusoidal signal from the local oscillator. The RF is the high-frequency port while the IF is the low-frequency port. An upconversion mixer is used in a transmitter for translating the information signal spectrum to a higher frequency for transmission. In a downconversion mixer, the received signal is downconverted, either to an intermediate frequency (IF) for further

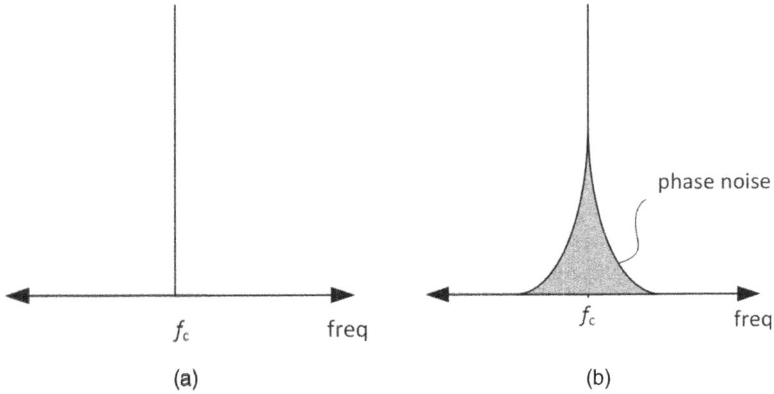

Figure 2.2 Output spectrum of (a) an ideal oscillator, and (b) a practical oscillator.

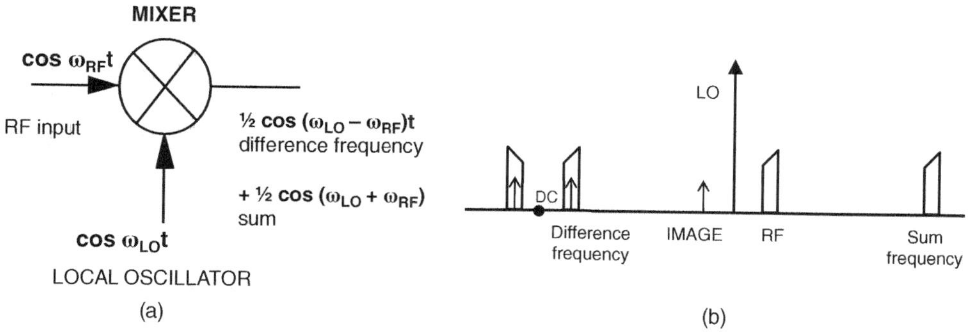

Figure 2.3 Mixer. (a) Symbol and function. (b) Simplified spectrum.

filtering and amplification, or directly to baseband. In an upconversion mixer, the RF port is the output port while the IF port is the input port. The ratio between the output power of mixer and the input power (not the LO power) is called the conversion gain or conversion loss, depending on whether the mixer is active or passive. While an ideal mixer can be simply modelled as a multiplication function, practical mixers generate a lot of unwanted output signals. To reduce the demand for filtering, various arrangements of balanced mixers, employing differential signals, can be used for partial cancellation of these unwanted signals.

Figure 2.3(a) shows the symbol for a mixer and indicates that the primary output signals are the sum and difference frequencies. In a downconversion application, the desired output signal is the difference frequency as illustrated in Figure 2.3(b), where the LO is shown below the RF frequency so that the spectrum of the downconverted IF output is not flipped over. It is common practice to draw the RF signal as a right trapezoid in the manner indicated to serve as a reminder that it contains information and the spectrum should be in the correct orientation before demodulation. Note that signals below the LO in this case are also downconverted to the same IF frequency (after flipping) and this is referred to as the image band. Normally, mixers are double-sideband, which means they are unable to discriminate between signals above and below the LO, referred to as the upper and lower sidebands, respectively. Techniques for realising single-sideband mixers are described in Chapter 12, but they are imperfect and there is nearly always a need for filters in a transmitter or receiver that deal with the problem.

2.2.4 Band-Pass Filter

As discussed in Chapter 1, the frequency spectrum is allocated for a very wide range of wireless communications systems and other applications. Unwanted signals from other systems and noise can interfere with the wanted signal. A band-pass filter (BPF) is normally required to eliminate the unwanted signals and allow the desired signal to pass through, as illustrated in Figure 2.4. A passive BPF is preferable to an active BPF because strong interfering signals can generate distortion in an active filter, leading to yet more undesirable signals. The frequency response of an ideal BPF is constant over the desired bandwidth and zero at other frequencies, as shown in Figure 2.4. The minimum bandwidth of the BPF is equal to the bandwidth of the desired signal. Theoretically, the signal at the output of this ideal filter output consists only of the signal components in the filter pass-band. Even then, these signal components include the

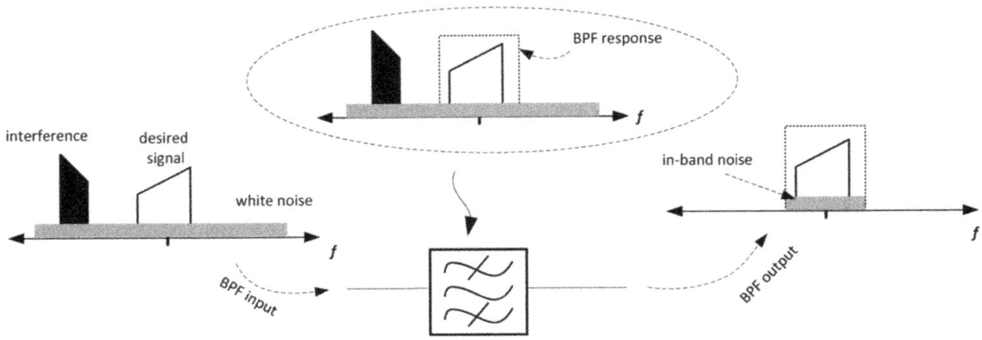

Figure 2.4 Band-pass filter function.

desired signal, noise and interference falling within the filter bandwidth. The noise in the filter bandwidth is called the in-band noise, and the actual noise power depends on the bandwidth of the filter.

2.2.5 *Modulator and Demodulator*

The modulator is an electronic circuit that maps the information signal onto a high-frequency carrier signal in some way, whilst the demodulator performs the reverse function of recovering the information signal. There are many different modulation techniques, the most basic of which is amplitude modulation; this can be described as a multiplication of the baseband information with the sinusoidal carrier wave, as illustrated in Figure 2.5 for a sinusoidal baseband information signal. As discussed later, the frequency or phase of the high-frequency carrier can also be modulated, and most modern wireless communications use an I-Q modulator, in which both the amplitude and phase of the carrier are modulated by separately amplitude-modulating orthogonal (0/90°) components of the carrier signal using two amplitude modulators, a 90° phase shifter and a combiner.

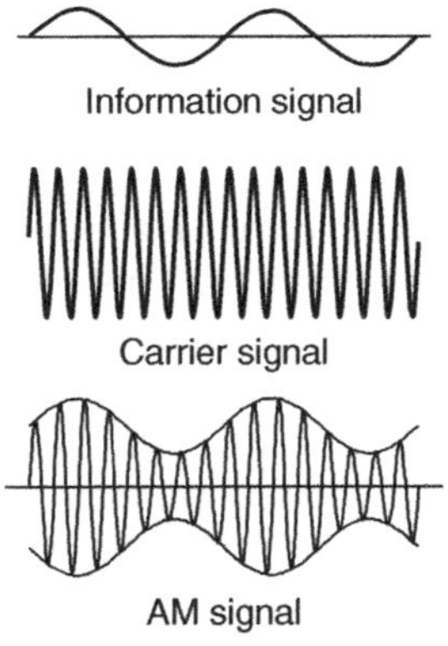

Figure 2.5 Basic amplitude modulation.

2.2.6 *Antenna*

The antenna is a component that converts electrical current into electromagnetic energy, or vice versa. The dimensions of an antenna are proportional to the carrier signal wavelength: for example, a monopole antenna has a physical length equal to a quarter-wavelength. One of the key parameters of an antenna is the antenna gain, which was introduced conceptually in Chapter 1. The gain, usually defined in dBi, is defined as the ratio of power transmitted by the antenna in the direction of peak radiation to that of an ideal isotropic radiator. The term *isotropic* derives from the Greek 'isos' (equal) and 'tropos' (way or manner), and an isotropic radiator is omnidirectional, with uniform radiated power in all directions. This ideal point source with power spreading equally over the surface of a sphere can only be approximated in practice, besides which it is preferable in many applications to direct the power in a certain direction.

2.2.7 *The Superheterodyne Receiver*

The superheterodyne architecture is a very important concept for modern radio systems. The technique employs a frequency mixer to convert a group of received signals to a lower frequency called the intermediate frequency (IF) before attempting to select the desired signal. The block diagram shown in Figure 2.6 brings together all the components described in the preceding sections. The LNA amplifies the received signal before any processing, while the RF filter selects the desired RF input signal band, rejecting the image band signals, but does not attempt to isolate the wanted signal from neighbouring signals as they are too close. The mixer is used to frequency-translate the RF signal by multiplying it with the local oscillator (LO) signal. The frequency-translated output signal of the mixer retains the information content of

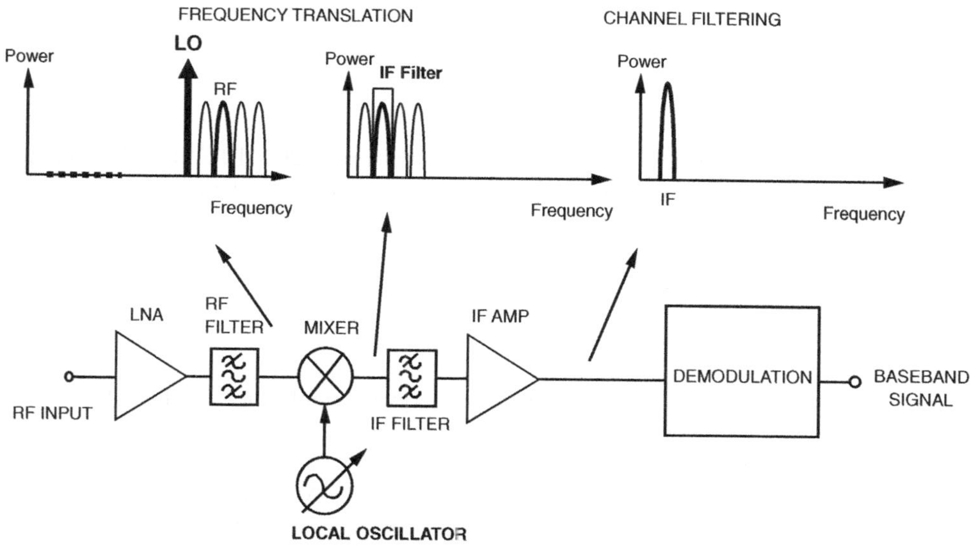

Figure 2.6 Superheterodyne ('superhet') receiver.

the RF input. Frequency shifting the RF to an IF enables a very selective IF band-pass filter to be implemented which selects the required signal and rejects the unwanted neighbouring signals. To tune the superheterodyne receiver from one channel to another, the LO frequency can be made variable, with the IF filter kept fixed. In this way, the IF filter can be extremely narrowband (selective) and the desired radio signal can be received cleanly even when the neighbouring signals are close together in frequency. A very important advantage of the very narrowband IF filter is that the amount of noise is minimised before the signal is demodulated. The relationship between noise power and bandwidth is very fundamental (details of this are introduced in the next section).

2.3 Noise and Interference

2.3.1 Thermal Noise

The proper understanding of noise dates back to 1918 when Walter Schottky, a German physicist, reported two main electronic noise sources, thermal and shot. Thermal noise, first described as the 'Warmeffekt', is caused by random motions of electrons agitated by thermal effects. Schottky deduced that the resulting noise power would be proportional to Boltzmann's constant and temperature. Thermal noise is usually referred to as Johnson or Nyquist noise to honour Johnson [5] and Nyquist [6], who first explained thermal noise in theory and experiment in 1920. In their studies, thermal noise was shown to be generated from thermal effects that cause randomly fluctuating electron flow in a conducting material. With an increase in temperature, these electrons obtain more energy and hence they collide more. The power density (S_N) of this noise (in W Hz^{-1}) is calculated from the average energy obtained from these electrons at the ambient temperature T_{in} in Kelvin (K) units [7] as follows:

$$S_N(f) = \frac{hf}{e^{\frac{hf}{k_B T}} - 1} \tag{2.1}$$

where k_{B} is Boltzmann's constant (1.381×10^{-23} J/K) and $h = 6.626 \times 10^{-34}$ J · s. Expanding Equation (2.1) with the Taylor's series and considering only the first-order term, this spectral density can be approximated as:

$$S_N(f) \approx k_B T \tag{2.2}$$

The term $k_{\text{B}}T$ is the same as for the average kinetic energy of a molecule, which confirms that thermal noise is a result of the thermal agitation of electrons.

A plot of exact power spectral density $S_N(f)$ from Equation (2.1), normalised with $k_{\text{B}}T$, is shown in Figure 2.7. The function plot is flat from 0 up to 83 GHz. At frequencies above 83 GHz, the spectral function gradually decreases with frequency. Consequently, the approximation for the noise power density defined in Equation (2.2) is only valid up to 83 GHz. The noise power density is found to be constant over several tens of GHz, and random noise with a constant

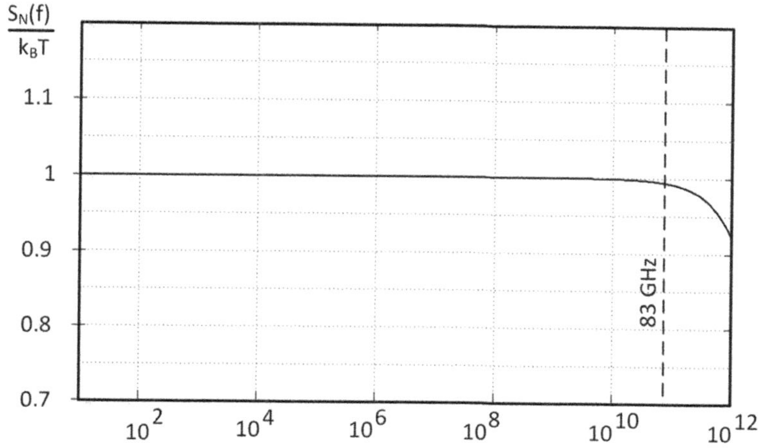

Figure 2.7 Normalised noise power density versus frequency.

power spectral density is referred to as 'white noise'. By expressing Equation (2.2) in dBm units and assuming room temperature (290 K), the noise power density will be:

$$S_N(f) = -173.83\,\text{dBm}\,\text{Hz}^{-1} \tag{2.3}$$

From statistical theory, the probability density function of the white noise is Gaussian. It is shown in Equation (2.2) that thermal noise power density N linearly increases with temperature, since fundamentally it is black body radiation within the conductor. The thermal noise power (in Watts) over a given bandwidth Δf(in Hertz) can be calculated from:

$$N = k_B T \Delta f \tag{2.4}$$

In dBm units, this is:

$$N = -173.83 + 10 \cdot \log_{10}(\Delta f) \tag{2.5}$$

In practice, a conductor has a limited conductivity and this is represented by a resistance, R. When measuring a thermal noise voltage across the conductor in the time domain, a randomly fluctuating signal is observed. The mean squared noise voltage signal $\overline{v_n^2}$ generated in a bandwidth Δf with ambient temperature T is:

$$\overline{v_n^2} = 4k_B TR \cdot \Delta f \tag{2.6}$$

The root-mean-square (r.m.s.) thermal noise voltage per $\sqrt{Hz}$ can be calculated by dividing with bandwidth and applying the square root operator in Equation (2.6). This leads to

$$\overline{v_n} = \sqrt{4k_B TR} \tag{2.7}$$

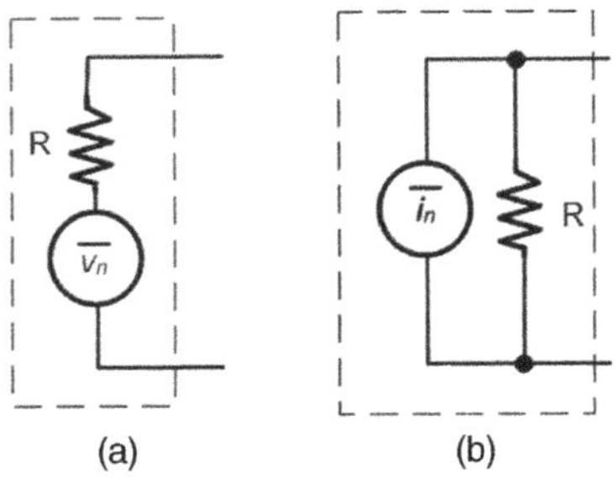

Figure 2.8 Thermal noise model using (a) voltage source, and (b) current source.

Equation (2.7) can be applied to model a thermal-noise voltage source with the series source resistance R. The noise voltage source has a magnitude relating to the square root of bandwidth, temperature and resistance. This noise voltage source is related to the current noise source ($\overline{i_n}$) per $\sqrt{Hz}$ by $\overline{i_n} = \frac{\overline{v_n}}{R}$. Hence

$$\overline{i_n} = \sqrt{\frac{4k_BT}{R}} \tag{2.8}$$

Two electrical models for thermal noise voltage and current sources are shown in Figure 2.8(a) and 2.8(b).The voltage model consists of a voltage source $\overline{v_n}$ and a series resistor R, while the current model has a current source of $\overline{i_n}$ shunted with the internal resistor R. These two models, which are electrically equivalent, generate the noise power of k_BT in a unit bandwidth. For a 1 $k\Omega$ resistor at 290 K, the current noise sources is $4pA/\sqrt{Hz}$ which is equivalent to the voltage noise source $4nV/\sqrt{Hz}$.

Figure 2.8 also suggests that thermal noise is generated by any lossy components. Every resistor in either lumped or parasitic form will generate thermal noise; hence, thermal noise always occurs in a circuit. In a practical communication system an increased thermal noise power, either due to elevated temperature or a wide bandwidth, will swamp the wanted signal, and if this happens the communication link will fail. With the introduction of noise, the range for communication is reduced and the receiver sensitivity is degraded. Using models shown in Figure 2.8(a) and (b), the total thermal noise power generated in the communication system can be analysed during the design process to ensure that the target performance is achieved.

2.3.2 Shot Noise

Schottky discovered shot noise in 1914 and published his work in 1918. He had claimed earlier than Johnson and Nyquist that shot noise is the second main form of electronic noise which significantly affects the performance of electronic circuits. The shot noise (i_s) is defined as the random fluctuation above the DC current (I_0), as shown in Figure 2.9. The DC part is calculated from the average number of electrons drifting. The variation around the mean value or the average quantity is the shot noise. Motion of individual electrons is random and the irregular arrival of electrons at the observation point can be statistically described with

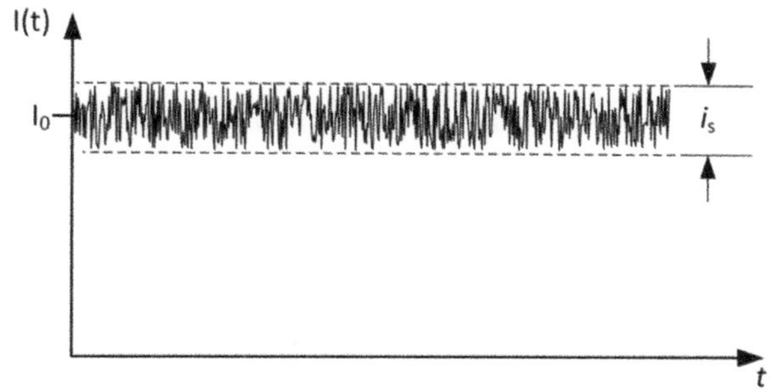

Figure 2.9 Shot noise.

a Poisson distribution. Schottky investigated the shot noise phenomenon in a temperature-limited thermionic diode under the assumption that the transit time of electrons travelling from cathode to anode is extremely short, and showed that the shot-noise mean square current in a unit bandwidth is defined by:

$$\overline{i_S^2} = 2q \cdot I_0 \tag{2.9}$$

where q is the charge of the electron (1.602×10^{-19} C) and I_0 is the DC current of shot noise generated device (in Amperes). Though the statistic time of arrival is described with a Poisson function, the amplitude distribution of shot noise has a Gaussian distribution. Hence, like the thermal noise case, the power spectral density of shot noise is flat up to several GHz.

The shot noise source can be modelled as a single current source of $\overline{i_s}$, the root-mean-square current. In practice, this noise is generally found in many devices where charges are carried across potential barriers such as a p-n junction diode, a Schottky diode, a PIN diode, a bipolar junction transistor as well as a field-effect transistor. In a p-n junction diode, the shot noise source is defined with the random flows of carriers across a potential barrier. For example, the current source for a p-n junction diode of 1 μA DC current is computed as:

$$\overline{i_s} = \sqrt{\overline{i_S^2}} = \sqrt{2 \times 1.602 \times 10^{-19} \times 10^{-6}} = 0.57pA/\sqrt{\text{Hz}}. \tag{2.10}$$

2.3.3 Flicker Noise

Flicker noise, which is a low-frequency phenomenon, derives its name from a flicker-type fluctuation that occurs in many nonequilibrium materials carrying a DC current (I_{DC}). Since flicker noise is found in various semiconductor devices, no single theory can be applied to completely describe the noise origin. Unlike the thermal and shot noise, the noise spectral density of flicker noise is inversely proportional to frequency. This distinct characteristic leads to the name '1/*f* noise', even though the exponent of the inverse term commonly deviates from unity. The mean-square noise current is given by:

$$\overline{i_f^2} = K\frac{I_{DC}^{\alpha}}{f^n} \tag{2.11}$$

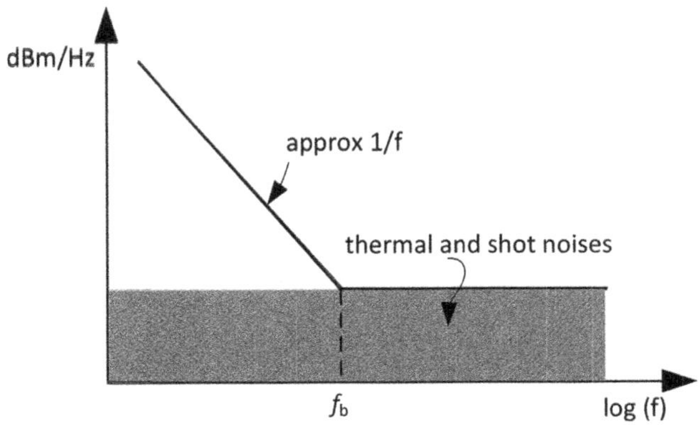

Figure 2.10 Spectral density of flicker (1/*f*) noise.

where K is a constant that depends on the material, n is the exponent which is close to unity and α is close to 2. Since it is inversely proportional to frequency it is swamped by thermal and shot noise at high frequency. The combination of flicker noise and white noise produces the noise spectrum shown in Figure 2.10. The corner frequency, f_b, is the frequency where the powers of white and flicker noises are equal. This corner frequency can be less than 1 kHz up to several tens of MHz depending on the material and device type. Microwave engineers may think that this flicker noise has little effect on high-frequency circuits because the corner frequency is well below the GHz region, but this is not true when there are nonlinear or linear time-variant effects present. Nonlinear and linear time-variant mechanisms generally exist in several key RF circuits, such as oscillators, mixers and power amplifiers. Low-frequency flicker noise in semiconductor devices is amplified and shifted to the carrier frequency where it appears as amplitude and phase noise in these RF circuits. The noise spectrum density is shown in Figure 2.11, where the slope is not necessarily the original slope due to the nonlinear or linear time-varying mechanism.

2.3.4 The Image Band and Noise

The image signal in a wireless receiver degrades the signal-to-noise ratio (SNR) performance considerably. The time-domain representation of the IF output of a downconversion mixer in a wireless receiver can be represented as:

$$v_{IF}(t) = 2\cos(2\pi f_{LO}t) \cdot v_{RF}(t) \tag{2.12}$$

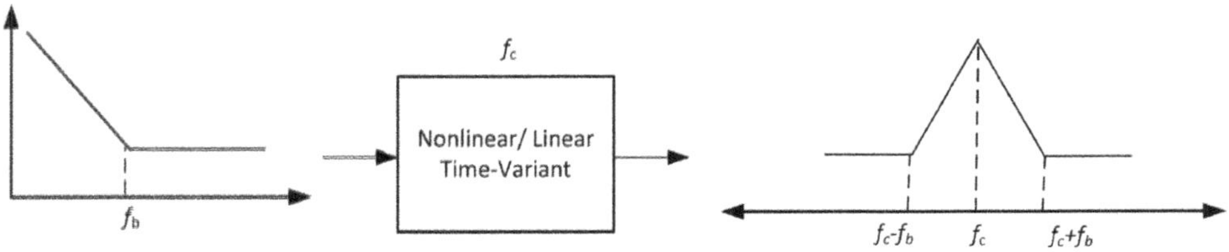

Figure 2.11 Flicker noise translation to high frequency in a nonlinear/linear time-variant system.

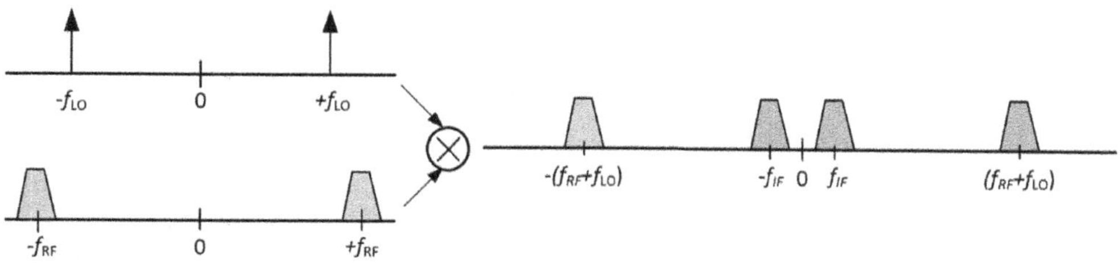

Figure 2.12 Spectrum representations for a downconversion mixer.

Using Fourier expansion, the mixing process shown in Equation (2.12) results in a frequency convolution between two signal spectra, the LO and RF. The IF frequency is the frequency difference between RF and LO, that is, $f_{IF} = f_{RF}$-f_{LO}. This is shown by:

$$V_{IF}(f) = [\delta(f + f_{LO}) + \delta(f - f_{LO})] \otimes [V_{RF}(f + f_{RF}) + V_{RF}(f - f_{RF})] \tag{2.13}$$

where $V_{RF}(f)$ denotes the spectrum of the downconverted signal, $v_{RF}(t)$, and $\delta(f)$ is the delta function at frequency f. Performing the convolution shown in Equation (2.13), the spectrum at mixer output is found to be:

$$V_{IF}(f) = [V_{RF}(f + f_{IF}) + V_{RF}(f - f_{IF}) + V_{RF}(f - f_{LO} - f_{RF}) + V_{RF}(f - f_{LO} + f_{RF})] \tag{2.14}$$

The spectrum plot of Equation (2.14) is shown in Figure 2.12. The spectrum has both positive and negative frequency components, as well explained with the Fourier theory. In Figure 2.12, only two spectrum components, f_{IF} and f_{RF}+f_{LO} (i.e., the difference and sum frequencies, respectively) are considered at the mixer output. These two spectrum components are symmetrical around f_{RF} with f_{LO} offset. The low-frequency spectrum components is the desired IF output from the downconversion mixer.

The problem occurs when there is an RF image, which has a frequency that is below f_{LO} by an amount equal to f_{IF}, as illustrated at the RF input of the mixer in Figure 2.13. The spectrum of image signal is given by:

$$IM(f) = [V_{IM}(f + f_{LO} - f_{IF}) + V_{IM}(f - f_{LO} + f_{IF})] \tag{2.15}$$

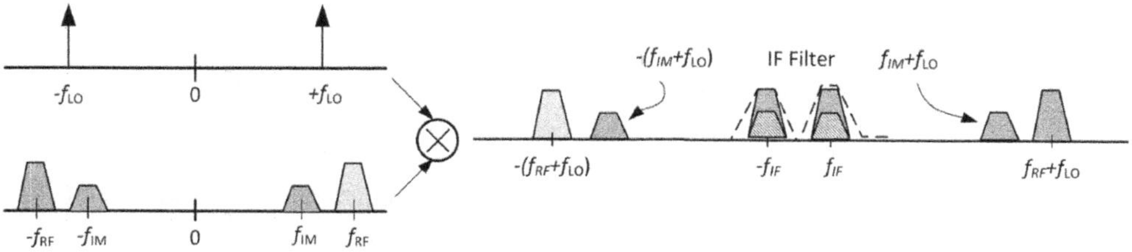

Figure 2.13 Spectrum showing image band signals at the mixer input and output.

The mixer output is,

$$V_{\text{IF}}(f) = [V_{\text{RF}}(f + f_{\text{IF}}) + V_{\text{RF}}(f - f_{\text{IF}}) + V_{\text{RF}}(f - f_{\text{LO}} - f_{\text{RF}}) + V_{\text{RF}}(f - f_{\text{LO}} + f_{\text{RF}})]$$
$$+ [V_{\text{IM}}(f - f_{\text{IF}}) + V_{\text{IM}}(f + f_{\text{IF}}) + V_{\text{IM}}(f - 2f_{\text{LO}} + f_{\text{IF}}) + V_{\text{IM}}(f + 2f_{\text{LO}} - f_{\text{IF}})] \tag{2.16}$$

To eliminate the undesired high-frequency sum signal in a downconversion situation, either a band-pass or low-pass filter is applied to the mixer output. Here, we assume that an IF band-pass filter with unity gain and bandwidth Δf is used. The filtered signal at the positive-frequency side is then given by:

$$V_{\text{IF,BPF}}(f) = V_{\text{RF}}(f - f_{\text{IF}}) + V_{\text{IM}}(f - f_{\text{IF}}) \tag{2.17}$$

The spectrum shown in Figure 2.13 clearly shows that the image signal appearing at the downconversion mixer input is frequency-translated into the bandwidth of IF filter. Hence, this downconverted image signal interferes with the wanted signal and degrades the receiver performance significantly. If we assume that the image signal is uncorrelated with the wanted signal, it can be considered as an in-band noise similar to a thermal noise. The SNR at the IF filter output is then given by:

$$SNR_{\text{IM}} = \frac{\int_{f_{\text{IF}}-\Delta f/2}^{f_{\text{IF}}+\Delta f/2} |V_{\text{RF}}|^2 \, df}{\int_{f_{\text{IF}}-\Delta f/2}^{f_{\text{IF}}+\Delta f/2} |V_{\text{IM}}|^2 \, df} \tag{2.18}$$

This shows that the image signal directly degrades the quality of the IF signal. Even if no interference signal appears at the image frequency, the noise in the image band is translated to the bandwidth of the desired IF output signal. To improve the SNR defined in Equation (2.18), the image term in the denominator should be reduced. A straightforward way is, in principle, to apply an RF band-pass filter before the mixer input in order to filter out the image signal prior to the mixing process. However, there is then a trade-off in the receiver design between the frequency of the IF and the Q-factor needed for this RF filter to reject the image signal. A high-Q RF band-pass filter is generally large and costly. A range of approaches for alternative receiver architectures which address this problem is provided in Section 2.6; these include the double-superheterodyne, zero-IF, low-IF, and image-rejection receiver, to name a few.

2.3.5 Interference

An interference signal is any unwanted signal that has different statistical characteristics to noise. Interference can be generated in the transmitter, communication channel, or receiver. Some of the common interference signals usually encountered are electromagnetic interference (EMI), intersymbol interference (ISI), co-channel interference (CCI), and adjacent channel interference (ACI). EMI is caused by two main mechanisms, conduction and radiation. Conduction EMI occurs at the physical contact of conductors, while radiation EMI results from electromagnetic induction. Proper system design techniques such as grounding, capacitive

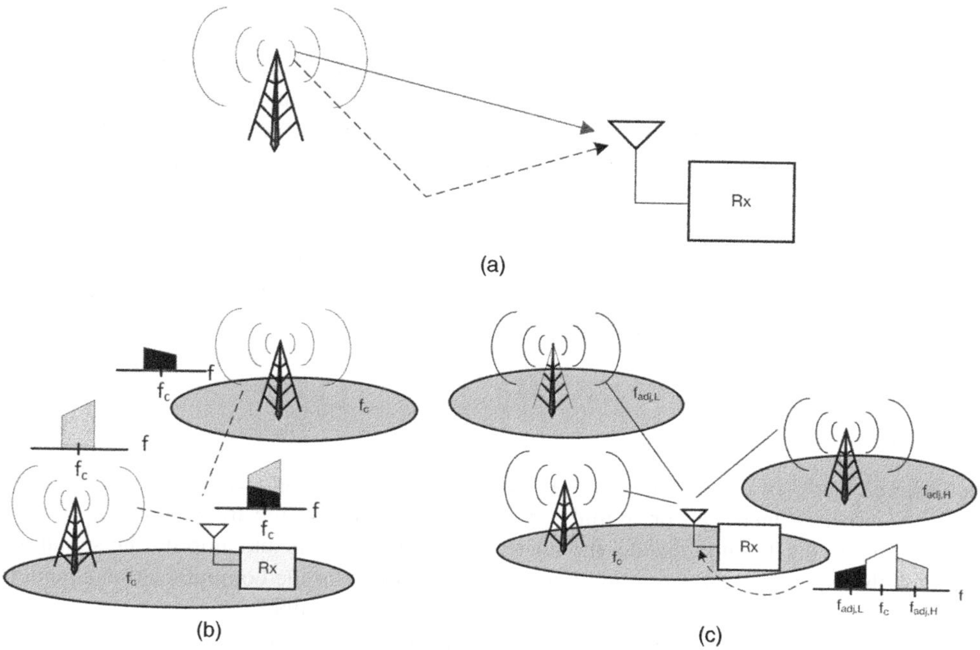

Figure 2.14 Examples of interference in communication systems. (a) Intersymbol interference (ISI). (b) Co-channel interference (CCI). (c) Adjacent channel interference (ACI).

coupling, as well as shielding, can reduce EMI effectively; however, this is a vast topic in its own right and is beyond the scope of the present text. ISI originates from multipath propagation effects occurring in the wireless communication or nonlinear channel.

Figure 2.14(a) shows the basic concept of ISI due to multipath propagation. A receiver collects a signal which is the sum of a symbol travelling from the direct path and other delayed symbols from other nondirect paths. These delayed symbols interfere with and distort the direct path symbol, degrading the signal quality. CCI is very common in mobile communications networks where frequency-reuse is used. A shown in Figure 2.14(b), the receiver (Rx) receives the signal at frequency f_c from the transmitter in the designated area or cell. However, there is another transmitter nearby transmitting at the same frequency f_c. The received signal will be the combination of the desired and undesired signals as shown in Figure 2.14(b), and if the power received from the undesired transmitter is large enough, compared to the desired signal, then severe interference will occur. It is not possible to reject this undesired signal since it is at the same frequency as the desired signal. Other, much more sophisticated techniques, such as those based on multiple antennas and beam-forming, must be adopted. Figure 2.14(c) shows the basic mechanism of ACI. A receiver seeks to receive the signal at frequency f_c from a transmitter in the same area. Unfortunately, there are two nearby transmitters transmitting signals at two adjacent frequencies $f_{adj,L}$ and $f_{adj,H}$. These two frequencies are very near to f_c and these signal powers will leak through in the receiver passband, which in turn corrupts the desired signal.

2.3.6 Atmospheric Attenuation

Radio waves propagating through the atmosphere are attenuated primarily by water vapour and oxygen. The attenuation level depends not only on atmospheric conditions such as pressure, temperature and humidity but is also dependent on the zenith angle. As discussed in Chapter 1, there are attenuation peaks at 22.3, 60, 118.74, 183.3, 323.8 and 380.2 GHz. The peaks around 60 and 120 GHz are the result of oxygen attenuation, while the others are due to absorption by water vapour. For a short-range communication such as IEEE 802.11ad, which gives 7 Gb s^{-1} communication within a 10-m range, this atmospheric attenuation in the 60 GHz region is around 12 dB km^{-1}. The atmospheric attenuation is only 0.12 dB for 10 m, which is negligibly small compared to other losses occurred.

2.3.7 Path Loss

The basic elements of a communication system are shown in Figure 2.15. The information sequence, either in analogue or digital form, is passed through the transmitter (Tx unit). The elements in the Tx unit encode, modulate, frequency-translate and amplify the signal to overcome the effect of loss, noise and interference occurring in the communication channel which can be in any form, including wireless, cabled and optical fibre.

The transmitted power (P_t) from the Tx unit is attenuated by loss in the communication channel. In a wireless channel, electrical power from the Tx unit is converted to an electromagnetic wave by the antenna in the transmitter. This wave propagates to the Rx unit and is converted back to an electrical signal by the antenna at the receiver unit (Rx). In the microwave and millimetre-wave frequency range, the direction of wave propagation is a straight line between the Tx and Rx units; this is called the direct wave or line-of-sight (LOS) mode. The wave of wavelength λ travelling from the transmitter to receiver at distance d is assumed to be a plane wave. This is valid if d is comparably large to the largest dimension of the antennas (a). That is:

$$d \geq \frac{2a^2}{\lambda} \tag{2.19}$$

The power that appears at the receiver (P_r) as shown in Figure 2.15 is calculated from Friis' formula as [8]:

$$P_r = \left(\frac{\lambda}{4\pi d}\right)^2 G_r G_t P_t \tag{2.20}$$

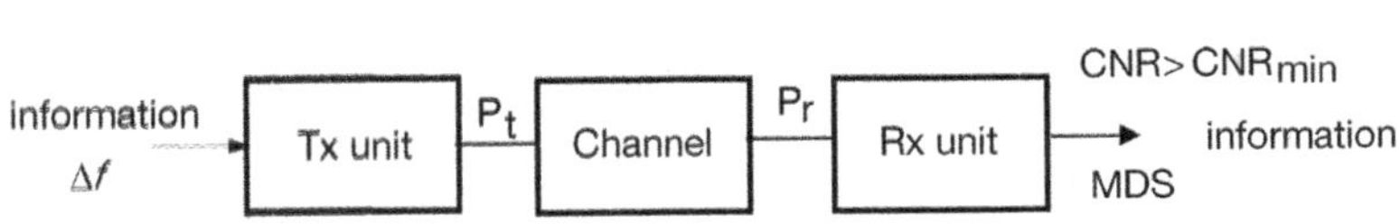

Figure 2.15 Communication channel.

Table 2.1 Path loss exponent for different environments.

Environment	Path loss exponent (n)
Free space	2.0
Urban area	2.7–3.5
Suburban area	3.0–5.0
Indoor (LOS)	1.7
Indoor (NLOS)	3.5

The antenna gains, denoted by G_t and G_r, respectively, are power amplification parameters. By inspection from Equation (2.20), P_r is attenuated relative to P_t by the factor $(\frac{\lambda}{4\pi d})^2$. This factor is called the free-space path loss, which increases with the square of distance d and f. However, the free space path loss is only valid for outdoor propagation and only if a direct path exists. This path loss model is too optimistic since, in a real situation, there is more than one propagation path due to earth ground reflection, diffraction around objects, scattering from large obstacles, and the dynamic characteristic of a mobile channel [9]. To simply take this effect into account, the second-order exponent in the free space loss model shown in Equation (2.20) is adjusted based on empirical data. The simplified path loss model is then:

$$P_r = \left(\frac{\lambda}{4\pi d}\right)^n G_r G_t P_t \tag{2.21}$$

where n is called the path loss exponent, depending on the environment as shown in Table 2.1. The path loss for indoor propagation has two modes, line-of-sight (LOS) and non-line-of-sight (NLOS). The values are obtained from averaging over the reported experimental data. It is noted that the path loss exponent in NLOS indoor propagation is rather large due to considerable absorption from office furniture, for example.

With the simplified path loss model, channel loss in different environments can be computed. Friis's formula then can be written in terms of effective isotropic radiated power (EIRP) as:

$$P_r\,(dBm) = EIRP\,(dBm) + G_r\,(dB) + 10n\log_{10}\left(\frac{\lambda}{4\pi d}\right) \tag{2.22}$$

where

$$EIRP\,(dBm) = P_t\,(dBm) - L_C\,(dB) + G_t\,(dB) \tag{2.23}$$

L_c, defined in Equation (2.23), is cable loss including antenna mismatch in decibel units. Effective isotropic radiated power is an important parameter that combines the transmitter power and peak transmitter antenna gain into a single figure: EIRP is the amount of power that a theoretical transmitter with an isotropic antenna would have to transmit to achieve the same peak power density.

Example: A100 mW 60 GHz indoor transmitter [10, 11] is transmitting a 5 Gb s^{-1} QPSK signal via an antenna with gain of 30 dBi to a receiver that is a distance of 10 m away. The antenna at the receiver has a gain of 20 dBi, the total cable and mismatch losses in the transmitter and receiver are 2 and 3 dB, respectively. Calculate the minimum possible power at the receiver.

The minimum power that can be received occurs for the maximum path loss for the indoor environment case, which is the NLOS case. From Table 2.1, the pass loss exponent n is 3.5. The wavelength at 60 GHz is:

$$\lambda = \frac{c}{f} = \frac{3 \times 10^8}{60 \times 10^9} = 5\,\text{mm}$$

At distance d equal to 10 m, the path loss (L_p) in decibels is:

$$L_p = 10\,(3.5)\log_{10}\frac{4\pi \times 10}{5 \times 10^{-3}} = 154.01$$

The EIRP from the 100 mW transmitter antenna is:

$$EIRP = 20 - 2 + 30 = 48\,\text{dBm}$$

Hence, the minimum received power at the receiver is:

$$P_r = EIRP + G_r - L_p = 48 + 20 - 154.01 = -86.01\,\text{dBm}$$

2.4 Introduction to Modulation

The baseband spectrum of information signals (voice, images, video, data, etc.) lies in the frequency range from DC to hundreds of MHz. It is impractical to directly transmit these signals wirelessly for many reasons, and in any case if information from several sources was transmitted simultaneously in a wireless environment, interference would occur. To prevent this, each information stream should be allocated a specific band of spectrum, spaced sufficiently apart that they do not interfere with one another and can be separated at the receiver. This can be done by shifting the original information band to a higher frequency band in the transmitter. The new spectrum is shifted to a carrier frequency, f_c, in the transmitter by a process called 'modulation'. At the receiver, a reverse – demodulation – process is needed to shift the modulated carrier back to a baseband information signal and recover the information. A wireless channel uses the radio spectrum which, according to the ITU, ranges from 3 kHz to 3 THz, and so a vast amount of spectrum is available for wireless communications, within the limits of government licensing rules and practical constraints such as the cost of hardware and availability of suitable devices for high-power sources and low-noise receivers.

Figure 2.16 illustrates how modulation allows information from various applications and users to be communicated over the same channel. Supposing the three users located in the same area are sending voice signals via a radio channel. The bandwidth of speech signals lies from DC to 4 kHz. The three speech signals are modulated with difference carrier frequencies (f_{c1}, f_{c2}, and f_{c3}) prior to transmission over the wireless channel, and the modulated spectra for user #A, #B and #C are centred at f_{c1}, f_{c2} and f_{c3}, respectively. The carrier frequencies are well chosen such that the modulated carriers are spaced well apart in the channel spectrum.

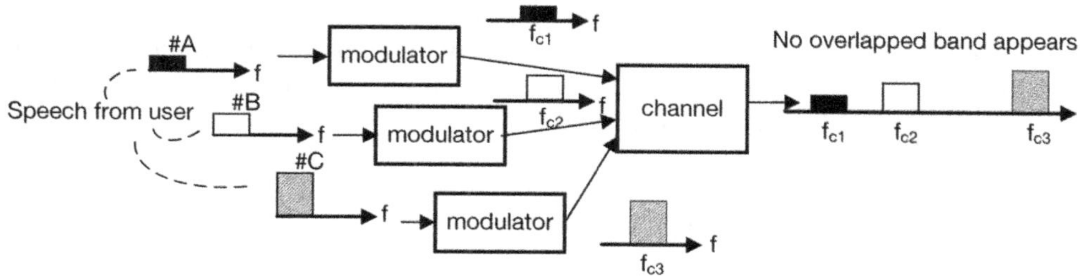

Figure 2.16 Modulation onto different carrier frequencies to allow multiple users to communicate over a single channel.

Inter-channel interference in this case does not occur. Modulation in wireless communication is really essential in any case since the antenna size is related to the wavelength. In practice, the carrier frequency is not arbitrary but is constrained with some specific communication guidelines, for example ITU regulations, industry standards and numerous theoretical and practical constraints. In addition, practicalities in the design of the modulator and demodulator hardware, plus other factors, mean that it is very common for a carrier signal to be modulated/demodulated at a lower frequency and then up/downconverted to the frequency used for transmission across the wireless channel.

Information is contained in the amplitude of the baseband signal, either in analogue or digital form. Many information signals, such as speech and music, are naturally analogue signals, and the analogue signal may be directly modulated onto the carrier. On the other hand, information is generally more robust in binary digital form, and digital data can represent any form of information with appropriate formatting, and so digital modulation is now far more common. Figure 2.17 shows the conceptual diagram of analogue and digital modulation for the simple case of varying the amplitude of the RF carrier in accordance with the information (baseband) signal amplitude.

2.4.1 Amplitude Modulation (AM)

The amplitude modulation technique gives a modulated carrier waveform whose amplitude is varied continuously in accordance with the information signal while the frequency is constant.

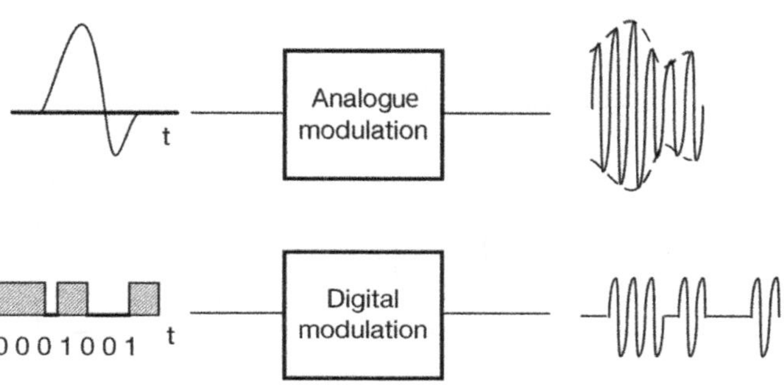

Figure 2.17 Analogue and digital modulation.

Therefore, in the time-domain, the envelope of the modulated carrier varies with the amplitude of the modulating signal. If B is the bandwidth of the information signal, the minimum bandwidth of an AM signal ranges from B to $2B$ depending on whether single-sideband (SSB) or double-sideband (DSB) techniques are used.

2.4.2 *Frequency Modulation (FM)*

Frequency modulation simply means that the carrier frequency changes with the amplitude of the information signal. Therefore, the envelope of the modulated signal is constant while the frequency is varied. In the frequency domain, the FM spectrum is expanded around the carrier frequency with the modulation index, m, and the bandwidth of the FM signal is given by:

$$BW = 2B\,(m + 1) \tag{2.24}$$

where m is the ratio between the maximum deviation frequency and the bandwidth of the modulating waveform. If m is selected to be less than 0.5, the modulated bandwidth is comparable to AM. This modulation case is called a narrowband FM technique. FM gained a lot of interest because it performs better in the presence of noise than AM.

2.5 Digital Modulation

Digital information can be obtained by sampling, quantising and encoding an analogue signal. The process of analogue-to-digital conversion is shown in Figure 2.18. The sampling block produces a sequence of samples of the input analogue signal at a sampling frequency (f_s).The output signal of the sampling block is called a sampled sequence. According to the sampling theorem of Claude E. Shannon, f_s should be chosen such that:

$$f_s > 2B \tag{2.25}$$

where B is the bandwidth of the input information. If an analogue information signal is a low-pass signal, B will be the maximum frequency of the analogue signal. However, if it is

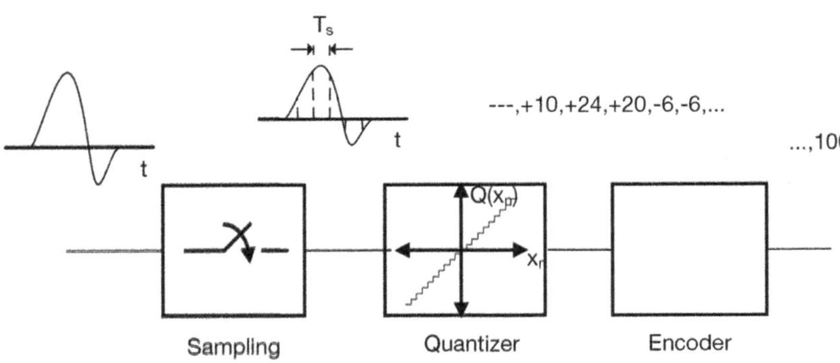

Figure 2.18 Analogue-to-digital conversion.

a band-pass signal, B is equal to the bandwidth of the input analogue signal. From Equation (2.25), f_s needs to be selected to fulfil this criterion in order to reconstruct the original signal from its digital representation, otherwise perfect reconstruction cannot be achieved.

A shown in Figure 2.18, the output of the sampling process is a sequence of impulse signals occurring at every sampling instant, spaced T_s apart in time ($T_s = 1/f_s$). The amplitude of each impulse matches the analogue waveform amplitude at the sampling instant. The sampled data is then input to the quantiser block, where the amplitude of each sample is mapped to one of a finite series of discrete values. The quantised discrete data is then input to the b-bit binary encoder to produce a binary digit number. In the quantisation process, a magnitude difference between the original value and the output discrete value occurs, referred to as 'quantisation error'. The scale of this error depends on the quantisation step (2^b) of the b-bit quantiser. This quantisation error occurs randomly and can be statistically described as a uniform distribution. It is regarded as quantisation noise and the SNR of the b-bit quantiser in decibels can be approximately by:

$$SNR = 10\log_{10}\frac{12S}{2^b} \tag{2.26}$$

The SNR can therefore be improved by increasing quantisation resolution or signal power.

2.5.1 Channel Capacity

In the early days of digital communication, many engineers believed that there was always a trade-off between transmission rate and error probability of data transmission, the so-called bit-error-rate (BER). During that period, engineers chose to reduce the bit rate when encountering a noisy channel. However, Claude E. Shannon developed a renowned formula called 'Channel capacity' (C) which defines the maximum rate (in bits per second) of transmission of error-free binary digits over a system when the signal is perturbed by noise. For binary digital transmission with average signal power S and white thermal noise of power N in the channel of bandwidth W, the data can be transmitted at a maximum rate of:

$$C = W\log_2\left(1 + \frac{S}{N}\right) \tag{2.27}$$

It is noted that term S/N means SNR, which should be measured at the receiver end of the channel. Hence, S means the average received signal power, which is a function of transmitted power and channel attenuation. The formula shows that if we increase channel bandwidth or S/N, we will get an increase in information rate. However, this does not mean that increasing bandwidth to infinity will allow the information rate to be infinite because the noise power will also be increased with the bandwidth.

Assuming that we are sending the binary digits at the information rate C, the bit duration is $1/C$ seconds and the energy per bit (E_b) will be equal to S/C. Given that the channel bandwidth W and the noise spectral density is denoted by N_0, the Shannon bound is:

$$\frac{E_b}{N_0} = \frac{(2^{\rho_{max}} - 1)}{\rho_{max}} \tag{2.28}$$

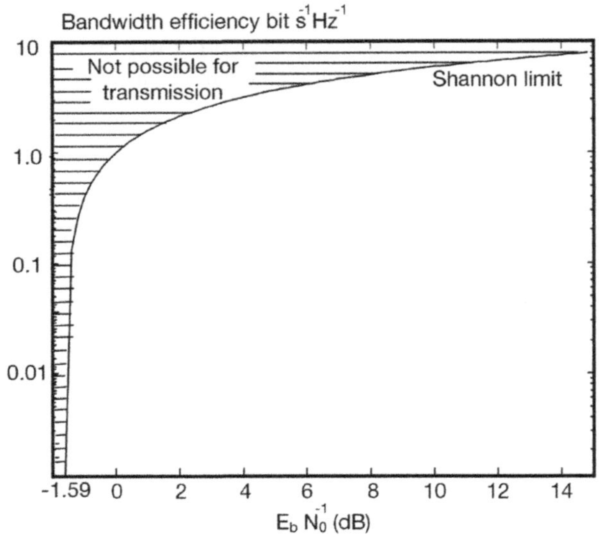

Figure 2.19 The Shannon limit in a digital communication system.

Note that in Equation (2.28) the term E_b/N_0 means the *energy-per-bit to noise-power-spectral-density ratio*. The term $\rho_{max} = \frac{C}{W}$ in unit of bits s^{-1} Hz^{-1} is the maximum bandwidth efficiency, which is the ratio of maximum transmitted information rate to the channel bandwidth B. An infinite bandwidth case means zero bandwidth efficiency. The minimum E_b/N_0 required to reliably transmit one bit of information is given by:

$$\lim_{\rho_{\max}\to 0} \frac{E_b}{N_0} = -1.59\, dB \tag{2.29}$$

If the energy-per-bit to noise-power-spectral-density ratio of any link is lower than the –1.59 dB given in Equation (2.29), that link will not be able to communicate. On the other hand, the maximum bandwidth efficiency can improve if the bit energy increases. The maximum rate shown in Figure 2.19 increases monotonically with E_b/N_0. The Shannon limit shown in Figure 2.19 suggests that, at a given bit energy, there might be some digital communication techniques that can be applied to allow transmission at the channel capacity. The Shannon theory does not say how to reach the channel capacity. Instead, it has inspired many engineers, for decades, to develop more sophisticated digital communication techniques. In the next section, the most common digital modulation techniques will be discussed and their performance in terms of bandwidth efficiency and BER will be described and compared.

2.5.2 Amplitude Shift Keying (ASK)

Amplitude shift keying (ASK) is one of the simplest forms of digital modulation. The amplitude of the digital information signal directly varies the envelope of the carrier signal. In the simplest form of ASK, the binary logic '0' produces no carrier signal and logic '1' produces carrier

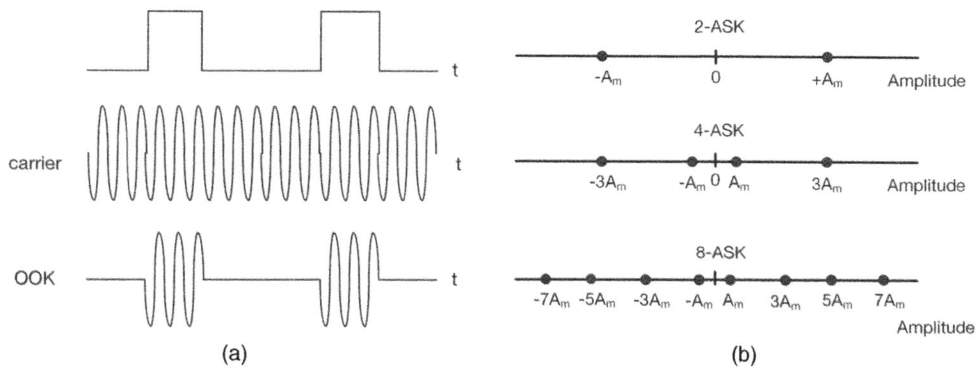

Figure 2.20 (a) Time-domain waveform of OOK. (b) Symbol diagram of M-ary ASK signals ($M = 2$, 4 and 8).

signal as shown in Figure 2.20(a). This simplest ASK is sometimes called On-Off-Keying (OOK) since the modulated waveform has just two levels which are on and off of the carrier. The more complicated form of ASK is called M-ary ASK, where M should be a power of 2, that is 4-ASK, 8-ASK, and so forth. In M-ASK modulation, each $k = \log_2(M)$ binary bits are grouped together to form a single *symbol* which is fed to the M-ary modulator to produce a carrier signal with M levels of amplitude. In this general form, the modulated waveform of each symbol may be simply represented in the following form:

$$s_{M-ASK}(t) = A_m g(t) \cos(2\pi f_c t), \quad A_m = \pm 1, \pm 3, \ldots, \pm(M-1) \tag{2.30}$$

The constellation diagrams of M-ary ASK for $M = 2$, 4, and 8 are shown in Figure 2.20(b). The constellation diagram is a convenient method of graphically showing the carrier signal's amplitude and phase states, without drawing the waveforms. This is the same as the phasor diagram used in power systems analysis – in that case, there is no need to keep drawing the 50/60 Hz sinusoidal signal, a simple vector in the complex plane shows any particular amplitude and phase.

The modulated signal for M-ary ASK has M amplitude levels, where the each amplitude level (symbol) is represented by the corresponding constellation point. With average energy per bit E_b, the symbol error probability for ASK using an optimal receiver is given by:

$$P_{e,M-ASK} = 2\left(1 - \frac{1}{M}\right) Q\left(\sqrt{\frac{E_b}{N_0}}\sqrt{\frac{6k}{M^2 - 1}}\right) \tag{2.31}$$

where the function $Q(x)$ is defined as:

$$Q(x) = \frac{1}{\sqrt{2\pi}} \int_x^{\infty} e^{-\frac{t^2}{2}} dt \tag{2.32}$$

The spectral bandwidth of the M-ary ASK signal depends on the spectral density of the pulse train $g(t)$ and the bit sequence. To limit the spectral bandwidth of the modulated waveform, a

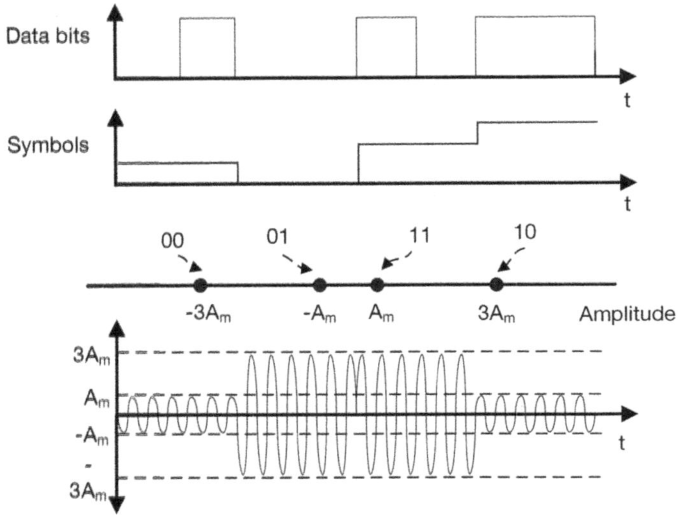

Figure 2.21 4-ASK modulation; binary digits, symbols and modulated carrier.

precoding technique can be used to provide a power spectral density that has a null at the rate $\frac{1}{2T}$, where T is the symbol interval.

An example of a 4-ASK signal is shown in Figure 2.21. The eight-bit data series {0 1 0 0 1 0 1 1} is converted to four symbols, which are {1, 0, 2, 3}. The four symbols are fed to an ASK modulator to produce a 4-ASK signal $\{-A_m, -3A_m, +3A_m, +A_m\}$. The mapping from the data bits to the symbols, the 4-ASK constellation diagram, and the 4-ASK modulated carrier waveform are shown. Note that the symbol rate is lower than the bit rate. Since it is the symbol rate that determines the bandwidth of the modulated signal, using a symbol containing n bits means that $n\times$ more data can be transmitted without requiring more bandwidth, all other things being equal.

2.5.3 Frequency Shift Keying (FSK)

The frequency shift keying (FSK) technique changes the frequency of the carrier in accordance with the binary digits of the information signal. The amplitude of the modulated signal is constant. Figure 2.22 shows binary FSK (BFSK), which is the simplest form of FSK. The signal frequency changes between two states with the logic input '0' or '1'. M-ary FSK is an FSK signal that sets the carrier frequency to one of M states according to the data input by grouping each $\log_2(M)$ bits into one of M symbols. The expression for an M-ary FSK signal is given by:

$$s_{m,FSK}(t) = A\cos(2\pi[f_c + m\Delta f]t); m = 1, 2, 3, \ldots, M \tag{2.33}$$

The bandwidth of the M-ary FSK signal is equal to the product of the frequency separation and M (i.e., $M\Delta f$). The frequency separation, Δf, should be suitably chosen to minimise

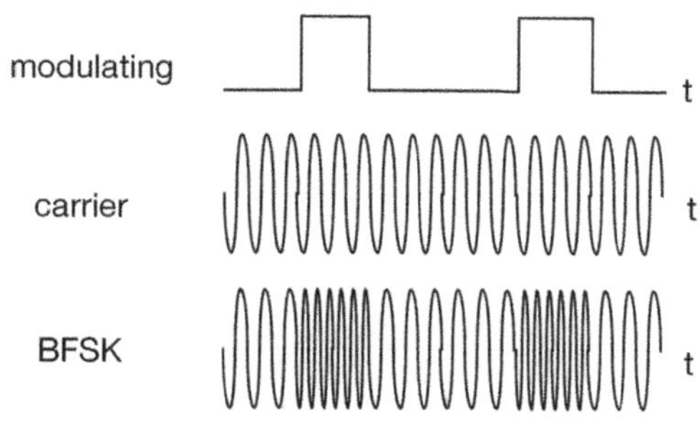

Figure 2.22 BFSK modulation.

interference between symbols. Given the symbol interval T, $\Delta f \geq \frac{1}{2T}$, it can be shown [12] that the minimum error probability for FSK will be obtained if $\Delta f = \frac{0.715}{T}$.
The error probability of M-ary FSK, when the minimum error probability case is chosen, is given by:

$$P_e = \frac{2^k - 1}{2^{k-1}} \frac{1}{\sqrt{2\pi}} \int_{-\infty}^{\infty} \left[1 - (1 - Q(x))^{M-1}\right] e^{-\frac{\left(x - \sqrt{\frac{2E_b}{N_0}}\right)^2}{2}} dx \tag{2.34}$$

2.5.4 *Phase Shift Keying (PSK)*

In the phase shift keying (PSK) technique, the phase of the carrier signal is varied in accordance with the data stream. Figure 2.23 shows the binary PSK (BPSK) signal in which the carrier phase switches between 0 and 180 degrees according to the information logic '0' and '1'. In a coherent receiver, the change in phase can be detected easily, making PSK relatively insensitive to noise. In M-ary PSK, the carrier has M different phase states that carry the binary data in symbol form. The M-ary PSK signal may be written as:

$$s_{M-PSK}(t) = g(t) \cos\left(2\pi f_c t + \frac{2\pi}{M}\{m - 1\}\right), m = 1, 2, \ldots, M \tag{2.35}$$

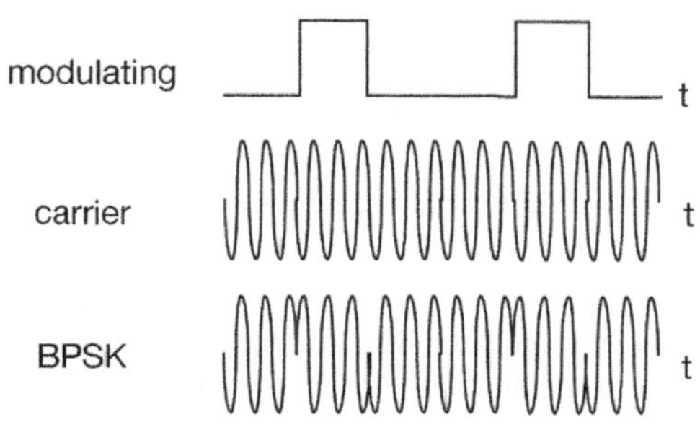

Figure 2.23 BPSK modulation.

The error probability for M-ary PSK is given by:

$$P_{e.M-PSK} = 1 - \int_{-\frac{\pi}{M}}^{\frac{\pi}{M}} \rho_{\Theta}(\theta)\, d\theta \tag{2.36}$$

where

$$\rho_{\Theta}(\theta) = \frac{1}{2\pi} e^{-\frac{kE_b}{N_0}\sin^2\theta} \int_0^{\infty} v e^{-\frac{1}{2}\left(v-\sqrt{2k\frac{E_b}{N_0}}\cos\theta\right)^2} \tag{2.37}$$

Once again, whilst it is feasible to draw the waveform of a BPSK signal, as in Figure 2.23, for greater values of M it is simply impractical to draw the waveform for M phase states, let alone base any useful analysis on this visual representation. In digital communications, the modulated signal is best represented symbolically in the complex domain, that is, on a constellation diagram.

2.5.5 *Constellation Diagram*

The constellation diagram is a signal space diagram for digitally modulated waveforms. Each symbol is characterised by a complex number which represents the amplitude and phase of the carrier signal for that symbol. Most often, Cartesian format is used, which means that the real part amplitude-modulates a cosine component of the signal carrier frequency, f_c, and the imaginary part modulates the sine component. The general form of modulated signals that can be represented on a constellation diagram is given by:

$$s_m(t) = A_{m,I}\, g(t) \cos(2\pi f_c t) - A_{m,Q}\, g(t) \sin(2\pi f_c t) \tag{2.38}$$

where the symbol amplitude of the cosine carrier component is $A_{\mathrm{m,I}}$ and $A_{\mathrm{m,Q}}$ is the symbol amplitude of the sine carrier component. The real and imaginary axes of the complex plane are called the in-phase (I) and quadrature (Q) axes, respectively. The whole complex plane with plotted symbols is called a constellation diagram. The constellation diagram is a vital tool for understanding the characteristics of a transmitter–receiver system. A constellation is not only useful to explain the different digital modulation techniques used in the communication system, but also gives a quick visual aid for diagnosing the system impairments that affect communication performance, such as white noise, thermal noise, distortion from nonlinearity, to name a few. Figure 2.24 shows four constellation diagrams which are an ideal quadrature-phase shift keying (QPSK) signal, QPSK with white thermal noise present, and QPSK with phase noise present. This clearly shows the usefulness of the constellation diagram as a tool for diagnosing systems performance and identifying problems.

To see how to interpret communication quality using a constellation diagram in more detail, consider the three constellation diagrams shown in Figure 2.24. QPSK has four constellation points that lie on a circle which means that QPSK provides a modulated signal with constant envelope. White thermal noise added to QPSK causes the constellation points to be randomly scattered around the ideal points. The radius of scattering from the ideal point is proportional

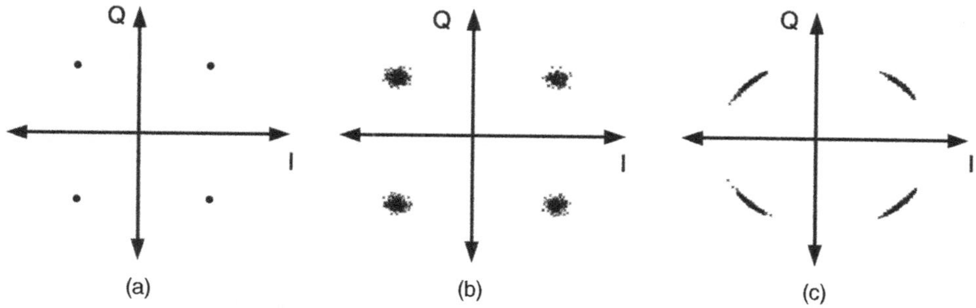

Figure 2.24 Constellation diagrams of (a) QPSK, (b) QPSK with added white noise, and (c) QPSK with added phase noise.

to the power of the thermal noise. The effect of phase noise on QPSK can be seen clearly on the constellation diagram (c): the constellation points are randomly located along an arc, the length of which is proportional to the phase noise power. The random deviation of constellation points increases the bit error rate when demodulating. Nonlinear effects in power amplifiers are also very easy to visualise on the constellation diagram, and this topic is covered in Chapter 10.

The manner in which bits are mapped onto symbols has a significant effect on the performance of the communication system. Figure 2.25 shows the constellation diagrams of Gray-encoded QPSK and 8-PSK. The Gray encoder is applied to the data bits to simplify the synchronisation task in the receiver. The mapping has a significant effect on the trajectory between symbols during modulation, which in turn affects the spectrum of the signal, the BER, and degree of envelope variation.

2.5.6 *Quadrature Amplitude Modulation (QAM)*

So far, we have seen that amplitude, frequency, and phase of the carrier signal can carry the binary digital information. High bandwidth efficiency can be achieved if multilevel amplitude, frequency or phase modulation is employed. The quadrature amplitude modulation (QAM) technique utilises both the amplitude and phase of the carrier to carry information. Baseband information is separated and simultaneously modulated with an in-phase (cosine channel) and

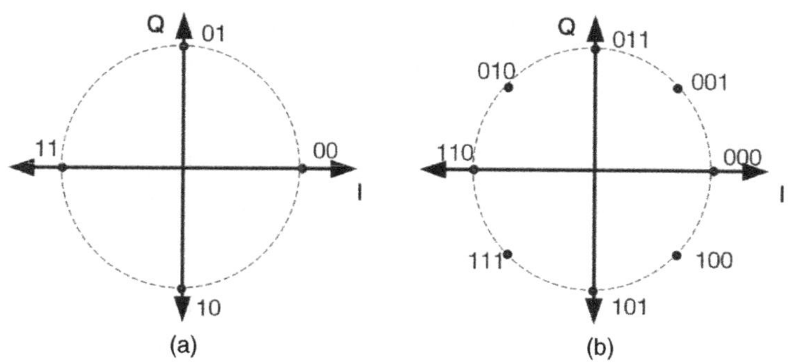

Figure 2.25 Constellation diagrams of (a) QPSK and (b) 8-PSK.

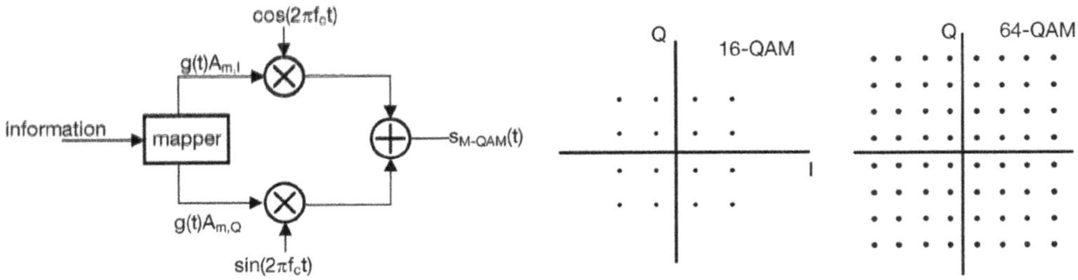

Figure 2.26 Quadrature modulator and 16- and 64-QAM constellation diagrams.

quadrature (sine channel) carrier signal pair. The quadrature carrier is used for its orthogonality with the in-phase carrier. According to information theory, the quadrature-phase difference prevents each channel from interfering with one another. Hence, the original bits can be perfectly retrieved when demodulating. The QAM output is obtained by combining two orthogonal modulated carriers, as shown in Figure 2.26, and can be written as:

$$s_{M\text{-}QAM}(t) = A_{m,I}g(t)\cos(2\pi f_c t) - A_{m,Q}g(t)\sin(2\pi f_c t) \tag{2.39}$$

From the equation above, the QAM technique can be visualised as a combination of two ASK signals from two orthogonal channels. This is obviously confirmed by the constellation diagram of 64-QAM in Figure 2.26.

The spectrum bandwidth of QAM is the same as for the M-ary ASK technique: the two orthogonal carriers share the same spectrum, which leads to a doubling of spectral efficiency. The error probability of QAM is given by:

$$P_{e,QAM} = 4\left(1-\frac{1}{\sqrt{M}}\right)Q\left(\sqrt{\frac{E_b}{N_0}}\sqrt{\frac{3k}{M-1}}\right)\left(1-\left(1-\frac{1}{\sqrt{M}}\right)Q\left(\sqrt{\frac{E_b}{N_0}}\sqrt{\frac{3k}{M-1}}\right)\right) \tag{2.40}$$

2.5.7 Orthogonal Frequency Division Multiplexing (OFDM) Modulation

Orthogonal frequency division multiplexing (OFDM) is a multi-carrier modulation technique, first proposed by Chang in 1966 [13]. The bandwidth of the transmitted signal is divided into many subcarriers that are closely spaced but orthogonal to one another. The data stream is applied to a serial-to-parallel converter to generate N parallel bit streams. Each parallel bit stream is then simultaneously modulated onto a separate narrowband orthogonal subcarrier. The modulated subcarriers are then combined and upconverted to high frequency. The primary advantage of OFDM is that the bit rate of each subcarrier is very low and this gives major advantages in a multi-path propagation environment, for example indoors. As a result, the OFDM technique is a core part of the IEEE 802.11 standards, including multi-Gigabit communication in the 60 GHz ISM band, as well as in being used in digital terrestrial TV and a host of other applications. The conceptual diagram of an OFDM modulator is shown in Figure 2.27. One of the main problems of the OFDM technique is the high degree of carrier

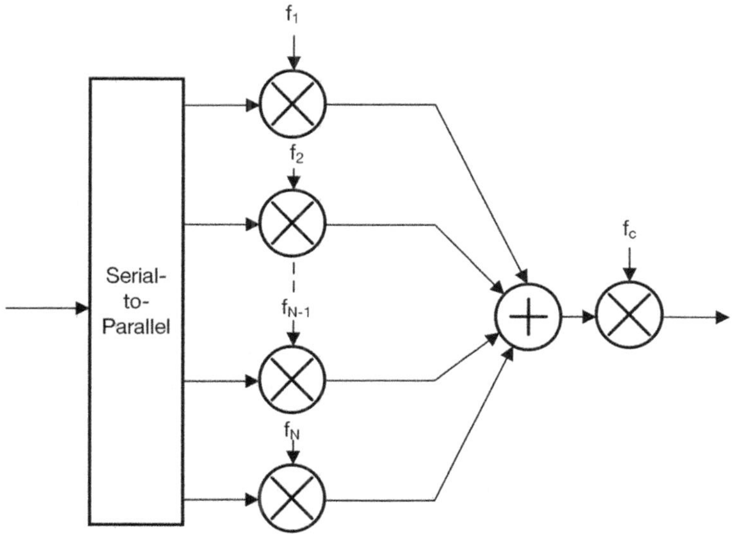

Figure 2.27 OFDM modulation technique.

signal envelope variation: the high peak-to-average power ratio problem (PAPR) puts stringent requirements onto the dynamic range of ADCs and DACs and makes efficient power amplifier design very challenging.

2.5.8 Comparison of E_b/N_0 versus BER for Various Modulation Schemes

The error probability equations for *M*-ary ASK, *M*-ary PSK and *M*-QAM as shown in Equations (2.34), (2.36) and (2.40) are very important for designing a wireless system with a systematic approach. These equations will help decide which digital modulation technique should be selected under the E_b/N_0 and the bandwidth constraints. Figure 2.28 shows the error probability plots of *M*-ary ASK ($M = 2, 4, 8, 16$), *M*-ary PSK ($M = 2, 4, 8, 16, 32$) and *M*-QAM ($M = 4, 16, 64$) versus the E_b/N_0. Note that the baseband signal modulated with *M*-ary ASK is an antipodal signal (i.e., the *M* levels are half-positive, half-negative, centred on zero). The minimum E_b/N_0 required for error-free communication is –1.59 dB, the fundamental limit given by Shannon's theory. This is shown in Figure 2.28. In a practical situation, a design has to comply with the communication standard. This means that the modulation and coding techniques are specified in advance. The error probability for a given modulation technique is specified and the required E_b/N_0 needs to be computed. For example, if the specification requires $P_e = 1 \times 10^{-5}$, meaning that there is single error per each transmission of 10 000 bits, the E_b/N_0 can be determined for each digital modulation techniques from Figure 2.28. If the transmit power is the primary concern, we should choose either binary-ASK or binary-PSK, which require approximately 9.0 dB E_b/N_0. However, these two techniques have a spectral efficiency of only 1 bit s^{-1} Hz^{-1}. Better spectral efficiency requires greater E_b/N_0: for example, QPSK achieves 2 bit s^{-1} Hz^{-1} but needs E_b/N_0 of 9.5 dB to achieve the required P_e. Pulse shaping and source coding

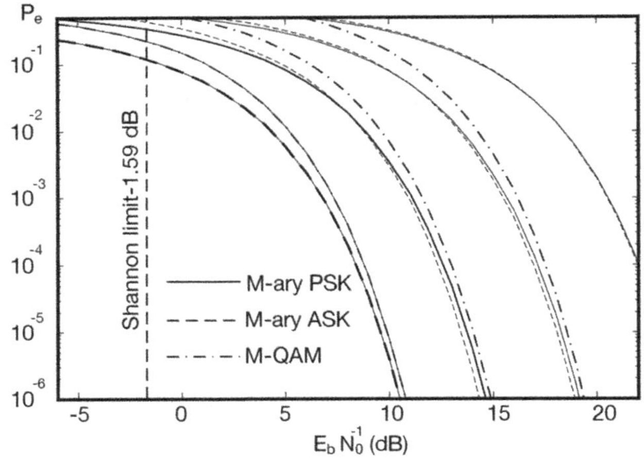

Figure 2.28 P_e of modulation techniques versus E_b/N_0 for various values of m.

techniques can be applied to improve spectral efficiency though this may add more digital hardware in the transceiver.

In a digital communication system, the E_b/N_0 can be reduced while the bandwidth utilisation efficiency is preserved by applying error-correcting codes. The encoder and decoder for a given error-correcting technique are implemented digitally at the transmitter and receiver. There are various error-correcting coding techniques, such as convolutional, Trellis and Turbo. The degree of E_b/N_0 reduction obtained from the error-correction coding technique is called the coding gain and depends on the method used. For convolutional coding with $\frac{1}{2}$ Viterbi decoding and eight-level quantised inputs, a coding gain of nearly 4 dB is obtained for $P_e = 1 \times 10^{-5}$. Hence, the coded QPSK requires only 5.5 dB E_b/N_0 to achieve the desired P_e. Considering that a 3 dB reduction means a halving of transmitter power, this is a very significant improvement.

2.5.9 Bandwidth Utilisation Efficiency and CNR

Since the radio spectrum is a very scarce resource, radio systems must be optimised so that the signal occupies a restricted amount of bandwidth. Consider that a channel must occupy a limited bandwidth of W Hz. The channel is called 'ideal' if a signal is transmitted or received through the channel without distortion occurring to either amplitude or phase. If a channel of bandwidth W Hz is ideal and a transmission rate R_T bits s^{-1} is desired, an appropriate modulation scheme should be chosen to meet this bandwidth constraint. For example, if R_T is less than W, FSK or BPSK can be chosen. However, if R_T approaches W, a more bandwidth-efficient modulation scheme such as QPSK should be used. The parameter of a digital modulation scheme that quantifies how many bits-per-second that can be transmitted in 1 Hz of bandwidth is called the spectral efficiency or bandwidth utilisation efficiency (BUE), with units of bits-per-second Hz^{-1}. The BUE of BPSK and BFSK is 1 bps Hz^{-1}, which is a low BUE. Higher BUE modulation techniques are, for example, QPSK, 8PSK or 16-QAM, which are M-ary modulation techniques. Simply choosing a high BUE modulation technique is not a simple and straightforward solution to providing high-speed communication, however, since in practice a general wireless channel has several impairments, including multipath propagation which causes inter-symbol interference (ISI). In such a case, a pulse-shaping technique is applied to shape the signal spectrum to roll off the signal bandwidth as desired.

The carrier-to-noise ratio (CNR) is defined as the ratio of the average power of the modulated carrier signal and the noise power. It is related to the E_b/N_0, channel bandwidth and bit rate as follows:

$$\frac{C}{N} = \frac{E_b}{N_0}\frac{R_b}{W} \tag{2.41}$$

where C is the high-frequency carrier signal power in watts, N is the noise power in watts, R_b is the channel bit rate in bits s^{-1} and W is the channel bandwidth in Hz.

2.6 Noise Analysis and Link Budget Calculation

2.6.1 Noise Equivalent Bandwidth

It has already been stated in the earlier sections that thermal and shot noise power spectral densities are flat over a large bandwidth. When these noise components pass through a filter having a band-pass response, some noise power is rejected by the filter. Figure 2.29 shows the amplitude response of a band-pass filter in frequency domain, $|H(f)|$, with the amplitude response of the filter, H_0, maximum at the centre frequency, f_c. The noise equivalent bandwidth, B_N, is defined by the bandwidth of an ideal filter (rectangular response) which passes the same noise power as the filter with response $H(f)$. This noise equivalent bandwidth in Hz is given by:

$$B_N = \frac{1}{H_0^2}\int_0^{\infty} |H(f)|^2\, df \tag{2.42}$$

2.6.2 Noise Analysis in a Circuit

To analyse noise in a circuit which has passive and active devices, source models for thermal and shot noise presented in the previous section can be used, along with the noise models for active devices, such as field-effect or bipolar transistors, that are available. For a passive circuit consisting of resistors only, each resistor is modelled with a thermal noise source, either as a voltage or current source. The total noise power is calculated by appropriate combining of

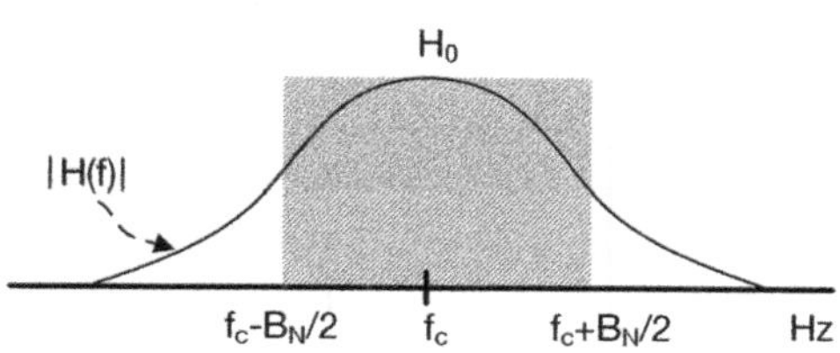

Figure 2.29 Noise equivalent bandwidth.

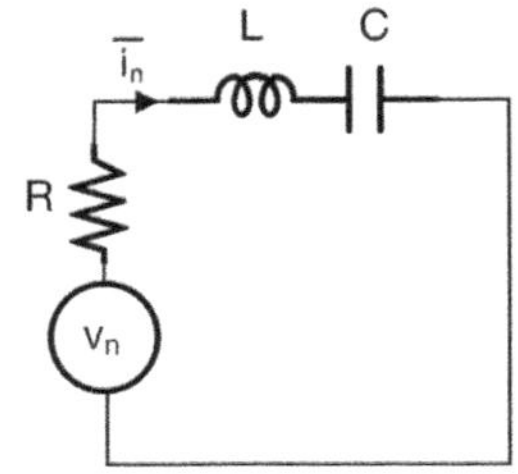

Figure 2.30 Noise analysis in series RLC circuit.

these mean-square noise sources. If there are other passive components such as a capacitor or inductor, noise analysis based on linear network theory is employed. Consider the simple series RLC circuit shown in Figure 2.30: Thermal noise naturally comes from the resistor, R, with the mean-square voltage $4k_BTR$. Using the linear network theory, the current from the noise source voltage is:

$$\overline{i_n^2} = \frac{4k_BTR}{\left|R + j\left(2\pi fL - \frac{1}{2\pi fC}\right)\right|^2} \tag{2.43}$$

Obviously, the noise current in Equation (2.43) is a frequency-dependent function. The total thermal noise power is obtained by integrating Equation (2.43) over the bandwidth of the RLC circuit, which gives:

$$\overline{i_n^2} = \frac{4k_BT}{R}\int_0^{\infty} \frac{df}{1 + \left(2\pi f\frac{L}{R} - \frac{R}{2\pi fC}\right)^2} \tag{2.44}$$

Comparing with Equation (2.43), the integrand term appearing in Equation (2.44) is the noise equivalent bandwidth of the series RLC circuit (B_N). The total thermal noise power is the product of the thermal white noise power in 1 Hz bandwidth, $4k_BT$, and B_N.

2.6.3 Figures of Merit

Noise figure (*NF*), noise factor (*F*) and noise temperature (T_e) are linear parameters to determine the noise performance of a system or a subsystem used in a receiver. Figure 2.31 shows a two-port linear network: an input is applied at port 1and the output port is port 2. The ratio of

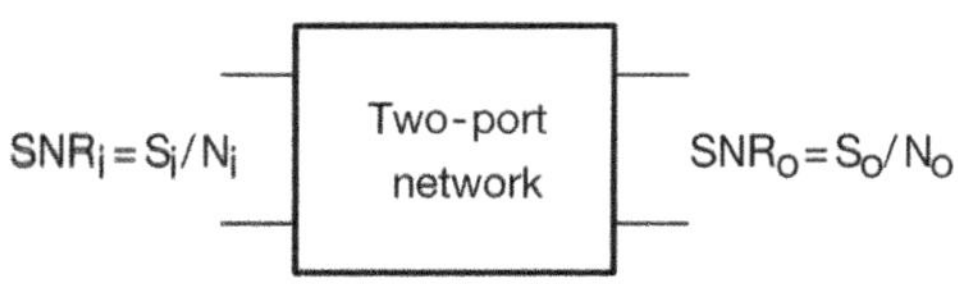

Figure 2.31 Input and output signal-to-noise ratios.

the desired signal and the noise is called a signal-to-noise ratio (SNR). The input and output SNRs are given by:

$$\mathrm{SNR}_i = \frac{S_i}{N_i} \tag{2.45}$$

$$\mathrm{SNR}_o = \frac{S_o}{N_o} \tag{2.46}$$

where S_i and S_o are the available desired signal powers at input and output ports, and N_i and N_o are available noise powers at the input and output ports, respectively. If a two-port linear network has an available power gain at specified input frequency, $G(f)$, and the available noise power contributed from the network to the output is N_A, the available output noise power at the output port will be produced from the sum of the amplified input noise and the noise added by the network, N_A:

$$N_O = G(f) \cdot N_i + N_A \tag{2.47}$$

It should be noted that Equation (2.47) is applied under the assumption that the input noise and the noise inherent in the two-port network are uncorrelated. Meanwhile the output signal is:

$$S_O = G(f) \cdot S_i \tag{2.48}$$

According to the IEEE standard definition, the noise factor (F) at a specified input frequency is defined as (a) the total noise power per unit bandwidth available at the output port when the noise temperature of the input termination is standard (290 K) to (b) that portion of (a) engendered at the input frequency by the input termination. The IEEE definition for F is strictly defined for a unit bandwidth and the specific temperature. The temperature 290 K is set as a room temperature. Hence the noise factor following this definition is called the *spot noise factor* and given by:

$$F = \frac{SNR_i}{SNR_o} \tag{2.49}$$

Applying Equations (2.47) and (2.48) into Equation (2.49) gives:

$$F = 1 + \frac{1}{G(f)} \frac{N_A}{N_i} \tag{2.50}$$

The second term of the right-hand-side term in Equation (2.50) is a contribution from the two-port network. F is always greater than 1 for a single-element receiver, meaning that the SNR at the output is worse than that at the input. In a front-end receiver, which normally has a linear amplifier to boost up the desired signal power, the noise factor of the receiver can be minimised by applying a linear amplifier with large $G(f)$ or/and small N_A. Especially in a radio receiver design, a front-end amplifier with very small N_A is mainly chosen and this amplifier is referred as a low-noise amplifier (LNA).

The noise performance of a network is most often stated in terms of its noise figure (*NF*), which is simply the noise factor in decibels:

$$NF = 10\log_{10} F \tag{2.51}$$

The spot noise figure of a two-port linear network at *f*, which is the other figure of merit is obtained by substituting Equation (2.50) to Equation (2.51):

$$NF = 10\log_{10}\left(1 + \frac{1}{G(f)}\frac{N_A}{k_B T}\right) \tag{2.52}$$

Sometimes, working with the dB units for *NF* gives an insufficient scale for specifying the noise performance of a receiver, especially for a very low-noise one. The effective input noise temperature (T_e) is another way to express the noise performance of a system. This parameter is defined in Ref. [14] as the equivalent temperature per unit bandwidth of the source impedance at the input of a noise-free device that would produce the same added noise, N_A. The available input noise N_i is referred at the reference temperature (T_0). From the definition of thermal noise power density, N_i and N_A per unit bandwidth are given by:

$$N_i = k_B T_0 \tag{2.53}$$
$$N_A = G(f)\, k_B T_e \tag{2.54}$$

Applying Equations (2.53) and (2.54) to Equation (2.50), T_e is related to the spot noise factor *F* by:

$$T_e = T_0(F - 1) \tag{2.55}$$

It is shown in Equation (2.55) that T_e is zero for a noise-free device. If a device generates noise, T_e will be increased with *F* (or *NF*). Figure 2.32 shows the plot of T_e with *NF* varied from 0.0 to 3.0 dB. The noise temperature increases from 0 to 289 K with 290 K T_0.

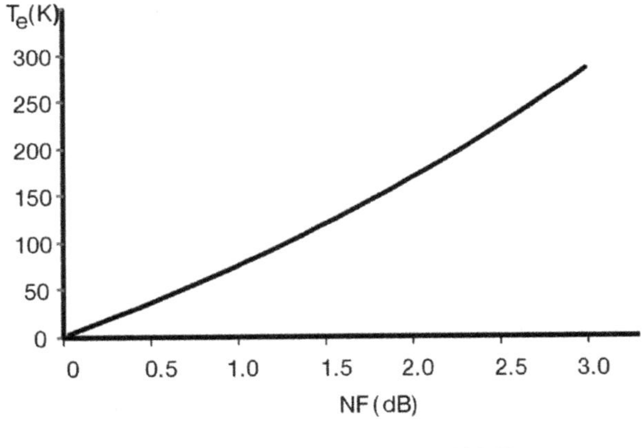

Figure 2.32 T_e and *NF* (T_0 = 290 K).

2.6.3.1 Noise Figure of Mixers

There are two definitions for the noise figure of a mixer. These are the single-sideband (SSB) and double-sideband (DSB) noise figure definitions, NF_{SSB} and NF_{DSB}. In a general mixer, the output signal spectrum is the frequency-translated spectrum of the input signal spectrum combined with an undesired white noise power contribution from both the desired signal band and the image band. The single-side band spot noise factor (F_{SSB}) calculated with both noise sidebands is defined as:

$$F_{SSB} = 2 + \frac{1}{G_{cv}(f)} \frac{N_A}{N_i} \tag{2.56}$$

where $G_{\mathrm{cv}}(f)$ is the conversion gain of the mixer. The single-sideband noise figure is normally the true reflection of the performance since the desired signal is only in one sideband at the mixer input, whilst unwanted noise appears in both upper and lower sidebands and is converted down to the same IF output band. Given a single-sideband noise figure, NF_{SSB}, in dB, normally NF_{DSB} will be 3 dB lower than NF_{SSB}. The distinction between double-sideband and single-sideband noise figures must be very carefully considered when measuring certain circuits, such as the image-rejection mixer and narrowband circuits, and it should not always be assumed that a 3 dB difference applies.

2.6.3.2 Noise Figure of a Lossy Device

There are several passive circuits in the front-end of a wireless transceiver, such as the antenna feed, filter, diplexer, isolator, to name a few, and the noise figure of the receiver is affected by the loss of these passive circuits. For a passive component on its own, it is widely presumed that the noise figure is equal to the insertion loss in dB. However, with proper analysis it can be shown that this is only true for $T = 290$ K, when the thermal noise added balances the signal dissipation, and when the component is perfectly matched. A much greater consideration is that the overall noise figure of the receiver depends strongly on where the lossy passive components are in the receiver chain. For example, a lossy cable placed before the LNA has a much greater effect than the same cable placed after the LNA; the next section will show why this is so.

2.6.4 *Noise Performance of Cascaded Networks*

In practice, a communication system normally consists of many networks connected in cascade. The noise figure defined for a single-stage, two-port network is not sufficient to determine the noise performance. Figure 2.33 shows a general cascade of N subcircuits. For the analysis it

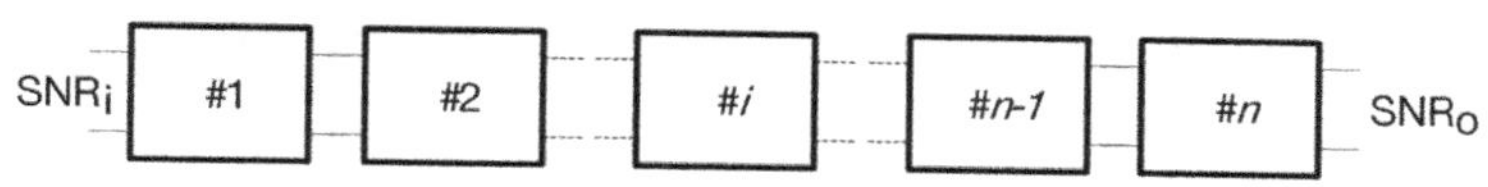

Figure 2.33 *N*-stage cascaded subsystem.

is assumed that they are unilateral two-port networks with the same bandwidth and operating at the same temperature.

The gain at frequency f of each subcircuit is denoted by G_i (not in dB, $i = 1, 2, 3,.N$) and the noise factor of the ith block is F_i. The total noise factor of the cascaded system shown in Figure 2.33 can be calculated using Friis's noise formula, and is given by:

$$F = F_1 + \frac{1}{G_1}(F_2 - 1) + \frac{1}{G_1G_2}(F_3 - 1) + \cdots + \frac{1}{G_1G_2 \cdots G_{n-1}}(F_n - 1) \qquad (2.57)$$

By inspection, it can be seen that F is dominated by the first two terms which are governed by F_1, F_2 and G_1. The contributions of following stages are reduced by the preceding gains. This shows that the first two stages of a cascaded system should be carefully designed. Therefore, the LNA, which is usually the first or second stage of a wireless front end, needs to have high gain and very low noise factor. Applying Equation (2.55) into (2.57), the noise temperature of a cascade of networks is:

$$T_e = T_1 + \frac{1}{G_1}(T_2 - 1) + \frac{1}{G_1G_2}(T_3 - 1) + \cdots + \frac{1}{G_1G_2 \cdots G_{n-1}}(T_n - 1) \qquad (2.58)$$

It should be noted that the equations of F and T_e in Equations (2.57) and (2.58) are only applicable for a unilateral device, meaning that there is no mismatch loss. The mismatch loss comes from the difference between input and the load impedances of the connected two-port devices. If this occurs, a complete analysis of the mismatch effect in the noise calculation should be considered. Fortunately, modern RF design software can take such effects into account.

2.6.4.1 Selectivity

Selectivity is the ability of a receiver to select the desired signal at a given frequency and reject interference signals at the adjacent channels. The receiver selectivity is very important for a multichannel communication system where frequency allocations for different systems and users are close together as the spectrum is scarce. The selectivity is determined primarily by the shape factor of the RF front-end circuits (the devices starting from receiver antenna to the output of first mixer) or the IF filter in a superheterodyne receiver. Figure 2.34 shows

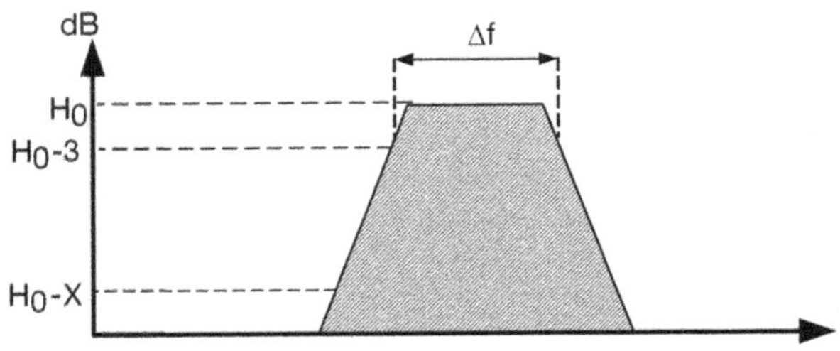

Figure 2.34 Selectivity factor.

the frequency response of a generic receiver for defining selectivity. The frequency response of the receiver is $H(f)$. The passband bandwidth, the 3-dB bandwidth, is denoted by Δf. The target rejection of X dB, and the rejection bandwidth, Δf_r, are determined from the difference at two frequencies for which the rejection level of X dB occurs. The selectivity of the receiver is given by:

$$\Omega = \frac{f_r}{f} - 1 \tag{2.59}$$

If the attenuation slope in Figure 2.34 is very large, Δf_rwill be nearly equal to Δf. This would mean Ω is very small. On the other hand, a receiver with poor selectivity will have large Ω since Δf_r is somewhat larger than Δf.

The rejection performance for interference determined by X is normally required by the standards specification: for example GSM, IEEE 802.11c, 3GPP, and so forth. If not specified, 60 to 90 dB is chosen for an excellent receiver and at least 45 dB amplitude attenuation is acceptable in most cases.

2.6.4.2 Sensitivity

Sensitivity is the ability of a receiver to detect the minimum required RF signal level for demodulating data at the operating frequency. This ability is defined by the minimum signal level required to yield a minimum carrier-to-noise ratio (CNR_{min}) for achieving a required bit-error-rate (BER) or frame-error-rate (FER), depending on the modulation and coding techniques in the communication link. The sensitivity depends on the NF of the receiver, hence this relates to noise generation in the receiver. If the signal power at the receiver input is smaller than the minimum detectable signal (MDS), it will be buried by the receiver noise. A receiver which has a very small MDS is a receiver with good sensitivity. Only thermal noise is assumed to appear in the system where other noises and interference are not defined. The sensitivity of a receiver in a given bandwidth $\Delta f$to obtain the minimum required SNR for obtaining the desired BER is given by:

$$P_{sense} = -173.83 + NF_{Rx} + 10\log_{10}\Delta f + CNR_{min} \tag{2.60}$$

NF_{Rx} shown in Equation (2.60) is the overall noise figure of the receiver. Generally, a radio receiver is a cascade structure of several subcircuits, and the formula for calculating the NF of cascaded networks is used to calculate NF_{rx}. The –173.83 dBm constant is the noise floor level at room temperature 290 K. The MDS, sometimes just called the receiver noise floor, is given by:

$$MDS = -173.83 + NF_{Rx} + 10\log_{10}\Delta f \tag{2.61}$$

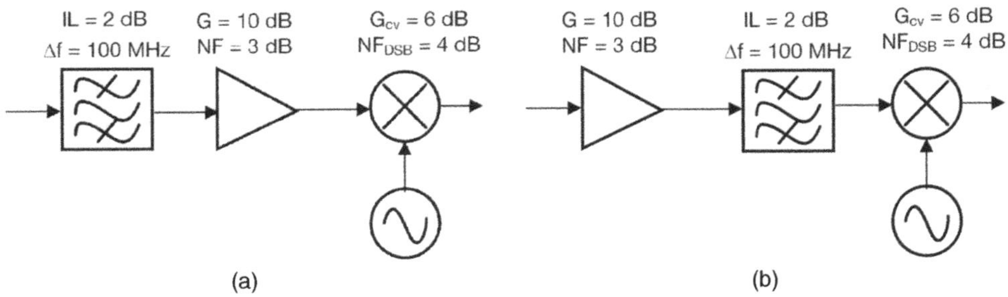

Figure 2.35 (a) Front-end type I. (b) Front-end type II.

2.6.5 Noise Calculation Examples

2.6.5.1 Noise figure cascade example

A high-frequency wireless receiver front-end should be carefully designed since it has a major influence on the overall performance. A simple front-end circuit is composed of a band-pass filter, LNA and a downconversion mixer. Let us assume that we have two front end design types (I and II) as shown in Figure 2.35. The received signal frequency is 950.30 MHz and the LO frequency is 950.00 MHz. The passive components and active devices used in these two front-end types are the same where their parameters are labelled. The parameters of each block are summarised in Table 2.2. The difference between these two front-ends is the location of the band-pass filter.

The mixer used is assumed to be a general mixer. Hence the calculation is based on NF_{SSB}, which is 7 dB. From Equation (2.57) the cascaded noise factors for type I and type II front-end circuits are:

$$F_I = 1.59 + \frac{1}{0.63}(2-1) + \frac{1}{0.63 \times 10}(5.01-1) = 3.80$$

$$F_{II} = 2 + \frac{1}{10}(1.59-1) + \frac{1}{0.63 \times 10}(5.01-1) = 2.70$$

As expected, type II has the better noise factor, since the lossy filter is after the LNA. However, since the purpose of the filter is to remove interfering signals, careful consideration must also be given to the possibility of type II suffering from nonlinear effects in the presence of strong interfering signals that fall within the LNA bandwidth. In practice, the optimum receiver chain

Table 2.2 Summary of design parameters of components for two front-end types.

Item	Gain (dB)	Gain	NF (dB)	F
BPF	−2	0.63	2	1.59
LNA	10	10	3	2
Mixer	6	4	7	5.01

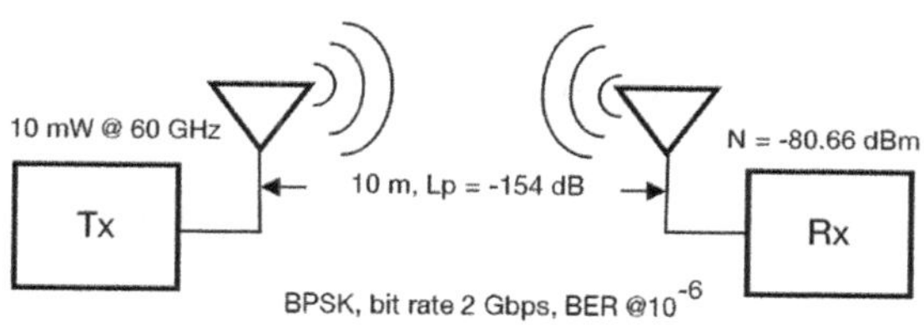

Figure 2.36 Example 60 GHz wireless transmission system.

requires careful consideration of all these effects, and since the receiver may have many tens of dB of gain, care must also be exercised in carefully considering the total noise power as the signal passes through the receiver chain. This is a reason why double- and triple-conversion superheterodyne receivers (introduced in the next section), give superlative performance; they offer the greatest flexibility, by choosing proper IF(s), for achieving step-by-step removal of interference by filtering, removal of noise power by using narrowband IF filters to reduce equivalent noise bandwidth, and spreading gain across the stages to improve linearity.

2.6.5.2 Link Budget Example

In this example, a 2 Gbps indoor wireless data transmission operating at 60 GHz is considered, as shown in Figure 2.36. The channel bandwidth for this communication link is 2.16 GHz. The targeted bit-error-rate is 1×10^{-6} and a BPSK modulation technique is proposed. For a 10 m LOS wireless link, the link budget is to be calculated given that the transmitter power is 10 mW and the noise figure of the receiver is 12 dB.

(a) **Minimum CNR**

From Figure 2.28, the required minimum E_b/N_0 is 9.5 dB, which is equal to 8.91. With the desired data rate $R_b = 2 \times 10^9$ bits s^{-1} and the available channel bandwidth $\Delta f = 2.16 \times 10^9$ Hz, the minimum CNR is given by:

$$CNR = \frac{E_b}{N_0}\frac{R_b}{\Delta f} = 8.91\left(\frac{2 \times 10^9}{2.16 \times 10^9}\right) = 8.25$$

The minimum CNR at the demodulator is 9.17 dB.

(b) **Noise Calculation at the Receiver**

The noise power for 1 Hz bandwidth at 290 K is approximately −174 dBm. With 2.16 GHz channel bandwidth, the total noise power is:

$$N \approx -174 + 10\log_{10}\left(2.16 \times 10^9\right) = -80.66\,\text{dBm}$$

The path loss exponent for indoor LOS model obtained from Table 2.1 is equal to 2. At 5 mm wavelength (60 GHz), the path loss for a 10 m indoor LOS link is calculated from:

$$L_p = 10\,(2)\log_{10}\frac{4\pi \times 10}{0.005} = 88.0\text{ dB}$$

Assuming that the antenna gains of the transmitter and receiver are equal, i.e. $G = G_t = G_r$, the power of received signal will be:

$$P_r = 10 + 2G - 88 = -78 + 2G\,\text{dBm}$$

Since the noise figure of the receiver front-end is 12 dB, the noise power appearing at the receiver output is:

$$N_{rx} = -80.66 + 12 = -68.66\,\text{dBm}$$

The required minimum signal power for the demodulator should be:

$$P_r = -68.66 + 9.17 = -59.49\,\text{dBm}$$

Therefore, to operate the communication link at 60 GHz, each antenna should have a gain given by:

$$G = \frac{P_r + 78}{2} = \frac{-59.49 + 78}{2} = 9.26\,\text{dBi}$$

The results of the link budget parameters obtained for this example are shown in Figure 2.37. Starting with the transmitter power at 10 dBm, the EIRP of the antenna output at the transmitter is 19.26 dBm. The received EIRP at the antenna input is attenuated to –68.74 dBm due to the path loss of 10 m wireless link at 60 GHz. The input power at the receiver is –59.49 dBm, while the receiver noise is –80.66 dBm within 2.16 GHz bandwidth. The noise in the receiver is added to the input noise for 12 dB, leading to the noise at the receiver output equal to –68.66 dBm. This finally gives an output CNR of 9.17 dB, which is the minimum CNR for demodulating BPSK data at 1×10^{-6} BER.

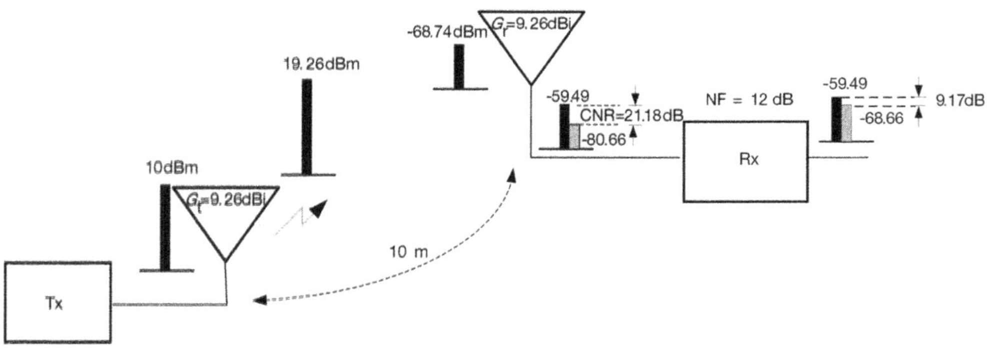

Figure 2.37 Diagram of wireless link budget calculation.

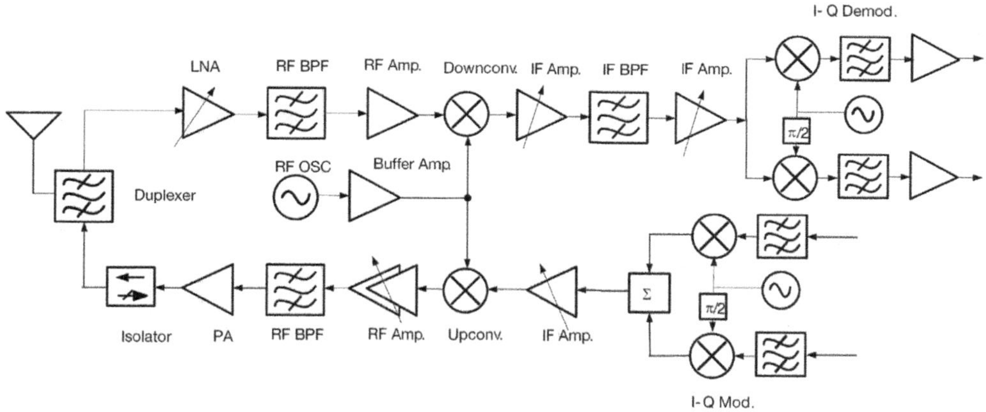

Figure 2.38 A full-duplex single-conversion superheterodyne transceiver.

2.7 Some Wireless Transceiver Architectures

The basic superheterodyne receiver was introduced earlier in order to give context to the sections on noise, interference, link budgets, and modulation theory. This traditional architecture generally employs a self-contained modulator and demodulator operating at an IF frequency, with frequency translation used to reach the RF carrier frequency. However, even the superheterodyne approach comes in several forms, including single-, double-, and triple-conversion types, since diverse applications may pose different challenges in terms of noise performance, interference rejection and selectivity. Silicon integrated circuit technology was the driving force behind the wireless internet revolution, and in the early years the single-chip transceiver was considered to be the dream solution for a low-cost wireless system. This forced engineers to develop a range of new transceiver architectures in order to minimise the need for off-chip components, such as IF filters, and in many cases some long-forgotten radio design techniques from the valve/tube era were revisited. The result is that there is now a very wide range of transceiver architectures in use, including zero-IF, low-IF and sampling-type transceivers, and the optimum solution depends very much on the application and the detailed specifications.

2.7.1 Superheterodyne

Figure 2.38 shows a full-duplex superheterodyne transceiver, which consists of a single-conversion transmitter and single-conversion receiver. In the receiver, a frequency mixer translates the incoming signal to the intermediate frequency (IF). In the transmitter, the modulated IF signal is upconverted to RF by a mixer/LO combination.

Figure 2.39 shows the simplified front end of a single-conversion superheterodyne receiver. The antenna receives a desired signal at f_c with bandwidth Δf. The LO signal is f_{LO}, which is the frequency difference between f_C and f_{IF1}, the intermediate frequency. The image band spectrum centred at f_{IM} also appears at the mixer RF input port. The frequency f_{IM} is the mirror image of f_c around f_{LO}. The downconversion mixer shifts the signal spectrum from f_c down to

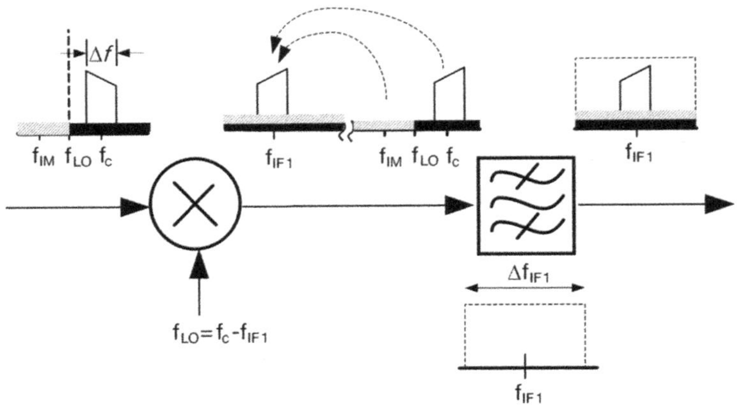

Figure 2.39 Simplified single conversion receiver front end.

f_{IF1}. The spectrum centred at f_{IM} also shifts to f_{IF1}. At the mixer IF output port, assuming there are no signals in the image band, the spectrum therefore consists of: (1) the desired signal downconverted from f_c to f_{IF1}; (2) the thermal white noise around f_c; and (3) the thermal white noise around f_{IM}. Hence, with the IF band-pass filter centred at f_{IF1} with bandwidth Δf_{IF1}, the desired and two noise components are in the IF filter passband. The SNR at the front end of the single-conversion receiver is therefore given by:

$$SNR_{single-conv} = \frac{P_S}{(N_0 + N_{o-IM})\Delta f_{IF1}} \tag{2.62}$$

where P_S is the power of the desired signal. The denominator term in Equation (2.62) is the total noise power from the image noise and in-band noise which are both in the passband of the IF filter. The equation shows that the SNR depends on the bandwidth of IF filter. If the filter bandwidth is narrow, the SNR will be enhanced. Theoretically, this filter bandwidth should be at least equal to half of the information bandwidth. However, a narrowband band-pass filter operating at high frequency with low loss and sharp rejection is generally known to be extremely difficult to realise.

A further consideration is the bandwidth of the RF preselector filter. The frequency separation between the image and the RF is equal to double the IF frequency, as shown in Figure 2.40.

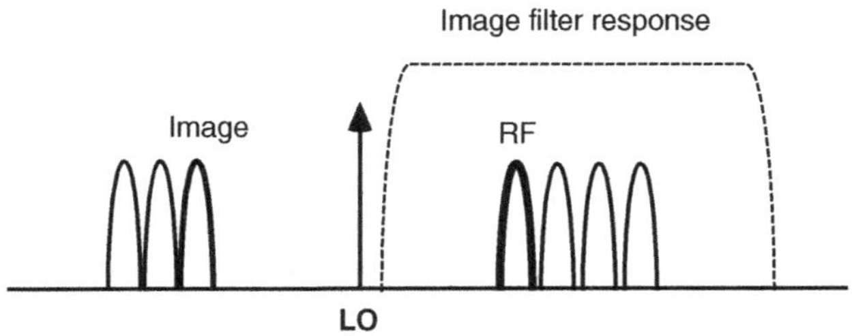

Figure 2.40 RF filter for image band rejection.

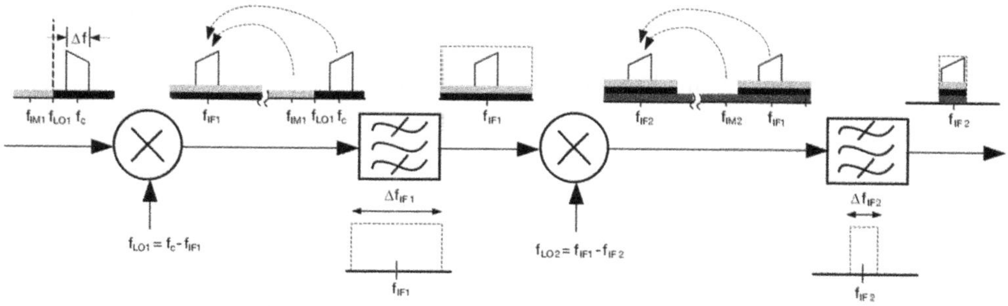

Figure 2.41 A simplified double-conversion receiver.

A highly selective RF band-pass filter will be bulky and expensive, so a high IF is preferred in order to make it easier to separate the image and RF signals before the mixer. However, if the IF is too high then designing a narrow-band IF filter may be difficult – and the bandwidth of the IF filter has been shown to be a crucial factor in achieving good SNR. So, there is a difficult trade-off which must be addressed in the receiver design. For many applications, this trade-off between the conflicting requirements on RF and IF filter specifications cannot be resolved satisfactorily.

2.7.2 *Double Conversion*

One solution to the problem is to use a double frequency conversion scheme. The first IF is chosen to be quite high, so that the image filtering is easy. The second IF is low to make the design and realisation of a narrowband channel selection filter easier. Often, three stages of frequency conversion may be necessary. In some applications, the first mixer may actually upconvert the signals. Hence, double- and triple-downconversion schemes are often used, leading to lower IF frequencies and narrow receiver bandwidth, and finally improved received SNR. There is a similar situation in the transmitter: the upconverting mixer will produce an upper and lower sideband at its output, separated by $2 \times f_{IF}$. Hence, if the IF is low, the sidebands will be very close together and it will be extremely difficult to remove the unwanted sideband by filtering.

Figure 2.41 shows a simplified double-conversion receiver. The second downconversion mixer shifts the spectrum output of the first IF filter from f_{IF1} to f_{IF2} through the use of a second LO signal at f_{LO2}. The downconverted signal is then passed through the second IF filter centred at f_{IF2}. Since the centre frequency of this IF filter is somewhat lower than f_{IF1}, the bandwidth of the second IF filter, Δf_{IF2}, can be designed to be much smaller than the first IF filter. The spectrum at the second IF filter output consists of: (1) the desired spectrum; (2) in-band noise; (3) image noise from the first IF; and (4) image noise from the second IF. Hence, the output SNR of the double-conversion receiver is given by:

$$SNR = \frac{P_S}{(N_0 + N_{0-IM1} + N_{0-IM2})\Delta f_{IF2}} \tag{2.63}$$

Although there are more noise terms in the double-conversion receiver, the total noise power at the output is smaller than a single-conversion one because the receiver bandwidth can be significantly reduced through the lower IF bandwidth; that is:

$$\Delta f_{IF2} \ll \Delta f_{IF1} \tag{2.64}$$

This double-conversion technique relaxes the need for a high-frequency IF filter with narrow bandwidth and sharp cut-off. Double downconverting the signal to a low IF makes it easier to realise a band-pass filter with sharp cut-off and narrow bandwidth. In fact, there are many IF filters available commercially and the choice of IF in a superheterodyne receiver should be determined as part of the systematic design process to ensure that the desired IF filters are available on the market. The most common technologies for IF filters are surface acoustic wave (SAW) filters, ceramic filters and crystal filters. Examples of IF frequencies commonly used are 455 kHz, 10.7 MHz, 70 MHz, 140 MHz, 325 MHz, 415 Hz, to name a few.

More IF stages can be added to achieve even better selectivity (sharp cut-off filter) and sensitivity (narrowband filter), but this would increase system complexity and power consumption. The multiple IF solution is not, therefore, an ideal choice for integrated circuit (IC) applications. In particular, since a very narrowband and low loss filter is extremely difficult to realise in IC technology, a transceiver IC using the superheterodyne architecture requires off-chip IF filters. This will typically require the use of buffer amplifiers that consume DC power and will make the packaging and system integration more challenging. For compact silicon IC implementations, it is virtually impossible to implement a receiver with all these filters, so numerous modern RFIC (radiofrequency integrated circuit) transceivers use direct conversion, or zero-IF architectures, that perform frequency translation directly to baseband.

2.7.3 Direct Conversion

Compared with the superheterodyne transceiver, the direct-conversion structure shown in Figure 2.42 is considerably simplified due to the removal of IF circuitry. It is reasonably straightforward to realise the direct conversion transmitter since there are no adjacent channels to worry about. However, we have already seen that the IF circuitry in a superheterodyne receiver plays a very important role in improving selectivity, selecting the desired signal and rejecting adjacent channels, interference and noise. A full-duplex direct-conversion transceiver is often called a zero-IF transceiver since the IF is DC, or close to DC. The main principle of the direct-conversion transceiver is to directly convert baseband signals to and from the RF carrier frequency. The image problem is eliminated since the IF is zero. By removing the chain of mixers, oscillators and filters from the double-conversion superheterodyne architecture, the direct conversion transceiver is ideally suited to RFIC implementation and has attracted a lot of attention from engineers as a low-cost transceiver solution [15, 16]. However, there are significant technical problems that need to be addressed, summarised in the following sections.

2.7.3.1 DC Offset Problem

Since the LO frequency is equal to the RF carrier and the RF signal is downconverted to DC, the LO cannot be distinguished from the RF and there are a number of causes for unwanted

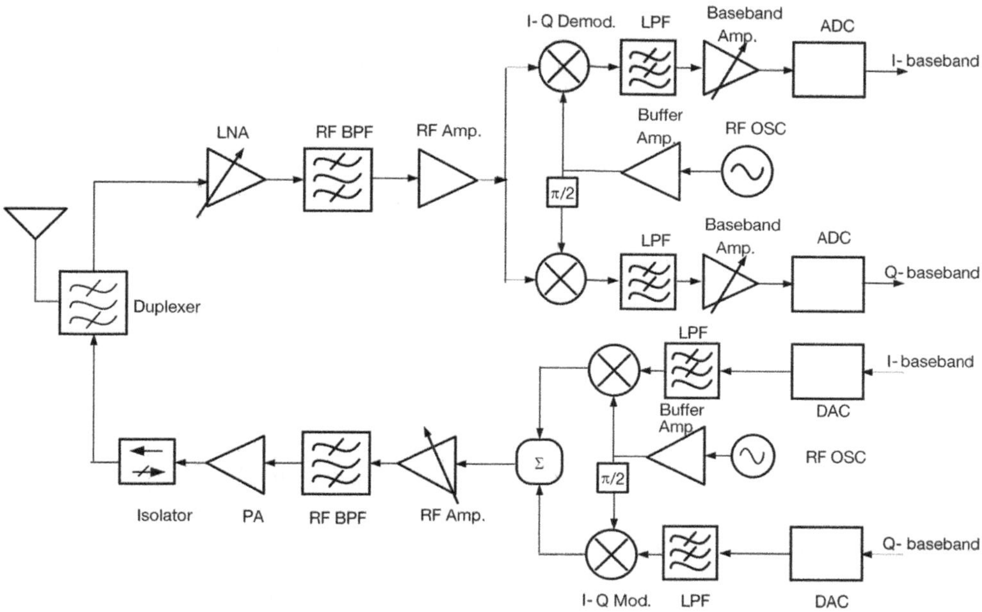

Figure 2.42 Full-duplex direct-conversion transceiver.

DC offsets at the mixer output which can degrade the BER. Removing the DC offset with a DC blocking capacitor is usually impractical because the digital modulation formats have a spectrum peak at DC. One major cause of DC offsets is LO self-mixing which can result from: (1) LO leakage which internally couples in the receiver; (2) a leakage from the transmitter in the full-duplex transceiver; (3) a radiated LO, due to limited LNA reverse isolation, which is reflected back to the receiver; and (4) a strong interferer. Figure 2.43 shows how the LO signal can leak to the RF port of the mixer and be reflected back, where it self-mixes and creates a DC offset at the demodulator output. The DC offset occurring in this case is static and could be eliminated prior to communication. By injecting an LO signal at the LO port of the demodulator, this static DC offset can be characterised and calibrated out.

In a full-duplex transceiver chip, the duplexer has finite isolation and some of the signal from the power amplifier will be mixed with the LO in the receiver, via several paths, as shown in Figure 2.44. To alleviate the DC offset from this self-mixing, the carrier frequency at the receiver is chosen to be twice the frequency of the transmitter carrier.

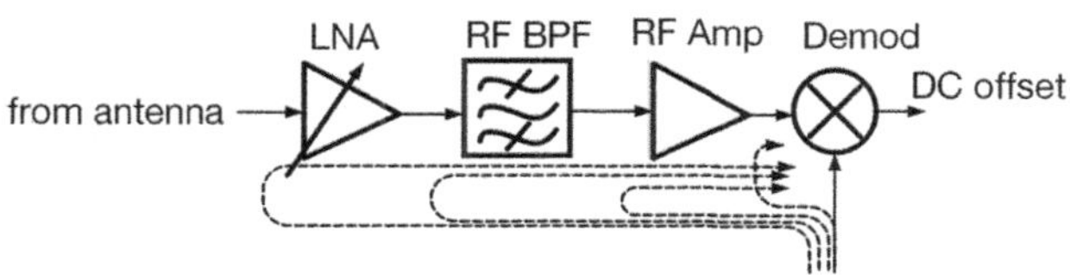

Figure 2.43 LO leakage causing self-mixing.

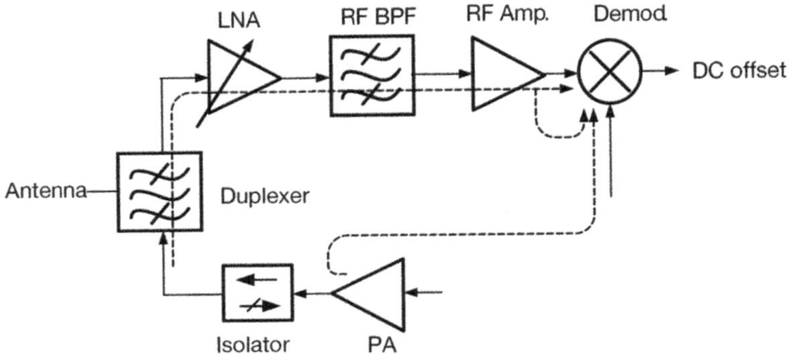

Figure 2.44 Leakage from the transmitter and self-mixing.

Whilst static DC offset problems can be overcome reasonably simply, the other causes of DC offset – that is, LO radiation from the antenna and reflection back into the receiver – and the strong interferer case caused by a nearby user in the same receiver band are more difficult to combat since the DC offsets produced are time-varying, as shown in Figure 2.45. The radiated LO signal reflects from nearby objects and is picked up by the front-end receiver and self-mixed in the demodulator. Another cause of DC offsets is the presence of a strong in-band interferer. The DC offsets caused by these two mechanisms are time-varying signals and are much more difficult to cancel out than for the static case. More sophisticated techniques, such as adaptive DC offset cancellation loops, are required to eliminate the offsets.

2.7.3.2 I-Q Mismatch

I-Q imbalance in the demodulator causes severe effects in a direct-conversion receiver since the received RF signal is directly demodulated to baseband. The imbalance is caused by different amplitude and phase responses in the I and Q baseband channels in the demodulator. An I-Q calibration technique either using analogue or digital techniques can alleviate this problem.

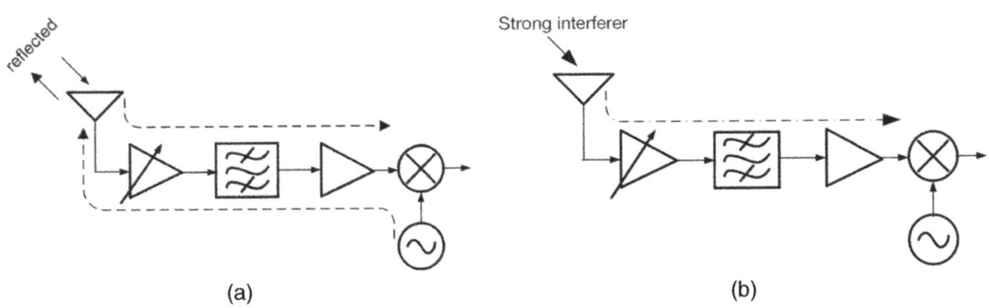

Figure 2.45 Time-varying DC offsets caused by (a) radiated LO or (b) strong interferer.

2.7.3.3 Second-Order Nonlinearity

Active devices are nonlinear and the nonlinearity can be simply described as a Volterra series. When the RF signal is applied in a direct-conversion receiver, the second-order nonlinear effects in the device generate DC components which subsequently distort the DC baseband signal. In fact, all even-order nonlinearity terms produce distortion components at DC, but the second-order term is more troublesome due to its stronger nonlinearity coefficient. The DC distortion generated from second-order nonlinear distortion in LNA is blocked by the RF BPF. The DC distortion signal is primarily produced from the I-Q demodulator. The distortion output from the demodulator depends on the degree of second-order nonlinearity of each modulator and the isolation between the LO and RF port. Consequently, the second-order intermodulation distortion occurring in the front-end receiver and the RF-LO isolation of the mixer as a demodulator must be given careful consideration in a direct-conversion architecture.

2.7.4 Low-IF Architecture

The DC offset problem in the direction-conversion transceiver brings many challenging design issues for RF engineers to overcome. On the other hand, a superheterodyne transceiver is not a low-cost IC solution due to the requirement of a high-selectivity IF filter. A low-IF receiver is a compromise between the superheterodyne and direct-conversion architectures. The DC offset problem and flicker noise from active devices are avoided by selecting the IF at low frequency. For a simple and low-cost solution, the IF frequency should not be as high as the IF required in a superheterodyne receiver. Normally, the IF frequency in a low IF receiver is not more than twice the information bandwidth. Figure 2.46 shows the block diagram of a low-IF receiver. The RF front-end of this receiver consists of an antenna, band-pass filter, LO, LNA and I-Q downconverter mixer. The low-frequency IF signals (I and Q) downconverted from the I-Q mixer are amplified, noise filtered by low-pass filters, digitised by analogue-to-digital converters (ADCs) to I and Q signals, and finally demodulated to baseband with either polyphase filters or digital dual I/Q downconverters. Thus, an important technical challenge in a low-IF receiver is the image suppression performance since the IF frequency is rather low so that the possible images are very close to the desired band. Either circuit, a polyphase filter

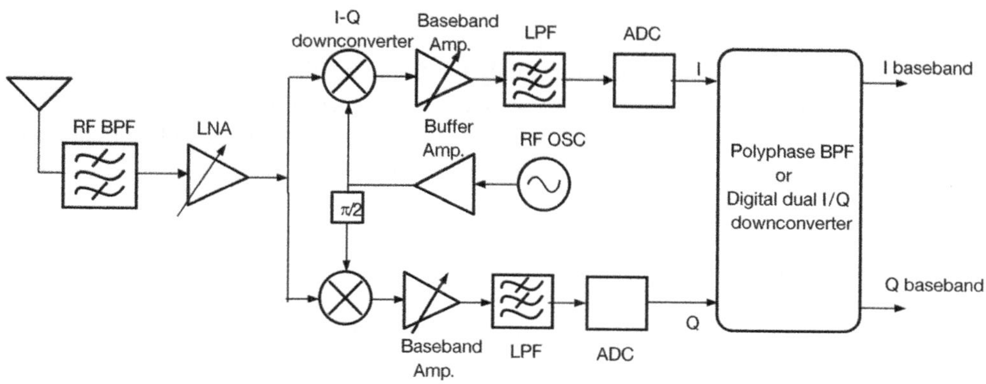

Figure 2.46 Low-IF receiver.

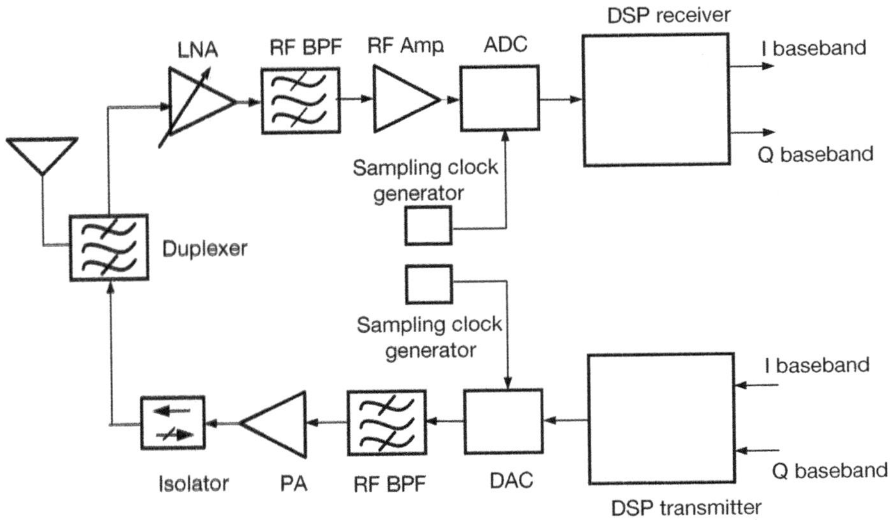

Figure 2.47 Band-pass sampling receiver.

or dual I/Q downconverter, are necessary to suppress unwanted image in a low-IF receiver. These two circuits require little DC power because they can be implemented with passive components.

2.7.5 *Band-Pass Sampling Receiver*

A band-pass sampling transceiver employs a subsampling technique that performs a frequency conversion using a sampling frequency at a rate lower than the highest frequency of the information signal that is frequency translated. The high-frequency signal is either directly digitised or combined with the superheterodyne technique to digitise an IF signal. A block diagram of the RF direct-digitised transceiver is shown in Figure 2.47. The frequency conversion in the transmitter and receiver are achieved using samplers in the ADC/DAC with the sampling clock signals fed from the sampling clock generators. The clock frequency is lower than the sampled signal frequency and this makes the technique, at first glance, more attractive. The architecture is considerably simplified, especially the front-end circuit, using ADC/DAC and DSP techniques but the architecture requires a high-performance ADC/DAC.

With a given band-pass analogue signal of bandwidth $\Delta f = f_H - f_L$, the sampling frequency (f_s) of the clock for no aliasing should be chosen such that:

$$\frac{2f_H}{n} \leq f_s \leq \frac{2f_L}{n-1} \tag{2.65}$$

n is a harmonic order number whose the maximum value is defined by:

$$n_{max} = \left\lfloor \frac{f_H}{\Delta f} \right\rfloor \tag{2.66}$$

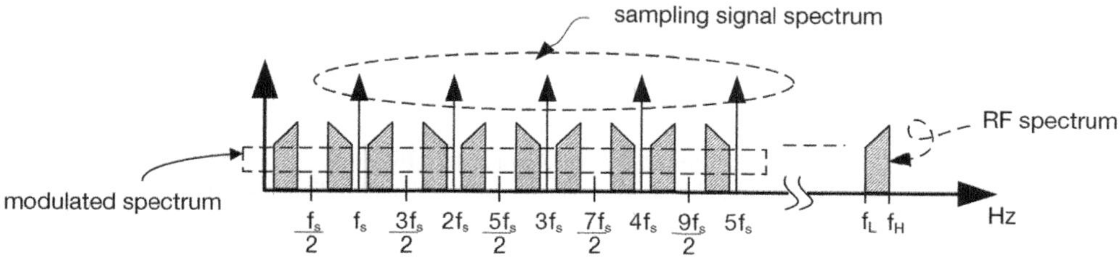

Figure 2.48 Sampling spectrum from a band-pass sampling receiver.

where $\lfloor \cdot \rfloor$ is the nearest integer number The signal spectrum obtained using a subsampling technique when used for frequency downconversion is shown in Figure 2.48. The clock spectrum of the sampling frequency, $f_{S,}$ used for sampling the modulated RF spectrum is a train of impulses of equal amplitude that occur at integer multiples of f_s. The information spectrum is duplicated from DC to nf_s and repeated at every f_S as shown in Figure 2.48. The relocated spectrum at low-frequency will be used and the binary digits will be demodulated by the DSP receiver. Compared with other receiver architectures, the SNR of the band-pass sampling receiver is degraded by the n-1 out-of-band noise signals and the in-band noise power ($P_{\text{N,inband}}$) by:

$$SNR_{SBP-sampling} = \frac{P_S}{P_{N,inband} + (n-1)\,P_{N,outofband}} \qquad (2.67)$$

Implementing an RF direct band-pass sampling receiver requires a high-dynamic range ADC. A less stringent requirement on the dynamic range of the ADC is to apply a subsampling technique to the IF signal. This is a combination of a single-conversion superheterodyne receiver and the subsampling receiver. For a band-pass sampling transmitter, a direct-digital-synthesiser is fed to a DAC to produce an analogue signal and upconvert it to high frequency.

2.8 Conclusion

This chapter has introduced a wide range of topics that are very important for designing a wireless communication transmitter or receiver. The trend of emerging wireless communication standards is to support ever-higher data rates. Some basic elements of a wireless transceiver have been described to provide a system background. In every communication link, noise and interference are unwanted signals which are unavoidable. Understanding their effects will help in designing a communication system effectively. The main digital modulation techniques commonly used nowadays have been summarised. Spectral efficiency (BUE), E_b/N_0 and BER, the key performance parameters of these digital modulation techniques, have been described. The concepts of path loss and noise figure analysis have been explained, with numerical examples. Finally, the key transceiver architectures have been introduced briefly. Understanding the topics presented in this chapter is essential for RF engineers to be able to contribute effectively to the design of microwave and millimetre-wave circuits and subsystems for communications applications.

References

1. http://www.thevalvepage.com/radtech/synchro/synchro.htm.
2. Tucker, D.G. (1954) The history of the homodyne and synchrodyne. *Journal of the British Institution of Radio Engineers*, **14** (4), 143–154.
3. Hogan, J.V.L. (1921) The heterodyne receiver. *Electric Journal*, **18**, 116.
4. http://frank.yueksel.org/other/RCA/Radiotron_Designers-Handbook_Fourth-Edition/34-Types-of-AM-Receivers.pdf.
5. Johnson, J.B. (1928) *Thermal Agitation of Electricity in Conductors*. The American Physical Society.
6. Nyquist, H. (1928) Thermal agitation of electric charge in conductors. *Physics Reviews*, **32**, 110.
7. Friis, H.T. (1944) Noise figures of radio receivers. *Proceedings of the IRE*, **32** (7), 419–422.
8. Friis, H.T. (1946) A note on a simple transmission formula. *Proceedings of the IRE*, **34** (5), 254–256.
9. Crane, R.K. (1981) Fundamental limitations caused by RF propagation. *Proceedings of the IEEE*, **69** (2), 196–209.
10. Smulders, P.F.M. (2009) Statistical characterization of 60-GHz indoor radio channels. *IEEE Transactions on Antennas and Propagations*, **57** (10), 2820–2829.
11. Valdes-Garcia, A., Nicolson, S.T., Jie-Wei Lai, Natarajan, A., *et al.* (2010) A fully-integrated 16-element phased-array receiver in SiGe BiCMOS for 60-GHz communications. *IEEE Journal of Solid-State Circuits*, **45** (12), 2757–2773.
12. Proakis, J.G. and Salehi, M. (2008) *Digital Communications*, 5th edn, McGraw-Hill.
13. Chang, R.W. (1966) Synthesis of band-limited orthogonal signals for multichannel data transmission. *Bell System Technical Journal*, **45** (10), 1775–1796.
14. Agilent/Keysight. *Fundamentals of RF and Microwave Noise Figure Measurements*, Application Note 57-1.
15. Svitek, R. and Raman, S. (2005) DC offsets in direct-conversion receivers: characterization and implications. *IEEE Microwave Magazine*, **6** (3), 76–86.
16. Razavi, B. (1997) Design considerations for direct conversion receivers. *IEEE Transactions on Circuits and Systems - II*, **44** (6), 428–435.

3

Scattering Parameters

3.1 Introduction

Before describing even lumped-element filters it is necessary to cover scattering parameters (S-parameters) because they are the standard way of representing high-frequency circuits. The need to employ S-parameters is linked to the use of transmission lines and the need to treat the signal as a wave. To introduce them, the standard two-port parameters are first introduced (Z-parameters, Y-parameters, G-parameters and $ABCD$ parameters). To measure these parameters, however, requires open- and short-circuit terminations, which makes them unsuitable for high frequencies: low-inductance short circuits are difficult to realise, and open circuits always have significant fringing capacitance. Furthermore, dealing with signals in terms of voltages at nodes and currents in branches rapidly becomes impractical at high frequencies. RF engineers are much more comfortable working with S-parameters, which are defined in terms of voltage waves in transmission lines – even if the transmission line is actually a conductor strip on an integrated circuit or PCB.

3.2 Z-Parameters (Open-Circuit Impedance Parameters)

Impedance is a very familiar concept for electronic engineers and a logical first step towards understanding the representation of a two-port network with a matrix of parameters. Figure 3.1 shows a two-port network with the voltages and currents labelled as per convention (the currents i_1 and i_2 flow *into* the ports). The advantage of the matrix two-port parameter representation is that any component or circuit can be treated as a 'black box' and the performance of the circuit or subsystem that it is part of can be determined by simple matrix transformations and reduction techniques to combine many other such 'black boxes' together.

Putting the parameters into matrix form gives:

$$\mathbf{V} = \begin{bmatrix} v_1 \\ v_2 \end{bmatrix} \quad \mathbf{Z} = \begin{bmatrix} z_{11} & z_{12} \\ z_{21} & z_{22} \end{bmatrix} \quad \mathbf{I} = \begin{bmatrix} i_1 \\ i_2 \end{bmatrix} \tag{3.1}$$

Microwave and Millimetre-Wave Design for Wireless Communications, First Edition.
Ian Robertson, Nutapong Somjit and Mitchai Chongcheawchamnan.
© 2016 John Wiley & Sons, Ltd. Published 2016 by John Wiley & Sons, Ltd.

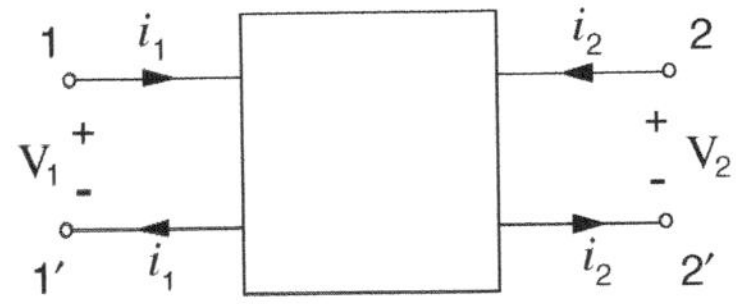

Figure 3.1 A two-port network with voltages and currents labelled.

$\mathbf{V} = \mathbf{Z} \cdot \mathbf{I}$ is Ohm's law in matrix form, and multiplying out yields:

$$v_1 = z_{11}i_1 + z_{12}i_2 \tag{3.2}$$

$$v_2 = z_{21}i_1 + z_{22}i_2 \tag{3.3}$$

These can be reduced by open-circuiting the input or the output, one at a time. The Z-parameters are then determined as follows:

$$z_{11} = \left.\frac{v_1}{i_1}\right|_{i_2=0} \qquad z_{12} = \left.\frac{v_1}{i_2}\right|_{i_1=0}$$
$$z_{21} = \left.\frac{v_2}{i_1}\right|_{i_2=0} \qquad z_{22} = \left.\frac{v_2}{i_2}\right|_{i_1=0} \tag{3.4}$$

Therefore, in words:

z_{11} is the open-circuit input impedance,

z_{12} is the open-circuit reverse transfer impedance,

z_{21} is the open-circuit forward transfer impedance, and

z_{22} is the open-circuit output impedance.

z_{11} and z_{22} have, therefore, very useful practical meanings, with some caveats, although 'forward/reverse voltage gain' would be a far more useful parameter in most situations, so z_{12} and z_{12} are less than ideal for day-to-day working.

3.3 *Y*-Parameters (Short-Circuit Admittance Parameters)

The *Y*-parameters of a network are defined in terms of terminal admittances under short-circuit conditions.

$\mathbf{Y} = \begin{bmatrix} y_{11} & y_{12} \\ y_{21} & y_{22} \end{bmatrix}$, and $\mathbf{I} = \mathbf{Y} \cdot \mathbf{V}$ is Ohm's law in matrix form, which yields the following:

$$i_1 = y_{11}v_1 + y_{12}v_2 \tag{3.5}$$

$$i_2 = y_{21}v_1 + y_{22}v_2 \tag{3.6}$$

The Y-parameters can be determined under the conditions of short circuit at either the input or output port. When the output is short circuit $v_2 = 0$, so:

$$y_{11} = \left.\frac{i_1}{v_1}\right|_{v_2=0} \qquad y_{21} = \left.\frac{i_2}{v_1}\right|_{v_2=0} \tag{3.7}$$

y_{11} is therefore defined as the short-circuit input admittance and y_{21} is defined as the forward transfer admittance with output short circuit. The parameters are referred to as the short-circuit input admittance and short-circuit forward transfer admittance, respectively. When $v_1 = 0$, the input is short circuit, giving:

$$y_{12} = \left.\frac{i_1}{v_2}\right|_{v_1=0} \qquad y_{22} = \left.\frac{i_2}{v_2}\right|_{v_1=0} \tag{3.8}$$

y_{12} is the short-circuit reverse transfer admittance, and y_{22} is the short-circuit output admittance.

3.4 *H*-Parameters (Hybrid Parameters)

Although it is not immediately obvious why they are useful, H-parameters relate (v_1, i_2) to (i_1, v_2):

$$\begin{bmatrix} v_1 \\ i_2 \end{bmatrix} = \mathbf{H} \cdot \begin{bmatrix} i_1 \\ v_2 \end{bmatrix}, \text{ where } \mathbf{H} = \begin{bmatrix} h_{11} & h_{12} \\ h_{21} & h_{22} \end{bmatrix}, \text{ giving:}$$

$$v_1 = h_{11}i_1 + h_{12}v_2 \tag{3.9}$$

$$i_2 = h_{21}i_1 + h_{22}v_2 \tag{3.10}$$

The H-parameters can be calculated or measured as follows:

$$h_{11} = \left.\frac{v_1}{i_1}\right|_{v_2=0} \qquad h_{12} = \left.\frac{v_1}{v_2}\right|_{i_1=0}$$
$$h_{21} = \left.\frac{i_2}{i_1}\right|_{v_2=0} \qquad h_{22} = \left.\frac{i_2}{v_2}\right|_{i_1=0} \tag{3.11}$$

In words,

h_{11} is the short-circuit input impedance,

h_{12} is the open-circuit reverse voltage gain,

h_{21} is the short-circuit forward current gain, and

h_{22} is the open-circuit output admittance.

It is found that H-parameters have a special relevance to bipolar transistors, and are commonly used at RF frequencies. For example, h_{21} is the transistor current gain, and by measuring it the f_t is easily found (the frequency at which the current gain is unity).

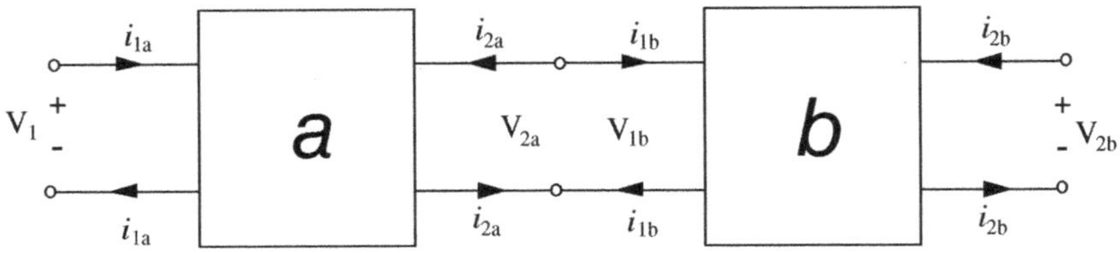

Figure 3.2 Illustration of two networks cascaded, showing the need for a sign change.

G-parameters (Inverse Hybrid Parameters) can also be used but will not be covered here for the sake of brevity.

3.5 *ABCD*-Parameters (Transmission or Chain Parameters)

The *Z*-, *Y*- and *H*-parameters are not useful in the case where two-ports are cascaded together. To facilitate such an interconnection of 'black boxes' requires parameters that relate the output variables (v_2, i_2) to the input ones (v_1, i_1). *ABCD*-parameters are designed to represent the relationship between the input variables (v_1, i_1) and the output (v_2, i_2). In matrix form, we write:

$$\begin{bmatrix} v_1 \\ i_1 \end{bmatrix} = \begin{bmatrix} a_{11} & a_{12} \\ a_{21} & a_{22} \end{bmatrix} \begin{bmatrix} v_2 \\ -i_2 \end{bmatrix} \tag{3.12}$$

which yields:

$$v_1 = a_{11}v_2 - a_{12}i_2 \tag{3.13}$$

$$i_1 = a_{21}v_2 - a_{22}i_2 \tag{3.14}$$

The reason for using -i_2 is that it simplifies things when two-ports are connected in cascade, because the port 2 current from the first black box flows directly into the input of the second one, as illustrated in Figure 3.2. However, this convention is not always followed. The *ABCD*-parameters are then defined by the following expressions:

$$a_{11} = \left.\frac{v_1}{v_2}\right|_{i_2=0} \quad a_{12} = \left.\frac{v_1}{-i_2}\right|_{v_2=0}$$

$$a_{21} = \left.\frac{i_1}{v_2}\right|_{i_2=0} \quad a_{22} = \left.\frac{i_1}{-i_2}\right|_{v_2=0} \tag{3.15}$$

In words,

a_{11} is the open-circuit reverse voltage gain,

a_{12} is the short-circuit reverse transfer impedance,

a_{21} is the open-circuit reverse transfer admittance, and

a_{22} is the short-circuit reverse current gain.

3.6 Summary of Two-Port Parameter Operations

Summarising, the *Z*-, *Y*-, *H*- and *ABCD*-parameter relationships are as follows:

Z-Parameters

$$\begin{bmatrix} v_1 \\ v_2 \end{bmatrix} = \begin{bmatrix} z_{11} & z_{12} \\ z_{21} & z_{22} \end{bmatrix} \begin{bmatrix} i_1 \\ i_2 \end{bmatrix}$$

Y-Parameters

$$\begin{bmatrix} i_1 \\ i_2 \end{bmatrix} = \begin{bmatrix} y_{11} & y_{12} \\ y_{21} & y_{22} \end{bmatrix} \begin{bmatrix} v_1 \\ v_2 \end{bmatrix}$$

H-Parameters

$$\begin{bmatrix} v_1 \\ i_2 \end{bmatrix} = \begin{bmatrix} h_{11} & h_{12} \\ h_{21} & h_{22} \end{bmatrix} \begin{bmatrix} i_1 \\ v_2 \end{bmatrix}$$

ABCD-Parameters

$$\begin{bmatrix} v_1 \\ i_1 \end{bmatrix} = \begin{bmatrix} a_{11} & a_{12} \\ a_{21} & a_{22} \end{bmatrix} \begin{bmatrix} v_2 \\ -i_2 \end{bmatrix}$$

It is when two ports are connected together in different ways that the usefulness of all these different parameters becomes apparent. Choosing the right parameters to represent the problem at hand is key. For example, Figure 3.3 shows the *ABCD*-parameters for a generalised series impedance and shunt admittance, with which a wide variety of circuits can be analysed. Figure 3.4 illustrates how the various parameters can be conveniently used to simplify a circuit, by combining particular arrangements of interconnected two-ports together. In order to analyse a circuit using these it will be necessary to transform between one parameter type and another: Table 3.1 shows how *Z*-, *Y*-, *H*- and *ABCD*-parameters can be converted from one type to another.

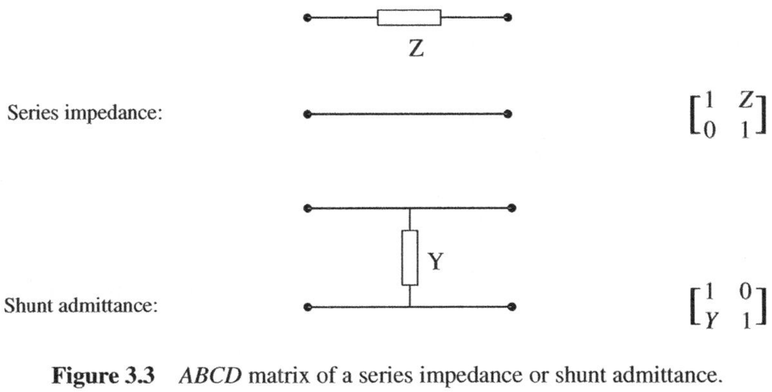

Figure 3.3 *ABCD* matrix of a series impedance or shunt admittance.

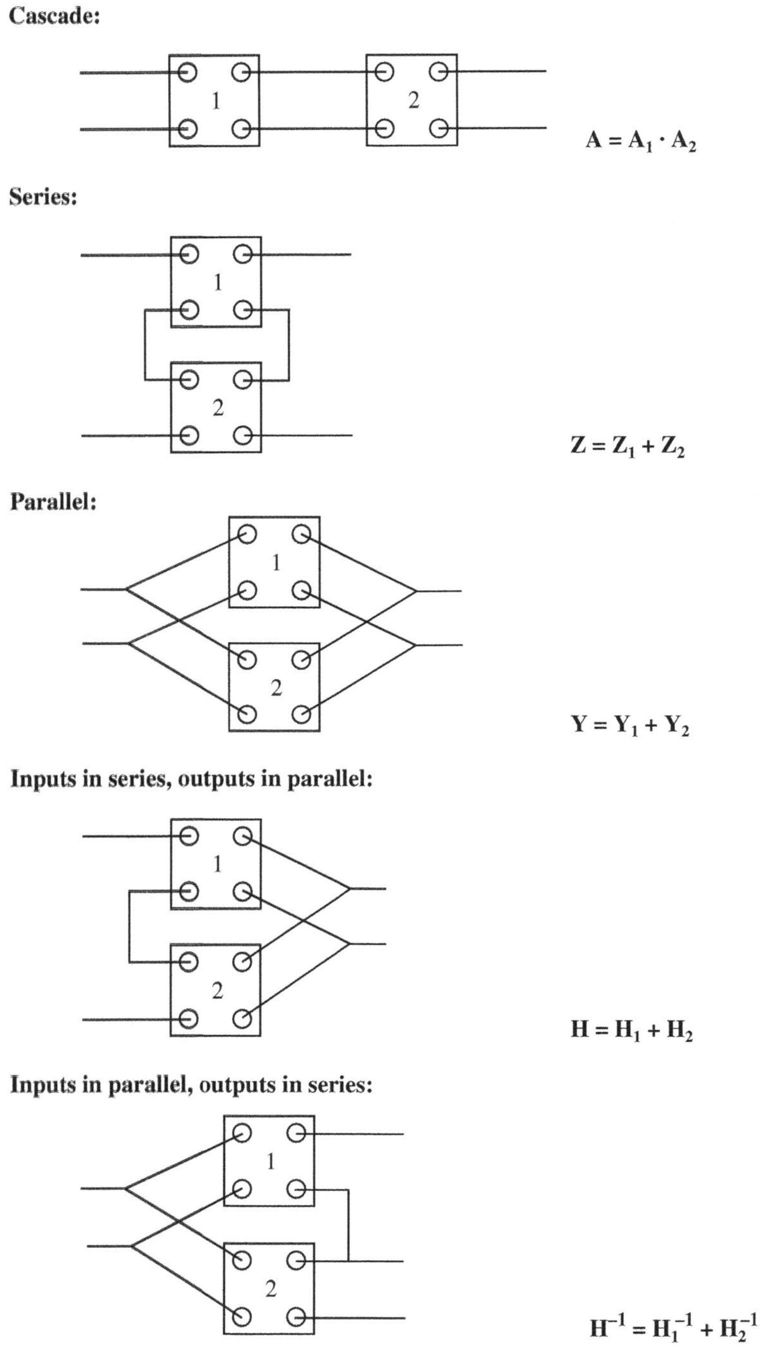

Figure 3.4 Interconnected two-port networks.

Table 3.1 Two-port parameter conversion.

	Z		Y		H		A	
Z	z_{11}	z_{12}	$\frac{y_{22}}{\Delta_y}$	$\frac{-y_{12}}{\Delta_y}$	$\frac{\Delta_h}{h_{22}}$	$\frac{h_{12}}{h_{22}}$	$\frac{a_{11}}{a_{21}}$	$\frac{\Delta_a}{a_{21}}$
	z_{21}	z_{22}	$\frac{-y_{21}}{\Delta_y}$	$\frac{y_{11}}{\Delta_y}$	$\frac{-h_{21}}{h_{22}}$	$\frac{1}{h_{22}}$	$\frac{1}{a_{21}}$	$\frac{a_{22}}{a_{21}}$
Y	$\frac{z_{22}}{\Delta_z}$	$\frac{-z_{12}}{\Delta_z}$	y_{11}	y_{12}	$\frac{1}{h_{11}}$	$\frac{-h_{12}}{h_{11}}$	$\frac{a_{22}}{a_{12}}$	$\frac{-\Delta_a}{a_{12}}$
	$\frac{-z_{21}}{\Delta_z}$	$\frac{z_{11}}{\Delta_z}$	y_{21}	y_{22}	$\frac{h_{21}}{h_{11}}$	$\frac{\Delta_h}{h_{11}}$	$\frac{-1}{a_{12}}$	$\frac{a_{11}}{a_{12}}$
H	$\frac{\Delta_z}{z_{22}}$	$\frac{z_{12}}{z_{22}}$	$\frac{1}{y_{11}}$	$\frac{-y_{12}}{y_{11}}$	h_{11}	h_{12}	$\frac{a_{12}}{a_{22}}$	$\frac{\Delta_a}{a_{22}}$
	$\frac{-z_{21}}{z_{22}}$	$\frac{1}{z_{22}}$	$\frac{y_{21}}{y_{11}}$	$\frac{\Delta_y}{y_{11}}$	h_{21}	h_{22}	$\frac{-1}{a_{22}}$	$\frac{a_{21}}{a_{22}}$
A	$\frac{z_{11}}{z_{21}}$	$\frac{\Delta_z}{z_{21}}$	$\frac{-y_{22}}{y_{21}}$	$\frac{-1}{y_{21}}$	$\frac{-\Delta_h}{h_{21}}$	$\frac{-h_{11}}{h_{21}}$	a_{11}	a_{12}
	$\frac{1}{z_{21}}$	$\frac{z_{22}}{z_{21}}$	$\frac{-\Delta_y}{y_{21}}$	$\frac{-y_{11}}{y_{21}}$	$\frac{-h_{22}}{h_{21}}$	$\frac{-1}{h_{21}}$	a_{21}	a_{22}

N.B. $\Delta_z = z_{11}z_{22} - z_{12}z_{21}$, etc.

3.7 Scattering Parameters

S-parameters treat the signal in terms of the flow of voltage waves, with the ports terminated with a reference impedance (the 'system impedance') which is most often 50 Ω, although not always. This immediately overcomes the need for open- and short-circuit terminations and means that a device-under-test (DUT) can be measured with matched loads. As discussed in Chapter 5, this is important in transmission lines because of the wave nature of the signal and because reflections, leading to forward- and backwards-travelling waves, must be considered. As illustrated in Figure 3.5, for a two-port network, the variables are the voltage waves going in to port 1 and port 2 (a_1, a_2, respectively) and the waves coming out (b_1, b_2). These parameters are vector voltage quantities with reference to the system impedance Z_0, such that:

$|a_1|^2$ = power of the incident wave at the network input (port 1)
$|a_2|^2$ = power of the incident wave at the network output (port 2)

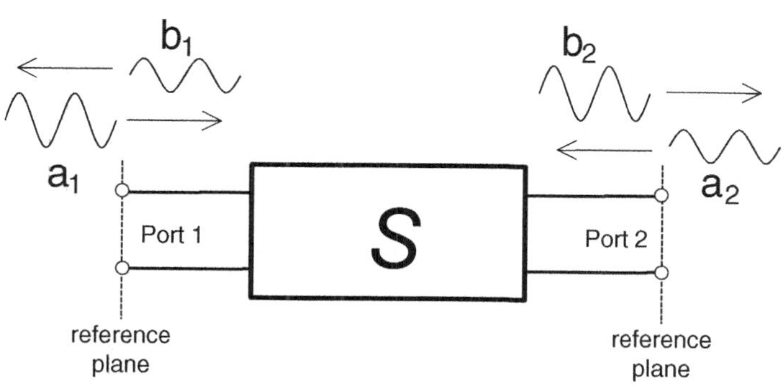

Figure 3.5 Two-port network represented in terms of voltage waves.

$|b_1|^2$ = power of the outgoing wave at port 1
$|b_2|^2$ = power of the outgoing wave at port 2

Note that the convention used here is that, for example a_l, the wave incident at port 1 is defined as $V_+/\sqrt{Z_o}$, where V_+ is the r.m.s voltage of the forwards-travelling wave. Some sources (mostly European, supposedly) use an alternate definition (peak voltage) that means the incident wave power is $\frac{1}{2}|a_1|^2$. Fortunately, *S*-parameters are ratios and so this difference is not usually a problem.

The four *S*-parameters represent the signal flow in terms of these voltage waves and relate the outgoing waves, b_1 and b_2 to the incident waves a_1, a_2, as follows [1]:

$$\begin{bmatrix} b_1 \\ b_2 \end{bmatrix} = \begin{bmatrix} S_{11} & S_{12} \\ S_{21} & S_{22} \end{bmatrix} \begin{bmatrix} a_1 \\ a_2 \end{bmatrix} \tag{3.16}$$

Multiplying out gives:

$$b_1 = S_{11}a_1 + S_{12}a_2 \tag{3.17}$$

$$b_2 = S_{21}a_1 + S_{22}a_2 \tag{3.18}$$

The phase of the waves is very important and *S*-parameters are vector quantities for which reference planes must always be defined. Each wave coming out of the two-port network can be seen to have two components: some signal is transferred from the other port and some is reflected from the port. Rearranging these equations yields the definitions of the *S*-parameters for a two-port network:

$$\begin{aligned} S_{11} &= \left.\frac{b_1}{a_1}\right|_{a_2=0} & S_{12} &= \left.\frac{b_1}{a_2}\right|_{a_1=0} \\ S_{21} &= \left.\frac{b_2}{a_1}\right|_{a_2=0} & S_{22} &= \left.\frac{b_2}{a_2}\right|_{a_1=0} \end{aligned} \tag{3.19}$$

S_{11} can now be seen to be the input reflection coefficient of the network when its output is terminated with a matched load. S_{11} in that case is determined by the input impedance, and three special cases are:

> Zero-input impedance (*short circuit*): $S_{11} = 1\angle 180^\circ$ because all of the signal is reflected, and the reflected voltage wave must be anti-phase in order to make the voltage at the short circuit zero.
>
> Infinite input impedance (*open circuit*): $S_{11} = 1\angle 0^\circ$ because all the signal power is reflected, and the voltage is a maximum at the open circuit.
>
> A *matched* input: This means that the input impedance is equal to the system impedance, and leads to $S_{11} = 0$ (no reflected signal, all the power goes into the two-port network).

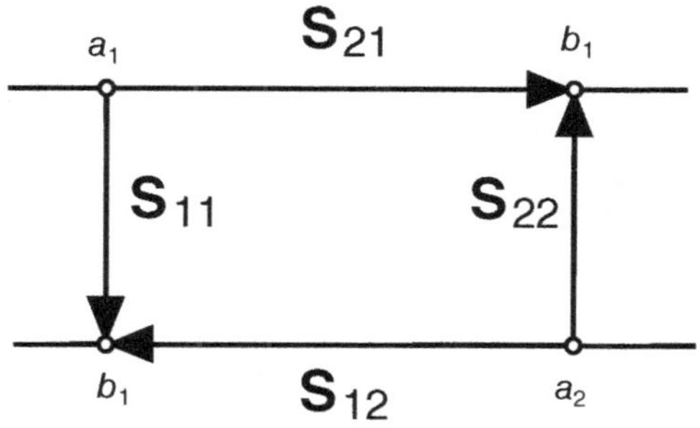

Figure 3.6 Flowgraph representation of a two-port S-parameter network.

There is a clear relationship between S_{11} and input impedance when the output is terminated in a matched load, but we need to be precise for the general case and flowgraphs are a useful tool for conducting a proper analysis of multiple signal paths.

3.7.1 Flowgraph Representation of Two-Port S-Parameters

The flowgraph representation of a two-port network is shown in Figure 3.6. As before, the parameter a_1 represents the voltage wave going into port 1, while b_1 is the voltage wave coming out from port 1. Similarly, a_2 is the wave going into port 2, and b_2 is the wave coming out of port 2. It is very important to recognise that this is a flowgraph, not a circuit diagram. The nodes are not the same as the nodes in a circuit, which were used in the low-frequency parameters earlier. a_1, a_2, b_1, b_2 are called nodes but they represent the voltage waves exactly at the input and output ports. This is important because as soon as you move away from the network's ports the voltage waves change in phase. The *S*-parameters S_{11}, S_{21}, S_{12}, S_{22} are branches and represent the signal flow, with reference planes defined since they are vector quantities.

From the flowgraph, the value of any node can be calculated by:

$$x_n = \sum_{\substack{\text{all branches pointing}\\ \textit{into}\text{ the node}}} \text{Branch value} \times \text{the value of the node at the other end of that branch} \tag{3.20}$$

Applying this to Figure 3.6 gives:

$$b_1 = S_{11}a_1 + S_{12}a_2 \tag{3.21}$$

$$b_2 = S_{21}a_1 + S_{22}a_2 \tag{3.22}$$

This confirms the earlier result, but the important advantage of the flowgraph representation is that complex networks can be analysed using a variety of tools to simplify the flowgraph, as described in Chapter 5.

3.7.2 Properties of S-Parameters

The properties of S-parameters will be more easily understood by studying some simple rules and examples of the S-parameter matrices of some simple ideal components. First, for a lossless, passive, two-port network the total power into the network must be equal to the total power out. If a signal a_1 is incident at port 1, then it can only be reflected (b_1) or transferred through the network (b_2) and therefore the lossless network requires:

$$|b_1|^2 + |b_2|^2 = |a_1|^2 \tag{3.23}$$

This is true in this case since $a_2 = 0$. A more general statement of the lossless condition for any network is that the sum of all the power going in (a waves) is equal to the sum of the power coming out (b waves), which for an network with N ports can be expressed as:

$$\sum_{1}^{N} |a_k|^2 = \sum_{1}^{N} |b_k|^2 \tag{3.24}$$

In S-parameter terms, since S_{11} and S_{21} give the reflected voltage wave and transmitted voltage wave relative to the input wave, Equation (3.23) can also be written as:

$$|S_{11}|^2 + |S_{21}|^2 = 1 \quad \text{for a lossless passive network}$$

Other important conditions and example S-parameters are as follows:

Matched at both ports:	$S_{11} = S_{22} = 0$
Reciprocal:	$S_{21} = S_{12}$
Symmetric:	$S_{21} = S_{12}$ and $S_{11} = S_{22}$
An ideal amplifier with voltage gain G:	$S = \begin{pmatrix} 0 & 0 \\ G & 0 \end{pmatrix}$ (n.b. G is not in dB)
A matched 3 dB attenuator:	$S = \begin{pmatrix} 0 & 0.707 \\ 0.707 & 0 \end{pmatrix}$
An ideal isolator:	$S = \begin{pmatrix} 0 & 0 \\ 1 & 0 \end{pmatrix}$
A section of matched transmission line:	$S = \begin{pmatrix} 0 & e^{-\gamma l} \\ e^{-\gamma l} & 0 \end{pmatrix}$

where γ is the propagation constant introduced in Chapter 5 and l is the physical length.

3.7.3 S-Parameters and Decibels

A simple low-pass filter is a useful example to give further insight into the meaning of S-parameter values and the wave behaviour of circuits, but first the decibel unit needs to be reviewed. The decibel (dB) is a unit based on the logarithmic ratio of powers, named in honour of Alexander Graham Bell. dB units are able to conveniently represent a huge range of signal levels and high gain/attenuation ratios; for example, hundreds of Watts at the transmitter to pW at the receiver are often encountered in a radio system.

For an amplifier, for example:

$$\text{Gain (dB)} = 10\log_{10}\frac{\text{Output Power}}{\text{Input Power}} \tag{3.25}$$

Note that this is a ratio and has no absolute units. The input and output power must have some absolute units, denoted by an extra suffix, such as:

$$\text{Power in dBW} = 10\log_{10}(\text{P in Watts})$$

$$\text{Power in dBm} = 10\log_{10}(\text{P in milliWatts})$$

Since P = V^2/R, it is important to remember to use $20\log_{10}$, not $10\log_{10}$, for voltages:

$$\text{Gain (dB)} = 20\log_{10}\frac{\text{Output Voltage}}{\text{Input Voltage}} \tag{3.26}$$

(Voltage can be expressed here as peak-to-peak, r.m.s. or amplitude; since it is a ratio the $\sqrt{2}$s cancel.)

Because S-parameters are voltage-based, to express their magnitude in dB you must, therefore, use $20\log_{10}|S_{mn}|$. For example, the power gain in dB of a matched amplifier is $20\log_{10}|S_{21}|$.

Going back to the three special cases of input impedance, for the open or short circuit $|S_{11}| = 1$, so $|S_{11}| = 0$ dB (only the phase is different, and in both the short- and open-circuit case all of the signal is reflected). For the matched case, $|S_{11}| = 0$, which gives $|S_{11}| = -\infty$ dB. This is slightly inconvenient, but in practice residual reflections from adapters/cables/etc. will mean that even –60 dB is rarely achieved for $|S_{11}|$. $|S_{11}|$ in dB is often referred to as the 'return loss' of the circuit.

Figure 3.7 shows the ADS™ simulated performance of a simple low-pass filter, duplicated from the design example described in Chapter 4 to aid understanding the nature of S-parameters. In the passband, $|S_{11}|$ is <–20 dB, while $|S_{21}| \approx 0$ dB. This indicates that nearly all the signal input to port 1 flows through the filter to port 2, and very little is reflected. In contrast, in the stopband, $|S_{11}| \approx 0$ dB, indicating that nearly all the input signal is reflected back, while $|S_{21}|$ is steadily falling and reaches <–40 dB quite quickly as the frequency increases.

3.7.4 N-Port S-Parameters

The matrix of scattering parameters can be extended to completely describe any N-port network. Linear algebra theory can be applied for analysis and design like other linear network parameters, such as impedance and admittance matrices. However, scattering parameters have received more attention than other linear network parameters since the concept is ideally suited to transmission line circuits. In addition, the admittance and impedance parameters at high frequency are rather difficult to determine due to the difficulty in measuring voltage and

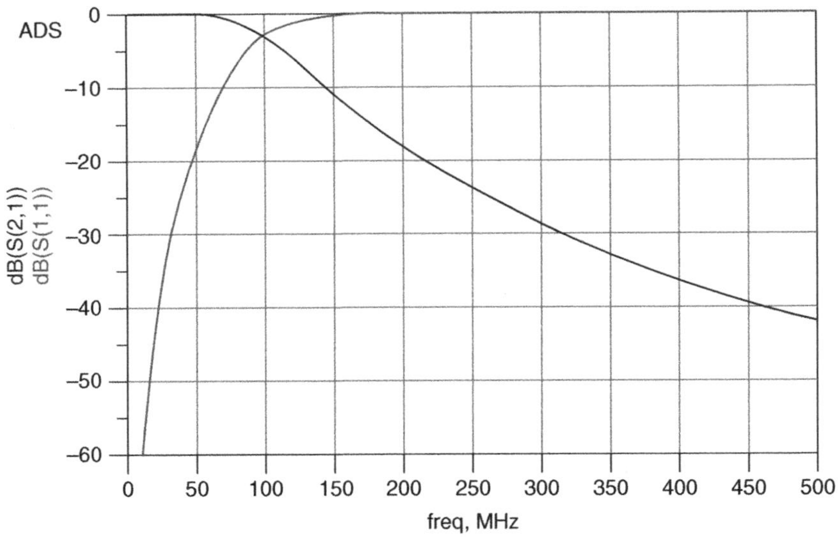

Figure 3.7 Simulated *S*-parameter plot for a simple low-pass filter example.

current at the network ports. Figure 3.8 shows the diagram and definition of variables for the scattering parameters of a generalised *N*-port. The concept of scattering parameters of an *N*-port network defines the incident waves (voltage or current) as the independent variables while the dependent variables are the reflected waves (again voltage or current). Let a_i and b_i denote the incident and reflected waves at port i and all network ports are referred the characteristic impedance Z_0. Therefore:

$$\begin{bmatrix} b_1 \\ b_2 \\ \vdots \\ b_N \end{bmatrix} = \begin{bmatrix} s_{11} & s_{12} & \cdots & s_{1N} \\ s_{21} & s_{22} & \cdots & s_{2N} \\ \vdots & \vdots & \ddots & \vdots \\ s_{N1} & s_{N2} & \cdots & s_{NN} \end{bmatrix} \begin{bmatrix} a_1 \\ a_2 \\ \vdots \\ a_N \end{bmatrix} \tag{3.27}$$

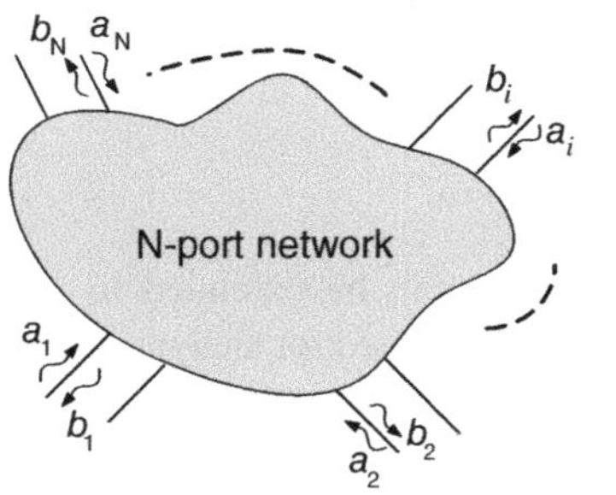

Figure 3.8 Scattering parameter definition for an *N*-port linear network.

where s_{ji} denotes the element of scattering matrix **S** which is given by

$$s_{ji} = \left.\frac{b_j}{a_i}\right|_{\forall a_k=0;k\neq i} \tag{3.28}$$

Each scattering matrix element, s_{ji}, relates to the transfer coefficient to port j where the wave is incident at port i and no incident wave is fed to other ports. The elements are related to the parameters defined in transmission theory. For example, s_{nn} is the reflection coefficient at port n or s_{ji} is the transfer coefficient from port i to port j. The terms a_k and b_k are in fact the incident and reflected voltage waves at port k normalised by the square root of Z_0. They are:

$$a_k = \frac{v_k^+}{\sqrt{Z_0}} \tag{3.29}$$

$$b_k = \frac{v_k^-}{\sqrt{Z_0}} \tag{3.30}$$

where v_k^+ is the rms voltage of the wave going into port k and v_k^- is the rms voltage of the wave coming out of port k. The voltage and current at port k are related by

$$v_k = v_k^+ + v_k^- \tag{3.31}$$

$$i_k = i_k^+ + i_k^- = \frac{v_k^+}{Z_0} - \frac{v_k^-}{Z_0} \tag{3.32}$$

One can write v_k^+ and v_k^- in terms of v_k and i_k as follows. From Equations (3.31) and (3.32) we obtain

$$v_k^+ = \frac{1}{2}\left(v_k + i_k Z_0\right) \tag{3.33}$$

$$v_k^- = \frac{1}{2}\left(v_k - i_k Z_0\right) \tag{3.34}$$

Substituting these incident and reflected voltage waves into the expressions for a_k and b_k given in Equations (3.29) and (3.30), we obtain these two waves in terms of the voltage and current at port k as follows:

$$a_k = \frac{1}{2}\left(\frac{v_k}{\sqrt{Z_0}} + i_k\sqrt{Z_0}\right) \tag{3.35}$$

$$b_k = \frac{1}{2}\left(\frac{v_k}{\sqrt{Z_0}} - i_k\sqrt{Z_0}\right) \tag{3.36}$$

The incident (P_k^+) and reflected (P_k^-) powers at port k are defined by:

$$P_k^+ = |a_k|^2 \tag{3.37}$$

$$P_k^- = |b_k|^2 \tag{3.38}$$

It is shown that the definition of scattering parameters is valid if the characteristic impedance is defined. This is very crucial whenever scattering parameters are applied to the analysis of a circuit.

3.7.5 Measuring S-Parameters

The instrument that has been developed for measuring the scattering parameters of a device, component or subsystem is called a vector network analyzer (VNA), an example of which is shown in Figure 3.9 (the principles of the VNA are described in detail in Chapter 18). A continuous-wave (CW) sinusoidal source is applied to the device-under-test (DUT) via a set of switches and couplers that are used to route the signal alternately to port 1 or 2 and sample the input and reflected waves. The high-frequency signals are downconverted using a reference oscillator so that accurate measurements of amplitude and phase can be made for a_1, b_1, a_2 and b_2. An on-board processor then calculates the ratioed measurements (a_1/b_1, etc.) and applies error-correction algorithms before the scattering parameters of the DUT are displayed.

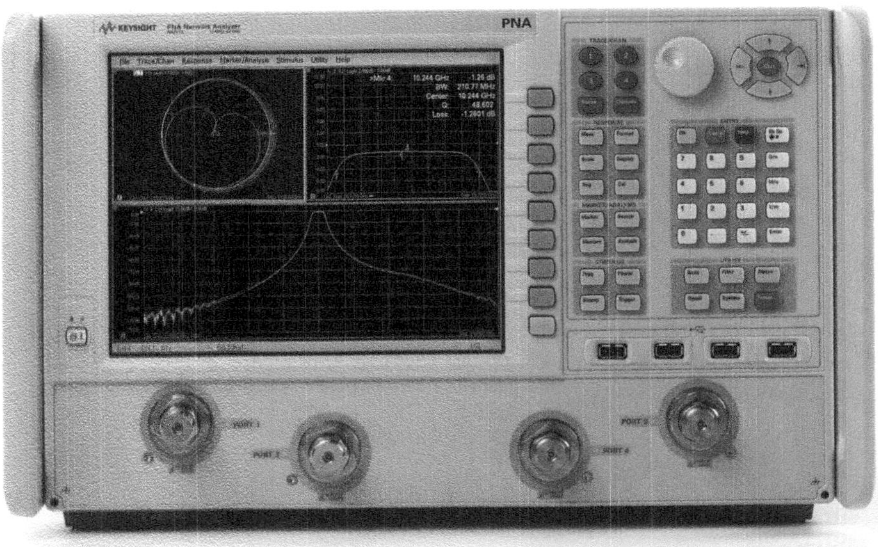

Figure 3.9 PNA Series Vector Network Analyzer. © 2016, Keysight Technologies, Inc. Reproduced with permission; all rights reserved.

3.8 Transmission Parameters

Frequently, a circuit or subsystem consists of many two-port networks (transmission lines, transistors, matching networks, filter, etc.) connected in a cascade arrangement. Analyzing cascaded networks with *S*-parameters is somewhat troublesome. What is required is a form that relates output variables to input variables, and transmission parameters (T-parameters) provide this. Analysis using *ABCD* parameters, as discussed earlier, can provide a simple approach to analyze cascaded networks, but the parameters are defined in terms of voltages and currents, not incident and reflected waves. For high-frequency circuit analysis, *T*-parameters are more suitable than the *ABCD* matrix since their definition is based on the concept of reflected and incident waves. Because of the similarity, in some texts *ABCD* parameters may be referred to as 'voltage-current transmission matrix parameters', and *T*-parameters as 'wave-amplitude transmission matrix parameters' or 'scattering transfer parameters' in order to clearly distinguish between them.

3.8.1 Definition

For a two-port network, the *T*-matrix is defined as the parameters determining how the waves at port 1 (a_1, b_1) transfer to the waves at port 2 (a_2, b_2) as follows (note there are alternative forms in use):

$$\begin{bmatrix} a_1 \\ b_1 \end{bmatrix} = \begin{bmatrix} T_{11} & T_{12} \\ T_{21} & T_{22} \end{bmatrix} \begin{bmatrix} b_2 \\ a_2 \end{bmatrix} \tag{3.39}$$

A conversion between *T*- and *S*-parameters for a two-port network can be obtained from,

$$\begin{bmatrix} T_{11} & T_{12} \\ T_{21} & T_{22} \end{bmatrix} = \frac{1}{s_{21}} \begin{bmatrix} 1 & -s_{22} \\ s_{11} & \Delta \mathbf{S} \end{bmatrix} \tag{3.40}$$

$$\begin{bmatrix} s_{11} & s_{12} \\ s_{21} & s_{22} \end{bmatrix} = \frac{1}{T_{11}} \begin{bmatrix} T_{21} & -\Delta \mathbf{T} \\ 1 & -T_{12} \end{bmatrix} \tag{3.41}$$

$\Delta \mathbf{S}$ and $\Delta \mathbf{T}$ are the determinants of the scattering matrix and transmission matrix, respectively. In fact, if we substitute the relation of waves to the voltage and current ports from Equations (3.35) and (3.36) into the matrix Equation (3.39), this leads to:

$$\begin{bmatrix} v_1 + i_1 Z_0 \\ v_1 - i_1 Z_0 \end{bmatrix} = \begin{bmatrix} T_{11} & T_{12} \\ T_{21} & T_{22} \end{bmatrix} \begin{bmatrix} v_2 - i_2 Z_0 \\ v_2 + i_2 Z_0 \end{bmatrix} \tag{3.42}$$

Expanding this, one obtains:

$$v_1 + i_1 Z_0 = T_{11}\left(v_2 - i_2 Z_0\right) + T_{12}\left(v_2 + i_2 Z_0\right) \tag{3.43}$$

$$v_1 - i_1 Z_0 = T_{21}\left(v_2 - i_2 Z_0\right) + T_{22}\left(v_2 + i_2 Z_0\right) \tag{3.44}$$

Combining Equations (3.43) and (3.44) and dividing by 2 gives:

$$v_1 = \frac{(T_{11} + T_{12} + T_{21} + T_{22})}{2} v_2 - \frac{Z_0 (T_{11} - T_{12} + T_{21} - T_{22})}{2} i_2 \tag{3.45}$$

If Equation (3.44) is now subtracted from Equation (3.33) and the result divided by $2Z_0$, then,

$$i_1 = \frac{(T_{11} + T_{12} - T_{21} - T_{22})}{2Z_0} v_2 - \frac{(T_{11} - T_{12} - T_{21} + T_{22})}{2} i_2 \tag{3.46}$$

Since the definition of *ABCD* parameters is given as

$$\begin{bmatrix} v_1 \\ i_1 \end{bmatrix} = \begin{bmatrix} A & B \\ C & D \end{bmatrix} \begin{bmatrix} v_2 \\ -i_2 \end{bmatrix} \tag{3.47}$$

we can conclude that the *ABCD* matrix is related to the *T*-matrix as follows:

$$\begin{bmatrix} A\ B \\ C\ D \end{bmatrix} = \frac{1}{2} \begin{bmatrix} T_{11} + T_{12} + T_{21} + T_{22} & Z_0 (T_{11} - T_{12} + T_{21} - T_{22}) \\ \dfrac{T_{11} + T_{12} - T_{21} - T_{22}}{Z_0} & T_{11} - T_{12} - T_{21} + T_{22} \end{bmatrix} \tag{3.48}$$

3.8.2 *Relationship Between T- and ABCD-Parameters*

From the *ABCD* to *T*-matrix relationship defined in Equation (3.48), four equations are obtained as follows:

$$A = \frac{1}{2} (T_{11} + T_{12} + T_{21} + T_{22}) \tag{3.49}$$

$$B = \frac{Z_0}{2} (T_{11} - T_{12} + T_{21} - T_{22}) \tag{3.50}$$

$$C = \frac{1}{2Z_0} (T_{11} + T_{12} - T_{21} - T_{22}) \tag{3.51}$$

$$D = \frac{1}{2} (T_{11} - T_{12} - T_{21} + T_{22}) \tag{3.52}$$

This leads to a 4 × 4 matrix relationship, as follows:

$$\begin{bmatrix} A \\ B \\ C \\ D \end{bmatrix} = \frac{1}{2} \begin{bmatrix} 1 & 1 & 1 & 1 \\ Z_0 & -Z_0 & Z_0 & -Z_0 \\ 1/Z_0 & 1/Z_0 & -1/Z_0 & -1/Z_0 \\ 1 & -1 & -1 & 1 \end{bmatrix} \begin{bmatrix} T_{11} \\ T_{12} \\ T_{21} \\ T_{22} \end{bmatrix} \tag{3.53}$$

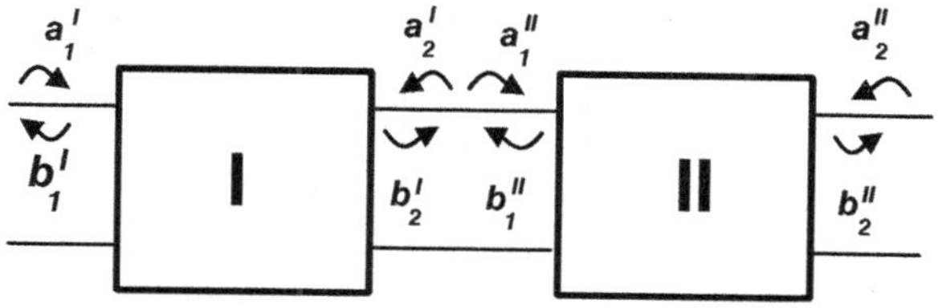

Figure 3.10 Waves in cascaded networks.

Inverting the matrix equation in Equation (3.53) one obtains:

$$\begin{bmatrix} T_{11} \\ T_{12} \\ T_{21} \\ T_{22} \end{bmatrix} = \frac{1}{2} \begin{bmatrix} 1 & 1 & 1 & 1 \\ 1/Z_0 & -1/Z_0 & 1/Z_0 & -1/Z_0 \\ Z_0 & Z_0 & -Z_0 & -Z_0 \\ 1 & -1 & -1 & 1 \end{bmatrix} \begin{bmatrix} A \\ B \\ C \\ D \end{bmatrix} \tag{3.54}$$

Finally, the T-matrix is related to the $ABCD$ matrix by:

$$\begin{bmatrix} T_{11} & T_{12} \\ T_{21} & T_{22} \end{bmatrix} = \frac{1}{2} \begin{bmatrix} A+B+C+D & \frac{1}{Z_0}(A-B+C-D) \\ Z_0(A+B-C-D) & A-B-C+D \end{bmatrix} \tag{3.55}$$

3.8.3 Application to Cascaded Networks

Consider the two networks (I and II) connected in cascade as shown in Figure 3.10. $\mathbf{T}^{\mathrm{I}}$ and $\mathbf{T}^{\mathrm{II}}$ are the T-parameters of network I and II, respectively. These two matrices are:

$$\begin{bmatrix} a_1^I \\ b_1^I \end{bmatrix} = \mathbf{T}^I \begin{bmatrix} b_2^I \\ a_2^I \end{bmatrix} \tag{3.56}$$

$$\begin{bmatrix} a_1^{II} \\ b_1^{II} \end{bmatrix} = \mathbf{T}^{II} \begin{bmatrix} b_2^{II} \\ a_2^{II} \end{bmatrix} \tag{3.57}$$

In order to rewrite these two matrix equations in terms of a_1^I, b_1^I, a_2^{II} and b_2^{II}, to derive the overall T-parameters, we note from Figure 3.10 that $b_1^{II} = a_2^I$ and $a_1^{II} = b_2^I$. Therefore, we can rewrite Equation (3.56) as:

$$\begin{bmatrix} a_1^I \\ b_1^I \end{bmatrix} = \mathbf{T}^I \begin{bmatrix} a_1^{II} \\ b_1^{II} \end{bmatrix} \tag{3.58}$$

Hence, the waves at port 1 of the cascade network are related to the waves at port 2 by:

$$\begin{bmatrix} a_1^I \\ b_1^I \end{bmatrix} = \mathbf{T}^I\mathbf{T}^{II} \begin{bmatrix} b_2^{II} \\ a_2^{II} \end{bmatrix} \tag{3.59}$$

It is evident, therefore, that the transmission matrix of the cascaded network is the matrix multiplication of each transmission matrix. Finally, if we have N networks cascaded, the transmission matrix of the cascaded network is easily determined as:

$$\mathbf{T} = \prod_{k=1}^{N} \mathbf{T}_k \tag{3.60}$$

where $\mathbf{T}_k$ is the transmission matrix of network k.

References

1. Hewlett Packard Test & Measurement, Application Note 95-1, S-Parameter Techniques, 1997.

4

Lumped-Element Filters

4.1 Introduction

From Chapter 2 on radio transmitters and receivers, it is clear that filters are a very crucial component for separating wanted signals from the unwanted signals of other users and systems. This chapter provides an introduction to the theory, design procedure and practical construction of RF filters based on lumped elements – inductors and capacitors. The purpose of most linear filters is to separate a wanted signal from a mixture of the wanted signal and one or more unwanted signals. As shown in Figure 4.1, filters that are required can be:

- Low-pass
- High-pass
- Band-pass
- Band-stop
- All-pass (but with some specific phase response).

The frequency band occupied by the wanted signal is referred to as the 'passband'. In this band the ideal requirement for the filter is to provide constant loss and linear phase so that the wanted signal(s) will be transmitted with no distortion. With a finite, lumped network one cannot achieve exactly constant loss over a finite band of frequencies, so it is customary to specify instead some acceptable upper and lower limits between which the loss can vary.

The 'stopband' is the part of the frequency response where unwanted signals are present. As far as the filter user is concerned, it is normally of no interest whether the actual filter loss in the stopband exceeds the amount required by 1 dB or 20 dB, and in this sense stopband specifications will specify a minimum level of loss for specific frequencies, in contrast to passbands where there are both upper and lower limits. The loss response of a filter is a continuous function of frequency and is thus unable to have sudden discontinuities. For this reason, there must always be some transition band between the passband and the stopband.

Microwave and Millimetre-Wave Design for Wireless Communications, First Edition.
Ian Robertson, Nutapong Somjit and Mitchai Chongcheawchamnan.
© 2016 John Wiley & Sons, Ltd. Published 2016 by John Wiley & Sons, Ltd.

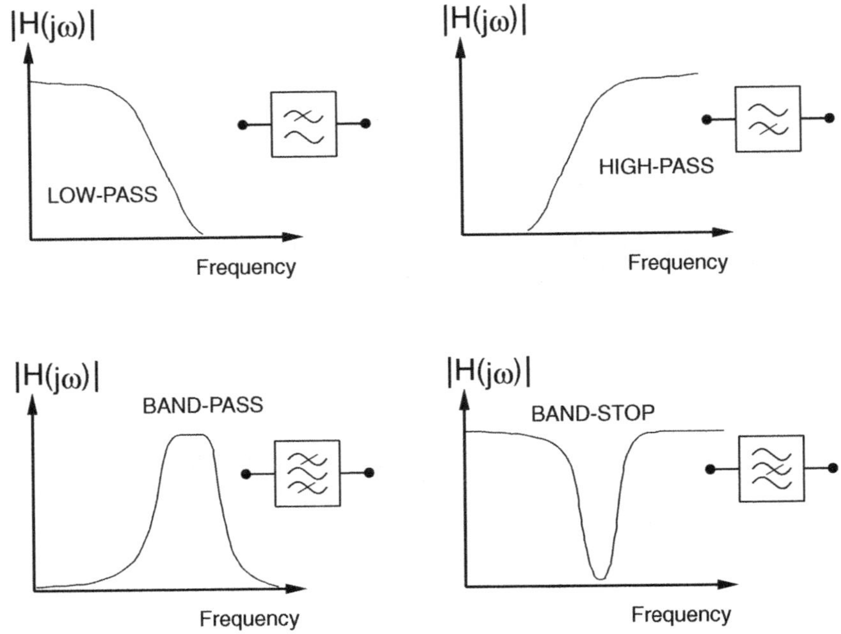

Figure 4.1 Low-pass, high-pass, band-pass and band-stop filters.

4.2 Filter Theory

4.2.1 Transfer Function Analysis

To start to understand the complexities of filter theory, it is necessary to briefly study transfer functions and Bode plots. Consider the simple RC low-pass filter shown in Figure 4.2. Using the ordinary potential divider expression, we can easily find that:

$$\frac{V_{out}}{V_{in}} = \frac{1/j\omega C}{R + 1/j\omega C} = \frac{1}{1 + j\omega CR} \tag{4.1}$$

Note that the source and load impedances are assumed to be zero and infinite, respectively. The transfer function in Equation (4.1) represents the frequency-domain response of the network.

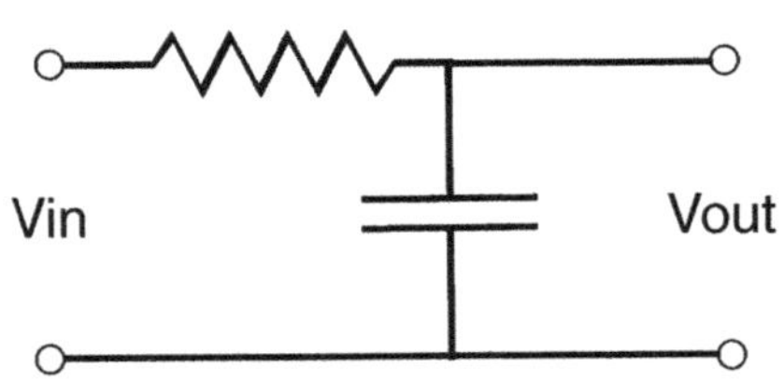

Figure 4.2 RC low-pass filter.

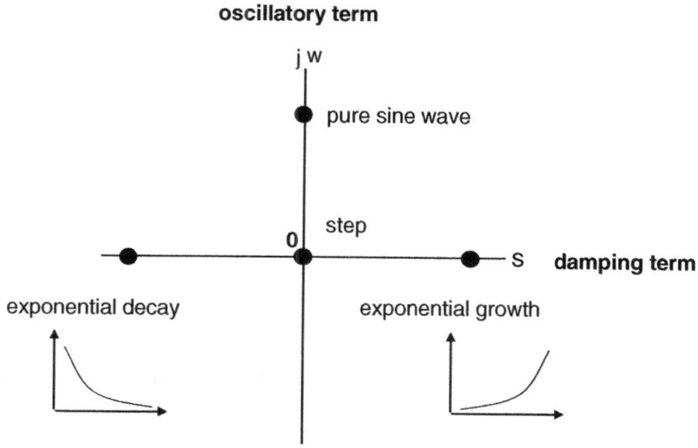

Figure 4.3 Example signals in the *s*-domain.

The circuit is a low-pass filter and the magnitude of the transfer function is $1/\sqrt{2}$ (the 3 dB point) when $\omega = 1/RC$.

4.2.2 *The s-Domain*

By using this analysis we have assumed that the signal is a pure sinusoid. In order to represent the filter more generally, the parameter '$j\omega$' is replaced by the complex 's' operator, $s = \sigma + j\omega$. The 's' operator is able to represent a wide range of signals, such as step functions, impulses, and sinusoids, as illustrated in Figure 4.3. For the same RC low-pass filter, we write the transfer function as:

$$H(s) = \frac{1/sC}{R + 1/sC} = \frac{1}{1 + sCR} \tag{4.2}$$

The transfer function $H(s)$ is related to the time-domain response of the network by the Laplace Transform. Specifically, the impulse response is the output, $r(t)$, of the network when an ideal unit impulse $\delta(t)$ is applied at the input at time $t = 0$ and is given by:

$$r(t) = \mathcal{L}^{-1}[H(s)] \tag{4.3}$$

For the simple RC low-pass filter this gives:

$$r(t) = u(t)\frac{1}{RC}e^{-t/RC} \tag{4.4}$$

where $u(t)$ is the unit step function and RC is recognisable as the time constant τ.

For a general network, the transfer function will have a polynomial for both the numerator $N(s)$ and for the denominator $D(s)$ of order m and n, respectively:

$$H(s) = \frac{N(s)}{D(s)} = \frac{a_m s^m + a_{m-1} s^{m-1} + \ldots + a_1 s + a_0}{b_n s^n + b_{n-1} s^{n-1} + \ldots + b_1 s + b_0} \tag{4.5}$$

Of course, these polynomials can be expressed in terms of the product of their factors:

$$H(s) = K.\frac{(s - z_1)(s - z_2) \ldots (s - z_m) \ldots}{(s - p_1)(s - p_2) \ldots (s - p_n) \ldots} \tag{4.6}$$

The quantities $z_1, z_2, \ldots, z_n$ are called the zeros, as the transfer function becomes zero at all these values.

The quantities $p_1, p_2, \ldots, p_n$ are called the poles, as the transfer function goes to infinity at these values.

The poles and zeros can be anywhere in the complex s-domain. If a network has complex poles or zeros, they must be in complex conjugate pairs. Figure 4.3 shows some examples of time-domain responses related to the pole positions.

4.2.3 Bode Plots

The Bode plot is graphical technique for representing transfer functions (both magnitude and phase) versus frequency. For the previous simple low-pass filter with $\boldsymbol{H(s) = 1/(1 + sCR)}$, and by normalising to $\boldsymbol{CR = 1}$ for convenience, the magnitude Bode plot shown in Figure 4.4 results. The beauty of the Bode plot stems from the fact that, for a generalised transfer function, each pole and zero can be treated individually and approximated by a straight line like that in Figure 4.4 when plotted on a log-log scale. This is because the transfer function consists of the products of factors in the numerator and denominator, as shown in Equation (4.6), and when the logarithm is taken these become separate terms, since log (A × B) = log A + log B and log (A ÷ B) = log A – log B. Each pole and zero is represented by a straight line with slope

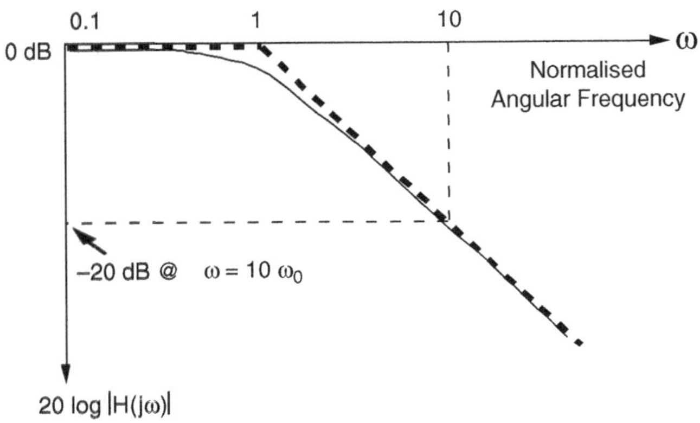

Figure 4.4 Bode plot for the RC low-pass filter.

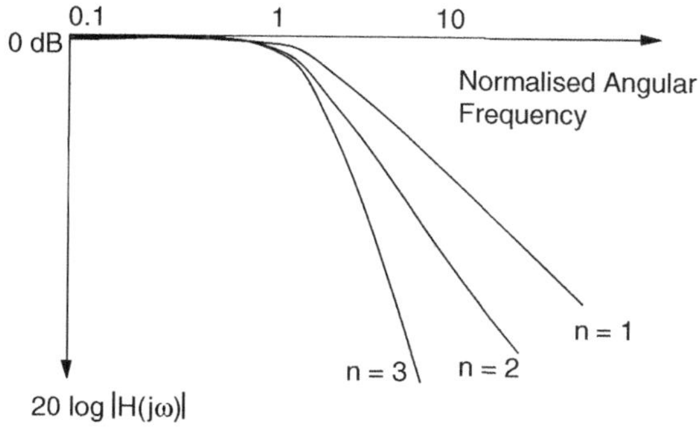

Figure 4.5 Achieving a sharper cut off by increasing the order of a Butterworth filter.

of 20 dB per decade, downwards for zeros and upwards for poles. So, by deliberately putting poles and zeros into the response at certain places, almost any desired frequency response can be obtained.

For example, the well-known Butterworth low-pass filter has a normalised transfer function given by:

$$H(s) = \frac{1}{(1+s^n)} \quad \text{or} \quad |H(j\omega)| = \frac{1}{\sqrt{(1+\omega^{2n})}} \tag{4.7}$$

At the normalised cut-off frequency, $\omega = 1$, clearly the magnitude of $H(j\omega)$ is $1/\sqrt{2}$ (the –3 dB point). The value of n, the order of the filter, determines how many poles are operating in the Bode plot, each giving a 20 dB per decade slope and therefore strongly affecting the steepness of the roll-off, as illustrated in Figure 4.5.

Poles in filters can be a source of considerable confusion because they are on the denominator, but when the signal frequency approaches a pole, the filter's transfer function starts to *fall*. It is very important to remember that the pole is in the complex s-domain. Real-world continuous-wave (sinusoidal) test signals are on the vertical axis in the s-domain. What matters then is how close the pole is to that vertical axis. With incorrect design of the low-pass filter, with a pole too close to the vertical axis, a very sharp peak in the frequency response is created. With proper design, the desired smooth roll-off is obtained. Figure 4.6 shows why a pole creates a low-pass response by plotting the magnitude of the transfer function in quasi-3D, to show the complex domain.

4.2.4 Impulse Response and Causality

Ideal characteristics such as perfectly sharp cut-offs, infinite attenuation in the stopband and zero attenuation in the passband (i.e., so-called 'brick wall' characteristics) cannot be realised in a practical circuit. These ideal characteristics for a low-pass filter, as illustrated in Figure 4.7, are defined as:

$$\begin{aligned} |H(j\omega)| &= A \quad 0 < \omega < \omega_0 \\ &= 0 \quad \omega > \omega_0 \end{aligned} \tag{4.8}$$

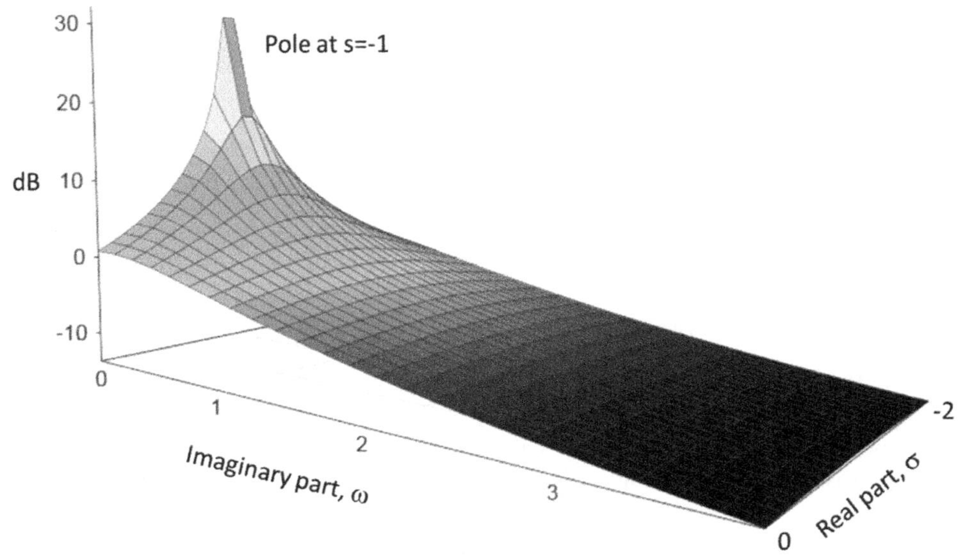

Figure 4.6 A pole plotted in quasi-3D to show the pole in a low-pass filter.

This ideal frequency response can be described as a *rect* function in the frequency domain, centred on $\omega = 0$. The inverse Laplace transform of this *rect* function, which gives the time-domain impulse response as shown in Equation (4.3), is a *sinc* function centred on $t = 0$. So, if there was a network that had this ideal brick wall response, it would need to be able to generate an output signal before the impulse at $t = 0$ was even applied. For a passive network this is clearly impossible, there is no energy in the system before the impulse is applied, and so such a filter is termed *non-causal*. This is summarised in Figure 4.8, which also shows an example of the type of impulse response that can be realised in practice from a passive network.

In the digital signal processing world, this problem can be partly tackled by putting a delay into the filter: the *sinc* function at the output is delayed so that the peak is at positive time and the ideal *sinc* function can be almost realised, albeit with some truncation of the left-hand side. DSP can do this by using digital memory devices. In the world of active analogue filters, there are limits on what can be achieved in approximating the ideal brick wall response without creating unstable circuits; the main advantage of active filters is that inductors can be

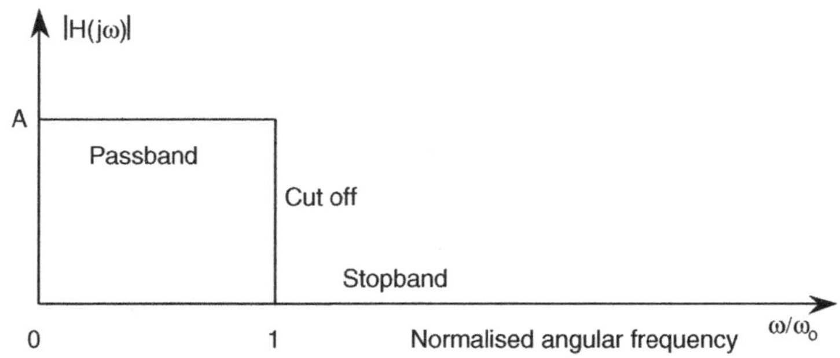

Figure 4.7 Ideal low-pass filter response.

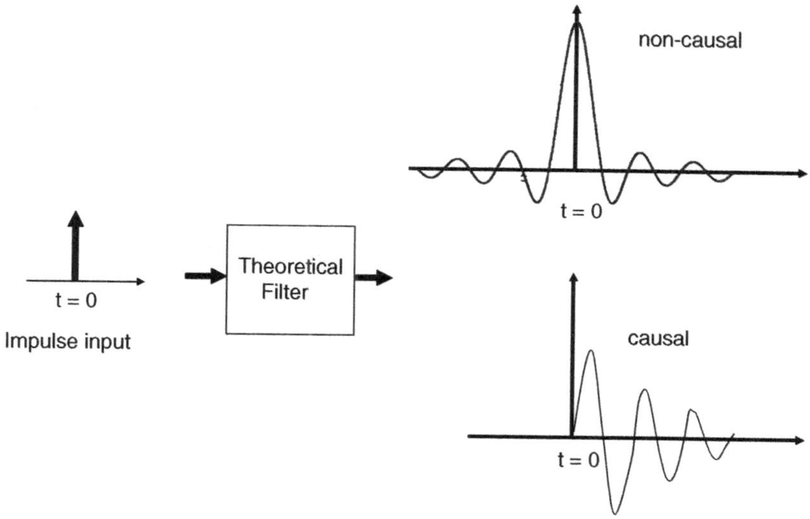

Figure 4.8 Impulse response of ideal brick wall low-pass filter.

avoided, making integrated circuit filters much more practical. For passive filters we can only approximate the ideal brick wall filter characteristics within a certain prescribed error.

4.2.5 The Need for a System Impedance

An important detail that we need to consider before the filter design method is introduced is the question of the desired input and output resistance. When working at low frequencies, where signals are generally treated in terms of voltages and currents, the most common situation is that sources have very low output resistance and loads have very high input resistance. For example, in the case of DC, when using a power supply to generate 5 V DC, an engineer wants the confidence that this will be a fixed 5 V, regardless of the amount of current drawn, within sensible limits. Hence, the power supply is carefully designed to have a low-output resistance and behaves very much like a perfect voltage source. A multimeter or oscilloscope, used to measure a voltage, will need a very large input resistance in order not to load the circuit that it is measuring. In the case of low-frequency AC signals, this approach holds true up to a certain point, although care needs to be exercised to take into account the input capacitance of the multimeter/oscilloscope and the capacitance of any cables used to connect the circuit to the measurement equipment. Even at a few MHz, it will be necessary to take measures such as using oscilloscope probes in order to avoid these loading effects.

Eventually, the low-frequency approach (low-output resistance/high-input resistance) has to be abandoned in favour of using a system impedance, such as 50 Ω. The need for a system impedance approach stems from the concept of the characteristic impedance of a transmission line – which may be a coaxial cable, an adapter or connector, or even a track on a circuit board. Characteristic impedance and the nature of transmission lines are explained in detail in Chapter 5. In the high-frequency scenario, the source output resistance and load resistance are both equal to the characteristic impedance of the transmission line(s) used to

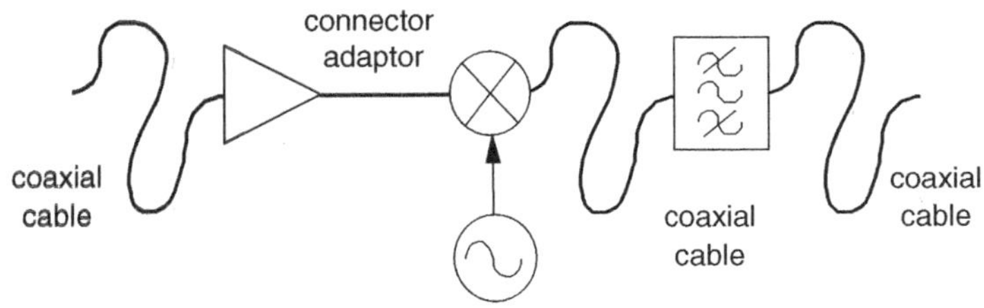

Figure 4.9 Example receiver constructed from modular components, showing the need for a system impedance.

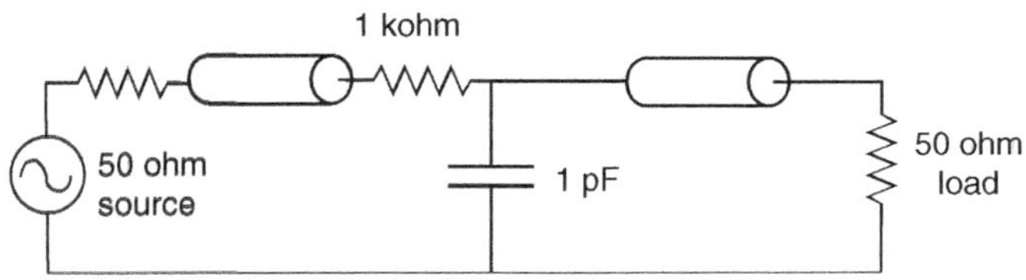

Figure 4.10 RC filter in a 50 Ω system.

connect up the system. This is illustrated in Figure 4.9, where a simple receiver made with modular components is connected together using coaxial cables and adapters. There is no exact rule that determines the frequency at which this approach takes over, since it very much depends on the length of the cables and connectors compared to the wavelength of the signal. The most important feature of this system impedance approach is that, with all components matched to the characteristic impedance of the transmission lines, it avoids reflections between components. Reflections would create standing waves and the frequency response would suffer from what is referred to as ripple: the reflected signals bounce backwards and forwards and add constructively and destructively at different frequencies, leading to peaks and troughs in the frequency response. Once introduced into the frequency response of a system, this gain ripple is extremely difficult to get rid of.

Now, if one were to imagine that the RC filter was one of these modular components in a 50 Ω system, as illustrated in Figure 4.10, it is very easy to see that there is now a problem – most of the signal voltage is dropped across the series resistor and the result is that the loss of the filter is very high when used in a 50 Ω system. The solution is to design filters that use only reactive components – inductors and capacitors (lumped elements) or some form of transmission line components (distributed elements). Figure 4.11 shows the circuit diagram of a fifth-order low-pass filter using only lumped reactive elements – series inductors and shunt

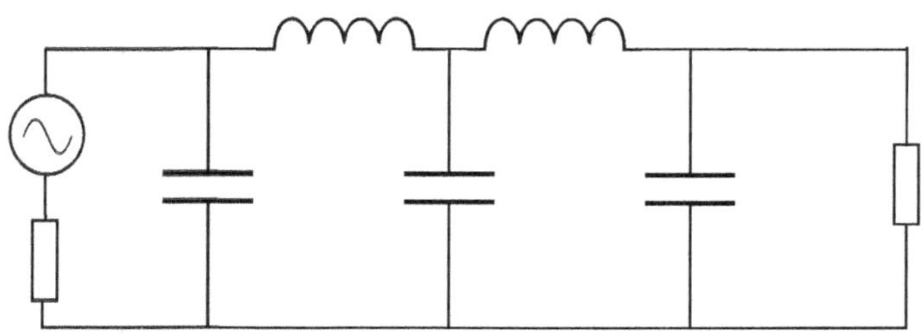

Figure 4.11 Fifth-order Cauer low-pass filter.

capacitors. This series-parallel ladder network is often called a Cauer topology in recognition of the pioneering work of Wilhelm Cauer in the field of network synthesis. A few quick derivations, or simulations, will show that the component values cannot be identical if this filter is to be well matched to 50 Ω over the passband, and network synthesis is required to yield the correct optimum component values.

4.3 Butterworth, Chebyshev and Elliptic Low-Pass Prototypes

This section describes three of the most common filter types for approximating the ideal filter characteristics, namely the Butterworth (or maximally flat), Chebyshev (or equi-ripple), and Elliptic filters. Band-pass, band-stop, and high-pass filter networks can be derived from a low-pass filter network, so treatment of these three filter types first focuses on the low-pass characteristics. The synthesis of filters is based on a low-pass filter, referred to as the low-pass prototype, shown in Figure 4.12, which is normalised to 1 Ω source and load impedance and 1 rad s^{-1} angular cut-off frequency. This keeps the analysis tidy and means that the component values for a wide variety of designs can be taken from published tables of normalised values, referred to as the g values, $g_1, \ldots, g_n$, where n is the order of the filter, which is equal to the number of elements in the low-pass prototype. For a low-pass filter, each series element in the prototype filter directly converts to an inductor, and each shunt element in the prototype is a capacitor. To obtain the required component values for a low-pass filter, these g values are simply de-normalised by scaling to the required cut-off frequency and impedance. For band-pass, band-stop and high-pass filters, the low-pass prototype network must be transformed before de-normalisation.

4.3.1 Butterworth Low-Pass Filter

The frequency response of the Butterworth approximation to the ideal brick wall filter is illustrated in Figure 4.13. Note the flatness of the response in the passband, with no ripples

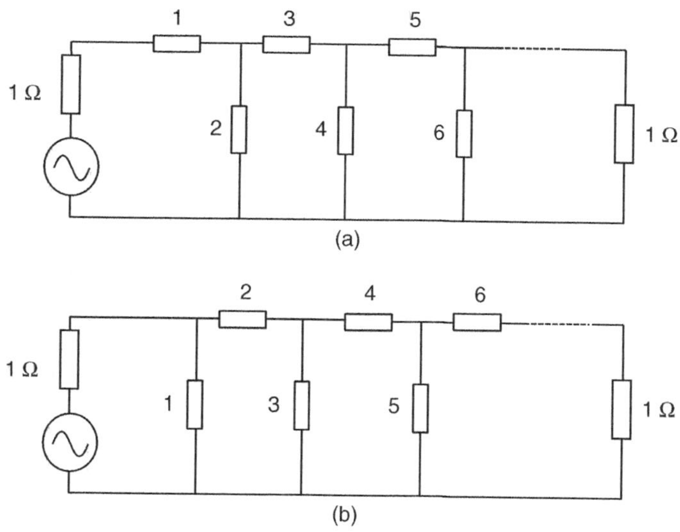

Figure 4.12 The prototype filter. (a) Series element first. (b) Shunt element first.

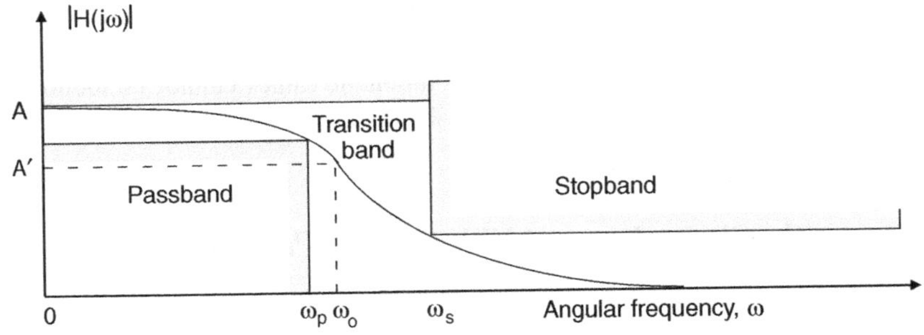

Figure 4.13 Butterworth (or maximally flat) filter response.

present. The component parameters for a certain order of Butterworth filter form a unique set; the only parameter that the circuit designer can specify for the Butterworth low-pass prototype is the order of the filter; a higher-order filter has a steeper cut-off as shown for the RC case. The amplitude response of the Butterworth, or maximally flat, filter is given by:

$$H(j\omega) = \frac{1}{\sqrt{1 + (\omega/\omega_0)^{2n}}} \tag{4.9}$$

The attenuation $\alpha_n(\omega)$ as a function of angular frequency is specified as:

$$\alpha_n(\omega) = 10\log_{10}\{1 + (\omega/\omega_0)^{2n}\} \tag{4.10}$$

where ω_0 is the frequency for which $\alpha = 3$ dB ('half-power point'), α_{max} is the maximum attenuation allowed in the passband, and α_{min} is the minimum attenuation allowed in the stopband. Hence,

$$\alpha_{max} = 10\log_{10}\{1 + (\omega_p/\omega_0)^{2n}\} \tag{4.11}$$

and

$$\alpha_{min} = 10\log_{10}\{1 + (\omega_s/\omega_0)^{2n}\} \tag{4.12}$$

The transfer function $H(j\omega)$ has n poles. To find them, consider the function $|H(j\omega)|^2$, and replace $j\omega$ by the complex frequency s. Since $|H(j\omega)|^2 = H(j\omega)H(-j\omega)$, replacing $j\omega$ by s results in $H(s)H(-s)$. If s_1 is a pole of $H(s)$ lying in the left-hand half-plane, then $-s_1$, is a pole of $H(-s)$, which is its mirror image in the right-hand half-plane. The function $H(s)H(-s)$ has $2n$ poles of which the n poles in the left-hand half-plane give rise to a stable filter. Hence, the desired poles are the values of s lying in the left half-plane. Hence, the function:

$$H(s)H(-s) = \frac{1}{1 + (-1)^n(s/\omega_0)^{2n}} \tag{4.13}$$

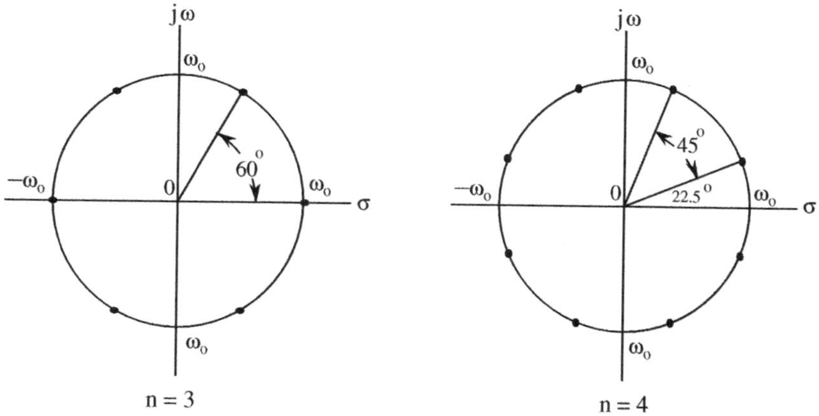

Figure 4.14 Poles in the s-domain for a Butterworth filter of order 3 and 4.

goes to infinity if s is one of the $2n$ roots of the equations:

$$(s/\omega_0)^{2n} = +1, \quad \text{for } n \text{ odd,}$$
$$= -1, \quad \text{for } n \text{ even.} \tag{4.14}$$

The roots are

$$s_i = \omega_0 e^{j\pi(2i/n)/2} \quad \text{for } n \text{ odd,}$$
$$s_i = \omega_0 e^{j\pi(2i-1/n)/2} \quad \text{for } n \text{ even} \tag{4.15}$$

where $i = 1, \dots, 2n$.

These roots are equally spaced points on a circle of radius ω_0. The poles of interest are the roots lying in the left-hand half-plane. Figure 4.14 shows the poles for Butterworth filters of order 3 and 4.

4.3.2 Chebyshev Filter

The Butterworth filter can only give a steeper transition from passband to stopband by increasing the order of the filter. Apart from simply requiring more components, with size and cost implications, this approach will introduce additional loss and difficulties in finding suitable component values. By allowing the magnitude of the filter response to have small ripples, one can achieve a sharper transition between the passband and the stopband. Here, the low-pass characteristic may be approximated in the passband by an equi-ripple response. This equi-ripple passband and maximally flat stopband response leads to the Chebyshev filter. The magnitude of the passband loss varies between two prescribed limits, and then drops monotonically outside the passband, as shown in Figure 4.15(a). The magnitude of the transfer function

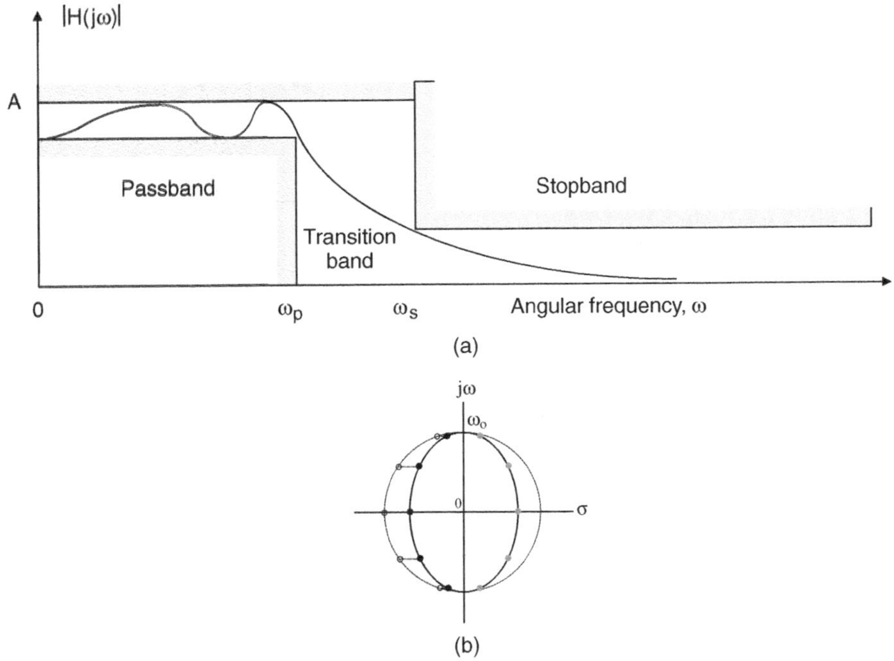

Figure 4.15 Chebyshev filter. (a) Frequency response. (b) Pole diagram for $n = 5$.

is given by:

$$|H(j\omega)|^2 = \frac{1}{1 + \varepsilon^2 T_n^2(\omega/\omega_0)} \tag{4.16}$$

where ε is a constant, and $T_n(\omega/\omega_0)$ is the Chebyshev polynomial, defined by:

$$T_n(\omega/\omega_0) = \cos[n \cos^{-1}(\omega/\omega_0)] \text{ for} |\omega/\omega_0| < 1 \tag{4.17}$$

Let $\cos^{-1}(\omega/\omega_0) = \theta$. Hence, this $T_n(\omega/\omega_0)$ is a polynomial because one can express $\cos n\theta$ as a polynomial in $\cos\theta$. For example, $\cos 2\theta = 2\cos^2\theta - 1$, and $\cos 3\theta = 4\cos^3\theta - 3\cos\theta$, etc. Observe that $-1 \leq T_n(\omega/\omega_0) \leq 1$ for $|\omega/\omega_0| \leq 1$ and therefore $|H(j\omega)|^2$ oscillates between $(1 + \varepsilon^2)^{-1}$ and 1 in the passband. At the cut-off frequency $(\omega/\omega_0) = 1$ and $T_n(\omega/\omega_0) = 1$. In terms of the poles in the s-domain, for a Chebyshev filter they are on an ellipse. For completeness, note than an inverse Chebyshev filter is one that has a monotonic frequency response in the passband and ripples in the stopband.

4.3.3 Elliptic Filter

The elliptic filter exhibits equi-ripple behaviour in both the passband and the stopband, as shown in Figure 4.16. The characteristic behaviour provides a more rapid transition from passband to stopband than does the Chebyshev or Butterworth filter, but at the expense of a

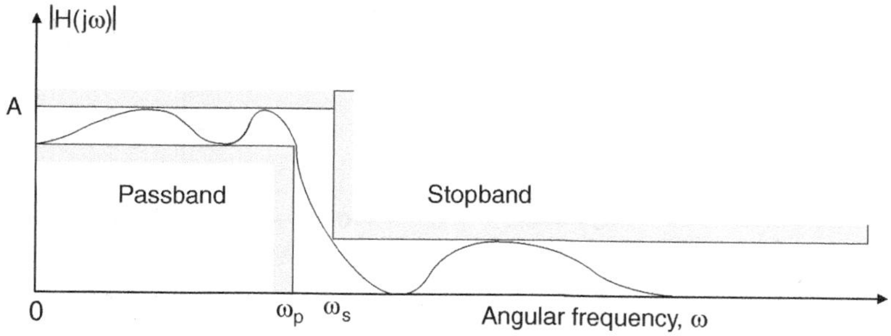

Figure 4.16 Elliptical filter response.

finite attenuation level in the stopband. Zeros are placed in the stopband in order to maintain sufficient stopband attenuation.

4.4 Filter Design Method

Filter design starts with the low-pass prototype filter which realises the required Butterworth, Chebyshev or elliptic transfer function, with the component values normalised to an angular cut-off frequency of 1 rad s^{-1} and impedance of 1 Ω. The normalised element values (g-values) can be obtained from standard tables, with a variety of computer tools, or with formulas found in deliberately rigorous tomes on filter design [1, 2]. For Butterworth filters, the number of elements in the prototype (the order of the filter) is the only parameter to be selected. For the other types, there are also parameters of maximum ripple and insertion loss to be decided. The g-values for Butterworth prototype filters up to $n = 6$ are given in Table 4.1. Since the poles of the Butterworth filter are on a circle, it can be shown that the g-values are given by:

$$g_k = 2\sin\left[\frac{(2k-1)}{2n}\pi\right] \quad k = 1, 2, 3 \ldots \text{n} \tag{4.18}$$

For any desired low-pass filter design, these g-values are simply scaled according to the cut-off frequency required and the impedance of the source and load. Remembering that $Z_c = 1/(j\omega C)$

Table 4.1 g-values for Butterworth filters.

	Element number					
Order	1	2	3	4	5	6
2	1.414	1.414				
3	1.000	2.000	1.000			
4	0.7654	1.848	1.848	0.7654		
5	0.6180	1.618	2.000	1.618	0.6180	
6	0.5176	1.414	1.932	1.932	1.414	0.5176

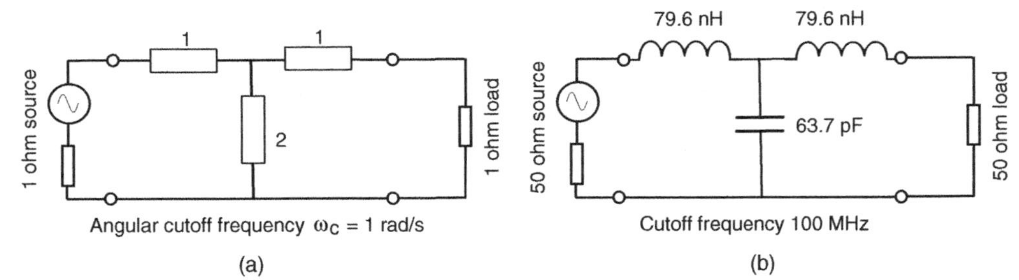

Figure 4.17 (a) Low-pass prototype for $n = 3$. (b) De-normalised values for a cut-off frequency of 100 MHz and 50 Ω impedance.

and $Z_L = j\omega L$, de-normalising for the low-pass filter is achieved as follows:

$$L_k = \frac{g_k \times R}{\omega_c} \tag{4.19}$$

$$C_k = \frac{g_k}{\omega_c R} \tag{4.20}$$

where R is the source and load resistance and the kth normalised value, while g_k is calculated from Equation (4.18) or read from the table.

Example: Design a 3rd-order Butterworth low-pass filter with cut-off frequency of 100 MHz and 50 Ω source and load impedance, using a series element first.

Figure 4.17(a) shows the schematic circuit diagram of the low-pass prototype, with $g_1 = g_3 = 1$ and $g_2 = 2$ from Table 4.1. For a low-pass filter, these simply need to be de-normalised using Equations (4.19) and (4.20):

$$L_1 = L_3 = \frac{1 \times 50}{2\pi \times 100 \times 10^6} = 79.6\ \text{nH} \qquad C_2 = \frac{2}{2\pi \times 100 \times 10^6 \times 50} = 63.7\ \text{pF}$$

The circuit is shown in Figure 4.17(b), while Figure 4.18 shows the simulated result using ADS^{TM} from Keysight Technologies.

It is useful at this point to remember that S-parameters deal with the signal as incident and reflected voltage waves, since it gives valuable insight into how filters work. Figure 4.19 shows the filter circuit with 50 Ω input and output transmission lines added – which might, for example, be coaxial cables connecting the circuit to a network analyser whilst the filter is being measured. In the pass-band, S_{21} is close to 0 dB, indicating that most of the input signal at port 1 is transferred to the output port, whilst S_{11} is of the order of –40 dB, representing the fact that only a very tiny proportion is reflected back, because the filter is well-matched to 50 in the pass-band.

In the stop-band, S_{21} drops steadily to –40 dB indicating that almost no signal transfers to port two, whilst S_{11} is close to 0 dB. This shows how the filter operates – in the stop-band the signal is reflected back to the input port. Given the properties of S-parameters described in Chapter 3, this is precisely the result to be expected: since this low-pass filter contains only

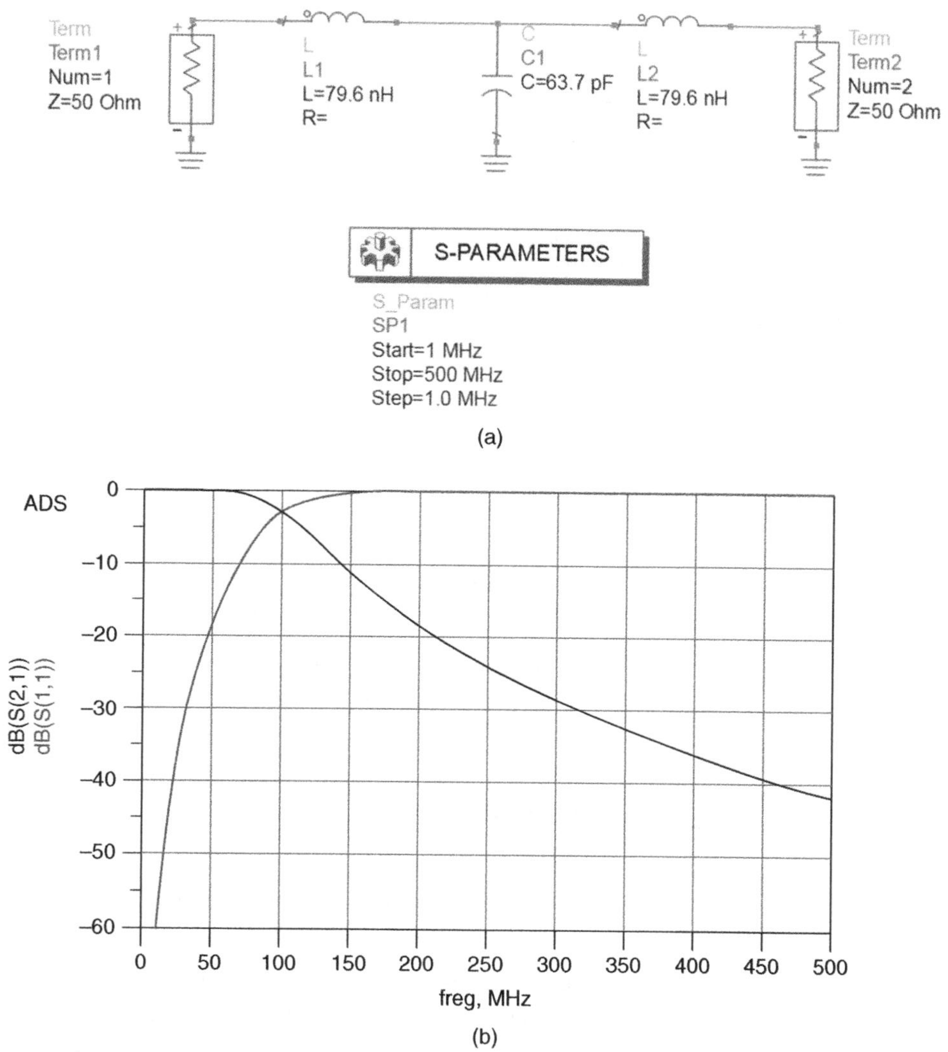

Figure 4.18 ADS™ circuit schematic and simulated response of the low-pass filter example.

lossless passive components, we know that $|S_{11}|^2 + |S_{21}|^2 = 1$. On this basis, it can be seen that a filter can be regarded as a component that selectively allows signals to pass through or be reflected back, based on their frequency.

4.4.1 High-Pass Transformation

For high-pass, band-pass, and band-stop filters, the prototype low-pass filter has each of its elements transformed into a new element, in order to achieve the desired response. The transformation of low-pass to high-pass requires ω to be mapped to $1/\omega$, as shown conceptually

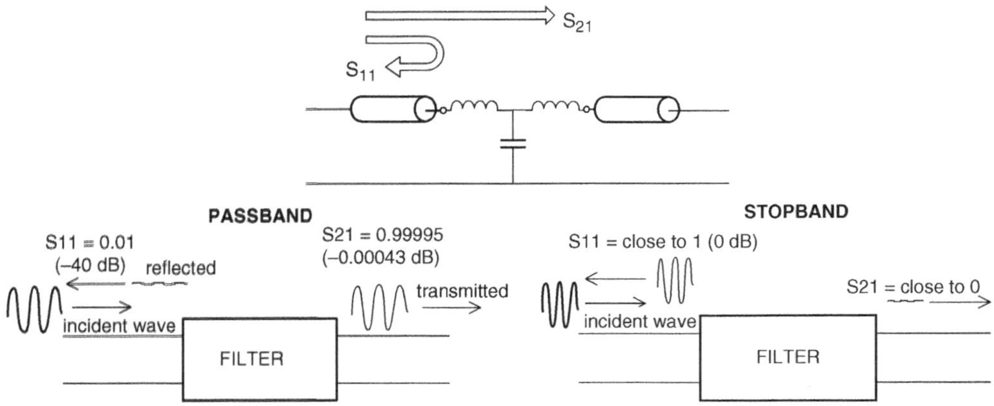

Figure 4.19 Filter behaviour in terms of S-parameters and incident/reflected waves.

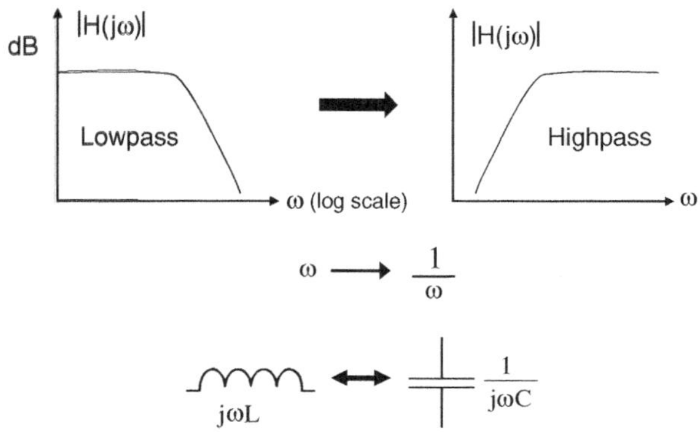

Figure 4.20 Low-pass to high-pass transformation on the Bode plot.

in Figure 4.20. To achieve this, the series inductors of the low-pass, with normalised value g_k, are replaced with shunt capacitors of normalised value $1/g_k$, while the shunt capacitors, with normalised value g_k, are replaced with series inductors of normalised value $1/g_k$, giving the topology shown in Figure 4.21. Having carried out the transformation, the component values are then de-normalised.

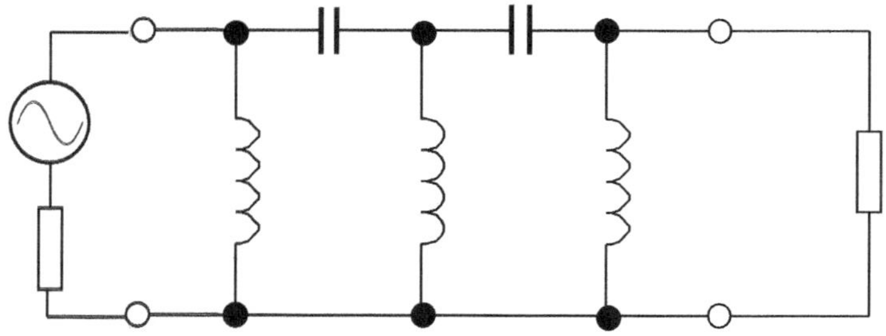

Figure 4.21 High-pass filter with $n = 5$.

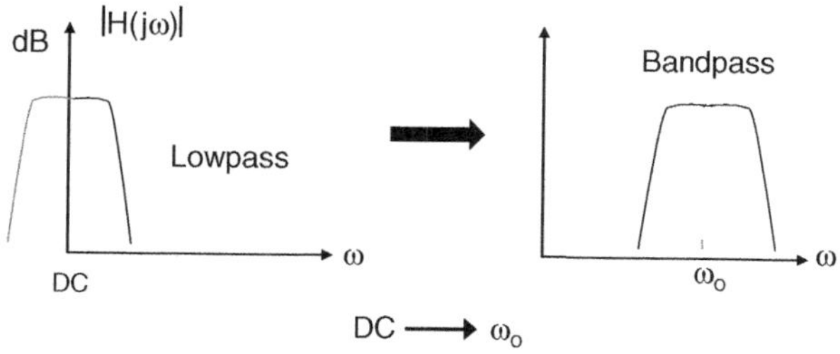

Figure 4.22 Low-pass to band-pass transformation.

4.4.2 Band-Pass Transformation

For a band-pass filter, with upper and lower cut-off frequencies ω_2 and ω_1, respectively, the following frequency transformation is used in order to shift $\omega = 0$ (DC) of the low-pass to the desired centre frequency, ω_0, as illustrated in Figure 4.22:

$$\omega \to \frac{1}{\Delta}\left(\frac{\omega}{\omega_o} - \frac{\omega_o}{\omega}\right) \tag{4.21}$$

where

$$\Delta = \frac{\omega_2 - \omega_1}{\omega_o} \tag{4.22}$$

and

$$\omega_o = \sqrt{\omega_1 \omega_2} \tag{4.23}$$

Note that ω_o is the *geometric* mean of the upper and lower angular cut-off frequencies; this is used instead of the arithmetic centre frequency for convenience in the mathematics. Δ is the fractional bandwidth, and becomes an additional term in the transformation. The element values for the *series* resonators, shown in Figure 4.23 for $n = 3$, are calculated from:

$$L_k = \frac{g_k R}{\Delta \times \omega_o} \tag{4.24}$$

$$C_k = \frac{\Delta}{\omega_o g_k R} \tag{4.25}$$

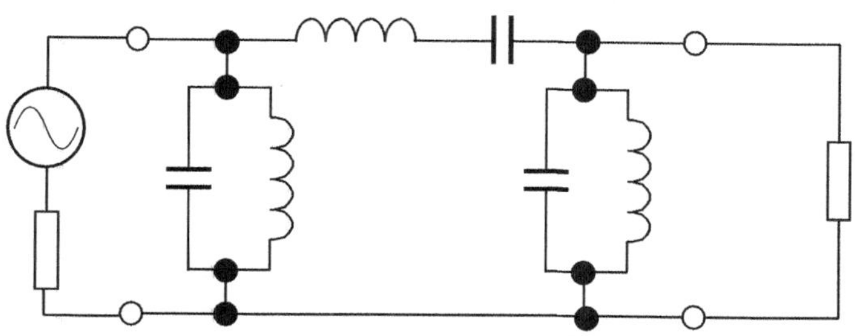

Figure 4.23 Band-pass filter with $n = 3$.

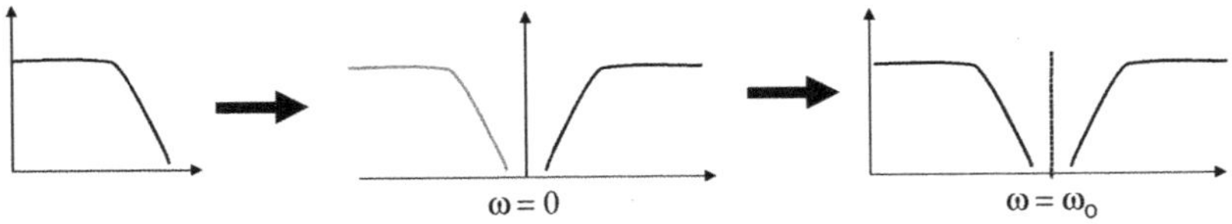

Figure 4.24 Low-pass to band-stop transformation.

The element values for the *parallel* resonators in Figure 4.23 are calculated from:

$$L_k = \frac{\Delta \times R}{\omega_o g_k} \tag{4.26}$$

$$C_k = \frac{g_k}{\Delta \times \omega_o R} \tag{4.27}$$

4.4.3 Band-Stop Transformation

For a band-stop filter, the transformation combines the high-pass and band-pass transformation of the low-pass frequency response, as shown in Figure 4.24, which requires:

$$\omega \to \frac{\Delta}{\left(\frac{\omega}{\omega_o} - \frac{\omega_o}{\omega}\right)} \tag{4.28}$$

The element values for the shunt-connected *series* resonators shown in Figure 4.25 are then calculated from:

$$L_k = \frac{R}{\omega_o g_k \Delta} \tag{4.29}$$

$$C_k = \frac{\Delta \times g_k}{\omega_o R} \tag{4.30}$$

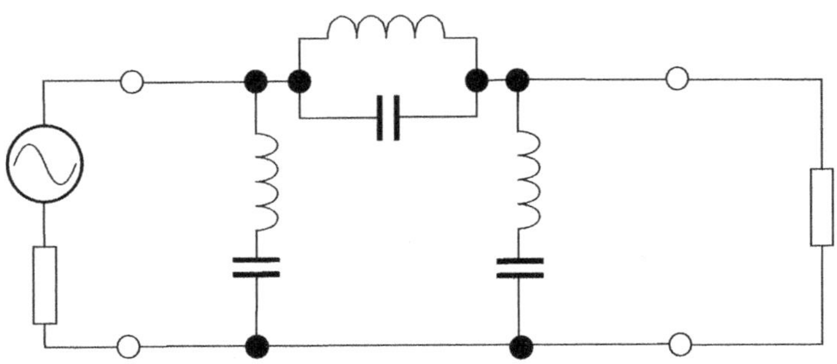

Figure 4.25 Band-stop filter with $n = 3$.

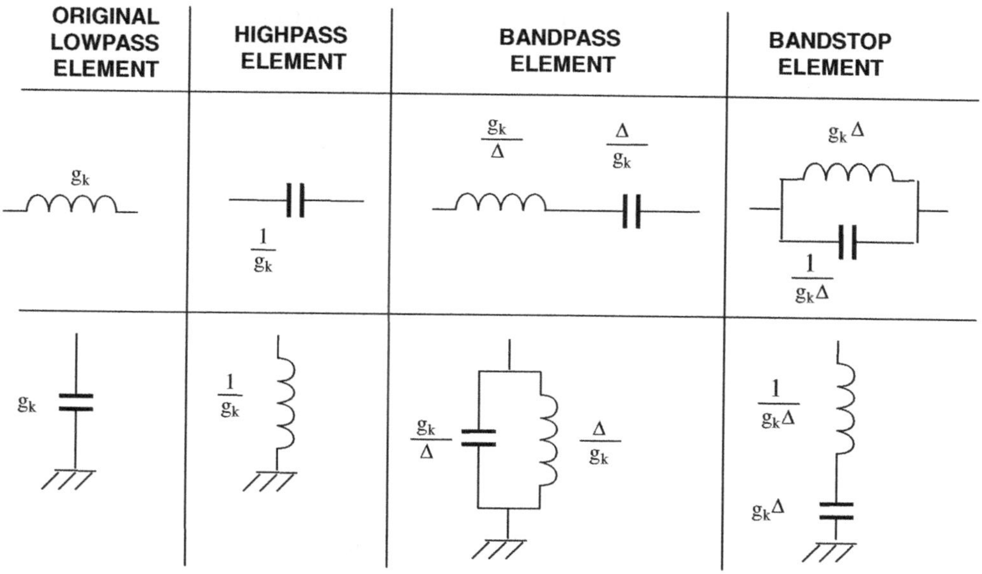

Figure 4.26 Summary of filter transformations.

The element values for the series-connected *parallel* resonators shown in Figure 4.25 are calculated from:

$$L_k = \frac{\Delta \times g_k R}{\omega_o} \tag{4.31}$$

$$C_k = \frac{1}{\Delta \times \omega_o g_k R} \tag{4.32}$$

4.4.4 Band-Pass Filter Design Example

Figure 4.26 summarises the filter transformations. To demonstrate the design method, a fifth-order Butterworth band-pass filter as shown in Figure 4.27 is designed for a passband from 65 to 75 MHz with 50 Ω source and load impedances using a series element first.

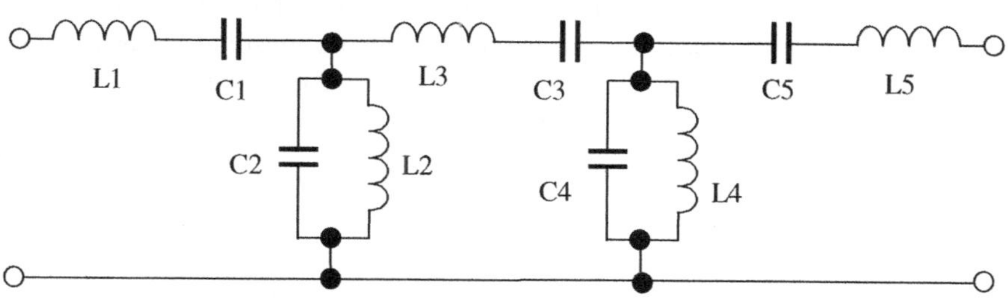

Figure 4.27 Fifth-order band-pass filter.

The low-pass prototype values are 0.618, 1.618, 2, 1.618 and 0.618.

$$\Delta = \frac{\omega_2 - \omega_1}{\omega_o} = \frac{f_2 - f_1}{\sqrt{f_1 f_2}} = 0.1432 \qquad \omega_o = 438699539 \text{ rad/s}$$

So

$$L_1 = \frac{0.618 \times 50}{0.1432 \times 438699539} = 492\text{nH}$$

$$C_1 = \frac{0.1432}{0.618 \times 50 \times 438699539} = 10.6\text{pF}$$

$$L_2 = \frac{0.1432 \times 50}{1.618 \times 438699539} = 10.1\text{nH}$$

$$C_2 = \frac{1.618}{0.1432 \times 50 \times 438699539} = 515\text{pF}$$

$$L_3 = \frac{2 \times 50}{0.1432 \times 438699539} = 1.59\mu\text{H}$$

$$C_3 = \frac{0.1432}{2 \times 50 \times 438699539} = 3.3\text{pF}$$

By symmetry, $C_5 = C_1$, $L_5 = L_1$, $C_4 = C_2$ and $L_4 = L_2$.

This design is checked with ADS™ as shown in Figure 4.28(a). Filter synthesis CAD programs make the design much easier and will be needed because more complicated filters are used in practice in order to achieve objectives such as improved group delay flatness. Note that the values in this example design have a large range, for example, inductors from 10.1 nH to 1.59 μH. Such a large range is difficult to realise in practice because of parasitics (introduced in the next section), and a different filter topology might be needed in reality.

4.4.5 Group Delay

So far, only the magnitude response of the filters has been discussed. For a two-port network, another parameter that is important to the systems engineer is the group delay response. What is important in a system is that, if an information-bearing signal (i.e., a modulated carrier) is applied to the input port, the information content should be preserved at the output. The modulated carrier has spectral components across its full bandwidth and, for the information to be reproduced at the output with no distortion, the network must have constant insertion loss and constant group delay across the signal's bandwidth. Group delay is the rate of change of the network's S_{21} phase response, given by:

$$\tau_g = -\frac{d\theta(\omega)}{d\omega} \tag{4.33}$$

A linear phase network would, therefore, have a constant group delay and all spectral components of the signal travel would through the network with the same time delay. Figure 4.28(b)

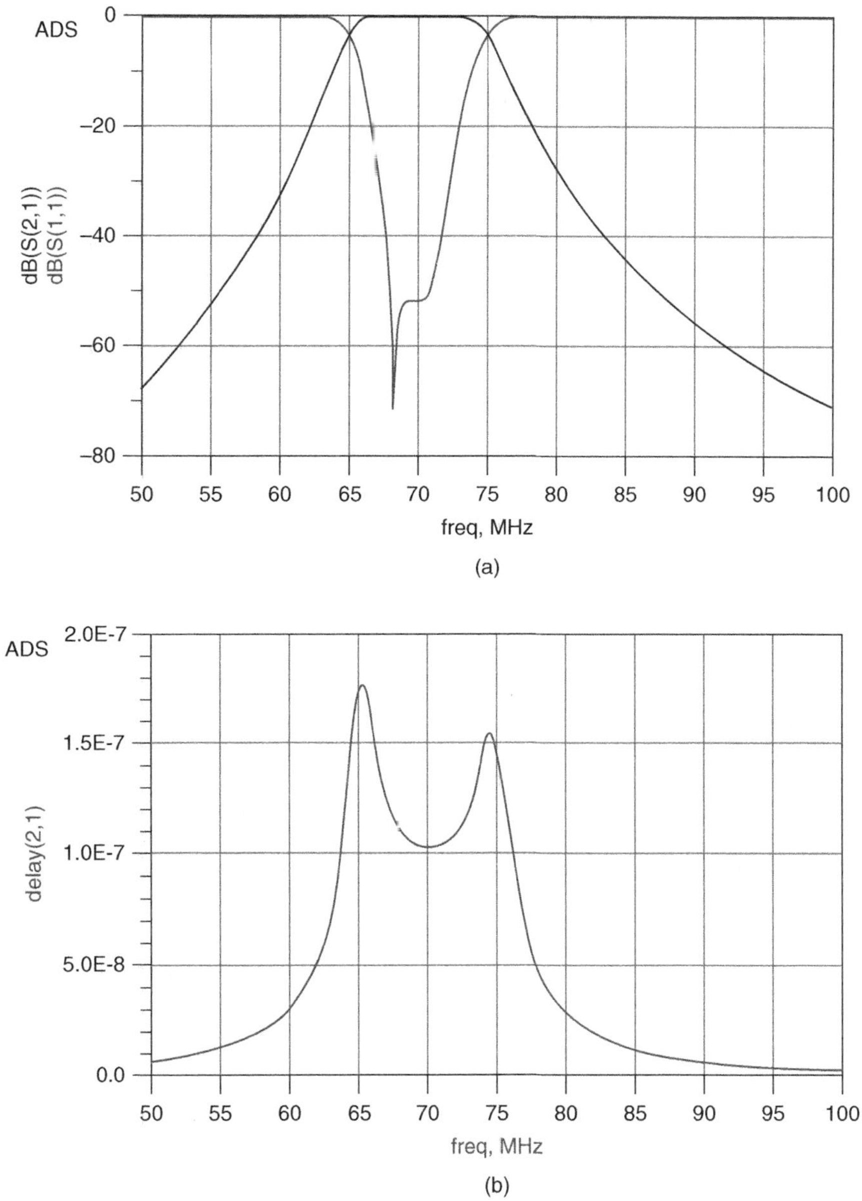

Figure 4.28 ADS simulation of the fifth-order band-pass filter design example. (a) Magnitude response (S_{21} and S_{11}). (b) Group delay response.

shows the group delay response of the fifth-order band-pass example filter. It can be seen that the group delay is constant in the centre of the band but increases quite significantly at the band edges. Filter specifications will typically impose rather stringent requirements on group delay flatness that necessitates the use of more sophisticated filter topologies.

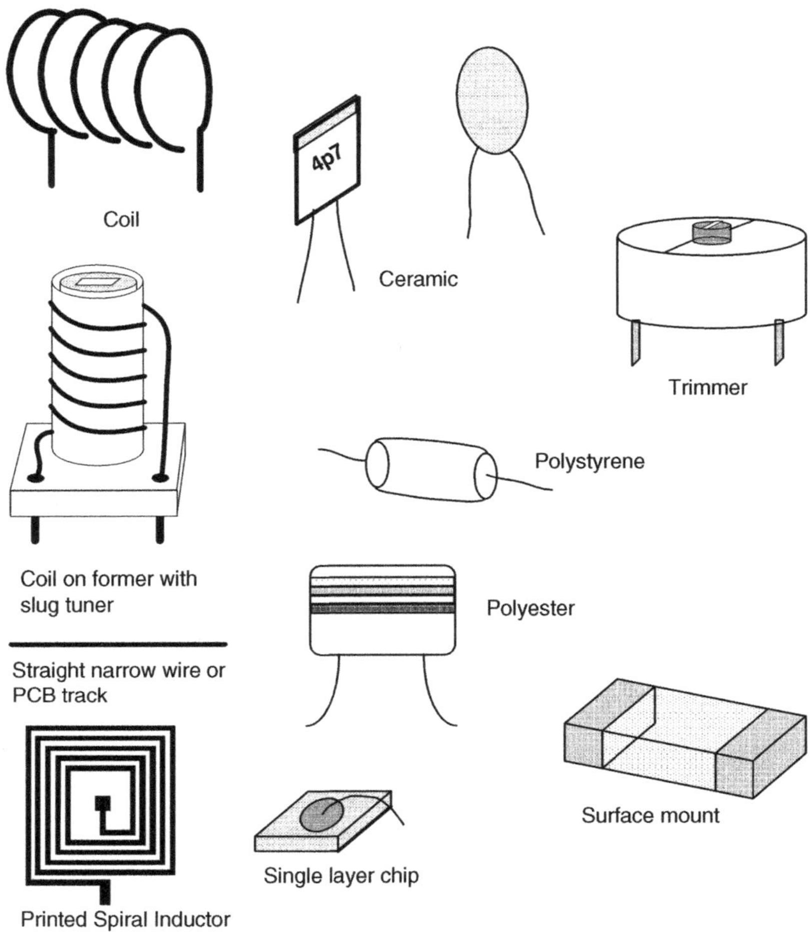

Figure 4.29 Typical discrete lumped components for RF filters.

4.5 Practical Lumped Elements

Having found the component values, the most suitable component types must be selected for construction of the filter. The best component type depends very much on the design frequency and on the desired value. Figure 4.29 shows examples of components that could be selected for lumped-element RF filters. Real components have 'parasitics' which include the inductance of leads, resistive losses in metals and dielectrics, fringing capacitance to ground, and capacitance between turns. These can be put into an 'equivalent circuit model' which approximates the true behaviour of the component. Figure 4.30 shows simple equivalent circuit models for a resistor, capacitor and inductor. The effects of these parasitics on the ideal filter include increased insertion loss in the passband, reduced rejection in the stop band, reduced sharpness of transitions and spurious behaviour at high frequency.

As Figure 4.30 shows, actually the inductor inter-turn feedback capacitance is in parallel with the inductance and this forms a resonant circuit. At low frequencies the feedback

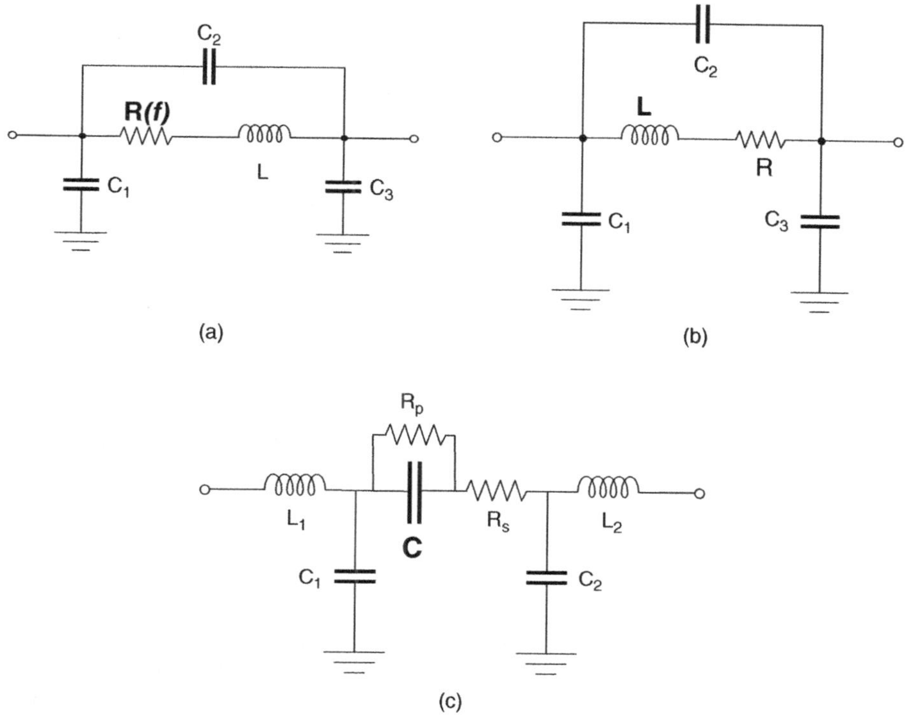

Figure 4.30 Simple equivalent circuit models showing parasitic. (a) Resistor. (b) Inductor. (c) Capacitor.

capacitance's effect may be negligible, but eventually the inductor resonates with its own feedback capacitance, and beyond that the capacitance dominates. Likewise, the capacitor model shows that there is series inductance (from the leads, package, and/or metallisation of the electrodes) and so the capacitor will resonate with its own parasitic inductance at some frequency. Choosing surface mount components helps to minimise parasitics, but every component has a self-resonant frequency (SRF), beyond which its behaviour can be unpredictable. Figures 4.31and 4.32 show typical impedance plots for SMD inductors and capacitors, respectively. Note that, because these are plotted on a log-log scale, the inductance behaviour is evident as a straight line with positive slope ($j\omega L$), while the capacitance gives a straight line with a negative slope ($1/j\omega C$). So, in Figure 4.31, for example, the 10 000 nH inductor characteristic clearly shows that inductance dominates up to only 30 MHz, while beyond the SRF the impedance starts to fall as the parasitic feedback capacitance dominates. To some extent, if proper models are used then the parasitics can be accommodated into the design using optimisation. However, lumped elements ultimately become unusable at a certain frequency and transmission lines become the only way of realising a useful filter.

4.6 Capacitively-Coupled Resonator Filter

The filter topologies described so far are sometimes referred to as the canonical form, with each element in the ladder prototype being directly implemented. However, this direct approach

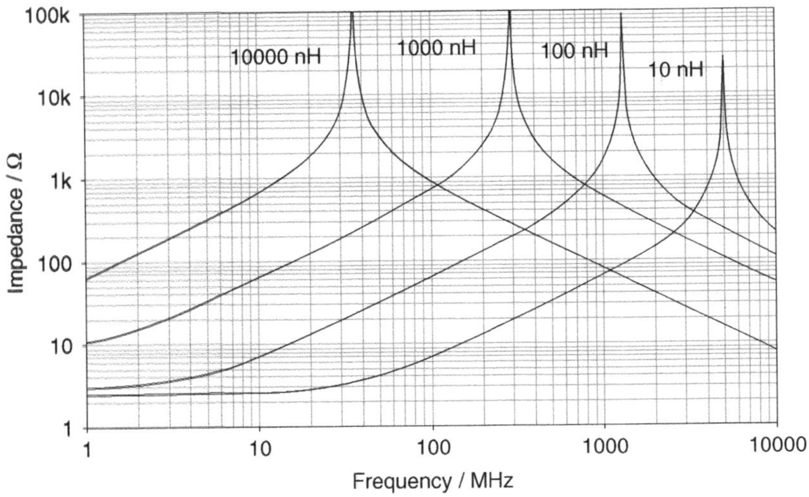

Figure 4.31 Plot of some SMD inductors' impedance versus frequency, showing typical SRF.

often leads to designs that are not practical because of the parasitics of real components. The problem is particularly evident in band-pass filters with extreme values of Δ, such as very narrow-band filters in a receiver. The capacitively-coupled resonator filter topology [3] (shown in Figure 4.33) is one of the design topologies that can solve this problem. Essentially, the series inductors are taken out of the circuit by replacing them with an element called a *K*-inverter (shown in Figure 4.34), with the series element in the *K*-inverter π–network being the inductor and the shunt components being absorbed by adjusting the adjacent component values.

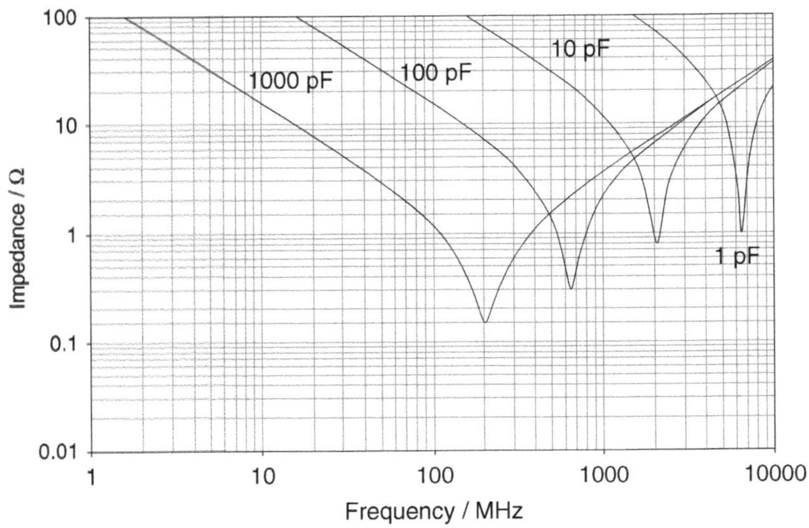

Figure 4.32 Plot of some SMD capacitors' impedance versus frequency, showing typical SRF.

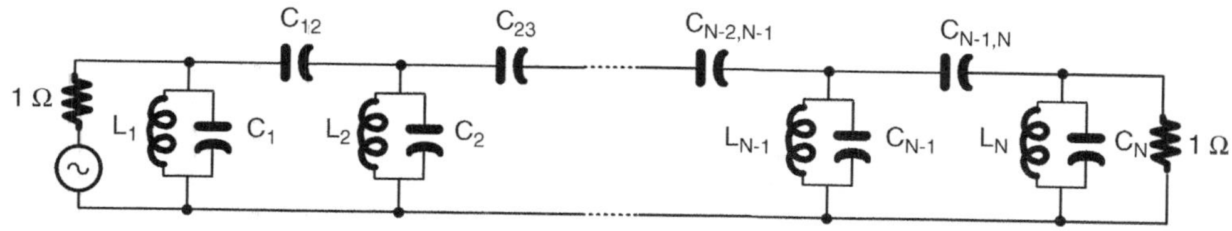

Figure 4.33 Capacitively-coupled band-pass filter.

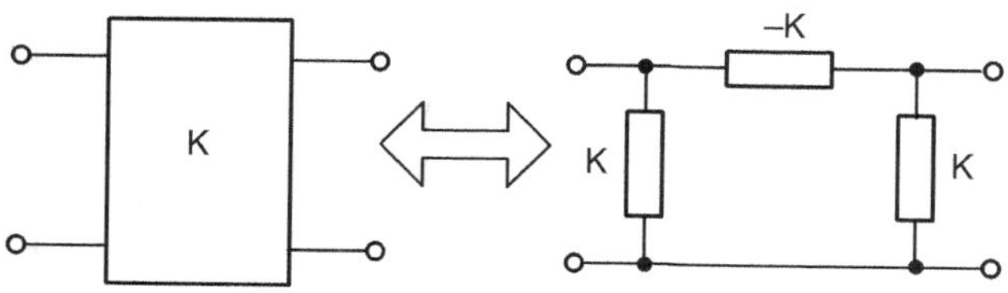

Figure 4.34 Realisation. A K-inverter with a π network.

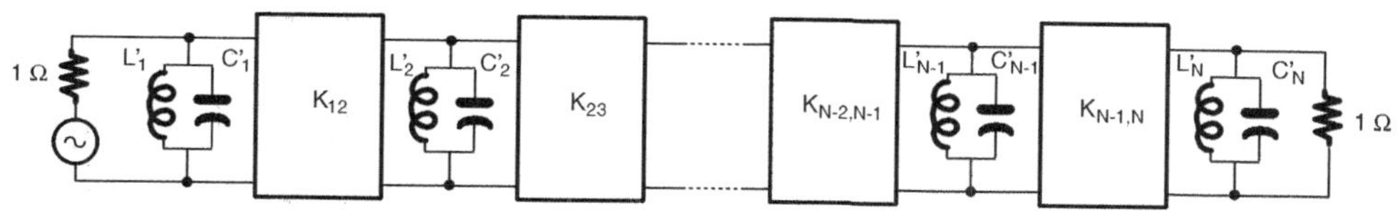

Figure 4.35 K-inverter-coupled band-pass filter.

The filter can be directly taken from the inverter-coupled topology shown in Figure 4.35, with shunt resonators and impedance inverters. The characteristic impedance of the impedance inverter is denoted by K_{ij}. The topology is developed from the impedance-inverter coupled low-pass prototype. The parameters of the inverters are kept unchanged while the band-pass transformation rule is applied to the capacitors in the low-pass prototype. The capacitively-coupled band-pass filter is obtained by realising the inverters using lumped elements in a π-network. For narrowband approximation, the K-inverter can be realised with capacitors as shown in Figure 4.36, where the capacitor value C is related to K by:

$$C = -\frac{K}{\omega} \tag{4.34}$$

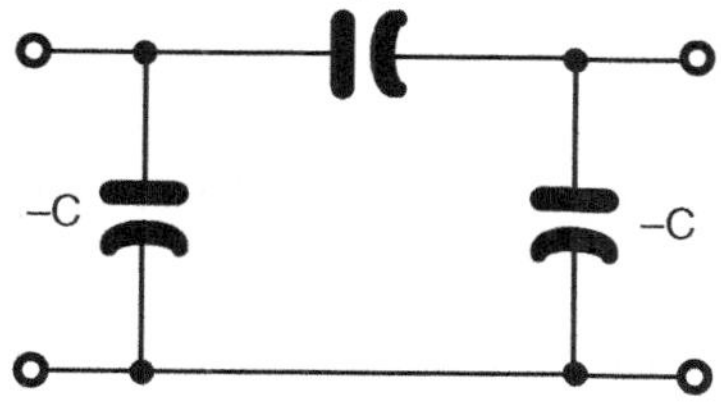

Figure 4.36 Narrow-band realisation approach for a K-inverter.

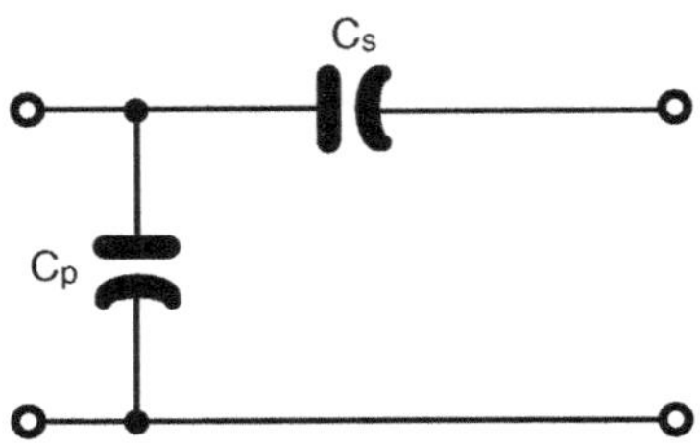

Figure 4.37 Series-shunt capacitive matching network for narrow-band matching.

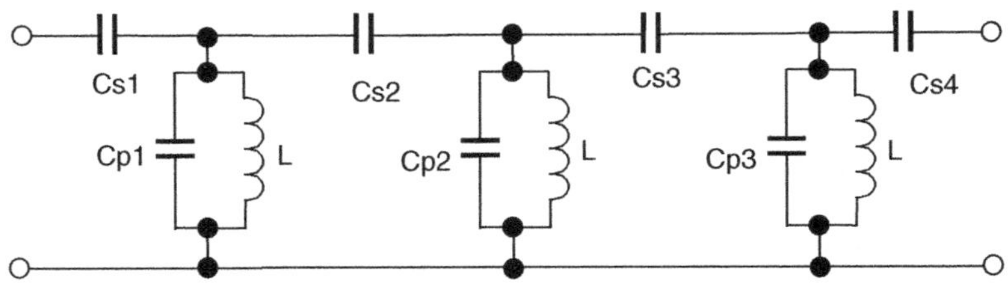

Figure 4.38 Modified capacitively-coupled resonator filter.

Hence, if all the *K*-inverters in the prototype shown in Figure 4.35 are replaced with the capacitive π-network shown in Figure 4.36, the capacitively-coupled band-pass filter shown in Figure 4.33 is obtained. The coupling capacitors directly implement the series capacitors in the *K*-inverters, whilst the negative capacitors are absorbed in the parallel LC resonators. This filter topology generally has quite poor upper stop-band rejection, however, which can be seen intuitively from the fact that there is direct capacitive-coupling across from the input to the output. The impedances of these series capacitors become small at high frequency, so the equivalent circuit of the filter at the upper stop-band is a single resonant parallel LC network, giving only a single transmission zero.

For a narrow-band filter, the capacitively-coupled filter topology shown in Figure 4.33 may not be appropriate due to the need for impractically large inductors. A simple way to solve this problem is to scale the source and load impedance and use a matching network at the input and output. There are several matching techniques (these are discussed in Chapter 9), but the simple series-shunt capacitive network shown in Figure 4.37 is suitable for the narrow-band matching required in a band-pass filter. Applying this matching network to Figure 4.33 at the source and load terminations gives the modified capacitively-coupled band-pass design shown in Figure 4.38.

References

1. Matthaei, G.L., Young L., and Jones, E.M.T. (1980) *Microwave Filters, Impedance-matching Networks and Coupling Structures*, New edition. Artech House Publishers. ISBN-10: 0890060991.
2. Hunter, I.C. (2001) *Theory and Design of Microwave Filters*. The Institution of Engineering and Technology, London, UK (reprinted 2006).
3. Cohn, S.B. (1957) Direct-coupled-resonator filter. *Proceedings IRE*, **45** (2), 187–196.

5

Transmission Line Theory

5.1 Introduction

This chapter describes three key approaches that are used for the analysis of transmission line problems – the Telegrapher's equation, the Smith chart, and flowgraph analysis using Mason's rule [1–4]. The purpose of a transmission line is to transfer a signal from a source over some distance to a remote load. In the case of electricity generation and transmission, it is just a continuous sine wave, and the aim is to transfer a great deal of power from the power station to the customers as efficiently as possible. High voltages are used in order to minimise the loss associated with the resistance of the conductors. In computer networking, the signal is a sequence of data pulses and the cable must be shielded from interference and terminated with an appropriate resistance in order to prevent reflections from corrupting the data. In radio systems, the signal is one or more modulated high-frequency carriers containing information. In the generic diagram of Figure 5.1, the source could be a cellular mobile transmitter in a cabinet on the ground, for example, connected by a coaxial cable to the load, which is an antenna mounted at the top of a tower. The same principles apply to all these examples and the most significant features of transmission lines are:

1. The transmission line must be properly designed to have low loss and be shielded in some way, to prevent the signal leaking away and to reject electrical interference from outside.
2. The line has electric and magnetic fields distributed along its length, representing local energy storage as the signal travels down the line.
3. There is a significant time delay from the signal leaving the source to it reaching the load.

In order to understand the basic theory of transmission lines it is helpful to focus attention initially on a coaxial cable (as shown in Figure 5.2), since these are well behaved and very familiar. Figure 5.2 shows the structure of a coaxial cable and a cross-section illustrating the electric and magnetic field patterns. The electric field, resulting from the difference in potential between the inner and outer conductors, is in the form of radial straight lines. The magnetic

Microwave and Millimetre-Wave Design for Wireless Communications, First Edition.
Ian Robertson, Nutapong Somjit and Mitchai Chongcheawchamnan.
© 2016 John Wiley & Sons, Ltd. Published 2016 by John Wiley & Sons, Ltd.

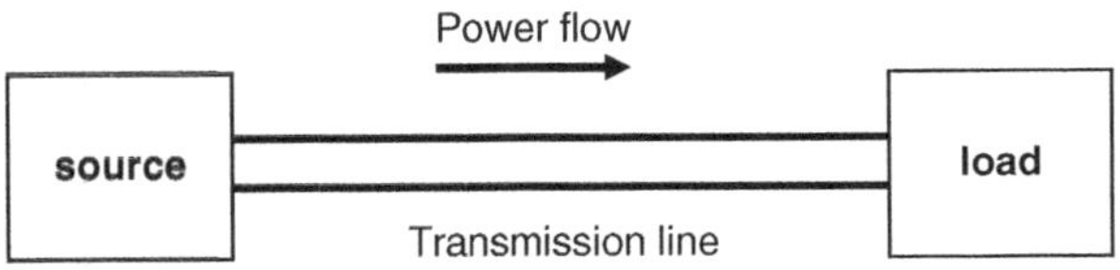

Figure 5.1 Generic transmission line situation.

field, resulting from the current in the conductors, has circular field lines. In circuit terms, the properties of the transmission line are represented by two main parameters:

Characteristic Impedance, Z_0: This is a resistive impedance for the lossless case, but there is no power dissipation associated with it. The inductance and capacitance of the line work in harmony so that, as the signal flows down the line, the energy is temporarily stored in the magnetic field of the line and the electric field of the line.

Propagation Constant, γ: This has a real part α, representing loss, and an imaginary part β, representing the time delay, $\gamma = \alpha + j\beta$. For sinusoidal signals the time delay is represented in terms of signal phase and β has units of radians per metre.

5.2 Reflections on Transmission Lines

Transmission lines obey Kirchhoff's laws and can be analysed with them provided that the time delay is taken into account. As a result of the time taken for the signal to travel down the transmission line, the transient behaviour of a transmission line system is very important and educational. If the voltage and current travelling down the line do not match the load impedance, a reflection occurs at the load end. A reflected signal is created, being essential in order that Ohm's law is obeyed at the load end. This reflected signal then travels backwards down the line towards the source. For simplicity, in this section it is assumed that there is no loss in the transmission lines.

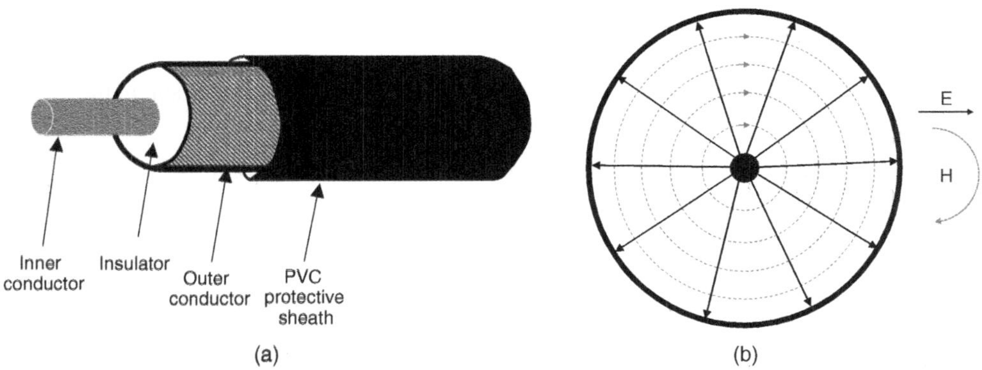

Figure 5.2 Coaxial cable. (a) Structure. (b) Cross-section illustrating the electric and magnetic field pattern.

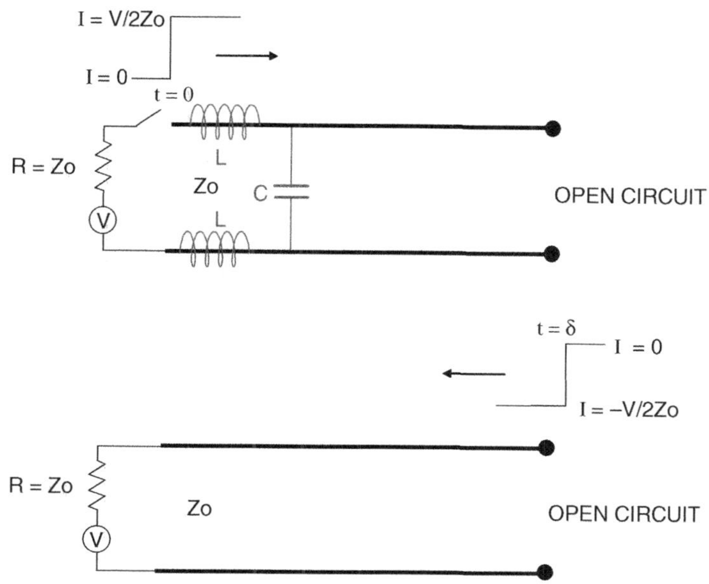

Figure 5.3 Reflection at the end of an open-circuited transmission line.

5.2.1 Open-Circuited Line

In Figure 5.3, a voltage source V with source resistance R is connected by a switch to the transmission line of characteristic impedance Z_0 and the switch is closed at time $t = 0$. To get maximum power from the source into the line, R is made equal to Z_0. The far end of the line, however, is open-circuited. Looking at the diagram, it should be obvious that the correct solution is that the current should be zero, although we only know that by cheating and looking immediately at the load end. The source cannot do that, so initially a current step is created at $t = 0$, with value $V/(2Z_0)$ (there is a potential divider effect between the source resistance and the Z_0 of the line, giving $\frac{1}{2}$V when $R = Z_0$). The step in current propagates along the transmission line, but when this current step arrives at the load the current has nowhere to go, so it is reflected and a reverse step is created at time $t = \delta$, where δ is the time taken to travel down the line. The value of the reverse step is $-V/(2Z_0)$ so that the two currents cancel out completely at the open circuit, satisfying Kirchhoff's and Ohm's laws. So, there is some interesting *transient* behaviour, but when everything settles down – in what is known as the 'steady state' – we get the expected result of no current flowing.

5.2.2 Short-Circuited Line

Now, consider what happens when the far end of the line is short-circuited, as shown in Figure 5.4. It should be obvious that the *voltage* at the far end will be zero in this case, but again we only know that by cheating and looking immediately at the load end. The source does not know what is connected at the end, so initially the voltage step starts to travel down the line, with value $V/2$. When this voltage step arrives at the load, Ohm's law is violated, so

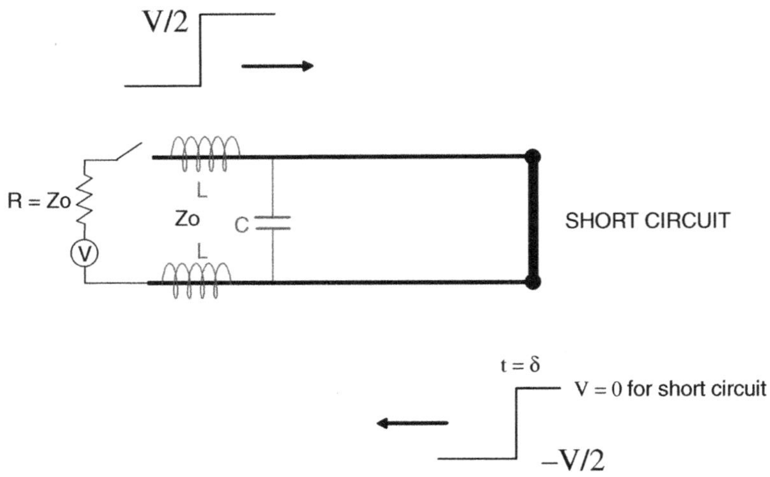

Figure 5.4 Reflection at the end of a short-circuited transmission line.

the step is reflected and a backwards-travelling step is created at time $t = \delta$, where δ is the time taken to travel down the line. The value of the reverse step is $-V/2$, and the two voltages cancel out at the short-circuit end, satisfying Ohm's law. The reflection coefficient is the ratio of the reflected and incident voltage waves. For the short-circuit load, its value is –1, or magnitude 1, phase 180 degrees.

The time-domain behaviour of the current in this transmission line with a short-circuit load is fascinating and demonstrates nicely why transmission line circuits need to be analysed rigorously with appropriate mathematical tools. Figure 5.5 shows the simulated current using QUCS (Quite Universal Circuit Simulator) for a $Z_0 = 50\ \Omega$ transmission line that is 20 km long and short-circuited at the end, with a 1 V step input, for three values of source impedance; $Z_s = 50$, 20 and 1 Ω. In the $Z_s = 50\ \Omega$ case, as described earlier, the voltage step initially travels along the line and is reflected back once it reaches the short-circuit load. When it has

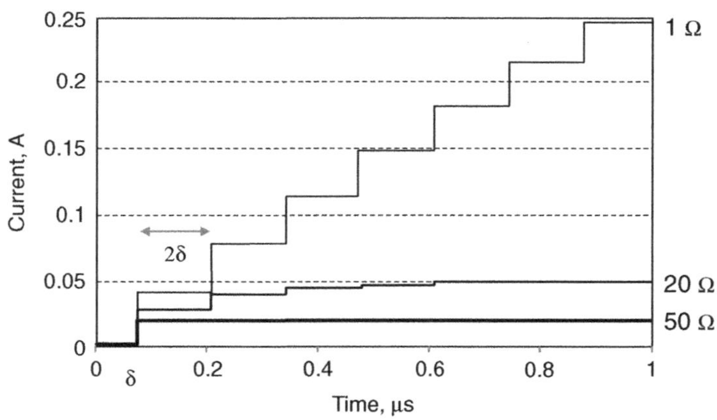

Figure 5.5 Simulated current at the short-circuit end of the line for $Z_s = 1$, 20 and 50 Ω.

made the complete round trip back to the source, the reflected step is then fully absorbed in the 50 Ω source output resistance, since it is matched to the Z_0 of the line. So, in this case there is a single reflection event and the current quickly reaches the expected steady state value of 0.02 A (1 V/50 Ω: the 1 V is dropped across Z_s).

If Z_s is set instead to 20 Ω, when the step that is reflected back from the short-circuit load returns to the source, it is again reflected, since the source output resistance is not matched to Z_0. Hence, there are multiple reflection events and it takes several round trips before the current at the short-circuit end reaches the expected steady state value of 0.05 A (1 V/20 Ω). In the $Z_s = 1$ Ω case, this transient behaviour is very easy to see, and the current at the short-circuit end goes up in steps as the reflections travel back-and-forth, until the steady state value of 1 A is eventually reached (it is not shown on the graph as it takes ≈25 μs). In all three cases, the first step in current occurs at $t = \delta$, which is the delay down the line (20×10^3 m ÷ 3×10^8 m s^{-1}) whilst the subsequent steps in Figure 5.5 are 133.33 μs in duration, which is the round trip delay of the line, 2δ.

Finally, consider the purely theoretical case of an ideal voltage source ($Z_s = 0$) feeding the short-circuited line. By inspection, we know that the current should be infinite (1 V across 0 Ω), but because of the transmission line behaviour the current actually increases step-by-step like the 1 Ω case, and the reflection bounces back-and-forth for ever, with the current not finally reaching infinity until $t = \infty$. It is not unusual for transmission line circuits to behave in unexpected ways and it is therefore important to rely on proper analysis tools and not just intuition, since concepts from regular lumped circuits often fail to work.

5.3 Transmission Line Theory

The previous section has introduced transmission line behaviour in terms of delay time and pulses, and the effects are important when the propagation delay of the line is comparable to the duration of the pulse. In most RF applications, the signal is sinusoidal and the issue is whether the period of the signal is comparable to the delay time. This is most easily expressed in terms of comparing the physical dimensions to the signal wavelength. A common rule of thumb is that transmission line effects are important for any component that has physical dimensions larger than around one twentieth of a wavelength. In such a case, the amplitude of the signal varies along the propagation direction. A circuit designer has to take these distributed effects into account and apply transmission line principles, either mathematically with the Smith chart, or with a computer-based tool. Engineers and researchers who are designing wireless communication systems operating at microwave or millimetre-wave frequencies must have a good background knowledge of transmission line theory. This section provides a brief introduction to transmission line theory and some related concepts. Transmission line theory can be described using either electromagnetic theory or a distributed circuit approach. Since the latter approach is more familiar to electronic and electrical engineers, a circuit approach describing the transmission line behaviour is adopted here.

5.3.1 Telegrapher's Equation

A transmission line is a set of conductors in a nonconducting space such that the conductors form a means of conveying the electrical signal with minimal loss and interference. The

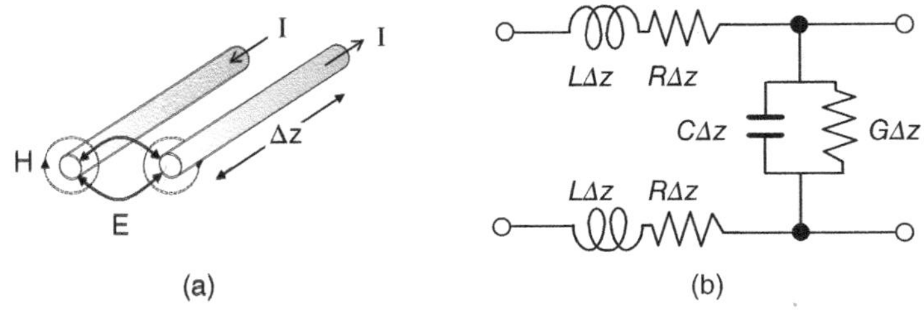

Figure 5.6 (a) Twin lead transmission line. (b) Lumped element equivalent circuit of a short section of line.

materials forming the conductor and insulating parts are assumed to be homogeneous and uniform in shape along the length. Therefore, when the electrical signal travels along the transmission line, the electrical parameters in the conductor and insulating materials are uniform. A transmission line can then be represented with a simple electrical circuit model. To show this, consider the 'twin lead' form of transmission line which consists of a pair of parallel wires, as illustrated in Figure 5.6(a). The transmission line is symmetrical and consists of two conductors. The signal travels down the line with voltage on the conductors in anti-phase, giving a virtual ground along the centre, and with the current in each conductor having the same magnitude but opposite direction. This leads to the magnetic and electric field patterns as shown, with a magnetic field being created due to the current in each conductor and an electric field due to the potential difference between them. For a small section of the line, of length Δz, the magnetic field can be represented in a circuit model by an inductance, and the electric field can be represented by a capacitance. The inductance and capacitance are spread uniformly all along the line. The conductors have some series resistance, and the dielectric medium between them will in general have some loss, represented by a shunt conductance between the conductors. Again, these are uniformly distributed all along the line. The parameters L, C, R and G are defined as follows:

- L represents the magnetic field around the transmission line as an inductance distributed all along its length (in units of inductance-per-unit-length, H/m),
- C represents electric field in the dielectric material (in units of capacitance-per-unit-length, F/m),
- R represents resistive loss in the conductor (in units of resistance-per-unit-length, Ω/m)
- G represents dielectric loss in the insulating material (in units of conductance-per-unit-length: $\Omega^{-1}/m \; or \; S/m$).

Hence, the resistance, inductance, conductance and capacitance for a short transmission line section of length Δz are $R\Delta z$, $L\Delta z$, $G\Delta z$, and $C\Delta z$; that is, the 'per unit length' value times the length Δz. Using these parameters, the short section with length Δz can be modelled with the electrical circuit shown in Figure 5.6(b). Using this model for a section of line is valid provided that the phase of the currents and voltages do not vary significantly along its length, which again relies on the rule of thumb that the length, Δz, should be less than one-twentieth of a wavelength. In order to model the whole transmission line of length, L, the line must be subdivided into many of these very short subsections sections.

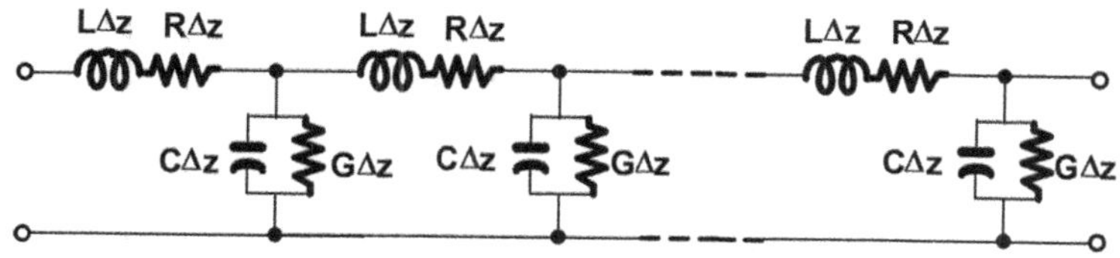

Figure 5.7 The equivalent circuit of a transmission line using a chain of small sections, each of length Δz.

Before attempting to use this equivalent circuit to develop an analysis of transmission line behaviour, note that the twin lead transmission line is referred to as a balanced line, since it consists of a pair of conductors operating in anti-phase. However, the most common transmission lines, such as coaxial cable and microstrip, are single-ended, which means that they consist of a single signal conductor and a ground. To aid the analysis then, we can represent the length of transmission line with a long chain of sections, each of length Δz, as illustrated in Figure 5.7, with a common ground. Indeed, noting that the twin lead line has a virtual ground along its centre line, this model can equally be applied.

Here, the transmission line is subdivided into N equal-length sections. Each section length is Δz, where Δz is infinitesimally small. Each section has the same electrical characteristics so a one-dimensional electrical model for the transmission line is formed by cascading N identical subnetworks. Each subnetwork is composed of a series RL and a shunt GC branch. It is noted that the electrical model of the subnetwork shown in Figure 5.7 is different to that of the model presented in Figure 5.6. There is no RL circuit for modelling the bottom conductor as the bottom conductor is now ground. This simplification greatly eases the analysis of the transmission line. However, it is important to remember the more general form of Figure 5.6 as this will often serve as an important reminder in circuit design that ground planes have current flowing and inductance.

To analyse the transmission line behaviour, the ith transmission line section, starting at distance z and ending at $z + \Delta z$, is considered, as illustrated in Figure 5.8. Let the node voltages at z and $z + \Delta z$ be $v(z, t)$ and $v(z + \Delta z, t)$, respectively, while the input and output currents of the transmission line section are denoted by $i(z, t)$ and $i(z + \Delta z, t)$. From Figure 5.8 and using Kirchhoff's current law, the input and output currents can be related as follows:

$$i(z,t) = i(z + \Delta z, t) + \left(C\frac{\partial}{\partial t}v(z + \Delta z, t) + G \cdot v(z + \Delta z, t) \right) \Delta z \tag{5.1}$$

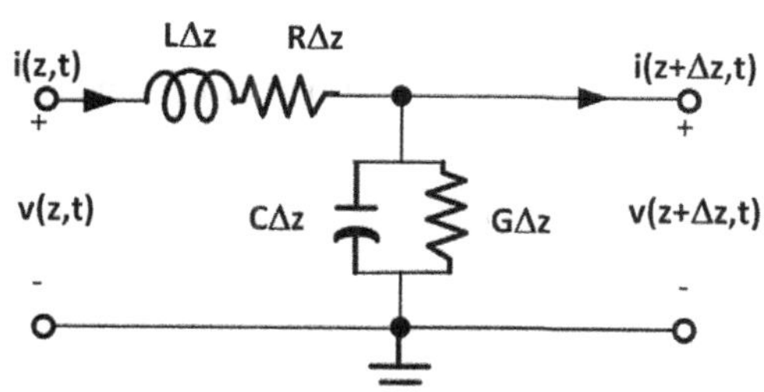

Figure 5.8 An equivalent circuit model for an infinitesimal transmission line section.

This simply states that the current going in is equal to the current going out plus the current down to ground through the capacitance and conductance. Similarly, applying Kirchhoff's voltage law, we obtain:

$$v(z,t) = v(z+\Delta z,t) + \left(L\frac{\partial}{\partial t}i(z,t) + R\cdot i(z)\right)\Delta z \tag{5.2}$$

In other words, the input voltage is equal to the output voltage plus the voltage drop across the series resistance and inductance. Rearranging these two equations, we obtain:

$$\frac{i(z+\Delta z,t) - i(z,t)}{\Delta z} = -\left(C\frac{\partial}{\partial t} + G\right)v(z+\Delta z,t) \tag{5.3}$$

$$\frac{v(z+\Delta z,t) - v(z,t)}{\Delta z} = -\left(L\frac{\partial}{\partial t} + R\right)i(z,t) \tag{5.4}$$

If we now apply the condition that each transmission line section is infinitesimally small, then $\Delta z \to 0$ and so Equations (5.3) and (5.4) are reduced to:

$$\lim_{\Delta z\to 0}\frac{i(z+\Delta z,t) - i(z,t)}{\Delta z} \equiv \frac{\partial i(z,t)}{\partial z} = -\left(C\frac{\partial}{\partial t} + G\right)v(z+\Delta z,t) \tag{5.5}$$

$$\lim_{\Delta z\to 0}\frac{v(z+\Delta z,t) - v(z,t)}{\Delta z} \equiv \frac{\partial v(z,t)}{\partial z} = -\left(L\frac{\partial}{\partial t} + R\right)i(z,t) \tag{5.6}$$

If Δz is sufficiently small, $v(z+\Delta z)$ is approximately equal to $v(z)$. Therefore the current Equation (5.5) can be written as:

$$\frac{\partial i(z,t)}{\partial z} = -\left(C\frac{\partial}{\partial t} + G\right)v(z,t) \tag{5.7}$$

Applying the Laplace transform and the sinusoidal steady-state condition, $s = j\omega$ to Equations (5.7) and (5.6), the partial differential Equations (5.6) and (5.7) are changed to first-order differential equations as follows:

$$\frac{dV(z)}{dz} = -(R + j\omega L)I(z) \tag{5.8}$$

$$\frac{dI(z)}{dz} = -(G + j\omega C)V(z) \tag{5.9}$$

where $V(z)$ and $I(z)$ are the phasor voltage and current of $v(t)$ and $i(t)$.

Equation (5.8) shows that the voltage is reduced with the rate controlled by the series impedance of the circuit model, $Z = R + j\omega L$. This is also similar to the current which depends on the the shunt conductance, $Y = G + j\omega C$. The propagation constant is defined by:

$$\gamma = \sqrt{Z\cdot Y} = \sqrt{(R + j\omega L)(G + j\omega C)} \tag{5.10}$$

where γ is the complex propagation constant referred to earlier.

This propagation constant can generally be written in a rectangular format as:

$$\gamma = \alpha + j\beta \tag{5.11}$$

where α is the attenuation constant of the transmission line in neper/m units and β is the phase constant of the transmission line (in radians m^{-1}).

The attenuation constant is the property of a transmission line that determines how the amplitude of the wave reduces as it travels along the line. The phase constant determines the phase delay of wave propagation. The effect of these constants on the waveform will be apparent in the following derivation of voltage and current wave equations. The second-order differential equation for the voltage and current obtained from Equations (5.8) and (5.9) are:

$$\frac{d^2V(z)}{dz^2} - \gamma^2 V(z) = 0 \tag{5.12}$$

$$\frac{d^2I(z)}{dz^2} - \gamma^2 I(z) = 0 \tag{5.13}$$

The solutions of the second-order differential equations for the voltage and current are given by:

$$V(z) = V^+e^{-\gamma z} + V^-e^{+\gamma z} \tag{5.14}$$

$$I(z) = I^+e^{-\gamma z} + I^-e^{+\gamma z} \tag{5.15}$$

These solutions for voltage and current show that there are forward and backward waves propagating along the transmission line. The forward wave, often referred as the incident wave, is the voltage (V^+) or current (I^+) wave travelling in the $+z$ direction. The backward wave, or reflected wave, is the voltage (V^-) or current (I^-) wave in the $-z$ direction. There are two propagating waves, the incident and reflected waves, as shown in Figure 5.9. The voltage and current Equations [(5.14) and (5.15)] are generalised equations for transmission lines modelled in Figure 5.7. If we substitute the voltage wave from Equation (5.14) into

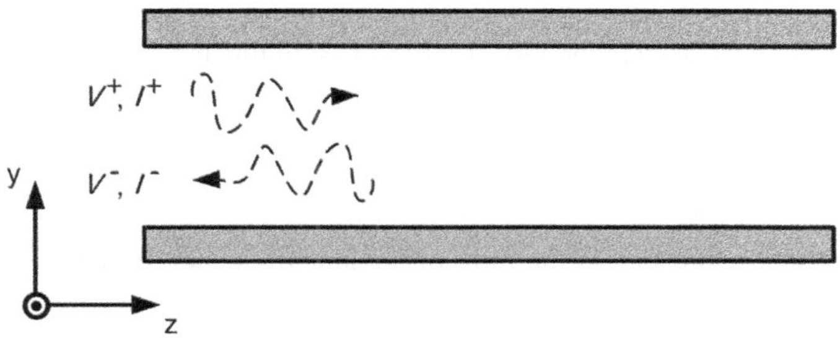

Figure 5.9 Forward (incident) and backward (reflected) waves in a transmission line.

Equation (5.8), we obtain the current equation in terms of incident and reflected voltage waves as follows:

$$I(z) = \frac{\gamma}{R + j\omega L}\left[V^{+}e^{-\gamma z} - V^{-}e^{+\gamma z}\right] \tag{5.16}$$

Hence, by comparing with the general solution of current in Equation (5.15), we find that the ratio of voltage and current waves in either direction is constant:

$$\frac{V^{+}}{I^{+}} = \frac{-V^{-}}{I^{-}} = Z_0 \tag{5.17}$$

where Z_0 is the characteristic impedance.

This parameter is the ratio of incident voltage and incident current or the ratio of the negative reverse voltage and reverse current. This means that in either direction of propagation, the ratio of voltage and current waves is always Z_0. This characteristic impedance, Z_0, is one of the important electrical parameters of a transmission line and is given by:

$$Z_0 = \sqrt{\frac{R + j\omega L}{G + j\omega C}} \tag{5.18}$$

The reflection coefficient at distance z, denoted by $\Gamma(z)$, is defined as the ratio of reflected voltage (current) wave and incident voltage (current) wave; that is

$$\Gamma(z) = \frac{V^{-}e^{+\gamma z}}{V^{+}e^{-\gamma z}} = \frac{V^{-}}{V^{+}}e^{+2\gamma z} \tag{5.19}$$

5.3.2 Lossless Transmission Line

A transmission line medium with very low loss is clearly desirable and this requires the conductor and dielectric losses (R and G) to be very small (i.e., R and $G \approx 0$). For this lossless transmission line, the propagation constant and characteristic impedance are given by:

$$\gamma = j\omega\sqrt{LC} \tag{5.20}$$

$$Z_0 = \sqrt{\frac{L}{C}} \tag{5.21}$$

It is shown in Equation (5.20) that the lossless transmission line has zero attenuation since the real part α is zero. Meanwhile, the phase constant is:

$$\beta = \omega\sqrt{LC} \tag{5.22}$$

Theoretically, the phase constant is the ratio of angular frequency in radians and the phase velocity, that is:

$$\beta = \frac{\omega}{v_p} \tag{5.23}$$

Therefore, the wave travels along a lossless transmission with a phase velocity given by:

$$v_p = \frac{1}{\sqrt{LC}} \tag{5.24}$$

Substituting the propagation constant for the lossless line case to the voltage and current Equations (5.14) and (5.15), gives:

$$V(z) = [V^+ + V^-]\cos\left(\frac{\omega}{v_p}z\right) - j[V^+ - V^-]\sin\left(\frac{\omega}{v_p}z\right) \tag{5.25}$$

$$I(z) = \left[\frac{V^+ - V^-}{Z_0}\right]\cos\left(\frac{\omega}{v_p}z\right) - j\left[\frac{V^+ + V^-}{Z_0}\right]\sin\left(\frac{\omega}{v_p}z\right) \tag{5.26}$$

Rewriting these equations in terms of a sinusoidal waveform yields:

$$V(z) = \sqrt{(V^+)^2 + (V^-)^2}\sin\left(\frac{\omega}{v_p}z + \phi\right) \tag{5.27}$$

$$I(z) = \frac{\sqrt{(V^+)^2 + (V^-)^2}}{Z_0}\sin\left(\frac{\omega}{v_p}z - \phi\right) \tag{5.28}$$

where

$$\phi = -\tan^{-1}\left(\frac{V^+ - V^-}{V^+ + V^-}\right) \tag{5.29}$$

5.4 Standing Waves on a Lossless Transmission Line with Mismatched Load

Consider a lossless transmission line terminated in a load that is not equal to the characteristic impedance, which is a situation often encountered in microwave and millimetre-wave wireless design. For example, a transmission line may be used as an impedance transformer to match two impedances, or used as a resonator, and so forth. Figure 5.10 shows the general case of a lossless transmission line length L terminated with a load impedance, Z_L.

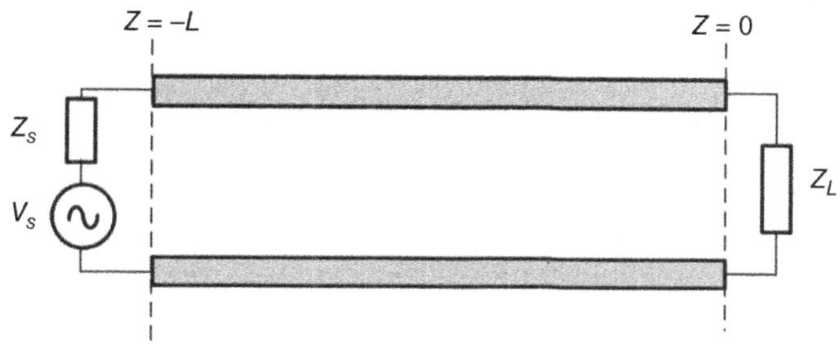

Figure 5.10 Terminated transmission line.

From the general voltage and current Equations (5.14) and (5.15), we apply the lossless condition, giving:

$$V(z) = V^+e^{-j\beta z} + V^-e^{+j\beta z} \tag{5.30}$$

$$I(z) = \frac{V^+}{Z_0}e^{-j\beta z} - \frac{V^-}{Z_0}e^{+j\beta z} \tag{5.31}$$

The impedance at $z = 0$ is the load impedance Z_L. This is equal to the ratio between voltage and current in Equations (5.30) and (5.31) at z = 0:

$$\frac{V(0)}{I(0)} = Z_L = \frac{V^+ + V^-}{\frac{V^+}{Z_0} - \frac{V^-}{Z_0}} \tag{5.32}$$

Applying the definition of reflection coefficient to Figure 5.10, the load reflection coefficient (Γ_L) is the reflection coefficient at $z = 0$. This is given by:

$$\Gamma_L \equiv \Gamma(0) = \frac{V^-}{V^+} \tag{5.33}$$

Substituting Equation (5.33) into (5.32) and simplifying, we obtain:

$$Z_L = Z_0\frac{1+\Gamma_L}{1-\Gamma_L} \tag{5.34}$$

Therefore, the reflection at the load is related to the load impedance and the characteristic impedance of the transmission line by:

$$\Gamma_L = \frac{Z_L - Z_0}{Z_L + Z_0} \tag{5.35}$$

With a similar approach, we can find the reflection coefficient at the source (Γ_S) to be:

$$\Gamma_S = \frac{Z_S - Z_0}{Z_S + Z_0} \tag{5.36}$$

From phasor theory, the power delivered to the load Z_L, denoted as P_L, is calculated by,

$$P_L = \frac{1}{2}\mathrm{Re}\{V(0)I(0)^*\} = \frac{1}{2}(V^+ + V^-)\left(\frac{V^+}{Z_0} - \frac{V^-}{Z_0}\right)^* \tag{5.37}$$

where $(\cdot)^*$ is the complex conjugate operator.

Simplifying Equation (5.37) further, we obtain:

$$P_L = \frac{|V^+|^2}{2Z_0}\left(1 - |\Gamma_L|^2\right) \tag{5.38}$$

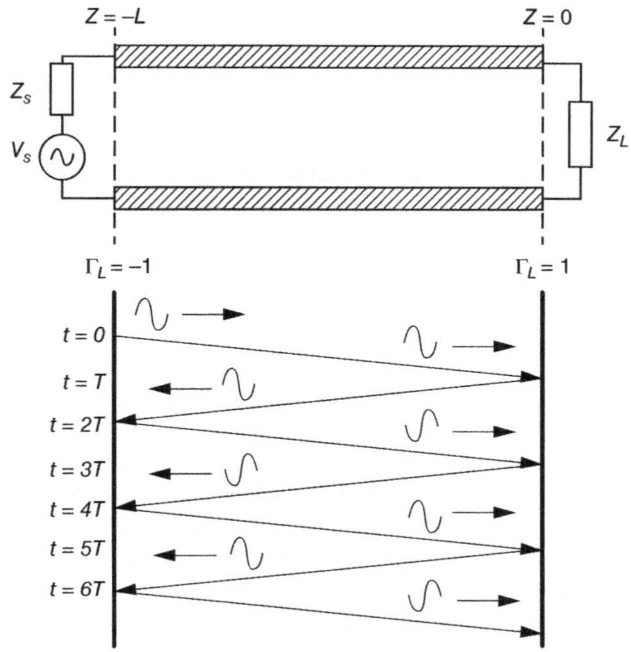

Figure 5.11 Reflections resulting from mismatching at source and load.

The reflected waves will always occur if there is a mismatched termination at the source, load or both. This is shown in Figure 5.11, which also shows graphically the multiple reflection effect that was discussed earlier for a pulse on a line. If the mismatch occurs at the load, the power delivered to the load is related to the reflection coefficient at the load. The maximum power at the load is obtained when $\Gamma_L = 0$. From the reflection coefficient definition, this means that there is no reflected voltage and current. A non-zero reflection coefficient means that there is some amount of backwards flowing power. This backwards power leads to the definition of a return loss (RL), which can be calculated in dB units from:

$$\mathrm{RL}\,(\mathrm{dB}) = 20\log_{10}|\Gamma_L| \tag{5.39}$$

The case of mismatch at the source can be described in a similar way. It can be concluded that the reflected wave is a result of unequal impedances in a transmission line. When impedance matching is achieved, there will be no reflected wave. This means that only the incident wave travels along the transmission line – from the source to the load, where it is dissipated. Therefore, for a matched transmission line, the travelling voltage and current in the transmission line consists of an incident wave only. This can be written as:

$$V(z) = V^{+}e^{-\gamma z} \tag{5.40}$$

$$I(z) = \frac{V^{+}}{Z_0}e^{-\gamma z} \tag{5.41}$$

From these two equations, it is obvious that the ratio between voltage and current at any point along the matched, lossless, transmission line is equal to Z_0. This is only the case for the matched situation, when the wave energy flows from the source along the transmission line to the matched load, propagating only in the forward direction and then being absorbed by the load.

Things are totally different when the transmission line is not matched. The reflection coefficient is non-zero and reflected waves do propagate along the transmission line. If the source and load are mismatched, then reflected waves travel in both forward and backward directions along the transmission line – bouncing back and forth forever in the case of a lossless line. To see this more clearly, consider the wave travelling along the lossless transmission line terminated with load and source impedances at $z = -L$ and $z = 0$, respectively as shown in Figure 5.11. At time $t = 0$, the wave starts to propagate from the voltage source at $z = -L$ to the load located at $z = 0$. The wave takes time T to travel along the line. Hence, at $t = T$, the incident wave arrives at the load where the reflection coefficient is +1. From the reflection coefficient definition, the reflected wave is equal in magnitude and in-phase with the incident wave. This reflected wave travels backwards down the line and arrives at the source at $t = 2T$. The reflection coefficient at the source interface is –1, so the wave is then reflected from the source with equal magnitude but opposite phase to the incident wave. This process is repeated and the wave travels backwards and forwards forever. The superposition of these forward and backward waves is constructive and destructive at different points along the transmission line, so the resulting wave varies in amplitude along the line. Furthermore, in the steady state there is no power flow along the line and what results is called a 'standing wave'. This standing wave concept can be defined in terms of either voltage or current standing waves. To understand the standing wave concept, we rewrite the voltage equation of a lossless transmission line from Equation (5.30) in terms of Γ_L:

$$V(z) = V^+ e^{-j\beta z}(1 + \Gamma_L e^{+j2\beta z}) \tag{5.42}$$

Since Γ_L is a complex number which can be written in polar form as:

$$\Gamma_L = |\Gamma_L| e^{j\emptyset_L} \tag{5.43}$$

So, the standing wave which is the magnitude of $V(z)$ can be written as:

$$|V(z)| = |V^+|\sqrt{1 + 2|\Gamma_L|\cos(\beta z + \emptyset_L) + |\Gamma_L|^2} \tag{5.44}$$

It is shown in Equation (5.44) that the standing wave is periodic. The standing wave has maximum and minimum voltages when $\cos(\beta z + \emptyset_L) = 1$ and $\cos(\beta z + \emptyset_L) = -1$ with the following magnitudes:

$$|V(z)|_{max} = |V^+|(1 + |\Gamma_L|) \tag{5.45}$$

$$|V(z)|_{min} = |V^+|(1 - |\Gamma_L|) \tag{5.46}$$

$|V(z)|_{max}$ and $|V(z)|_{min}$ (the maxima and minima of the standing wave) occur at half-wavelength intervals (π radians). The voltage standing wave patterns for $|\Gamma_L| = 0$, 1and less

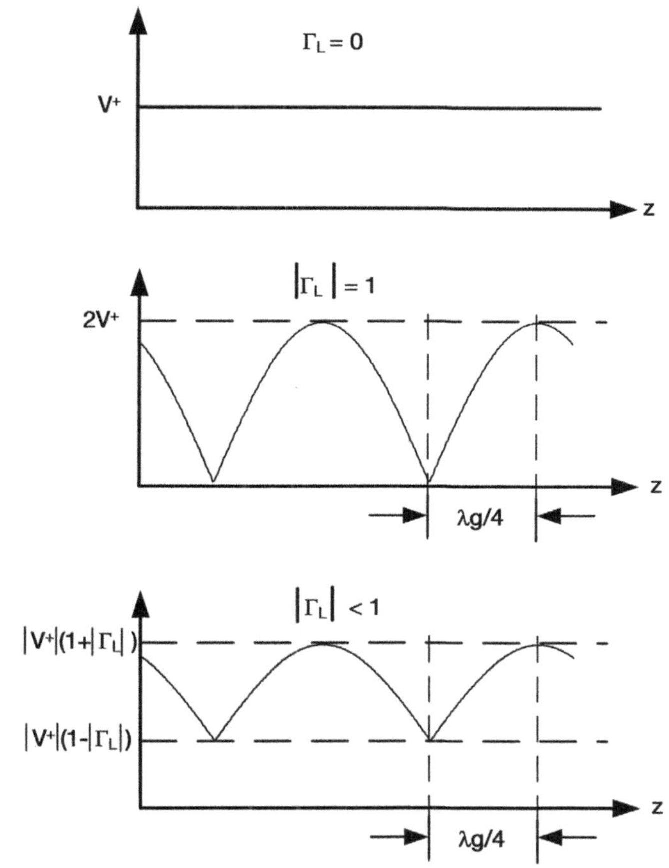

Figure 5.12 Standing wave patterns for three different reflection coefficients.

than 1 are shown in Figure 5.12. For a matched load ($\Gamma_L = 0$), the wave amplitude is independent of distance, z, because there is no standing wave as the wave travels in a single direction only. On the other hand, there is a standing wave for an unmatched load ($\Gamma_L \neq 0$). The largest swing occurs when $|\Gamma_L| = 1$. The voltage amplitude swings from $2V^+$ to 0 volts. The amplitude swing reduces with a lower $|\Gamma_L|$. From these three cases, we can conclude that the degree of amplitude variation depends strongly on the magnitude of Γ_L.

The defining parameter for a standing wave is the relative magnitude of the maxima and minima, which is called the voltage standing wave ratio (VSWR) or current standing wave ratio (CSWR), depending on which parameter is of interest. The standing wave ratio indicates the amplitude variation of the standing wave along the transmission line. The VSWR is defined by the ratio of maximum and minimum voltage:

$$VSWR = \frac{|V_{max}|}{|V_{min}|} \tag{5.47}$$

From Equations (5.45) and (5.46), the VSWR becomes:

$$VSWR = \frac{1 + |\Gamma_L|}{1 - |\Gamma_L|} \tag{5.48}$$

The value of VSWR depends on the magnitude of reflection coefficient. From Figure 5.12, for the extremely mis-matched case, that is $|\Gamma_L| = 1$, the VSWR in infinity. For the perfectly matched case, the VSWR is 1. From the standing wave pattern, $|\Gamma_L|$ can be calculated from:

$$|\Gamma_L| = \frac{VSWR - 1}{VSWR + 1} \tag{5.49}$$

Obviously, the derivation of Equation (5.49) is obtained from Equation (5.48). Using these relationships, it is possible to derive the impedance of a load by measuring the standing wave pattern with a slotted line, using an open- or short-circuit load as a reference to determine the phase of Γ_L.

5.4.1 Impedance Variation along a Transmission Line

One of the most common applications of a transmission line is to use one as an impedance transformer to transform a certain load impedance (say a transistor's input impedance) to a desired value (e.g., 50 Ω). Consider the circuit shown in Figure 5.13, where the transmission line of characteristic impedance Z_0 is terminated with a load impedance Z_L and fed with a source impedance Z_s. The input, or driving-point, impedance at the source, Z_{in}, is equal to the ratio of the voltage and current at $z = -L$:

$$Z_{in} = \frac{V(-L)}{I(-L)} \tag{5.50}$$

Substituting the voltage and current Equations (5.14) and (5.15) to (5.50), this gives:

$$Z_{in} = Z_0 \frac{\left(1 + \Gamma_L e^{-2\gamma L}\right)}{\left(1 - \Gamma_L e^{-2\gamma L}\right)} \tag{5.51}$$

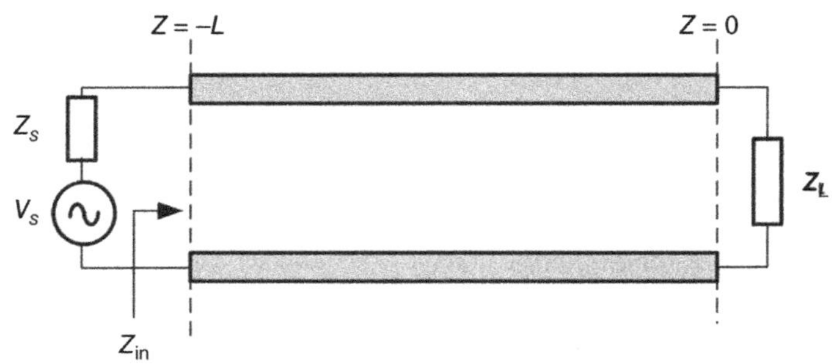

Figure 5.13 Input impedance of a terminated transmission line.

But:

$$\Gamma_L = \frac{Z_L - Z_0}{Z_L + Z_0} \tag{5.52}$$

Hence:

$$Z_{in} = Z_0 \frac{Z_L + Z_0 \frac{1-e^{-2\gamma L}}{1+e^{-2\gamma L}}}{Z_0 + Z_L \frac{1-e^{-2\gamma L}}{1+e^{-2\gamma L}}} \tag{5.53}$$

There is a mathematical identity that says:

$$\tanh \gamma L \equiv \frac{1 - e^{-2\gamma L}}{1 + e^{-2\gamma L}} \tag{5.54}$$

Therefore Equation (5.53) can be written as:

$$Z_{in} = Z_0 \frac{Z_L + Z_0 \tanh \gamma L}{Z_0 + Z_L \tanh \gamma L} \tag{5.55}$$

Assuming the transmission line is lossless, $\gamma = +j\beta$, giving:

$$Z_{in} = Z_0 \frac{Z_L + jZ_0 \tan \beta L}{Z_0 + jZ_L \tan \beta L} \tag{5.56}$$

This is a very important result and shows that the input, or driving-point, impedance actually depends more strongly on L than it does on the load impedance, because of the tangent function. There are three special cases that illustrate how important, and useful, the standing wave effect can be for a terminated lossless transmission line.

5.4.1.1 Short-Circuited Line

Substituting $Z_L = 0$ in Equation (5.56) and simplifying, the input impedance of a shorted transmission line is:

$$Z_{in,short} = jZ_0 \tan \beta L \tag{5.57}$$

The input impedance, normalised to Z_0, is a tangent function and is periodic, repeating every 2π radians. If L is selected such that $L < \frac{\lambda_g}{4}$, then $Z_{in,short}$ will be positive and purely imaginary. Therefore, this shorted transmission line can be used as an inductive element in a circuit such as a matching network or filter.

5.4.1.2 Open-Circuited Line

In this case, the end of a transmission line is left open circuit. Substituting $Z_L = \infty$ in Equation (5.56) and simplifying, the driving-point impedance of the open-ended transmission line is:

$$Z_{in,open} = -jZ_0 \cot \beta L \tag{5.58}$$

Similarly, $Z_{in,open}$ normalised to Z_0 is a cotangent function and this transmission line can be used as a capacitive element in a circuit if $L < \frac{\lambda_g}{4}$.

5.4.1.3 Matched Termination ($Z_L = Z_0$)

When the transmission line is terminated with a matched load, the driving-point impedance becomes:

$$Z_{in,Z_L=Z_0} = Z_0 \tag{5.59}$$

The driving-point impedance is equal to Z_0 regardless of the length of the line. The wave propagates along the transmission line in a forward direction from source to load only. This is usually the desired situation in a *systems* context, as was discussed earlier for the example of a transmitter connected via a coaxial cable to an antenna. However, in a *circuit* context, it is very convenient to be able to deliberately use the impedance transforming effect of transmission lines, and Equation (5.56) is tremendously important. The Smith chart, described in the next section, in effect implements this equation in graphical form.

5.5 The Smith Chart

In the past, before calculators and personal computers were available, RF engineers had to use a slide rule and mathematical tables for these mathematical formulas used in transmission line problems. During the 1930s, P.H. Smith introduced a chart that allows engineers to avoid this tedious calculation task and solve transmission line problems graphically. Since then, the chart has become an icon of RF engineering and is widely used to solve problems and visualise data. Even though computers are now used, the Smith chart remains in use as a way of understanding transmission line problems and designing circuits. Consequently, the chart is extensively used as a teaching aid in hundreds of courses relating to a transmission lines and RF engineering. Its important role in the academic world it witnessed by the publication of several standard microwave textbooks on its use. In addition, the chart is included in vector network analysers and in commercial design software such as Advanced Design System (ADS™), Microwave Office™, etc.

The Smith chart is a polar diagram on which the reflection coefficient of the load can be conveniently plotted. Since the reflection coefficient has a maximum magnitude of unity, the chart is circular with a radius of one. It can be shown that circles of constant load resistance and reactance can be added to the chart: this makes the chart immediately useful as a calculator for converting from Z (= $R + jX$) to Γ, and back again. To achieve this, the contours are all labelled with normalised values of resistance, r (= R/Z_0), and reactance, x (= X/Z_0). Normalising, which means simply dividing by the system reference impedance Z_0, means the chart has

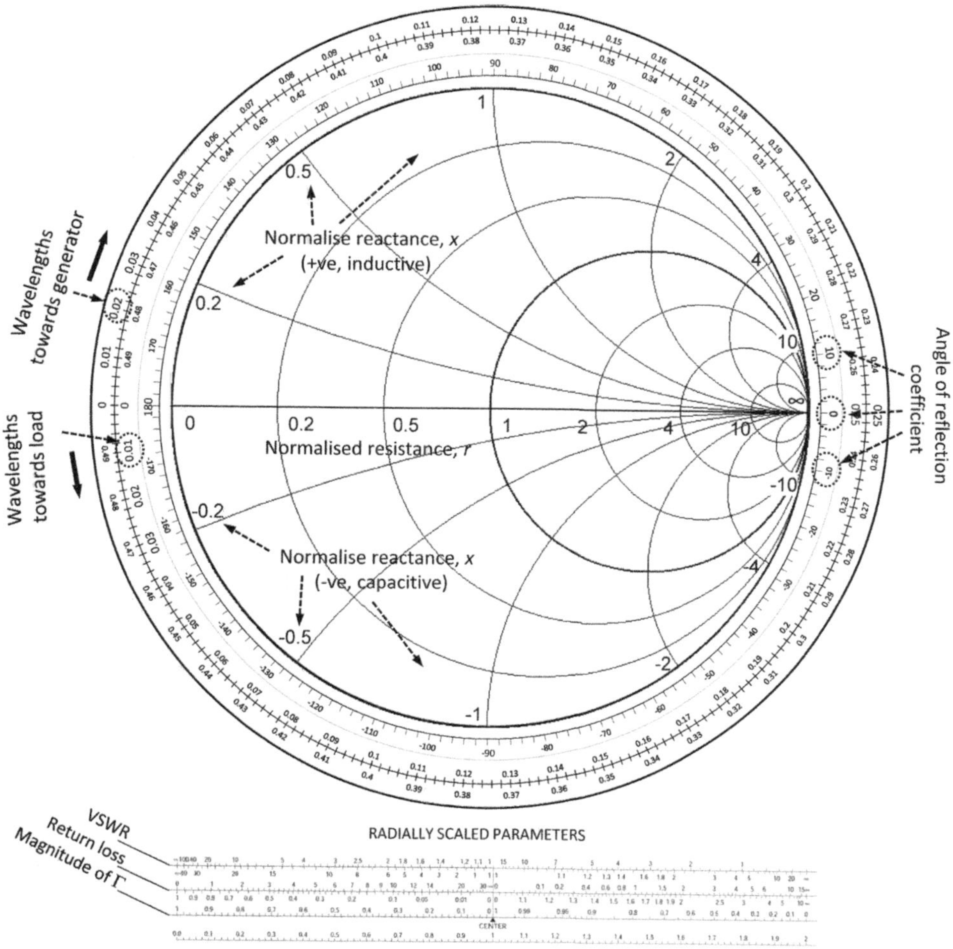

Figure 5.14 The Smith chart.

much wider applicability. The key feature that makes the Smith chart so valuable is that, since it is based around Γ, it is very easy to cope with transmission line problems: as you move along a transmission line, the magnitude of Γ is fixed, so you are actually rotating around the centre of the chart with fixed radius. By convention, if you move away from the load you rotate clockwise.

The phase of the reflection coefficient is given by the angle labelled around the outside edge of Figure 5.14 (from −180° to +180°). The Smith chart also has a wavelength scale of 0 to 0.5 λ, labelled around the edge. This scale enables the designer to perform rotations around the chart that are the effect of transmission lines introduced into the circuit. On a paper chart, two scales are given: one (clockwise) represents movement towards the generator, labelled 'wavelengths towards generator', while the other (anti-clockwise) represents movement towards the load, labelled 'wavelengths towards load'.

As shown in Figure 5.14, the Smith chart basically consists of two sets of orthogonal circles. Given that a complex load impedance, Z_L, is connected at the end of a lossless transmission line with characteristic impedance Z_0, the normalised complex load z_L can be written as:

$$z_L = Z_L/Z_0 = r_L + jx_L \tag{5.60}$$

where r_L and x_L are the load resistance and reactance normalised to Z_0. The load reflection coefficient Γ_L can be written in the Cartesian form as:

$$\Gamma_L = \Gamma_r + j\Gamma_i \tag{5.61}$$

where Γ_r is the real part and Γ_i is the imaginary part.

From the definition of Γ_L, Γ_L is related to z_L by:

$$\Gamma_L = \frac{z_L - 1}{z_L + 1} = \frac{r_L - 1 + jx_L}{r_L + 1 + jx_L} \tag{5.62}$$

Equating (5.61) and (5.62), two equations for circles which lie in the Γ_r, Γ_i complex plane are obtained as follows:

$$\left(\Gamma_r - \frac{r_L}{1 + r_L}\right)^2 + \Gamma_i^2 = \frac{1}{(1 + r_L)^2} \tag{5.63}$$

$$(\Gamma_r - 1)^2 + \left(\Gamma_i - \frac{1}{x_L}\right)^2 = \frac{1}{x_L^2} \tag{5.64}$$

The centre of the first set of circles are at the coordinates $(\frac{r_L}{1+r_L}, 0)$. This set of circles forms the constant resistance circles. The radius of each circle depends on the normalised load resistance. For a normalised load resistance, $r_L = 1$ the circle has a centre at $(0.5, 0)$ and radius 0.25 (the '50 Ω' circle). If the load resistance is infinite, the circle will be centred at the right most side of the plot, that is at (1, 0), and the radius is zero: the open circuit point. If r_L is zero, the circle is centred at (0, 0) and has a unit radius (i.e., the outer edge of the chart). Figure 5.15 shows some examples of constant resistance circles ($r_L = 0, 0.5, 1, 2, 5$ and ∞). It should be noted that a circle for a negative r_L is located outside the unit circle, and this situation can be encountered when active devices are present.

Another set of circles is obtained from the second equation and this is the set of constant reactance circles. The normalised reactance, x_L, can be either negative or positive, depending on whether the load is capacitive or inductive. By inspection, it can be seen that the constant reactance circles have their centres located at $(1, 1/x_L)$ and their radius is $\frac{1}{|x_L|}$. For $|x_L| \to 0$, the reactance circles are centred at $(1, \pm\infty)$ and the radius tends to infinity. This gives the horizontal axis, referred to as the real axis since Z_L is purely real. For $|x_L| \to \infty$, the centre moves to (1,0) and its radius tends to zero. This is the open-circuit point on the right-hand side, where the circles are bunched up. Figure 5.16 shows the reactance circles for $|x_L| = 0.5, 1, 2$, and 5.

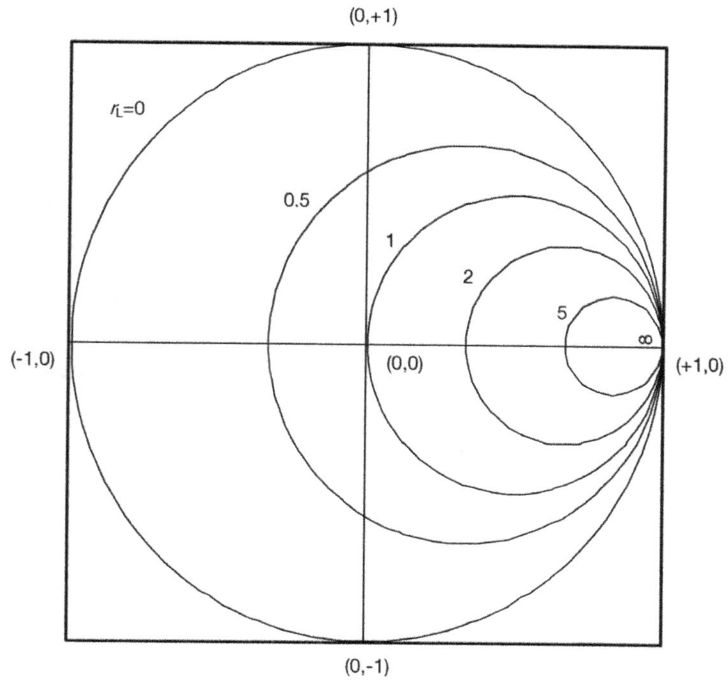

Figure 5.15 Constant resistance circles.

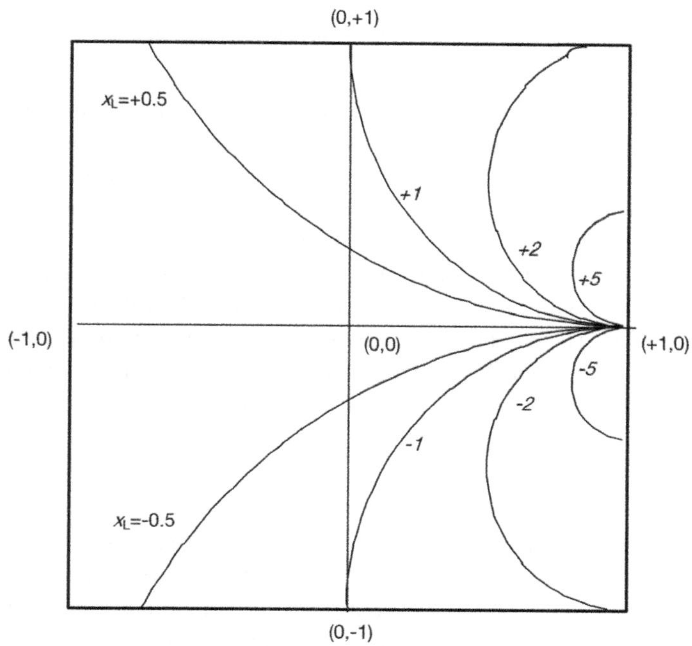

Figure 5.16 Constant reactance circles.

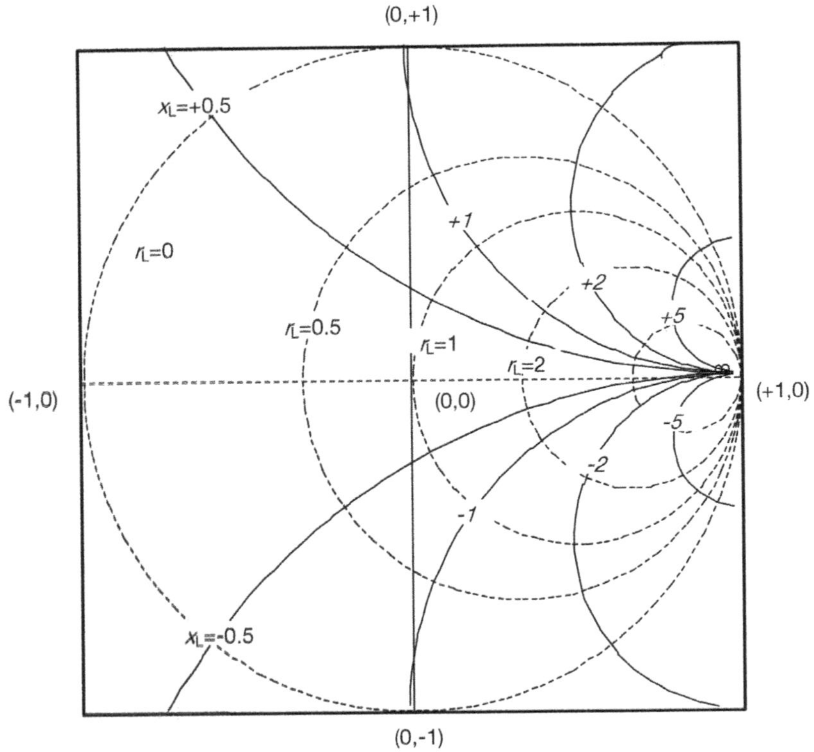

Figure 5.17 Smith chart formed by combining the constant resistance and constant reactance circles.

With these two sets of circles, the Smith chart is simply formed by combining the constant resistance circles and constant reactance circles onto one graph, as shown in Figure 5.17. This chart is referred to as the 'impedance Smith chart'. The short-circuit load is located at the left end of the real axis. The open circuit is located at the right-hand end of the real axis, and the matched point, $Z_L = Z_0$, is located at the centre of the chart.

Figure 5.18 shows the admittance Smith chart. Equations for the circles can be derived in a similar manner to before, with $Y_L = 1/Z_L = G_L + jB_L$, where G_L is the load conductance and B_L the load susceptance. The chart can also be seen to be the result of rotating the impedance Smith chart by 180 degrees. For this chart, each reading is a normalised admittance, that is, $y_L = 1/z_L$. The normalised admittance can be written in the Cartesian form as:

$$y_L = g_L + jb_L \tag{5.65}$$

where g_L and b_L are the normalised load conductance and susceptance.

The admittance chart is vital for matching networks that include shunt elements, such as stub matching networks (these will be covered later). It is important to note that the Smith chart remains at its heart a plot of Γ_L, so when moving from impedance to admittance, Γ_L has not changed, but the contours have flipped over. Unfortunately, text books deal with the admittance chart in different ways, which can cause confusion and sometimes errors. It is

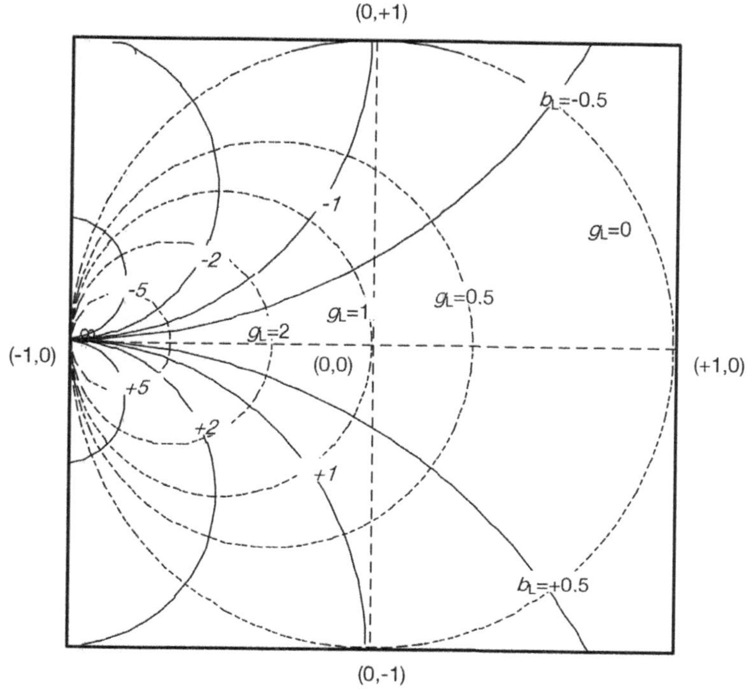

Figure 5.18 Admittance chart resulting from rotating the impedance chart by180°.

possible to use the same Smith chart for both impedance and admittance, and to convert from one to the other by mirror-imaging Γ through the centre (i.e., rotating 180°). This has the same effect as flipping the contours, but is highly risky because the 'anchor' points, such as the open and short circuit points, and all other features, are also flipped, potentially causing major confusion. A safer alternative is to use two charts, one for impedance and one for admittance, or a single chart with both admittance and impedance contours (often referred to as a double Smith chart but more precisely a dual impedance and admittance chart). This chart, obtained by combining impedance and admittance contours onto one chart, is shown in Figure 5.19. This impedance-admittance Smith chart is particularly useful when designing multielement matching networks.

5.5.1 Standing Wave Ratio Circles

So far, we have seen four kinds of constant circles which are resistance, reactance, conductance and admittance. The fifth kind of the circle in this section is called a standing wave ratio (SWR) circle. All SWR circles are centred on the origin (0, 0), and the radius represents a certain SWR whose value is the same as the resistance value read from the right-hand set of resistance contours, where the SWR circle crosses the real axis. Figure 5.20 shows examples of constant SWR circles of SWR = 2 and 5 on the impedance Smith chart. For each constant SWR circle, all points along the perimeter of the circle have the same SWR, and the physical meaning is

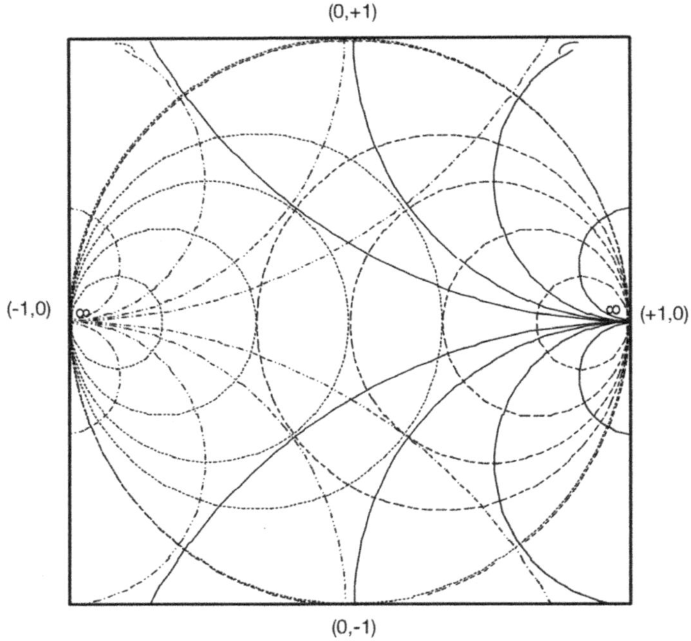

Figure 5.19 Dual impedance-admittance Smith chart.

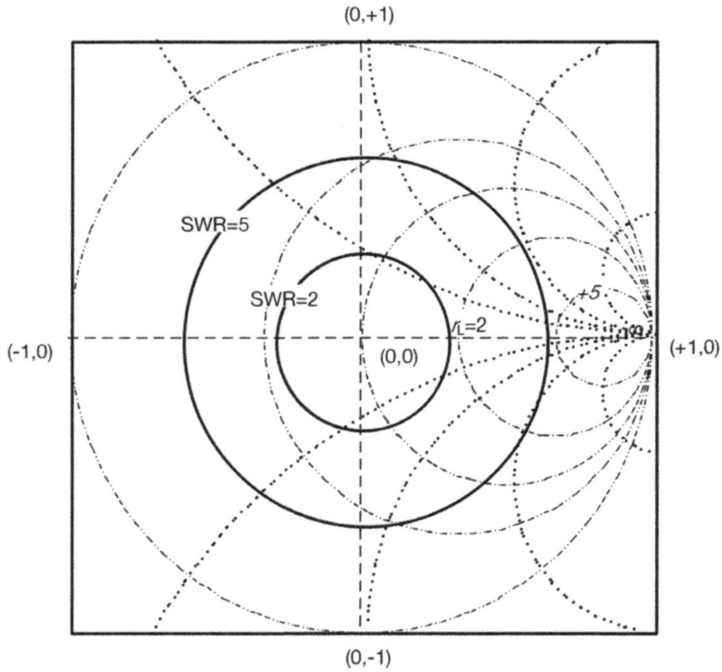

Figure 5.20 Constant SWR circles for SWR = 2 and 5.

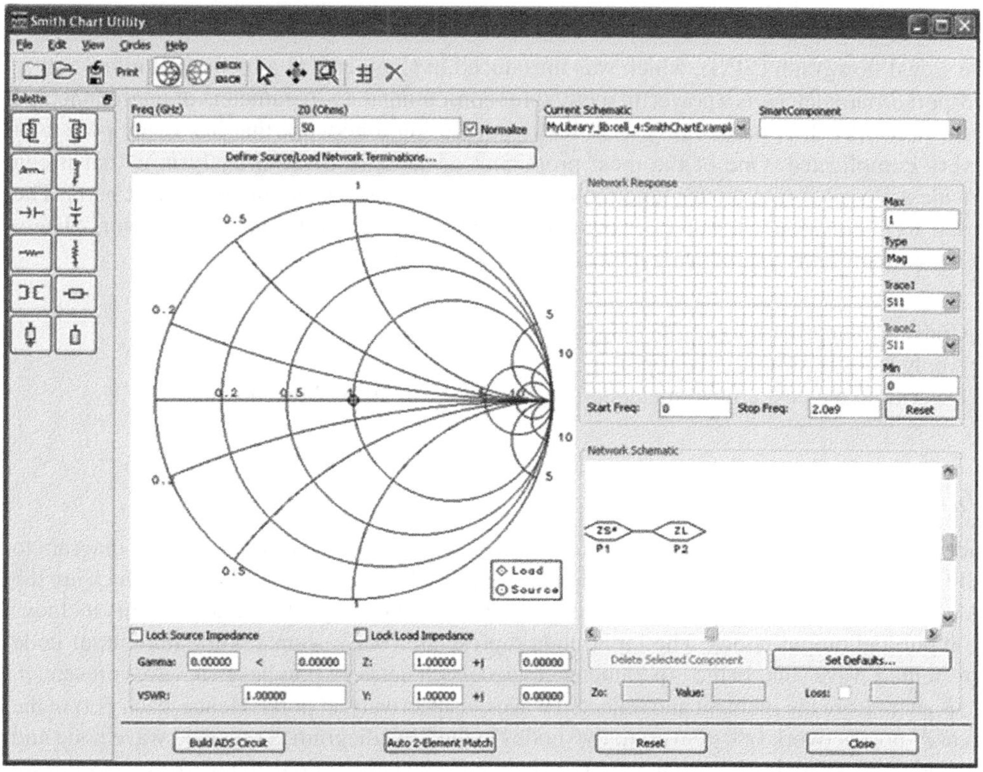

Figure 5.21 The Keysight Technologies Smith chart tool in ADS.

that as you move along the transmission line, away from the load, you move along the standing wave pattern and the impedance varies. This gives one of the key features of the Smith chart: as you move along a transmission line, the impedance at any point along the line can be found by simple rotation about the origin using a compass.

5.5.2 Smith Chart Computer Tools

There are a number of software tools that provide a Smith chart utility, both commercially available and free or shareware. One such is the Smith chart utility in the Advanced Design System (ADS™) from Keysight Technologies, shown in Figure 5.21. The Smith chart tool is integrated with other design tools to provide an automatic solution for circuit design. There are at least two other notable tools: one is the 'Iowa Hills Smith chart' available for download from the Iowa Hills website, and another is Smith Version 3.1 (or later) by Fritz Dellsperger. These tools are easy to use, but most importantly they calculate parameters automatically and allow the designer to try ideas quickly and sweep parameters. The ability to see how Γ moves around the chart using a mouse to tune a parameter gives exceptional insight into the Smith chart and how to use it for circuit design.

5.6 The Signal Flow Graph

The signal flowgraph (SFG), which was introduced in Chapter 3 as a method of representing two-port S-parameters, is a powerful method for computing the S-parameters of interconnected linear networks. It is particularly useful when analysing a network that has many ports and is very complicated. One of the most prominent advantages of SFG analysis is that it can provide a closed-form transfer function in terms of the S-parameters of constituent parts. The closed-form equation brings proper insight into the behaviour of the analysed network. There are three main steps in analysing a linear network with the SFG tool:

1. Convert the network to a SFG diagram.
2. Apply rules to simplify the topology.
3. Determine the desired parameters by inspection, or by using Mason's rule.

5.6.1 *SFG Diagram*

The first step in applying SFG analysis to a linear network is to convert the network diagram to a SFG diagram. To do this, we must understand the two basic elements of a SFG, which are the node and the path. The node in a SFG diagram is represented by a black dot and this includes the input and output nodes. The input node represents a wave source while the output node represents a wave sink. In the S-parameter context, each network port has two waves present, a_i and b_i, which are the incident and reflected waves, respectively, at port i. Hence, each port in the linear N-port network will convert to two nodes in the SFG diagram, a reflected wave node and an incident wave node. A branch in a SFG diagram indicates the flow of a wave along a certain path, with an arrow indicating the direction of travel. Each branch is defined by a parameter value, which is the scattering parameter between the destination and the origin node (e.g., S_{21} represents flow from port 1 to 2) and is, in general a complex, frequency-dependent, quantity.

Figure 5.22 shows the example of converting a one-port network to a SFG diagram. This one-port network could, for example, represent a load resistor, R_A. The reference plane, defined

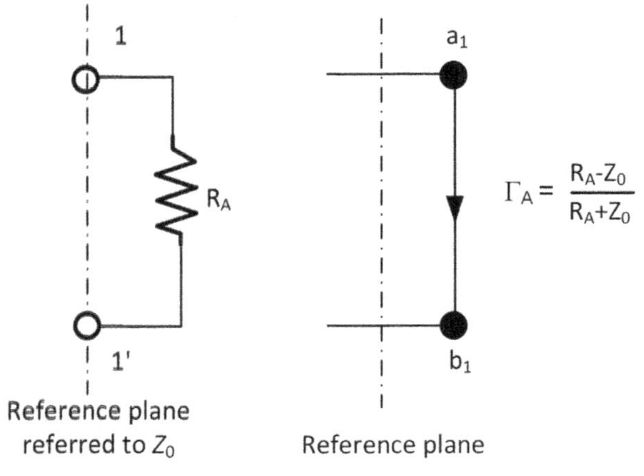

Figure 5.22 One-port network and its SFG diagram.

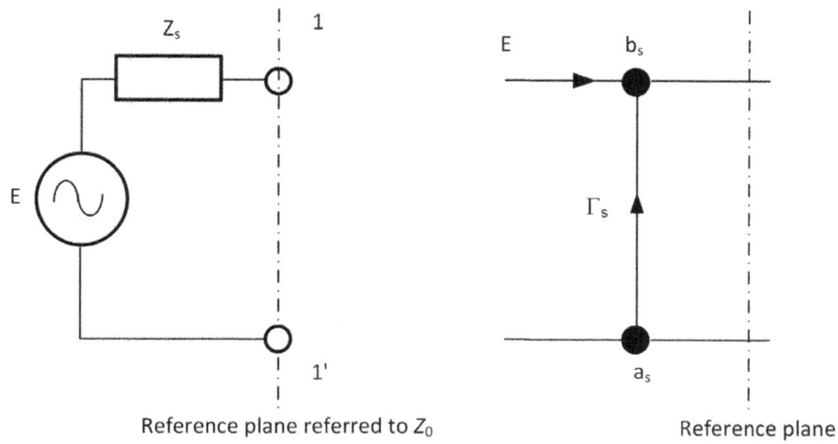

Figure 5.23 Signal source and its SFG diagram.

by the dashed line 1-1‘, is the plane interfacing the network and the rest of the circuit. The characteristic impedance is assumed to be Z_0. From Figure 5.22, the SFG diagram has two nodes, a_1 and b_1, representing the incident and reflected nodes. There is a branch a_1 to b_1 connecting between these two nodes. The arrow of the branch a_1 to b_1 is in the direction originating from node a_1 towards node b_1. Hence, the transfer coefficient is the ratio of the reflected wave b_1 to the incident wave a_1 which is the reflection coefficient referred to Z_0. This has the value $\Gamma_A = (R_A - Z_0)/(R_A + Z_0)$.

A generator can also be converted to a SFG diagram. A generator consists of a source E and impedance Z_s, as shown in Figure 5.23. The SFG diagram of a generator has two nodes, the incident node a_s and the reflected node b_s. The wave generated from the source E ends at node b_s. With the source resistance Z_s, there is a reflected wave travelling from node a_s and ending at node b_s with the ratio Γ_S which is defined by, $(Z_S - Z_0)/(Z_S + Z_0)$. Note that the node arrangement of the SFG source from top to bottom begins with node b_s and ends with node a_s. The energy from the source is travelling from the node b_s.

A two-port linear network can be converted to a SFG diagram if the scattering parameters of the considering network are defined. Using the same rules, there are four nodes; two for port 1 (a_1, b_1) and two for port 2 (a_2, b_2) in the diagram. The wave from node a_1 travels to node b_1 (reflection; a contribution to the wave coming out of port 1) and to b_2(forward transmission; a contribution to the wave coming out of port 2) with coefficients s_{11} and s_{21}, respectively. Similarly, the wave travels from node a_2 to node b_1 and b_2 with coefficients s_{12} and s_{22}. These are shown as the lines with arrows and parameters s_{nm} denoted in the SFG diagram shown in Figure 5.24.

Figure 5.25 shows the SFG diagram for a two-port network with a source connected to port 1, with output impedance Z_s, and a load R_L connected to port 2. From the SFG diagrams of the individual parts, shown in Figures 5.22, 5.23 and 5.24, the complete SFG diagram of this two-port network can easily be constructed and is shown in Figure 5.25. For clarity, the connections between two cascaded subnetworks are represented with branches of transfer coefficient ‘1’ linking between the two nodes.

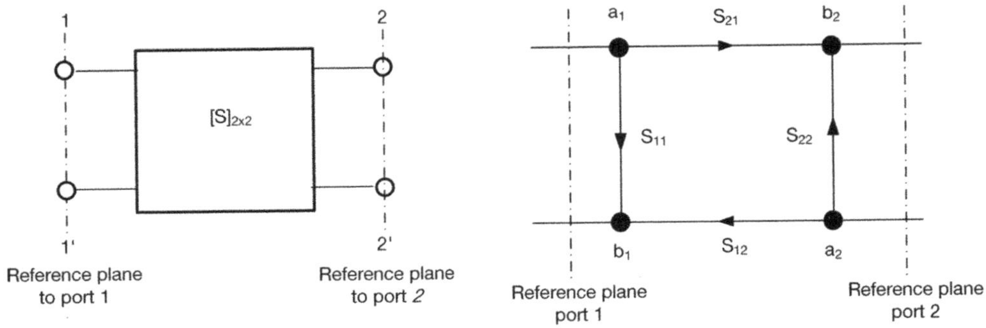

Figure 5.24 Two-port network and its SFG diagram.

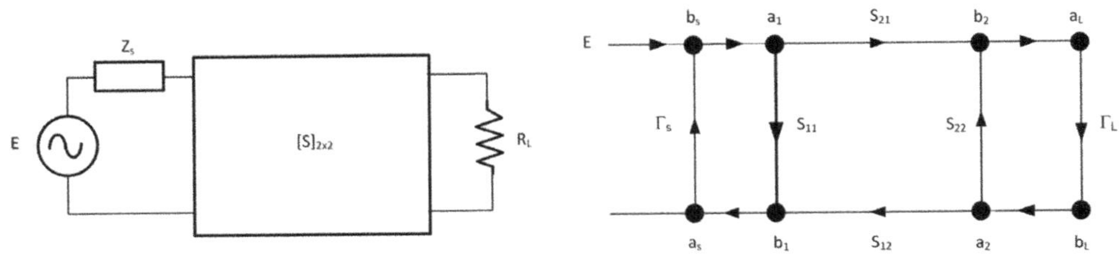

Figure 5.25 The SFG diagram for a two-port network with a source connected to port 1 with output impedance Z_s and a load R_L connected to port 2.

5.6.2 SFG Simplification

There are two further elements that must be defined for analysing SFG diagrams. From the definitions of input and output nodes, a forward path is defined as a path that starts from an input node and ends at an output node. Meanwhile, a loop in the SFG is defined as a path that starts and ends at the same node without passing other nodes more than once along the travelled path. In order to solve a SFG, for example to find the relationships between a certain output and input node, two methods are employed. The first method is to employ a set of four topological rules for simplifying the diagram step-by-step. The second method is Mason's non-touching loop rule.

5.6.2.1 The Four Topological Rules

Rule 1: The middle node B of the three-node SFG (with nodes A, B, C) shown in Figure 5.26 can be removed. The simplified SFG diagram has only two nodes (A and C). The gain of path AC in the new SFG diagram is the product of the gains of path AB and path BC.

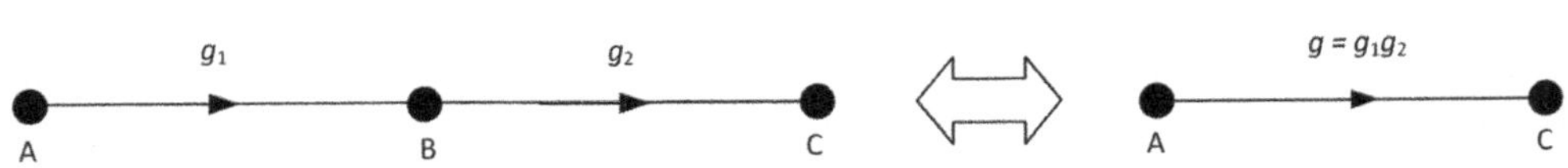

Figure 5.26 Middle node removal.

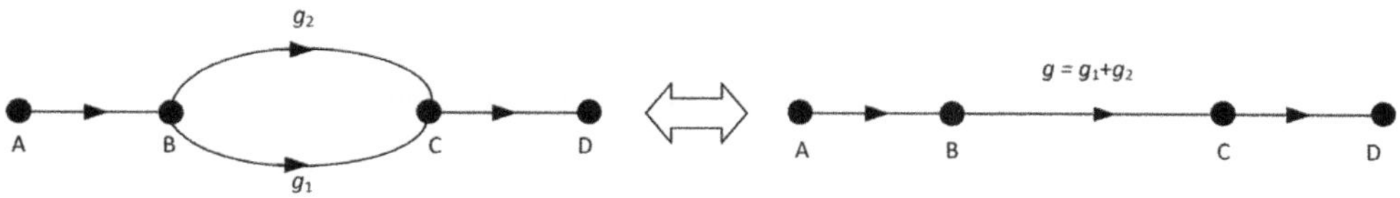

Figure 5.27 Parallel path simplification.

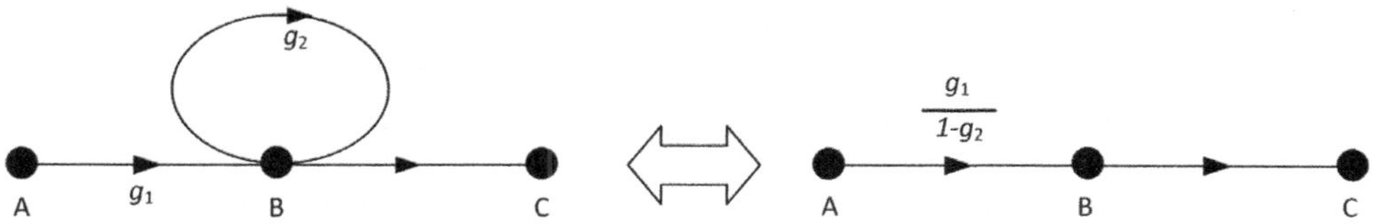

Figure 5.28 Self-loop simplification.

Rule 2: Two parallel paths with path gains g_1 and g_2, as shown in Figure 5.27 can be reduced to a single path with the new path gain $g = g_1 + g_2$.

Rule 3: A self-loop of gain g_2, as shown in Figure 5.28, can be removed by dividing the path gain of each path entering the node B by $1 - g_2$.

Rule 4: A node can be split into several separated nodes and parallel paths as shown in Figure 5.29.

5.6.2.2 Mason's Rule

The other important concept is Mason's rule. This rule, sometimes called Mason's non-touching loop rule, can be applied to compute the transfer function ratio (T) between two nodes of quite complex networks by following a step-by-step procedure to identify all the paths between the two nodes and any loops in the network, and then applying the formula:

$$T = \frac{P_1\left(1 - \sum L(1)^{(1)} + \sum L(2)^{(1)} - \sum L(3)^{(1)} + \ldots\right) + P_2\left(1 - \sum L(1)^{(2)} + \sum L(2)^{(2)} - \sum L(3)^{(2)} \ldots\right) + \ldots}{1 - \sum L(1) + \sum L(2) - \sum L(3) + \sum L(4) - \ldots} \tag{5.66}$$

where P_k is the gain of the kth forward path,

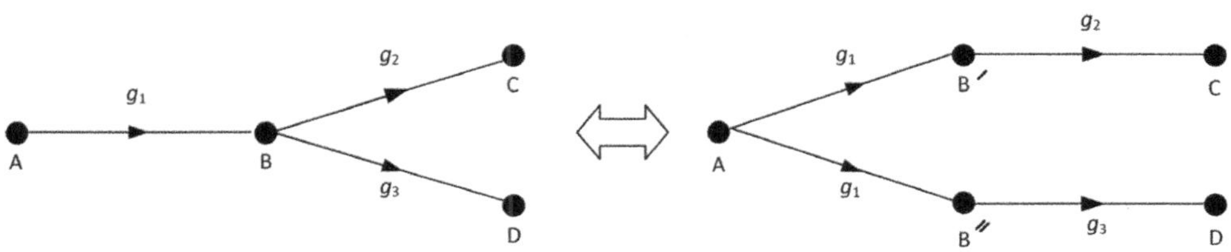

Figure 5.29 Splitting a node.

$\sum L(1)$ denotes for the sum of all first-order loop gains;

$\sum L(2)$ denotes for the sum of all second-order loop gains; and

$\sum L(n)$ denotes for the sum of all n^{th}-order loop gains.

The first-order loop gain is the gain of any loop in the network under consideration. The second-order loop gain is the product of any pair of first-order loops that do not touch at any point. Touching means sharing any branch. The third-order loop gain is the product of any three first-order loops that do not touch at any point, and so on for $L(3)$, $L(4)$, and so forth.

Having identified all the loop gain products, the following are determined:

$\sum L(1)^{(i)}$ stands for the sum of the first-order loop gains that do not touch path P_i;

$\sum L(2)^{(i)}$ stands for the sum of the second-order loop gains that do not touch path P_i;

$\sum L(n)^{(i)}$ stands for the sum of the nth-order loop gains that do not touch path P_i.

Mason's rule is applied to the simple example of a transistor amplifier in Chapter 10.

5.7 Conclusion

This chapter has introduced the characteristics of transmission lines, starting with considering how the time delay creates surprising effects when step functions and pulses are sent down a line. By developing a simple inductor-capacitor ladder network representation of the transmission line, the wave behaviour and standing waves have been introduced. The characteristic impedance and propagation constant are used in practice to represent the electrical behaviour of a transmission line. A most important effect is that impedance is found to be a function of position along the transmission line, leading the use of the Smith chart to perform calculations. For more complex networks, the signal flowgraph can be used, with reduction techniques and Mason's rule being used to determine a closed-form expression for the characteristics of an arbitrary network.

References

1. Ramo, S., Whinnery, J.R., and Van Duzer, T. (1994) *Fields and Waves in Communication Electronics*, 3rd edition. John Wiley & Sons. ISBN-10: 0471585513.
2. Pozar, D. (2011) *Microwave Engineering*, 4th edition. John Wiley & Sons. ISBN-10: 0470631554.
3. Chang, K. (2000) *RF and Microwave Wireless Systems*. John Wiley & Sons, Inc.
4. Schmitt, R. (2002) *Electromagnetics Explained*. Newnes.

6

Transmission Line Components

6.1 Introduction

Coaxial line was introduced briefly in Chapter 5 so that the parameters of characteristic impedance and propagation constant could be used for the analysis of transmission line behaviour, which underpins the concept of distributed circuits. This chapter now describes the extensive range of transmission line media that are used for circuits and introduces some of the key distributed passive components that are used in wireless communications systems. After a brief description of coaxial lines and waveguides, the emphasis thereafter is on transmission lines that can be fabricated with photolithographic techniques, since that is absolutely fundamental to manufacturing circuits in large quantities. Microstrip is the default choice of medium for designing hybrid and monolithic integrated circuits since it is well behaved and there is a wealth of design information available, with accurate modelling techniques that have been developed over several decades. However, there are many alternatives that have been developed with the aim of improving performance, reducing size, and so forth, and the most important of these alternatives, such as coplanar waveguide, substrate integrated waveguide and stripline, will be briefly summarised.

6.2 Coaxial Components

Coaxial cable is often used as a transmission line medium for low-loss signal transmission in communications and measurement systems. A coaxial cable consists of two concentric cylindrical conductors. The outer ground conductor can be formed from a solid or braided conductor, usually copper and sometimes silver-plated. The dielectric material separating the conductors is commonly made of solid polyethylene (PE) or Teflon material, but may be in the form of a foam, or microstructured in some way in order to create hollow regions to reduce dielectric loss. The outer jacket insulates and protects the coaxial cable. If the outer diameter of the centre conductor is d, the inside diameter of the outer conductor is D, and the relative permittivity of the dielectric material is ε_r, it can be shown that the characteristic impedance

Microwave and Millimetre-Wave Design for Wireless Communications, First Edition.
Ian Robertson, Nutapong Somjit and Mitchai Chongcheawchamnan.
© 2016 John Wiley & Sons, Ltd. Published 2016 by John Wiley & Sons, Ltd.

of the coxial line is given by:

$$Z_o = \frac{377}{2\pi\sqrt{\varepsilon_r}} \ln\left(\frac{D}{d}\right) \tag{6.1}$$

The dominant mode of propagation in the coaxial cable transmission line is transverse electromagnetic (TEM) because there are no components of E or H field along the z-axis: the electric field lines are radial and the magnetic field lines form circles around the centre conductor. In the TEM mode, the EM field propagates along the z-direction and the electric and magnetic field lines are orthogonal to the propagation direction. The field pattern is the same all along the line and TEM propagation means that coaxial line has constant electrical properties over an extremely wide frequency range, which is why it is used extensively for measurement equipment. On close inspection, some dispersion occurs at very low frequency, where the conductor resistance is no longer negligible compared to $j\omega L$, while at high frequencies the coaxial cable starts to suffer from unwanted non-TEM modes of propagation. One of the key frequency limitations is that, at very high frequency, the outer conductor can start to support modes similar to those in a hollow waveguide (this is discussed shortly). For this reason, in order to create coaxial cables and connectors that work at higher and higher frequencies it is necessary to make the outer conductor's inside diameter smaller and smaller. This trend can be observed easily in the common microwave coaxial connector types, which are listed in Table 6.1. Some examples are shown in Figure 6.1.

Table 6.1 Common coaxial connector types.

Name/outer conductor inside diameter (mm)	Maximum frequency of operation (GHz)	Note 1	Note 2
SMA, 4.2	18 (typical), 24 GHz (possible)	Sub-miniature type A	Not a precision connector, quality and maximum frequency varies greatly
N type, 7	11, 18 for precision versions	75 Ω variants exist with a narrower centre conductor	
APC7	18	Amphenol Precision Connector	Sexless
3.5	33	Compatible with SMA	
2.4	50		
K, 2.92	40	Compatible with APC3.5 and SMA	
V, 1.85	65	Compatible with APC2.4	Also known as the V connector
W1, 1.0	110		
0.8	145	Anritsu	

K, V and W1 connector are trademarks of Anritsu.

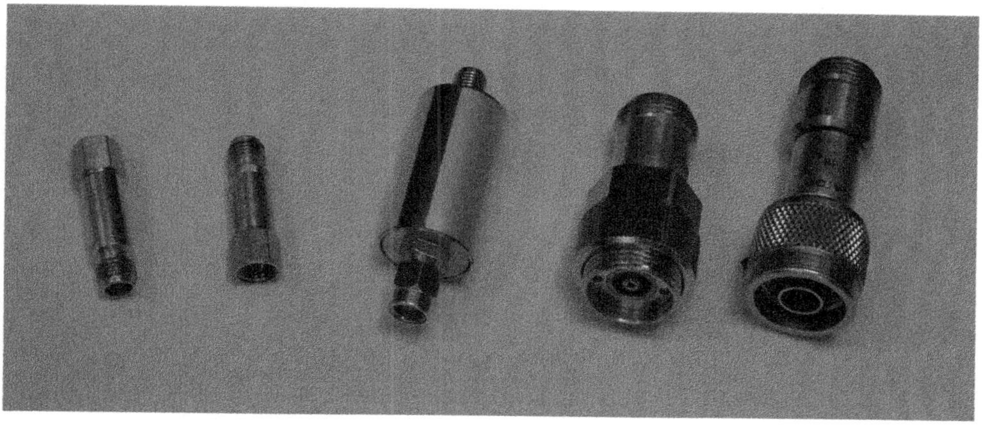

Figure 6.1 An assortment of coaxial connector adapters and attenuators (from left to right: APC2.4, APC2.4, SMA, APC7, N-type).

The BNC (Bayonet Neill–Concelman) connector is a quick-fit twist-on connector that is used very widely at low frequencies, and high-grade versions can be used to 4 GHz. BNC connectors and cables are available in 50 and 75 Ω types, so care should be exercised. A few connectors, such as the APC7, are 'sexless' but the vast majority come in male/female pairs, where the connector sex is determined by the centre conductor. In some countries, the terminology of plugs and jacks is often used, where the plug has the protruding pin. Table 6.1 covers only mainstream microwave and millimetre-wave connectors and there is a vast range of other coaxial connectors available for more specialist requirements such as weatherproof, high-power and military-grade connectors and, more recently, connectors for coaxial/ribbon cables, and surface mount snap-on coaxial connectors for mounting on printed circuit boards.

Coaxial components can benefit significantly from the TEM propagation, which leads to well-behaved performance with wide spurious-free operation. With its cylindrical profile, it is not convenient to design circuits in coaxial form, although various filter structures can be realised in coaxial form. Many coaxial components, for example attenuators and power detectors, will solve this issue by using a small substrate inserted into the cylinder. Another common approach is to employ a structure with square or rectangular conductor cross-sections [1] which makes components easier to manufacture. It is especially attractive to use this form to realise components using photolithographic fabrication [2], including micro-machined structures operating at very high frequencies (these are described in Chapter 13).

6.3 Twisted Pairs and Twin-Lead

Figure 6.2 shows the form of a twisted pair and a twin-lead transmission line. Twisted pairs started off life in telephony and were generally regarded as a cheap and simple means of achieving signal confinement for a low-frequency transmission line. Nowadays, they find widespread use in computer networking: 'UTP' stands for unshielded twisted pair, and these cables are routinely used to 100 Mb s^{-1}. Twin-lead has particular relevance to antennas, where the balanced transmission line is compatible with a number of important structures, such as the dipole and folded-dipole antenna. A balun (balanced-unbalanced transition) is then

Figure 6.2 Twisted pair line and 300 Ω FM twin-lead ribbon.

required to connect the balanced line to a coaxial cable. Some of these balun structures give the impression of requiring 'black art' techniques in the design, but actually the principles outlined in Chapter 5 all apply and the most important thing to remember is that the concept of a ground plane is quite dangerous if it is not recognised that there is a reverse current flowing in any transmission line structure. Twin-lead is a powerful reminder of this fact, and there are many practical considerations that can be applied to the planar equivalent, called coplanar strips (CPS) (see later in this chapter) which is of increasing importance in RFIC design.

6.4 Rectangular Waveguide

Rectangular waveguide is used in many applications due to its low loss and high power handling. It is also often used in millimetre-wave and terahertz systems where planar transmission lines may have too much loss to be practical. Some examples (S-band, X-band and V-band) of hollow rectangular waveguide are shown in Figure 6.3(a). Hollow waveguide is used most often, but dielectric materials can be used to reduce the waveguide size, with careful consideration needed to the trade-offs in power handling capability, weight and cost. The waveguide with dielectric is called a dielectric-loaded waveguide. There is a range of other waveguide

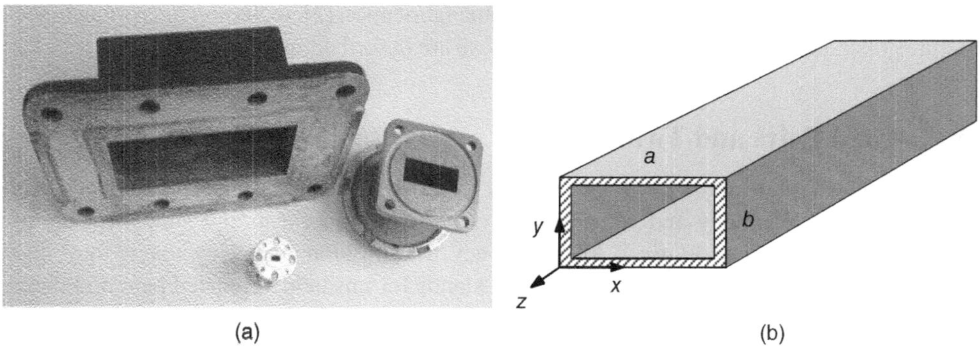

Figure 6.3 Rectangular waveguide. (a) Types of waveguide. (b) Rectangular waveguide structure.

cross-sections, such as circular waveguide and ridged waveguide. There are also many forms of dielectric waveguide, which can be advantageous for millimetre-wave and terahertz applications since conductor loss can become a serious limitation. The optical fibre is at the very far end of this 'spectrum' of wave-guiding structures. It is simply not possible to do all of these structures justice in this chapter, which is focused on planar transmission lines for circuit design; hence, the reader is referred to one of the very many texts on traditional waveguide components, millimetre-wave waveguides and dielectric guides for further information [3–5].

The rectangular waveguide is mostly chosen due to its simpler manufacture than the circular waveguide or other more sophisticated structure. The propagation modes in a waveguide can be either Transverse Electric (TE) or Transverse Magnetic (TM), depending on the waveguide's dimensions, the signal frequency and the nature of the input/output coupling. The rectangular waveguide shown in Figure 6.3(b) lies along the z-axis and the cross-section is in the x–y plane. For TE wave propagation, also called the H-mode, no electric field components exist along the z-axis ($E_z = 0$). Likewise, in a TM or E-mode, there are no magnetic field components along the z-axis, that is, $H_z = 0$. In a certain waveguide, there are a number of field patterns, or propagation modes, which can propagate independently. For a lossless rectangular waveguide, for the TE_{mn} modes where m and n are any positive integers, the electric and magnetic field intensities are given by:

$$E_x = \frac{j\omega\mu n\pi}{k_c^2 b} A_{mn} \cos\frac{m\pi x}{a} \sin\frac{n\pi y}{b} e^{-j\beta z} \tag{6.2}$$

$$E_y = \frac{-j\omega\mu m\pi}{k_c^2 a} A_{mn} \sin\frac{m\pi x}{a} \cos\frac{n\pi y}{b} e^{-j\beta z} \tag{6.3}$$

$$H_x = \frac{j\beta m\pi}{k_c^2 a} A_{mn} \sin\frac{m\pi x}{a} \cos\frac{n\pi y}{b} e^{-j\beta z} \tag{6.4}$$

$$H_y = \frac{j\beta n\pi}{k_c^2 b} A_{mn} \cos\frac{m\pi x}{a} \sin\frac{n\pi y}{b} e^{-j\beta z} \tag{6.5}$$

$$H_z = A_{mn} \cos\frac{m\pi x}{a} \sin\frac{n\pi y}{b} e^{-j\beta z} \tag{6.6}$$

For TM_{mn} modes, the electric and magnetic field intensities are:

$$E_x = \frac{-j\beta m\pi}{k_c^2 a} B_{mn} \cos\frac{m\pi x}{a} \sin\frac{n\pi y}{b} e^{-j\beta z} \tag{6.7}$$

$$E_y = \frac{-j\beta n\pi}{k_c^2 b} B_{mn} \sin\frac{m\pi x}{a} \cos\frac{n\pi y}{b} e^{-j\beta z} \tag{6.8}$$

$$E_z = B_{mn} \sin\frac{m\pi x}{a} \sin\frac{n\pi y}{b} e^{-j\beta z} \tag{6.9}$$

$$H_x = \frac{j\omega\varepsilon n\pi}{k_c^2 b} B_{mn} \sin\frac{m\pi x}{a} \cos\frac{n\pi y}{b} e^{-j\beta z} \tag{6.10}$$

$$H_y = \frac{-j\omega\varepsilon m\pi}{k_c^2 a} B_{mn} \cos\frac{m\pi x}{a} \sin\frac{n\pi y}{b} e^{-j\beta z} \tag{6.11}$$

where β is the imaginary part of the propagation constant introduced in Chapter 5, which in this case is given by:

$$\beta = \sqrt{k^2 - k_c^2} \tag{6.12}$$

In Equation (6.12), k_c is the cut-off wave number which is given by:

$$k_c = \sqrt{\left(\frac{m\pi}{a}\right)^2 + \left(\frac{n\pi}{b}\right)^2} \tag{6.13}$$

The propagation constant, β, from Equation (6.12) will be an imaginary number if $k_c > k$. In such a case, the wave amplitude along the z-axis of the rectangular waveguide decays exponentially. The critical point occurs when $k = k_c$. The frequency at this critical point is called the cut-off frequency of the waveguide (f_c) and is given by:

$$f_c = \frac{1}{2\pi\sqrt{\mu\varepsilon}}\sqrt{\left(\frac{m\pi}{a}\right)^2 + \left(\frac{n\pi}{b}\right)^2} \tag{6.14}$$

Any signal whose frequency is greater than f_c is able to propagate along the waveguide. The characteristic of wave propagation mode (m, n) can be observed from the number of half wavelengths along the x- and y-axis, respectively. For example, as shown in Figure 6.4, the TE_{10} and TE_{20} modes have one and two half-wavelengths of electric field, respectively, across the width of the guide. Even if a single mode is propagating, Equation (6.12) shows that rectangular waveguides exhibit group delay dispersion. Furthermore, the combination of the cut-off frequency at lower frequency and multiple modes of propagation at high frequencies means that hollow waveguides have a limited useful bandwidth. Waveguides are, therefore, available in a range of bands, such as Ka-band (26.5 to 40 GHz, WR-28), V-band (50–75 GHz, WR-15), E-band (60–90 GHz, WR-12) and W-band (75–110 GHz, WR-10). The WR-*number* designation is from standards set by the former Electronics Industry Association in the USA [6], and the number is the inner dimension of the waveguide in mils divided by 10. It must be noted that *mils* are the Imperial unit of one-thousandth of an inch, and should not be confused with millimetres in the metric system, although sometimes millimetres is shortened colloquially to 'mils', to add further confusion. Also, mils are often the default unit in various EDA software and equipment that originates from the USA, so great care must be exercised.

Whilst a rectangular waveguide has distinct advantages of low-loss and high-power handling capability, and is often required for filters due to the very high-Q factor that can be achieved, because of the 3-D machine-shop fabrication requirement much research has been conducted

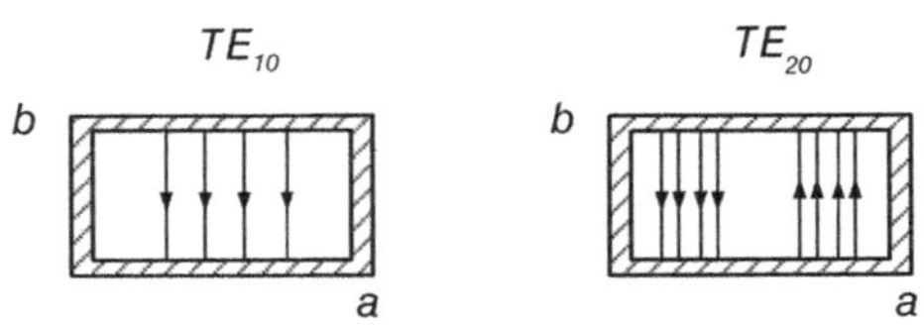

Figure 6.4 TE_{10} and TE_{20} propagation modes in a rectangular waveguide (E-field).

into methods to replace hollow waveguides with similar structures that can be fabricated photolithographically. Some of these, including substrate integrated waveguide (SIW) and micromachined waveguides, are described later in this chapter.

6.5 Microstrip

Microstrip consists of a signal conductor strip on the upper surface of a dielectric substrate, with the lower surface forming a solid ground plane, as shown in Figure 6.5. The dielectric material has a dielectric constant ε_r, loss tangent δ and height h, while the conductor strip has width w and thickness t. The loss tangent, the metal conductivity and surface roughness all contribute to the loss of the line. The field energy is mainly confined in the substrate with

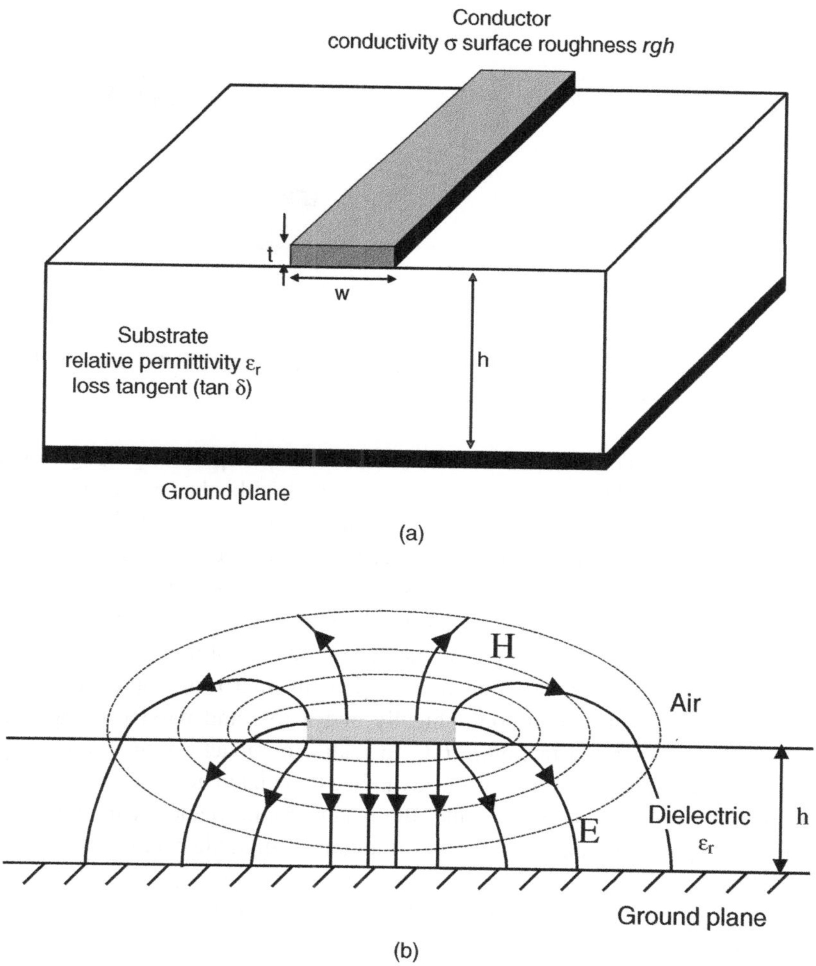

Figure 6.5 (a) Microstrip transmission line, with the key parameters labelled. (b) Approximate E- and H-field patterns in cross-section.

Figure 6.6 Microstrip 5W X-band power amplifier circuit using discrete GaN transistor. Illustration courtesy of Plextek RFI. Used with permission; all rights reserved.

some part in the air above. This means there is a difference in the speed of propagation and the field pattern cannot be pure TEM (transverse electromagnetic) like the coaxial case, where the dielectric is homogeneous. However, with the assumption that the ground plane is effectively infinite, the dominant propagation mode in a microstrip line is called a quasi-TEM mode, which means the longitudinal field components are very much smaller than the transverse ones. This is generally valid provided that h is small compared to the signal wavelength. The degree of energy confinement depends on the dielectric constant of the substrate.

Microstrip is suitable for use as a transmission line in microwave and millimetre-wave circuits since it can be fabricated lithographically and active and passive components can easily be attached, as illustrated in Figure 6.6, to realise compact circuits and subsystems. Compared to coplanar structures, it has the disadvantage that plated holes through the substrate are needed for grounding. On the other hand, the lower ground plane means microstrip can easily be mounted in packages and test fixtures without unpredictable behaviour and a heatsink can be attached.

Broadly speaking, there are three main technologies for fabricating microstrip circuits, which are described in more detail in Chapter 16:

- Copper-clad boards, including FR-4, PTFE/ceramic composites (such as Duroid^{TM}), and liquid crystal polymer (LCP).

- Thick-film fabrication on ceramic substrates, including low temperature co-fired ceramic (LTCC).
- Thin-film fabrication, on a variety of substrates (e.g., alumina, quartz, glass, aluminium nitride), and on integrated circuits (GaAs, InP, Si, etc).

6.5.1 *Analysis and Synthesis*

The most basic function of EDA software for microstrip design is to calculate the electrical parameters for a line of given physical dimensions (analysis) or to provide the physical width and length required for the desired characteristic impedance and electrical length (synthesis). Tools such as LineCalc from Keysight Technologies perform this function. With some software (e.g., freeware) it may be the case that simplified models are used with 'fitting factors' so the accuracy may be quite poor. The software may not be suitable for data outside a certain range, but may not necessarily warn you if you go outside it. Because of the inhomogeneous dielectric, the fields in microstrip are too complicated to solve for directly. One traditional technique is to use conformal mapping to map the structure into an ideal parallel-plate waveguide, and this is the origin of the parameters effective width, w_{eff}, and effective dielectric constant, $\varepsilon_{\mathrm{eff}}$. The latter is constrained to be in the range:

$$\frac{\varepsilon_r + 1}{2} \leq \varepsilon_{eff} \leq \varepsilon_r \tag{6.15}$$

In other words, for narrow lines the field tends towards being equally distributed in the substrate and the air, while wide lines are similar to an ideal parallel-plate waveguide, with the fringing fields having less effect. In fact, $\varepsilon_{\mathrm{eff}}$ also varies with frequency because the field profile changes. From the effective width and effective dielectric constant, the characteristic impedance can be calculated as [7]:

$$Z_0(f) = \frac{h\eta}{w_{eff}(f)\sqrt{\varepsilon_{eff}(f)}} \tag{6.16}$$

This highlights the fact that Z_0 changes slightly with frequency. η is the impedance of free space, $\sqrt{\mu_o/\varepsilon_o} \approx 377\ \Omega$.

6.5.2 *Microstrip Frequency Limitations*

For wide lines (low characteristic impedance) the maximum frequency of use is usually determined by a transverse resonance effect, which occurs when the width of the line (plus a fringing field component) approaches a half-wavelength in the dielectric:

$$f_{CT} = \frac{C}{\sqrt{\varepsilon_r}(2W + 0.8h)} \tag{6.17}$$

Transverse resonances can be avoided by careful choice of Z_0, but unwanted TM modes can be supported by the substrate itself, acting as a grounded dielectric slab waveguide. TM mode propagation is not width-dependent and the lower cut-off frequency is given by:

$$f_{TM} = \frac{C\tan^{-1}\varepsilon_r}{\sqrt{2}\pi h\sqrt{\varepsilon_r - 1}} \tag{6.18}$$

TM modes, often referred to as surface-wave modes, can be avoided by choosing a thin substrate and so this frequency limitation is usually specified as:

$$h_{\max} = \frac{0.354\lambda_o}{\sqrt{\varepsilon_r - 1}} \tag{6.19}$$

For higher and higher frequencies of operation, the substrate must be made thinner and thinner. This means that the ultimate limit on the applicability of microstrip lines is dictated by practical considerations – the fabrication tolerances and the high losses due to narrow conductors. Microstrip can be used to well over 100 GHz on GaAs monolithic microwave integrated circuits (MMICs).

6.5.3 *Microstrip Discontinuities and EDA Software*

Microstrip circuit design primarily uses schematic capture to create a circuit schematic diagram using symbols. This makes it is easy to visualise the circuit and prevent mistakes, but compared to regular circuit simulations the difference is that the physical layout of the design is crucial to its behaviour. So, modern microwave EDA software, such as ADS^{TM} by Keysight Technologies, will have a separate layout window. In most cases, microstrip circuits are modelled by breaking the circuit down into its constituent transmission lines, but discontinuities such as bends, steps and T-junctions have to be manually inserted into the schematic so that their effect on the performance is taken into account. Figure 6.7 shows some examples of the discontinuity elements available in ADS^{TM}. An MSub data block defines the substrate parameters. It is increasingly common to use electromagnetic modelling techniques (as described in Chapter 17), as these can be more accurate when circuits are miniaturised. This is because the building-block approach to microstrip modelling requires that individual elements are sufficiently far apart that there is no unintended coupling and the input and output wave into every element is single mode: within a discontinuity, unwanted modes of propagation may exist, but they are evanescent and decay away exponentially with distance. If elements are placed too close together, the validity of models may be violated.

6.5.4 *Test Fixtures and Housings*

A test fixture or housing is required as a protective enclosure and as a transition between the microstrip circuit and the coaxial connectors or rectangular waveguide flange required for connecting to the outside world. Microstrip launchers are available with a wide range of connectors, such as SMA, APC3.5, APC2.4, K and V, but the design of the transition is

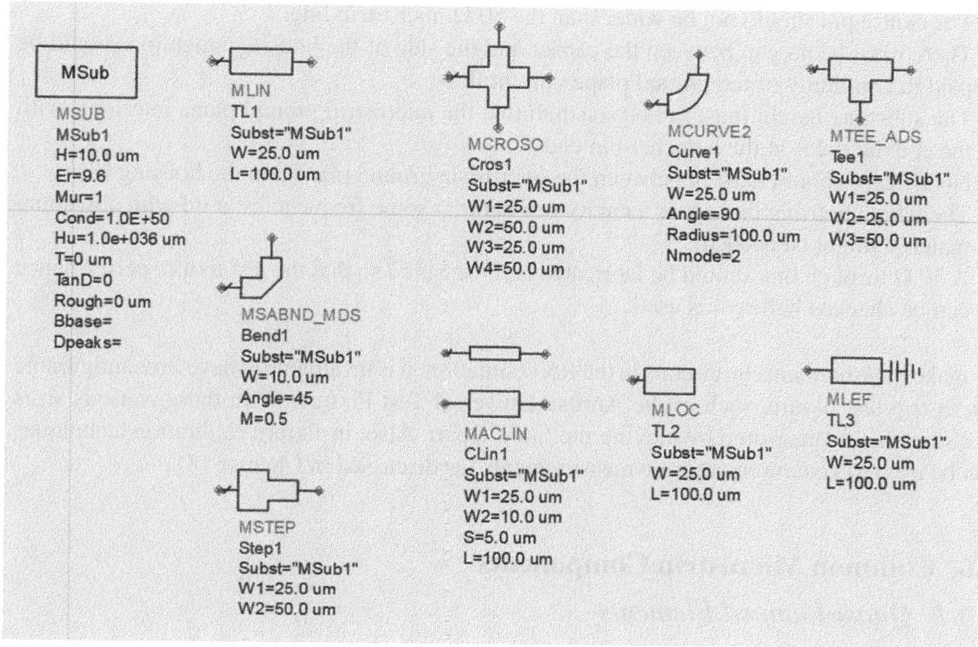

Figure 6.7 Examples of microstrip components in ADS™.

crucial. To operate at high frequencies with low VSWR, the coaxial fields must be smoothly converted to the microstrip field. Launchers might connect directly to the microstrip, as shown in Figure 6.8, or a 'bead' might be soldered into the housing wall. A typical launcher cross-section is shown in Figure 6.8. The main design principles are:

1. The housing wall/centre pin should be selected to give a 50 Ω coaxial line, and care should be taken to avoid eccentricity.

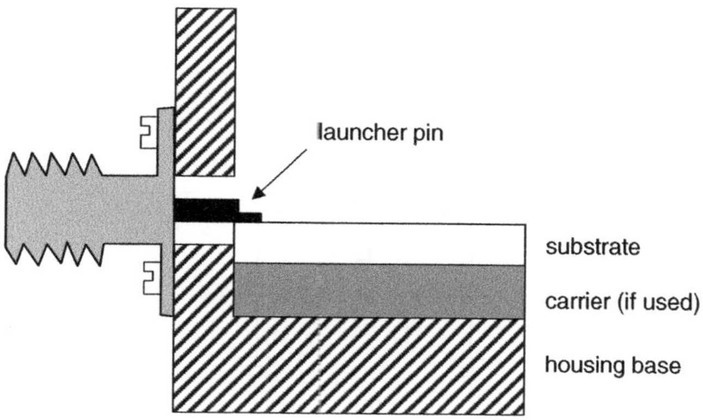

Figure 6.8 A typical coax-microstrip launcher cross-section.

2. The centre pin should not be wider than the 50 Ω microstrip line.
3. There must be no gap between the carrier and the side of the housing; attention should be paid to continuity of the ground plane current flow.
4. The substrate height must not be so small that the microstrip ground plane interferes with the coaxial fields at the launcher pin end.
5. No air gaps should exist in between the microstrip ground plane and the housing base.
6. The whole housing can act as a cavity resonator at some frequencies; a lid with absorbing material might be needed.
7. A 50 Ω through line should be fabricated and measured so that the test fixture performance can be checked before it is used.

To make measurements on circuits in the R&D situation, it is invaluable to have a reconfigurable microstrip test fixture, such as the Anritsu Universal Test Fixture. With these, various sizes of circuit can be measured by moving the 'jaws' apart. Also, in-fixture calibration techniques can be used to give more accurate measurements (as discussed in Chapter 18).

6.6 Common Microstrip Components

6.6.1 *Quasi-Lumped Elements*

Whilst surface mount passive components can readily be soldered onto a microstrip circuit, there will be additional parasitic components (e.g., the capacitance of the solder pads), plus additional assembly work and component costs. It is therefore attractive to be able to realise integrated quasi-lumped elements, for example in bias networks, alongside the distributed microstrip elements.

6.6.1.1 Ribbon Inductors

Depending on the inductance required, inductors can be realised in a microstrip circuit as straight narrow tracks (ribbon inductors), as single loop inductors, or as multi-turn spiral inductors. A microstrip ribbon inductor and its equivalent circuit are shown in Figure 6.9. For

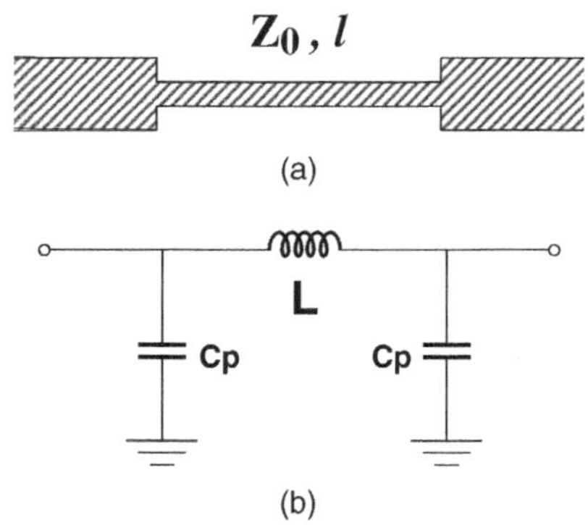

Figure 6.9 Microstrip ribbon inductor. (a) Layout. (b) Equivalent circuit.

short lengths ($<\lambda g/4$), the inductance and shunt end-capacitances can be calculated from the following well-known equations [7]:

$$L = \frac{Z_0}{2\pi f}\sin\left(\frac{2\pi l}{\lambda_g}\right) \tag{6.20}$$

and

$$C = \frac{1}{2\pi f Z_0}\tan\left(\frac{\pi l}{\lambda_g}\right) \tag{6.21}$$

A narrow track with high Z_0 is needed to achieve high inductance with low parasitic capacitance. However, in practice the choice of track width is determined by fabrication limits, by the DC current carrying capacity, and by the high resistance of very narrow tracks.

6.6.1.2 Spiral inductors

Spiral inductors are essential for values above approximately 1 nH. The mutual inductance between turns yields a significant increase in the spiral inductor's overall inductance and higher Q. A bond-wire or a second metal layer is required to connect to the centre.

6.6.1.3 Interdigital capacitors

These consist of a number of interleaved microstrip fingers coupled together, as shown in Figure 6.10. The maximum value of an interdigital capacitor is limited by its physical size, and its maximum usable operating frequency is limited by the distributed nature of the fingers. They are ideal as tuning, coupling and matching elements, where small capacitor values are required and precise values are necessary.

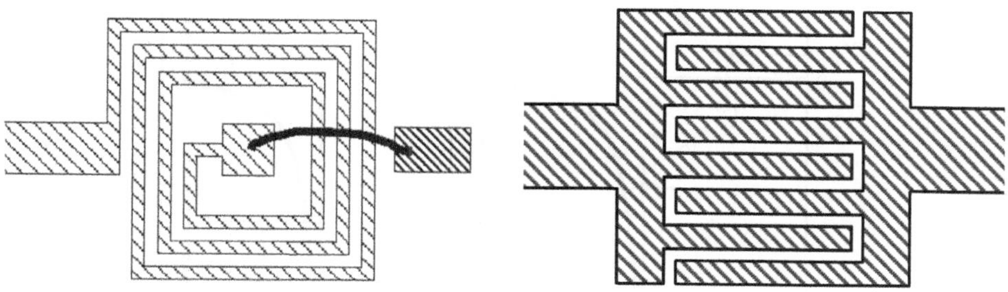

Figure 6.10 Microstrip spiral inductor and interdigital capacitor.

6.6.1.4 Resistors

Resistors use a deposited resistive layer, fabricated either with thin-film techniques (sputtering) or thick-film printing. The most commonly used materials are tantalum nitride, cermet and nickel chrome. Since the layer or film thickness is fixed, resistivity is usually given in terms of an ohms-per-square figure. Hence, the value of the resistor is chosen by selecting a suitable aspect ratio.

6.6.2 Distributed Elements

6.6.2.1 Parallel Stub

Since microstrip comprises a signal conductor placed over a solid ground plane, series stubs are impractical. Open- and short-circuited parallel stubs are often used in microstrip circuits, including matching networks, bias circuits and filters. As parallel elements, it is useful to work with admittance and the input admittance of a short-circuited parallel stub is given by:

$$Y_{IN} = Y_0 \frac{0 + jY_0 tan\beta d}{Y_0 + j0} = jY_0 \tan\beta d \qquad (6.22)$$

This is plotted in Figure 6.11 over the range 0 to 360 degrees of electrical length (βd). Any value of positive or negative susceptance can be obtained, so by choosing a suitable length any required equivalent L or C value can be obtained in principle. Up to a quarter-wavelength the admittance is capacitive, whilst between a quarter and half-wavelength it is inductive, although it is more likely that a short-circuited stub would be used if inductance was required. This is partly because it is physically shorter, and also because the cyclical nature of the frequency response, which is very evident in Figure 6.11, can create problems such as instability in active devices and longer lines make it worse.

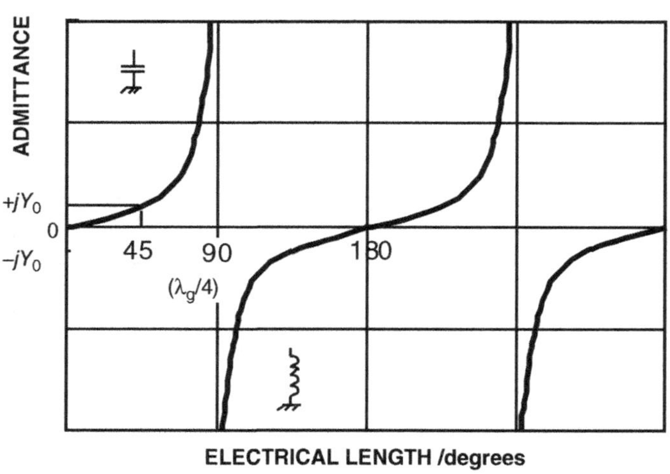

Figure 6.11 Input admittance of an open-circuited parallel stub versus electrical length.

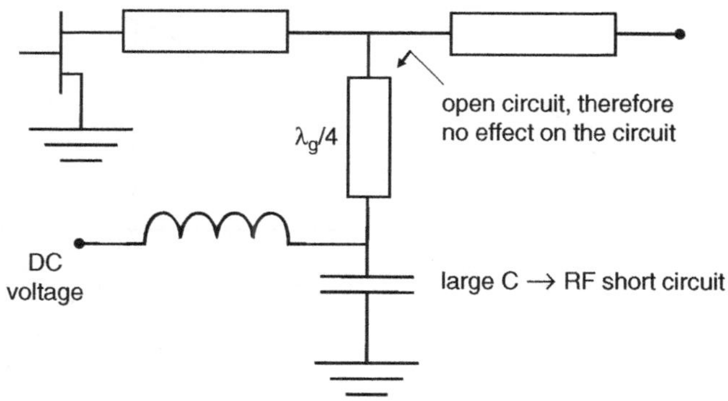

Figure 6.12 Field effect transistor (FET) biasing using a short-circuited parallel stub.

Short-circuited microstrip stubs are often used as a means of applying DC bias to an active device because of the problem in finding bias choke inductors with a sufficiently high self-resonant frequency (SRF). As shown in Figure 6.12, the short-circuit end is grounded with a decoupling capacitor so that DC can be applied. If the bias stub is a quarter-wavelength long, the RF short of the decoupling capacitor is transformed to an open circuit at the transistor end. However, this is a very frequency-dependent element. The decoupling at the end of these stubs is very critical, and care must be taken to avoid oscillations. Since the SRF of a high-value capacitor is low, in practice a small capacitor (e.g., a 10 pF single-layer capacitor) would be placed at the end of the stub, backed up by larger values further away, with additional inductors and resistors to aid the DC/RF separation and improve stability by removing the chance for feedback to occur via the bias circuitry and power supply rails.

6.6.2.2 Radial and Butterfly Stub

Continuing the theme of using an open-circuited stub as a shunt capacitance, it can be observed from Equation (6.22) that a low-impedance line (large Y_0) can realise a larger shunt capacitance for a given length. On the other hand, the wider stub is less convenient to connect in to the circuit, since a large discontinuity is introduced. Hence, a radial stub is a common component, where the line is tapered in width: narrow at the end connected into the circuit, but very wide at the open-circuit end. As illustrated in Figure 6.13, a butterfly stub is simply the same structure connected on both sides of a signal line in order to achieve double the capacitance or improve symmetry. Again, noting the cyclical nature of the frequency response, it can be very beneficial to keep the physical length short, and the radial and butterfly stubs can have relatively broadband operation.

6.6.2.3 Quarter-wave transformer

This well-known technique can match a load resistance, R, to Z_0 by simply using a transmission line one quarter-wavelength long (90° or $\pi/2$). Using the general formula for impedance

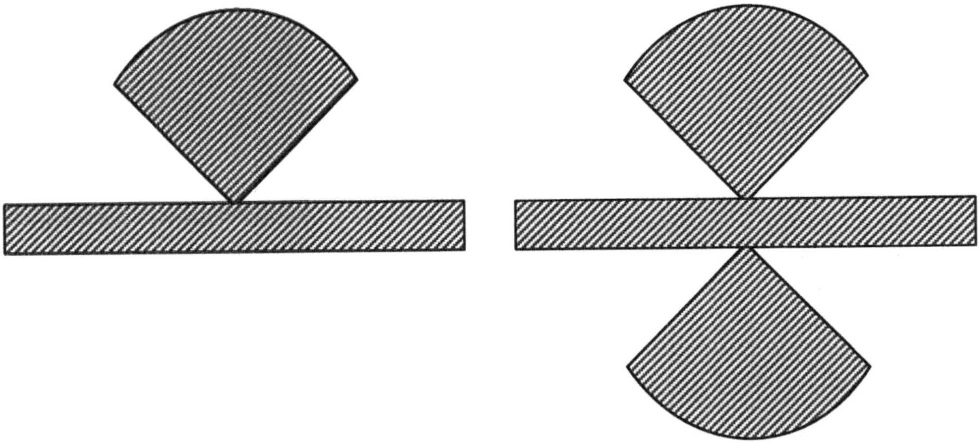

Figure 6.13 Radial and butterfly stubs.

transformation along a line, with the length set to $\pi/2$ gives:

$$Z_{in} = Z_T \frac{Z_L + jZ_T \tan\frac{\pi}{2}}{Z_T + jZ_L \tan\frac{\pi}{2}} = \frac{Z_T^2}{Z_L} \tag{6.23}$$

For a resistive load, $Z_L = R$, then Z_{in} is equal to Z_0 if:

$$Z_T = \sqrt{Z_0 \times R} \tag{6.24}$$

Since a device's impedance is rarely purely real, first a reactive element must be used to resonate out the imaginary part. This can be achieved with either a series or shunt inductor/capacitor or with a transmission line stub. Since Z_{in} is proportional to $1/Z_L$, the quarter-wave transformer is often referred to as an impedance inverter (this will be examined in detail in Chapter 7, on filters). Whilst they are less well-known in microwave circles, there are many other important forms of transmission line transformer, including the Guanella balun and Ruthroff transformer [8–10] that have a discrete range of impedance transformation ratios but are very broadband. These have received renewed attention for MMIC and RFIC applications, where multilayer techniques can be used to realise the interconnected transmission lines [11].

6.6.3 Wilkinson Splitter/Combiner

It is not possible to achieve power division in RF and microwave circuits just by using a T-junction because it does not maintain 50 Ω matching. Indeed, it can be shown that it is theoretically impossible for a lossless three-port circuit to be matched at all ports. The Wilkinson power splitter is a convenient solution. It consists of two quarter-wave matching sections, with 70.7 Ω characteristic impedance (for a 50 Ω system), and a 100 Ω isolation resistor connected across the outputs, as shown in Figure 6.14(a). Note that the Wilkinson splitter can also be used for power combining; if two equal-phase, equal-amplitude signals are

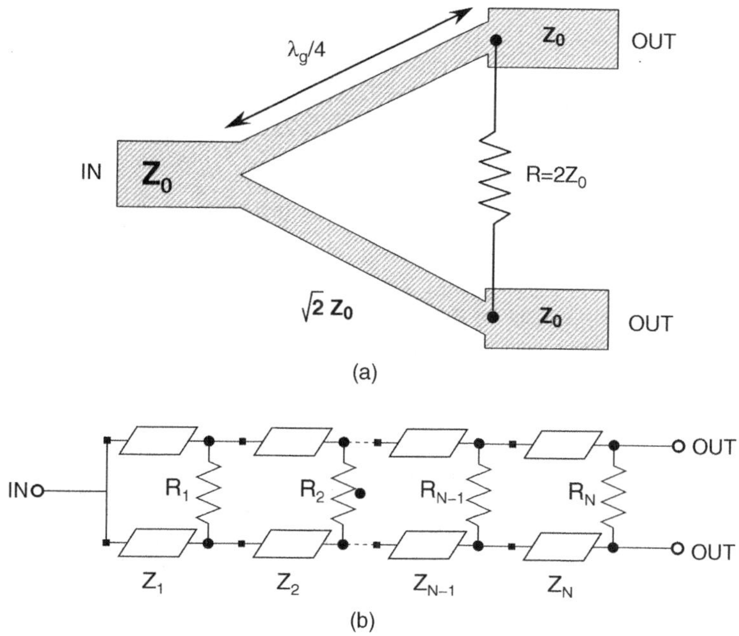

Figure 6.14 (a) Microstrip layout of the Wilkinson power splitter. (b) Multisection broadband circuit schematic.

input to the two arms, then there is no voltage across the isolation resistor and no loss in the combining process (except, of course, the normal transmission line losses). If the two inputs are not identical, there is a combining loss. For broadband applications a multisection design can be used [12], as illustrated in Figure 6.14(b).

6.6.4 Parallel-Coupled Lines

There are many situations that require a directional coupler: for example, to separate waves according to their direction of travel for measuring reflection coefficient or to sample the output signal of a power amplifier for power monitoring or linearisation purposes. A directional coupler can be realised in transmission line form by placing two signal conductors in close proximity so that the signal couples from one to the other. In microstrip form, this is shown in Figure 6.15 in cross-section and plan view.

If the directional coupler is chosen to have a length of one quarter-wavelength, it can be shown that maximum power is transferred in the reverse direction to the coupled arm, whilst very little is coupled in the forward direction to the isolated port. The remaining power travels straight out of the direct port. The coupler is often called a backward-wave coupler because of its directional property. A remarkable feature of the coupler is that the phase difference between the signals from the coupled and direct ports is exactly 90° over a huge frequency range – although the amplitudes vary with frequency.

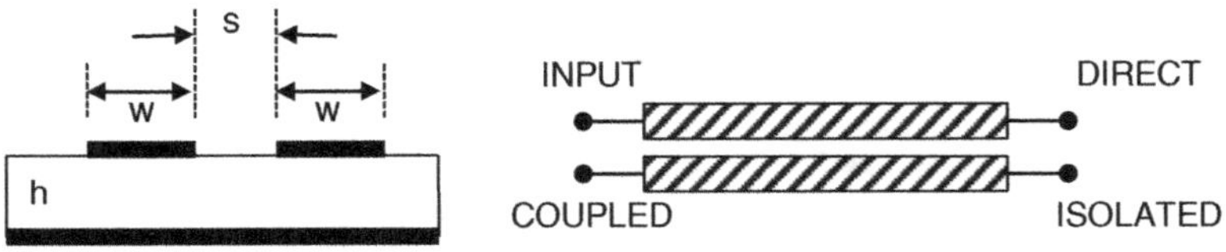

Figure 6.15 Symmetrical parallel-coupled microstrip lines, indicating key design parameters and the port definitions.

The backward-wave coupler can be explained using standing-wave principles. It involves the superposition of two modes of propagation in the coupled line structure, as shown in Figure 6.16. In the even mode, the two conductors are at equal potentials, and there are no electric field lines from one to the other. In the odd mode, the conductor potentials have opposite polarity and electric field lines run direct from one to the other. In microstrip, the different field patterns of the two modes mean that different proportions are in the air and dielectric. So, the two modes have slightly different velocities as well as different characteristic impedances (Z_{0e} and Z_{0o}). The differing phase velocities have a big effect on the isolation that is achieved: to achieve high isolation (e.g., 40 dB) a pure TEM transmission line medium is best. In coaxial line and stripline (as discussed later) the dielectric is homogeneous (and the fields pure TEM) and the two modes have identical phase velocities.

For coupling factors up to 10 dB, approximately, the values of Z_{0e} and Z_{0o} must satisfy the following conditions:

$$C' = 20\log\left|\frac{Z_{0e} - Z_{0o}}{Z_{0e} + Z_{0o}}\right| \tag{6.25}$$

where C is the coupling factor in dB, and:

$$Z_0^2 \approx Z_{0e} Z_{0o} \tag{6.26}$$

This yields:

$$Z_{0e} \approx Z_0\sqrt{\frac{1 + 10^{C'/20}}{1 - 10^{C'/20}}} \text{ and } Z_{0o} \approx Z_0\sqrt{\frac{1 - 10^{C'/20}}{1 + 10^{C'/20}}} \tag{6.27}$$

Design curves for Z_{0e} and Z_{0o} are available, or EDA software can be used. The parallel-coupled line structure also finds application in filters; then Z_{0e} and Z_{0o} are the parameters

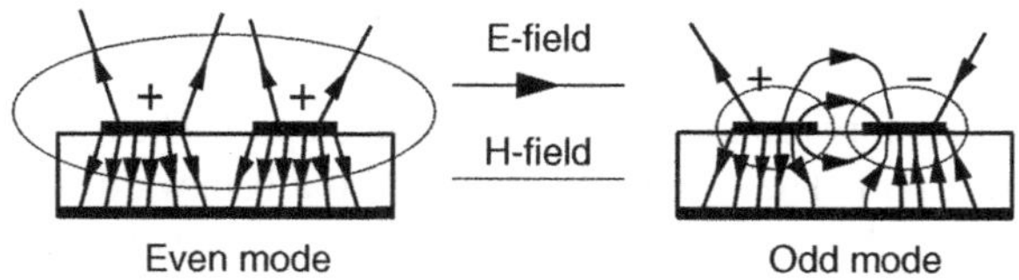

Figure 6.16 Approximate field patterns for even and odd mode in parallel-coupled microstrip.

which are synthesised from filter theory, in a similar manner to the *L*s and *C*s in the lumped-element band-pass filters. Other applications include DC blocks, phase shifters, and matching networks.

6.6.5 Lange Coupler

The parallel-coupled microstrip structure cannot be used to realise an equal power split (a 3 dB coupler); even with a conductor spacing close to zero, the coupling is limited to approximately 6 dB. In order to overcome this limit, the two lines must overlap. This can be done easily with a multilayer structure, but with ordinary microstrip it has to be done by interleaving the two conductors. Each arm consists of a number of 'fingers' which alternate, and this was first proposed by Julius Lange [13]. Depending on the achievable conductor dimensions, and on the substrate height and ε_r, either four-, six- or eight-finger Lange couplers are used. To keep the fingers of a particular arm at the same potential, straps must be connected across using bond-wires, air-bridges, or underpasses. In the Lange coupler shown in Figure 6.17, the structure is flipped at the centre, so that the Coupled and Direct ports are on the same side for convenience. Multisection couplers can be designed in order to increase the bandwidth [14].

6.6.6 Branch-Line Coupler

Coupled-line and Lange couplers can be difficult to realise because they require narrow tracks and/or gaps. So, for many applications the branch-line coupler is used when a 90°, 3 dB split is required. In fact, this can also be analysed in terms of odd- and even-modes, but more in circuit terms than as fields. The standard layout is shown in Figure 6.18; the ports can be interchanged to remove the step in width. A single-section branch-line coupler is rather narrowband, but multisection versions can be designed for more bandwidth [12].

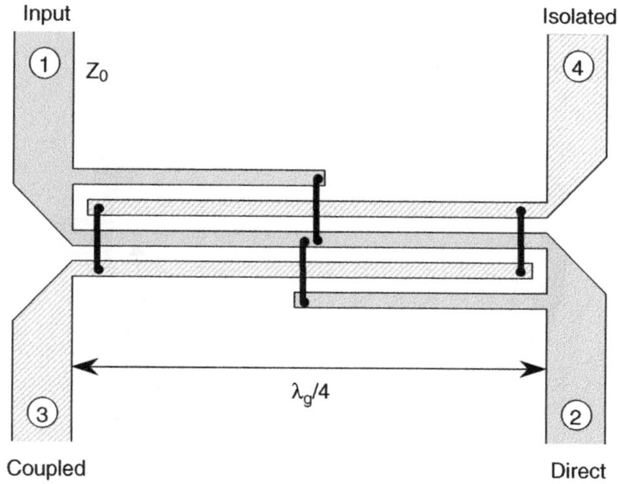

Figure 6.17 Layout of a four-finger Lange coupler.

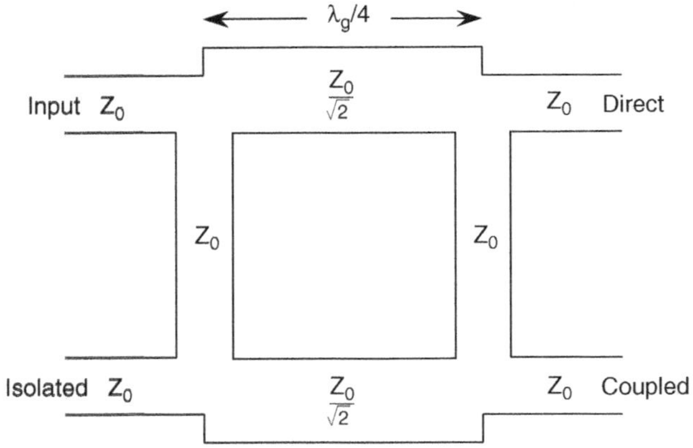

Figure 6.18 Microstrip branch-line coupler.

6.6.7 *Rat-Race Hybrid*

When a 180°, 3 dB split is required, which is often the case in mixers and modulators, the rat-race coupler can be used. As illustrated in Figure 6.19, the coupler is traditionally circular, and by creating two paths between any input or output port pair, constructive or destructive interference occurs so that some ports are isolated, whilst signals appear as in-phase or

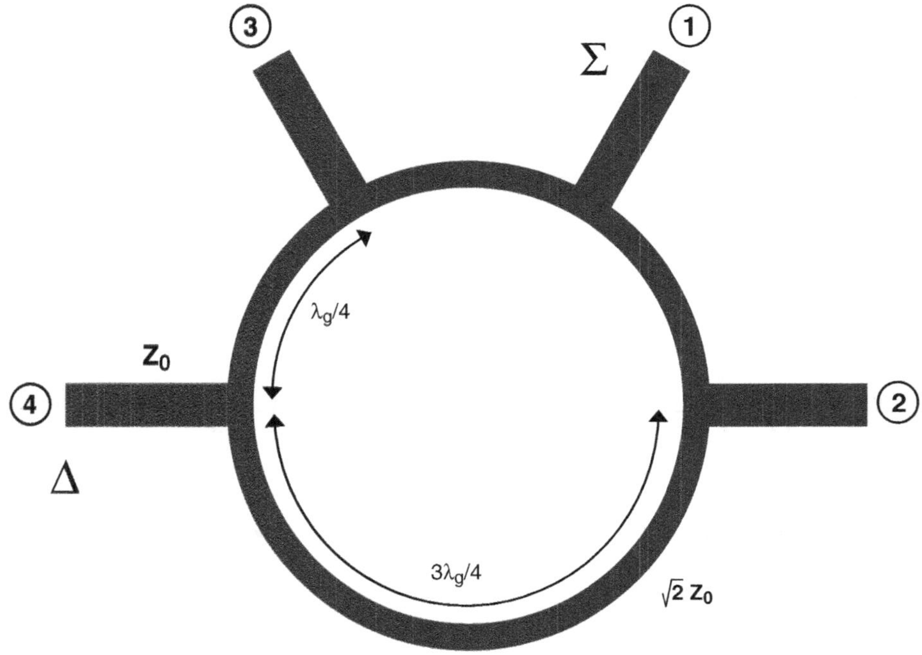

Figure 6.19 Microstrip rat-race coupler.

anti-phase pairs at others. The S-parameter matrix of the ideal rat-race coupler, with port numbers as labelled in Figure 6.18 is:

$$S = -\frac{j}{\sqrt{2}}\begin{pmatrix} 0 & 1 & 1 & 0 \\ 1 & 0 & 0 & -1 \\ 1 & 0 & 0 & 1 \\ 0 & -1 & 1 & 0 \end{pmatrix} \tag{6.28}$$

This shows that the rat-race coupler is matched at all ports (the diagonal elements are all zero). A signal input to port 1 will split 3 dB, 0° to ports 2 and 3, while a signal input to port 4 will split 3 dB, 180° to ports 2 and 3. One simple application using this behaviour is the single-balanced mixer where the LO and RF are input to the Δ and Σ ports and the individual mixers are connected at ports 2 and 3. By reciprocity, the rat-race can also be used as a signal combiner: if a pair of signals input at ports 2 and 3 then the sum appears at port 1 and the difference at port 4 (because of the 180° phase shift). Ports 1 and 4 are often referred to as the Σ and Δ ports, respectively, for this reason.

6.6.8 *Planar Marchand Balun*

In hybrid mixers, the main forms of balun are wire-wound ones on ferrite cores, or suspended microstrip structures. The wire-wound type is widely used for drop-in modular mixers, most notably from Mini-Circuits™, who founded a very successful business on double-balanced mixers (DBMs), with the original 'SBL-1' being perhaps of historical importance. With ferrite cores or suspended microstrip, there is some intricate manual assembly required and various planar balun configurations have been proposed, most notably the planar version of the Marchand balun which has wideband performance. It consists of two coupled sections, connected as shown in Figure 6.20, which may be realised using microstrip coupled lines, Lange couplers, multilayer coupled structures or spiral coils. The coupling factor of the couplers can be adjusted to realise an impedance-transforming balun [15]. Many commercial mixer products use a form of this planar Marchand balun, fabricated in LTCC or other multilayer technology. The balun using spiral transformers is a compact form of the planar Marchand balun. The frequencies that this technique can be used at are limited by the size and DC resistance at low frequency, and the inter-spiral capacitances at higher frequencies.

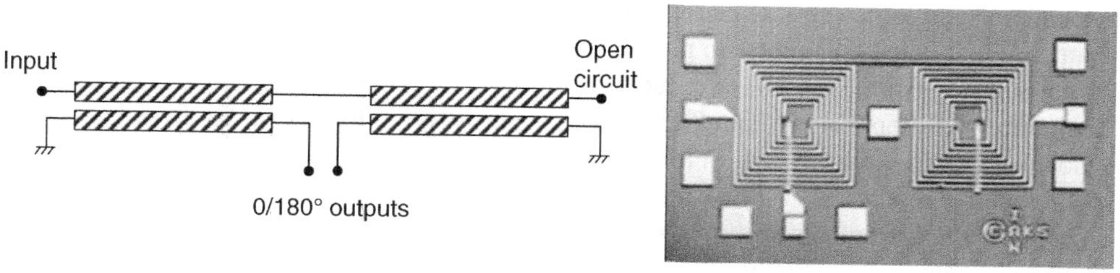

Figure 6.20 Planar Marchand balun and GaAs spiral coupler realisation.

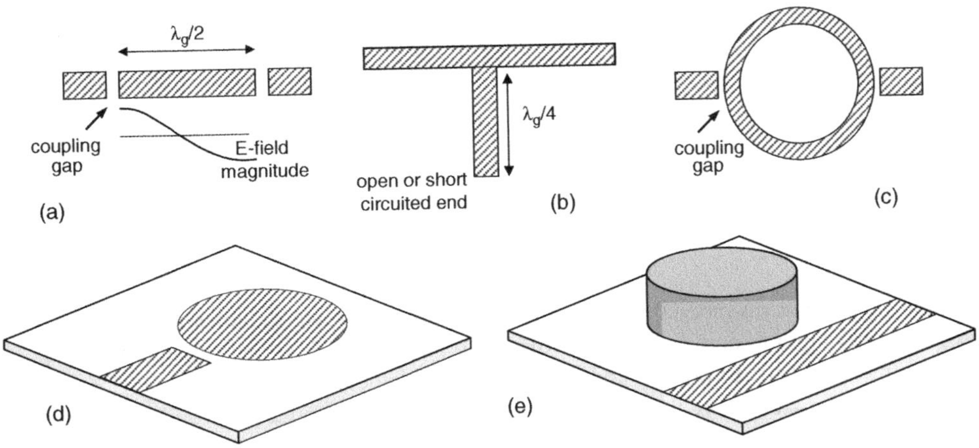

Figure 6.21 Common forms of microstrip resonator. (a) Half-wavelength resonator. (b) Quarter-wavelength resonator. (c) Ring resonator. (d) Disc resonator. (e) Dielectric puck.

6.6.9 Microstrip Resonators

Transmission line filters are described in detail in Chapter 7, and the main types of microstrip resonator will be described here first. The half-wavelength resonator with open ends is shown in Figure 6.21(a). This is highly suited to microstrip because it does not need any connection to ground, which would require a via-hole, but the open ends have a maximum electric field that leads to radiation loss. The quarter-wavelength resonator, shown in (b) has one end open-circuit and one short-circuited. The ring resonator shown in (c) is one wavelength in circumference and has quite low radiation loss as there is no open end. The disc resonator in (d) can operate in a range of modes [16] and is representative of a range of patch-style microstrip resonators. Microstrip resonators have rather low unloaded Q-factors, the maximum typically achieved is on the order of a few hundreds, and this depends very much on the substrate parameters and the choice of structure. Absolutely thousands of research papers have been published on novel types of microstrip resonator. Popular techniques that have been reported include:

- Different shapes that give higher Q, change the harmonic response, or give multiple resonances.
- Notches, slots, stubs and other perturbations to the geometry to obtain more than one resonance from a single resonator.
- Meandered and folded structures for miniaturisation.
- Lumped-distributed techniques to reduce size.
- Metamaterial-inspired circuits to reduce size or improve performance.
- Techniques to suppress harmonic resonances (microstrip is dispersive, so these harmonic resonances can be troublesome in filters).
- Techniques to reduce radiation loss.
- Micromachined structures to reduce dielectric loss.
- Defected ground plane structures.
- Slotted ground plane structures based on E/H field duality.

Fundamentally, the Q-factor of microstrip resonators is much lower (a few hundred maximum) than can be achieved with coaxial lines and waveguides. Much excitement was generated by the prospects for high-temperature superconductors for high-Q microstrip resonators, but room temperature operation was not achieved and the need for cooling is a major drawback. Whilst there are planar transmission line structures with better shielding and lower loss than microstrip (some of which are described later), generally these are far less suitable for the integration of active components. Hence, a very popular solution is to mount dielectric resonator pucks onto the surface of the substrate and couple these from a microstrip line magnetically, as shown in Figure 6.21(e). Dielectric resonators are the dual of a waveguide cavity, using total internal reflection at a high ε_r/air interface, and high ε_r (e.g., 20 to 90) ceramic materials are widely available, giving miniature resonators. This approach can yield unloaded Q-factors that are sufficiently high (1000s) to enable low phase-noise oscillators to be realised in microstrip form. The mounting, coupling and production line tuning become critical, and the temperature coefficient of the ceramic material must be carefully considered in the design.

6.7 Uniplanar Transmission Lines

The major attraction of uniplanar transmission lines is that through-substrate via-holes are not required for grounding purposes. The major types are coplanar waveguide (CPW), coplanar strips (CPS) and slotline, although there are many derivatives of these basic structures.

6.7.1 Coplanar Waveguide

This transmission line type was first proposed by C.P. Wen in 1969 [17], but attracted renewed interest during the 1990s for MMIC design as it does not need via-hole grounds. Figure 6.22 shows the structure of a CPW transmission line. CPW consists of a signal conductor placed between two ground planes. The dominant mode of CPW is quasi-TEM, and there is no low frequency cut-off. The principal advantages of CPW are:

- No via-holes needed for grounding.
- Low dispersion.

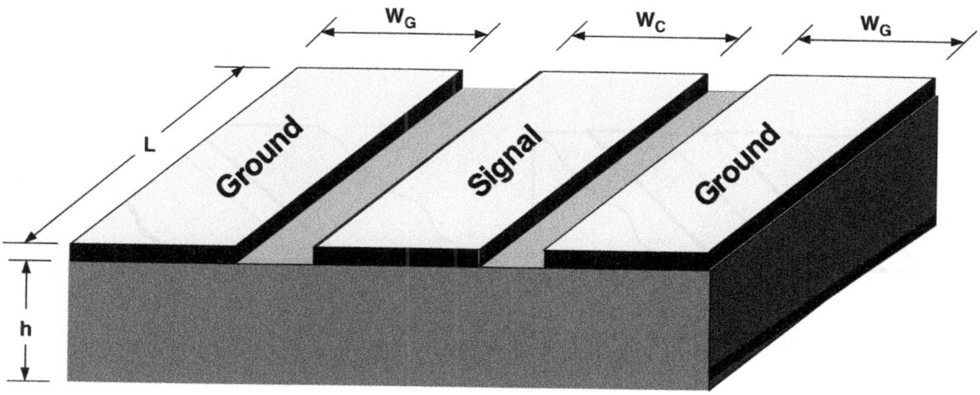

Figure 6.22 Coplanar waveguide structure.

- Tapers with constant characteristic impedance can be realised by varying the track width and gap combination.
- The ground planes can provide some shielding between lines.
- Circuit components may exhibit less parasitic capacitance as there is no ground plane underneath.

Many EDA software packages now have CPW elements in their libraries. These models include bends, steps, T-junctions and many other common elements. One of the major problems with CPW is that the mode of propagation can easily degenerate from quasi-TEM into a balanced coupled-slotline mode. This happens very often at discontinuities, but can be avoided by incorporating grounding straps between the ground planes, using either bond-wires, air-bridges or a second metal layer. The characteristic impedance of a CPW line depends on several parameters, ground width (W_G), signal width (W_s), the gap (G), substrate height (h) and dielectric constant. CPW is sometimes used with a back conductor (conductor-backed CPW) or with narrow ground plane strips (finite ground CPW, FGCPW).

6.7.2 Coplanar Strips

The coplanar strip (CPS) transmission medium consists of a pair of closely coupled parallel strips, as shown in Figure 6.23(a). It is essentially a planar equivalent to the twisted pair or twin-lead line. As a balanced transmission line, it is ideally suited to balanced mixers and push–pull amplifiers. The lack of design information restricted its use for many years, but it now finds widespread application in RFICs, where balanced signals are common and where its relatively high characteristic impedance may be advantageous.

6.7.3 Slotline

Slotline consists of a pair of ground planes with a narrow slot between them, as shown in Figure 6.23(b), with the signal propagating in a TE (transverse electric) mode. This means that slotline is not an ideal general purpose transmission line medium, but this mode makes it very useful for circuits such as balanced mixers and amplifiers, where push–pull operation is required. Also, since CPW has the quasi-TEM mode and slotline has the TE mode, a number of useful hybrid junctions and transition circuits can be produced. The most popular of these is

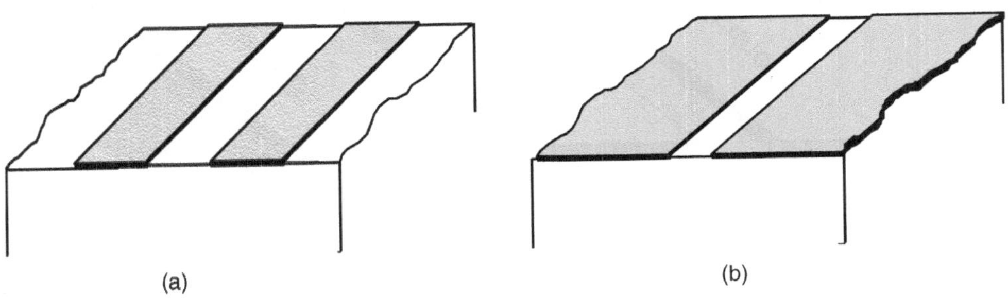

Figure 6.23 (a) Coplanar strip transmission line (CPS). (b) Slotline.

the CPW-to-slotline transition. This type of transition has demonstrated balun operation over more than two octaves of bandwidth, and has been used in miniaturised uniplanar mixers and amplifiers [18].

6.8 Other Transmission Line Types

Microstrip is the most popular transmission line medium for the design of microwave and millimetre-wave circuits, but its performance is found wanting in some applications. It is quite lossy, which may restrict its use in filters and power-combining networks. The quasi-TEM behaviour leads to dispersion, which degrades the performance in some applications, including directional couplers, where the isolation is reduced by the difference in even- and odd-mode phase velocities. In filters, this dispersion creates spurious harmonic behaviour that degrades the stop-band rejection. Hence, it is often necessary to use other transmission line media, which are still fabricated using photolithography but have a structure that leads to better performance.

6.8.1 Stripline

Stripline, sometimes called Tri-plate, consists of a signal conductor track sandwiched between two substrates, both of which have a ground plane, as shown in Figure 6.24(a). Since the dielectric is homogeneous, stripline has a pure-TEM mode of propagation. This means that coupled lines have equal even- and odd-mode phase velocities and enables high-directivity couplers to be realised. It also improves the stop-band response of filters, which also benefit from the reduced radiation loss. Of course, it is impractical to attach active devices and passive components, so stripline is most often encountered in high-performance connectorised passive components and subsystems. However, in multilayer technologies such as LTCC it is possible to integrate stripline into a module very easily for those components that need the high performance.

6.8.2 Suspended Substrate Techniques

Figure 6.24(b) shows a suspended stripline, (c) shows a suspended microstrip line, and (d) shows an inverted microstrip line. A key aim of this approach is to maintain the photolithographic fabrication and related advantages of microstrip, but to reduce loss by having air as the primary dielectric. As well as reducing dielectric loss, the conductor strip will be wider and, therefore, conductor loss is also reduced. Furthermore, when chip components are attached there is less parasitic capacitance to ground. Hence, these transmission line types are

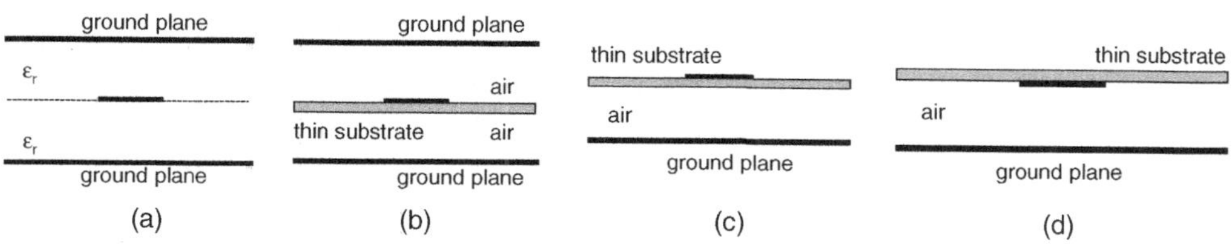

Figure 6.24 (a) Stripline. (b) Suspended stripline. (c) Suspended microstrip. (d) Inverted microstrip.

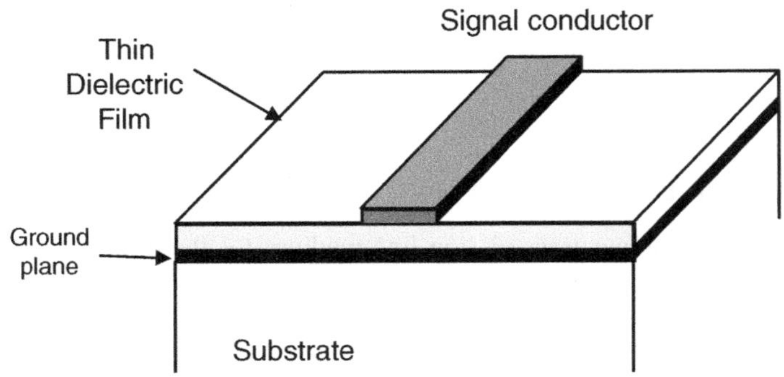

Figure 6.25 Thin-film microstrip transmission line medium.

often used in specialist applications such as ultrawideband bias-Tees and DC blocks, where a large value of series capacitor is required to have very low parasitic shunt capacitance in order to operate at very high frequencies.

6.8.3 Thin-Film Microstrip

Multilayer techniques are now widely used on integrated circuits, and a simple example is the use of thin-film microstrip (TFMS) where miniature lines can be fabricated by using a thin added dielectric on top of the main substrate, as illustrated in Figure 6.25. This is valuable in silicon ICs where the substrate loss is high, but if low ε_r materials are used the length of the lines will be longer, and the width is narrow because of the low h, and both contribute to high loss.

6.8.4 Lumped and Lumped-Distributed Components

Where distributed elements are too long, such as on integrated circuits, many familiar transmission line components can be miniaturised by replacing individual transmission line sections with lumped inductor-capacitor networks or a shorter, high-impedance line loaded with a lumped capacitance at each end [19, 20]. For the purely lumped case, a quarter-wave line of impedance Z_0 can be replaced with a C-L-C pi-network if the following values are used:

$$L = {}^{Z_o}\!/_{2\pi f} \tag{6.29}$$

$$C = {}^{1}\!/_{2\pi f Z_o} \tag{6.30}$$

For the lumped-distributed case, a quarter-wave line of impedance Z_0 can be replaced with a short section of line with a higher impedance (Z_{LD}) that has electrical length θ, loaded with a capacitance C_{LD} at each end, providing:

$$Z_{LD} = {}^{Z_o}\!/_{\sin\theta} \tag{6.31}$$

$$C_{LD} = {}^{\cos\theta}\!/_{2\pi f Z_o} \tag{6.32}$$

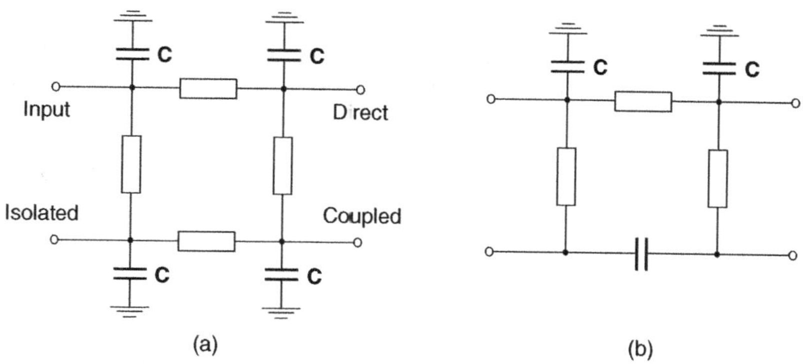

Figure 6.26 Schematic diagrams of (a) a lumped-distributed branch-line coupler, and (b) a rat race coupler.

Figure 6.26(a) shows the schematic diagram of a lumped-distributed branch-line coupler, while Figure 6.26(b) shows a lumped-distributed rat-race coupler using $\lambda_g/8$ line sections, which leads to significant miniaturisation.

6.8.5 Substrate Integrated Waveguide

Substrate integrated waveguide (SIW) is a rectangular waveguide-like structure, embedded in a substrate for a low-cost and high-volume production [21]. Figure 6.27(a) illustrates an SIW implemented with PCB technology, while the detail of the structure is shown in Figure 6.27(b). The top and bottom conductor layers form the upper and lower broadwalls of the waveguide. Two rows of via-holes with specific uniform spacing form the sidewalls. It has been shown that the two rows of metal posts give performance similar to that of a waveguide with solid metal walls. The field leaking from the gaps will be evanescent, provided that suitable design guidelines are adhered to. The important design parameters are the hole diameter (d), via-hole

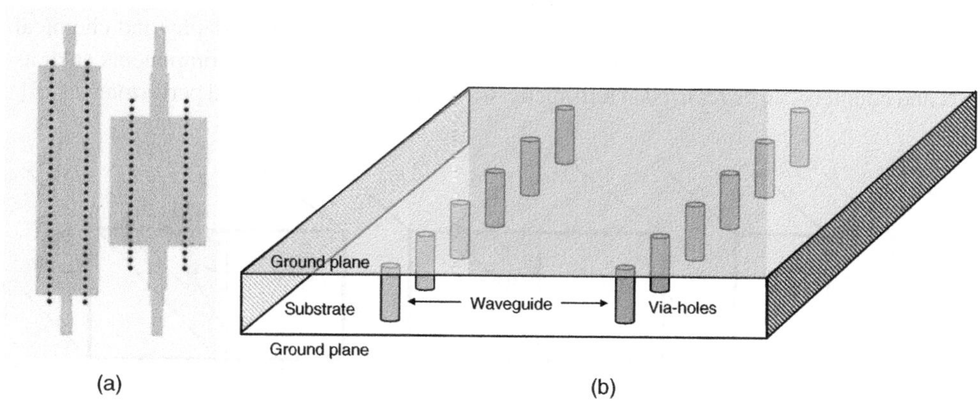

Figure 6.27 Substrate integrated waveguide. (a) SIWs in PCB technology, with transitions to microstrip. (b) Detail of the waveguide structure.

separation (s) and the separation of the rows (W). The via-hole separation is measured from the centres of the via-holes (i.e., the pitch). To obtain a low radiation loss, Ke Wu, the key pioneer of SIW, proposed the following design rules:

$$d < \frac{\lambda_g}{5} \tag{6.33}$$

and

$$b \leq 2d \tag{6.34}$$

The major advantage of SIW is that low-loss waveguides, for example for power combining or for an antenna feed network, can easily be integrated with other transmission lines such as microstrip and CPW, as well as with active devices and ICs. Many circuits have been designed with SIW technology, and there is still a huge opportunity for SIW technology to have a major impact in microwave and millimetre-wave wireless communication systems. For reduced loss, air-filled or hollow substrate integrated waveguide (HSIW) has been demonstrated in both soft-board and LTCC technologies [22, 23].

6.8.6 Ridge Waveguide and Finline

A ridge waveguide consists of a hollow rectangular waveguide with one or more ridges in the broadwalls [24]. Figure 6.28(a) shows the cross-section of a ridge waveguide. The advantage of ridge waveguides is that the TE_{10} cut-off frequency can be significantly lowered, while the cut-off frequency of the TE_{20} and TE_{30} modes is increased. The result is a waveguide that has a much wider useable bandwidth than the standard rectangular waveguide, albeit with some increase in loss and reduction in maximum power handling. The reason that ridge waveguide attracts special attention here is that it is a technique which provides a valuable link between the domain of waveguides, with the advantages of low loss and high power handling, and the world of planar circuits where photolithography is used to fabricate intricate low-cost circuits. Figure 6.28(b) shows the cross-section of an E-plane circuit using a metal insert in a split-block waveguide. With this method, a metal insert is fabricated by photolithography and chemical etching, or by laser machining, which means that a very wide range of components such as filters and couplers can be realised at millimetre-wave frequencies with good performance [25].

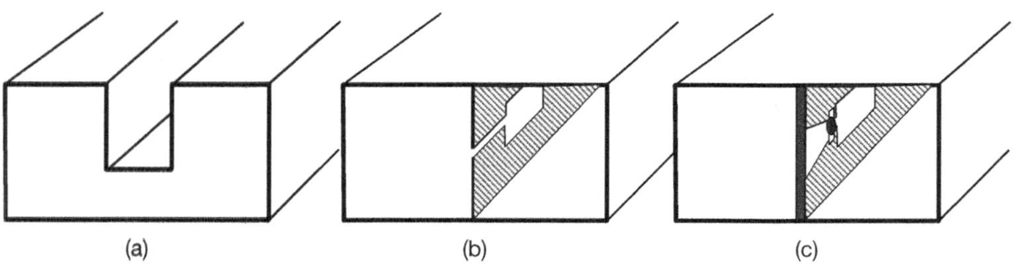

Figure 6.28 Cross-section of (a) ridged waveguide, (b) an E-plane circuit using an all-metal insert and (b) finline.

The all-metal insert is a single piece of sheet metal, which causes some geometry limitations. Figure 6.28(c) shows the structure of finline, in which a substrate with a conductor pattern is inserted between the blocks, and this overcomes that restriction [26].

Waveguide-to-coax and waveguide-to-microstrip transitions often employ a single-ridge waveguide in which the ridge height is increased in steps and the ridge brought down to touch the coax centre pin or microstrip signal conductor. Alternatively, transitions from microstrip and other planar media to finline can be used as a way of integrating transmitter/receiver circuits with high-performance waveguide filters and antennas or antenna feeds. Furthermore, ridge waveguides can be integrated into multilayer modules and ridge waveguide bandpass filters have been demonstrated in LTCC Technology [27].

6.8.7 Dielectric Waveguides

Conductor loss is a major factor that limits the performance of microwave and millimetre-wave circuits and components. The extensive use of gold and silver plating and the need for shiny polished conductors with low surface roughness are evidence of the importance of conductor loss at high frequency. This is because of the skin effect, in which the current decays exponentially with distance into the conductor, with a skin depth given by:

$$\delta = \sqrt{\frac{1}{\pi f \sigma \mu_o \mu_r}} \tag{6.35}$$

where σ = metal conductivity, f = frequency, μ_o = permeability of free space, and μ_r = relative magnetic permeability of the conductor.

The skin depth is the distance that it takes for the current density to drop to $1/e$ of its value at the surface, and in practice the current can be assumed to be confined to a few skin depths of the conductor. As the signal frequency gets higher and higher, the surface roughness becomes a limiting factor and mirror-finish is required.

An alternative approach is to employ dielectric waveguides of various forms, in which the signal is confined by reflection at the interface between two dielectric media. For example, the dielectric rod or ribbon waveguide is a solid dielectric structure with air outside. Dielectric guides are well-known from the optical domain, and most optical fibres for telecommunications are cylindrical guides that use a narrow silica glass core inside a lower-refractive index cladding. For millimetre-wave and terahertz applications, there is a wide range of dielectric guide structures that can be used [28], some of which are illustrated in Figure 6.29. Generally, these have a lower loss than metal waveguides, provided that very low-loss dielectrics are

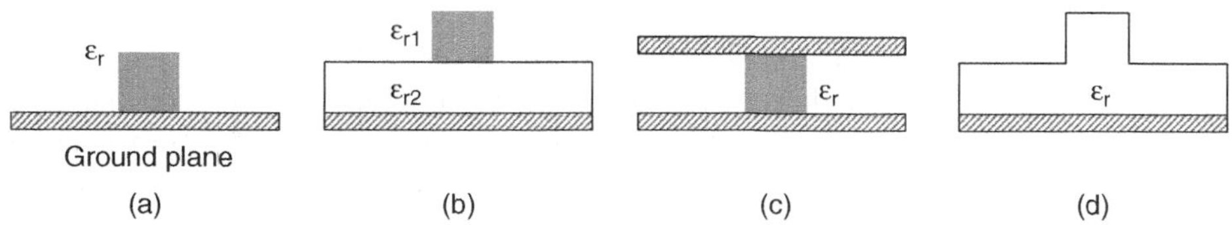

Figure 6.29 Cross-section of (a) dielectric image guide, (b) dielectric insular image guide, (c) nonradiative dielectric guide (NRD), and (d) rib image guide.

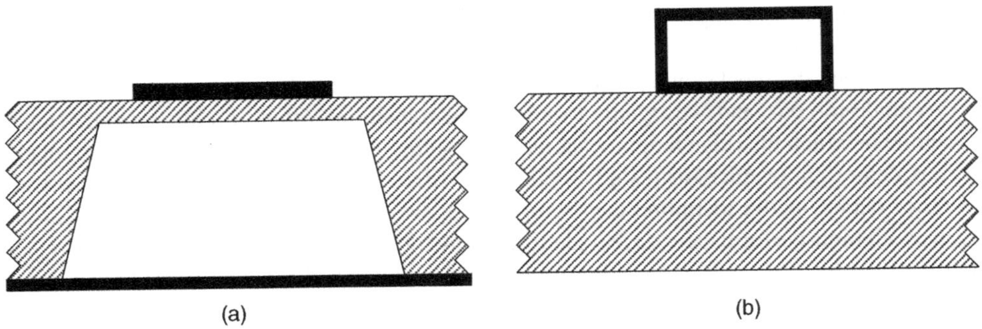

Figure 6.30 (a) Micromachined suspended microstrip. (b) Micromachined rectangular waveguide.

employed, although the fields can be less confined and it is sometimes difficult to mount or package them without disturbing the fields. Also, it is not easy to integrate active and passive components.

6.8.8 Micromachined Lines

Micromachined silicon components, using selective crystallographic etching techniques, have been widely developed for high-volume commercial markets such as air-bag sensors, displays, disk drives and print-heads. These miniature components are classified as microelectromechanical systems (MEMS) or microsystems. For microwave circuits, MEMS technology has the important feature of being able to realise moving parts for switching, tuning and steering as well as structures using air as the main dielectric, leading to low loss. The thick photoresist 'SU-8' has been widely applied to the fabrication of air-based transmission lines. Figure 6.30 shows two types of transmission line that have been demonstrated using micromachining techniques. Further examples are given in Chapter 13.

6.9 Conclusion

This chapter has followed on from the chapter on transmission line theory by describing the wide range of practical transmission line media that are used in microwave and millimetre-wave systems. Whilst microstrip remains the workhorse, it has been shown that there are many other media that can be used and the designer needs to be familiar with these in order to be able to use creative design techniques that meet the sometimes conflicting requirements of cost, performance and size for the various subsystems in a wireless communication system. For example, in a base station filter the specification will be very much more stringent than in a cellular handset because of the higher power level. In a power amplifier, transmission line loss in the combining network is very damaging to power efficiency and it may be necessary to design a microstrip module that interfaces with a waveguide or coaxial power-combining network, with substrate integrated waveguides providing an elegant solution in some circumstances. Lastly, the antenna may have very different requirements and the transceiver module may need to be cleverly integrated with a transition to another transmission line medium that forms the feed network or antenna itself. Antennas are described in Chapter 14, while multichip modules and system-in-package techniques are described in Chapter 16.

References

1. Metcalf, W.S. (1968) Why Not Use Rectangular Coax? *Microwaves*, **April**, 52–56.
2. http://www.nuvotronics.com/.
3. Marcuvitz, N. (1986) *Waveguide Handbook*, New edition. IET Publishers. ISBN-10: 0863410588.
4. Cronin, N.J. (1995) *Microwave and Optical Waveguides*. CRC Press. ISBN-10: 075030216X.
5. Pozar, D. (2011) *Microwave Engineering*, 4th edition. John Wiley & Sons. ISBN-10: 0470631554.
6. EIA RS-261-B (1979) Rectangular Waveguides (WR3 to WR2300); Standard of the Electronic Industries Association of the United States of America.
7. Edwards, T.C. and Steer, M.B. (2000) *Foundations of Interconnect and Microstrip Design*, 3rd edition. John Wiley & Sons. ISBN: 9780471607014.
8. Sevick, J. (2001) *Transmission Line Transformers*. Noble Publishing. ISBN 1-884932-18-5.
9. Ruthroff, C.L. (1959) Some broad-band transformers. *Proc. IRE*, **August**, 1337–1342.
10. Guanella, G. (1944) New Method of Impedance Matching in Radio-Frequency Circuits. The Brown Boveri Review, September, pp. 327–329.
11. Sobrany, R.F. and Robertson, I.D. (2003) Ruthroff Transmission Line Transformers using Multilayer Technology. 33rd European Microwave Conference, October, pp. 559–562. doi: 10.1109/EUMA.2003.341014.
12. Matthaei, G., Young, L. and Jones, E.M.T. (1980) *Microwave Filters, Impedance-matching Networks and Coupling Structures*, New edition. Artech House Publishers. ISBN-10: 0890060991.
13. Lange, J. (1969) Interdigitated Strip-Line Quadrature Hybrid. G-MTT International Microwave Symposium, 5–7 May, pp. 10–13. doi: 10.1109/GMTT.1969.1122649.
14. Uysal, S. and Aghvami, H. (1989) Synthesis, design, and construction of ultra-wide-band nonuniform quadrature directional couplers in inhomogeneous media. *IEEE Transactions on Microwave Theory and Techniques*, **37** (6), 969–976. doi: 10.1109/22.25398.
15. Ang, K.S. and Robertson, I.D. (2001) Analysis and design of impedance-transforming Marchand baluns. *IEEE Transactions on Microwave Theory and Techniques*, **49** (2), 402–406.
16. Watkins, J. (1969) Circular resonant structures in microstrip. *Electronics Letters*, **5** (21), 524–525. doi: 10.1049/el:19690393.
17. Wen, C.P. (1969) Coplanar Waveguide, a Surface Strip Transmission Line Suitable for Nonreciprocal Gyromagnetic Device Applications. G-MTT International Microwave Symposium, 5–7 May, pp. 110–115. doi: 10.1109/GMTT.1969.1122668.
18. Chen, T.-H., Ton, T.-N., Dow, G.S., Nakano, K., Liu, L.C.T., and Berenz, J. (1991) A Q-band monolithic balance resistive HEMT mixer using CPW/slotline balun. *IEEE Journal of Solid-State Circuits*, **26** (10), 1389–1394. doi: 10.1109/4.90092.
19. Vogel, R.W. (1992) Analysis and design of lumped- and lumped-distributed-element directional couplers for MIC and MMIC applications. *IEEE Transactions on Microwave Theory and Techniques*, **40** (2), 253–262. doi: 10.1109/22.120097.
20. Chongcheawchamnan, M., Siripon N., and Robertson, I,D. (2001) Design and performance of improved lumped-distributed Wilkinson divider topology. *Electronics Letters*, **37** (8), 501–503.
21. Deslandes, D. and Wu, Ke (2003) Single-substrate integration technique of planar circuits and waveguide filters. *IEEE Transactions on Microwave Theory and Techniques*, **51** (2), 593–596.
22. Parment, F., Ghiotto, A., Tan-Phu Vuong, Duchamp, J.-M., and Wu, Ke (2015) Air-filled substrate integrated waveguide for low-loss and high power-handling millimeter-wave substrate integrated circuits. *IEEE Transactions on Microwave Theory and Techniques*, **63** (4), 1228–1238. doi: 10.1109/TMTT.2015.2408593.
23. Jin, Lukui, Lee, R.M.A., and Robertson, I.D. (2014) Analysis and design of a novel low-loss hollow substrate integrated waveguide. *IEEE Transactions on Microwave Theory and Techniques*, **62** (8), 1616–1624. doi: 10.1109/TMTT.2014.2328555.
24. Helszajn, J. (2000) *Ridge Waveguides and Passive Microwave Components*. IEE Electromagnetic Waves Series. ISBN: 0852967942.
25. Bhartia, P. and Pramanick, P. (1987) *E-Plane Integrated Circuits*. Artech House Publishers. ISBN: 0890062420.

26. Meier, P.J. (1978) Millimeter integrated circuits suspended in the E-plane of rectangular waveguide. *IEEE Transactions on Microwave Theory and Techniques*, **26** (10), 726–733. doi: 10.1109/TMTT.1978.1129477.
27. Rong, Yu, Zaki, K.A., Hageman, M., Stevens, D., and Gipprich, J. (1999) Low-temperature cofired ceramic (LTCC) ridge waveguide bandpass chip filters. *IEEE Transactions on Microwave Theory and Techniques*, **47** (12), 2317–2324. doi: 10.1109/22.808977.
28. Koul, S.K. (1997) *Millimeter Wave Optical Dielectric Integrated Guides and Circuits*. Wiley Interscience. ISBN: 0471168416.

7

Transmission Line Filters

7.1 Introduction

As discussed in Chapter 2, filters play an absolutely key role in communication systems as the spectrum is crowded and there are many undesired signals that must be rejected. From a practical viewpoint, active filters are noisy, power-hungry and have poor linearity. Whilst they have can have an important role to play in signal processing, it is simply impractical to use an active filter to remove an interfering signal that is stronger than the desired signal(s) as nonlinear effects will create many unwanted intermodulation products. Passive filters are, therefore, preferable but the filter loss is then very important as it degrades the noise performance of the receiver and dissipates valuable transmitter power. Passive filters can be realised either with lumped or transmission line components. While a filter realised with lumped components can be attractive due to its compact size, at high operating frequency the parasitic effects in lumped components become impossible to overcome. The loss from parasitic effects is unavoidable and hence the unloaded quality factor of lumped resonators at high frequency is very low. Normally, the electrical lengths of transmission line resonators for a filter are in the range from $0.125\lambda_g$ to $0.50\lambda_g$. Hence, at sufficiently high frequency, the length of transmission line resonators is small enough to be practical in most applications. One exception is the mobile handset, where the operating frequency is only a few GHz typically but the size of the filters must be extremely small. For this kind of application, miniature filters are realised using piezoelectric materials, in which the wavelength is much smaller as the wave is acoustic and not electromagnetic and therefore travels much more slowly.

Designing transmission line filters requires several significant steps that extend from the lumped design steps described in Chapter 4. This extension requires extensive mathematical treatment in order to synthesise realistic filter structures starting from first principles. Some theories were developed quite some years ago and yet are still the standard techniques in use, while many design new techniques have been proposed as the demand for smaller filters grows. The topics presented in this chapter cannot possibly cover the full range of modern filter design techniques; hence, coverage is focused on those mainstream techniques that are most-often used in microwave and millimetre-wave applications.

Microwave and Millimetre-Wave Design for Wireless Communications, First Edition.
Ian Robertson, Nutapong Somjit and Mitchai Chongcheawchamnan.
© 2016 John Wiley & Sons, Ltd. Published 2016 by John Wiley & Sons, Ltd.

Table 7.1 Unloaded Qs of transmission line resonators compared with a dielectric resonator

Transmission line technology	Unloaded Q
Microstrip/stripline/coplanar waveguide	100–600
Coaxial cavity	1000–6000
Waveguide	4000–15 000
Dielectric resonator	5000–50 000

7.2 Unloaded Q of a Transmission Line Resonator

The design of a transmission line filter should start by investigating the filter specifications and choosing an appropriate solution for implementing the resonators. The quality factor of a transmission line resonator depends strongly on the implementation technology. The most common technologies for microwave and millimetre-wave applications are microstrip, coaxial cavity and waveguide. Table 7.1 lists the typical unloaded Q of these transmission line resonators compared with a high-Q dielectric resonator used as a benchmark. The Q-values in a specific technology depend on the dimensions and the material used. It is noted that the values listed in Table 7.1 are typical values that can be accomplished in practice, and readers may find in some research reports that the values are higher. In a band-pass filter design, the unloaded Q of a resonator is a key parameter to determine the loss in the filter pass-band. If the coupling mechanisms at the filter input and output are identical, the insertion loss (IL) is related to the loaded (Q_L) and unloaded Qs (Q_U) by:

$$\mathrm{IL}\,(dB) = 20\log_{10}\left(\frac{Q_U}{Q_U - Q_L}\right) \tag{7.1}$$

where Q_L is determined from the specifications of the band-pass filter. This Q_L can be determined from the ratio between f_0 and the filter bandwidth. Using the unloaded Q guidelines shown in Table 7.1, one can choose a suitable transmission line technology to design a band-pass filter to meet the filter specifications. The following example will explain this statement.

Example: Using the transmission line guideline shown in Table 7.1, choose the appropriate transmission line technology to design a filter which has 3.44 GHz centre frequency, 70 MHz bandwidth and 0.8 dB insertion loss.

From the problem, the desired Q_L for this filter is calculated as:

$$\mathrm{Q_L} = \frac{f_0}{BW} = \frac{3440}{70} = 491 \tag{7.2}$$

The unloaded Q is computed from the insertion loss defined in Equation (7.1). Rearranging Equation (7.1), one obtains:

$$\mathrm{Q_U} = \mathrm{Q_L}\left(\frac{10^{\frac{\mathrm{IL}(dB)}{20}}}{10^{\frac{\mathrm{IL}(dB)}{20}} - 1}\right) \tag{7.3}$$

Substituting the insertion loss and Q_L, this leads to:

$$Q_U = 5585 \tag{7.4}$$

If the transmission line resonator solution is preferred, either coaxial cavity or waveguide resonator is suggested. A condition to select the technology for implementation may take other issues into account such as power handling or fabrication cost. Planar transmission lines such as microstrip, coplanar waveguide or stripline, are not suitable because the maximum achievable unloaded quality factor is only 600, which is much smaller than the desired unloaded Q in this example.

7.3 Lumped-to-Distributed Conversion

7.3.1 Shunt Stub

Realising a transmission line filter requires a step replacing all lumped elements in the prototype filter (inductors and capacitors) with distributed elements that have equivalent electrical characteristics. Generally, a prototype filter synthesised from either an image or insertion loss approach is a ladder network that consists of series and shunt elements. Shunt elements can be realised using a short- or open-circuited transmission line (shunt stub). Assuming that the transmission line is lossless, a shunt capacitor and inductor can be realised, as shown in Figure 7.1.

The shunt capacitor (C) can be replaced with an open-circuited transmission line. The electrical length of the transmission line (θ_C) is less than 90°. By equating the one-port input admittance of these two circuits, the characteristic admittance of the open transmission line ($Y_{0,C}$) with a given electrical length θ_C operating at frequency ω_0 can be calculated from:

$$Y_{0,C} = \omega_0 C \tan \theta_C \tag{7.5}$$

For the shunt inductor (L), a short-circuited transmission line with electrical length less than 90° can be used to replace the shunt inductor. The characteristic impedance of the short transmission line ($Z_{0,L}$) is given by:

$$Z_{0,L} = \omega_0 L \tan \theta_L \tag{7.6}$$

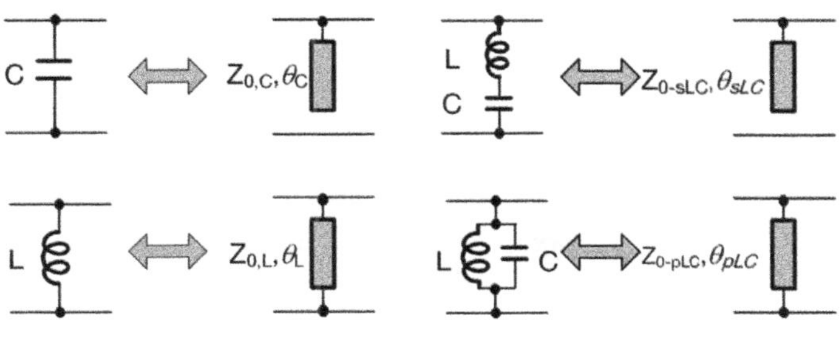

Figure 7.1 Lumped elements converted to shunt stubs.

Parallel and series LC resonators can be realised using either a quarter-wave short- or open-circuited transmission line. The total Q of a transmission line at ω_0 is affected by both conductor and the dielectric losses:

$$\frac{1}{Q} = \frac{1}{Q_c} + \frac{1}{Q_d} \tag{7.7}$$

where Q_c and Q_d are the quality factors due to the conductor and dielectric losses. These are defined by:

$$Q_c = \frac{\beta}{2\alpha_c} \text{ and } Q_d = \frac{\beta}{2\tan\delta} \tag{7.8}$$

Substituting Equations (7.8) into Equation (7.7), gives:

$$Q(\omega_0) = \frac{\omega_0}{2v_p\left(\tan\delta + \alpha_c\right)} \tag{7.9}$$

This shows that the quality factor at ω_0 is inversely proportional to the phase velocity and the attenuation coefficient resulting from the conductor loss (α_c) and the dielectric loss (α_d). The latter loss is represented by the tangent loss term ($\tan\delta$).

7.3.2 Series Line

It is shown in Figure 7.1 that open- and short-circuited shunt stubs are able to replace parallel or series *L-C* shunt resonators. A series line can be designed to function as a series inductor as shown in Figure 7.2, but is limited to a small inductance. Assuming a lossless transmission line, the ABCD matrix of a series inductor and a two-port lossless transmission line of length θ and characteristic impedance Z_0 are:

$$[ABCD]_L = \begin{bmatrix} 1 & j\omega L \\ 0 & 1 \end{bmatrix} \tag{7.10}$$

$$[ABCD]_{Tx} = \begin{bmatrix} \cos\theta & jZ_0\sin\theta \\ jY_0\sin\theta & \cos\theta \end{bmatrix} \tag{7.11}$$

At first, these two matrix equations seem totally different. However, they can be approximately equal if the transmission line is sufficiently short such that the following condition is met, that is:

$$\omega L = Z_0 \sin\theta \tag{7.12}$$

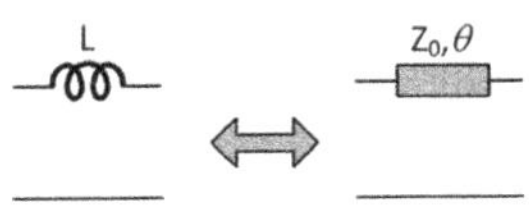

Figure 7.2 A series lines used as a series inductor.

Therefore, the value of Z_0 that provides the electrical characteristics of a series inductor at frequency ω is determined from:

$$Z_0 = \frac{\omega L}{\sin\theta} \tag{7.13}$$

Equation (7.13) is valid if and only if the transmission line is sufficiently short. For a series capacitor, the ABCD matrix suggests that this is not possible unless the characteristic impedance is not a real positive value.

7.4 Impedance and Admittance Inverters

7.4.1 Definition

Transforming a lumped-element prototype to a distributed form poses a difficult challenge for series capacitors. An impedance or admittance inverter (which were introduced briefly in Chapter 4) can solve this problem by converting a prototype network formed with inductors and capacitors into one with only inductors or only capacitors. The ABCD matrices of an ideal impedance inverter (called a K-inverter) and admittance inverter (called J-inverter) are given by:

$$\mathbf{K} = \begin{bmatrix} 0 & \mp jK \\ \pm\frac{1}{jK} & 0 \end{bmatrix} \tag{7.14}$$

$$\mathbf{J} = \begin{bmatrix} 0 & \pm\frac{1}{jJ} \\ \mp jJ & 0 \end{bmatrix} \tag{7.15}$$

As indicated in the ABCD matrices, there is ± 90° transfer phase from port 1 to port 2. The names of 'impedance' and 'admittance' inverters are derived from the network applications shown in Figure 7.3. In Figure 7.3(a), a K-inverter is terminated with a load Z_b at one port, and the input impedance at the other port (Z_a) is:

$$Z_a = \frac{K^2}{Z_b} \tag{7.16}$$

Similarly, for the admittance (J-) inverter terminated with an admittance Y_b as shown in Figure 7.3(b), the input admittance Y_a can be written as:

$$Y_a = \frac{J^2}{Y_b} \tag{7.17}$$

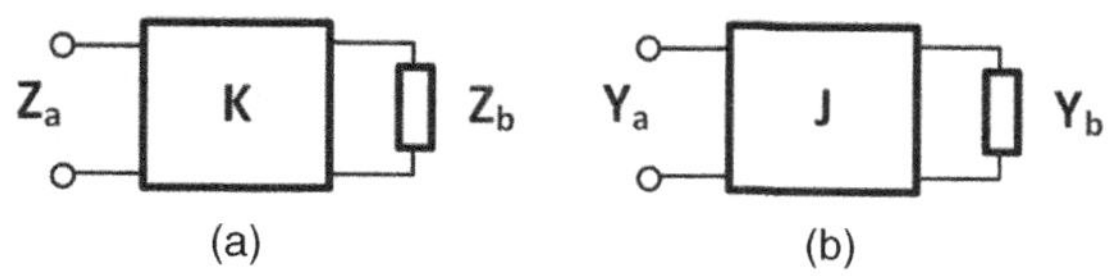

Figure 7.3 Input impedance and admittance of terminated K- and J-inverters.

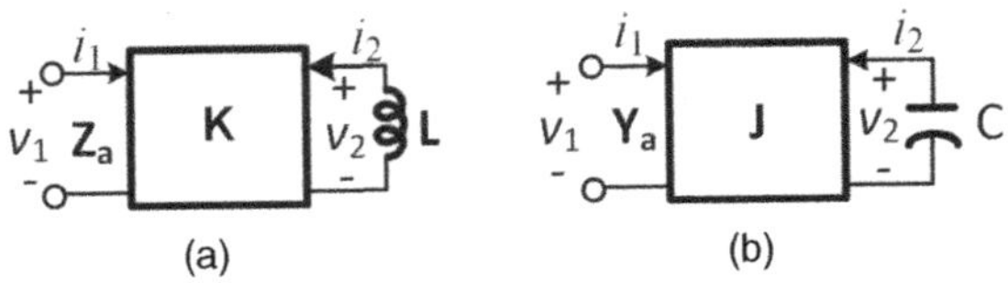

Figure 7.4 (a) K-inverter terminated with and inductor, and (b) J-inverter terminated with a capacitor.

From Equations (7.16) and (7.17), it can be seen that the function of either K- or J-inverter is equivalent to a quarter-wave transformer of characteristic impedance K or characteristic admittance J, respectively. The only difference is the operating bandwidth which is infinite for ideal K and J inverters. Using a K- or J-inverter can transform an inductor to a capacitor and vice versa. The following example illustrates this.

Example: Derive the input impedance and admittance of the circuits shown in Figure 7.4(a) and (b).

From the ABCD matrix of the K-inverter, defined in Equation (7.14), the voltage and current at port 1 of the network in Figure 7.4(a) can be written as:

$$v_1 = \mp jKi_2 \tag{7.18}$$

$$i_1 = \pm \frac{1}{jK} v_2 \tag{7.19}$$

The voltage at port 2 is related to the current at port 2 by:

$$v_2 = -j\omega L i_2 \tag{7.20}$$

Substituting Equation (7.20) into Equation (7.19), this leads to:

$$i_1 = \mp \frac{\omega L}{K} i_2 \tag{7.21}$$

Hence, the input impedance at port 1 is obtained by the ratio of v_1 and i_1 as:

$$Z_a = \frac{v_1}{i_1} = \frac{K^2}{j\omega L} \tag{7.22}$$

It is shown that the driving-point impedance of network in Figure 7.4(a) is capacitive which has a capacitance equal to:

$$C_a = \frac{L}{K^2} \tag{7.23}$$

Similarly, the voltage and current at port 1 of the network shown in Figure 7.4(b) are:

$$v_1 = \pm \frac{1}{jJ} i_2 \tag{7.24}$$

$$i_1 = \mp jJv_2 \tag{7.25}$$

The impedance at port 2 of the network in Figure 7.4(b) is related with the terminating capacitor C, i.e.:

$$v_2 = \frac{1}{j\omega C} i_2 \tag{7.26}$$

Therefore,

$$i_1 = \mp \frac{J}{\omega C} i_2 \tag{7.27}$$

Finally, the input impedance of the J-inverter terminated with a capacitor is given by:

$$Z_a = \frac{v_1}{i_1} = \frac{j\omega C}{J^2} \tag{7.28}$$

Obviously, the driving impedance of the network in Figure 7.14(b) is inductive and its inductance is:

$$L_a = \frac{C}{J^2} \tag{7.29}$$

The results shown in Equations (7.23) and (7.29) demonstrate that a using K- and J-inverter allows an inductor or capacitor to be replaced with each other.

Example: Prove the equivalence of the networks shown in Figure 7.5(a) and (b).

As shown in Chapter 3, the ABCD matrix of a two-port series impedance, z, is:

$$\mathbf{A}_{Series} = \begin{bmatrix} 1 & z \\ 0 & 1 \end{bmatrix} \tag{7.30}$$

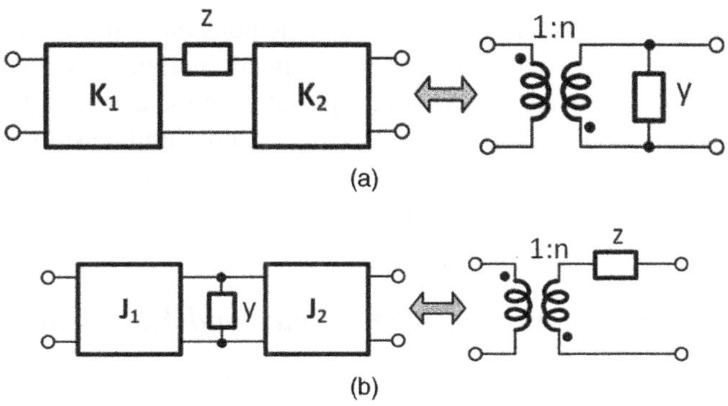

Figure 7.5 Equivalent networks using (a) K-inverter and (b) J-inverter.

If we cascade the two K-inverters with this network, the ABCD matrix of the cascaded network is:

$$\mathbf{K}_1 \cdot \mathbf{A}_{Series} \cdot \mathbf{K}_2 = \begin{bmatrix} -\frac{K_1}{K_2} & 0 \\ -\frac{z}{K_1 K_2} & -\frac{K_2}{K_1} \end{bmatrix} \tag{7.31}$$

Rewriting the ABCD matrix in terms of the multiplication of two ABCD matrices, this leads to:

$$\mathbf{K}_1 \cdot \mathbf{A}_{Series} \cdot \mathbf{K}_2 = \begin{bmatrix} -\frac{K_1}{K_2} & 0 \\ 0 & -\frac{K_2}{K_1} \end{bmatrix} \begin{bmatrix} 1 & 0 \\ \frac{z}{K_1 K_2} & 1 \end{bmatrix} \tag{7.32}$$

From Equation (7.32), the first matrix in the right-hand-side (RHS) is equivalent to a phase-reversion ideal transformer with a turns ratio $\frac{K_1}{K_2}$. The second matrix of the RHS term is a shunt admittance (as per Chapter 3) with value $\frac{z}{K_1 K_2}$.

Considering the network shown in Figure 7.5(b), if two J-inverters (J_1 and J_2) surround the shunt admittance Y, so the ACBD matrix of the cascade network is

$$\mathbf{J}_1 \cdot \mathbf{A}_{Shunt} \cdot \mathbf{J}_2 = \begin{bmatrix} -\frac{J_2}{J_1} & -\frac{y}{J_1 J_2} \\ 0 & -\frac{J_1}{J_2} \end{bmatrix} \tag{7.33}$$

Similarly, we can rewrite this in the form of two cascaded matrices as:

$$\mathbf{J}_1 \cdot \mathbf{A}_{Shunt} \cdot \mathbf{J}_2 = \begin{bmatrix} -\frac{J_2}{J_1} & 0 \\ 0 & -\frac{J_1}{J_2} \end{bmatrix} \begin{bmatrix} 1 & \frac{y}{J_1 J_2} \\ 0 & 1 \end{bmatrix} \tag{7.34}$$

Again, the first matrix is a phase-reversal transformer, with turns ratio $\frac{J_2}{J_1}$, but now the second matrix is a series impedance with value $\frac{y}{J_1 J_2}$.The phase reversion in the transformer has no effect to the magnitude of frequency response of the transformed network, while the transformer ratio should be designed appropriately. This example demonstrates the application of using a K-inverter to transform a series network to a shunt network and vice-versa with a J-inverter. These network transformations are very useful, especially for the design of transmission line filters in a medium such as microstrip where a series transmission line stub is rather difficult to implement. The detail of deriving common transmission line filter topologies using these network transformations is discussed in the following section.

7.4.2 Network Transformation with Impedance and Admittance Inverters

It was shown in Chapter 4 that the canonical forms of the L-C band-pass filter required a mixture of series and shunt resonators, and even in lumped form these are impractical when parasitics are taken into account. In a transmission line medium it is very much preferred to realise a band-pass filter with high-Q resonators of one type coupled together. To obtain a

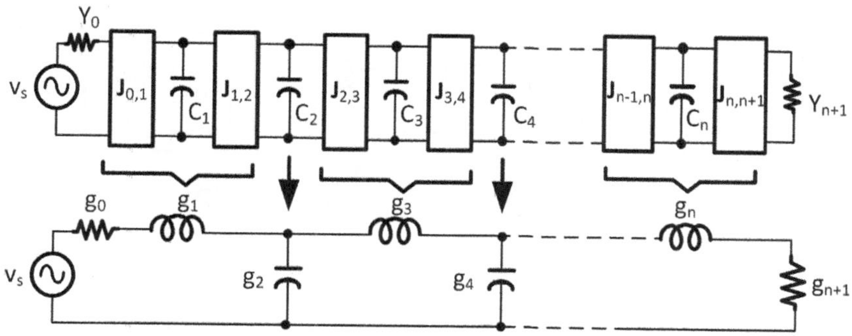

Figure 7.6 Converting a low-pass filter to a purely-capacitive topology using J-inverters.

topology that is more amenable to realisation in transmission line form, K-inverters can be used to transform unwanted shunt sections to series ones, and vice-versa using J-inverters. In practice, K- or J-inverters can be simply realised with a quarter-wavelength transmission line transformer.

Series filter network sections should especially be avoided to achieve low insertion loss in the filter pass-band. Starting with the low-pass prototype shown in Figure 7.6, all the series inductors are replaced with a shunt capacitor J-inverter combination. There are various degrees of design freedom. One of the design approaches is to set the shunt capacitors C_i arbitrarily and compute the parameters $J_{i,i+1}$ from:

$$J_{0,1} = \sqrt{\frac{Y_0 C_1}{g_0 g_1}} \tag{7.35}$$

$$J_{i,i+1} = \sqrt{\frac{C_i C_{i+1}}{g_i g_{i+1}}} \tag{7.36}$$

$$J_{n,n+1} = \sqrt{\frac{C_n Y_{n+1}}{g_n g_{n+1}}} \tag{7.37}$$

Using Equations (7.35) to (7.37), all the network parameters including J-inverters can be determined. The resulting network provides the same frequency response as the original low-pass filter.

Next, the J- and K-inverters will be used to convert the canonical band-pass filter with series and parallel resonators into a topology with only shunt parallel-resonators. As shown in Figure 7.7, each of the series L-C resonators connected in series is first replaced with a parallel L-C resonator connected in shunt, along with a pair of J-inverters. This network transformation is carried out because a shunt parallel L-C resonator can be realised using a short-circuited quarter-wave transmission line stub (as shown in Figure 7.1), whereas realising a series resonant circuit in series is not possible. The impedance of a series L-C resonator can be written as:

$$Z_i(\omega) = j\left(\omega L_i - \frac{1}{\omega C_i}\right) \tag{7.38}$$

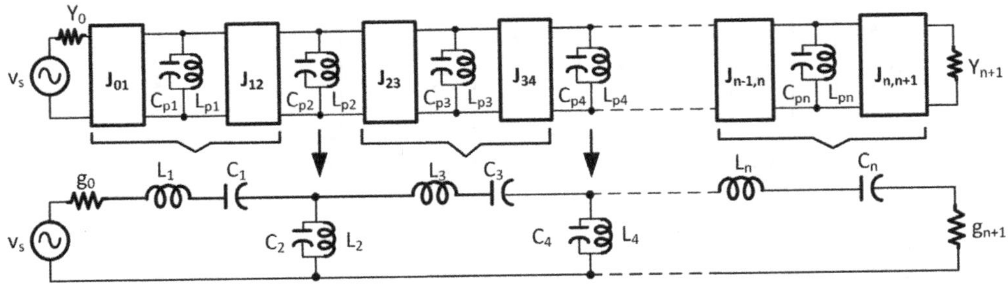

Figure 7.7 Converting a band-pass filter prototype to a J-inverter based band-pass filter.

Converting this series network with the technique presented in Figure 7.5(b), the admittance of the shunt element $Y_i(\omega)$ is related to $Z_i(\omega)$ by:

$$Y_i(\omega) = Z_i(\omega) J_{i-1,i} J_{i+1,i} = j\left(\omega L_i - \frac{1}{\omega C_i}\right) J_{i-1,i} J_{i+1,i} \tag{7.39}$$

where $J_{i-1,i}$ and $J_{i+1,i}$ are the inverter parameters of the two J-inverters to the left and right of the shunt element $Y_i(\omega)$. From Equation (7.39), the new shunt inductors and capacitors of the transformed network are:

$$L_{pi} = \left(\frac{1}{J_{i-1,i} J_{i+1,i}}\right) C_i \tag{7.40}$$

$$C_{pi} = \left(J_{i-1,i} J_{i+1,i}\right) L_i \tag{7.41}$$

Given an arbitrary parallel capacitor, C_{pi}, the other design parameters of the J-inverter based band-pass filter shown in Figure 7.7 are computed as follows:

$$L_{pi} = \frac{1}{\omega_0^2 C_{pi}} \tag{7.42}$$

$$J_{0,1} = \sqrt{\frac{Y_0 \omega_{3-dB} C_{p1}}{g_0 g_1}} \tag{7.43}$$

$$J_{i,i+1} = \omega_{3-dB} \sqrt{\frac{C_{pi} C_{pi+1}}{g_i g_{i+1}}} \tag{7.44}$$

$$J_{n,n+1} = \sqrt{\frac{Y_{n+1} \omega_{3-dB} C_{pn}}{g_n g_{n+1}}} \tag{7.45}$$

where ω_{3-dB} is the 3 dB bandwidth in rad s^{-1}.

The design obtained from applying J-inverters provides a topology that can be simply realised using transmission line elements. The J-inverter can be implemented with a quarter-wave transformer and a shunt L-C parallel resonator can be designed with a quarter-wave short-circuited transmission line. Figure 7.8 shows the generalised K- and J-inverter band-pass filter realisations with distributed elements described in terms of series reactances, $X_j(\omega)$,

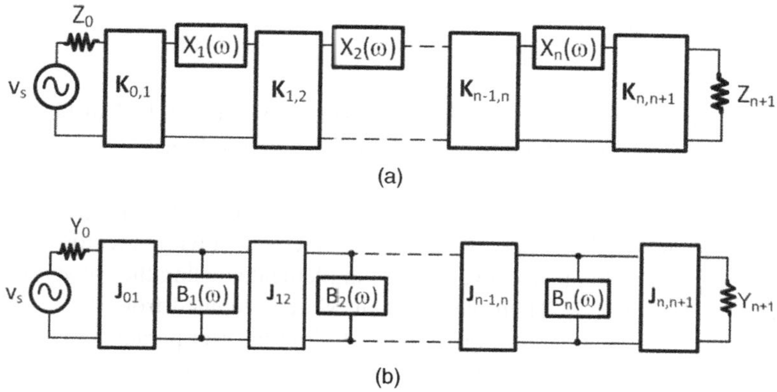

Figure 7.8 Generalised band-pass filters based on (a) K- and (b) J-inverters.

or shunt susceptances, $B_j(\omega)$. The fractional bandwidth relative to centre frequency ω_0 is denoted by $\Delta\omega$. Let a_j and b_j be the slope parameters of X_j and B_j, they are defined at ω_0 by:

$$a_j = \frac{\omega_0}{2} \left. \frac{dX_j(\omega)}{d\omega} \right|_{\omega=\omega_0} \tag{7.46}$$

$$b_j = \frac{\omega_0}{2} \left. \frac{dB_j(\omega)}{d\omega} \right|_{\omega=\omega_0} \tag{7.47}$$

If the K-inverter topology shown in Figure 7.8(a) is chosen, the parameter $K_{j,j+1}$ can be calculated from:

$$K_{0,1} = \sqrt{\frac{Z_0 a_1 \Delta\omega}{g_0 g_1 \Omega_c}} \tag{7.48}$$

$$K_{j,j+1} = \frac{\Delta\omega}{\Omega_c} \sqrt{\frac{a_j a_{j+1}}{g_j g_{j+1}}} j = 1, 2, \ldots, n-1 \tag{7.49}$$

$$K_{n,n+1} = \sqrt{\frac{Z_{n+1} a_n \Delta\omega}{g_n g_{n+1} \Omega_c}} \tag{7.50}$$

Similarly, the parameter $J_{j,j+1}$ in the J-inverter topology shown in Figure 7.8(b) can be calculated from:

$$J_{0,1} = \sqrt{\frac{Y_0 b_1 \Delta\omega}{g_0 g_1 \Omega_c}} \tag{7.51}$$

$$J_{j,j+1} = \frac{\Delta\omega}{\Omega_c} \sqrt{\frac{b_j b_{j+1}}{g_j g_{j+1}}} j = 1, 2, \ldots, n-1 \tag{7.52}$$

$$J_{n,n+1} = \sqrt{\frac{Y_{n+1} b_n \Delta\omega}{g_n g_{n+1} \Omega_c}} \tag{7.53}$$

7.4.3 Inverter Realisation

7.4.3.1 Lumped Inverter

As well as being realised with a quarter-wave transformer which is purely distributed, inverters can be realised using lumped and lumped-distributed topologies. Figure 7.9 shows the J- and K-inverters using lumped elements, along with their respective transmission matrices. The inductive and capacitive Π or T networks can realise both inverter types if negative capacitors and inductors are allowed. The lumped K-inverters implemented with inductors are shown in Figure 7.9(a) and (b). The lumped J-inverters are implemented using the capacitor networks shown in Figure 7.9(c) and (d). Their element values are computed from the corresponding transmission matrices in Figure 7.9 to provide the desired inverter parameter (J or K value). The lumped inverters are more narrowband than is the quarter-wave transformer.

A crucial application using the lumped element inverter realisation is using a J-inverter topology to realise a capacitively-coupled band-pass filter. To illustrate this, Figure 7.10(a) shows a J-inverter based band-pass topology with two resonators and three J-inverters whose parameters are J_{01}, J_{12} and J_{23}. The input and output ports of the filter are terminated with

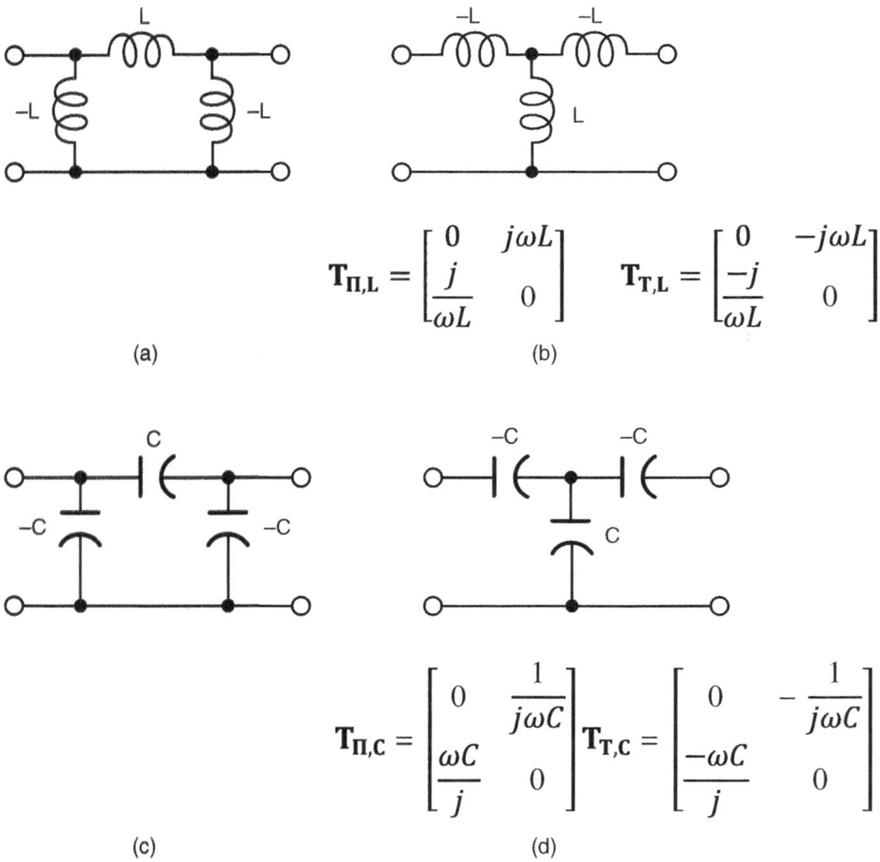

$$\mathbf{T}_{\Pi,L} = \begin{bmatrix} 0 & j\omega L \\ \dfrac{j}{\omega L} & 0 \end{bmatrix} \qquad \mathbf{T}_{T,L} = \begin{bmatrix} 0 & -j\omega L \\ \dfrac{-j}{\omega L} & 0 \end{bmatrix}$$

$$\mathbf{T}_{\Pi,C} = \begin{bmatrix} 0 & \dfrac{1}{j\omega C} \\ \dfrac{\omega C}{j} & 0 \end{bmatrix} \qquad \mathbf{T}_{T,C} = \begin{bmatrix} 0 & -\dfrac{1}{j\omega C} \\ \dfrac{-\omega C}{j} & 0 \end{bmatrix}$$

Figure 7.9 Lumped element inverters.

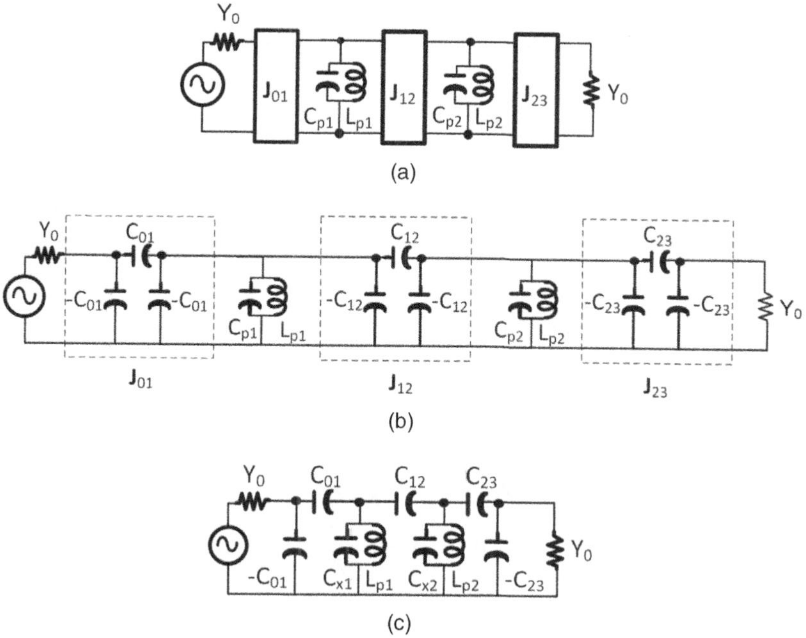

Figure 7.10 (a) J-inverter band-pass filter. (b) Lumped-inverter based band-pass filter. (c) Capacitive coupling band-pass filter.

admittance Y_0. Figure 7.10(b) is obtained by replacing all the J-inverters with the Π-network realisation, as shown in Figure 7.9(c). It is evident that four of the negative capacitors are in parallel with the capacitors of the parallel L-C resonators, so they can be realised very elegantly by reducing the values of the shunt capacitors, C_{x1} and C_{x2}, of the resonators as follows:

$$C_{x1} = C_{p1} - C_{01} - C_{12} \tag{7.54}$$

$$C_{x2} = C_{p2} - C_{12} - C_{23} \tag{7.55}$$

The final band-pass filter, referred to as the capacitively-coupled resonator band-pass filter, is shown in Figure 7.10(c). To realise this filter with transmission lines, quarter-wave transmission lines are implemented for the parallel L-C resonators. An inductively coupled band-pass filter can also be obtained using the lumped inverter realisation in a similar way.

7.4.3.2 Lumped-Distributed Inverter

There is an alternative way to realise K- and J-inverters using a lumped-distributed network topology. There are two topologies for K-inverter implementation and two for J-inverter implementation. The lumped-distributed forms for realising impedance inverters are shown in Figure 7.11. The realisation for a K-inverter can be accomplished using a negative or positive length transmission line. From Figure 7.11, the parameters of the lumped-distributed inverters

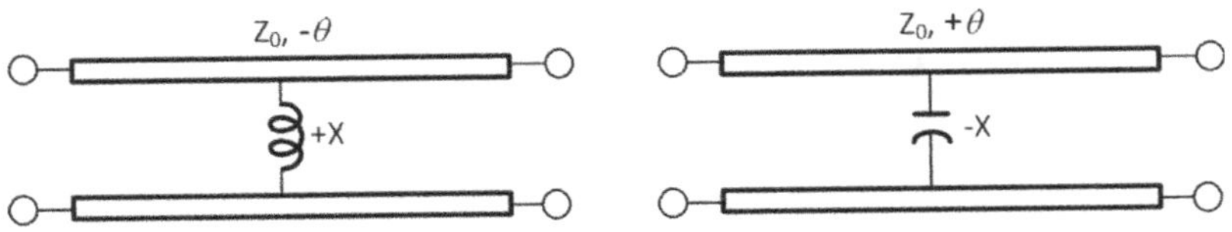

Figure 7.11 Lumped-distributed topologies for K-inverter realisation using (a) a negative, and (b) a positive length of transmission line.

are related to the parameter K by:

$$K = Z_0 \tan\left|\frac{\theta}{2}\right| \tag{7.56}$$

$$\left|\frac{X}{Z_0}\right| = \frac{K/Z_0}{1-(K/Z_0)^2} \tag{7.57}$$

$$\theta = -\tan^{-1}\frac{2X}{Z_0} \tag{7.58}$$

The lumped-distributed topologies for a J-inverter are shown in Figure 7.12. Similar to the K-inverter case, two forms using either negative- or positive-length transmission line can be used. To realise the negative transmission line length topology, the adjacent transmission lines with equal characteristic impedance can simply be shortened by the requisite amount. The parameters of the lumped-distributed J-inverter are determined from:

$$J = Y_0 \tan\left|\frac{\theta}{2}\right| \tag{7.59}$$

$$\left|\frac{B}{Z_0}\right| = \frac{J/Y_0}{1-(J/Y_0)^2} \tag{7.60}$$

$$\theta = -\tan^{-1}\frac{2B}{Y_0} \tag{7.61}$$

There are two other topologies which are in lumped-distributed form. The distributed network consists of two transmission lines of length $\theta/2$. The lumped networks in these topologies are arranged in the form of T and Π networks, as shown in Figure 7.13. These two topologies

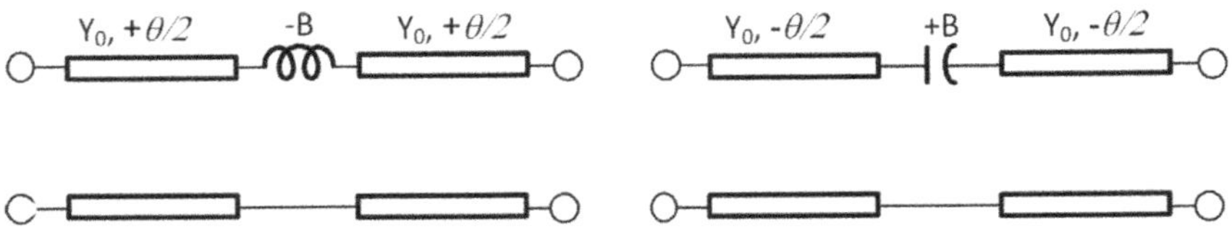

Figure 7.12 Lumped-distributed topologies for J-inverter realisation using (a) negative- and (b) positive-length transmission line.

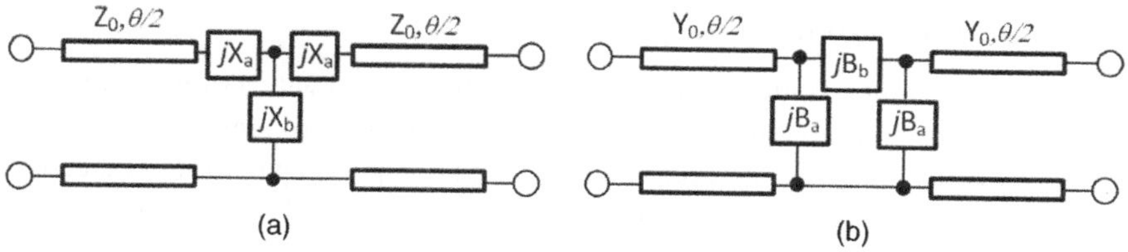

Figure 7.13 Lumped-distributed implementations for (a) K-inverter using T-network and (b) J-inverter using Π-network.

are useful for a transmission line filter designed with discontinuities such as a capacitive-gap coupled transmission line filter or a shunt-inductively coupled waveguide filter. The circuit parameters for the T-lumped network are computed from:

$$\frac{K}{Z_0} = \left|\tan\left(\frac{\theta}{2} + \tan^{-1}\frac{X_a}{Z_0}\right)\right| \tag{7.62}$$

$$\theta = -\tan^{-1}\left(\frac{2X_b}{Z_0} + \frac{X_a}{Z_0}\right) - \tan^{-1}\frac{X_a}{Z_0} \tag{7.63}$$

The parameters of the Π-lumped network are computed from:

$$\frac{J}{Y_0} = \left|\tan\left(\frac{\theta}{2} + \tan^{-1}\frac{B_a}{Y_0}\right)\right| \tag{7.64}$$

$$\theta = -\tan^{-1}\left(\frac{2B_b}{Y_0} + \frac{B_a}{Y_0}\right) - \tan^{-1}\frac{B_a}{Y_0} \tag{7.65}$$

7.4.3.3 Distributed Inverter

Finally, another important implementation of an admittance inverter is shown in Figure 7.14. The network is composed of two negative-length and one positive-length transmission lines. The negative transmission lines can be absorbed to their adjacent positive-length transmission lines. This distributed topology is very useful for realising a transmission line band-pass filter. The parameter J is related to the parameters of transmission lines by:

$$J = |Y_b \cot\theta| \tag{7.66}$$

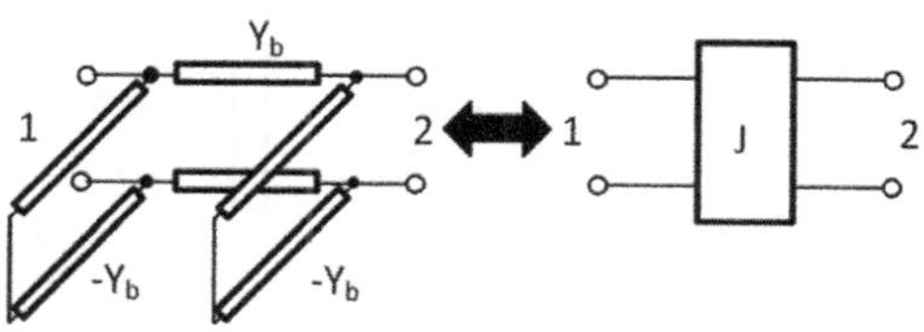

Figure 7.14 J-inverter realisation with distributed Π-network.

7.5 Richards Transformation

We have shown that a lumped inductor or capacitor can be realised with a short- or open-circuit stub. For a lossless transmission line, short- and open-circuit stubs of length l can be designed to be electrically equivalent to a one-port inductor L and one-port capacitor C, respectively, at a frequency ω. Let Z_{SC} and Z_{OC} be the impedances of the short- and open-stubs, then:

$$j\omega L \equiv Z_{SC} = jZ_{0,SC} \tan\left(\frac{\omega}{v_p} l\right) \tag{7.67}$$

$$\frac{-j}{\omega C} \equiv Z_{OC} = -j\frac{Z_{0,OC}}{\tan\left(\frac{\omega}{v_p} l\right)} \tag{7.68}$$

If l is chosen to be equal to $\frac{\lambda_g}{8}$, one obtains:

$$\tan\left(\frac{\omega}{v_p} l\right) = 1 \tag{7.69}$$

Hence Equations (7.67) and (7.68) are simplified to:

$$Z_{0,SC} = \omega L \tag{7.70}$$

$$Z_{0,OC} = \frac{1}{\omega C} \tag{7.71}$$

Example: Convert the lumped element low-pass filter in Figure 7.15(a) to a transmission line equivalent network. The low-pass filter has a cut-off frequency at 1 GHz when terminated at both ports with 50 Ω impedance.

Open- and short-circuited transmission lines with length $\frac{\lambda_g}{8}$ are used. The characteristic impedances of the short ($Z_{0,SC}$) and open stubs ($Z_{0,OC}$) are computed from Equations (7.70) and (7.71) to be:

$$Z_{0,SC} = 2\pi \times 10^9 \times 3.80 \times 10^{-9} = 23.88\,\Omega \tag{7.72}$$

$$Z_{0,OC} = \frac{1}{2\pi \times 10^9 \times 3.65 \times 10^{-12}} = 43.60\,\Omega \tag{7.73}$$

The transmission line topology for the filter is shown in Figure 7.15(b).

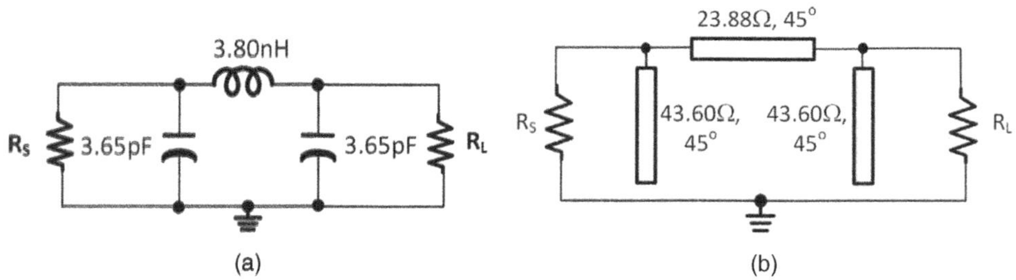

Figure 7.15 Schematics of (a) lumped element low-pass filter at 1 GHz, and (b) transmission line counterpart.

The example above provides a quick design technique for a $\frac{\lambda_g}{8}$ transmission line filter. There is a mathematical technique called the Richards transformation that can be generally applied to transform a lumped network to a commensurate distributed network. The transformation provides a distributed network whose frequency response is mapped to that of the original lumped filter network. The transformation is a mapping of the complex variable $p = \sigma + j\omega$ to the new complex variable t called the Richards variable, which is given by:

$$t = \tanh\left(\frac{pl}{v_p}\right) \tag{7.74}$$

The real part of the complex variable p (i.e., σ) results from any loss in the lumped network, and ω is the frequency variable. In Equation (7.74), l and v_p, the parameters of the transmission line, are the length and phase velocity of the wave propagating through the line, respectively. Applying the Richards transformation, a one-port lumped inductor L whose impedance $Z_L = pL$ corresponds to a shorted stub with impedance $Z_{0,SC}$ providing:

$$Z_L = t \cdot Z_{0,SC} \tag{7.75}$$

The one-port admittance of a lumped capacitor $Y_C = pC$ is mapped to an open-circuit stub of admittance $Y_{0,OC}$ providing:

$$Y_C = t \cdot Y_{0,OC} \tag{7.76}$$

There is a simplified case when a lumped network is lossless. The complex variable p is reduced to $j\omega$. Hence, the Richards variable t can be written as:

$$t = j\tan\left(\frac{\omega}{v_p}l\right) \tag{7.77}$$

while the network transformation rules are still governed by Equations (7.75) and (7.76).

7.6 Unit Element, Kuroda's Identity and Coupled-Lines Section

7.6.1 Unit Element

The distributed filter designed with transmission lines of equal length is called a 'commensurate' design. An example of this is given in the example presented in Section 7.4. For this design approach, a lossless transmission line is a basic building block, so it is referred to as a unit element (UE) in the structure. The transfer matrix of the UE element in terms of the Richards variable can be written as:

$$\mathbf{T_{UE}} = \frac{1}{\sqrt{1-t^2}}\begin{bmatrix} 1 & Z_u t \\ Y_u t & 1 \end{bmatrix} \tag{7.78}$$

where $Y_u = (Z_u)^{-1}$.

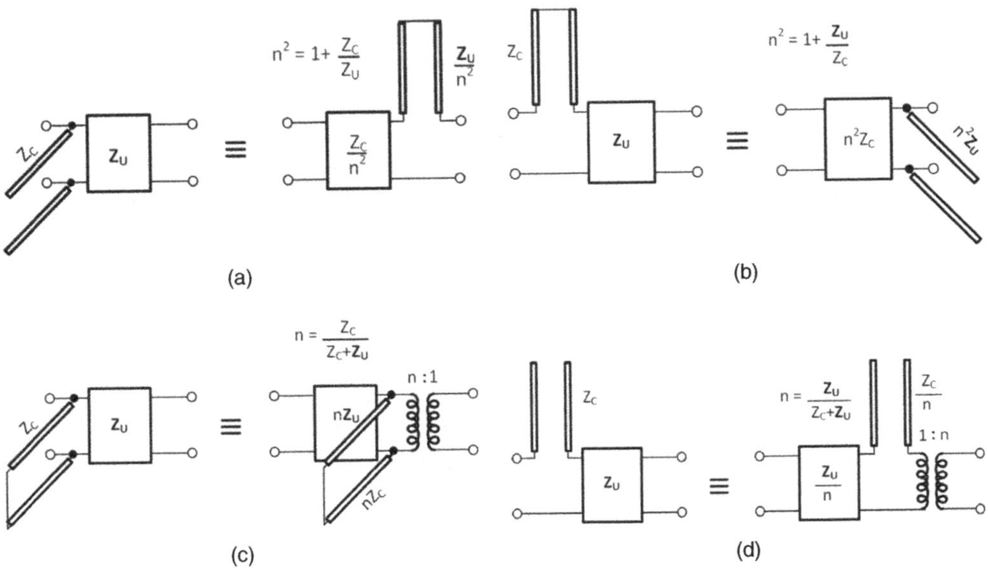

Figure 7.16 Kuroda's identities.

The parameter Z_u is the basic electrical parameter of the UE. It is indeed the characteristic impedance of a lossless transmission line.

7.6.2 *Kuroda's Identity*

Kuroda's identity provides four pairs of equivalent two-port networks. Each pair of two-port networks has a different topology but the same electrical characteristics. Using Kuroda's identity to replace one two-port network with its equivalent form, a more practical implementation of a filter can be obtained; for example, changing a transmission line series stub to a shunt stub. The four Kuroda's identities formed with UEs are shown in Figure 7.16.

Figure 7.16(a) and (b) show that a two-port network using a series stub can convert to a two-port network using a shunt stub with the same length. The characteristic impedance and the parameter of the UE of the converted two-port network are calculated from the equations. The mathematical proof of the equivalence shown in Figure 7.16(a) is given in the following example.

Example: Prove that two networks in Figure 7.16(a) are electrically equivalent.

Generally, the transmission matrices of the shunt ($\mathbf{T}_{shunt}$) and series stubs ($\mathbf{T}_{series}$) illustrated in Figure 7.17 can be written in the form of the input impedance Z_{in} as:

$$\mathbf{T}_{shunt} = \begin{bmatrix} 1 & 0 \\ \frac{1}{Z_{in}} & 1 \end{bmatrix} \tag{7.79}$$

$$\mathbf{T}_{series} = \begin{bmatrix} 1 & Z_{in} \\ 0 & 1 \end{bmatrix} \tag{7.80}$$

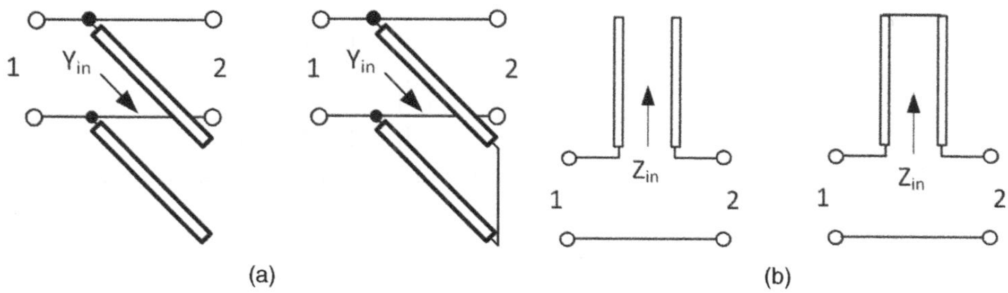

Figure 7.17 Transmission matrix definitions of (a) shunt stubs and (b) series stubs.

The input impedances of open ($Z_{in,oc}$) and short stubs ($Z_{in,sc}$) are related to the characteristic impedance Z_0 of the stub by:

$$Z_{in,oc} = \frac{Z_0}{t} \tag{7.81}$$

$$Z_{in,sc} = Z_0 t \tag{7.82}$$

The transmission matrix of the left-hand-side network shown in Figure 7.16(a) ($\mathbf{T}_{LHS}$) is obtained by multiplying the transmission matrix of the shunt open stub with the transmission matrix of the UE, giving:

$$\mathbf{T}_{LHS} = \begin{bmatrix} 1 & 0 \\ \frac{t}{Z_c} & 1 \end{bmatrix} \frac{1}{\sqrt{1-t^2}} \begin{bmatrix} 1 & Z_u t \\ \frac{t}{Z_u} & 1 \end{bmatrix} = \frac{1}{\sqrt{1-t^2}} \begin{bmatrix} 1 & Z_u t \\ t\left(\frac{1}{Z_c}+\frac{1}{Z_u}\right) & t^2\frac{Z_u}{Z_c}+1 \end{bmatrix} \tag{7.83}$$

For the right-hand-side network, its transmission matrix is obtained by multiplying the transmission matrix of the UE and the series short stub. This is given by

$$\mathbf{T}_{RHS} = \frac{1}{\sqrt{1-t^2}} \begin{bmatrix} 1 & \frac{Z_c}{n^2}t \\ \frac{n^2 t}{Z_c} & 1 \end{bmatrix} \begin{bmatrix} 1 & \frac{Z_u t}{n^2} \\ 0 & 1 \end{bmatrix} = \frac{1}{\sqrt{1-t^2}} \begin{bmatrix} 1 & \frac{t}{n^2}(Z_c+Z_u) \\ \frac{n^2 t}{Z_c} & t^2\frac{Z_u}{Z_c}+1 \end{bmatrix} \tag{7.84}$$

It is shown in Equations (7.83) and (7.84) that $\mathbf{T}_{LHS}$ will be equal to $\mathbf{T}_{RHS}$ if and only if

$$n^2 = 1 + \frac{Z_c}{Z_u} \tag{7.85}$$

This finishes the proof of the equivalence shown in Figure 7.16(a).

The Kuroda's identities shown in Figure 7.16 are very useful for filter implementation because they can be applied to transform some filter topologies which are difficult to implement into other realisable topologies. The following example explains how to apply the Kuroda's identities for this application.

Example: Figure 7.18 shows a low-pass filter designed for a system impedance Z_0 with two series short-circuit stubs of characteristic impedance Z_1 and a shunt open-circuit stub of characteristic impedance Z_2. All the stubs have electrical length equal to $\frac{\lambda_g}{8}$. This low-pass

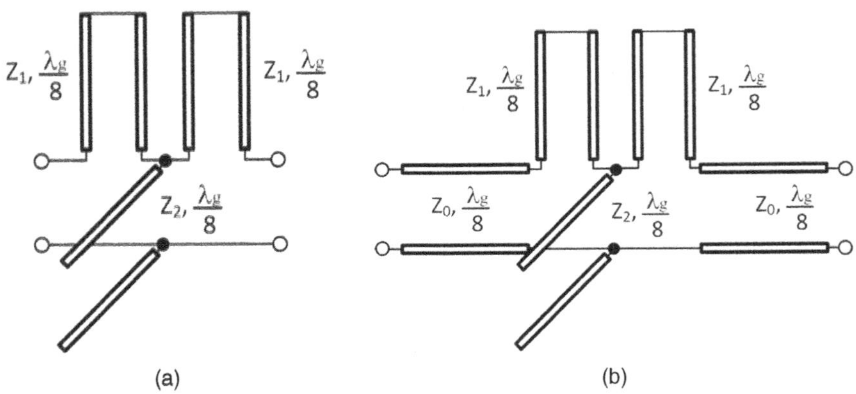

Figure 7.18 (a) A low-pass filter, and (b) its modified topology by adding lines at the input and output ports.

transmission line filter faces some difficulties in implementation because there is no physical space between the transmission lines. The filter response will change from the original response drastically if the filter topology is not modified.

The first step is to connect two identical transmission lines at the input and output ports of the filter. To preserve the magnitude response of the filter, the characteristic impedance of these transmission lines should be equal to the system impedance, Z_0. To maintain a commensurate design, the length of each inserted transmission line is $\frac{\lambda_g}{8}$, the same as the original lines. The circuit with these two added transmission lines is shown in Figure 7.18(b). We then apply the second identity to convert the transmission line section Z_0 and Z_1. This results in the new topology shown in Figure 7.19, where $n^2 = 1 + \frac{Z_0}{Z_1}$. The topology is now simpler to implement than the original structure because there are no series stubs and the three shunt stubs are now physically separated. In addition, if the characteristic impedance of some transmission lines in the original filter is small, the new characteristic impedance, increased by the factor n^2, may relax the design, layout and/or fabrication.

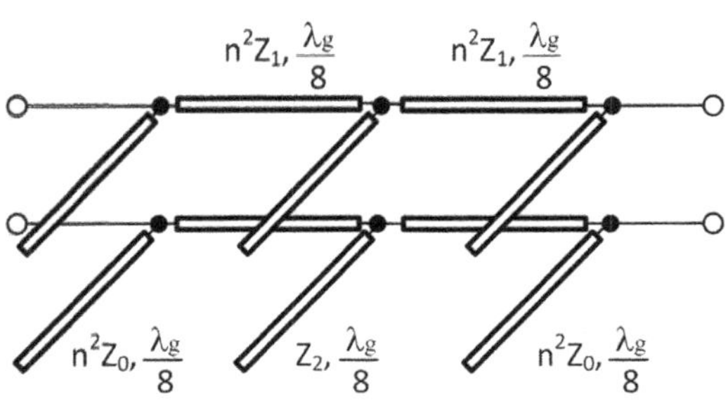

Figure 7.19 Filter design obtained by applying Kuroda's identities to the filter in Figure 7.18(b).

7.6.3 Coupled-Lines Section

Parallel-coupled lines are a very common element in microwave and millimetre-wave circuits. They were introduced in Chapter 6 as directional couplers. With some specific coupled-line configurations, various filter responses – that is, band-pass, band-stop, low-pass and high-pass – can be obtained. Figure 7.20 shows five coupled-line configurations applied to microwave filter applications. The electrical parameters of the coupled lines are the even-mode impedance (Z_{0e}), odd-mode impedance (Z_{0o}) and electrical length (θ).

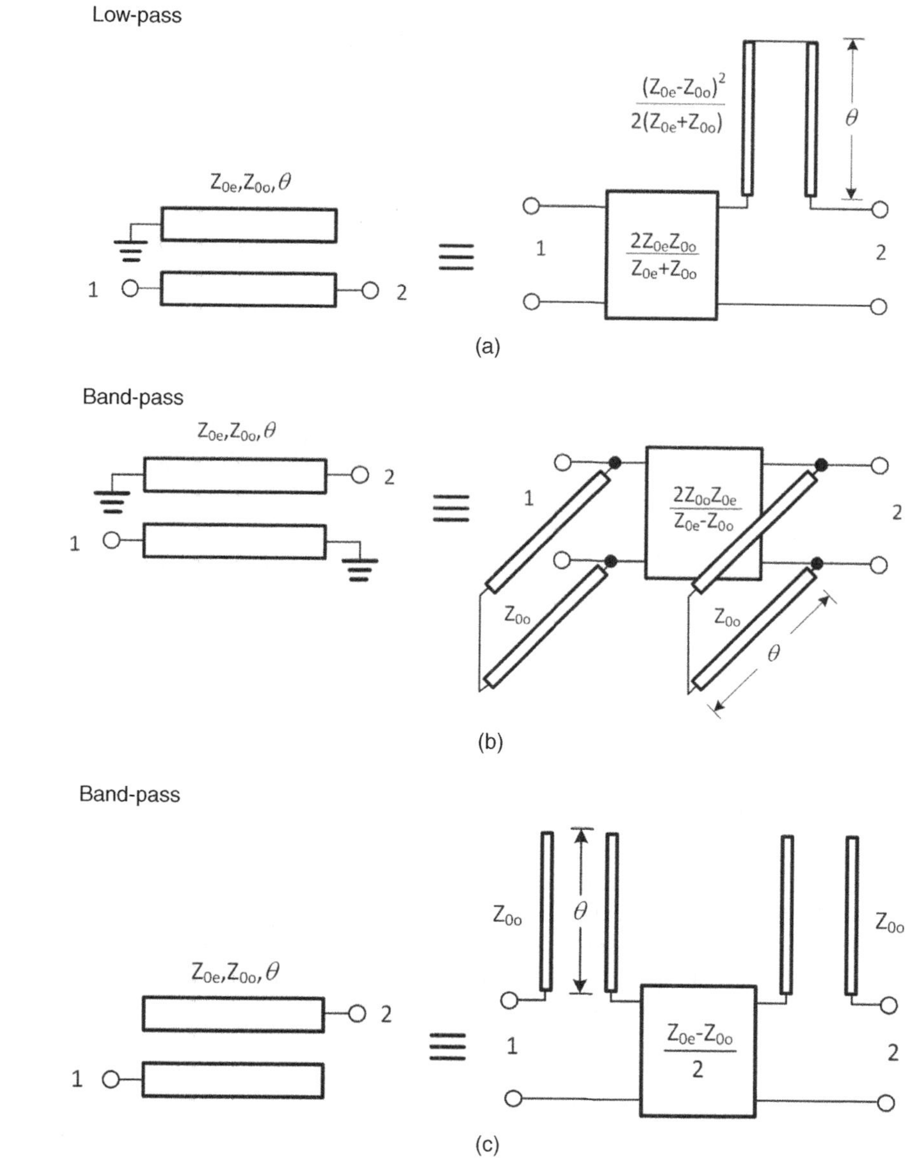

Figure 7.20 Coupled-lines section as a filter element.

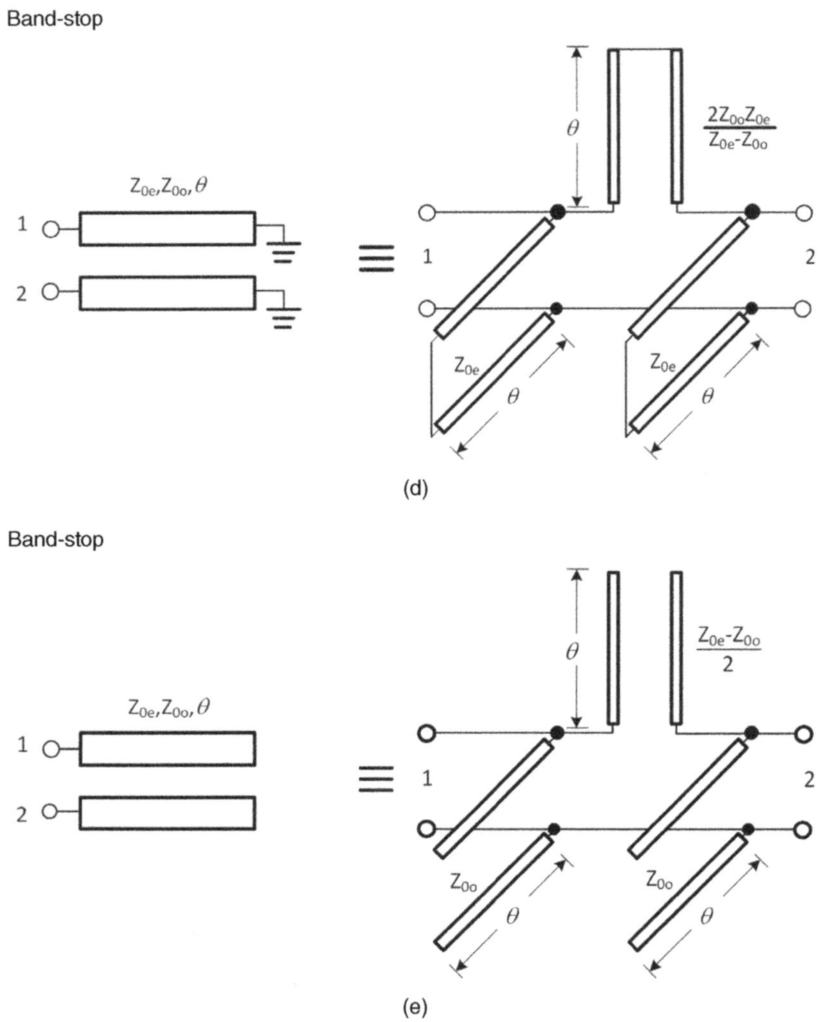

Figure 7.20 (*Continued*)

For each configuration, the electrically equivalent model using UEs and stubs is also shown in Figure 7.20. All the UEs have electrical length of θ and their characteristic impedances are shown. The models show commensurate distributed design, indicating that all transmission lines (including UEs) are of equal length. The desired filter response can be achieved using parallel-coupled lines as the main filter network element.

7.7 Stepped-Impedance Low-Pass Filter

A stepped-impedance low-pass filter consists of a cascade of low-impedance transmission line sections (Z_{low}) alternately connected with high-impedance transmission line sections (Z_{high}).

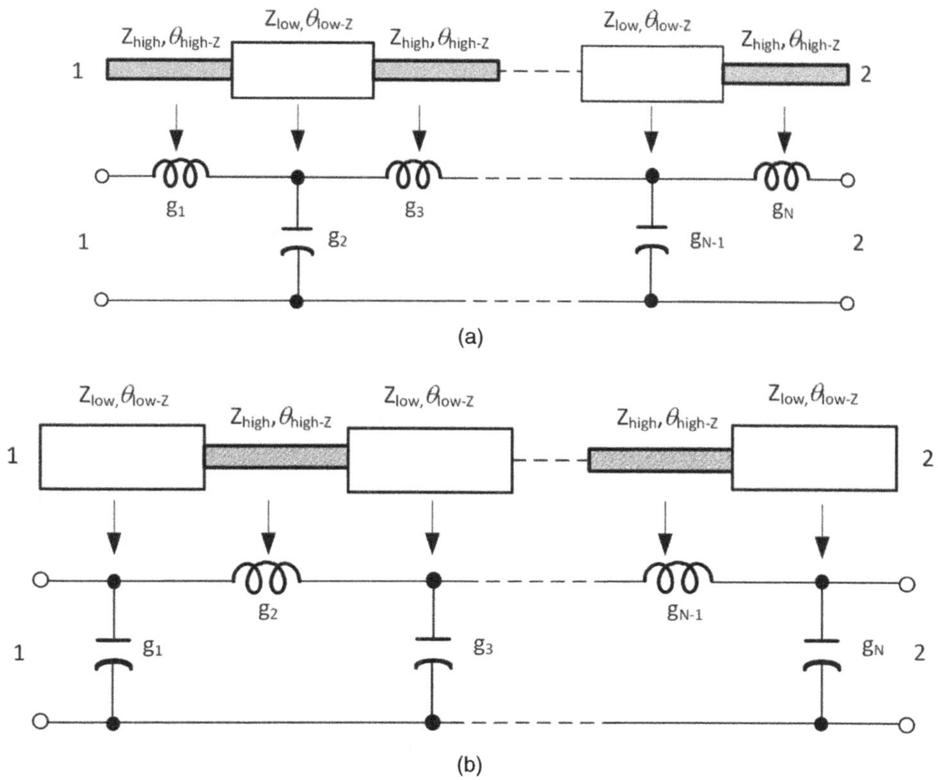

Figure 7.21 Stepped-impedance topologies and lumped-element prototypes.

The low-pass prototype can start with a series inductor or shunt capacitor, giving two topologies as shown in Figure 7.21. The prototype parameters are denoted by g_k, where $k = 1, 2, 3, \cdots, N$. The electrical lengths of the low- and high-impedance transmission line sections are denoted by θ_{low-Z} and θ_{high-Z}, respectively. The high-impedance transmission line functions as a series inductor, while the low-impedance transmission behaves like a shunt capacitor. Since the width of the low-impedance transmission line is wider than that of a high-impedance transmission line, a field discontinuity does occur and this needs to be taken into account during the design and optimisation process.

A simple design approach for this filter is to assume that the electrical length of each transmission line is short, that is, $\theta_k \leq \frac{\pi}{8}$. This assumption considerably simplifies the design task by converting each lumped element in the low-pass prototype to a short transmission line section. The first design step is to investigate the maximum and minimum impedance of the transmission lines which can be implemented. These maximum and minimum impedances have a significant effect on the stop-band performance of the filter and are constrained by details of the transmission line technology, substrate parameters and fabrication limitation. The maximum and minimum widths for microstrip lines were discussed in Chapter 6. If the topology shown in Figure 7.21(a) is employed, the lengths of the high- and low-impedance transmission lines can be calculated from:

$$\theta_{high-Z,k} = \frac{g_k Z_0}{Z_{high}} k = 1, 3, 5, \ldots \tag{7.86}$$

$$\theta_{low-Z,k} = \frac{g_k Z_{low}}{Z_0} k = 2, 4, 6, \ldots \tag{7.87}$$

On the other hand, if the topology in Figure 7.21(b) is chosen, the lengths of transmission lines are calculated from:

$$\theta_{high-Z,k} = \frac{g_k Z_0}{Z_{high}} k = 2, 4, 6, \ldots \tag{7.88}$$

$$\theta_{low-Z,k} = \frac{g_k Z_{low}}{Z_0} k = 1, 3, 5, \ldots \tag{7.89}$$

It is very important after this design step to check that the lengths of transmission lines obtained meet the condition:

$$\theta_i \leq \frac{\pi}{4} \tag{7.90}$$

If this is not the case, a different choice of impedance values should be made until the electrical length of each transmission line meet this condition.

Example: Design a stepped-impedance low-pass filter with a cut-off frequency at 20 GHz for 50 Ω system impedance. The desired filter has a third-order Chebyshev response with 0.1 dB ripple. The design employs microstrip transmission lines fabricated on a substrate with dielectric constant of 10.5 and thickness of 1.27mm. The loss tangent of the substrate is 0.02 at 20 GHz.

According to the substrate detail and the possible impedance range for microstrip, the low- and high-characteristic impedances are chosen to be 30 and 90 Ω. The g-values g_k for a third-order Chebyshev low-pass filter with 0.10 dB ripple are $g_1 = g_3 = 1.0316$ and $g_2 = 1.1474$. If the topology in Figure 7.21(a) is used, the electrical lengths of three transmission line sections are computed from Equations (7.86) and (7.87) to be:

$$\theta_{high-Z,1} = \theta_{high-Z,3} = \frac{g_1 Z_0}{Z_{high}} = \frac{1.0316 \times 50}{90} = 0.5731 \equiv 32.8362^\circ \tag{7.91}$$

$$\theta_{low-Z,2} = \frac{g_2 Z_{low}}{Z_0} = \frac{1.1474 \times 30}{50} = 0.6884 \equiv 39.4424^\circ \tag{7.92}$$

On the other hand, applying the topology in Figure 7.21(b) and computing the electrical lengths of transmission lines from Equations (7.88) and (7.89), one obtains

$$\theta_{low-Z,1} = \theta_{low-Z,5} = \frac{g_1 Z_{low}}{Z_0} = \frac{1.0316 \times 30}{50} = 0.6190 \equiv 35.4661^\circ \tag{7.93}$$

$$\theta_{high-Z,2} = \frac{g_2 Z_0}{Z_{high}} = \frac{1.1474 \times 50}{90} = 0.6374 \equiv 37.5203^\circ \tag{7.94}$$

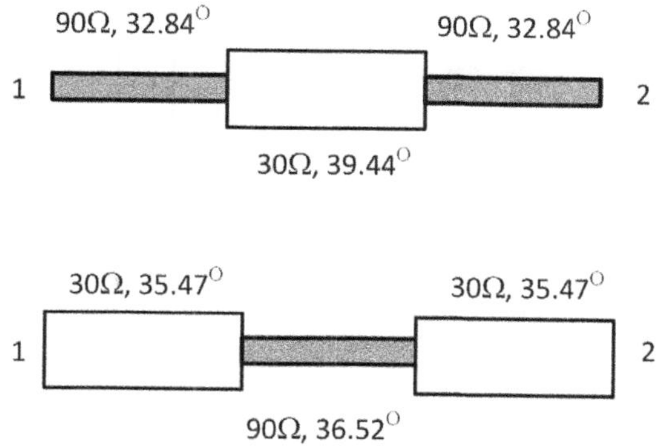

Figure 7.22 Final designs for the stepped-impedance low-pass filter.

It is noted that the lengths of all transmission lines satisfy the condition given in Equation (7.90). Hence, both topologies are valid and the designs for the stepped-impedance low-pass filters are shown in Figure 7.22.

7.8 Parallel-Coupled Line Filter

Due to its simplicity, the parallel-coupled line filter is one of the most commonly used band-pass filters for microwave and millimetre-wave communication systems. The basic cell used in this filter is either an open- or short-circuit half-wavelength resonator, parallel-coupled to its neighbours. Generally, an open-circuit resonator is preferred since it does not need any via-holes. The structure then comprises a number of open-ended parallel-coupled line sections in cascade, as shown in Figure 7.23. Each subsection has electrical length of 90° at the centre frequency of the pass-band and is type (c) in Figure 7.20, with a band-pass response. The frequency response of N cascaded sections allows a band-pass response with improved selectivity. The equivalent circuit model of a single section of open-ended coupled-lines is electrically equivalent to a network composed of a J-inverter and two UEs, as shown in Figure 7.24. Assuming that the coupled lines are lossless, the transmission matrix of the

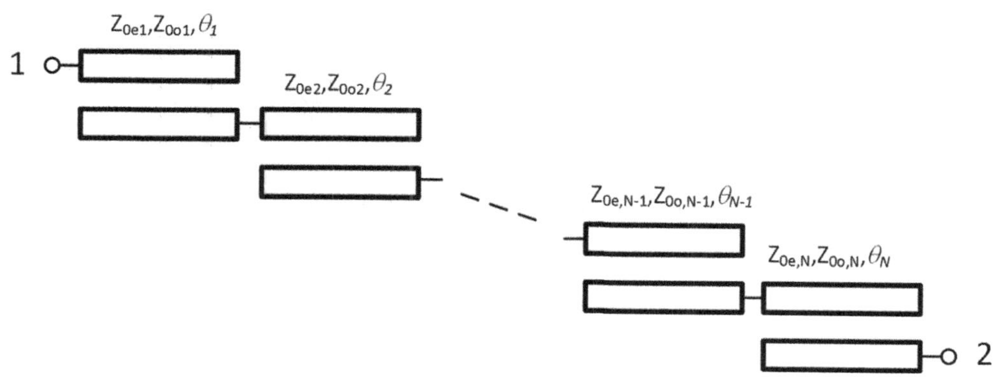

Figure 7.23 Parallel-coupled line filter with open-circuit resonators.

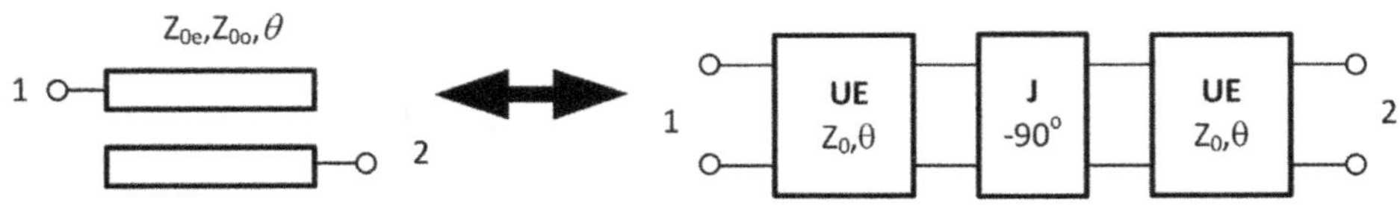

Figure 7.24 Open-ended coupled-line section and its equivalent circuit.

open-ended coupled line section is given by:

$$\mathbf{T}_{CL} = \begin{bmatrix} \cos\theta & jZ_0\sin\theta \\ jY_0\sin\theta & \cos\theta \end{bmatrix} \begin{bmatrix} 0 & -\frac{j}{J} \\ -jJ & 0 \end{bmatrix} \begin{bmatrix} \cos\theta & jZ_0\sin\theta \\ jY_0\sin\theta & \cos\theta \end{bmatrix} \tag{7.95}$$

Multiplying out the matrices in Equation (7.95), one obtains:

$$\mathbf{T}_{CL} = \begin{bmatrix} \left(Z_0J + \frac{Y_0}{J}\right)\sin\theta\cos\theta & j\left(JZ_0^2\sin^2\theta - \frac{1}{J}\cos^2\theta\right) \\ j\left(\frac{Y_0^2\sin^2\theta}{J} - J\cos^2\theta\right) & \left(Z_0J + \frac{Y_0}{J}\right)\sin\theta\cos\theta \end{bmatrix} \tag{7.96}$$

Setting the length of the coupled-line sections, θ, to $\frac{\pi}{2}$, one obtains:

$$\mathbf{T}_{CL} = \begin{bmatrix} 0 & jJZ_0^2 \\ j\frac{Y_0^2}{J} & 0 \end{bmatrix} \tag{7.97}$$

Meanwhile, the transmission matrix of an open-ended quarter-wave coupled-line section can be derived using the circuit model presented in Figure 7.20(c). This is given by:

$$\mathbf{T}_{CL} = \begin{bmatrix} 1 & jZ_{0o}\cot\theta \\ 0 & 1 \end{bmatrix} \begin{bmatrix} \cos\theta & j\frac{Z_{0e}-Z_{0o}}{2}\sin\theta \\ j\frac{2}{Z_{0e}-Z_{0o}}\sin\theta & \cos\theta \end{bmatrix} \begin{bmatrix} 1 & jZ_{0o}\cot\theta \\ 0 & 1 \end{bmatrix} \tag{7.98}$$

Multiplying out the matrices in Equation (7.98), gives:

$$\mathbf{T}_{CL} = \begin{bmatrix} \cos\theta\left(1 - \frac{2Z_{0o}}{Z_{0e}-Z_{0o}}\right) & j\left(\frac{Z_{0e}-Z_{0o}}{2}\sin\theta + Z_{0o}\frac{\cos^2\theta}{\sin\theta} + Z_{0o}\cot\theta\left[\cos\theta\left(1 - \frac{2Z_{0o}}{Z_{0e}-Z_{0o}}\right)\right]\right) \\ j\frac{2\sin\theta}{Z_{0e}-Z_{0o}} & \cos\theta\left(1 - \frac{2Z_{0o}}{Z_{0e}-Z_{0o}}\right) \end{bmatrix} \tag{7.99}$$

Substituting $\theta = \frac{\pi}{2}$ in Equation (7.99) and equating with Equation (7.97), gives the following conditions:

$$Z_{0e} - Z_{0o} = 2JZ_0^2 \tag{7.100}$$

$$\frac{Z_{0e} + Z_{0e}}{Z_{0e} - Z_{0o}} = Z_0J + \frac{1}{JZ_0} \tag{7.101}$$

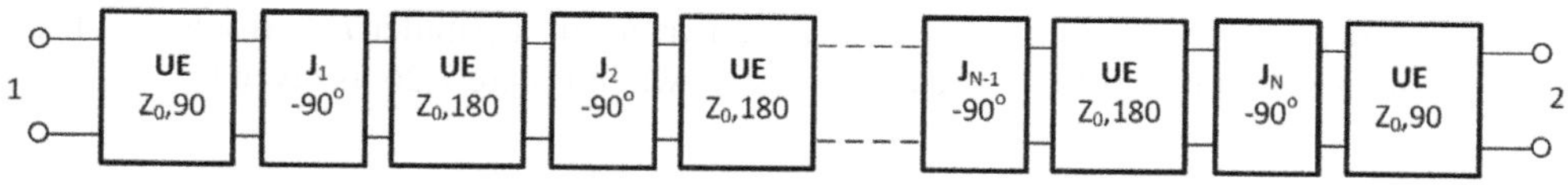

Figure 7.25 Circuit model for a cascaded coupled-lines filter based on *J*-inverters and UEs.

Solving these two conditions simultaneously, the parameters of each quarter-wave open-ended coupled-line section in the band-pass filter can be determined from:

$$Z_{0e} = Z_0\left(1 + JZ_0 + J^2Z_0^2\right) \tag{7.102}$$

$$Z_{0o} = Z_0\left(1 - JZ_0 - J^2Z_0^2\right) \tag{7.103}$$

A single coupled-line section is not sufficient to meet the specifications dictated by most wireless communication standards. Hence, a number, N, of quarter-wave coupled-line sections are cascaded to meet the requirement. The equivalent circuit model for a cascaded design is shown in Figure 7.25. The UE elements between two J-inverters have an electrical length equal to 180° at the centre frequency (ω_0). The transmission matrix of UE_{180} can be written in terms of frequency as:

$$\mathbf{T}_{UE,\theta=180^o}(\omega) = \begin{bmatrix} \cos\left(\frac{\omega}{v_p}l\right) & jZ_0\sin\left(\frac{\omega}{v_p}l\right) \\ j\frac{1}{Z_0}\sin\left(\frac{\omega}{v_p}l\right) & \cos\left(\frac{\omega}{v_p}l\right) \end{bmatrix} \tag{7.104}$$

At frequencies around ω_0 (i.e., $\omega_0 + \Delta\omega$), the term $\frac{\omega}{v_p}\ l$ is equal to $\frac{\omega_0+\Delta\omega}{\omega_0}\pi$. The transmission matrix of the half-wave UE can therefore be approximated by:

$$\mathbf{T}_{UE,\theta=180^o}\left(\omega_0 + \Delta\omega\right) = \begin{bmatrix} -1 & -j\pi Z_0\frac{\Delta\omega}{\omega_0} \\ -j\frac{\pi}{Z_0}\frac{\Delta\omega}{\omega_0} & -1 \end{bmatrix} \tag{7.105}$$

The resulting transmission matrix is equal to the transmission matrix of the shunt network cascaded with a $1 : -1$ transformer, as shown in the circuit in Figure 7.26. The shunt network will have an admittance given by:

$$Y_{in,UE} = j\frac{\pi}{Z_0}\frac{\Delta\omega}{\omega_0} \tag{7.106}$$

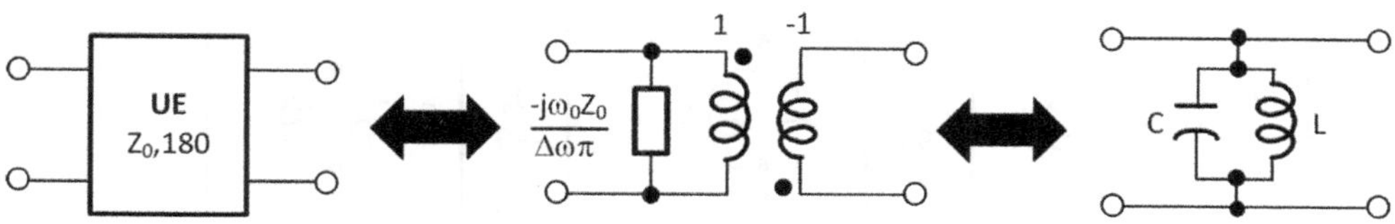

Figure 7.26 Half-wavelength UE as a parallel resonator.

It can be shown that the response of UE_{180} is equivalent to a parallel L-C resonator. The transmission matrix of a parallel L-C resonator, as shown in Figure 7.26, is given by:

$$\mathbf{T}_{shunt-LC} = \begin{bmatrix} 1 & 0 \\ \frac{1-\omega^2 LC}{j\omega L} & 1 \end{bmatrix} \tag{7.107}$$

The matrix element $\mathbf{T}_{shunt-LC}(2,1)$ is the input admittance of the L-C resonator network. At the frequency $\omega_0 + \Delta\omega$, the input admittance of L-C resonator can be approximated by:

$$Z_{in} = -j\frac{\omega_0\left(\omega_0 + \Delta\omega\right)}{2\Delta\omega}L \approx -j\frac{\omega_0^2}{2\Delta\omega}L \tag{7.108}$$

Hence, from Equations (7.106) and (7.108), the inductor and capacitor of the shunt L-C network which is equivalent to the half-wave UE of parameter Z_0 are given by:

$$L = \frac{2Z_0}{\pi\omega_0} \tag{7.109}$$

$$C = \frac{\pi}{2Z_0\omega_0} \tag{7.110}$$

Looking at Figure 7.25, there are still two quarter-wave UEs connected at the input and output ports. We have already shown that a quarter-wave UE is equivalent to a quarter-wave transmission line of Z_0. Therefore, this has no effect on the magnitude response of the filter. Hence, the parallel-coupled band-pass filter modelled by the J-inverter network shown in Figure 7.25 can be converted to the model shown in Figure 7.27. According to the design equations for a lumped-element band-pass filter, all design parameters of a parallel coupled-line band-pass filter can be computed from:

$$J_1 = \frac{1}{Z_0}\sqrt{\frac{\pi\Delta f}{2g_0g_1f_0}} \tag{7.111}$$

$$J_i = \frac{\pi\Delta f}{2Z_0f_0\sqrt{g_ig_{i-1}}}, i = 1,2,3,\ldots,n \tag{7.112}$$

$$J_{n+1} = \frac{1}{Z_0}\sqrt{\frac{\pi\Delta f}{2g_ng_{n+1}f_0}} \tag{7.113}$$

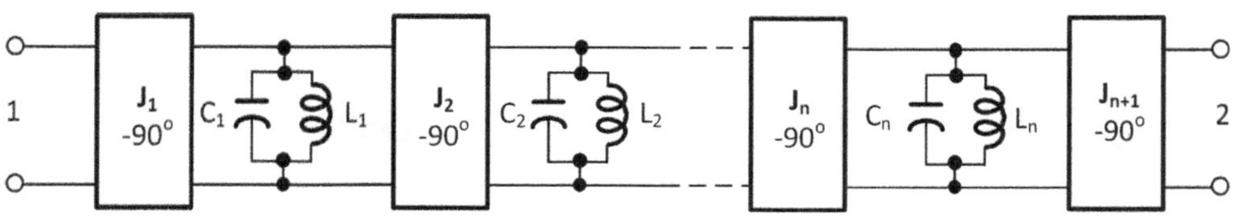

Figure 7.27 J-inverter band-pass model for the parallel-coupled band-pass filter.

From Equations (7.102) and (7.103), the parameters of each coupled lines section can be computed from

$$Z_{0e,i} = Z_0\left(1 + J_i Z_0 + J_i^2 Z_0^2\right) \quad (7.114)$$

$$Z_{0o,i} = Z_0\left(1 - J_i Z_0 - J_i^2 Z_0^2\right) \quad (7.115)$$

Example: Design a parallel-coupled band-pass filter with 0.1 dB ripple and fifth-order Chebyshev response at a centre frequency of 10 GHz with 1 GHz bandwidth and 50 Ω system impedance.

The bandwidth $\Delta\omega$ is 1 GHz and the center frequency ω_0 is equal to 10 GHz. The coefficients for 0.1 dB ripple and fifth-order Chebyshev response ($n = 5$) are:

$$g_0 = 1, g_1 = 1.1468,\ g_2 = 1.3712,\ g_3 = 1.9750,\ g_4 = 1.3712,\ g_5 = 1.1468,\ g_6 = 1$$

From Equations (7.111) to (7.113) we obtain,

$$J_1 = J_6 = \frac{1}{Z_0}\sqrt{\frac{\pi\Delta f}{2g_0 g_1 f_0}} = \frac{1}{50}\sqrt{\frac{\pi\times 1\times 10^9}{2\times 1\times 1.1468\times 10\times 10^9}} = 0.0074 \quad (7.116)$$

$$J_2 = J_5 = \frac{\pi\Delta f}{2Z_0 f_0\sqrt{g_2 g_1}} = \frac{\pi\times 1\times 10^9}{2\times 50\times 10\times 10^9\sqrt{1.3712\times 1.1468}} = 0.0025 \quad (7.117)$$

$$J_3 = J_4 = \frac{\pi\Delta f}{2Z_0 f_0\sqrt{g_3 g_2}} = \frac{\pi\times 1\times 10^9}{2\times 50\times 10\times 10^9\sqrt{1.975\times 1.3712}} = 0.0019 \quad (7.118)$$

Using Equations (7.114) and (7.115), the even- and odd-mode impedances of each coupled-line are:

$$Z_{0e,1} = Z_{0e,6} = Z_0\left(1 + J_1 Z_0 + J_1^2 Z_0^2\right) = 50\left(1 + 0.0074\times 50 + (0.0074\times 50)^2\right) = 75.3450\,\Omega \quad (7.119)$$

$$Z_{0o,1} = Z_{0o,6} = Z_0\left(1 - J_1 Z_0 - J_1^2 Z_0^2\right) = 50\left(1 - 0.0074\times 50 - (0.0074\times 50)^2\right) = 24.6550\,\Omega \quad (7.120)$$

$$Z_{0e,2} = Z_{0e,5} = Z_0\left(1 + J_2 Z_0 + J_2^2 Z_0^2\right) = 50\left(1 + 0.0025\times 50 + (0.0025\times 50)^2\right) = 57.0313\,\Omega \quad (7.121)$$

$$Z_{0o,2} = Z_{0o,5} = Z_0\left(1 - J_2 Z_0 - J_2^2 Z_0^2\right) = 50\left(1 - 0.0025\times 50 - (0.0025\times 50)^2\right) = 42.9688\,\Omega \quad (7.122)$$

$$Z_{0e,3} = Z_{0e,4} = Z_0\left(1 + J_3 Z_0 + J_3^2 Z_0^2\right) = 50\left(1 + 0.0019\times 50 + (0.0019\times 50)^2\right) = 55.2013\,\Omega \quad (7.123)$$

$$Z_{0o,3} = Z_{0o,4} = Z_0\left(1 - J_3 Z_0 - J_3^2 Z_0^2\right) = 50\left(1 - 0.0019\times 50 - (0.0019\times 50)^2\right) = 44.7988\,\Omega \quad (7.124)$$

The design equations for the impedance $Z_{0,i}$ and coupling coefficient (C_i) of the coupled lines are:

$$Z_{0,i} = \sqrt{Z_{0o,i} Z_{0e,i}} \tag{7.125}$$

$$C_i = \frac{Z_{0e,i} - Z_{0o,i}}{Z_{0e,i} + Z_{0o,i}} \tag{7.126}$$

For coupled-lines 1 and 6 this gives:

$$Z_{0,1} = Z_{0,6} = \sqrt{Z_{0o,1} Z_{0e,1}} = 43.1002\,\Omega \tag{7.127}$$

$$C_1 = C_6 = 20\log_{10}\left(\frac{Z_{0e,1} - Z_{0o,1}}{Z_{0e,1} + Z_{0o,1}}\right) = -5.9016\,\text{dB} \tag{7.128}$$

For coupled-lines 2 and 5:

$$Z_{0,2} = Z_{0,5} = \sqrt{Z_{0o,2} Z_{0e,2}} = 49.5032\,\Omega \tag{7.129}$$

$$C_2 = C_5 = 20\log_{10}\left(\frac{Z_{0e,2} - Z_{0o,2}}{Z_{0e,2} + Z_{0o,2}}\right) = -17.0388\,\text{dB} \tag{7.130}$$

And for coupled-lines 3 and 4:

$$Z_{0,3} = Z_{0,3} = \sqrt{Z_{0o,3} Z_{0e,3}} = 49.7288\,\Omega \tag{7.131}$$

$$C_3 = C_4 = 20\log_{10}\left(\frac{Z_{0e,3} - Z_{0o,3}}{Z_{0e,3} + Z_{0o,3}}\right) = -19.6573\,\text{dB} \tag{7.132}$$

The electrical length of each coupled-line section is 90° at 10 GHz. The filter topology for this example is shown in Figure 7.28. Figure 7.29 shows a typical implementation of this kind of filter in a subsystem.

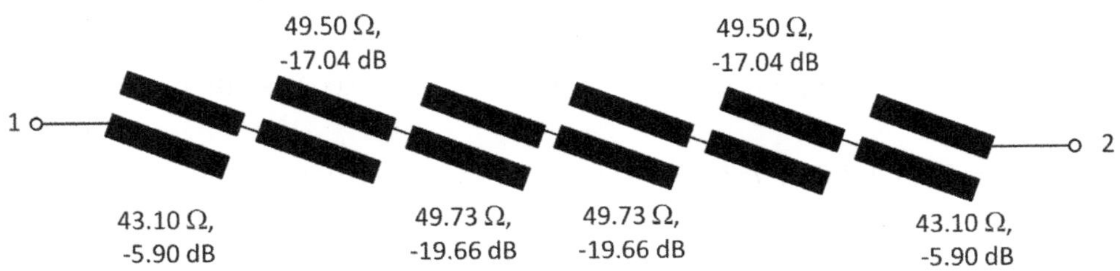

Figure 7.28 Coupled-line band-pass filter.

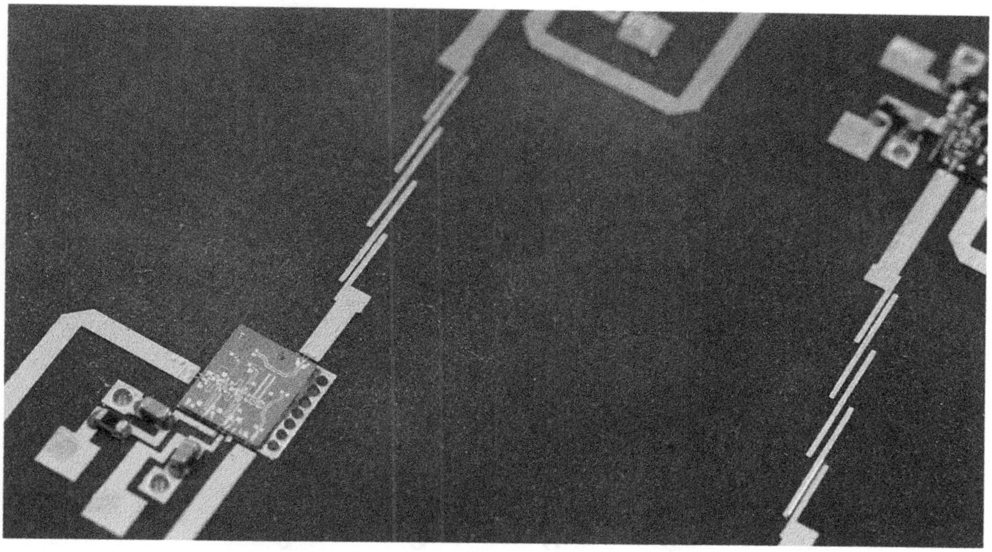

Figure 7.29 Photograph of microstrip parallel-coupled bandpass filters integrated with MMICs on Duroid™. Illustration courtesy of Plextek RFI. Used with permission; all rights reserved.

7.9 Interdigital Filter

An interdigital filter network designed for microwave applications is shown in Figure 7.30. The filter is formed with an array of quarter-wave transmission line resonators. Each resonator is shorted to ground at one end and open-circuited at the other end. The transmission line resonators are placed side-by-side and oriented alternately in order to give the correct sense of coupling. An n-section interdigital filter, as shown in Figure 7.30, can be approximately described using short-circuit coupled-line elements, as shown in Figure 7.31(a). As shown in Figure 7.20(b), each short-circuited coupled-line element has a band-pass response. Consequently, the interdigital filter shown in Figure 7.31(a) can be modelled by its equivalent circuit, as shown in Figure 7.31(b) with the model in Figure 7.20(b). This model comprises n

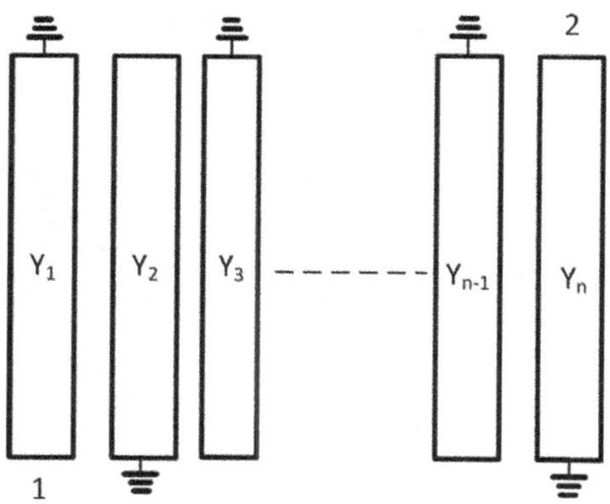

Figure 7.30 Interdigital filter network.

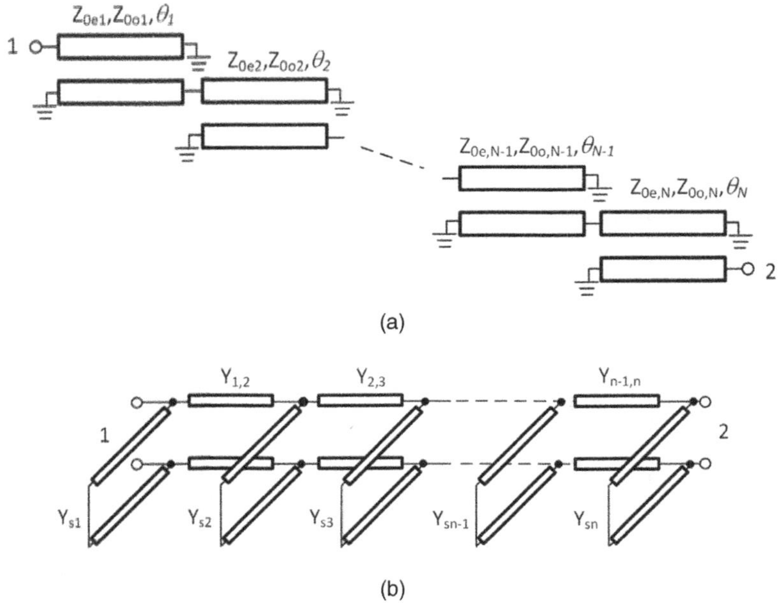

Figure 7.31 Circuit model for an interdigital filter using (a) short-circuited coupled lines, and (b) quarter-wave transmission lines and short-circuited stubs.

quarter-wave short-circuited stubs with characteristic admittance Y_{si} connected with n quarter-wave transmission lines with characteristic admittance $Y_{i,i+1}$. Generally, a J-inverter can be converted to the Π-network Y_b as shown in Figure 7.14, where the inverter parameter J can be written as a function of Y_b as:

$$J = \frac{Y_b}{sin\theta} \tag{7.133}$$

Hence, each transmission line section $Y_{i,i+1}$ in Figure 7.31(b) can be converted to the equivalent J-inverter and two shorted stubs of characteristic admittance $Y_{i,i+1}$. This model is shown in Figure 7.32. It is composed of $(n-1)$ J-inverters and n shunt stubs. Obviously, the

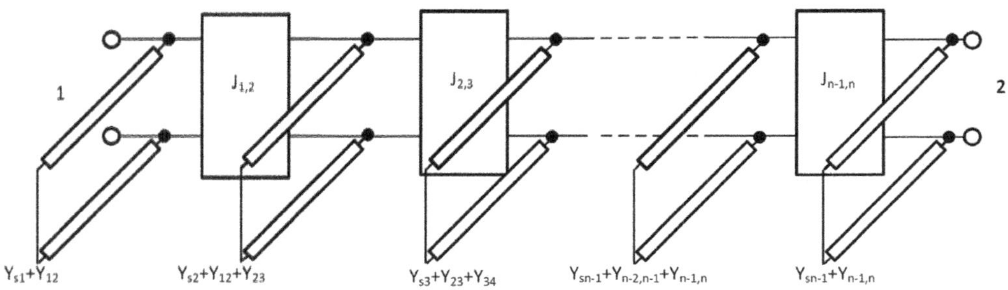

Figure 7.32 Circuit model for an interdigital band-pass filter using J-inverters and short-circuited stubs.

model is a cascade of N band-pass sections where each section consists of a J-inverter and two shunt stubs.

The physical structure of an interdigital filter shown in Figure 7.30 can be considered as an array of conductors. The electric field to ground of each conductor can be modelled as a capacitance connected to ground and is referred to as the self-capacitance, denoted by C_i for the ith conductor. There are also electric-field couplings between conductors. As a first-order approximation, only electric field coupling between adjacent conductors is considered. This is modelled by a capacitance denoted by C_{ij}, which represents the mutual coupling between conductor i and j. For simplicity, an interdigital filter with five conductors as shown in Figure 7.33(a) is considered here. There are five self-capacitances from five conductors denoted by C_1, C_2, C_3, C_4 and C_5. Four mutual capacitances denoted by C_{12}, C_{23}, C_{34} and C_{45} represent field coupling between each pair of adjacent conductors, 1–2, 2–3, 3–4 and 4–5, respectively. Each self-capacitance can be separated into two self-capacitances, as shown in Figure 7.33(b). Using this approach, the three inner conductors can be split into six conductors, with each conductor being split into two. Consequently, the five-conductor interdigital filter can be modelled as four pairs of parallel coupled lines, as shown in Figure 7.34. Each

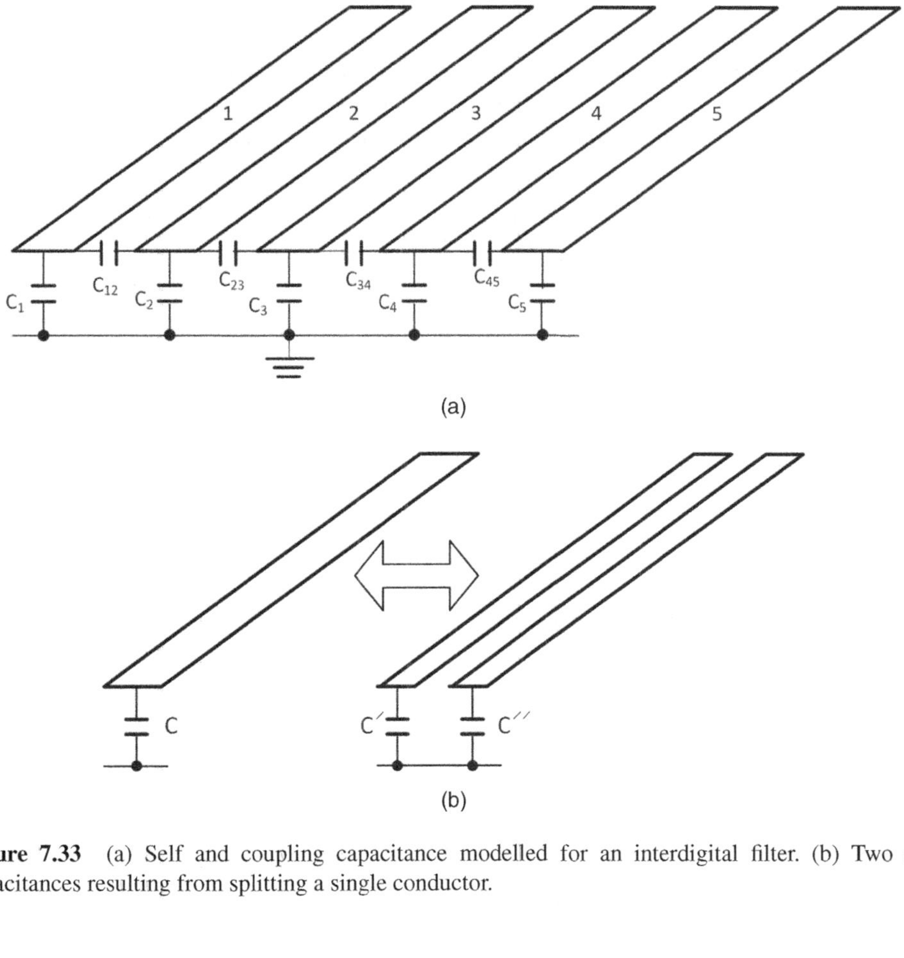

Figure 7.33 (a) Self and coupling capacitance modelled for an interdigital filter. (b) Two self-capacitances resulting from splitting a single conductor.

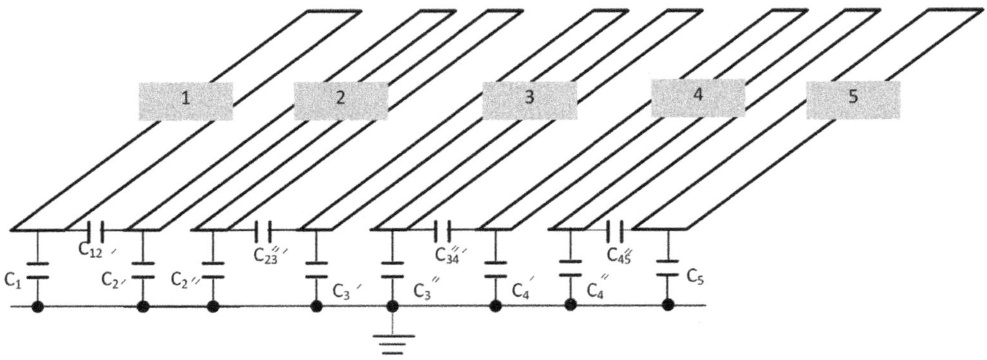

Figure 7.34 Asymmetrical coupled-line model for a five-conductor interdigital filter.

self-capacitance C_i is split into two self-capacitances, C_i' and C_i'' with the rule $C_i = C_i' + C_i''$. Hence, a five-conductor interdigital filter can be modelled as four pairs of asymmetrical coupled lines and the design of this filter can be performed if the parameters of each pair of coupled lines are determined. As shown in Figure 7.34, there are eight parameters of coupled-line pairs, that is $Z_{0oi,i+1}$ and $Z_{0ei,i+1}$, ($i = 1, 2, 3$ and 4), which need to be found.

The design procedure for the interdigital filter based on this asymmetrical coupled-lines model begins by determining the parameters of the J-inverters appearing in the model of Figure 7.31. The input admittance of the short-circuit stub is $-j\frac{Y_0}{\tan\theta}$, so the inverter parameter in the equivalent circuit of the distributed J-inverter should first be computed from the parameters of the prototype filter using:

$$J_{i,i+1} = \frac{Y_1}{\tan\theta\sqrt{g_i g_{i+1}}}; i = 1, \ldots, n-1 \tag{7.134}$$

$$\theta = \frac{\pi}{2}\left(1 - \frac{\Delta f}{2f_0}\right) \tag{7.135}$$

where Y_1 is chosen arbitrarily.

As already described, an interdigital filter can be modelled as multiconductor coupled lines. Given that $i = 2$ to $n - 2$, the parameters of each coupled line section can be computed from:

$$Z_{0e1,2} = \frac{1}{Y_1 - Y_{1,2}} \tag{7.136}$$

$$Z_{0o1,2} = \frac{1}{Y_1 + Y_{1,2}} \tag{7.137}$$

$$Z_{0ei,i+1} = \frac{1}{2Y_1 - Y_{0ei-1,i} - Y_{i,i+1} - Y_{i-1,i}} \tag{7.138}$$

$$Z_{0oi,i+1} = \frac{1}{2Y_{i,i+1} + Y_{0ei,i+1}} \tag{7.139}$$

$$Z_{0en-1,n} = \frac{1}{Y_1 - Y_{n-1,n}} \tag{7.140}$$

$$Z_{0on-1,n} = \frac{1}{Y_1 + Y_{n-1,n}} \tag{7.141}$$

where

$$Y_{i,i+1} = J_{i,i+1} \sin\theta \tag{7.142}$$

The input and output matching of the interdigital filter is achieved by feeding a transmission line of characteristic admittance Y_t at the input and output conductors at a point distance θ_t from the shorted end of the conductor, where:

$$\theta_t = \frac{\sin^{-1}\left(\sqrt{\frac{Y_1 \sin\theta \cos\theta}{Y_0 g_0 g_1}}\right)}{1 - \frac{\Delta f}{2f_0}} \tag{7.143}$$

The admittance Y_t is computed from:

$$Y_t = Y_1 - \frac{Y_{1,2}^2}{Y_1} \tag{7.144}$$

Using this tapped-feed on the input and output conductors does shift the resonant frequency. To shift the resonant frequency back to the original one, a compensation capacitor (C_t) is needed to load the input and output conductors. The value of C_t is given by:

$$C_t = \frac{Z_t \cos\theta_t \sin^3\theta_t}{\omega_0\left(Z_0^2 + Z_t^2 \cos^2\theta_t \sin^2\theta_t\right)} \tag{7.145}$$

With all the electrical parameters derived from the design equations above, the next step is the physical realisation which depends on the transmission line medium to be used. If a microstrip or stripline technology is chosen, readers are suggested to consult some excellent textbooks [1–4].

Example: Design a fifth-order interdigital filter for 50 Ω system impedance with a centre frequency of 20 GHz and bandwidth of 4 GHz, giving a Chebyshev magnitude response with 0.1 dB ripple.

The g-values for the fifth Chebyshev filter with 0.1 dB ripple are:

$$g_0 = 1,\ g_1 = 1.1468,\ g_2 = 1.3712,\ g_3 = 1.9750,\ g_4 = 1.3712,\ g_5 = 1.1468,\ g_6 = 1$$

The fractional bandwidth in this example is $\frac{\Delta f}{f_0} = 4/20 = 0.20$ where the electrical length of each transmission line resonator is:

$$\theta = \frac{\pi}{2}\left(1 - \frac{0.20}{2}\right) = 81^o \tag{7.146}$$

We choose Y_t such that its characteristic impedance is 50 Ω, hence $Y_t = 0.02$. The characteristic admittance (Y_1) is chosen to be equal to 0.025. Hence, we can calculate inverter parameters as follows:

$$J_{1,2} = \frac{Y_1}{\tan\theta\sqrt{g_1 g_2}} = \frac{0.025}{\tan 81^o\sqrt{1.1468\times 1.3712}} = 0.0032 \quad (7.147)$$

$$J_{2,3} = \frac{Y_1}{\tan\theta\sqrt{g_2 g_3}} = \frac{0.025}{\tan 81^o\sqrt{1.3712\times 1.9750}} = 0.0024 \quad (7.148)$$

$$J_{3,4} = \frac{Y_1}{\tan\theta\sqrt{g_3 g_4}} = \frac{0.025}{\tan 81^o\sqrt{1.3712\times 1.9750}} = 0.0024 \quad (7.149)$$

$$J_{4,5} = \frac{Y_1}{\tan\theta\sqrt{g_4 g_5}} = \frac{0.025}{\tan 81^o\sqrt{1.1468\times 1.3712}} = 0.0032 \quad (7.150)$$

Thus, from Equation (7.142) we obtain:

$$Y_{1,2} = J_{1,2}\sin\theta = 0.0032\times\sin 81^o = 0.00316 \quad (7.151)$$

$$Y_{2,3} = J_{2,3}\sin\theta = 0.0024\times\sin 81^o = 0.00237 \quad (7.152)$$

$$Y_{3,4} = J_{3,4}\sin\theta = 0.0024\times\sin 81^o = 0.00237 \quad (7.153)$$

$$Y_{4,5} = J_{4,5}\sin\theta = 0.0032\times\sin 81^o = 0.00316 \quad (7.154)$$

The even- and odd-mode impedances of the coupled lines are computed from Equations (7.136) to (7.141), yielding:

$$Z_{0e1,2} = Z_{0e4,5} = \frac{1}{0.025 - 0.00316} = 45.8750\,\Omega \quad (7.155)$$

$$Z_{0o1,2} = Z_{0o4,5} = \frac{1}{0.025 + 0.00316} = 35.5114\,\Omega \quad (7.156)$$

$$Z_{0e2,3} = Z_{0e3,4} = \frac{1}{2\times 0.025 - 1/45.8750 - 0.00237 - 0.00316} = 44.1112\,\Omega \quad (7.157)$$

$$Z_{0o2,3} = Z_{0o3,4} = \frac{1}{2Y_{2,3} + Y_{0e2,3}} = \frac{1}{2\times 0.00237 + 1/44.1112} = 37.4830\,\Omega \quad (7.158)$$

The tapped length and the characteristic admittance of the tapped line are computed from Equations (7.143) to (7.145) as:

$$\theta_t = \frac{\sin^{-1}\left(\sqrt{\frac{0.025\sin 81^o\cos\theta 81^o}{0.02\times 1\times 1.1468}}\right)}{1-\frac{0.2}{2}} = 27.9209^o \quad (7.159)$$

$$Y_t = Y_1 - \frac{Y_{1,2}^2}{Y_1} = 0.025 - \frac{(0.00316)^2}{0.025} = 0.0246 \quad (7.160)$$

Figure 7.35 shows the parameters of the interdigital line filter for this design example.

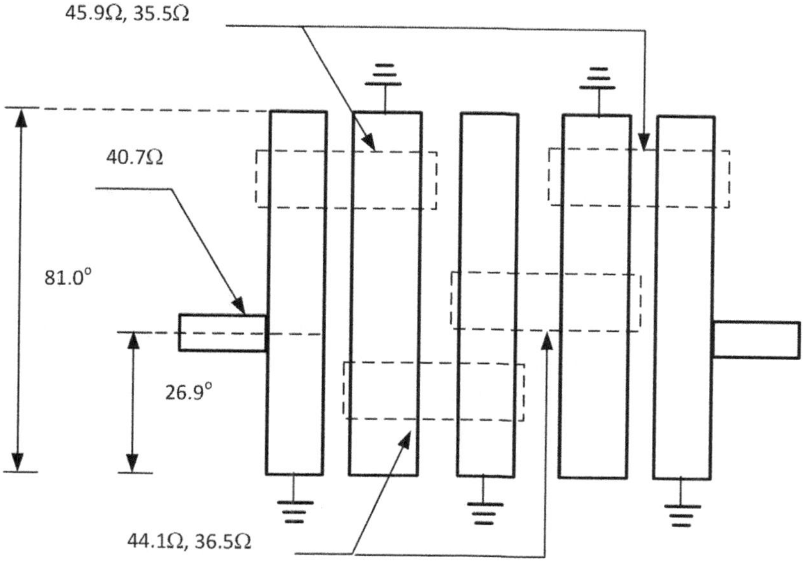

Figure 7.35 Final design for the fifth order interdigital filter.

7.10 Combline Filter

Compared with an interdigital filter, the combline filter shown in Figure 7.36 provides a more compact band-pass design. For an nth order combline filter, there are $n + 2$ transmission lines. The transmission lines Y_0 and Y_{n+1} match the input and output impedance. In Figure 7.36, there are n lumped-distributed resonators in a row between the transmission lines Y_0 and Y_{n+1}. The open end of each resonator is capacitively loaded with a lumped capacitor, while the other end is connected to ground. All resonators have equal length, which is less than

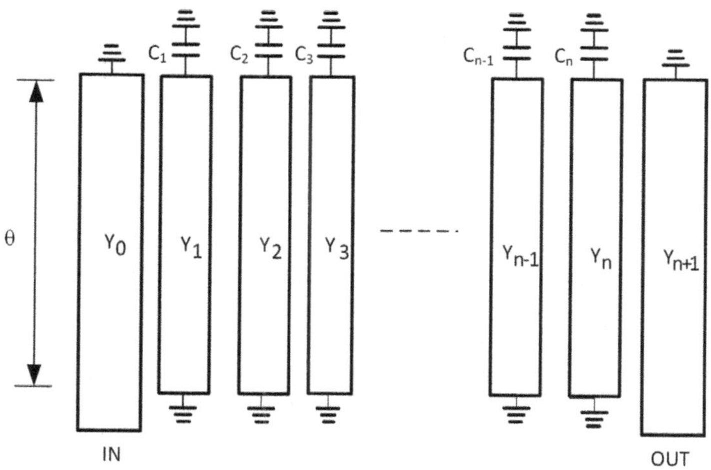

Figure 7.36 Combline filter network.

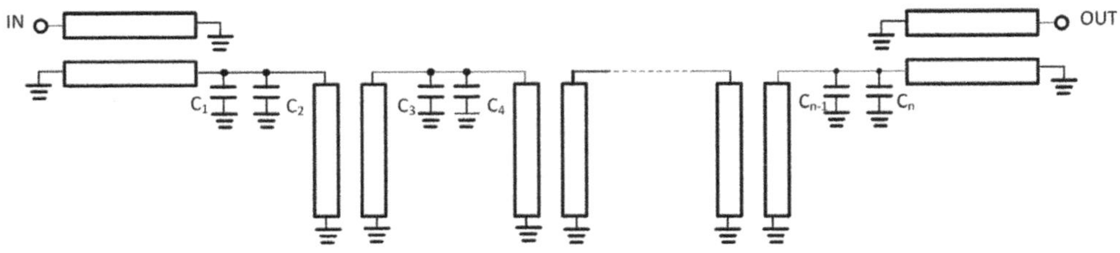

Figure 7.37 Coupled-line model for a combline filter.

90° at the centre frequency, f_0, because of the loading capacitor C_i. By applying the lumped-distributed implementation approach, a wider stopband response is achieved comparing with purely distributed resonators. If these capacitors are removed, the lengths of all transmission line resonators will be 90°. A larger loading capacitor offers a shorter transmission line. This reduces the dimensions of the combline filter and provides a wider stopband. However, a minimum of transmission line length exists since the insertion loss of the filter is a combination of the transmission line resonators as well as the lumped capacitors. Theoretically, this loss depends on the quality factors of the loading capacitors and the transmission lines. Figure 7.37 shows an approximate model of the combline filter based on coupled lines. The model is obtained by arranging the topology in a specific manner. Some couplings between adjacent resonators are neglected. For example, the coupling between resonator pair $Y_1 - Y_2$, $Y_3 - Y_4$, and so on.

Applying the equivalent model shown in Figure 7.20(b), the input (Y_0, Y_1) and output coupled-line sections (Y_n, Y_{n+1}) can be modelled with a UE plus two shorted stubs. The inner coupled-line sections $(Y_i, Y_{i+1};\ i = 2, 3, \ldots, n-1)$ in the combline filter are modelled with series stubs sandwiched with two shunt stubs, as shown in Figure 7.20(d). The loading capacitors are connected in parallel with all shunt stubs except those at the input and output ports. Since a UE represents the lossless transmission line model as previously described, the equivalent circuit using shorted stub and transmission line model for a combline filter is shown in Figure 7.38.

To simplify the combline filter model in Figure 7.38 further, the distributed Π-model for a J-inverter in Figure 7.39(a) is considered. Based on this model, a series stub can be modelled with a J-inverter sandwiched between two shunt stubs, as shown in Figure 7.39(b). The parameter of the admittance-inverter (J) is related to the characteristic admittance (Y) of the series stub length θ by:

$$J = Y \cot\theta \tag{7.161}$$

Based on the equivalent circuit in Figure 7.39(b), each series stub of length θ in Figure 7.38 can be modelled with a J-inverter and two shunt stubs of the same length. In the coupled-line model shown in Figure 7.37, the input and output networks of the combline filter are composed of shorted asymmetrical coupled-line networks. This part can be modelled using an ideal transformer and a shunt shorted stub. Compared with the input and output networks formed with the Π network, as shown in Figure 7.38, we can conclude that the two networks are electrically equivalent. Figure 7.40 shows these two networks which are electrical equivalent with the shorted-end asymmetrical coupled lines $(Y^a_{0o}, Y^a_{0e}, Y^b_{0o}, Y^b_{0e})$. The coupled-line network

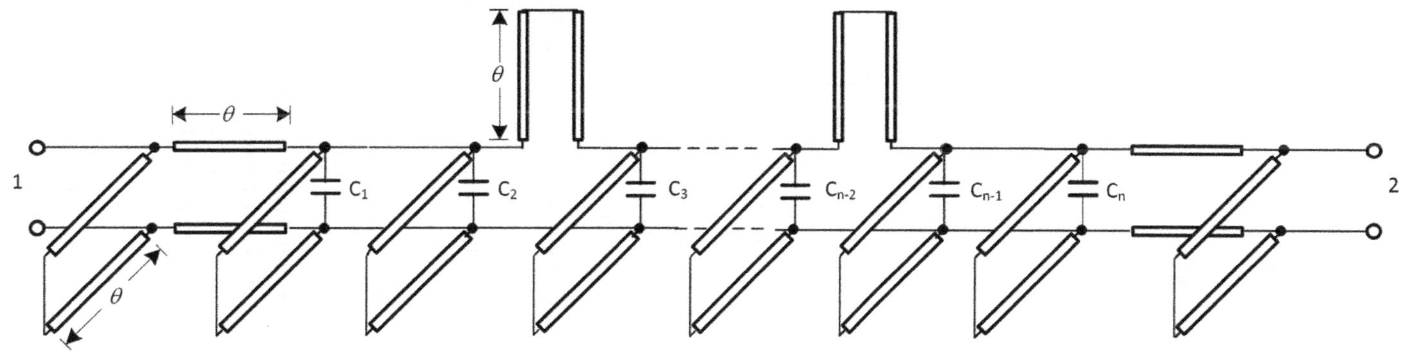

Figure 7.38 Equivalent circuit model for a combline filter using shunt and series stubs.

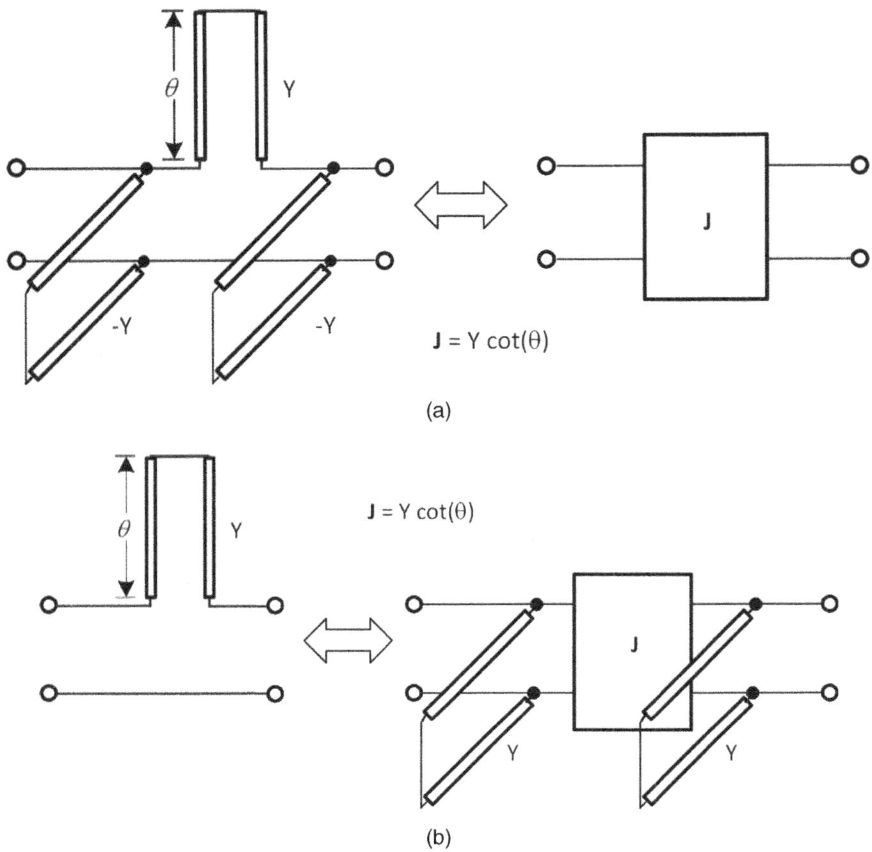

Figure 7.39 Equivalent circuit for (a) J-inverter using a series stub and (b) series stub using a J-inverter.

is terminated at one port by Y_0, which is either source or load admittance. It is related to the parameters of asymmetrical coupled-lines by:

$$\frac{Y^a_{0o} + Y^a_{0e}}{2} = Y_0 \tag{7.162}$$

The transformer model is chosen to model the input and output matching networks due to its simpler form than the Π network. The transformer model consists of a 1 : N ideal transformer and a shunt shorted stub. The electrical length of the shorted stub is equal to that of the asymmetrical coupled lines where other design parameters are related to the parameters of the asymmetrical coupled lines by:

$$Y_s = Y_0 \left(\frac{n^2 - 1}{N^2} \right) + Y^b_{0e} - Y^a_{0e} \tag{7.163}$$

$$N = \frac{2Y_0}{Y^a_{0o} - Y^a_{0e}} \tag{7.164}$$

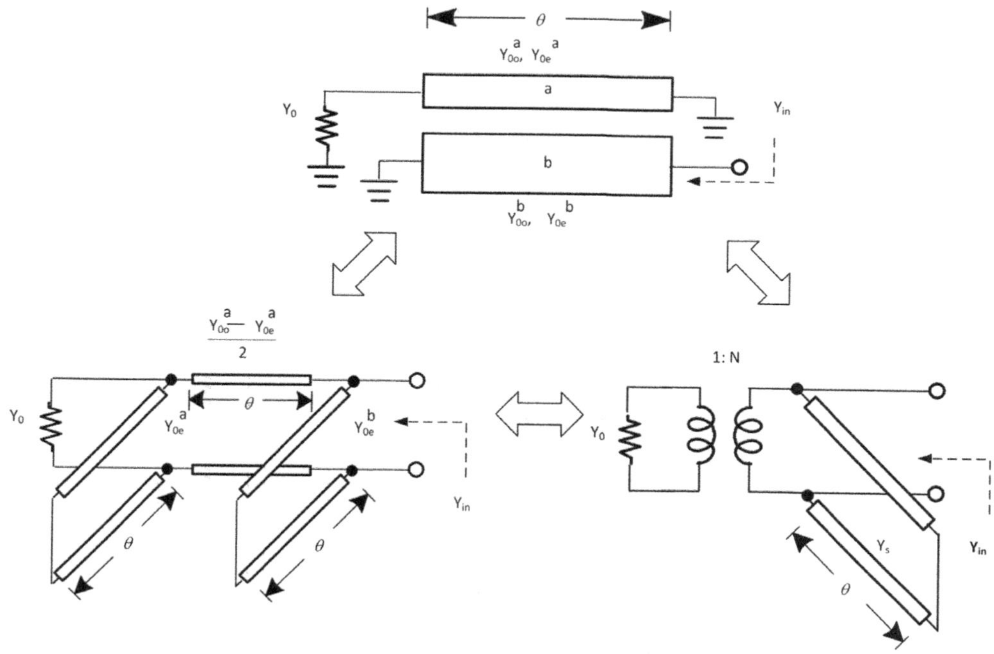

Figure 7.40 Equivalent circuit for input and output matching networks of a combline filter.

The final simplification is done by combining several shunt elements, either shorted stubs or lumped capacitors. Based on the equivalent networks shown in Figure 7.39(b) and 7.40, the model for a combline filter after applying these models is shown in Figure 7.41.

As shown in Figure 7.41, the network section 1A at the input and output ports comes from the equivalent circuit of the shorted asymmetrical coupled lines in Figure 7.40. Meanwhile, the network section 1B is obtained by replacing the shorted series stub with the equivalent network in Figure 7.39(b). If we assume that the stubs and loading capacitors are lossless, the adjacent shunted stubs and capacitors can be combined to a single reactive element of susceptance B_i appearing between each J-inverter. Hence, the final model of the combline filer will be reduced to the model presented in Figure 7.42.

The model for designing a combline filter shown in Figure 7.42 is composed of J-inverters $J_{k,k+1}$, where $k = 1, 2, \ldots, n-2$. It is obvious that the final model for a combline filter is very similar to a generalised band-pass filter designed using J-inverters presented in Figure 7.8(b). The only difference is the input and output transformers with turn ratio N that appeared in Figure 7.42. Hence, the design equations for a combline filter can be taken from a generalised band-pass filter using J-inverters. The design procedure for a combline filter is now explained in detail.

Step I: Design the prototype filter which meets the desired filter specifications.

Step II: Select the characteristic admittance Y_{ai} of each transmission line resonator, i. For simple design and implementation, the characteristic impedance of each resonator should be equal. Since Y_{ai} determines the cross-section of the resonator, which in turn affects the

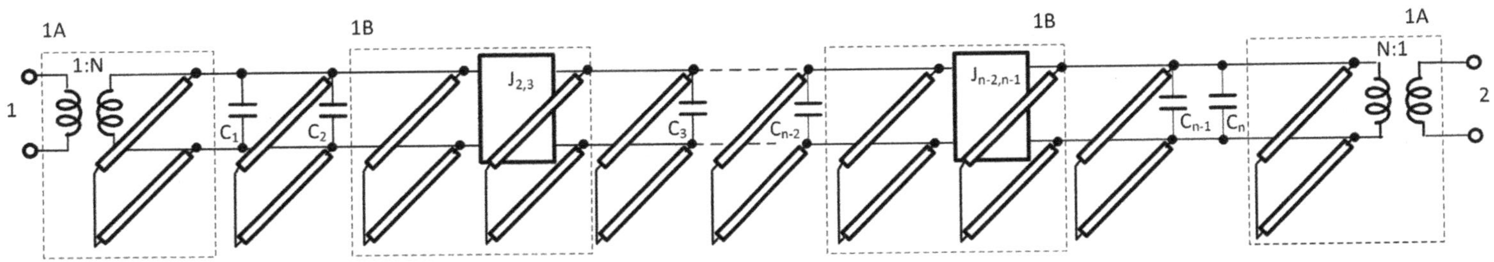

Figure 7.41 A model for a combline filter obtained by replacing input/output network and shorted series stubs with the networks in Figures 7.39 and 7.40.

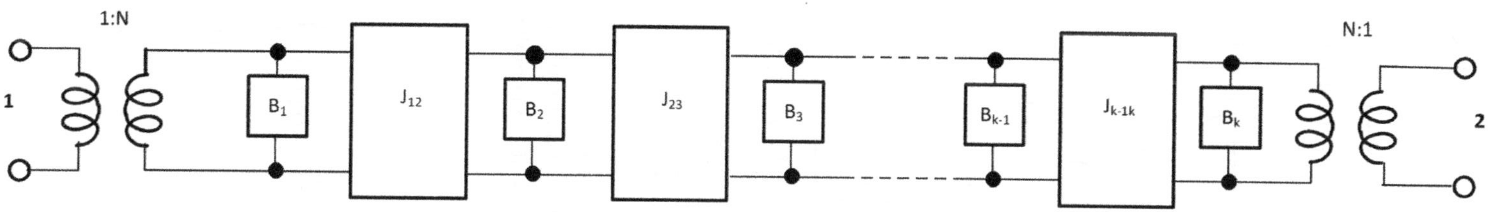

Figure 7.42 Model for designing a combline filter.

unloaded quality factor, an optimum of Y_{ai} should be chosen in the design. It is suggested [1] that the value of Y_{ai} should be 0.0143 or $1/70$.

Step III: Choose the electrical length of the resonator θ.

Step IV: Calculate the slope parameter b_i of resonator i from:

$$b_i = Y_{ai}\left(\frac{\cot\theta + \theta\csc^2\theta}{2}\right);\ i = 1, 2, \ldots, n-1 \quad (7.165)$$

Step V: Calculate the loading capacitors, C_i, from:

$$C_i = \frac{Y_{ai}\cot\theta}{2\pi f_0} \quad (7.166)$$

Step VI: Calculate the parameters of the admittance inverters from:

$$J_{01} = \sqrt{\frac{\Delta f}{f_0}\frac{b_1 Y_0}{g_0 g_1}} \quad (7.167)$$

$$J_{ii+1} = \frac{\Delta f}{f_0}\sqrt{\frac{b_i b_{i+1}}{g_i g_{i+1}}};\ i = 1, 2, \ldots, n-1 \quad (7.168)$$

$$J_{nn+1} = \sqrt{\frac{\Delta f}{f_0}\frac{b_n Y_0}{g_n g_{n+1}}} \quad (7.169)$$

Step VII: Asymmetrical coupled lines can be modelled with the self- and mutual-capacitance model. This step is to calculate the self-(C_i^s) and mutual (C_{ij}^M) capacitances which will be used to synthesise the physical dimensions of the combline filter. Each self-capacitance can be calculated from:

$$\frac{C_0^s}{\varepsilon_0\sqrt{\varepsilon_r}} = 377\left(Y_0 - J_{01}\right) \quad (7.170)$$

$$\frac{C_1^s}{\varepsilon_0\sqrt{\varepsilon_r}} = 377\left(Y_{a1} - Y_0 + \frac{J_{01}^2}{Y_0} - J_{12}\tan\theta\right) + \frac{C_0^s}{\varepsilon_0\sqrt{\varepsilon_r}} \quad (7.171)$$

$$\frac{C_i^s}{\varepsilon_0\sqrt{\varepsilon_r}} = 377\left(Y_{ai} - J_{i-1,i}\tan\theta_0 - J_{i,i+1}\tan\theta\right);\ i = 2, 3, \ldots, n-2 \quad (7.172)$$

$$\frac{C_n^s}{\varepsilon_0\sqrt{\varepsilon_r}} = 377\left(Y_{an} - Y_0 + \frac{J_{n,n+1}^2}{Y_0} - J_{n-1,n}\tan\theta\right) + \frac{C_{n+1}^s}{\varepsilon_0\sqrt{\varepsilon_r}} \quad (7.173)$$

$$\frac{C_{n+1}^s}{\varepsilon_0\sqrt{\varepsilon_r}} = 377\left(Y_0 - J_{n,n+1}\right) \quad (7.174)$$

while each mutual capacitance can be calculated from:

$$\frac{C_{01}^{M}}{\varepsilon_0\sqrt{\varepsilon_r}} = 377Y_0 - \frac{C_0^s}{\varepsilon_0\sqrt{\varepsilon_r}} \tag{7.175}$$

$$\frac{C_{i,i+1}^{M}}{\varepsilon_0\sqrt{\varepsilon_r}} = 377J_{i,i+1}\tan\theta \tag{7.176}$$

$$\frac{C_{n,n+1}^{M}}{\varepsilon_0\sqrt{\varepsilon_r}} = 377Y_0 - \frac{C_{n+1}^s}{\varepsilon_0\sqrt{\varepsilon_r}} \tag{7.177}$$

Step VIII: Realising all the physical dimensions of the combline filter from the self- and mutual-capacitances obtained from Step VII.

Example: Design a 50 Ω, seventh-order, combline filter operating at 5 GHz with 0.25 GHz bandwidth that has a Chebyshev response with 0.01 dB ripple.

The fractional bandwidth is $\frac{f}{f_0} = 0.05$ where $f_0 = 5\times10^9$ Hz and $n = 7$.

Step I: The parameters of the low-pass filter prototype that provides the desired response magnitude response are:

$$g_0 = g_8 = 1,\ g_1 = g_7 = 0.7970,\ g_2 = g_6 = 1.3924,\ g_3 = g_5 = 1.7481,\ g_4 = 1.6331$$

Step II: Choose Y_{ai}: In this example, $Y_{ai} = 0.0143$ as recommended in Ref. [1].

Step III: Choose θ: A compact design and wide stopband is required. In this example, we choose θ to be $\frac{\pi}{4}$.

Step IV: Using Equation (7.165) to calculate the slope parameter, we obtain:

$$b_i = 0.0143\left(\frac{\cot\frac{\pi}{4} + \frac{\pi}{4}\csc^2\frac{\pi}{4}}{2}\right) = 0.0112 \tag{7.178}$$

Step V: Compute the loading capacitor, C_i, from Equation (7.166):

$$C_i = \frac{0.0143\times\cot\frac{\pi}{4}}{2\pi\times5\times10^9} = 0.091n\text{F} \tag{7.179}$$

Step VI: Compute the parameters of all admittance inverters from Equations (7.167) to (7.169):

$$J_{01} = \sqrt{\frac{\Delta f}{f_0}\times\frac{0.0112\times0.02}{1\times0.7970}} = 0.0037 \tag{7.180}$$

$$J_{1,2} = \frac{\Delta f}{f_0}\sqrt{\frac{b_1 b_2}{g_1 g_2}} = 0.05\sqrt{\frac{0.0112 \times 0.0112}{0.7970 \times 1.3924}} = 7.4225 \times 10^{-4} \tag{7.181}$$

$$J_{2,3} = \frac{\Delta f}{f_0}\sqrt{\frac{b_2 b_3}{g_2 g_3}} = 0.05\sqrt{\frac{0.0112 \times 0.0112}{1.3924 \times 1.7481}} = 5.0118 \times 10^{-4} \tag{7.182}$$

$$J_{3,4} = \frac{\Delta f}{f_0}\sqrt{\frac{b_3 b_4}{g_3 g_4}} = 0.05\sqrt{\frac{0.0112 \times 0.0112}{1.7481 \times 1.6331}} = 5.4277 \times 10^{-4} \tag{7.183}$$

$$J_{4,5} = \frac{\Delta f}{f_0}\sqrt{\frac{b_4 b_5}{g_4 g_5}} = 0.05\sqrt{\frac{0.0112 \times 0.0112}{1.6331 \times 1.7481}} = 5.4277 \times 10^{-4} \tag{7.184}$$

$$J_{5,6} = \frac{\Delta f}{f_0}\sqrt{\frac{b_5 b_6}{g_5 g_6}} = 0.05\sqrt{\frac{0.0112 \times 0.0112}{1.7481 \times 1.3924}} = 5.0118 \times 10^{-4} \tag{7.185}$$

$$J_{6,7} = \frac{\Delta f}{f_0}\sqrt{\frac{b_6 b_7}{g_6 g_7}} = 0.05\sqrt{\frac{0.0112 \times 0.0112}{1.3924 \times 0.7970}} = 7.4225 \times 10^{-4} \tag{7.186}$$

$$J_{7,8} = \sqrt{\frac{\Delta f}{f_0}\frac{b_7 Y_0}{g_7 g_8}} = \sqrt{0.05\frac{0.0112 \times 0.02}{0.7970 \times 1}} = 0.0037 \tag{7.187}$$

Step VII: The self-capacitances normalised with the term $\varepsilon_0\sqrt{\varepsilon_r}$ are computed from Equations (7.170) to (7.174) as follows:

$$\frac{C_0^s}{\varepsilon_0\sqrt{\varepsilon_r}} = 377\,(0.02 - 0.0037) = 7.1451 \tag{7.188}$$

$$\frac{C_1^s}{\varepsilon_0\sqrt{\varepsilon_r}} = 377\left(0.0143 - 0.02 + \frac{(0.0037)^2}{0.02} - 7.4225 \times 10^{-4}\tan 45^o\right) + 7.1451 = 3.9744 \tag{7.189}$$

$$\frac{C_2^s}{\varepsilon_0\sqrt{\varepsilon_r}} = 377\left(0.0143 - 7.4225 \times 10^{-4}\tan 45^o - 5.0118 \times 10^{-4}\tan 45^o\right) = 4.9223 \tag{7.190}$$

$$\frac{C_3^s}{\varepsilon_0\sqrt{\varepsilon_r}} = 377\left(0.0143 - 5.0118 \times 10^{-4}\tan 45^o - 5.4277 \times 10^{-4}\tan 45^o\right) = 4.9975 \tag{7.191}$$

$$\frac{C_4^s}{\varepsilon_0\sqrt{\varepsilon_r}} = 377\left(0.0143 - 5.4277 \times 10^{-4}\tan 45^o - 5.4277 \times 10^{-4}\tan 45^o\right) = 4.9819 \tag{7.192}$$

$$\frac{C_5^s}{\varepsilon_0\sqrt{\varepsilon_r}} = 377\left(0.0143 - 5.4277 \times 10^{-4}\tan 45^o - 5.0118 \times 10^{-4}\tan 45^o\right) = 4.9975 \tag{7.193}$$

$$\frac{C_6^s}{\varepsilon_0\sqrt{\varepsilon_r}} = 377\left(0.0143 - 5.0118\times10^{-4}\tan 45^o - 7.4225\times10^{-4}\tan 45^o\right) = 4.9223 \tag{7.194}$$

$$\frac{C_8^s}{\varepsilon_0\sqrt{\varepsilon_r}} = 377\,(0.02 - 0.0037) = 7.1451 \tag{7.195}$$

$$\frac{C_7^s}{\varepsilon_0\sqrt{\varepsilon_r}} = 377\left(0.0143 - 0.02 + \frac{(0.0037)^2}{0.02} - 7.4225\times10^{-4}\tan 45^o\right) + 7.1451 = 3.9744 \tag{7.196}$$

The mutual capacitors are computed from Equations (7.175) to (7.177) as follows:

$$\frac{C_{01}^M}{\varepsilon_0\sqrt{\varepsilon_r}} = 377Y_0 - \frac{C_0^s}{\varepsilon_0\sqrt{\varepsilon_r}} = 377\times0.02 - 7.1451 = 1.3949 \tag{7.197}$$

$$\frac{C_{1,2}^M}{\varepsilon_0\sqrt{\varepsilon_r}} = 377J_{1,2}\tan\theta_0 = 377\times7.4225\times10^{-4}\tan 45^o = 0.2798 \tag{7.198}$$

$$\frac{C_{2,3}^M}{\varepsilon_0\sqrt{\varepsilon_r}} = 377J_{2,3}\tan\theta_0 = 377\times5.0118\times10^{-4}\tan 45^o = 0.1889 \tag{7.199}$$

$$\frac{C_{3,4}^M}{\varepsilon_0\sqrt{\varepsilon_r}} = 377J_{3,4}\tan\theta_0 = 377\times5.4277\times10^{-4}\tan 45^o = 0.2046 \tag{7.200}$$

$$\frac{C_{4,5}^M}{\varepsilon_0\sqrt{\varepsilon_r}} = 377J_{4,5}\tan\theta_0 = 377\times5.4277\times10^{-4}\tan 45^o = 0.2046 \tag{7.201}$$

$$\frac{C_{5,6}^M}{\varepsilon_0\sqrt{\varepsilon_r}} = 377J_{5,6}\tan\theta_0 = 377\times5.0118\times10^{-4}\tan 45^o = 0.1889 \tag{7.202}$$

$$\frac{C_{6,7}^M}{\varepsilon_0\sqrt{\varepsilon_r}} = 377J_{6,7}\tan\theta_0 = 377\times7.4225\times10^{-4}\tan 45^o = 0.2798 \tag{7.203}$$

$$\frac{C_{7,8}^M}{\varepsilon_0\sqrt{\varepsilon_r}} = 377Y_0 - \frac{C_8^s}{\varepsilon_0\sqrt{\varepsilon_r}} = 377\times0.02 - 7.1451 = 1.3949 \tag{7.204}$$

Based on the mutual and self-capacitances obtained, the last step is to realise the physical circuit. Since this step depends very much on the technology used, it will be left as an exercise for the reader.

7.11 Hairpin Filter

The hairpin filter is a modified form of the parallel-coupled band-pass filter. The filter is formed with U-shaped half-wavelength resonators to provide a compact design. An array of hairpin

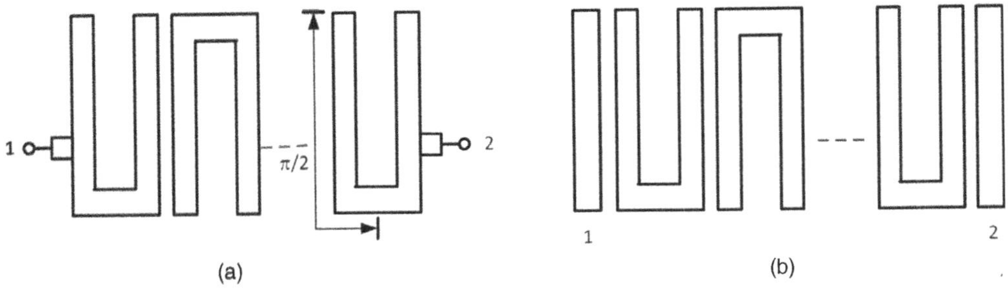

Figure 7.43 (a) Tapped-line and (b) coupled-line matched hairpin filters.

resonators is formed as shown in Figure 7.43(a) and (b), which shows two configurations for input/output matching. Figure 7.43(a) shows the matching configuration using tapped lines, while Figure 7.43(b) shows a coupled-line matching technique. The hairpin filter can be used effectively in several planar transmission line technologies. However, the performance is somewhat different from a parallel-coupled line filter since there are two main features in the topology which are:

- The coupling of each half-wave resonator is discontinuous: there is no coupling at the bend section, and there are additional couplings between adjacent U-shape resonators.
- Discontinuity effects from the bends.

The first effect reduces the coupling compared with parallel-coupled line structure while the bends affect the response of filter. With these two effects, the design equations are derived based on the loaded-Q resonator. The coupling coefficient between adjacent U-shape resonators (resonator i and $i+1$) can be calculated from:

$$M_{i,i+1} = \frac{\Delta f}{f_0\sqrt{g_i g_{i+1}}} \tag{7.205}$$

The external loaded-Q factors of the resonators at the input ($Q_{e,1}$) and output ($Q_{e,n}$) at the resonant frequency (f_0) for bandwidth Δf are related to the prototype parameters, g_i, by:

$$Q_{e,1} = \frac{g_0 g_1}{\Delta f} f_0 \tag{7.206}$$

$$Q_{e,n} = \frac{g_n g_{n+1}}{\Delta f} f_0 \tag{7.207}$$

From Equations (7.205) to (7.207), $M_{i,i+1}$ and external loaded-Q factors Q_e are applied to synthesise the dimensions of the filter.

In the tapped U-shaped resonator shown in Figure 7.43(a), t denotes the distance from the tapped input point to the centre of the U-shaped resonator. The equivalent circuit of the tapped-line matching used at either the input or output resonator is equivalent to a parallel resonant network loaded with an external admittance Y_{ex}. The input admittance Y_{in} is:

$$Y_{in} = jB = j\left(\omega L - \frac{1}{\omega C}\right) \tag{7.208}$$

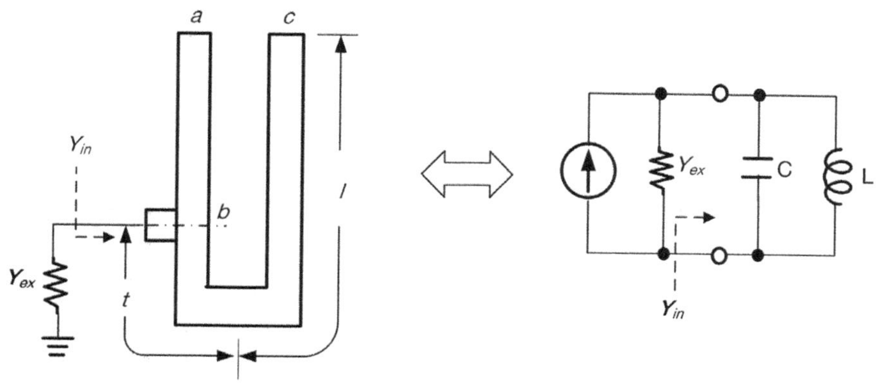

Figure 7.44 Equivalent circuit for a tapped U-shaped resonator.

The slope factor b at resonant frequency ω_0 is given by:

$$b = \frac{\omega_0}{2} \frac{\partial B}{\partial \omega}\bigg|_{\omega=\omega_0} = \omega_0 C \tag{7.209}$$

It is generally known that the external quality factor of a parallel resonant network at ω_0 is computed from:

$$Q = \frac{\omega_0 C}{Y_{ex}} = \frac{b}{Y_{ex}} \tag{7.210}$$

Next, we derive the slope factor of the tapped U-shaped resonator connected with Y_{ex}, as shown in Figure 7.44. At the input, Y_{in} is formed by two open-ended transmission lines (line b-a and line b-c) connected in parallel. This equivalence is sufficient if the coupling between two transmission lines is negligibly small. Assuming that all transmission lines are lossless and have a characteristic admittance Y_0, the open-ended transmission line b-a and b-c have lengths $l - t$ and $l + t$, respectively. Hence, Y_{in} can be written as:

$$Y_{in} = jB_t = jY_0 \tan(\beta[l - t]) + jY_0 \tan(\beta[l + t]) \tag{7.211}$$

With the condition $\beta l = \frac{\pi}{2}$, the slope factor at ω_0 is obtained as:

$$b_t = \frac{\omega_0}{2} \frac{\partial B_t}{\partial \omega}\bigg|_{\omega=\omega_0} = \frac{\pi}{2} \frac{Y_0}{\sin^2\left(\frac{\pi t}{2l}\right)} \tag{7.212}$$

Hence, the external quality factor at ω_0 of this tapped U-shaped resonator is:

$$Q_e = \frac{b_t}{Y_{ex}} = \frac{\pi}{2} \frac{Y_0}{Y_{ex} \sin^2\left(\frac{\pi t}{2l}\right)} \tag{7.213}$$

Therefore, with the given Y_{ex} and Q_e, the position for the tap t can be determined from:

$$t = \frac{2l}{\pi}\sin^{-1}\sqrt{\frac{\pi}{2}\frac{Y_t}{Y_{ex}Q_e}} \tag{7.214}$$

To clearly demonstrate how to design a hairpin filter, we will use the following example to illustrate.

Example: Design a hairpin band-pass filter operating at 10 GHz with 1.5 GHz bandwidth. The desired response of the band-pass filter is seventh-order Chebyshev function with 0.01 dB ripple. The microstrip filter is to be designed for 50 Ω system impedance on a substrate of thickness of 1.29 mm and dielectric constant of 7.2.

The seventh-order Chebyshev prototype function has design parameters as follows:

$$g_0 = g_8 = 1,\ g_1 = g_7 = 0.7970,\ g_2 = g_6 = 1.3924,\ g_3 = g_5 = 1.7481,\ g_4 = 1.6331$$

The fractional bandwidth, $\frac{\Delta f}{f_0}$, is 0.15. Hence the external quality factors are:

$$Q_{e,1} = Q_{e,7} = \frac{g_0 g_1}{\Delta f / f_0} = \frac{1 \times 0.7970}{0.15} = 5.3133 \tag{7.215}$$

The coupling coefficients between adjacent U-shaped resonators are computed from:

$$M_{1,2} = \frac{\Delta f}{f_0\sqrt{g_1 g_2}} = \frac{0.15}{\sqrt{0.7970 \times 1.3924}} = 0.1424 \tag{7.216}$$

$$M_{2,3} = \frac{\Delta f}{f_0\sqrt{g_2 g_3}} = \frac{0.15}{\sqrt{1.3924 \times 1.7481}} = 0.0961 \tag{7.217}$$

$$M_{3,4} = \frac{\Delta f}{f_0\sqrt{g_3 g_4}} = \frac{0.15}{\sqrt{1.7481 \times 1.6331}} = 0.0888 \tag{7.218}$$

$$M_{4,5} = \frac{\Delta f}{f_0\sqrt{g_4 g_5}} = \frac{0.15}{\sqrt{1.6331 \times 1.7481}} = 0.0888 \tag{7.219}$$

$$M_{5,6} = \frac{\Delta f}{f_0\sqrt{g_5 g_6}} = \frac{0.15}{\sqrt{1.7481 \times 1.3924}} = 0.0961 \tag{7.220}$$

$$M_{6,7} = \frac{\Delta f}{f_0\sqrt{g_6 g_7}} = \frac{0.15}{\sqrt{1.3924 \times 0.7970}} = 0.1424 \tag{7.221}$$

The free space wavelength λ_0 at 10 GHz is 3 cm, hence the guide wavelength is around 13.82 mm. Therefore, the length l in millimetres which is a quarter-wavelength is:

$$l = \frac{\lambda_g}{4} = 3.45 \tag{7.222}$$

Given that $Y_{ex} = 0.02$ and choosing $Y_0 = 1/90$, the tap position t (in millimetres) computed from Equation (7.214) is:

$$t = \frac{2l}{\pi}\sin^{-1}\sqrt{\frac{\pi}{2}\frac{Y_0}{Y_{ex}Q_e}} = \frac{2\times 3.45\times 10^{-3}}{\pi}\sin^{-1}\sqrt{\frac{\pi}{2}\times\frac{1}{1.8\times 5.3133}} \approx 0.92 \quad (7.223)$$

The microstrip line is designed on the substrate of thickness 1.29 mm and dielectric constant of 7.2, which gives Z_0 as 90 Ω for a width of 0.54 mm. Therefore, the width of each U-shaped resonator is 0.54 mm. From the calculation, the short-arm length of each U-shape is 2.2 mm. Hence, the length of each long arm (in mm) is $3.45 - \frac{2.2}{2} = 2.35$. Finally, the coupling coefficient of each resonator can be used to determine the gap between adjacent U-shaped resonators. The coupling coefficients with various gaps can be investigated with full-wave EM simulation software.

7.12 Cross-Coupled Filters

The term cross-coupling refers to a generic method of modifying the transfer function of a filter network by introducing coupling between nonadjacent resonators. The elliptic filter response is commonly realised with a cross-coupled filter and can provide a filter response with several transmission zeros close to the edge of the pass-band. This results in a band-pass design with very steep skirts in the response. For example, Figure 7.45 shows the frequency response of a seventh-order elliptic band-pass filter with 0.5 dB pass-band ripple and 60-dB target stopband attenuation, compared with the frequency response of a Chebyshev filter with 0.5 dB ripple of the same order. The elliptic function with these extra transmission zeros requires additional

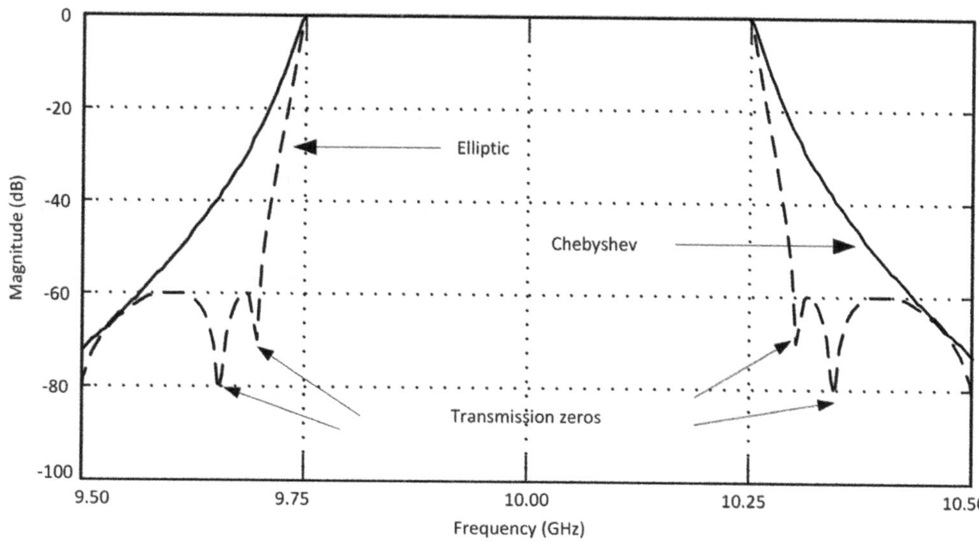

Figure 7.45 Comparison of the frequency responses of seventh-order elliptic and Chebyshev filters with the same degree of pass-band ripple.

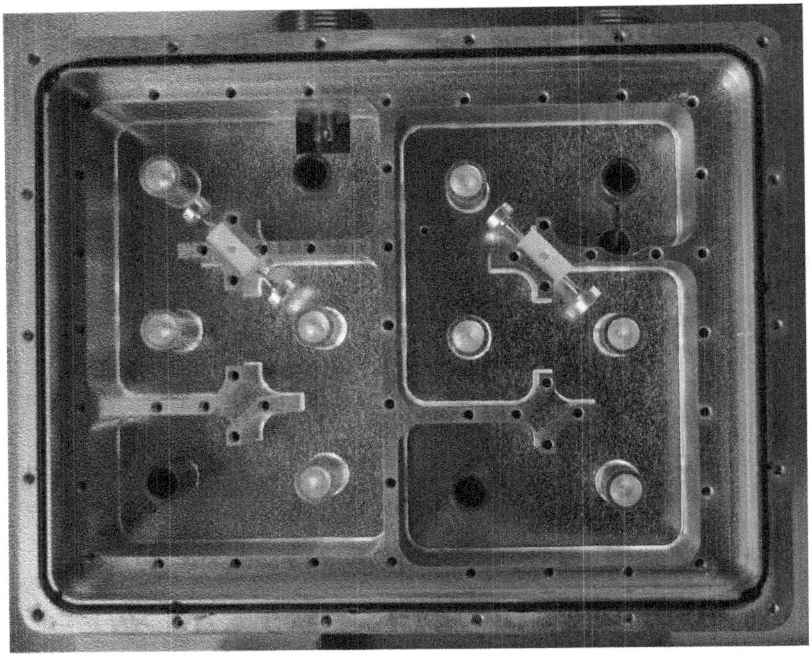

Figure 7.46 Coaxial cellular base-station filter using cross-couplings. Illustration © Phase2 Microwave Ltd. Used with permission; all rights reserved.

cross-couplings in the filter structure. The filter response obtained by the elliptic function shown in Figure 7.45 not only has an improved roll-off, but the group delay flatness can also be enhanced by adding poles which can be achieved using a cross-coupling structure. A flat group delay filter response is generally desired for digital transmission applications.

A cross-coupled filter can be implemented by folding the design in a U-shape and placing a coupling element, such as a probe feed, between two normally nonadjacent resonators, as shown in Figure 7.46. To synthesise the filter, a coupling matrix should be formed. A coupling matrix is a square $N \times N$ matrix where N is equal to the number of resonators. Each nondiagonal element in the coupling matrix represents the coupling coefficient between the resonators. The diagonal elements relate to the self-coupling occurring at each resonator. There are five steps in the design procedure of this cross-coupled filter, which are now described.

<u>Step I</u>: Polynomial Synthesis

For a two-port filter network, the reflection and transfer functions can be written as polynomial functions of ω, $F_n(\omega)$, $P_n(\omega)$ and $E_n(\omega)$ as follows:

$$s_{11}(\omega) = \frac{F_n(\omega)}{E_n(\omega)} \tag{7.224}$$

$$s_{21}(\omega) = \frac{P_n(\omega)}{\varepsilon E_n(\omega)} \tag{7.225}$$

where ε, the ripple in the pass-band, is related to normalised frequency Ω and the return loss (RL) (in dB) by:

$$\varepsilon = \frac{1}{\sqrt{10^{\frac{RL}{10}} - 1}} \left. \frac{P_n(\omega)}{F_n(\omega)} \right|_{\omega=1} \tag{7.226}$$

For a lossless network, the transfer function magnitude is given by:

$$|s_{21}|^2 = \frac{1}{\left|1 + \varepsilon^2 \frac{F_n(\omega)}{P_n(\omega)}\right|^2} = \frac{1}{\left[1 + j\varepsilon \frac{F_n(\omega)}{P_n(\omega)}\right]\left[1 - j\varepsilon \frac{F_n(\omega)}{P_n(\omega)}\right]} \tag{7.227}$$

The polynomial function $F(\omega)$ with transmission zeros ω_i can be written in the following form:

$$F_n(\omega) = \frac{1}{2}\left[G_n(\omega) + G_n'(\omega)\right] \tag{7.228}$$

where

$$G_n(\omega) = \prod_{i=1}^{n}\left[c_n + d_n\right] = \prod_{i=1}^{n}\left[\left(\omega - \omega_i^{-1}\right) + \sqrt{\left(1 - \omega_i^{-2}\right)\left(\omega^2 - 1\right)}\right] \tag{7.229}$$

$$G_n'(\omega) = \prod_{i=1}^{n}\left[c_n - d_n\right] = \prod_{i=1}^{n}\left[\left(\omega - \omega_i^{-1}\right) - \sqrt{\left(1 - \omega_i^{-2}\right)\left(\omega^2 - 1\right)}\right] \tag{7.230}$$

The polynomial $P_n(\omega)$ is given by:

$$P_n(\omega) = \prod_{i=1}^{n}\left(1 - \omega \cdot \omega_i^{-1}\right) \tag{7.231}$$

The formulation of the polynomial $F_n(\omega)$ can be obtained from a recursive technique because its n-degree solution can be computed from its $(n-1)$-degree polynomial $F_{n-1}(\omega)$. If we rearrange the polynomial $G_n(\omega)$ in terms of two polynomials, $U_n(\omega)$ and $V_n(\omega)$, as follows:

$$G_n(\omega) = U_n(\omega) + V_n(\omega) \tag{7.232}$$

where

$$U_n(\omega) = u_0 + u_1\omega + \ldots + u_n\omega^n \tag{7.233}$$

$$V_n(\omega) = \sqrt{\omega^2 - 1}\left(v_0 + v_1\omega + \ldots + v_n\omega^n\right) \tag{7.234}$$

For $n = 1$, the polynomial function $F_1(\omega)$ is related to the first transmission zero ω_1 by:

$$G_1(\omega) = U_1(\omega) + V_1(\omega) = \left(\omega - \omega_1^{-1}\right) + \sqrt{\omega^2 - 1}\sqrt{1 - \omega_1^2} \tag{7.235}$$

Applying the prescribed second transmission zero, ω_2 for $n = 2$, the polynomial $G_2(\omega)$ is obtained as:

$$G_2(\omega) = U_2(\omega) + V_2(\omega) = \left[U_1(\omega) + V_1(\omega)\right]\left[\left(\omega - \omega_2^{-1}\right) + \sqrt{\omega^2 - 1}\sqrt{1 - \omega_2^2}\right] \quad (7.236)$$

The process to calculate the polynomial function $G_n(\omega)$ is performed until all n prescribed zeros are computed. Similarly, the calculation process is applied for computing $G_i'(\omega)$ which is obtained from:

$$G_n'(\omega) = U_n(\omega) - V_n(\omega) \quad (7.237)$$

Hence the numerator of $F_n(\omega)$ is equal to $U_n(\omega)$.

The polynomial $E_n(\omega)$ can be determined using the properties of Hurwitz polynomials. In addition, to be able to realise the filter, the number of prescribed transmission zeros should not be greater than $n - 2$. Furthermore, all zeros must be located symmetrical around the imaginary axis of the s-plane to provide real polynomial functions $P_n(\omega)$ and $F_n(\omega)$.

Step II: Admittance and impedance functions synthesis

Theoretically, the input impedance of a two-port network is related to s_{11} by

$$Z_{in}(\omega) = \frac{1 - s_{11}(\omega)}{1 + s_{11}(\omega)} \quad (7.238)$$

From Equation (7.224), one obtains:

$$Z_{in}(\omega) = \frac{E_n(\omega) + F_n(\omega)}{E_n(\omega) - F_n(\omega)} = \frac{m_1 + n_1}{m_2 + n_2} \quad (7.239)$$

where m_1, n_1 are the even and odd parts of the numerator polynomial and m_2, n_2 are the even and odd parts of the denominator polynomial.

Next, the admittance parameters of the filter are derived and their values used to compute the coupling matrix. For the even-order case, the short-circuit output admittance of a two-port doubly-terminated network, $y_{22}(\omega)$, is given by:

$$y_{22}(\omega) = \frac{n_1}{m_1} \quad (7.240)$$

The short-circuit transfer admittance $y_{22}(\omega)$ is:

$$y_{21}(\omega) = \frac{\mathrm{P}_n(\omega)}{\varepsilon \cdot \mathrm{m}_1} \quad (7.241)$$

For the odd-order case:

$$y_{22}(\omega) = \frac{m}{n_1} \quad (7.242)$$

$$y_{21}(\omega) = \frac{\mathrm{P}_n(\omega)}{\varepsilon \cdot \mathrm{n}_1} \quad (7.243)$$

Step III: Coupling matrix synthesis

A general prototype cross-coupled filter structure is shown in Figure 7.47. There are n resonators composed of a 1F lumped capacitor and two 0.5 H lumped inductors. All resonators have resonant angular frequency normalised to 1rads^{-1}. The resonator number i is coupled to its adjacent resonators $i-1$ and $i+1$ by coefficients $M_{i-1,i}$ and $M_{i+1,i}$, respectively. Coupling between nonadjacent resonator numbers j and i is denoted with $M_{i,j}$ as shown in Figure 7.47. If a voltage source e_s with source resistance R_s is applied at the first resonator and there is a load resistor R_L terminating the last resonator, the loop equation matrix of the circuit shown in Figure 7.47 is:

$$\begin{bmatrix} -je_s \\ 0 \\ 0 \\ \vdots \\ 0 \\ 0 \end{bmatrix} = \begin{bmatrix} \lambda - jR_s & j\omega M_{1,2} & j\omega M_{1,3} & \cdots & j\omega M_{1,n-1} & j\omega M_{1,n} \\ j\omega M_{1,2} & \lambda & j\omega M_{2,3} & \cdots & j\omega M_{2,n-1} & j\omega M_{2,n} \\ j\omega M_{1,3} & j\omega M_{2,3} & \lambda & \cdots & j\omega M_{3,n-1} & j\omega M_{3,n} \\ \vdots & \vdots & \vdots & \ddots & \vdots & \vdots \\ j\omega M_{1,n-1} & j\omega M_{2,n-1} & j\omega M_{3,n-1} & \cdots & j\omega M_{n-1,n-1} & j\omega M_{n-1,n} \\ j\omega M_{1,n} & j\omega M_{2,n} & j\omega M_{3,n} & \cdots & j\omega M_{n-1,n} & \lambda + R_L \end{bmatrix} \begin{bmatrix} i_1 \\ i_2 \\ i_3 \\ \vdots \\ i_{n-1} \\ i_n \end{bmatrix} \tag{7.244}$$

where i_k is the current flows in the resonator k and

$$\lambda = \omega - \frac{1}{\omega} \tag{7.245}$$

This equation can be written in matrix form by relating the excitation voltage vector e and the loop current vector i as follows:

$$-je = (\lambda \cdot \mathbf{I} + \mathbf{M} - j\mathbf{R}) \cdot i \tag{7.246}$$

where $\mathbf{I}$ is the identity matrix and $\mathbf{R}$ is a matrix which mostly contains zero elements. The only nonzero elements are:

$$\mathbf{R}(1,1) = R_s \tag{7.247}$$

$$\mathbf{R}(n,n) = R_L \tag{7.248}$$

The $n \times n$ coupling matrix $\mathbf{M}$ must be determined to design a cross-coupled filter. This coupling matrix has zero elements in the diagonal line. The elements of matrix $\mathbf{M}$ can be either positive or negative, which can be implemented using either electric or magnetic coupling techniques. From the loop equation matrix in Equation (7.244), $y_{21}(\omega)$ is obtained by the element row n and column 1 of the inverse matrix using:

$$y_{21}(\omega) = \left.\frac{i_n}{e_s}\right|_{R_1=R_n=0} = j[-\mathbf{M} - \omega\mathbf{I}]_{n1}^{-1} \tag{7.249}$$

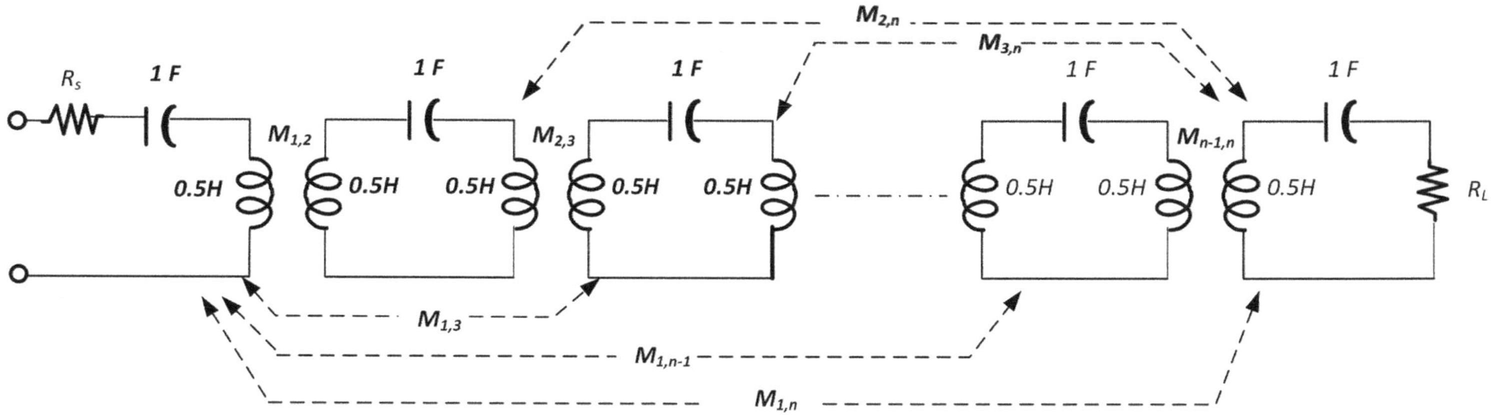

Figure 7.47 General cross-coupled filter.

Similarly $y_{22}(\omega)$ is determined from:

$$y_{22}(\omega) = \left.\frac{i_n}{e_n}\right|_{R_1=R_n=0} = j\left[-\mathbf{M} - \omega\mathbf{I}\right]^{-1}_{nn} \quad (7.250)$$

Since **M** is a real symmetrical matrix and all of its eigenvalues are real, there exists an orthogonal matrix, **T**, that satisfies:

$$-\mathbf{M} = \mathbf{T}\cdot\mathbf{\Lambda}\cdot\mathbf{T}^T \quad (7.251)$$

where

$$\mathbf{\Lambda} = \mathrm{diag}\left(\lambda_1, \lambda_2, \ldots, \lambda_n\right) \quad (7.252)$$

is the $n \times n$ matrix of eigenvalues of $-\mathbf{M}$

It is noted in Equation (7.252) that $\mathbf{T}^T$ is the transposed matrix of **T** which satisfies the condition $\mathbf{TT}^T = \mathbf{I}$. The elements of the first and last rows of matrix **T** and the eigenvalues are related to the admittance parameters of the filter by:

$$y_{21}(\omega) = j\sum_{k=1}^{n} \frac{T_{n,k}T_{1,k}}{\omega - \lambda_k} \quad (7.253)$$

$$y_{22}(\omega) = j\sum_{k=1}^{n} \frac{T_{n,k}^2}{\omega - \lambda_k} \quad (7.254)$$

The poles of the admittance functions given in Equations (7.253) and (7.254) are calculated by discarding the term 'j' in both equations. Identifying these poles, the first and the final rows of matrix **T** are obtained.

Let $r_{22,k}$ and $r_{21,k}$ are the kth residues (poles) of y_{22} and y_{21}, the elements of matrix **T** are calculated from:

$$T_{n,k} = \sqrt{r_{22,k}} \quad (7.255)$$

$$T_{1,k} = \frac{r_{21,k}}{\sqrt{r_{22,k}}} \quad (7.256)$$

Noting that the elements of the first and last row of **T** should be normalised with the transformer turns ratios N_1 and N_2 given by:

$$N_1^2 = R_s = \sum_{k=1}^{n} T_{1,k}^2 \quad (7.257)$$

$$N_2^2 = R_L = \sum_{k=1}^{n} T_{n,k}^2 \quad (7.258)$$

Hence,

$$\breve{T}_{1,k} = T_{1,k}/N_1 \tag{7.259}$$

$$\breve{T}_{n,k} = T_{n,k}/N_2 \tag{7.260}$$

The matrix elements of other rows, $T_{j,k}$, can be determined using the Gram Schmidt orthonormalisation process, which is a procedure to construct a set of orthogonal vectors $(\boldsymbol{u}_1, \dots, \boldsymbol{u}_n)$ from a set of linearly independent vectors $(\boldsymbol{v}_1, \dots, \boldsymbol{v}_n)$. The process is defined by:

$$\boldsymbol{u}_i = \boldsymbol{v}_i - \sum_{j=1}^{i-1} \text{proj}_{\boldsymbol{u}_j}(\boldsymbol{v}_i) \tag{7.261}$$

The term $\text{proj}_{u_j}(\boldsymbol{v}_i)$ is the projection operator of vector $\boldsymbol{v}_i$ to the line spanned by vector $\boldsymbol{u}_j$. This projection operator is defined by the inner product between $\boldsymbol{u}_j$ and $\boldsymbol{v}_i$. The inner product is normalised by the norm of orthogonal vector $\boldsymbol{u}_j$ given by:

$$\text{proj}_{\boldsymbol{u}_j}(\boldsymbol{v}_i) = \frac{\langle \boldsymbol{u}_j, \boldsymbol{v}_i \rangle}{\langle \boldsymbol{u}_j, \boldsymbol{u}_j \rangle} \boldsymbol{u}_j \tag{7.262}$$

Step IV: Coupling matrix reduction

The coupling matrix is obtained from the previous steps, with each element representing the coupling between two resonators. For a passive, reciprocal, symmetrical filter, $M_{i,j} = M_{j,i}$ where i is the row and j is the column of the matrix **M**. Each coupling coefficient can be either zero, positive or negative. A zero value means that no coupling exists between two resonators. Meanwhile, a positive coefficient requires inductive coupling and negative coefficient requires a capacitive coupling structure. Hence, the design complexity of a band-pass filter depends on the number of nonzero elements in the matrix **M**.

In general, the obtained matrix **M** has only a few zero matrix elements. If we realise a filter directly from **M** without matrix modification, the design will be very complicated and the manufacturing cost high. Moreover, the performance of the filter is very sensitive and extensive tuning may be required. Consequently, the nonzero elements should be eliminated as much as possible while the reflection and transfer responses of the filter are kept unchanged. Fortunately, this is possible by applying a mathematical transformation technique. A technique to serve this purpose was proposed by Cameron [5]. The technique is efficient to eliminate the nonzero elements such that the final filter structure realising from the converted coupling matrix is in a fold form. This technique transforms a $n \times n$ matrix $\mathbf{M}_{s-1}$ with a square rotation matrix $\mathbf{R}_s$:

$$\mathbf{M}_s = \mathbf{R}_s \cdot \mathbf{M}_{s-1} \cdot \mathbf{R}_s^T \tag{7.263}$$

where $s = 1, 2, 3, \dots, r$.

The number of transformations for the nth order filter can be calculated from

$$r = \sum_{t=1}^{n-3} t \tag{7.264}$$

From Equation (7.263), the eigenvalues of $\mathbf{M}_s$ are equal to the original matrix $\mathbf{M}_{s-1}$. Therefore the original response is preserved.

The rotation matrix $\mathbf{R}_s$ is specially designed to eliminate the matrix elements, $\mathbf{M}_{s-1}\,(i,k)$ and $\mathbf{M}_{s-1}\,(k,i)$. If the coupling matrix is rotated around the pivot (i,j) where $i \neq j$, the elements of rotation matrix are:

$$\mathbf{R}\,[i,i] = \mathbf{R}\,[j,j] = \cos\theta_r \tag{7.265}$$

$$\mathbf{R}\,[i,j] = -\mathbf{R}\,[j,i] = -\sin\theta_r \tag{7.266}$$

$$\mathbf{R}\,[p,q] = 1; p,q \neq i,j \tag{7.267}$$

$$\mathbf{R}\,[u,v] = 0; u \neq i, v \neq j \text{ and } u \neq v \tag{7.268}$$

The elements of the transformed matrix $\mathbf{M}_s$ are related to the elements of $\mathbf{R}_s$ and $\mathbf{M}_{s-1}$ as follows:

Element in row i:

$$\mathbf{M}_s\,[i,k] = \cos\theta_r \mathbf{M}_{s-1}\,[i,k] - \sin\theta_r \mathbf{M}_{s-1}\,[j,k] \tag{7.269}$$

Element in row j:

$$\mathbf{M}_s\,[j,k] = \sin\theta_r \mathbf{M}_{s-1}\,[i,k] + \cos\theta_r \mathbf{M}_{s-1}\,[j,k] \tag{7.270}$$

Element in column i:

$$\mathbf{M}_s\,[k,i] = \cos\theta_r \mathbf{M}_{s-1}\,[k,i] - \sin\theta_r \mathbf{M}_{s-1}\,[k,j] \tag{7.271}$$

Element in column j:

$$\mathbf{M}_s\,[k,j] = \sin\theta_r \mathbf{M}_{s-1}\,[k,i] + \cos\theta_r \mathbf{M}_{s-1}\,[k,j] \tag{7.272}$$

where $k \neq i,j$.

These four equations are used to determine the rotation angle θ_r. By selecting the pivot (i,j) and the target element to eliminate, the rotation angle can be calculated from either one of four equations depending on the location of the element that it is desired to eliminate:

Element in row i,

$$\tan\theta_r = \frac{\mathbf{M}_{s-1}\,[i,k]}{\mathbf{M}_{s-1}\,[j,k]} \tag{7.273}$$

Element in row j,

$$\tan\theta_r = -\frac{\mathbf{M}_{s-1}[j,k]}{\mathbf{M}_{s-1}[i,k]} \tag{7.274}$$

Element in column i,

$$\tan\theta_r = \frac{\mathbf{M}_{s-1}[k,j]}{\mathbf{M}_{s-1}[k,i]} \tag{7.275}$$

Element in column j,

$$\tan\theta_r = -\frac{\mathbf{M}_{s-1}[k,j]}{\mathbf{M}_{s-1}[k,i]} \tag{7.276}$$

The computation for matrix rotation is performed for r times as calculated from Equation (7.264). The result of coupling matrix is applied to the final design step for filter realisation.

Step V: Filter realisation

Each element of the coupling matrix **M** is converted into the filter elements. The J-inverters at the input and output coupling are related to the input and output transformer ratios. They are calculated from:

$$J_{0,1} = \frac{N_1}{Z_0}\sqrt{\frac{\pi\Delta f}{2f_0}} \tag{7.277}$$

$$J_{n,n+1} = \frac{N_2}{Z_0}\sqrt{\frac{\pi\Delta f}{2f_0}} \tag{7.278}$$

The J-inverters obtained from the inner coupling are obtained by:

$$J_{i,j} = \frac{\pi\Delta f}{2Z_0 f_0} M_{i,j} \tag{7.279}$$

With these design steps, a cross-coupled filter with the desired response can be implemented.

7.13 Conclusion

This chapter has explored some of the most important concepts that are used in the design of transmission line filters for microwave and millimetre-wave wireless communication systems. Several design examples of low-pass and band-pass filters have been provided. The examples given show that it is important to start the design procedure with a network prototype and to derive the electrical parameters of the ideal transmission lines in a step-by-step fashion. The final step of determining the physical dimensions of the transmission lines depends strongly on the transmission line medium and fabrication technology. As discussed in Chapter 6, there is an exceedingly wide variety of transmission line media in use in modern systems. EDA techniques are required in order to properly take account of practical considerations such as discontinuities in the practical design. Furthermore, the very highest performance filters, such as those required in cellular base-stations, will use sophisticated techniques that are beyond the scope of this chapter. Consequently, the reader is urged to carry out extensive further reading of the references for further information.

References

1. Matthaei, G.L., Young L., and Jones, E.M.T. (1980) *Microwave Filters, Impedance-Matching Networks, and Coupling Structures.* Artech House Publishers.
2. Hong, J.S. (2011) *Microstrip Filters for RF/Microwave Applications*, 2nd edition. John Wiley & Sons. ISBN-10: 0470408774.
3. Cameron, R.J., Mansour, R., and Kudsia, C.M. (2007) *Microwave Filters for Communication Systems: Fundamentals, Design and Applications*, 1st edition. Wiley-Interscience. ISBN-10: 0471450227.
4. Hunter, I. (2001) *Theory and Design of Microwave Filters.* The Institution of Engineering and Technology, London, UK (reprinted 2006).
5. Cameron, R.J. (1999) General coupling matrix synthesis methods for Chebyshev filtering functions. *IEEE Transactions on Microwave Theory Techniques*, **47** (4), 433–442.

8

Semiconductor Devices

8.1 Introduction

Whilst silicon is the dominant semiconductor technology for most applications, for high power and low noise at microwave and millimetre-wave frequencies there are considerable advantages in using compound semiconductors such as gallium arsenide (GaAs), gallium nitride (GaN) and indium phosphide (InP). This means there is an extremely varied range of transistors available for microwave designers to choose from, as well as a range of special-purpose two-terminal devices such as Schottky, Gunn and PIN diodes. The most important transistor technologies for wireless communications systems include:

- GaAs HEMT or HBT
- GaN HEMT
- Indium phosphide HEMT or HBT
- Silicon CMOS, BiCMOS, LDMOS
- SiGe HBT and BiCMOS

There are many other technologies and devices on the horizon, particularly as applications in the terahertz region develop, with graphene, antimonide-based and plasmonic devices looking particularly promising [1, 2]. The GaAs-based pseudomorphic high-electron mobility transistor (pHEMT) offers high performance and, with gate lengths as short as 30 nm, circuits operating to 100 GHz have been reported [3] while InP HEMT MMIC amplifiers operating to 1 THz have been demonstrated [4]. The GaAs heterojunction bipolar transistor (HBT) is a vertical structure with critical device dimensions defined by the material growth (layer thicknesses) and doping rather than by lithography. An HBT with an emitter width of 2 μm can offer good microwave performance. Many manufacturers are now offering HBT foundry processes, and there is a wide range of material systems that have been employed. InP HBTs with f_{max} of 800 GHz have been reported [5].

GaAs completely dominated the first 20 years of MMIC development, with a lot of research and development targeted at aerospace and defence applications. Electronically steered phased-array antennas were, and remain, a particularly important application, in which hundreds of

Microwave and Millimetre-Wave Design for Wireless Communications, First Edition.
Ian Robertson, Nutapong Somjit and Mitchai Chongcheawchamnan.
© 2016 John Wiley & Sons, Ltd. Published 2016 by John Wiley & Sons, Ltd.

transmit/receive modules are required and hybrid MICs are too bulky, expensive and difficult to mass produce. One of the many significant milestones in GaAs MMIC development [6] was the demonstration of 2.45 GHz spread spectrum communications [7, 8], an early forerunner of Wi-Fi. However, improvements to silicon devices, including deep submicron technology, enabled silicon RFICs to reach the frequencies needed for cellular mobile communication and indoor networking, and it was this that really fuelled the consumer wireless revolution during the 1990s. Silicon bipolar devices have achieved f_t values of well over 20 GHz, with silicon-germanium HBTs having f_t values of well over 100 GHz [9]. The f_t of bulk CMOS devices depends very strongly on the technology 'node' with 32 nm NMOS devices pushing beyond 400 GHz, while multi-gate devices, FinFET technology and 3-D integration have extended Moore's law into the 2020s [10, 11] and allowed silicon to challenge III–Vs in terms of operating frequency.

For microwave and millimetre-wave applications, GaAs and InP still offer higher performance (noise figure, RF power efficiency, maximum operating frequency), but silicon has the advantage of low cost in high volumes and the ability to integrate digital and analogue baseband circuits. For high power, GaN (gallium nitride) has emerged as the key wide bandgap semiconductor technology for microwave systems and continues to replace GaAs and LDMOS in many applications.

8.2 Fabrication Technology

To fabricate a semiconductor device or integrated circuit requires many processing steps in order to pattern the semiconductor material, the metal layers and dielectrics. Photolithography is the key method of patterning, as in PCB technology. However, the layers are much thinner, the dimensions are very much smaller, and there is a much greater variety of processing steps. To successfully fabricate devices and integrated circuits with high yield requires an ultraclean environment and carefully controlled temperature and humidity. Figure 8.1 shows part of a cleanroom for semiconductor fabrication. Cleanrooms are rated in terms of how many particles per ft^3 are present in the air. For example, a Class 1000 cleanroom has less than 1000 particles of 0.5 μm or larger per cubic foot of air. With the push to smaller and smaller devices, this terminology has started to be replaced with the more exacting ISO cleanroom standards definition which includes 'ISO 1' with no more than 10 particles of 0.1 μm per cubic metre and no more than two of 0.2 μm. In some clean rooms, the technologists wear special suits with a fully enclosed hood so that particulates exhaled on the breath are filtered out. In areas where photolithography is performed, cleanrooms will have yellow lighting – this safe lighting uses special bulbs or filters so that ultraviolet (UV) light is removed.

The fabrication starts with ultrapure semiconductor wafers. For silicon, the process starts with very pure sand – usually from the beaches of Australia – which goes through various treatment steps to create metallic-grade silicon. Semiconductor-grade silicon is then fabricated in the form of ingots (or boules), which are large round bars of silicon, using the Czochralski process: the silicon is melted at 1600 °C in a crucible and a seed crystal is used to slowly draw out a solid bar, rotating as it goes, until an ingot up to 1 or 2 meters in length is made. The ingot is then sliced up like salami with a diamond band saw to produce round wafers with thicknesses of between about 0.5 and 0.75 mm and diameters between 100 and 675 mm. A notch or flat is made on the wafer so that the orientation of the crystal is known. The wafers

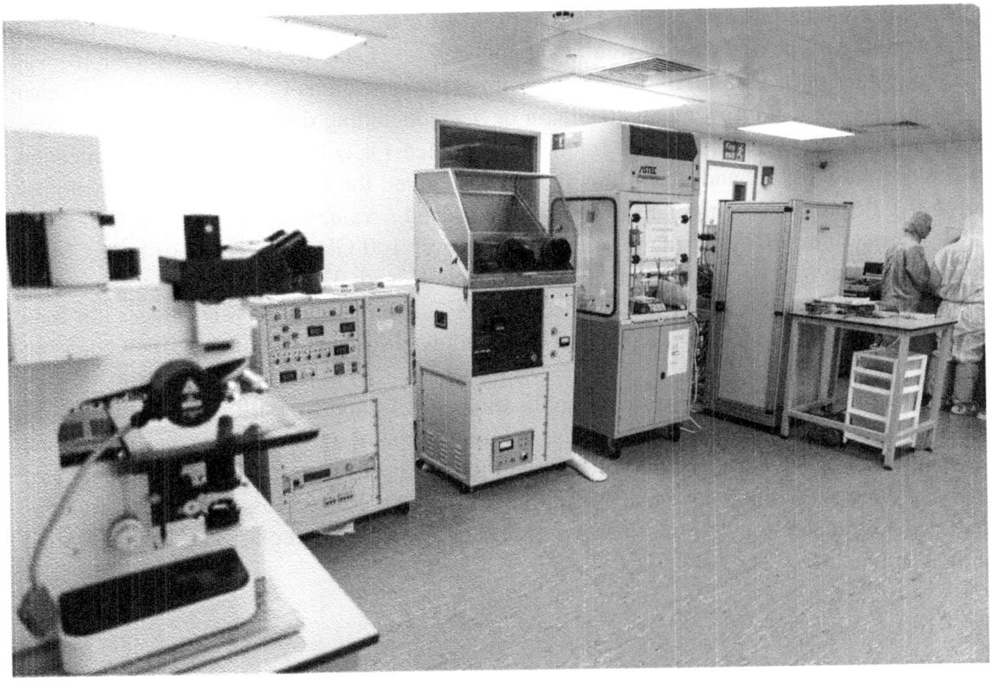

Figure 8.1 Part of a cleanroom suite used for semiconductor chip fabrication. Illustration courtesy of University of Leeds. Reproduced with permission; all rights reserved.

are then polished to a mirror-like finish with a combination of mechanical and chemical polishing. At the time of writing, the two biggest suppliers of silicon wafers were Shin-Etsu Chemical Co. and SUMCO (Sumitomo Mitsubishi Silicon Corp.), both based in Japan but with factories worldwide. According to SEMI, the global industry association for the micro- and nano-electronics supply chain, more than 10 billion square inches of silicon were produced in 2014 (not including solar panel applications) [12]. To put that in perspective, this equates to approximately 900 football (soccer) pitches.

8.2.1 Processing Steps

Having procured suitable wafers, the next step is to produce the required patterned layers of doped material in the surface to make active devices, then deposit metal electrodes to form the device terminals, followed by depositing and patterning additional conductor and dielectric layers to realise passive components and interconnects, or just bond pads for a discrete device. Figure 8.2 shows the simplified processing steps for etching a metal conductor pattern using the so-called 'etch-back' method. Photoresist is applied in liquid form to flood the wafer surface, after which the wafer is spun at high speed (e.g., 3000 rpm) so that the photoresist spreads out very evenly to form an ultra-thin film. It is then dried (soft-baked) before being exposed to UV light through a patterned photomask. After exposure, the photoresist is developed, leaving the required pattern, and the photoresist is then hard-baked to make it chemically resistant to the

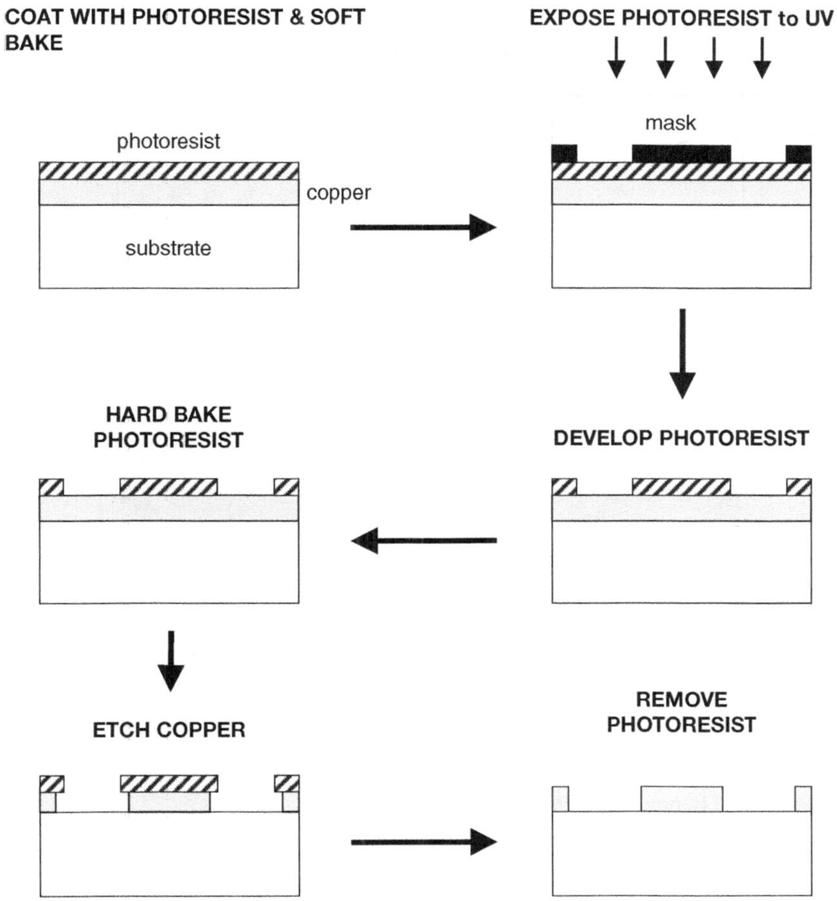

Figure 8.2 Photolithographic patterning of a conductor layer.

etchant, with etching being carried out to create the conductor pattern. Finally, the remaining photoresist is washed away with a suitable solvent.

For completeness, note that in Figure 8.2, the areas of photoresist that were exposed to UV light are shown as being removed by the developer, which is the case for positive photoresist. With positive photoresist, the exposed photoresist becomes soluble in the developer and the mask looks the same as the desired pattern. If a negative photoresist were used, the reverse would occur – the exposed photoresist is polymerised and becomes resistant to the developer, so the mask needs to be the negative image of the desired pattern. Also, note that one major potential disadvantage of the etch-back process is that the whole substrate is exposed to the etchant, which might cause damage to any previously deposited layers. An alternative patterning technique that avoids this is the so-called 'lift-off' process in which the photoresist is deposited and patterned first, after which the conductor is deposited on top. The photoresist is then removed, leaving the conductor pattern behind, with any conductor that was deposited on top of the photoresist being removed.

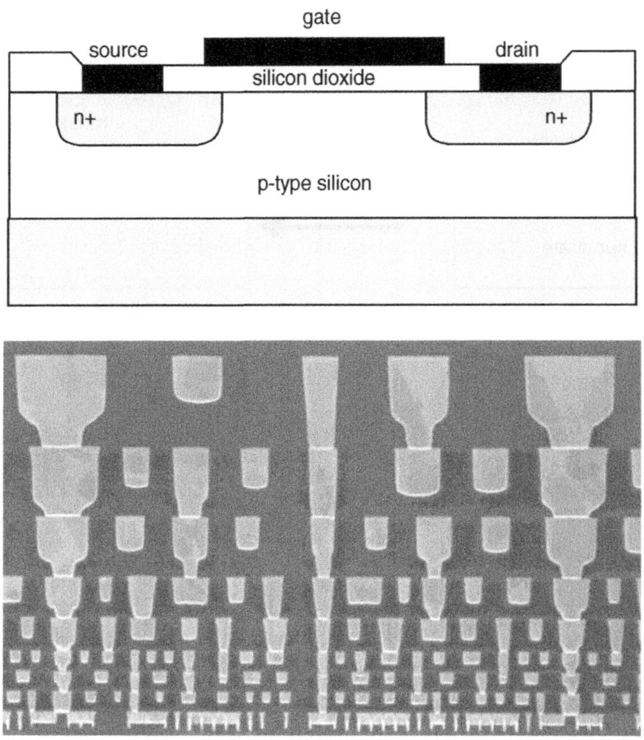

Figure 8.3 Cross-section of NMOS device and photographic image of an interconnect stack. Source: Brain, R. *et al.* (2009) Reproduced with permission of IEEE.

Photolithography is fundamental to all forms of microfabrication. The fabrication steps required to make semiconductor devices are very varied and depend on the size of the features and the materials used, making it very hard to generalise. Figure 8.3 shows the structure of a typical NMOS device and a cross-sectional photographic image through an interconnect stack [13], which serve to illustrate just how many process steps are involved. The main processing steps, involving photolithography at every stage, that may be used in the various device technologies include:

Ion implantation, in which ions are accelerated using voltages up to 2 MV and blasted into the wafer surface in order to create the doped layers needed to create active devices.

Epitaxial growth, including vapour-phase, liquid-phase and molecular beam epitaxy (MBE), where semiconductor layers are deposited directly onto the surface of the wafer.

Diffusion, in which material diffuses into the surface by flowing suitable gases over the surface at high temperature; unwanted thermal and lateral diffusion may also occur, leading to special low temperature techniques being used for other process steps.

Annealing, in which a material is processed by the application of intense heat, using either a furnace or a very powerful lamp. For example, the crystal damage caused by ion implantation is repaired by annealing, and low-resistance ohmic contacts are often created by annealing a metal-semiconductor interface.

Oxidation, in which thin layers of insulating silicon dioxide are created by heating a silicon substrate to very high temperatures in an oxygen-rich atmosphere.

Vacuum deposition, including chemical vapour deposition (CVD) and physical vapour deposition (PVD), in which carefully chosen gases flow over the wafer surface and deposit a thin layer of the required material evenly on the surface. This system is used for depositing silicon nitride dielectric.

Spinning, to deposit liquid photoresist or thick dielectric materials such as polyimide.

Baking, both soft baking to dry material prior to exposure, or hard baking to thoroughly harden the material prior to etching.

Exposure, using a mask aligner or stepper to expose photoresist or other photosensitive materials using UV, deep UV light (optical lithography) or using an electron-beam (E-beam lithography).

Etching, using wet chemicals or a plasma technique (plasma etching, reactive ion etching) to pattern layers.

Metallisation by evaporation or sputtering, in which metal layers are deposited, starting with the devices' own electrodes. A very pure sample of the metal is held in a vacuum chamber and either heated intensely or bombarded with ions so that it evaporates onto the wafer.

Electroplating, in which evaporated metals are plated up to increase the thickness and reduce resistance.

All of the process steps require a very thorough optimisation of the 'recipe' to work properly. The processing times, temperatures, chemical compositions and so forth must all be carefully selected, and machines often require calibration so that the desired layer thicknesses and material properties are achieved with a high degree of precision. For this reason, a full-scale IC manufacturing plant will rarely be used for a new or experimental process. Getting a process up to full production quality requires a huge effort and, once it is achieved, the machines must be disturbed as little as possible. Figure 8.4 shows a photograph of a finished 6-inch diameter wafer of GaAs MMICs.

8.2.2 Extending Moore's Law

Figure 8.5 shows how the feature size of transistors has reduced, from 0.5 μm for the first Pentium® at 60 MHz in 1993, to 14 nm for the 2015 Intel® Skylake processor architecture. The photolithographic process uses UV light; for many years a wavelength of 365 nm was employed – the so-called 'I-line', but device sizes have been smaller than that wavelength for

Figure 8.4 A 6-inch diameter wafer of GaAs PHEMT MMICs. Illustration courtesy of Plextek RFI. Used with permission; all rights reserved.

many years. For a given wavelength, as the feature size becomes smaller and smaller, diffraction around the edges of the photomask pattern (as illustrated in Figure 8.6) becomes more and more problematic. Hence, the wavelength of light used for advanced photolithography has been reduced progressively, down to 193 nm for 45 nm technology. To push even further down this path, extreme UV (EUV) with wavelengths down to 13.5 nm has been developed, and it has been shown that diffraction can be reduced by using immersion lithography. The end of Moore's law has been the subject of much debate in the industry and in academia: the trend clearly cannot go on for ever because the dimensions of transistors would eventually become comparable to the size of the atoms themselves (ca. 0.24 nm for silicon).

There are economic arguments too; even for the 45 nm processes, there are typically 22–25 mask layers, leading to a CAD layout file size on the order of 200 GB. The mask-making facility cost is on the order of US$ 250 million and, taking equipment depreciation into account,

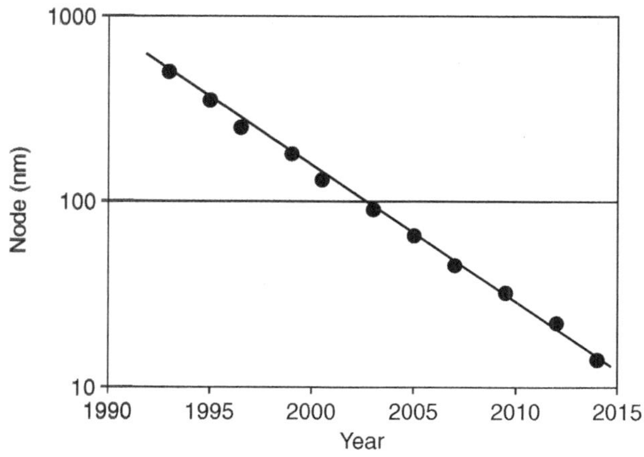

Figure 8.5 Feature size reduction versus Year.

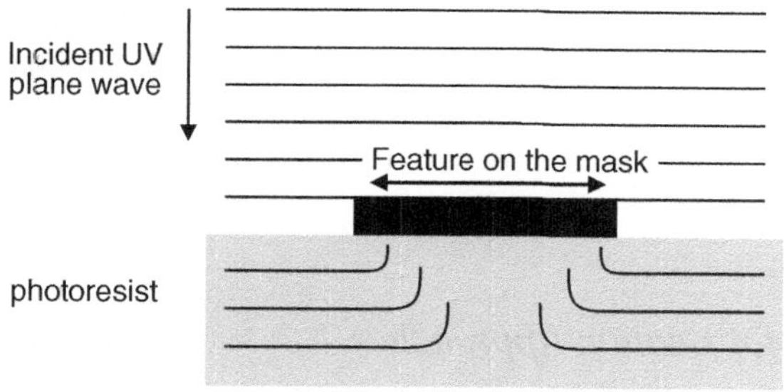

Figure 8.6 Diffraction effect in photolithography.

the resulting cost per set of masks is more than US$ 1 million. The fabrication facilities ('fabs') already cost approximately US$ 3 billion, and EUV tools are particularly costly. As a result, many companies and research groups are pursuing different agendas. One approach is to move to self-assembling structures and new 2-D materials such as graphene; another approach is to investigate ways of adding to the capabilities of the CMOS process.

8.2.3 More than Moore

In the 'More than Moore' paradigm, technology development and design activity is focussed on finding ways of complementing the CMOS technology's capabilities by adding special components. These include optical, mechanical and bioelectronic devices, as illustrated in Figure 8.7. MEMS (microelectromechanical systems) technology uses photolithography to fabricate miniature mechanical structures such as actuators, sensors and motors on the silicon

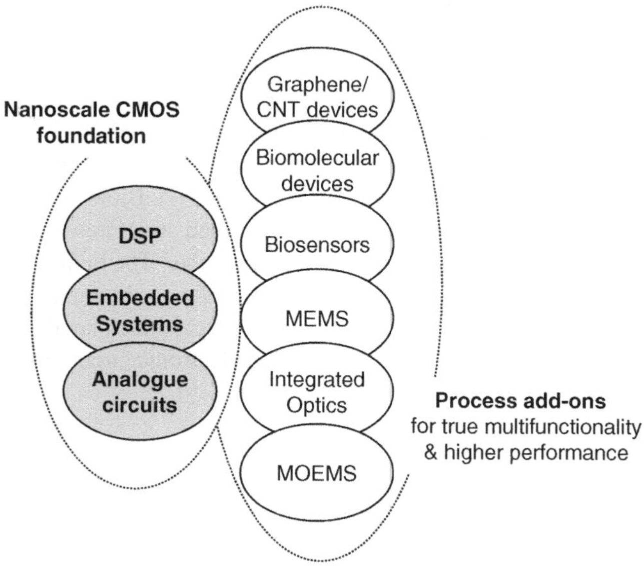

Figure 8.7 Diagrammatic representation of the *More than Moore* paradigm.

chip. Another example is the 'lab-on-chip', in which microfluidic pumps and actuators are combined with biosensors and sophisticated electronics to create small-scale devices for detecting medical conditions. Taking this further leads to the concept of 'nanobots', which would have many applications, such as delivering drugs directly to certain parts of the body in medicine.

8.2.4 Other Nanofabrication Approaches

Whilst the end of Moore's law is topical, there is no doubt that the limit of established silicon-based system-on-chip technology has yet to be reached: the roadmap has shifted its emphasis to how many cores can be realised on a single processor chip. Nevertheless, there are many technologies being developed in research laboratories that could replace silicon in key applications, or at least complement it in others. Plastic-based circuits take several forms; plastic can be used essentially as a flexible PCB, silicon devices can be deposited onto the surface with low-temperature processing, or organic molecules can be used to make electronic devices. The advantage of plastic circuits is, naturally, flexibility and this feature has made displays a key target for the technology. Printed electronics includes a range of technologies for using established inkjet printing, screen printing or lithographic processes to deposit devices and interconnects onto plastic, paper, textiles, and so forth. Such processes are unlikely to beat silicon in terms of circuit complexity or performance, but for key applications such as displays, wearable electronics and internet of things they offer the prospect of large areas and low cost by employing techniques from the printing industry, leading to reel-to-reel manufacture.

Nanotechnology offers a radically different approach for circuit manufacture. Established silicon technology generally uses what is known as *top-down* fabrication: that is, the process starts with a bare silicon wafer and devices are etched or deposited onto its surface, using sophisticated patterning techniques. Taking inspiration from Nature, a *bottom-up* process is one where structures are formed by making use of the directional and selective bonding properties of atoms or molecules, as illustrated conceptually in Figure 8.8. For electronics, the bucky-ball, carbon nanotube and graphene have attracted great interest because carbon's atomic structure allows structures to be grown with a very regular structure, for use as either conducting or semi-conducting devices and interconnections. In biology, it is well-known that DNA has a double-helix structure that employs pairs of just four key building blocks called bases: adenine (A), thymine (T), cytosine (C), and guanine (G). Biological molecules can be coded with sequences of these pairs, creating self-assembled structures through the powerful ability of molecules to recognise one another and bind together. The human genome has three billion base pairs, and a strand of human DNA is 1.8 m long when uncoiled. This field of molecular electronics provides a means of producing devices much smaller than is possible with photolithography, giving the prospect of extending electronic circuit densities beyond the foreseen limit of Moore's law.

8.3 Field-Effect Transistors

The earliest microwave transistors were GaAs metal–semiconductor field-effect transistors (MESFETs), which use a reverse-biased Schottky junction between the gate and channel. These were pioneered during the late 1960s and early 1970s by research groups at Fairchild

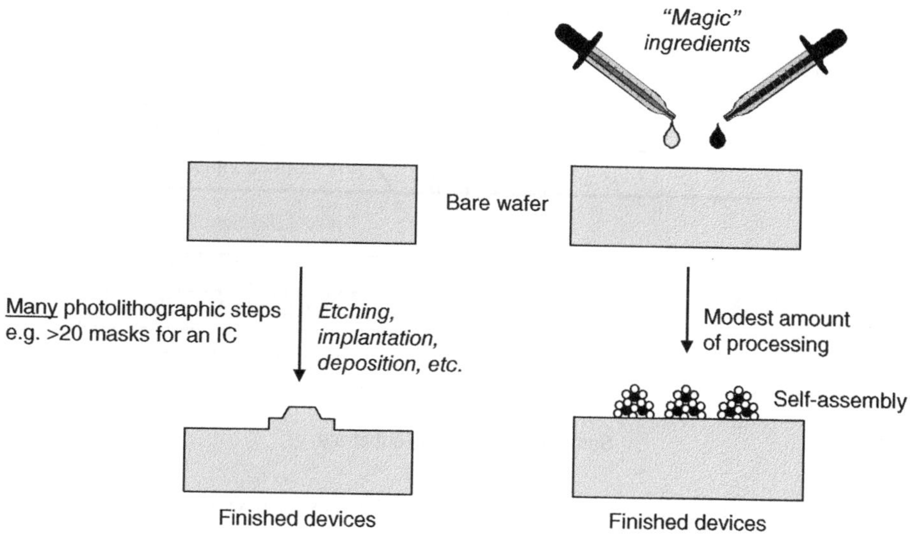

Figure 8.8 Simplified representations comparing 'top-down' (left) and 'bottom-up' nanofabrication (right).

[14], Plessey Research Caswell [15], Philips, RCA and IBM Zurich, amongst others. This device had the double benefit of being a majority carrier device, similar to a junction gate field-effect transistor (JFET) in nature, and making use of the high electron mobility and drift velocity offered by GaAs. The additional advantage of the GaAs semi-insulating substrate led to the demonstration of the world's first MMIC amplifier by Turner and Pengelly in 1976 [16]. Ever since then, monolithic integration of active devices in a range of technologies has been the key to the exploitation of the microwave and millimetre-wave spectrum.

8.3.1 GaAs and InP HEMTs

The GaAs pHEMT (pseudomorphic high-electron mobility transistor) superseded the MESFET, using a more sophisticated semiconductor layer structure that yields a very significant improvement in performance by confining electrons with a heterojunction, separating the electrons from the scattering centres of the doped layers. The cross-section of a GaAs pHEMT is shown in Figure 8.9. The source and drain are ohmic contacts onto the GaAs, this being achieved by annealing a suitable metal (typically gold/germanium) with the highly doped n+ layer. The gate metal is different (typically titanium/platinum/gold) and not annealed, so that there is a Schottky barrier at the GaAs/gate interface. The channel is the conducting path between the source and drain. Without the gate, this would act as a linear resistor, up until the point where the carriers reach maximum possible velocity; after that, the I–V characteristic levels off and the device is in the saturation region. With the gate finger in place above the channel, the Schottky barrier underneath controls the thickness of the channel, since a depletion region is formed. The depletion region depth is controlled by the gate-source voltage, and the current can thus be varied: A family of I–V curves results, as shown in Figure 8.10. In the

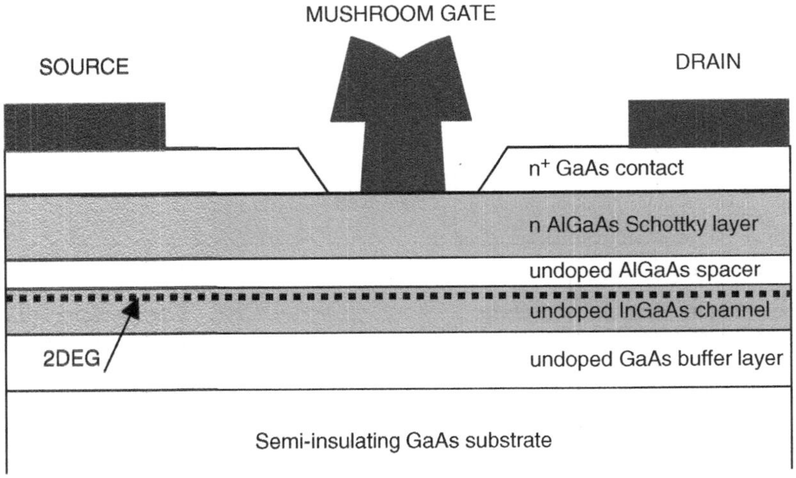

Figure 8.9 Cross-section of a typical GaAs pHEMT.

saturation region, a small change in gate-source voltage yields a large change in I_{ds}, and this is the gain mechanism of the FET. The transconductance, g_m, is given by:

$$g_m = \left. \frac{\partial I_{ds}}{\partial V_{gs}} \right|_{V_{ds}=\text{constant}} \tag{8.1}$$

The inclusion of indium gives high electron mobility, but the addition of indium to the channel causes lattice mismatch. In the pseudomorphic HEMT, this mismatch is accommodated if the channel is kept sufficiently thin; it takes on the lattice constant of the bulk. By constraining the carriers to a thin 'sheet' in undoped material, referred to as a two-dimensional electron gas (2DEG), the transconductance and carrier mobility are much higher than for a MESFET.

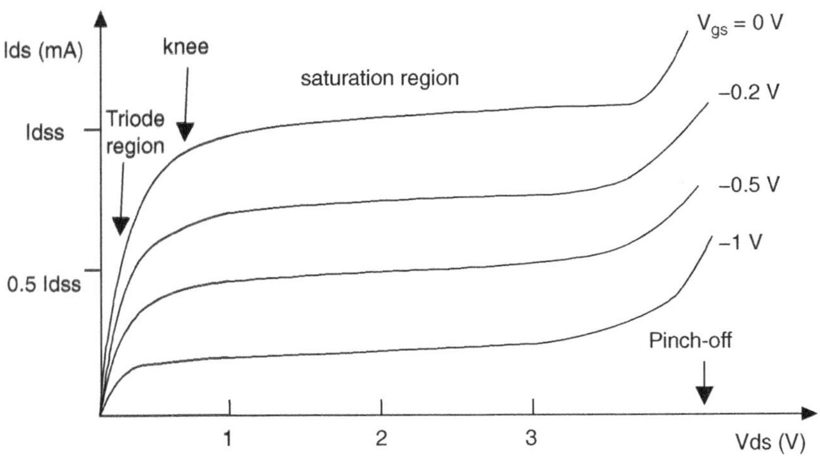

Figure 8.10 Typical low-noise pHEMT I–V curves.

The intricate material layers must be fabricated with either molecular beam epitaxy (MBE) or metal organic chemical vapour deposition (MOCVD), both of which are time-consuming and expensive processes. The gates must be defined with electron beam lithography, and for good noise figures the mushroom gate is needed for low gate resistance, which must be produced using a precise multilayer photoresist technique. pHEMT foundries commonly offer 0.25 μm gate-length devices, and 0.15 μm devices are available from some foundries.

Other devices of importance are the double-heterojunction pHEMT, the lattice-matched HEMT, and the metamorphic HEMT. The double-heterojunction is used for power devices; it adds a second sheet of '2DEG' to increase current handling and also increases the breakdown voltage. The lattice-matched HEMT is a device fabricated on an InP substrate; here, the lattice mismatch is gone and so the In concentration can be further increased to improve performance. InP HEMTs have produced the absolute best high-frequency performance. The fragility of InP wafers causes significant difficulties in mass production, however. To achieve InP-type performance on GaAs substrates, in order to aid mass production, the metamorphic device (mHEMT) has been developed and is available commercially from a few companies.

8.3.2 GaN HEMTs

There has been intensive research into wide bandgap devices since they can be operated at higher supply voltages than GaAs and Si ones, yielding more output power. The main contender is a gallium nitride (GaN) HEMT on either a silicon or silicon carbide (SiC) substrate. GaN devices fabricated on a 4H–SiC substrate can run at very high power without serious thermal problems since the thermal conductivity of 4H–SiC is more than double that of GaN and more than sixfold that of GaAs. The advantage of running the HEMT at a high supply voltage is that the current is less and a higher Watts mm^{-1} gate periphery is achieved. This, in turn means that power combining can sometimes be avoided, leading to a significant increase in power efficiency. At the time of writing, a packaged GaN-on-SiC device running at 28 V could supply 180 W up to 3 GHz with up to 70% efficiency. This shows that GaN devices are of tremendous importance for communications and radar applications. GaN-on-silicon devices are a more cost-effective alternative for high-volume applications, but the silicon substrate is more lossy and has lower thermal conductivity than SiC.

8.3.3 HEMT Small-Signal Equivalent Circuit

Inevitably, the HEMT device has undesirable parasitic resistances and capacitances that degrade its inherent gain mechanism. At a particular DC bias point in the saturation region, the HEMT can be represented by the generic FET small-signal equivalent circuit model shown in Figure 8.11. The following are the main elements in the model and the typical values for a low-noise GaAs HEMT are shown in Table 8.1:

- L_d, R_d, R_g, L_g, R_s, L_s are contact and electrode resistances/inductances.
- C_{gs} is the capacitance across the gate-source depletion region.
- R_i is the resistance of the semiconductor material from source to C_{gs}.
- C_{dg} and C_{ds} are mostly fringing capacitances between contacts.
- R_{ds} is the output resistance; the slope of the I–V curve in the saturation region.
- g_m is the transconductance of the voltage-controlled current source.

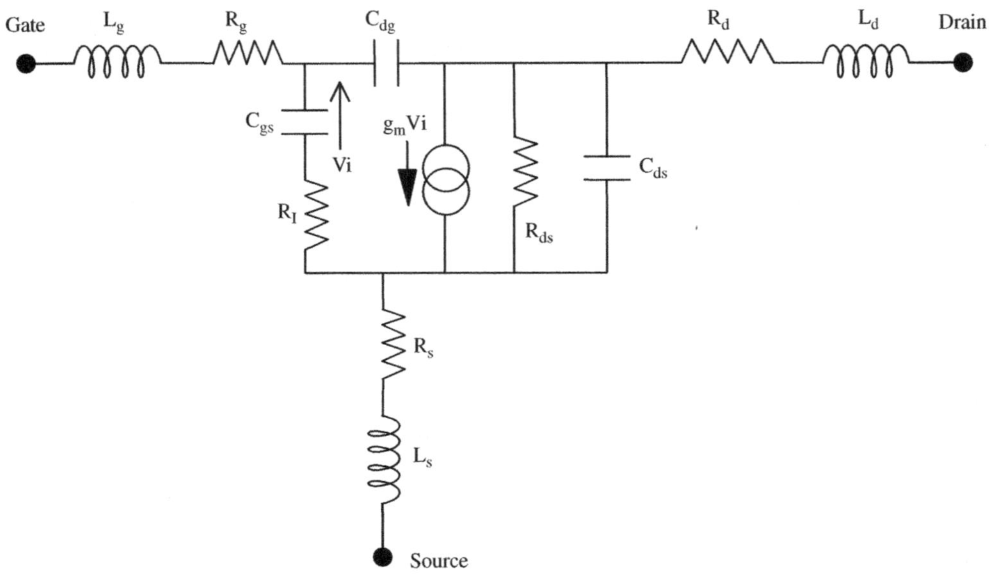

Figure 8.11 HEMT small-signal equivalent circuit.

It can be shown that the maximum frequency of operation, which is determined by these various unwanted model parameters, is given by [17]:

$$f_T = \frac{g_m}{2\pi\left[(C_{gs}+C_{gd})\left(1+\dfrac{R_s+R_d}{R_{ds}}\right)+g_m C_{gd}(R_s+R_d)\right]} \tag{8.2}$$

and

$$f_{\max} = \frac{f_T}{2\left(\dfrac{Rg+R_i+R_s}{R_{ds}}+(2\pi f_T R_g C_{gd})\right)^{\frac{1}{2}}} \tag{8.3}$$

f_T is the frequency where the current gain drops to unity, which is more relevant to bipolar transistors. The $f_{\max}$ is the frequency where the maximum available gain drops to unity (0 dB). This is, therefore, also the maximum frequency of oscillation (in theory). The device is conditionally stable at low frequency; because of the feedback through C_{dg} and R_s/L_s,

Table 8.1 Typical low-noise HEMT equivalent circuit parameter values.

L_g	0.02 nH	g_m	40 mS	C_{gs}	0.2 pF	R_{ds}	400 Ω
R_g	2 Ω	C_{ds}	0.05 pF	R_I	4 Ω	R_s	2 Ω
C_{dg}	0.02 pF	L_s	0.05 nH	R_d	2 Ω		
		V_{ds}	3 V	V_{gs}	–0.5 V	I_{ds}	10 mA

the device will oscillate if conjugately matched. At higher frequencies, the gain drops and it reaches a point beyond which the device is unconditionally stable. This is illustrated in Figure 8.13. The stability and maximum gain formulas are described in Chapter 10. As well as depending on the basic device structure (i.e., the semiconductor layers, doping densities, etc.), the FET RF characteristics depend on the layout geometry of the device:

- The **gate length** is the short dimension of the gate finger; this is the directional of carrier travel, and the transit-time under the gate is one of the most important factors determining f_T. Hence, short gates are needed for high frequencies.
- The **unit gate width** is the 'length' of each individual finger in the layout.
- The **gate periphery** or total gate width is the *number of fingers × the unit gate width.* This determines the maximum output power and is used in the important **Watts mm^{-1}** figure of merit for a given transistor technology.

A '4 × 75' pi-gate device, for example, has four gate fingers, each 75 μm long, as illustrated in Figure 8.12. The equivalent circuit model parameters scale almost linearly with the device total gate width: the capacitances double if the gate width is doubled, while the resistances are halved. The exceptions are the more layout-dependent electrode parasitics. In particular, the gate resistance depends on both the number of fingers, *n*, and their individual length (the unit gate width). For a given total gate width, R_g is inversely proportional to n^2. However, with a large number of fingers other parameters may be degraded, so a balanced optimum must be found. R_g is very important because in the FET the R_g/C_{gs} pair forms a potential divider and the low-pass characteristic will dramatically affect f_{max}. In the MSG/MAG graph shown in Figure 8.13, the 4 × 75 device has a lower R_g but the same C_{gs} as the 2 × 150 one, and thus operates to higher frequencies. This illustrates that the unit gate width is very important in determining f_{max}.

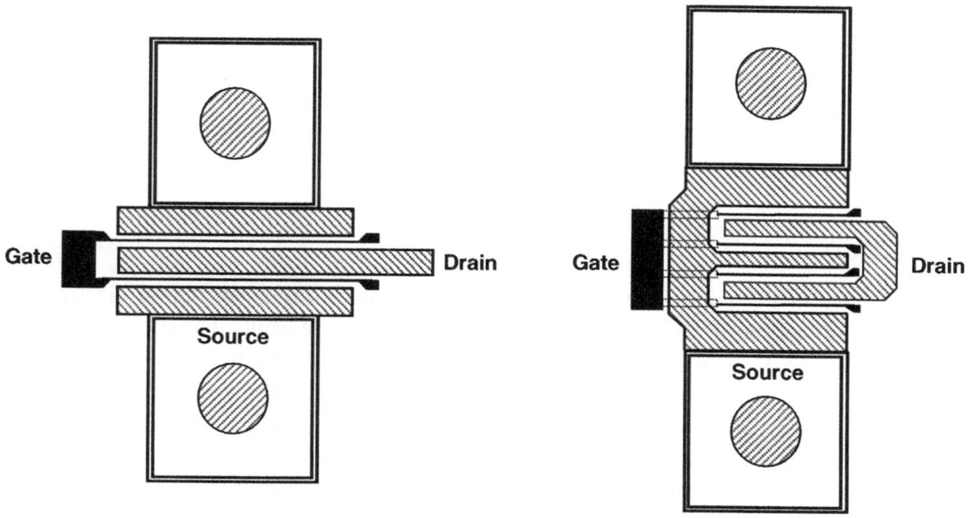

Figure 8.12 Left: 2 × 150 HEMT. Right: 4 × 75 HEMT.

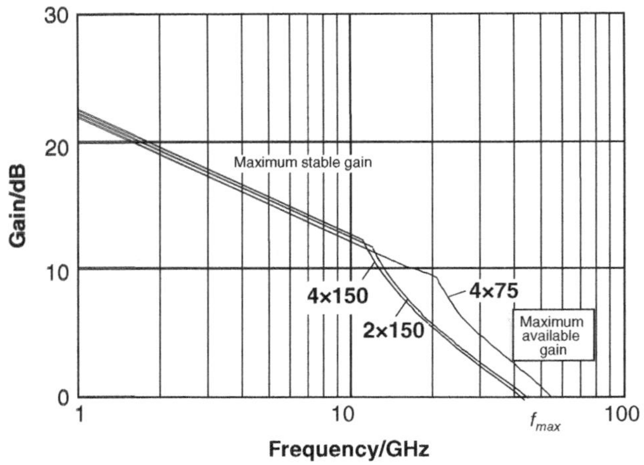

Figure 8.13 The effect of HEMT geometry on maximum stable/available gain versus frequency.

The 4 × 150 has almost the same response as the 2 × 150 (C_{gs} is doubled in the 4 × 150, but R_g is halved). The 4 × 150 has 600 μm total gate width and a higher drain current and gives more power than both of the 300 μm devices. The total gate width determines the maximum output power. Extending this argument, it can be shown that the maximum output power of a transistor technology has a $1/f^2$ dependency, and in fact a $1/f^3$ dependency is often observed. This is why even the most esoteric transistor technologies struggle to provide significant output powers above 100 GHz.

8.3.4 Dual-Gate FETs

For certain applications, such as mixers and gain-control circuits, it is useful to use a dual-gate FET (DGFET). This enables one gate to be used as an RF input, and the other as a gain control or local oscillator input. The arrangement can be represented as a cascode of individual FETs that has surprisingly complex behaviour. The cascode arrangement means the DGFET has a higher gain than a single device, but almost the same physical size. Therefore, DGFETs are used sometimes as a standard device, although their stability may be very poor, depending on the performance of the basic device.

8.3.5 CMOS

Complementary metal-oxide-semiconductor technology has its roots in digital electronics, where the combination of n-channel and p-channel metal oxide–semiconductor field-effect transistors (MOSFETs) can be used to create logic circuits with extremely low power consumption. The FETs only draw current momentarily when the logic gates switch state. This leads to a DC power requirement that is proportional to the voltage squared × switching speed, and reducing the power consumption of GHz processors has been one of the major challenges for silicon technology. Traditionally, CMOS used a polysilicon gate but in recent years the drive to deep submicron devices has led to the use of metal gate devices with high-k dielectrics [18] to reduce gate leakage and improve performance. Whilst digital circuits have been the

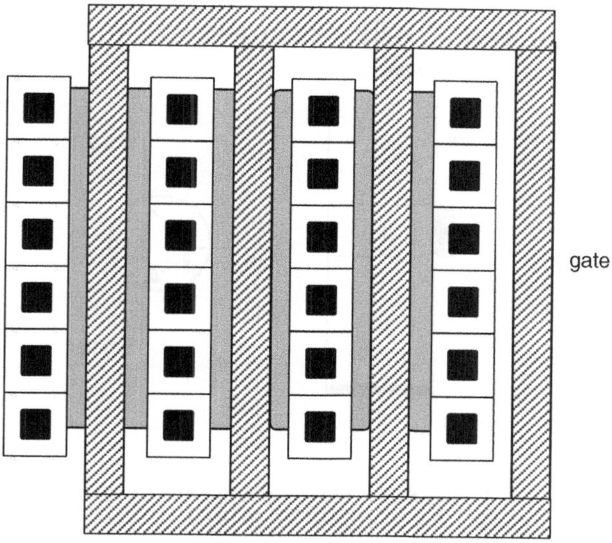

Figure 8.14 Layout of a typical MOSFET for increased high-frequency power output with low gate resistance.

main driving force behind the industry roadmap, as device feature sizes have steadily been reduced it is found that the noise performance, including $1/f$ noise, has improved considerably, whilst f_t is in the >500 GHz regime for state-of-the-art technologies.

The basic digital device is optimised for very low current or charge transfer levels, and for analogue design the layout of the transistor has to be carefully optimised. Figure 8.14 shows the layout and model of a typical NMOS transistor for use in an RF amplifier [19]. Like the HEMT described earlier, a multifingered device is used to reduce gate resistance, but the gates are connected at both ends to further reduce resistance (note that this multifinger layout feature is different to that used at nanoscale within the device structure itself). A guard band is placed around the transistor in order to isolate it from the substrate and reduce noise.

8.3.6 LDMOS

For high power, the LDMOS (lateral diffused MOS) device has been very successful. LDMOS can operate to higher frequencies than a vertical device (VDMOS) as it has a reduced source inductance and feedback capacitance. NXP, a major supplier of such devices, offers 400 Watt, 2 GHz devices operating with a 28 V supply. Push-pull operation is often used with these devices, which are already starting to be used in domestic microwave ovens, where the greater control of power compared to a magnetron gives a number of advantages.

8.4 Bipolar Transistors

A bipolar transistor consists of a pair of p-n junctions placed back-to-back. In the n-p-n transistor, the base is doped p-type, the emitter is heavily doped n-type, and the collector is

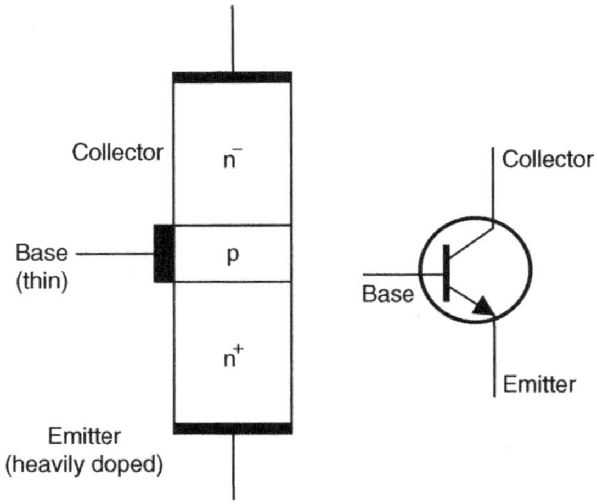

Figure 8.15 Basic n-p-n transistor.

doped n-type. In common-emitter configuration, as shown in Figure 8.15, the base-emitter junction is forward-biased while the base-collector junction is reverse-biased. Electrons trying to flow from the emitter to the base will actually diffuse into the collector, where the large electric field in the reverse-biased junction sweeps them to the collector contact. Thus, the emitter-collector current is controlled by the base-emitter bias. In terms of current gain, the ratio I_c/I_b (beta, β) is proportional to the ratio of the emitter and base doping concentrations, which is why the emitter is heavily doped. In terms of transconductance, the base current is exponentially dependent on the base-emitter voltage, and so large values of g_m can be achieved. The cross-section of a basic device is shown in Figure 8.16. This is a mesa-type device, fabricated by etching; it has the advantage that there is no material around the edges, which would otherwise add unwanted capacitance. In most silicon IC technologies, the device is integrated into the bulk of the substrate and isolated by (for example) proton bombardment. In silicon-on-insulator (SOI) technology, the device parasitics are reduced by mounting it on an insulating layer such as silicon dioxide or sapphire.

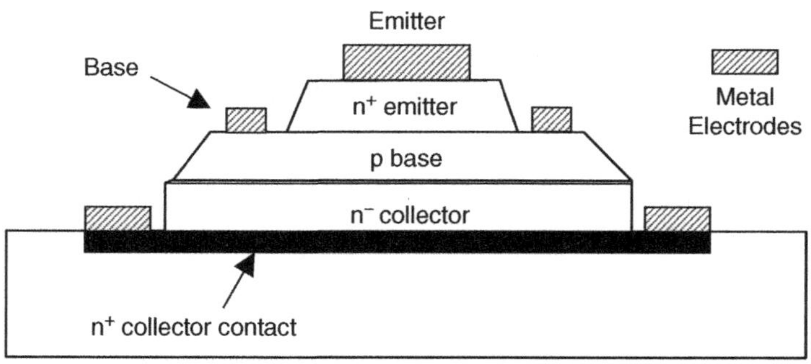

Figure 8.16 Basic n-p-n transistor in cross-section.

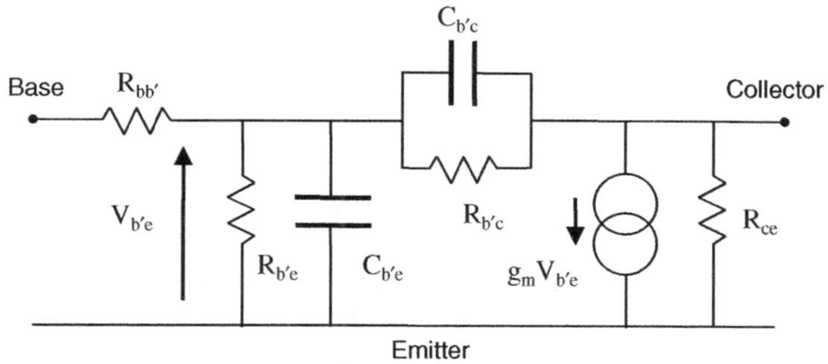

Figure 8.17 High-frequency, small-signal π-model for a bipolar transistor.

Various equivalent circuit models, such as current-based β or α control versions and the famous Gummel–Poon version may be used. The transconductance-based small-signal RF equivalent circuit of a bipolar transistor is shown in Figure 8.17. From this model, it can be shown that [17]:

$$f_T \cong \frac{g_m}{2\pi\left(C_{b'e} + C_{b'c}\right)} \tag{8.4}$$

and

$$f_{\max} = \sqrt{\frac{f_T}{8\pi C_c r_b}} \tag{8.5}$$

The base resistance of bipolar devices is often high, which gives a higher noise figure than HEMTs, and also means that the $f_{\max}$ is not as good as f_T, which is the reverse of the FET situation earlier. Careful device optimisation helps, but ultimately the BJT has a number of parameters which restrict the high-frequency performance. Standard silicon BJT technology is typically capable of $f_T \sim 50$ GHz. Unlike the FET case, this f_T gives an exaggerated indication of the useable frequency limit of the device, which might be more like 20 GHz.

8.4.1 SiGe Heterojunction Bipolar Transistors (HBTs)

An important difference between HEMTs and bipolar transistors is that HEMTs have lateral current flow, whereas bipolars have vertical current flow. So, HEMTs need ultrashort gate lengths and demanding electron-beam lithography is required. Bipolars, however, need thin layers, which are relatively easy to produce. The critical device dimensions are defined by the material growth and doping rather than by lithography. However, in the standard bipolar transistor, if the base is made very thin then the base resistance is very high and the $f_{\max}$ is reduced. The SiGe heterojunction bipolar transistor (HBT) uses silicon-germanium for the base, which has a narrower bandgap than silicon itself and the heterojunction means that the barrier for electrons is significantly smaller than for holes, as shown in Figure 8.18. The key advantage is that the base can be much more highly doped than in a standard BJT, so a thin

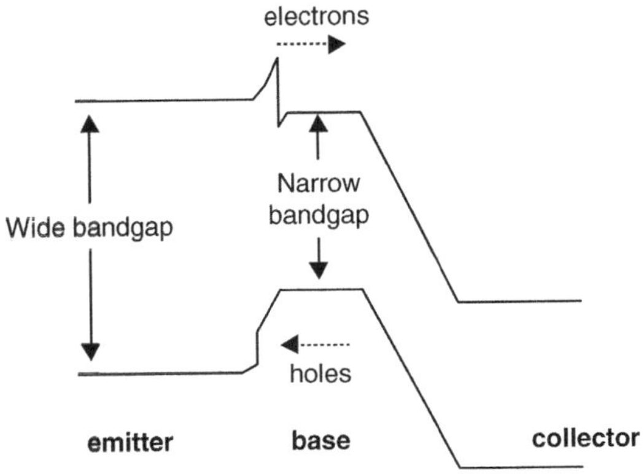

Figure 8.18 HBT band diagram.

base (5–10 nm) can be used without causing high base resistance. The thin base then leads to a much shorter transit time. Since Ge and Si atoms are different sizes, there is a limit to the maximum concentration of Ge that can be used before strain in the lattice produces excessive defects. There are various device parameters that need to be traded-off against each other for an optimum device. Foundry processes are available that offer $f_T > 300$ GHz and $f_{max} >$ 500 GHz with optical lithography and integration with CMOS. This technology is widely recognised as key for commercial broadband wireless communications and radar systems in the millimetre-wave range.

8.4.2 GaAs and InP HBTs

HBTs were first realised in GaAs, partly because GaAs device engineers already had extensive experience of growing thin layers using MBE and MOCVD for HEMTs. A cross-section of a basic InGaP/GaAs HBT is shown in Figure 8.19. Again, a key feature is that the bandgap

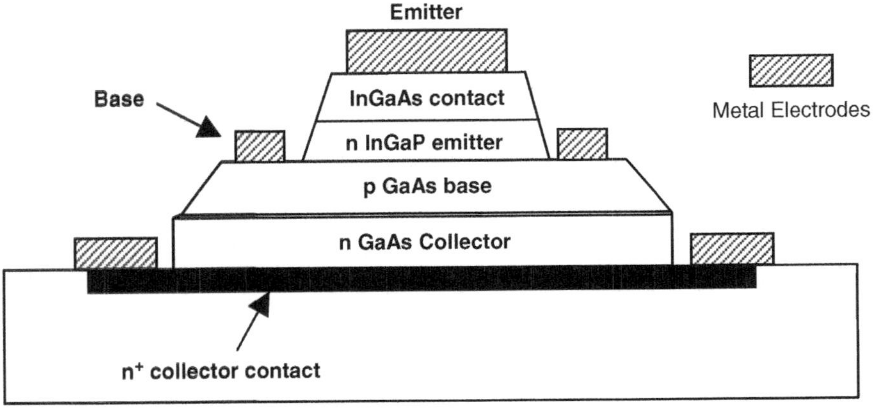

Figure 8.19 InGaP/GaAs HBT cross-section.

of the emitter is wider than that of the base. This means that the current gain is greatly increased, because the barrier to holes in the emitter-base region is higher. For an optimum device design the increased current gain can be traded-off to improve other parameters. For example, the base doping can be increased to reduce base resistance, and the emitter doping can be reduced to reduce C_{be}. The HBT can offer exceedingly high power-density and efficiency and gives much lower phase noise than a HEMT. A number of manufacturers offer InGaP/GaAs HBT HBT foundry processes, and HBTs were extensively developed for mobile handset power amplifier modules since they have the advantage that they can operate off a single power supply rail and with high efficiency. A major advantage of III–V HBTs over HEMTs is that they are far better suited to digital circuits; hence, devices such as 100 Gb s^{-1} multiplexers for optical communications have been realised in HBT technology. InP HBTs are amongst the highest frequency transistors currently available, with f_{max} > 800 GHz having been reported [5]. InP devices have the advantage that the bandgap matches that required for long-haul telecommunications systems, and so lasers, modulators, photodiodes and microwave circuits can be realised on one chip (in principle). As of 2014, an InP HBT MMIC was pushing the limits of high-frequency transistor-based power, with 10 mW at 300 GHz being demonstrated [20].

8.5 Package Styles

Discrete transistors are available from many manufacturers, each offering a range of different devices and package styles. The manufacturer will have made careful design choices (total device size, number of fingers, contact pad size, etc.) and will offer a range of device types for different applications. The package style will depend largely on the operating frequency and power level. Chip-on-board is a technique where a bare-chip device or integrated circuit is directly attached to the substrate and wire-bonded into the circuit. Often, the chip is then covered with epoxy or similar substance to protect the bond-wires and chip. Sometimes this is referred to as ‘glob-top’ packaging and has an extremely low cost, which means it is used in many consumer products. For microwave applications, there are a number of limitations to this approach: first, the epoxy will degrade the performance of the device; second, mounting the chip proud of the surface means longer bondwires; and third, there may be too much inductance to ground. As indicated in Figure 8.20, in microstrip design it is possible to mount a bare chip transistor on a chip carrier or stud. The advantage of the bare chip is that package parasitics are minimised. However, the device bondpads on the chip have some capacitance, the bond-wires have inductance (ca. 0.8 nH mm^{-1} length), and there is also a fringing capacitance of the open end of the microstrip line. Even the carrier or stud itself has inductance, and the thickness of the substrate needs to be considered carefully. These parasitics can seriously degrade the device’s high-frequency capability, and certainly make it impossible to construct complicated circuits without running into stability problems.

Whilst bare chip assembly does have some performance advantages, the need for cleanroom conditions and specialist die-attaching and wire-bonding equipment means that this approach is avoided if at all possible. Prepackaged devices can be conveniently soldered directly into a hybrid circuit and are much more convenient to use. Figure 8.21 shows typical packages for a low-noise device and power device: packages can be plastic, for low-power, non-hermetic applications, or ceramic for high-power or hermetic applications. High-power devices often

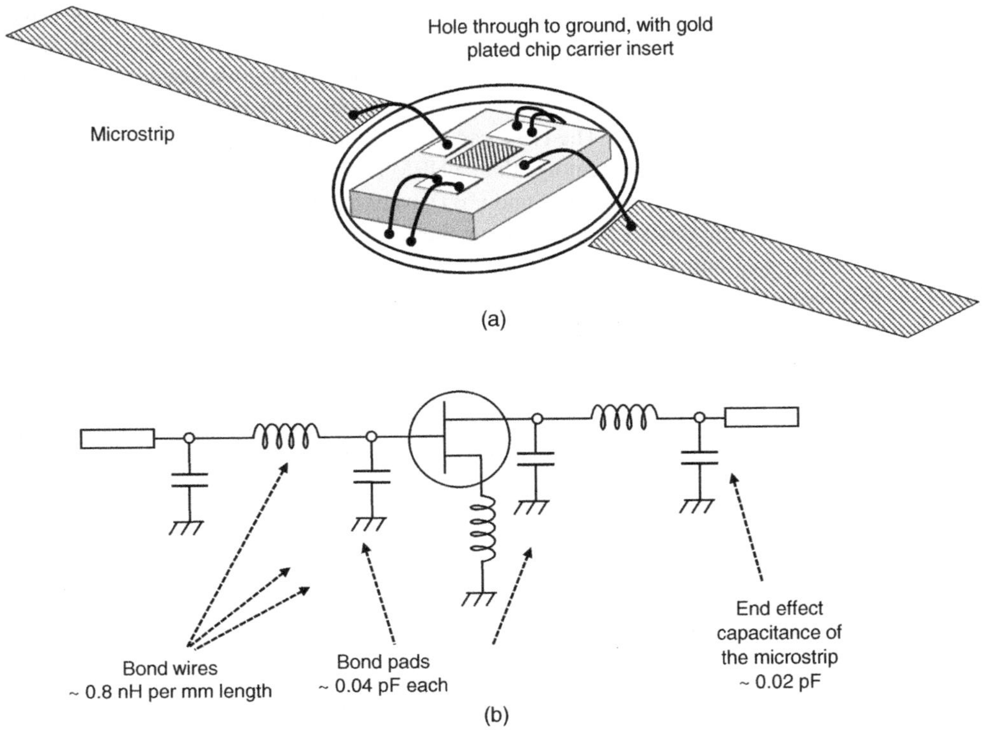

Figure 8.20 (a) A bare chip FET assembled into a microstrip circuit. (b) Typical parasitics of a bare-chip device when in a circuit.

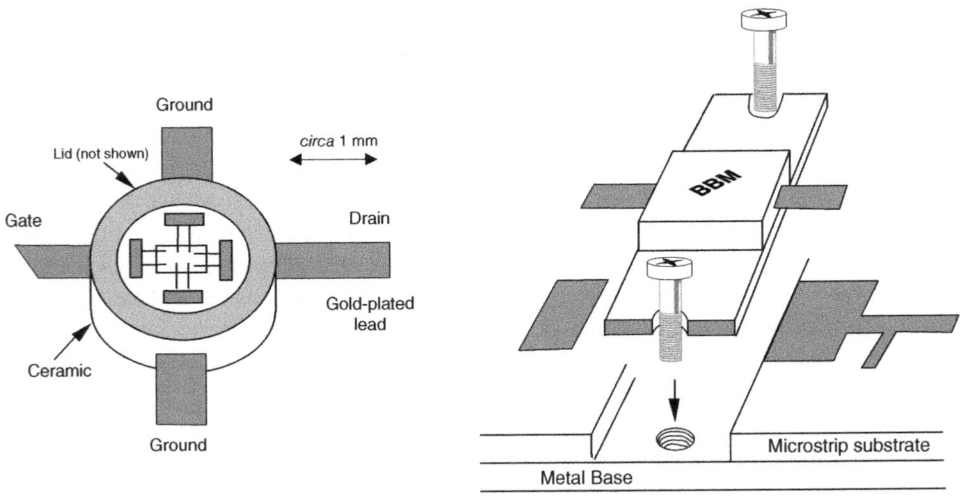

Figure 8.21 Typical ceramic packages for a low-noise device (left) and power device (right).

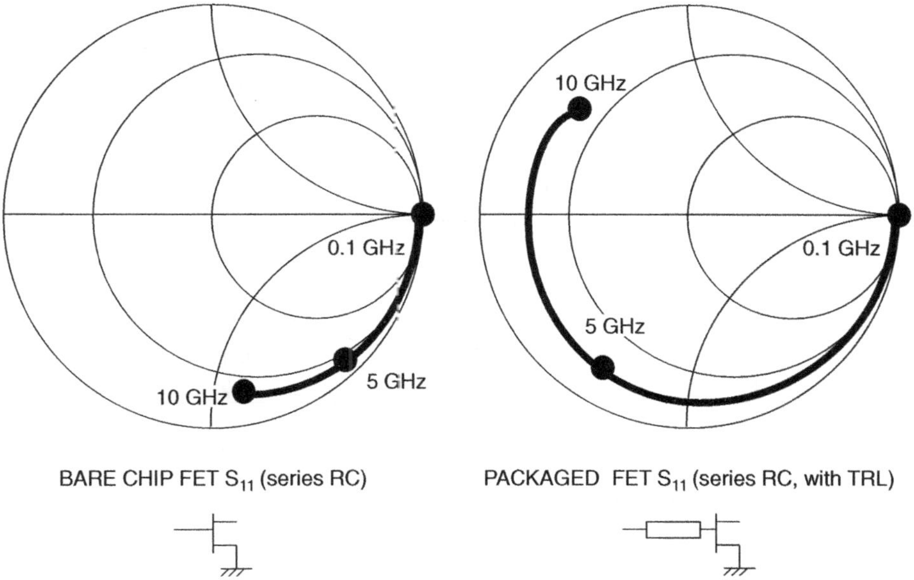

Figure 8.22 Smith chart plot of S_{11} for a bare-chip FET, compared with a packaged FET.

have flanges, so that they can be tightly screwed down onto the ground plane and heat sink. This is to reduce inductance and improve thermal conduction.

A packaged transistor has not only the bond-wire and pad parasitics, but also significant electrical length in the leads of the package. The ground lead length will have a very serious effect on the high-frequency gain and stability. The input/output lead lengths make it difficult to design a wideband circuit because the electrical length of the lead increases with frequency. For example, Figure 8.22 illustrates how the S_{11} of a FET is affected by the packaging: S_{11} is 'stretched out' on the Smith chart, and it is more difficult to match over a wide frequency range. Problems like these are largely overcome when designing RFICs and MMICs: the freedom over the device geometry, as discussed earlier, allied with the lack of package parasitics, enables the designer to be creative and design complex circuits that extract the best high-frequency performance from a given fabrication process.

8.6 High-Power Transistors

In power amplifier design, the overriding concern is to ensure that the load impedance presented to the transistor gives the optimum load-line on the I–V characteristic, in order to optimise the voltage and current swings. The load line for a FET operating in class A, is illustrated in Figure 8.23. At the input, the gate voltage swing is limited by the cut-off voltage and the forward barrier potential. If the transistor has low gain it is this maximum swing that determines the output power. Otherwise, the drain voltage swing becomes the limitation, since this is limited by the knee voltage and the gate-drain breakdown voltage. Similar limits occur in the bipolar transistor. If the transistor structure and material parameters are optimised for

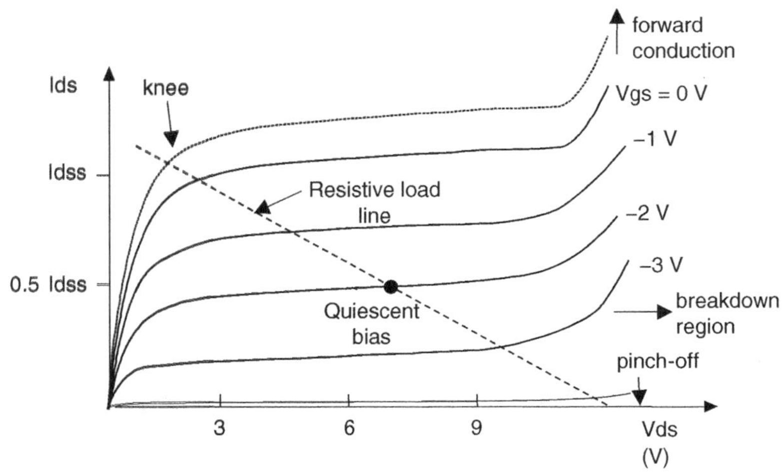

Figure 8.23 FET I–V curves with Class A resistive load line.

maximum power, then it is the breakdown voltage and barrier potential that ultimately limit the output power of a solid-state device. This is why wide bandgap semiconductors such as GaN are so important.

The bipolar transistor has similar limits, such as the collector-base breakdown voltage. For low-frequency devices, the V_{BR} can be very high, but as f_T is improved by device engineering, so the V_{BR} is reduced. Emitter crowding, whereby current concentrates along the edges of the emitter, means that interdigitated structures are required. With the multifinger device, however, there is a danger of thermal runaway; as soon as one finger draws more current than another it gets hotter, leading to a further increase in current. Emitter ballast resistors are used so that the extra current will decrease V_{be} and compensate.

8.6.1 High-Power Device Geometries

In order to overcome the limited input and output voltage swing, a power transistor is a large periphery device which can handle a large current. For FETs, this means using the largest gate-width possible, while for bipolar transistors it means using the largest device area possible. In order to maintain regular grounding points across the device, a power device might often be realised by paralleling-up identical cells, as shown in Figure 8.24. With paralleling-up, the total impedance drops. In a reactively matched power amplifier, the transistor operates in a low-impedance environment (less than a few ohms), and matching transformers/networks are required to bring this up to the desired 50 Ω level.

However, there are four major limitations to this approach of paralleling-up cells to make a single large power transistor:

- **The thermal problem** is addressed by using a thin substrate for good heat dissipation, or flip-chip mounting the device onto a heat sink or high thermal conductivity substrate. Selective substrate thinning can also be used, giving the advantage of lower bond-pad capacitances,

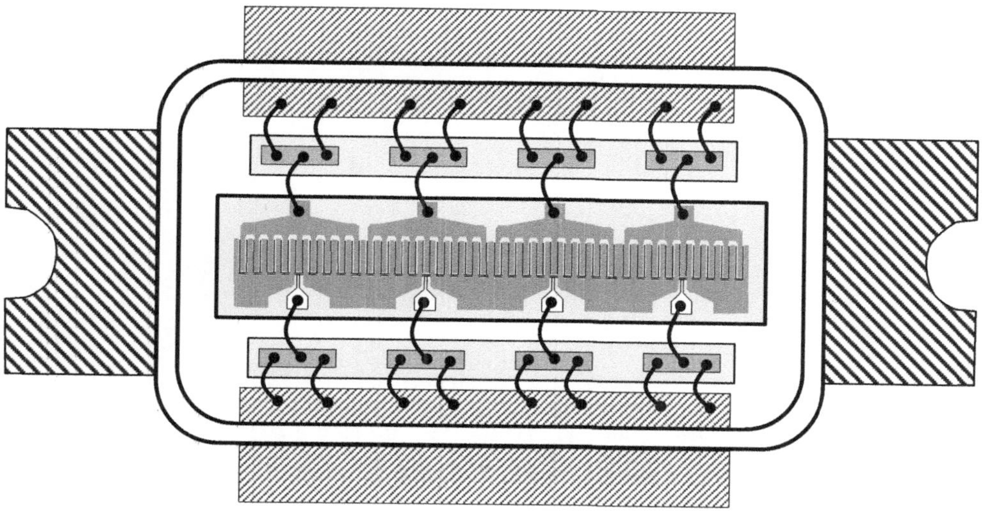

Figure 8.24 Packaged power device. Single-layer capacitors are integrated inside the package and combine with the bond-wire inductances to ease matching.

a more manageable wafer thickness and lower loss for on-chip passive components, if applicable.

- **The matching problem** is caused by the very low input/output impedances: a hybrid MIC power amplifier can readily use (say) a 10 Ω microstrip transformer, even though it may be 8 mm wide, although it is difficult to achieve broadband matching to impedances so close to the edge of the Smith chart. For discrete power transistors, pre-matching may often be employed, using series bond-wire inductances and shunt single-layer capacitors inside the package, as illustrated in Figure 8.24. On an integrated circuit, more sophisticated design approaches have been developed such as enhanced cluster matching (as discussed in Chapter 10).
- **The yield problem** is simply due to the need to successfully produce devices with a very large periphery.
- **Distributed effects** are caused by the device's long input and output interconnections. This significant electrical length can mean that different parts of the device are not fed in-phase, resulting in a nonuniform power distribution within the device. Thus, some cells run at more RF power than others, reducing the output power as not all cells are operating at their maximum power. There may also be stability problems that need careful investigation [21,22].

8.7 RFICs and MMICs

In radiofrequency integrated circuits (RFICs) and monolithic microwave integrated circuits (MMICs), active devices and passive components are fabricated on the same semiconductor substrate. Figure 8.25 shows a photograph of a balanced three-stage GaAs MMIC 60 GHz low-noise amplifier. A distinctive feature of design at microwave and millimetre-wave frequencies

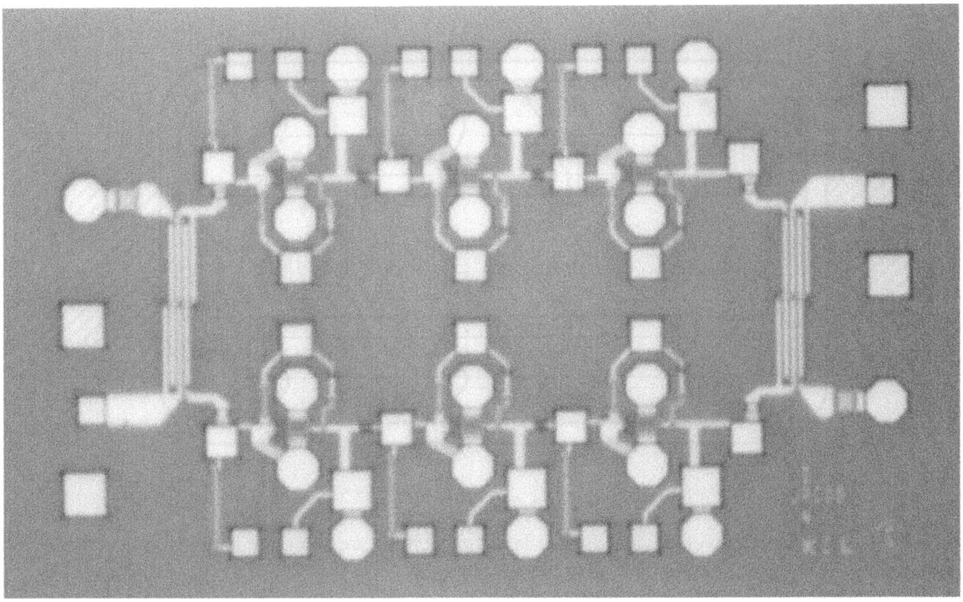

Figure 8.25 GaAs MMIC 60 GHz balanced three-stage amplifier.

is the extensive use of passive components and transmission lines for matching networks, DC biasing, phase shifting, and so forth. Microstrip is the most common form of transmission line in technologies where the semiconductor substrate is high resistivity, with coplanar waveguide and coplanar strips and also being used where appropriate. For silicon RFICs, a considerable effort is made to use multilayer transmission lines based on added dielectric layers, in order to reduce loss from the substrate and aid miniaturisation. Figure 8.26 shows a 2.4 GHz RFIC receiver [23] which clearly has a higher level of integration and uses less passive components.

8.7.1 On-Chip Lumped Passive Components

It is the extensive use of inductors, capacitors and transmission lines that makes microwave and millimetre-wave ICs so different to conventional integrated circuits. In RFIC and MMIC design, passive components are used for impedance matching, DC biasing, phase-shifting, resonators, filters, and many other functions. The transmission line components can realised in various transmission line media, including microstrip, coplanar waveguide (CPW), coplanar strips (CPS), slotline, and multilayer thin-film microstrip. The choice of transmission line medium depends on the type of semiconductor substrate used and the level of performance required. There is a well-known relationship from filter design that the Q-factor tends to vary as the volume of a resonator, so it is fundamentally difficult to realise high-Q miniature passive components on integrated circuits. For this reason there has been a concerted effort to develop new transmission line types such as substrate integrated waveguide (SIW), hollow SIW (HSIW), and various forms of micromachined transmission line, as described in Chapter 13.

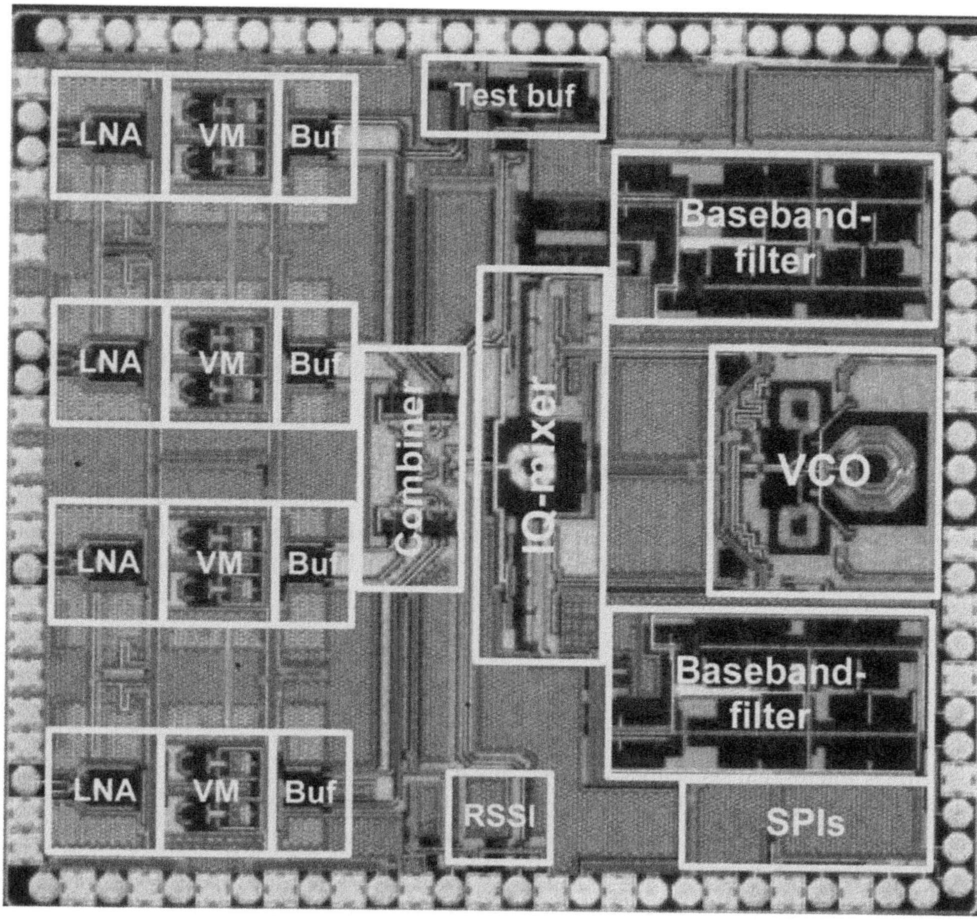

Figure 8.26 RFIC receiver. Source: Wickert, M. *et al.* (2012) Reproduced with permission of IEEE.

8.7.1.1 Inductors

Inductors can be realised either as straight narrow tracks, as single loop inductors, as planar multiturn spiral inductors, or as multilayer quasi-3D structures. Spiral inductors are essential for values above approximately 1 nH, and the connection to the centre turn is usually made on a different metal layer with a spacer dielectric, as shown in Figure 8.27(a), although sometimes air-bridges are used. The basic inductor equivalent circuit consists of a prime inductance along with a series resistance, inter-turn and crossover feedback capacitance, and some capacitance to ground. This lumped element model is shown in Figure 8.28. A multilayer metal/dielectric interconnect stack enables the use of stacked spiral inductors, which were pioneered in the 1990s at Plessy Caswell [24]. These consist of a number of interwound spirals placed on separate metal layers, as shown in Figure 8.27(b), which gives a high inductance because of the mutual coupling of turns, but may also have significant series resistance as the lower metal layers may be thin due to processing limitations. The layout of the stacked spiral, as shown in Figure 8.27(b) may look a little inelegant, but the turns need to be configured so that the mutual

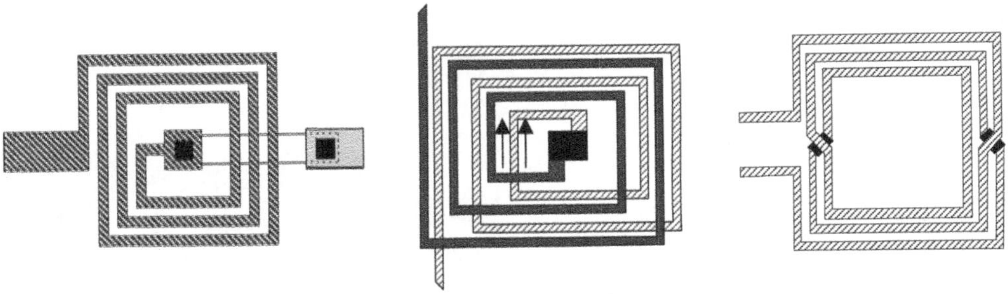

Figure 8.27 Spiral inductors. (a) Standard spiral with an underpass. (b) Single-ended stacked spiral. (c) Interwound windings.

inductance is of the correct sense. Figure 8.27(c) shows an example of an inductor using a pair of interwound windings for application in differential circuits. This structure can be extended to form baluns and transformers [25]. Many research groups have investigated techniques for realising micromachined inductors with a higher *Q*-factor and higher self-resonant frequency, and some of these techniques are described in Chapter 13.

8.7.1.2 Capacitors

Metal-insulator-metal (MIM) capacitors and interdigital capacitors can be used on RFICs and MMICs. Interdigital capacitors are large, and distributed effects will be significant. MIM capacitors are used for matching, filtering, DC blocking and decoupling, where pF-range capacitor values are required. The most common dielectrics are silicon nitride, silicon dioxide, polyimide and benzocyclobutene (BCB). Where polyimide or BCB is used as a spacer dielectric for interconnects and spiral inductors, small-value capacitors can use this dielectric for applications such as matching networks and filter coupling. Other capacitor structures include the woven structure, which has less inductance than an interdigital one, and those using vertical interconnect conductors side-by-side. Also, in CMOS it is possible to use MOS capacitors. By using a combination of inductors and capacitors, a wide range of traditional microwave components, such as a branch-line coupler, can be realised in lumped element form for reduced size [26, 27].

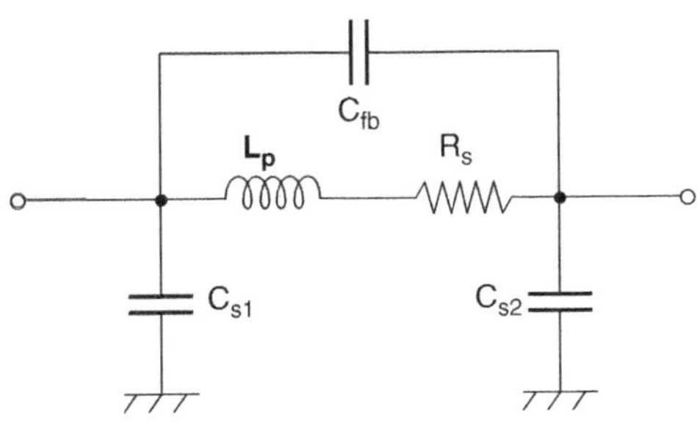

Figure 8.28 Inductor model.

8.7.1.3 Resistors

Resistors on ICs can either use the doped semiconductor layer (mesa resistors or implanted planar resistors), a deposited thin-film resistive layer, or polysilicon. Resistivity is normally quoted in terms of an ohms-per-square figure, and the value of the resistor is chosen by selecting a suitable aspect ratio. The monolithic resistor has a frequency-dependent resistance resulting from the skin effect, which can be accounted for at the design stage in most cases. The parasitic shunt capacitance of the resistor and its contact pads, and the small end-to-end feedback capacitance of the resistor have to be included in any simulation. The distributed nature of the resistor can be taken into account with the series inductance, or for greater accuracy it may be necessary to model the resistor as a distributed RLC ladder.

8.8 Two-Terminal Devices

8.8.1 Schottky Diodes

Schottky diodes are widely used for mixers and detectors in microwave and millimetre-wave applications. The Schottky diode is a majority carrier device and can operate at higher frequencies than PN junction diodes, which are limited by the minority carrier lifetime. GaAs devices give higher cut-off frequencies than silicon because the high low-field mobility reduces the series resistance. GaAs devices can be made with cut-off frequencies over 3 THz, and breakdown voltages over 6 V. To achieve low resistance in silicon devices, a very thin epitaxial layer is needed and this lowers the breakdown voltage and makes silicon devices delicate. However, GaAs diodes have a larger knee voltage, and this may make it necessary to operate with higher LO power or with DC bias. Figure 8.29 shows the common physical

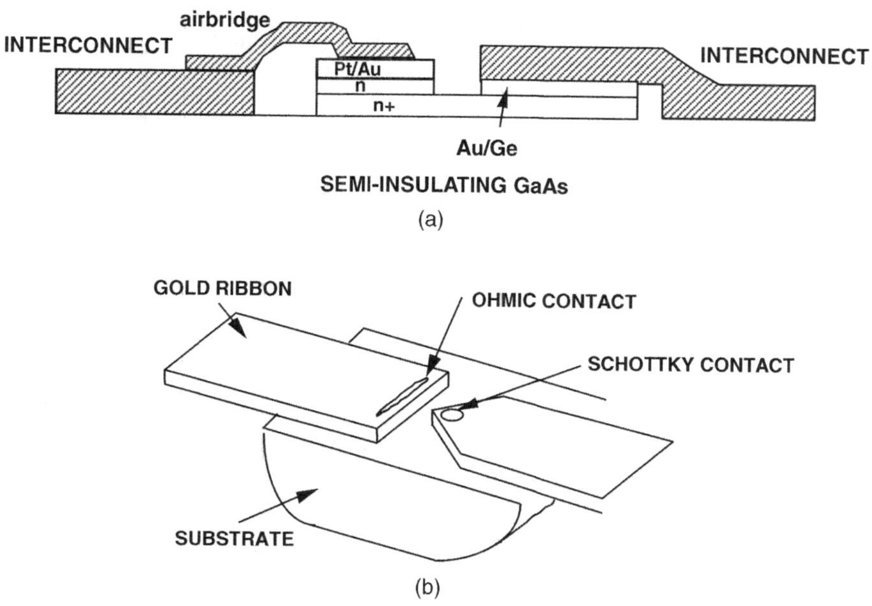

Figure 8.29 Schottky diode. (a) Bare chip air bridge mesa diode. (b) Beam-lead device.

form of practical Schottky diodes. In Figure 8.29(a) the device is isolated by etching and an air bridge is used to access the anode. The bare chip is, inevitably, extremely difficult to handle and assemble into circuits and the beam-lead device, shown in Figure 8.29(b), is a variant where the diode can be more easily mounted into circuit. However, the added parasitics of this form restrict the maximum frequency of operation. Various other forms of diode have been reported in the literature. Schottky diodes remain one of the very best devices for mixers, working into the terahertz regime [28].

The Schottky diode can be used as a square-law detector, with silicon diodes preferred for their low barrier voltage. There are also zero-bias detector diodes and other optimised detectors, such as the planar doped barrier diode, that offer improved noise, sensitivity, temperature sensitivity, and so forth. Microwave varactor diodes are usually reverse-biased Schottky barrier devices in which the device structure has been optimised to operate as a voltage-controlled capacitance. These devices are widely used in oscillators, for realising a voltage-tuned L-C resonant circuit, and phase shifters, to realise a variable-reactance reflection termination. Abrupt and hyperbrupt junctions are used for varactors. The minimum device capacitance will often be determined by package effects.

8.8.2 PIN Diodes

PIN diodes are very widely used for microwave switching, and often also as variable resistance devices for circuits such as modulators. A normal PN junction diode can be used as a switch, but the PIN is one which has been optimised for low off-state capacitance, and a forward bias series resistance which is almost linear with bias current. This is achieved by incorporating a thin layer of intrinsic semiconductor between the P and N regions. PIN diodes typically require up to 50 mA of forward bias current, but can switch quickly (~5 ns) and are solid state without moving parts. For high-isolation switches, combinations of series and shunt PIN diodes can be used. The bias applied then needs to be complementary.

8.8.3 Sources

Two-terminal devices can be important signal sources in some specialist millimetre-wave and sub-millimetre-wave applications. The Gunn diode is a transferred electron device (TED) that relies on the nature of the band structure in certain III–V semiconductors to allow electrons to transfer from one conduction band to another. With a direct band-gap semiconductor, the electrons start to transfer to the upper conduction band as the electric field in the semiconductor is increased. However, their mobility and velocity drops, so the I–V curve exhibits a region of negative resistance. This effect was first discovered experimentally by J.P. Gunn, who observed an unexpected output from a GaAs sample as he ramped up the bias applied. The negative resistance in this bias region can be used for signal generation: the diode must be used with a high-Q cavity to give a spurious-free output. Gunn diodes are low noise, but have low efficiency. For example, a typical 77 GHz Gunn diode might require 7 V bias, with 250 mA current, and generate 80 mW.

IMPact ionisation Avalanche Transit Time (IMPATT) diodes rely on the avalanche current at breakdown in a reverse-biased PN junction. This was pioneered by W.T. Read in the late 1950s and early 1960s. The diode is biased just below the breakdown voltage, so that an applied RF signal just pushes it into avalanche during part of the cycle. The avalanche effect greatly increases the electron flow, and by matching the transit time to the RF frequency, a continuous effect occurs. This causes the RF signal to build up until, in equilibrium, the device's negative resistance matches the total load resistance. The IMPATT gives higher power and efficiency than the Gunn diode, but is more noisy. Variants such as the TRAPATT, BARRITT and double-drift region diode exist [29].

Harmonic generation is often used in microwave and millimetre-wave mixers and oscillators because of the severe difficulty in realising stable low-noise oscillators at very high frequencies. Whilst any active device can generate harmonics, the efficiency of conversion becomes very important when there is no amplifier available (e.g., >100 GHz). Hence, diodes which are optimised for efficient harmonic generation are often encountered in microwaves. The key is to make a device with an abrupt nonlinearity, so that the current waveform is rich in harmonic content. The step-recovery diode (SRD) yields a current waveform that is close to an ideal impulse train, giving a highly efficient conversion to high-order harmonics. SRDs employ a number of structures, making use of charge storage during the forward cycle, and sharp step-function discharge during the reverse cycle.

Tunnelling devices use heavily-doped p+ and n+ semiconductors, brought together to create a slightly extreme band diagram in which a very thin layer of forbidden energy gap is created. It is found that carriers can traverse this gap by tunnelling. The speed of the tunnelling carriers is much higher than the speed of electrons in the normal semiconductor bulk, and so diodes with a very high intrinsic cut-off frequency can be fabricated. Resonant tunnelling diodes (RTDs) have found applications as >100 GHz sources and as detectors. As a source, the power generated is far lower than that of Gunns and IMPATTs, but ≫100 GHz operation is achievable.

References

1. Koswatta, S.O., Valdes-Garcia, A, Steiner, M.B., Yu-Ming Lin, and Avouris, P. (2011) Ultimate RF performance potential of carbon electronics. *IEEE Transactions on Microwave Theory and Techniques*, **59** (10), 2739–2750. doi: 10.1109/TMTT.2011.2150241.
2. Hacker, J.B., Bergman, J., Nagy, G., *et al.* (2005) An ultra-low power InAs/AlSb HEMT W-band low-noise amplifier. IEEE MTT-S International Microwave Symposium, June, pp. 1029–1032.
3. Niehenke, E.C. (2015) The evolution of transistors for power amplifiers: 1947 to today. IEEE MTT-S International Microwave Symposium, pp. 1–4. doi: 10.1109/MWSYM.2015.7166768.
4. Xiaobing Mei, Yoshida, W., Lange, M., Lee, J., Zhou, J., Po-Hsin Liu, Leong, K., Zamora, A., Padilla, J., Sarkozy, S., Lai, R., and Deal, W.R. (2015) First demonstration of amplification at 1 THz using 25-nm InP high electron mobility transistor process. *IEEE Electron Device Letters*, **36** (4), 327–329. doi: 10.1109/LED.2015.2407193.
5. Hacker, J., Seo, M., Young, A., Griffith, Z., Urteaga, M., Reed, T., and Rodwell, M. (2010) THz MMICs based on InP HBT technology. IEEE MTT-S International Microwave Symposium Digest, pp. 1126–1129.
6. Lucyszyn, S. and Robertson, I.D. (2001) *RFIC & MMIC Design and Technology*. IEE Publishers.
7. Devlin, L.M., Buck, B.J, Clifton, J.C., Dearn, A.W., and Long, A.P. (1993) A 2.4 GHz Single Chip Transceiver. IEEE Microwave and Millimeter-Wave Monolithic Circuits Symposium Digest, June, pp. 23–26.
8. Williams, D.A. (1993) A Frequency Hopping Microwave Radio System for Local Area Network Communication. IEEE International Microwave Symposium Digest, June, pp. 685–690.
9. Malz, S., Heinemann, B., and Pfeiffer, U.R. (2014) A 233-GHz low noise amplifier with 22.5 dB gain in 0.13 μm SiGe. Ninth European Microwave Integrated Circuit Conference (EuMIC), pp. 190–193. doi: 10.1109/EuMIC.2014.6997824.

10. International Technology Roadmap for Silicon (www.itrs.net).
11. Jan, C.-H, *et al.* (2015) A 14 nm SoC platform technology featuring 2nd generation Tri-Gate transistors, 70 nm gate pitch, 52 nm metal pitch, and 0.0499 μm^2 SRAM cells, optimized for low-power, high-performance and high-density SoC products. Symposium on VLSI Technology (VLSI Technology), pp. T12–T13. doi: 10.1109/VLSIT.2015.7223683.
12. SEMI global industry association (www.semi.org).
13. Brain, R., *et al.* (2009) Low-k interconnect stack with a novel self-aligned via patterning process for 32 nm high-volume manufacturing. IEEE International Interconnect Technology Conference (IITC), pp. 249–251. doi: 10.1109/IITC.2009.5090400.
14. Hooper, W.W. and Hower, P.L. (1967) A microwave GaAs field-effect transistor. International Electron Devices Meeting, vol. 13, pp. 38–38. doi: 10.1109/IEDM.1967.187800.
15. Turner, J.A. and Waller, A.J. (1969) An S-band gallium arsenide field effect transistor. 1st European Microwave Conference, 8–12 September, pp. 480–482. doi: 10.1109/EUMA.1969.331950.
16. Pengelly, R.S. and Turner, J.A. (1976) Monolithic broadband GaAs f.e.t. amplifiers. *Electronics Letters*, **12** (10), 251–252. doi: 10.1049/el:19760193.
17. Thayne I.G., Elgaid, K., and Ternent, G. (2001) Devices and Fabrication Technology. In: *RFIC & MMIC Design and Technology*, Chapter 2. IEE Publishers.
18. Ouhachi, R., Pottrain, A., Ducatteau, D., Okada, E., Gaquiere, C., and Gloria, D. (2013) RF power potential of high-k metal gate 28 nm CMOS technology. International Semiconductor Conference (CAS), vol. 2, pp. 181–184. doi: 10.1109/SMICND.2013.6688649.
19. Fritsche, D., Tretter, G., Carta, C., and Ellinger, F. (2015) Millimeter-wave low-noise amplifier design in 28-nm low-power digital CMOS. *IEEE Transactions on Microwave Theory and Techniques*, **63** (6), 1910–1922. doi: 10.1109/TMTT.2015.2427794.
20. Yu, H.G., Choi, S.H., Jeon, S., and Kim, M. (2014) 300 GHz InP HBT amplifier with 10 mW output power. *Electronics Letters*, **50** (5), 377–379. doi: 10.1049/el.2013.3288.
21. Freitag, R.G. (1992) A unified analysis of MMIC power amplifier stability. IEEE MTT-S International Microwave Symposium Digest, pp. 297–300.
22. Ohtomo, M. (1993) Stability analysis and numerical simulation of multidevice amplifiers. *IEEE Transactions on Microwave Theory and Techniques*, **MTT-41** (6/7), 983–991.
23. Wickert, M., Mayer, U., and Ellinger, F. (2012) 802.11a Compliant Spatial Diversity Receiver IC in 0.25 μm BiCMOS. *IEEE Transactions on Microwave Theory and Techniques*, **60** (4), 1097–1104. doi: 10.1109/TMTT.2012.2184297.
24. Geen, M.W., Green, G.J., Arnold, R.G., Jenkins, J.A., and Jansen, R.H. (1989) Miniature multilayer spiral inductors for GaAs MMICs. IEEE Gallium Arsenide Integrated Circuit (GaAs IC) Symposium, pp. 303–306. doi: 10.1109/GAAS.1989.69348.
25. Ferguson, D., *et al.* (1984) Transformer Coupled High-Density Circuit Technique for MMICs. IEEE Microwave and Millimetre-wave Monolithic Circuits Symposium Digest, pp. 34–36.
26. Gupta, R.K. and Getsinger, W.J. (1984) Quasi-lumped-element 3- and 4-port networks for MIC and MMIC applications. IEEE International Microwave Symposium Digest, pp. 409–411.
27. Vogel, R.W. (1992) Analysis and design of lumped- and lumped-distributed-element directional couplers for MIC and MMIC applications. *IEEE Transactions on Microwave Theory and Techniques*, **MTT-40**, 253–262.
28. Retzloff, S.A., Young, A., and Hesler, J.L. (2014) A 1.46 THz Schottky receiver at cryogenic temperatures. 39th International Conference on Infrared, Millimeter, and Terahertz waves (IRMMW-THz), pp. 1–2. doi: 10.1109/IRMMW-THz.2014.6956176.
29. Haddad, G.I. and Trew, R.J. (2002) Microwave solid-state active devices. *IEEE Transactions on Microwave Theory and Techniques*, **50** (3), 760–779. doi: 10.1109/22.989960.

9

Impedance Matching

9.1 Introduction

Matching networks are normally used in RF circuits to ensure the best possible power transfer from input to output port. For example, if a power amplifier output is not matched to the characteristic impedance of the duplexer and antenna, there is a signal reflection at the point where the impedance mismatch occurs. The signal reflection leads to a standing wave, with constructive and destructive interference at different points in the system. The mismatch reduces the efficiency of the power amplifier and it can even be damaged because the standing wave leads to a high peak voltage and current.

In this chapter, some of the different types of networks used to perform impedance matching are described. The graphical way of designing matching networks with the Smith chart is studied. Different types of impedance matching, using lumped or distributed elements, are analysed. The relationship between the bandwidth of the network with the type of network is also studied, and the design requirements for an impedance matching network are discussed. The use of transmission lines as impedance transformers in a number of topologies is covered.

9.2 The Purpose of Impedance Matching

Impedance matching is used to ensure the transfer of maximum power from a source to the load. It is particularly important when dealing with high-frequency signals for two main reasons. First, active devices have relatively low gain and output power, so matching is essential in order to achieve useful performance. Second, since RF systems are based on transmission lines as interconnections between components, there is usually a need to match all the individual components to a system impedance to avoid unwanted reflections. For example, consider the amplifier chain shown in Figure 9.1. Unless the amplifiers are matched to the characteristic impedance of the cables and connectors, reflections are created between components. The reflected signals interfere constructively and destructively with the incoming signal at different frequencies because the difference in phase between the incoming and reflected signals varies with frequency. At a certain point on the line, at some frequencies the incident and reflected

Microwave and Millimetre-Wave Design for Wireless Communications, First Edition.
Ian Robertson, Nutapong Somjit and Mitchai Chongcheawchamnan.
© 2016 John Wiley & Sons, Ltd. Published 2016 by John Wiley & Sons, Ltd.

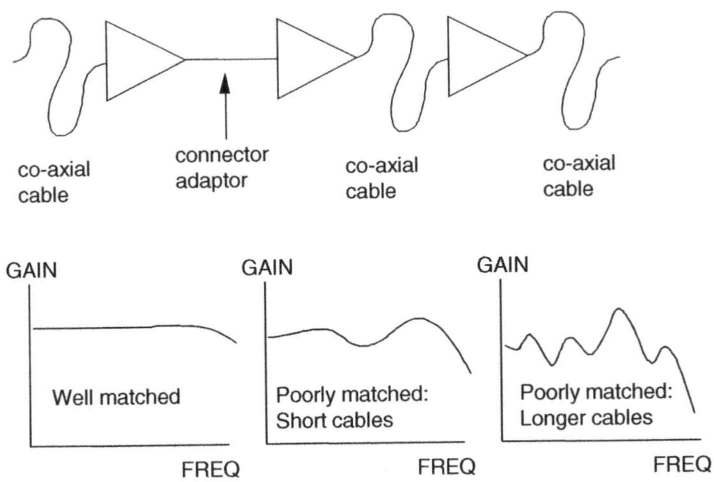

Figure 9.1 Amplifier chain illustrating the need for a system impedance.

waves may be in-phase and add constructively, while at other frequencies they become anti-phase and partially cancel each other out. The end result is gain ripple, a phenomenon where the gain changes with frequency, going up and down. Depending on the physical length of the cables, relative to the signal guided wavelength, these ripples can be quite gentle (short cables) or very dramatic (long cables), with the gain going up and down rapidly with frequency. Furthermore, without matching, a standing wave is created on the transmission line as the forward and reflected waves superimpose constructively and destructively along the line. This can lead to high values of voltage and current on the transmission line and these values can cause damage to electronic devices in the circuit.

In low-frequency design, even a 1-metre long cable will be short relative to the wavelength, which is why these problems are not generally considered in low-frequency design. However, in the tens of GHz regime, even a short microstrip line on a board or integrated circuit will have significant electrical length and the reflections must be avoided by making sure components are matched to the characteristic impedance of the line. Today, 50 Ω is the most common choice of system impedance for historical reasons that relate to the value of coaxial cable Z_0 that gives the best trade-off between low loss and high power handling. However, what matters is that the source and load are matched to the characteristic impedance of the line that connects them. For on-chip design, 50 Ω may not be the best choice since it is rather low, requiring larger transistors in general in an active circuit and wide lines (microstrip) or small gaps (CPW/CPS).

Figure 9.2 shows the basic circuit representation of a source feeding a load. For DC systems, the maximum power transfer theorem shows that maximum power is delivered to the load when the source and load have the same the resistance. The main thrust of this theorem is that when the source resistance (R_s) matches with the load resistance (R_L), the amount of power received at the load will be equal to the dissipated power at the source and, therefore, the efficiency will be 50%. For this case, the source should generate double the desired power to deliver maximum power to the load; this phenomenon is shown in Figure 9.3.

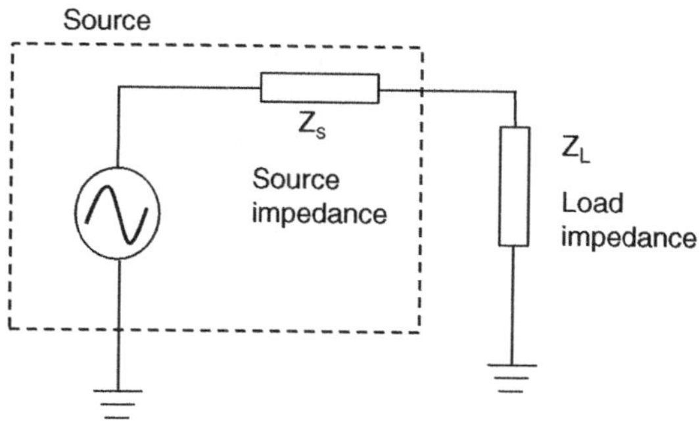

Figure 9.2 Basic source and load circuit.

In the case of AC and RF circuits, the complex impedance must be used because inductors and capacitors have significant reactance. In this scenario, the requirement for impedance matching is that the real parts of Z_S and Z_L should be equal, while the imaginary parts should be equal in magnitude but opposite in sign. For example, if the source impedance (Z_s) is $R + jX$, then load impedance (Z_L) should be $R - jX$ for perfect matching. More generally, this is referred to as conjugate matching and $Z_S = Z_L^*$.

In RF applications, the source output is connected to the load through a transmission line, for example a coaxial cable, as shown in Figure 9.4. All TEM transmission lines have a characteristic impedance (Z_0), that is formed from the inductance and capacitance distributed all along the line. To obtain maximum power at the load side, the impedance of the transmission line should be equal to the source and load impedance. If it is not, maximum power will not be transferred to the load and a standing wave will be created.

The reflection coefficient (Γ) at the load is given by:

$$\Gamma = \frac{Z_L - Z_O}{Z_L + Z_O} \tag{9.1}$$

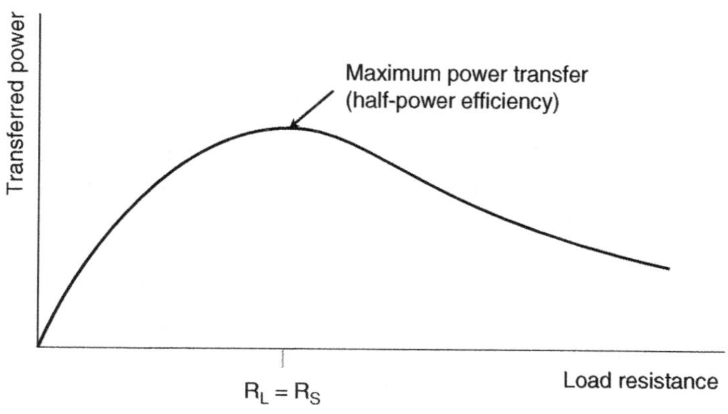

Figure 9.3 Graph of load power versus load resistance.

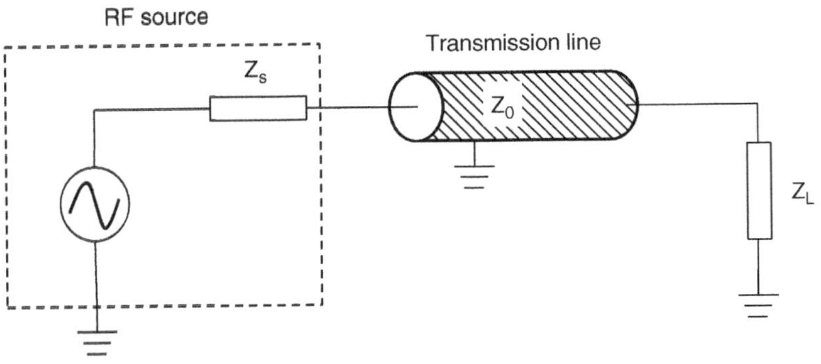

Figure 9.4 Source connected to load through a transmission line.

If $Z_L = Z_0$ then the reflection coefficient = 0, the system is perfectly matched, there is no standing wave, and maximum power transfer is achieved. Generally, Z_L is very different to Z_0 and it has significant reactance, for example a transistor's input capacitance, so a matching network must be inserted between the source and load. There are many methods to solve this problem and some of these will be explained in detail in the following sections.

When matching complex impedances, it will be shown that resonators play a key role. The quality factor (Q) is an important parameter that quantifies the resonant bandwidth of a circuit compared to the centre frequency. If the Q-factor is high, then only a small amount of energy is lost compared to the total energy in the system. In designing any system, it is better to have high-Q components in order to have good control of the performance. Quality factor is defined as the ratio of total energy of the system to the energy loss in the same system:

$$Q = \frac{2\pi \times \textit{total energy stored}}{\textit{energy loss per cycle}} \tag{9.2}$$

Normally, Q-factor is defined at the resonant frequency and takes into account the effective total resistance of the circuit. The stored energy is the sum of the energy stored in the inductors and capacitors, whereas the energy lost is the sum of the energy dissipated in the resistors, in one cycle.

9.3 Lumped-Element Matching Networks

The loss in a matching network must be as low as possible, and so reactive components (inductors, capacitors and transmission lines) should be used. As discussed in Chapter 4, lumped elements have a self-resonant frequency (SRF) which depends very much on the practical form of the component. Lumped-element matching networks are typically used for those circuits that have operating frequencies of approximately 20 GHz on an integrated circuit, or a few GHz if SMD components are used. A Smith chart can be used to design lumped-element matching networks, but a matching circuit using lumped elements will require the use of components connected both in series and parallel with the load impedance, and so some confidence in working in both impedance and admittance is required. Figure 9.5 shows four examples of how series and parallel inductors or capacitors move a load on the Smith chart.

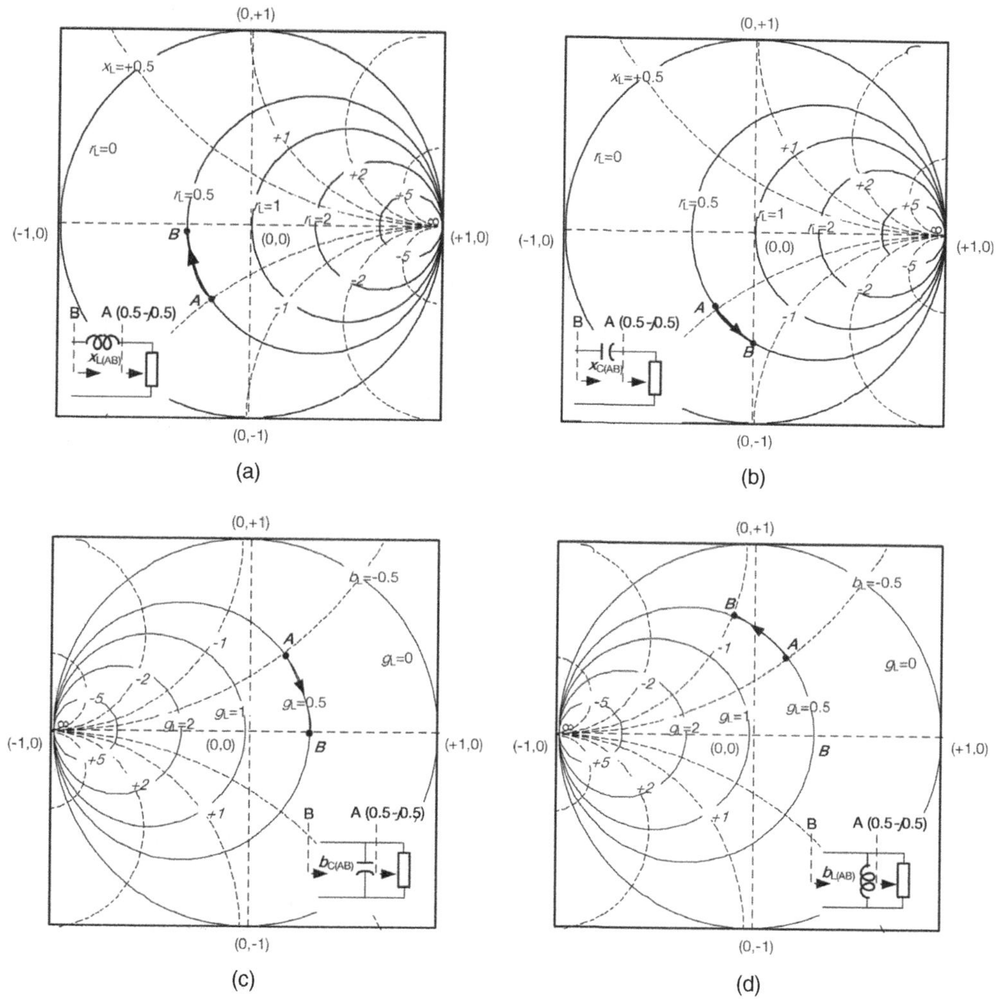

Figure 9.5 Smith chart representations for connecting a circuit with (a) a series inductor, (b) a series capacitor, (c) a parallel capacitor, and (d) a parallel inductor.

In Figure 9.5(a) and (b), the impedance Smith chart is used for a series-reactive component to transform the load impedance at point A (normalised load impedance, $z = 0.5 - j0.5$) to the impedance at point B. The point $z = 0.5 - j0.5$ is found from the intersection between the resistance constant circle $r = 0.5$ and reactance constant circle $x = -0.5$. A series L or C moves the impedance around the constant resistance circle of $r = 0.5$ since only the reactance can be changed. A series inductor moves the impedance clockwise, as shown in Figure 9.5(a), while a series capacitor moves the impedance counter-clockwise, as shown in Figure 9.5(b).

When a parallel reactive component is used to match a load, the admittance Smith chart should be used. In Figure 9.5(c) and (d), the admittance chart is shown with a load at point A, with normalised admittance $y = 0.5 - j0.5$. Now, by adding a parallel L or C, only the imaginary part of the admittance (i.e., the susceptance) can be changed and so the trace follows the constant conductance circle, $g = 0.5$. In this case, a parallel capacitor moves the load clockwise around the constant conductance circle and a parallel inductor moves it counterclockwise. Note that the magnitude of the change in imaginary part, either in the parallel or series case, is determined by the length of the arc A–B, read off using the reactance/susceptance contours as appropriate. In Figure 9.5(c), for example, the length of the arc A–B is 0.5, which indicates that the parallel capacitor requires a normalised susceptance value +0.5.

9.3.1 Lumped-Element Matching with L-Networks

The four ways of moving a load on the Smith chart shown in Figure 9.5 provide the basis for designing a matching network using a pair of lumped elements, one in series and one in parallel. These simple *L*-networks will only achieve narrowband matching, but this gives a sufficiently flat frequency response for many communications systems. There are eight different combinations, and the usage is dependent on the load impedance (Z_L) and source impedance (Z_0), along with some practical constraints. The load impedance, Z_L, with real and imaginary parts (resistance and reactance) is assumed to be given by:

$$Z_L = R_L + jX_L \tag{9.3}$$

whilst the load admittance has real and imaginary parts (conductance and susceptance) of:

$$\mathrm{Y_L} = {}^{1}/_{Z_L} = G_L + \mathrm{j}B_L = (R_L - jX_L)/\left(R_L^2 + X_L^2\right) \tag{9.4}$$

There are two groups of network that can be used, which are applicable depending on whether $R_L < Z_o$ or $G_L < 1/Z_o$, as shown in Figure 9.6. They are called *L*-networks because the structure resembles the letter 'L'. One group is implemented with a series element first (closest to the load), followed by a shunt element. The other group has a shunt element first, followed by a series one. The elements can be either inductors or capacitors, so in total the two groups each have four combinations, giving eight circuit topologies of *L*-network that can be used, as shown in Figures 9.7 and 9.8. In general, any impedance can be matched with at least two

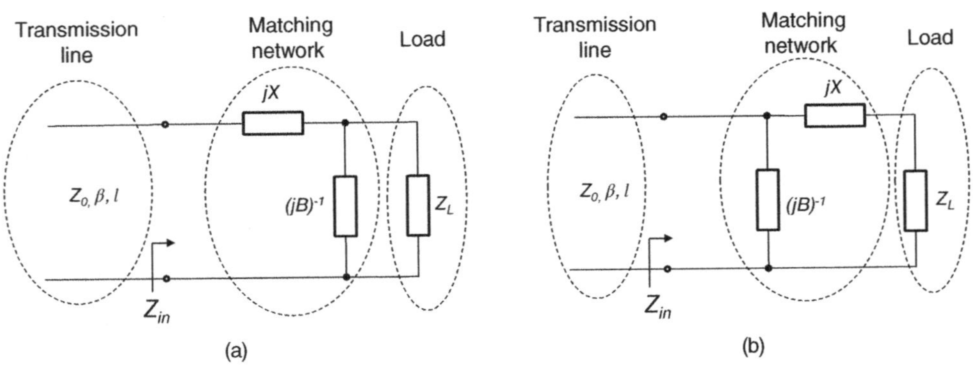

Figure 9.6 *L*-network for two conditions. (a) $R_L < Z_0$. (b) $G_L < 1/Z_0$.

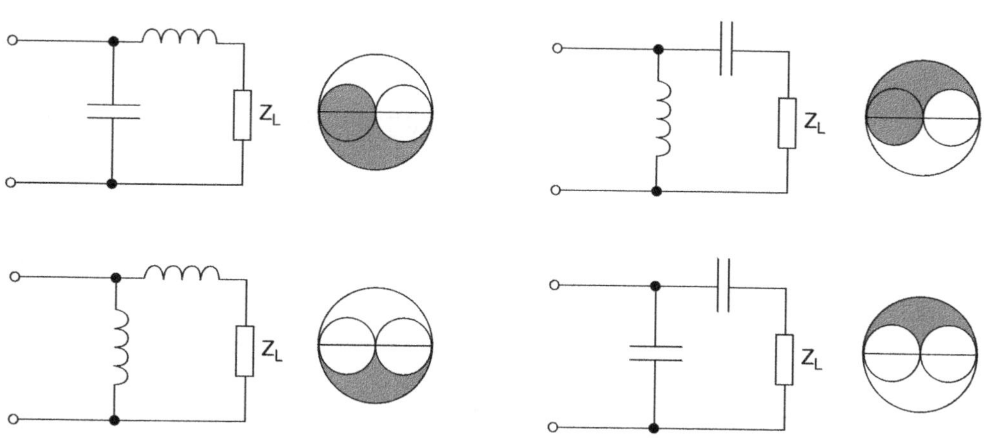

Figure 9.7 L-network circuit topologies for $R_L < Z_0$, showing the regions of applicability for Z_L on the Smith chart (each network can be used only for Z_L in the respective shaded area).

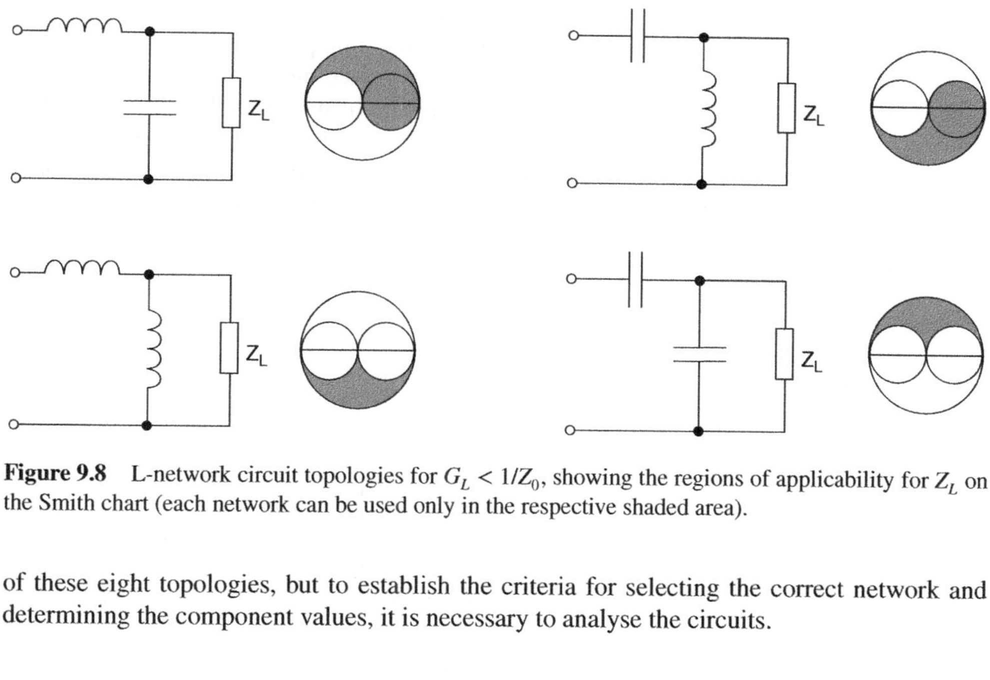

Figure 9.8 L-network circuit topologies for $G_L < 1/Z_0$, showing the regions of applicability for Z_L on the Smith chart (each network can be used only in the respective shaded area).

of these eight topologies, but to establish the criteria for selecting the correct network and determining the component values, it is necessary to analyse the circuits.

9.3.1.1 Solution for $R_L < Z_0$

For the first topology group, shown in Figure 9.6(a), the input impedance is:

$$Z_{in} = \left[\frac{1}{jB + \dfrac{1}{R_L + jX_L + jX}} \right] \tag{9.5}$$

Setting this equal to Z_0 to achieve a matched condition, gives:

$$Z_0 = \frac{R_L + jX_L + jX}{jBR_L - BX_L - BX + 1} \tag{9.6}$$

By separately equating the real and imaginary parts, the expressions for the required values of B and X are found to be:

$$B = \pm\frac{1}{Z_0}\sqrt{\frac{Z_0 - R_L}{R_L}} \tag{9.7}$$

$$X = \pm\sqrt{R_L(Z_0 - R_L)} - X_L \tag{9.8}$$

In this case, if $R_L < Z_0$, the values of B and X will be real. The $\pm$ indicates that inductors or capacitors may be selected, giving the four networks in Figure 9.7, which also shows the region of the Smith chart that each one can cover. On the Smith chart, the series component is first moving z_L to the $g = 1$ circle, so that the parallel component can then cancel the remaining susceptance. However, the $g = 1$ circle is not plotted on the impedance chart and so either the extra circle can be drawn on, or the double Smith chart used. The best method is to use a computer Smith chart tool, which has a full range of display options, avoids mistakes, calculates component values automatically, and allows a try it-and-see approach. The worst option is to randomly try matching networks in a circuit simulator and use optimisation to obtain the values. This will, quite likely, give results that are not optimum, for example very narrow bandwidth.

9.3.1.2 Solution for $G_L < 1/Z_0$

The input impedance of network in Figure 9.6(b) is given by:

$$Z_{in} = jX + \left[\frac{1}{jB + \dfrac{1}{R_L + jX_L}}\right] \tag{9.9}$$

Simplifying and setting this equal to Z_0, to achieve matching to the source, gives:

$$jX + \frac{R_L + jX_L}{jBR_L - BX_L + 1} = Z_0 \tag{9.10}$$

To solve the above equation, it is rearranged as:

$$jZ_0BR_L - Z_0BX_L + Z_0 = R_L - XBR_L + j(X + X_L - XBX_L) \tag{9.11}$$

Equating the real and imaginary parts, gives two equations:

$$B(XR_L - Z_0X_L) = R_L - Z_0 \tag{9.12}$$

and

$$BZ_0R_L - X_L = X(1 - BX_L) \tag{9.13}$$

The second equation gives an expression for the series component as follows:

$$X = \frac{BZ_0R_L - X_L}{1 - BX_L} \tag{9.14}$$

Now, putting this expression for X into Equation 9.12, gives the susceptance of the shunt component:

$$B = \frac{X_L \pm \sqrt{\frac{R_L}{Z_0}\left(R_L^2 + X_L^2\right) - R_L^2}}{R_L^2 + X_L^2} \tag{9.15}$$

The condition that $R_L^2 + X_L^2 \geq Z_0R_L$ is now evident, which is the same as $G_L < 1/Z_O$. Two values of B are obtained, and when these are put into the expression for X, two values of X are obtained, yielding the four networks of Figure 9.8. The sign of the X and B obtained obviously will determine whether the component is a capacitor or inductor.

In this case, on the Smith chart the parallel component is first moving y_L to the $r = 1$ circle, so that the series component can then cancel out the remaining reactance. The $r = 1$ circle is not plotted on the admittance chart and so, again, it can be drawn on, or a double Smith chart used.

In some cases, there are two locations on the $g = 1$ or $r = 1$ circle that can be reached, and generally it is the nearer one that yields smaller component values. The choice of network that is used to match the load to the source will be dictated by factors such as bandwidth, convenience of applying DC bias, availability of components, SRF, and so forth. Note that a drawback of L-networks is that the quality factor is dictated by the source and load. The quality factor can be calculated with:

$$Q = \sqrt{\frac{R_{source}}{R_{load}} - 1} \tag{9.16}$$

9.3.2 *Matching with π- and T-Networks*

Multiple reactance matching is used to achieve a high quality factor and to match a large source impedance to a small load impedance, or to match a small source impedance with a large load impedance. To give better control over Q-factor, a pair of L-networks can be used to choose an intermediate resistance value before matching. The pair of networks is selected such that the central two components are either both in series, or both in parallel, which means

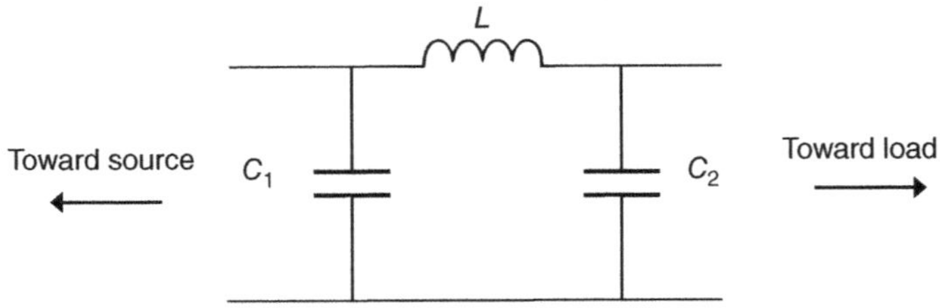

Figure 9.9 *C-L-C* π-network for impedance matching.

they can be merged into a single component. This means that there are then three components and the two types of multiple reactance matching are the π-network and the T-network.

9.3.2.1 π-network

The π-network in capacitor-inductor-capacitor form is shown in Figure 9.9. In this example, two L-network series components are combined into a single series inductor. Before calculating the component values, the desired Q-factor should be decided first because it affects harmonic attenuation and bandwidth. After setting the Q-factor, the input impedance should be converted to parallel impedance. So, if the load impedance is:

$$Z_D = R_D + jX_D \tag{9.17}$$

Its parallel impedance is:

$$R_P = \frac{R_D^2 + X_D^2}{R_D} \quad \text{and} \quad X_P = \frac{R_D^2 + X_D^2}{X_D} \tag{9.18}$$

The components values of the π-network can be calculated from:

$$Y_{C1} = \frac{1}{X_{C1}} = \frac{Q}{R_P} + \frac{1}{X_P} \tag{9.19}$$

$$X_{C1} = R_L \sqrt{\frac{R_P/R_L}{(Q^2+1) - R_P/R_L}} \tag{9.20}$$

$$X_L = \frac{QR_P + \left(\dfrac{R_P R_L}{X_{C2}}\right)}{Q^2+1} \tag{9.21}$$

9.3.2.2 T-network

The T-network circuit topology in inductor-capacitor-inductor form is shown in Figure 9.10. As seen in the figure, the two L-networks that it is derived from are merged, yielding a single

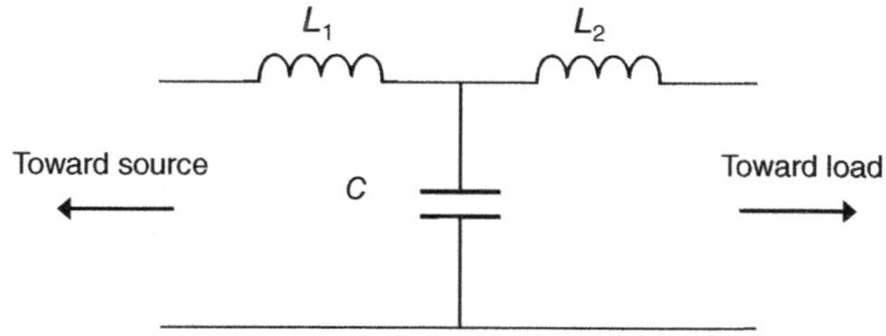

Figure 9.10 *L-C-L T*-network circuit diagram.

shunt capacitor at the centre. Again, if the device impedance is:

$$Z_D = R_D + jX_D \tag{9.22}$$

Then it can be shown that the three component values can be calculated from:

$$X_{L1} = QR_D - X_D \quad X_{L2} = R_L A \quad X_C = \frac{B}{Q + A} \tag{9.23}$$

where,

$$A = \sqrt{\frac{\{R_D(1 + Q^2)\}}{R_L} - 1} \quad \text{and} \quad B = R_D(1 + Q^2) \tag{9.24}$$

9.4 Distributed Matching Networks

Distributed elements are transmission line elements, and are used instead of lumped elements at high frequencies because there are many cases where lumped elements are not practical. The reasons for using distributed element matching networks are:

- They work at high frequency with low loss.
- Lumped elements may not be available with high enough SRF.
- Lumped elements are found to have unfeasibly small or large values.

The Smith chart was originally introduced by Phillip H. Smith in 1940, while working at Bell Laboratories, to perform the calculations associated with transmission line problems, and it is an essential aid to impedance matching using distributed elements. For example, as shown in Chapter 5, impedance varies along a lossless transmission line according to the formula:

$$Z(d) = Z_0 \frac{Z_L + jZ_0 \tan \beta d}{Z_0 + jZ_L \tan \beta d} \tag{9.25}$$

where Z_L is the impedance of the load and d is the distance moved away from the load, β is the propagation phase constant and Z_0 is the characteristic impedance of the line. It is rather tedious to put numerical values into this formula using a slide rule and book of mathematical tables. Using a Smith chart, the same calculation can be performed graphically. Moving along a transmission line of characteristic impedance Z_0 by an electrical length θ (= βd) is achieved on the chart by plotting the load Z_L normalised to Z_0, and then rotating an angle $2 \times \theta$, with a compass centred on the origin. With the circles of resistance and reactance overlaid on the chart, the value of $Z(d)$ can be directly read off. Since the Smith chart is a plot of reflection coefficient, a full rotation around the chart represents a half-wavelength, or 180° of electrical length. To determine the length of the transmission line in wavelength units, the direction of rotation must be taken into account, and there are two cases; that is, wavelengths to generator and wavelengths to load, often labelled on the edge of the chart. By convention, transmission line lengths are determined in the counterclockwise direction for the *wavelengths toward load*, and in the clockwise direction for *wavelengths toward generator*. There are many methods of using transmission lines for matching, but only the stub matching and quarter-wave transformer technique will be described here in detail.

9.4.1 Stub Matching

Two types of stubs are used in stub matching, namely shunt and series connected stubs, as shown in Figure 9.11. In shunt stub matching, the input impedance of the stub is connected in parallel with the circuit. Hence, it is convenient to work with admittance and the far end of the stub is either short- or open-circuited depending on the required stub susceptance. The series stub requires a twin-conductor transmission line medium, not one with a ground plane, so that the impedance looking into the stub can be connected in series with the circuit. More than one stub can be used according to specific requirements of systems, for example to obtain more bandwidth.

The Smith chart can be used for matching the load with a short-circuited parallel stub placed at a certain distance from the load. This is, in fact, the classic technique used to introduce distributed matching circuits and has probably featured in hundreds of University examination papers. The length of the stub and its distance from load are easily calculated using the Smith chart. An example of this single stub matching technique is shown in Figure 9.12, where d is the distance of the stub from the load and d_1 is the length of the stub. As the stub is connected in parallel, it is convenient to work in admittance.

The admittance of the stub will be purely imaginary, that is, a susceptance. Hence, in order to match to impedance Z_0, the first step is to rotate the load on the Smith chart to a point

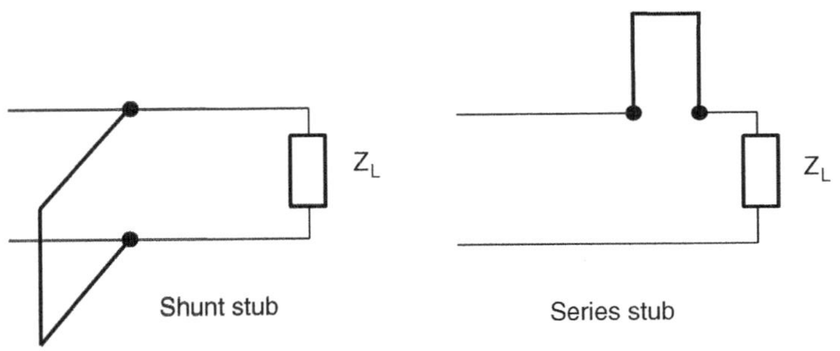

Figure 9.11 Stub matching with a shunt or series stub.

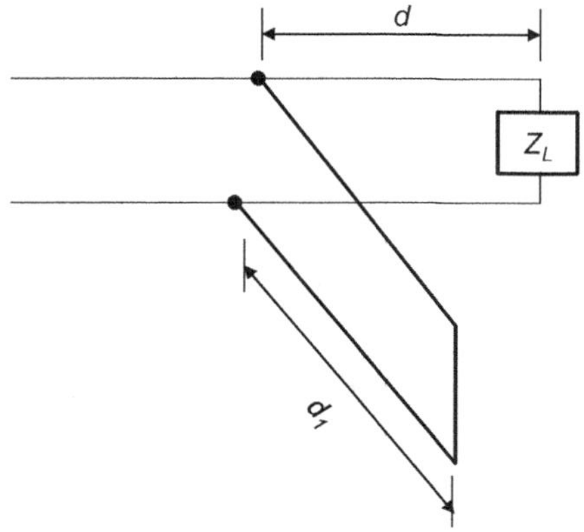

Figure 9.12 Single stub matching.

where the admittance has a real part (conductance, G) equal to $1/Z_0$ (i.e., Y_0). This rotation is achieved by the line, of length d, between the stub and the load. With $G = 1/Z_0$, the stub length, d_1, is then chosen so that the stub's susceptance has the same magnitude but opposite sign as the susceptance remaining at the point where the stub is connected to the circuit. Then, when it is added with the admittance at the intersection of the stub and line, the resultant input admittance will be only the real part, $1/Z_0$, and matching to the source has been achieved. Once matching to the source has been achieved, the remaining length of transmission line, to the left of the stub in the figure, has no effect on the impedance provided that it has characteristic impedance Z_0.

Example: Match a load with impedance of $105 + 40j\Omega$ to $50\ \Omega$ using a shunt (parallel) stub connected a suitable distance from the load.

Step I: First, Z_L is normalised to $z = 2.1 + 0.8j$ and the normalised admittance, y, is calculated:

$$y = \frac{1}{2.1 + 0.8j} = 0.41 - 0.16j$$

Step II: The admittance can be plotted directly on the admittance chart, shown as point 'A' in Figure 9.13.

Step III: Since the parallel stub can only change the susceptance, the load is rotated to the point 'B', where the normalised conductance is equal to unity. In other words, the load must be moved to the $g = 1$ circle by choosing the correct length of the series line before the stub (the 'suitable distance from the load'). In Figure 9.13, this rotation is labelled φ and for this example the angle on the chart is 136°. The electrical length of the series line is therefore $\varphi/2 = 68°$.

Step IV: Read off the remaining susceptance. In Figure 9.13, the susceptance contour is read off at point 'C'. For this example, the susceptance remaining at point 'B' is therefore 0.935 (still normalised).

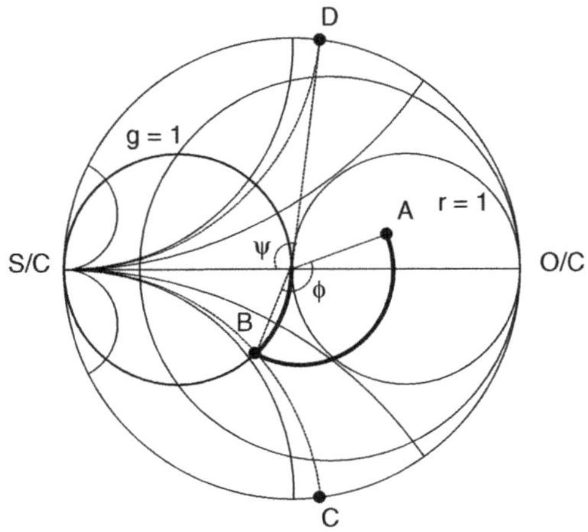

Figure 9.13 Stub matching example on the Smith chart.

Step V: The required stub type (open or circuited) and electrical length are now determined. The stub must have susceptance –0.935, in order to cancel the remaining susceptance of the load. $-0.935j$ is labelled as point 'D' on Figure 9.13. Since rotations are clockwise, the short-circuit point is closest to 'D' and so a short-circuited stub is selected. The electrical length is half the angle ψ from the short-circuit point to 'D', giving an electrical length 47°. Alternatively, on a paper Smith chart, the wavelengths scale around the outer edge can be used again, giving 0.13λ.

So, the solution is that the series line has electrical length 68° and a short-circuited parallel stub of length 47° is required.

Note that an open-circuited stub would also work, but to rotate from the open-circuit point to 'D' requires more than a quarter wavelength (you must rotate in the correct direction) and in fact an open-circuit stub would be 0.38λ and the likely effect of a much longer line is lower useful bandwidth. Likewise, the load could have been rotated even further by the series line, to the other side of the $g = 1$ circle, and that would have meant a long series line and less bandwidth, although it may help in other ways in some cases.

Finally, note that there are many other ways in which this same problem could be solved with a paper Smith chart, and these other methods may be seen in other text books [1–3] for the various steps, including:

1. Plot z directly, then rotate z by 180° to get y on the same chart (this is quite brave since the conductance/susceptance circles are on the wrong side now).
2. Plot z, mirror it through the origin (which is actually the same as a 180° rotation) to get y, then rotate the chart itself by 180° (this means the conductance/susceptance circles are on the correct side).
3. Use separate charts for working in impedance and admittance.
4. Use the dual impedance-admittance (double) Smith chart.
5. Use the equation form to get the length of the stub.

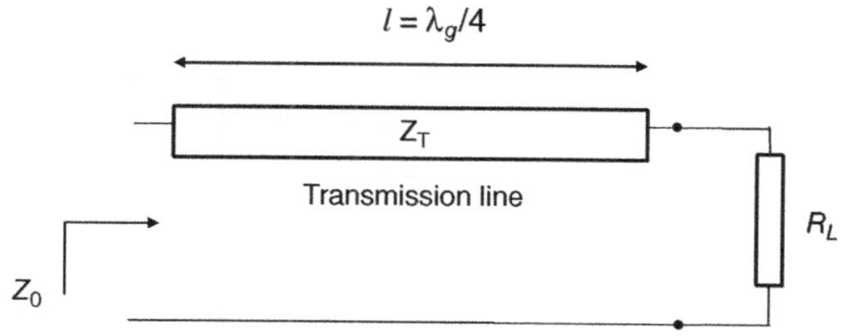

Figure 9.14 Quarter-wave transmission line matching.

One of the biggest risks in the 180° rotation or mirror-image method is that the open- and short-circuit points also flip over, so great care must be taken. Hence, the double Smith chart is safest, but since it is difficult to distinguish the many contours and values, the computer tool is ideal, combining the visual aid with accurate, error-free, calculations.

9.4.2 Single-Section Quarter-Wave Transformer

A quarter-wave transformer can be used between resistive source and load impedances and consists of a quarter-wavelength transmission line as shown in Figure 9.14. The characteristic impedance, Z_T, of the transmission line transformer can be calculated by:

$$Z_T = \sqrt{Z_0 \times R_L} \tag{9.26}$$

where Z_0 is the system impedance (typically 50 Ω) that is to be matched to, and R_L is the load resistance. This can easily be derived as a special case of Equation (9.25). The single quarter-wave transmission line can only provide perfect impedance match at one frequency (ignoring harmonics thereof, discussed later). The frequency response for this case is shown in Figure 9.15. If the load is not resistive, one of the many techniques described in this chapter can be first used to cancel out the imaginary part.

To match a load with impedance of 100 Ω to a 50 Ω source using a quarter-wave transformer, the transformer impedance is $Z_T = \sqrt{50 \times 100} = 70.71$ Ω. This is why 70.7 Ω lines appear in many applications, such as the Wilkinson power divider. To match 25 Ω to 50 Ω requires

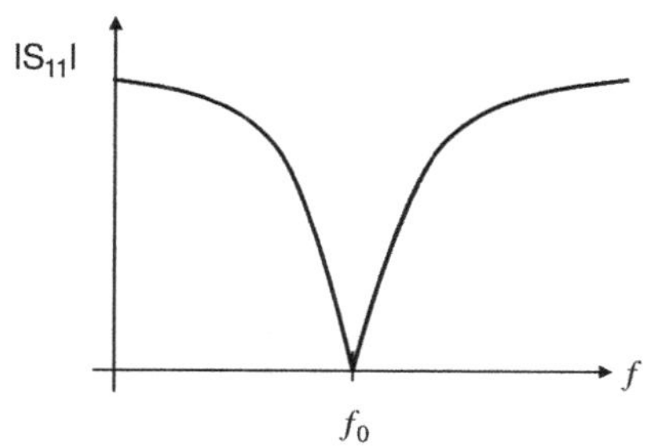

Figure 9.15 Frequency response with single-section quarter-wave transformer.

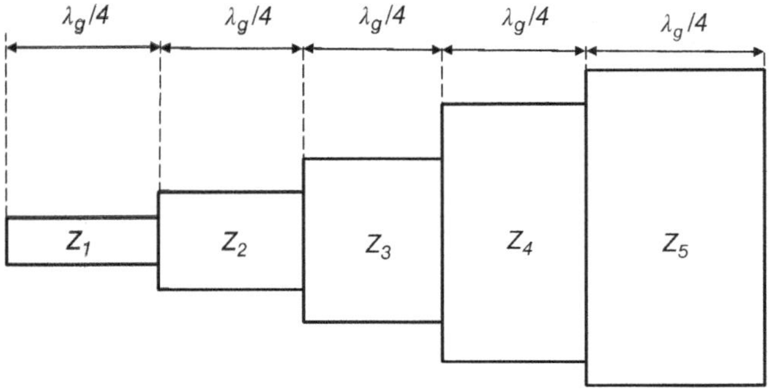

Figure 9.16 Multisection quarter-wave transformer.

a transformer characteristic impedance of $Z_T = \sqrt{50 \times 25} = 35.4\ \Omega$, which is why this value appears in the branch-line coupler, for example.

9.4.3 Multisection Quarter-Wave Transformer

To achieve impedance matching over a wide frequency range, multiple quarter-wave transformer sections can be used, gradually changing the load impedance. In this scenario, it is not possible to achieve zero reflection coefficient at all frequencies. The topology and typical frequency response are given in Figures 9.16 and 9.17, respectively. The multisection transformer can be designed with the characteristics of Butterworth or Chebyshev filters, and others, and the design procedure is available in Matthaei, Young and Jones [4]. There are also various design tools available on the internet. Note that in microstrip form the physical lengths of the sections will not be equal because ε_{eff} varies with the conductor width.

9.4.4 Other Types of Transmission Line Transformer

It was shown in Chapter 7 on transmission line filters that quarter-wavelength coupled-line sections can be configured in various ways to act as impedance or admittance inverters. The coupled-line element, therefore, forms the basis of a very wide range of matching techniques based on a transmission line transformer approach [5]. The Ruthroff and Guanella approaches

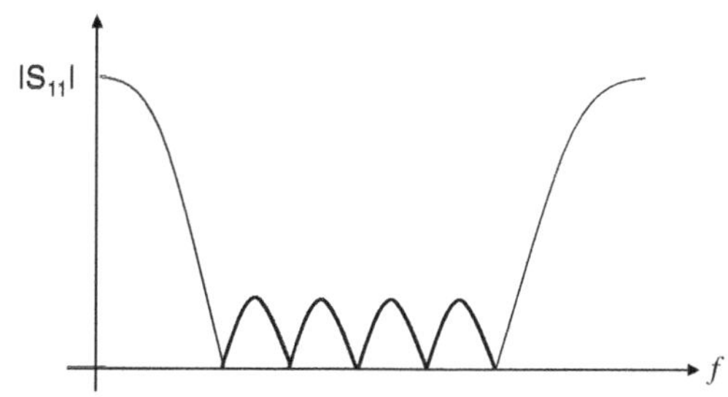

Figure 9.17 Multisection quarter-wave transformer frequency response.

[6, 7] are generally limited to certain discrete impedance transformation rations (such as 4:1) but offer very wide bandwidth. More recently, further investigations have been made of a range of developments, including broadband and multiband techniques [8, 9] and various multilayer and integrated circuit implementations for miniaturisation [10–12].

9.5 The Cyclic Nature of Distributed Circuits

It is important to be aware that distributed matching elements have a cyclic nature when their frequency response is considered over a wide frequency range, and this can cause significant problems. For example, even if it is designed with great care at the desired operating frequency, an amplifier may be completely unstable at some other frequency because the electrical length of the transmission line matching elements changes continuously with frequency.

As a simple design example, suppose we wish to match a load resistance of 20 Ω using a quarter-wave transformer, at a design frequency of 900 MHz. For the quarter-wave transformer, $Z_T = \sqrt{Z_o R}$ gives $Z_T = 31.6\ \Omega$ and the electrical length is 90°. The electrical parameters of the quarter-wave transformer must be converted to physical dimensions. In microstrip form, the transformer is a track with a certain width (which determines the characteristic impedance) and physical length, l. The width is first found for the chosen substrate parameters, for example using LineCalc™. The guided wavelength at 900 MHz is then calculated using the value of ε_{eff} for that width of microstrip line. The transformer physical length is, of course, one-quarter of that guided wavelength, to give 90° electrical length at 900 MHz (the design frequency). However, at other frequencies, the electrical length is different. The electrical length is βl, and remembering that $\beta = \omega/v$ it can be seen that the electrical length will increase linearly with frequency (for a TEM transmission line). To investigate how this changing electrical length affects the circuit behaviour, first consider what happens in the same 900 MHz quarter-wavelength line at 20× and 1/20th of the design frequency. This is illustrated in Figure 9.18, which shows that there is very little electrical length at 1/20th of the design frequency, whereas five whole wavelengths can fit into the same physical length of line at 20 × 900 MHz.

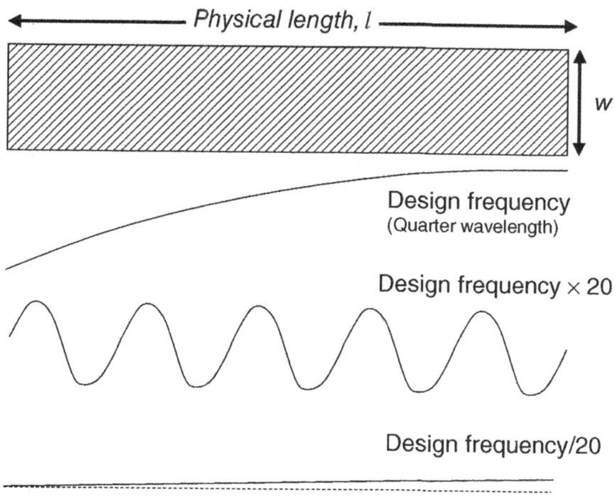

Figure 9.18 Illustration of change in electrical length with signal frequency.

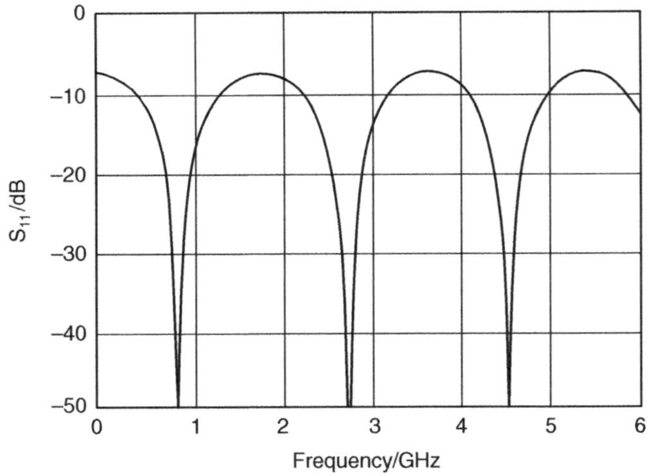

Figure 9.19 S_{11} plot showing cyclical behaviour of the matching network.

At 20× the frequency, the wavelength is shorter and five whole cycles fit in the same physical length of transmission line. The electrical length is '5 × 360 degrees' or 'five wavelengths'. At low frequency, the line length is very small compared with the wavelength. Now, consider what happens on the Smith chart in the 20× frequency case. The load resistance is rotated around and around the centre, since every half wavelength of line rotates the impedance right around the chart by a full rotation. This leads to the cyclical behaviour of transmission line circuits. As illustrated in Figure 9.19, therefore, the quarter-wave transformer actually works at an infinite number of regularly spaced frequencies (in the theoretical case).

9.6 Conclusion

The purpose of this chapter has been to study the different types of impedance matching networks. Matching is important in high-frequency systems for many reasons. For example, it is needed in amplifiers in order to achieve useful gain from a transistor, and when connecting circuits together to form a system, it is important that they are well matched so that ripple in the frequency response can be avoided. The Smith chart is an essential tool for the design of both lumped and distributed matching networks. Simple lumped-element *L*-networks and stub matching networks can easily be designed in this way, but when wide bandwidth is required then it is often necessary to use the matching network synthesis tools provided in EDA software.

References

1. Radmanesh, M.M. (2001) *Radio Frequency and Microwave Electronics Illustrated.* Prentice-Hall. ISBN-10: 0130279587.
2. Pozar, D. (2011) *Microwave Engineering*, 4th edition. John Wiley & Sons. ISBN-10: 0470631554.

3. Ludwig, R. and Bogdanov, G. (2008) *RF Circuit Design: Theory & Applications*, 2nd edition. Prentice-Hall. ISBN-10: 0131471376.
4. Matthaei, G.L., Young L., and Jones, E.M.T. (1980) *Microwave Filters, Impedance-matching Networks and Coupling Structures*, New edition. Artech House Publishers. ISBN-10: 0890060991.
5. Sevick, J. (2001) *Transmission Line Transformers*. Noble Publishing. ISBN: 1-884932-18-5.
6. Guanella, G. (1944) New Method of Impedance Matching in Radio-Frequency Circuits. The Brown Boveri Review, September, pp. 327–329.
7. Ruthroff, C.L. (1959) Some broad-band transformers. *Proceedings of the IRE*, **47** (8), 1337–1342. doi: 10.1109/JRPROC.1959.287200.
8. Jensen, T., Zhurbenko, V., Krozer, V., and Meincke, P. (2007) Coupled transmission lines as impedance transformer. *IEEE Transactions on Microwave Theory and Techniques*, **55** (12), 2957–2965. doi: 10.1109/TMTT.2007.909617.
9. Chongcheawchamnan, M., Patisang, S., Srisathit, S., Phromloungsri, R., and Bunnjaweht, S. (2005) Analysis and design of a three-section transmission-line transformer. *IEEE Transactions on Microwave Theory and Techniques*, **53** (7), 2458–2462. doi: 10.1109/TMTT.2005.850408.
10. Sobrany, R.F. and Robertson, I.D. (2003) Ruthroff transmission line transformers using multilayer technology. 33rd European Microwave Conference, October, pp. 559–562. doi: 10.1109/EUMA.2003.341014.
11. Bahl, I.J. (2006) Broadband and compact impedance transformers for microwave circuits. *IEEE Microwave Magazine*, **7** (4), 56–62. doi: 10.1109/MMW.2006.1663990.
12. LaRocca, T., Liu, J.Y.-C., and Chang, M.-C.F. (2009) 60 GHz CMOS amplifiers using transformer-coupling and artificial dielectric differential transmission lines for compact design. *IEEE Journal of Solid-State Circuits*, **44** (5), 1425–1435. doi: 10.1109/JSSC.2009.2015794.

10

Amplifiers

10.1 Introduction

Amplifying a signal may seem like a simple task, but amplifier design has to take into account a very wide range of parameters including gain, operating frequency, bandwidth, gain flatness, matching, stability, DC bias, efficiency, noise figure, output power at 1 dB gain compression and third-order intercept point. Clearly, amplifier design is far more than matching the source and load impedance to the transistor, although understanding the classical fundamentals of high-frequency amplifier design are, nevertheless, essential. In receivers, RFIC amplifiers may be required to operate with very low current, requiring the adoption of a raft of analogue design techniques from traditional low-frequency applications. The design of power amplifiers ranges from the order of just 10 dBm for short-range wireless systems, through to large output powers requiring power-combining of many modules. The high reliability of solid-state devices has led to a lot of interest in GaN and other amplifier technologies as replacements for vacuum devices. A number of new classes of operation have been developed to push the limits on power efficiency, in many cases because of the large peak-to-average power ratio of modern digital signals, such as OFDM, that have high spectral efficiency and good resilience to multipath. For frequencies beyond 100 GHz, the available power from transistors remains modest, but techniques such as spatial power combining and active antennas offer exciting opportunities for addressing this longstanding challenge.

10.1.1 Gain Definitions

In radiofrequency amplifiers, power gain is generally the most important gain parameter since designers are dealing with components that are used in a transmission line environment, with source and load impedance equal to the characteristic impedance, Z_0, such as 50 Ω. This leads to the traditional techniques of impedance matching to the transistors, with associated stability and gain circles. However, amplifiers are increasingly realised on an integrated circuit and so, especially in RFIC design, there is no longer any need to focus on traditional matching techniques and a wide range of traditional analogue design techniques have been adopted and optimised for high-frequency applications.

Microwave and Millimetre-Wave Design for Wireless Communications, First Edition.
Ian Robertson, Nutapong Somjit and Mitchai Chongcheawchamnan.
© 2016 John Wiley & Sons, Ltd. Published 2016 by John Wiley & Sons, Ltd.

The operating, or actual, power gain of an amplifier is the ratio of the output power to the input power, given by:

$$\text{Power Gain} = \frac{P_{load}}{P_{in}} \tag{10.1}$$

where P_{load} is the power transferred to the load from the amplifier and P_{in} is the power input to the amplifier.

Gain is normally expressed in decibel units, which have the advantage that they can simply be added and subtracted to calculate the overall gain of a cascade of active and passive networks:

$$\text{Power Gain in dB} = 10\log\frac{P_{load}}{P_{in}} \tag{10.2}$$

Note that this is a ratio and has no absolute units. However, the input and output power must have some absolute units:

$$\text{Power in dBm} = 10\log(P\ in\ mW)\ \text{or Power in dBW} = 10\log(P\ in\ Watts)$$

where the ‘m’ or ‘W’ suffix indicates the absolute power reference units.

It is often useful to remember that, despite various complications within the active devices, for resistive source and load impedances, the input power can be calculated by:

$$P_{in} = \frac{|V_i|^2}{R_i} \tag{10.3}$$

where V_i is the input voltage and R_i is the input resistance, and the output power is given by:

$$P_{load} = \frac{|V_L|^2}{R_L} \tag{10.4}$$

where V_L is the load voltage and R_L is the load resistance. The power gain in dB is given by:

$$G(\text{dB}) = 20\log\frac{|V_L|}{|V_i|} \tag{10.5}$$

For a linear, *matched*, amplifier, the power gain in dB is also equal to 20 log $|S_{21}|$.

A high-frequency power amplifier may have quite low gain, so the input power is significant compared to the DC power consumed, P_{DC}. In high-frequency power amplifiers it is common, therefore, to specify the power-added efficiency, given by:

$$PAE = \frac{P_{load} - P_{in}}{P_{DC}} \times 100\% \tag{10.6}$$

although it has been reported that this is not, strictly speaking, thermodynamically correct and can give odd results in some unusual cases [1].

Source and load mismatch will clearly affect both the input and output power. That is why there are other gain parameter definitions for a two-port. For the *transducer power gain* definition, the input power is the power that would be delivered into a load that was the conjugate of the specified source impedance. The *maximum available power gain* is the gain that would be achieved with the source and load conjugately matched to the input and output impedance of the amplifier or device. The *insertion gain* is the ratio of the power delivered to the load with and without the amplifier in place: in other words, the input power is defined as that which would be delivered by the source to the load if the amplifier were not present and the same source and load were connected directly together.

10.2 Transistor Configurations

Field-effect (MOSFET, MESFET or HEMT) and bipolar (BJT or HBT) transistors can be used in a range of configurations, such as common-source or common-base, as described in many excellent texts on RF analogue electronics [2–4]. Some configurations have high voltage gain, some high current gain, some are inverting, some non-inverting, and so forth. For example, one well-known configuration is the emitter-follower, which has low voltage gain but very low output impedance and is often used as a buffer stage. In high-frequency design, there are a number of additional constraints, however. One is that some configurations may exhibit very poor stability. Another is that the device may be required to operate very close to its f_t/f_{max}, where the gain may be at a premium. Many high-performance devices have very low breakdown voltage. Finally, a fundamental constraint is the Miller effect, which means that the feedback capacitance in an inverting amplifier appears effectively as an additional input capacitance, with its value multiplied by the gain. The following sections describe the main configurations used in microwave and millimetre-wave design along with their advantages and disadvantages.

10.2.1 FET Configurations

Figure 10.1 shows the simplified small-signal equivalent circuit of a FET. This is a model of the intrinsic device and does not include parasitics related to the physical layout. This is useful since it can therefore be scaled according to the device total periphery (gate width) and the physical

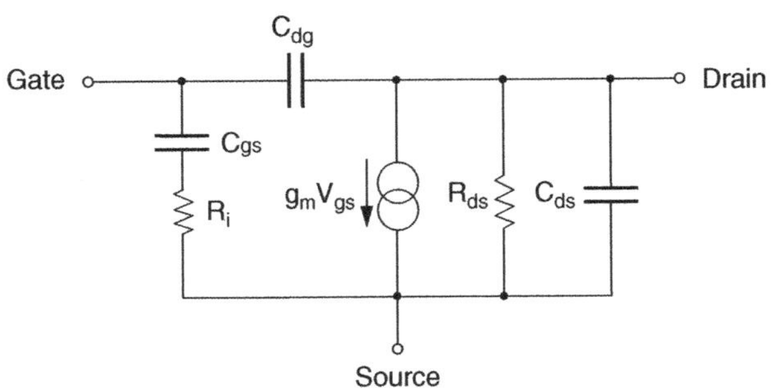

Figure 10.1 Intrinsic FET linear model.

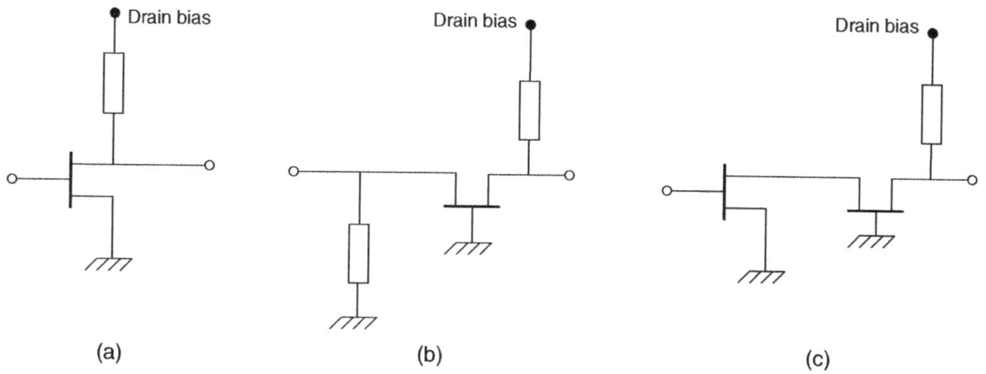

Figure 10.2 FET configurations. (a) Common source. (b) Common gate. (c) Cascode.

layout parameters modelled separately. The key gain mechanism is a transconductance, with the gate voltage controlling the drain-source current. For high-frequency amplifiers, FETs are most often used in common-source configuration, as illustrated in Figure 10.2(a). In this configuration the amplifier will be inverting and achieves the best absolute high-frequency gain and usually the best output power or noise performance.

The common-gate stage, shown in Figure 10.2(b) can be useful because it has an input resistance approximately equal to $1/g_m$ and by careful choice of device geometry a single device, or devices in parallel (e.g., in an active power divider), can give 50 Ω or other desired input impedance over a wide bandwidth. The cascode arrangement, shown in Figure 10.2(c) has a significantly increased f_t/f_{max} because of a reduced Miller effect, and it provides high output resistance and higher breakdown voltage than a single common-source device. These advantages make it particularly attractive in the distributed amplifier (discussed later), where the underlying device performance is very important in determining the gain and bandwidth achievable.

10.2.2 Bipolar Transistor Configurations

Heterojunction bipolar transistors are of increasing importance compared to HEMTs for high-frequency amplifiers because a high f_t/f_{max} is achieved through the thin semiconductor layers, grown by methods such as MBE and MOCVD, rather than using electron beam lithography to define very short gates in HEMTs. III–V HBTs are excellent for high linearity and power-efficient mobile handset power amplifiers and for the broadband amplifiers needed for optoelectronic applications, including OEICs where optical devices are integrated with electronic circuits. However, the lower cost of volume manufacture makes Si/SiGe much more attractive and there has been extensive research and development, leading to $f_{max} > 500$ GHz.

From a designer's perspective, a significant difference compared to FETs is that the base-emitter junction is forward-biased and bipolar transistors do not have the very high input impedance that FETs provide. This can be a disadvantage in some applications, but on the other hand if matching to 50 Ω is required it can be an advantage, although the noise performance is not as good usually. Another difference is in the models used for bipolar transistors. Often, these are hybrid-π models with a transconductance similar to a FET but with the base-emitter

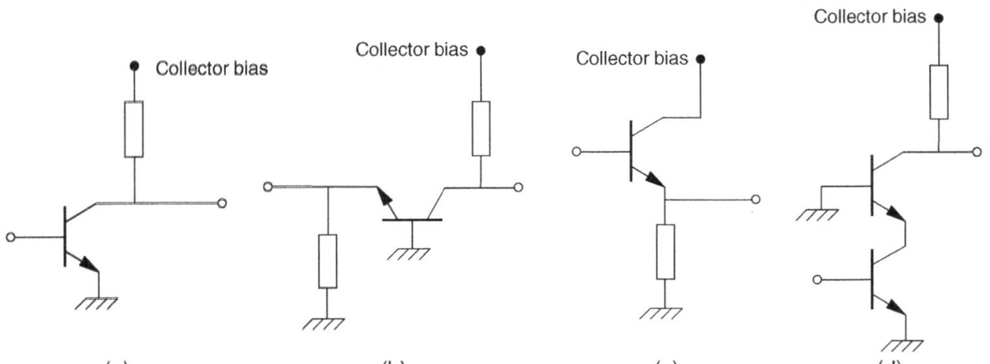

Figure 10.3 HBT configurations. (a) Common emitter. (b) Common base. (c) Common collector. (d) Cascode.

resistance in parallel with the base-emitter capacitance. Sometimes, a model based on a current-controlled current source is used. Furthermore, there are three current gains used for amplifiers: alpha (α), beta (β) and gamma (γ) for common-base, common-emitter and common-collector configurations, as shown in Figure 10.3(a), (b) and (c), respectively. H-parameters are popular for design with bipolar transistors since the individual H-parameters are closely related to the parameters of the intrinsic transistor. For example, h_{21} is the current gain, β or h_{fe}, so a plot of h_{21} versus frequency gives an immediate indication of the f_t of the transistor.

A high-f_t SiGe device will typically have a collector-emitter breakdown voltage of only a few volts. The collector-base breakdown voltage is typically much larger (ca. 8 V), but the gain will be low in common-base configuration. Therefore, the cascode shown in Figure 10.3(d) is an extremely important configuration in high-frequency design, and the common-emitter/common-base device geometry can be separately chosen for high gain and high breakdown voltage, respectively.

10.3 Classical Analysis of Gain and Stability

A perfect transistor would be a unilateral device, which means that the signal passes from its input to the output with no reverse transmission. This would mean that the S-parameter S_{12} would be zero at all frequencies. In practice, of course, this is not possible and reverse transmission in the transistor can easily lead to a positive feedback effect and subsequently instability. In the FET equivalent circuit of Figure 10.1, the forwards gain is the transconductance g_m, while the reverse transmission is via C_{dg}, which forms a backwards potential divider with C_{gs} that gives nearly constant S_{12} versus frequency over most of the operating range. In practice, the FET will also have source resistance and inductance, which contributes further to feedback. The effect of this feedback loop is directly analogous to the familiar example of an audio public address system, as shown in Figure 10.4. When the gain of the amplifier is too high, the sound from the loudspeaker feeds back into the microphone. At a frequency where the round-trip delay is such that the signal feeds back with exactly $n \times 360$ degrees phase delay, a positive feedback loop is created. The signal grows in amplitude until either:

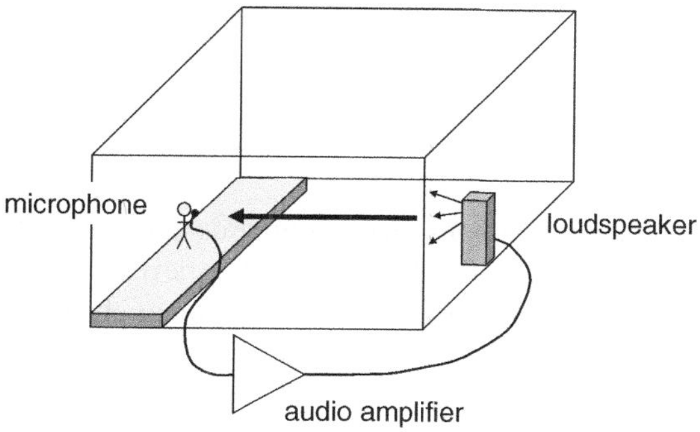

Figure 10.4 Illustration of positive feedback in a public address system.

(i) someone turns down the gain (volume); or (ii) the power amplifier and/or loudspeaker are destroyed by the very large amplitude signal. Exactly the same phenomenon occurs in a microwave amplifier if not designed properly. It might be noted, however, that such is the power of DSP that in the audio case many modern public address systems have the ability to measure the frequency response in real time and dynamically compensate for any positive feedback through digital filtering.

The stability problem in a microwave amplifier can be analysed using Mason's rule. Figure 10.5 shows a flowgraph representation of a transistor, represented by its S-parameters, with a load connected to its output that has a reflection coefficient Γ_L. We can see that there is a stability condition by determining an expression for the input reflection coefficient, b_1/a_1.

Following the steps from Mason's rule in Chapter 5 gives the following paths:

Path no. 1, $P_1 = S_{11}$ Path no. 2, $P_2 = S_{21} \cdot \Gamma_{\mathrm{L}} \cdot S_{12}$

The only loop is $S_{22} \cdot \Gamma_{\mathrm{L}}$ which touches P_2, and there are no second- or third-order loops. Therefore, from the rule:

$$\Gamma_{in} = \frac{b_1}{a_1} = \frac{[S_{11}(1 - S_{22}\Gamma_L) + S_{21}\Gamma_L S_{12}(1-0)]}{1 - S_{22}\Gamma_L} = S_{11} + \frac{S_{21}S_{12}\Gamma_L}{1 - S_{22}\Gamma_L} \tag{10.7}$$

This immediately illustrates an important basic consideration in amplifier design – that what is connected at the output affects the INPUT reflection coefficient. In a similar manner, it can

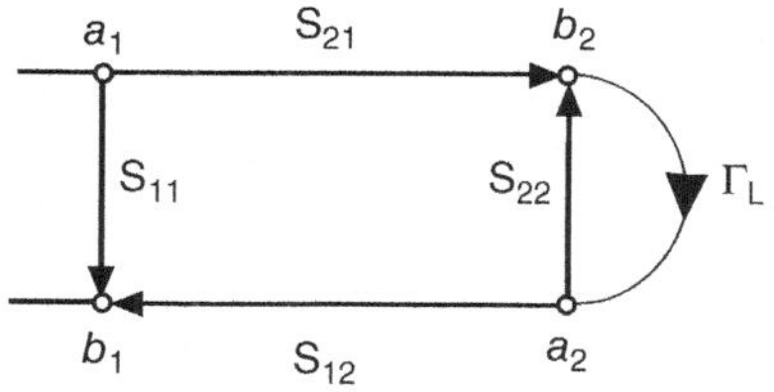

Figure 10.5 Flowgraph representation of a transistor with a load on its output.

be shown that what is connected at the input affects the OUTPUT reflection coefficient:

$$\Gamma_{out} = S_{22} + \frac{S_{21}S_{12}\Gamma_S}{1 - S_{11}\Gamma_S} \tag{10.8}$$

Because of this, it is immediately obvious that amplifier design is more than simply matching the transistor S_{11} and S_{22} to 50 Ω. Indeed, in fact that is never done since when the output matching network is connected, the input reflection coefficient changes and is no longer S_{11}, and vice versa.

10.3.1 Stability Circles

Like the audio public address system, whether the device is stable or not depends on the gain and the amount of feedback, and therefore the stability condition varies with frequency. We can develop a stability boundary condition by setting the input reflection coefficient equal to unity ($\Gamma > 1$ means negative resistance) when the load is connected at the output, as follows:

$$\left| S_{11} + \frac{S_{21}S_{12}\Gamma_L}{1 - S_{22}\Gamma_L} \right| = 1 \tag{10.9}$$

This gives the equation for a circle, with radius r_L, and centre, Γ_{CL}, given by:

$$r_L = \frac{|S_{12}S_{21}|}{\left||\Delta|^2 - |S_{22}|^2\right|} \text{ and } \Gamma_{C_L} = \frac{\left(S_{22} - \Delta S_{11}^*\right)^*}{|S_{22}|^2 - |\Delta|^2} \tag{10.10}$$

where

$$\Delta = S_{11}S_{22} - S_{12}S_{21}$$

The S-parameters vary with frequency, of course, so the input (source) and output (load) stability circles move with frequency. At a given frequency, the source/load stability circle on the Smith chart shows the range of source/load impedances that can be connected to the transistor output without causing the input reflection coefficient to go above unity. In most cases, each stability circle is either partially overlapping the Smith chart, as shown in Figure 10.6(a), or it is completely outside the Smith chart, as shown in (b). This is not the full set of possible situations but is sufficient for understanding the principles. In case (a), whether the device is stable or not depends on the source/load impedance and this is referred to as 'conditional stability'. Unfortunately, it is found that the point of conjugate matching, and maximum gain, is inside the unstable region. Hence, conjugate matching is not possible in the case of conditional stability. Either the device must be stabilised in some way or the device is left mismatched and some other approach is used to deal with the reflections, such as the balanced amplifier (as described later). In case (b), there are no values of source/load impedance on the Smith chart that can cause the device to be unstable, and this case is referred to as 'unconditional stability'.

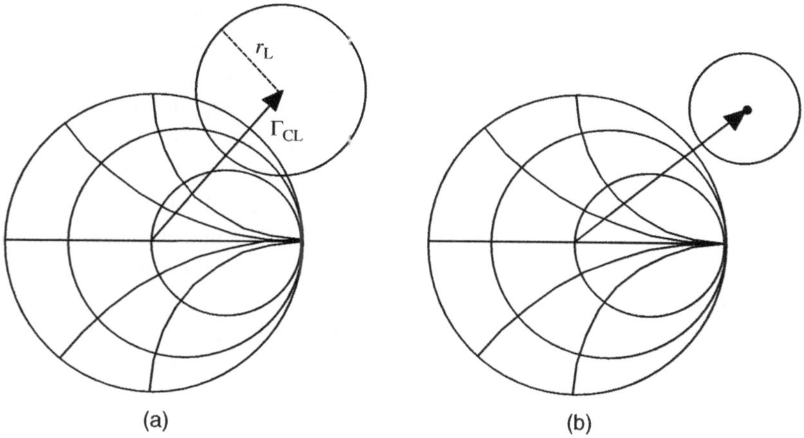

Figure 10.6 Stability circles on the Smith chart. The two most common situations.

10.3.2 Stability Factor

Early in the design process, an indication of the stability of the transistor is required. There are various key stability factors that are used to indicate the conditional/unconditional stability condition at a particular frequency. The most common is Rollett's stability factor, K, which is given by [5]:

$$K = \frac{1 - |S_{11}|^2 - |S_{22}|^2 + |\Delta|^2}{2|S_{12}S_{21}|} \tag{10.11}$$

where

$$\Delta = S_{11}S_{22} - S_{12}S_{21}$$

If this stability factor is greater than unity and $|\Delta| < 1$ then the device is unconditionally stable. The other stability factors are:

$$B_1 = 1 + |S_{11}|^2 - |S_{22}|^2 - |\Delta|^2 \tag{10.12}$$

and,

$$B_2 = 1 + |S_{22}|^2 - |S_{11}|^2 - |\Delta|^2 \tag{10.13}$$

If the amplifier circuit is stable, Δ should be less than unity and B_1 and B_2 should be positive.

The Edwards–Sinsky stability factors [6], μ_1 and μ_2, for the source and load side are calculated from:

$$\mu_1 = \frac{1 - |S_{11}|^2}{|S_{22} - \Delta S_{11}^*| + |S_{21}S_{12}|} \tag{10.14}$$

$$\mu_2 = \frac{1 - |S_{22}|^2}{|S_{11} - \Delta S_{22}^*| + |S_{21}S_{12}|} \tag{10.15}$$

The advantage of these factors over K is that an indication is given of which port is causing stability problems. If either one is equal or greater than unity, stability is guaranteed.

Considering the full set of combinations of stability factor, in fact there are six possible stability circle arrangements on the Smith chart. In Figure 10.6(a) the stability circle is partially inside the Smith chart and it is the region within the stability circle that is unstable, whereas in Figure 10.6(b) the stability circle is completely outside the Smith chart. However, the circle can be partially inside and encompassing the 50 Ω point, or the stability circle can completely encompass the Smith chart, or it can be completely inside but not encompassing the 50 Ω point, or completely inside and encompassing the 50 Ω point. If the circle encompasses the 50 Ω point then it is normally the region inside the stability circle that is stable. Lombardi and Neri have given an up-to-date analysis of two-port stability criteria [7]. Clearly, it is not practical to calculate the stability factor manually at many frequency points but these are given as standard outputs in most EDA software.

10.3.3 Maximum Gain

Having carefully examined the stability of the device across the full frequency range and at both the input and output, the next step is to investigate how much gain is available. When $K < 1$ the device is conditionally stable and the maximum stable gain (MSG) that can be achieved is given by:

$$\mathrm{MSG} = \left|\frac{S_{21}}{S_{12}}\right| \tag{10.16}$$

When $K > 1$ and $|\Delta| < 1$ the device can be conjugately matched at both input and output and the maximum available gain (MAG) is achieved, given by:

$$\mathrm{MAG} = \left|\frac{S_{21}}{S_{12}}\right|\left(K - \sqrt{K^2 - 1}\right) \tag{10.17}$$

In this case, the transistor can be conjugately matched at both input and output, but the designer must not attempt to match to S_{11} or S_{22}. The parameters called the simultaneous conjugate match reflection coefficients, gm_1 and gm_2 (gm = gamma matched) must be used since these include the loading effect described earlier.

'Typically' a transistor will be conditionally stable at low frequencies, with the stability circle partially overlapping the Smith chart as in Figure 10.6(a). The maximum stable gain drops at 3 dB per octave as shown in Figure 10.7. As the gain falls, the stability improves and eventually K reaches unity and the stability circle moves outside the Smith chart. Beyond that frequency, $K > 1$ and the maximum available gain is achievable, but it falls at 6 dB per octave. As shown in Figure 10.7, the MAG can be extrapolated to lower frequencies and clearly illustrates the problem of stability being the major limiting factor in amplifier design in many cases.

Of course, 'typically' is a risky assumption and it should be noted that the behaviour of a transistor will be greatly affected by its size and also by its packaging. For example, significant

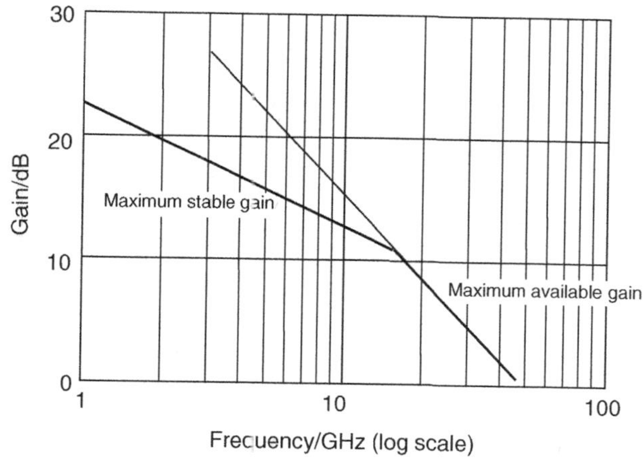

Figure 10.7 Idealised maximum gain response of a transistor.

common lead inductance in a packaged transistor will lead to poor stability and the device may go through several bands of frequencies where it is conditionally and unconditionally stable.

In order to investigate how the transistor gain varies with input and output loading, it is necessary to go back to the flowgraph analysis but now assume that $S_{12} = 0$ in order to solve for the gain of the amplifier. This leads to the parameter called 'unilateral transducer gain' (*GTU*), which is given by:

$$G_{TU} = \frac{|S_{21}|^2 \left(1 - |\Gamma_S|^2\right)\left(1 - |\Gamma_L|^2\right)}{|(1 - S_{11}\Gamma_S)(1 - S_{22}\Gamma_L)|^2} \quad (10.18)$$

It is evident that this can be separated into three factors:

$$G_{TU} = G_S \cdot G_0 \cdot G_L \quad (10.19)$$

where

$$G_S = \frac{1 - |\Gamma_S|^2}{|1 - S_{11}\Gamma_S|^2}, \quad G_0 = |S_{21}|^2 \text{ and } G_L = \frac{1 - |\Gamma_L|^2}{|1 - S_{22}\Gamma_L|^2} \quad (10.20)$$

What this form clearly illustrates is that the amplifier has three gain elements: G_0 is the raw unmatched transistor power gain ($|S_{21}|^2$because S_{21} is a voltage gain), G_S defines how much signal is lost in mismatch at the input (source), and G_L defines how much gain is lost due to mismatch at the output (the load). This leads to the concept of gain circles, where for the source and load side the gain is constant for certain values of source/load impedance that lie on a circle on the Smith chart. The conjugate match point lies at the epicentre of these circles, and when $K < 1$ this epicentre is, unfortunately in the unstable region, as illustrated in Figure 10.8. This is why conjugate matching is not possible in the case of conditional stability, which leads on to techniques for stabilising the device.

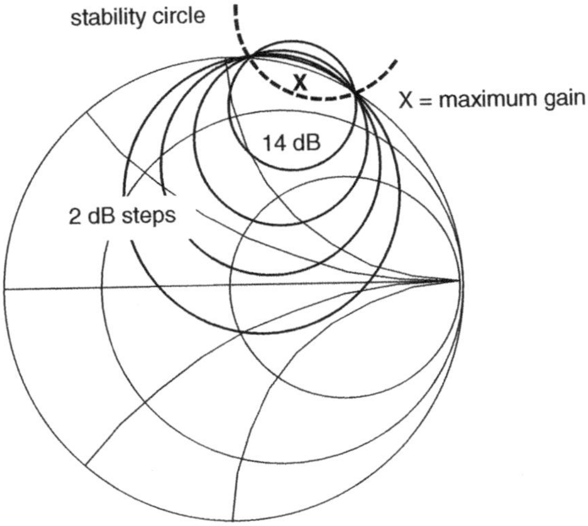

Figure 10.8 Example of gain and stability circles ($K < 1$).

10.3.4 Stabilisation Techniques

In the simplest approach, small series resistors or large shunt resistors are connected to the device to limit the gain and move the stability circles outside the Smith chart. Another way of looking at this is that the resistor limits the coverage of the matching network on the Smith chart so that the source/load cannot enter the unstable region. Using this approach, it becomes clearer that a series or shunt resistor will be required, depending on the location of the stability circle on the Smith chart. If the unstable region is on the left-hand side of the impedance chart, then a small series resistor is enough to ensure that the source/load impedance cannot enter that region. If the unstable region is on the right-hand side of the chart, then a large shunt resistor (a few hundred ohms) is usually enough to prevent the source/load entering that region, leading to unconditional stability. Resistors will degrade the noise figure and also reduce the gain at lower frequencies where stability may not be a problem. The latter can be addressed by using a resistor-inductor combination, which leads to the lossy match amplifier topology.

Microwave transistors are often used in common-source or common-emitter configuration, which means they are inverting and shunt-resistive feedback can be used to stabilise the device over a very wide bandwidth. This leads to a range of feedback amplifier topologies. In the series feedback approach, a small resistor is inserted into the common lead of the transistor. Alternatively, an inductor can be used if great care is taken to evaluate its effect on stability across the full frequency range of the transistor. Transformer feedback, widely used at lower frequencies, can be employed in microwave amplifiers provided that the parasitics of the transformer are carefully modelled. This approach, often referred to as lossless feedback, was pioneered by Norton [8].

Finally, note from Figure 10.7 that the MAG of the transistor (extrapolated backwards) is much higher than the MSG at the lower frequency end and, if the stability problem could be solved, a much higher gain could be achieved. Since it is typically a feedback capacitance that is causing this reduction in gain due to instability, a number of techniques have been reported for cancelling out the reverse transmission by creating an equal but opposite signal path round the transistor. Traditionally, this is referred to as 'neutralisation' and has been demonstrated with RC networks, inductors, transformers and active circuit techniques.

10.4 DC Biasing

The transistor must have DC bias applied to set the operating point on its I–V characteristic about which the AC high-frequency signal swings. It is important to appreciate that a DC voltage source is a perfect short for AC signals, so the DC power supplies must be carefully isolated from the RF signals in the amplifier. Furthermore, RF transistors generally have extremely high gain at low frequencies and the designer must ensure there is no unwanted feedback through the bias circuits that leads to instability. The simple aim of connecting a DC bias point to ground at AC (referred to as decoupling) is, therefore, nontrivial. A very large capacitance is required to satisfactorily short-circuit the low frequencies, but a large capacitance has a very low SRF due to parasitics. On the other hand, a very high SRF capacitor, used to decouple the high frequencies, will be limited in value to the order of tens of pF (depending on many factors) and cannot decouple the low frequencies well. Hence, for an IC, a careful design of bias network is required that is focussed on decoupling the highest frequencies on-chip, followed by decoupling the next band of signals in a package, followed by larger inductors/capacitors next to the package, and so forth, leading finally to some very large capacitors across the DC supply inputs. All of that ought to be modelled carefully at the design stage since some simple on-chip circuitry (such as some negative feedback) may considerable simplify the bias circuitry/packaging and so forth and prevent stability problems.

If the terminal being biased does not draw current, such as the gate of a FET, then usually a high-value resistor can be used to connect the DC bias without loading the RF signal. However, when there is any significant current drawn, it is necessary to find a method that does not give any significant DC voltage drop. The three common methods are bias chokes, bias stubs and active loads.

10.4.1 Inductors as Bias Chokes

An inductor has $Z = j\omega L$ which means zero impedance at DC but a high impedance at RF. DC blocking capacitors are used at the input and output to isolate the bias from other circuits, and with decoupling capacitors to prevent the leakage of RF signals into the power supplies. It is evident from Figure 10.9(a) that the *L-C* bias network forms a low-pass filter, to separate the RF and DC, but it also a resonator and so considerable effort must be taken to study the effect of the bias circuits on the low-frequency stability of the active device. Indeed, this part of the design and modelling must include careful consideration of package parasitics and off-chip bias components since these will also have an important effect.

10.4.2 Bias Stubs

Inductors suffer significant parasitics as discussed in Chapter 4, and so there is often a need to use a transmission line technique to isolate the DC and RF. A short-circuit stub can be used for this purpose, where the far end is grounded with a decoupling capacitor so that DC can be

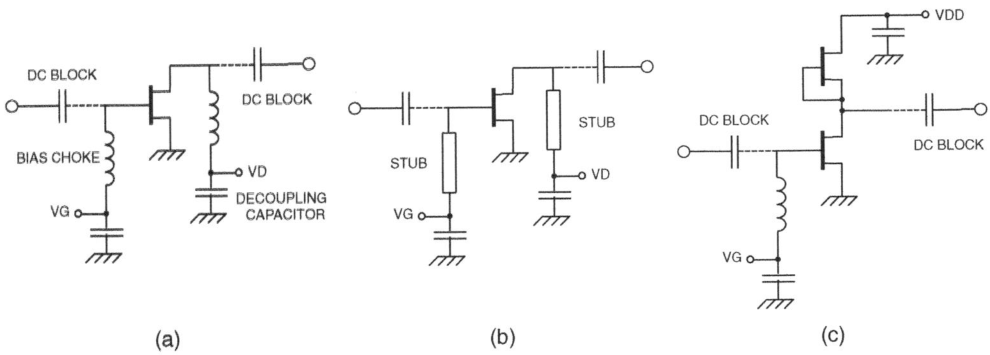

Figure 10.9 Basic bias circuits. (a) Bias chokes. (b) Short-circuit stubs. (c) An active load.

applied, as shown in Figure 10.9(b). If the bias stub is a quarter-wavelength long, the RF short of the decoupling capacitor is transformed to an open circuit at the transistor end. Alternatively, the stub may form part of a matching network. However, it is important to remember that this is a very frequency-dependent element and to be aware that the decoupling at the RF-shorted end of the stub is very critical, and care must be taken to avoid oscillations.

10.4.3 Active Loads

Bias inductors and stubs will require a very large amount of chip area, especially at low frequencies. Many digital IC processes have no resistor layer and, besides, a resistor as a DC bias element and load will have significant voltage drop and be able to feed very little current due to power dissipation. The active load is a transistor-based method of separating the DC voltage drop from the AC load impedance on the circuit. This is easy to achieve by making use of the nonlinear I–V characteristics. For the simple case shown in Figure 10.9(c), the gate is strapped to the source of the upper transistor, meaning that $V_{gs} = 0$ V. This then constrains the upper FET to operate in the saturation region, where the AC output resistance (the slope resistance) is high. The bias on the low FET in this simple circuit must be set rather precisely with this simple arrangement, and various techniques have been reported for making the active load more practical, including reducing its sensitivity to temperature. If the FET has a small gate width, then the slope resistance will be very high and the active load is treated as a constant current source in many circuits, such as the long-tailed pair and many current mirror circuits.

10.5 Common Amplifier Topologies

Whilst many University courses in microwave amplifier design still describe the design of amplifiers with stub matching, this traditional design technique is rarely used nowadays. Where the absolute best noise figure or power output is required from a transistor, there may be some merit in adopting such a traditional approach, but such a simple design is very unlikely to meet modern systems requirements. For example, it will be narrowband and it will be very difficult to achieve unconditional stability at all frequencies without additional circuit

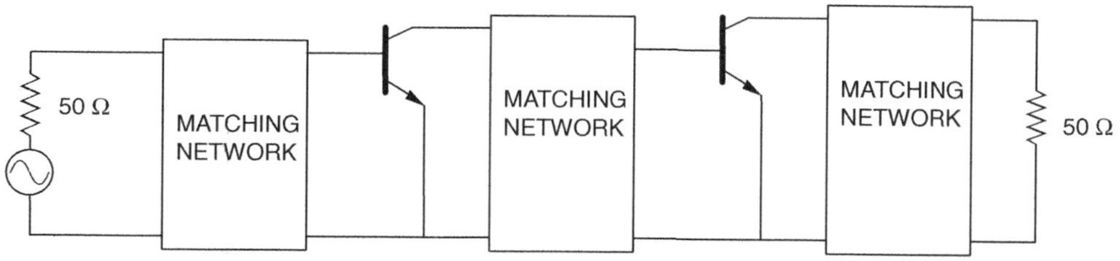

Figure 10.10 A generic two-stage reactively matched amplifier.

elements. This section describes just a few of the most common circuit topologies that are used for microwave and millimetre-wave amplifier design. The advent of MMICs and RFICs has enabled circuit designers to continue to develop increasingly better circuit techniques that improve the performance of amplifiers. These developments are particularly important for power amplifier design and linearisation, which are described in later dedicated sections.

10.5.1 Reactively Matched Amplifiers

In a reactively matched amplifier, the matching techniques described in Chapter 9 are used to transform the input and output impedances of the transistors to match the source and load. Figure 10.10 shows a two-stage reactively matched amplifier, chosen to illustrate some of the challenges facing the circuit designer in this approach. First, conjugate matching of the input and output may not be possible due to stability, as discussed earlier. However, let us assume that the devices are unconditionally stable over the frequency range of the design. This actually means they are in the MAG region, and their gain is falling at 6 dB per octave. The input and output matching networks must be optimised for good matching, so there is no scope there for combating this gain slope. Instead, the inter-stage matching network must be designed in such a way that it achieves a 12 dB per octave positive slope, which balances the two devices' 6 dB per octave and gives a flat gain response. Such an inter-stage circuit can be designed using matching network synthesis tools. However, great care must be exercised as one of the transistors can present a reflection coefficient to the other that is outside the Smith chart, so the normal boundary condition for stability does not apply – the situation can be much worse. Furthermore, any such issues in the inter-stage matching network may not be observable with an overall two-port *S*-parameter simulation of the whole two-stage amplifier. Various simulators allow the designer to insert a special block in the network that allows any problems to be observed separately, however.

The reactively matched amplifier is very important for microwave and millimetre-wave design because it extracts the very best performance from the devices, for low noise figure and high power at very high frequencies. A purely reactively matched design is rarely possible because of the need for unconditional stability and good gain flatness. Some degree of frequency-dependent loss can be incorporated into the matching networks to facilitate this, and this technique has been referred to as a lossy-match amplifier [9]. Low-frequency stability is often very challenging to achieve due to the very high low-frequency gain of high-performance devices. The bias networks are a very important part of the complete circuit and can be designed in various ways to aid low-frequency stability.

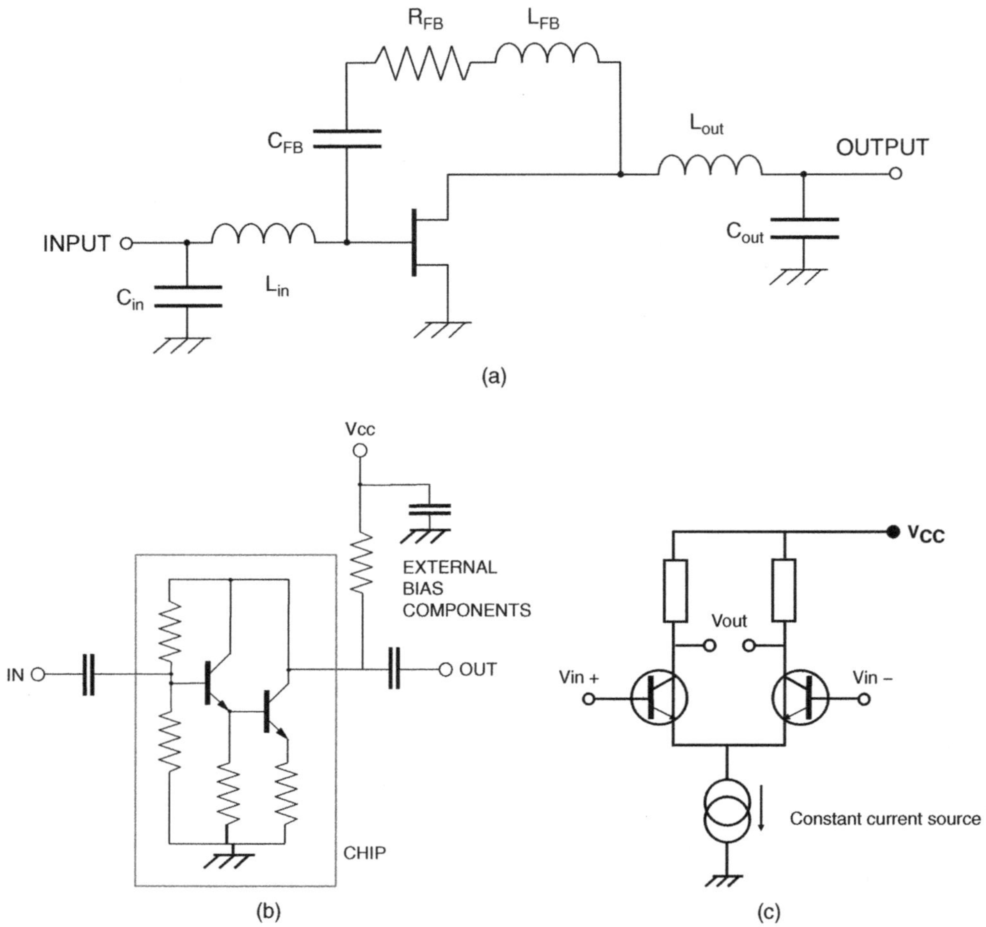

Figure 10.11 (a) FET feedback amplifier. (b) Darlington pair broadband amplifier. (c) Long-tailed pair.

10.5.2 FET Feedback Amplifier

The common-source FET is inverting and resistor feedback between the drain and gate can be used to realise a simple broadband amplifier with minimal matching network complexity and good stability. The circuit diagram of such an amplifier is illustrated in Figure 10.11(a). The resistor R_{FB} provides the negative feedback whilst L_{FB} flattens the gain by reducing the amount of feedback at high frequency. C_{FB} isolates the positive drain bias from the negative gate bias. C_{in} and L_{in} form a simple L-network for matching the input, while L_{out} and C_{out} do the same at the output.

10.5.3 Darlington Pair

The Darlington pair broadband amplifier has become a very familiar amplifier configuration through its use in the ubiquitous silicon BJT modular monolithic amplifier, available packaged for 99 cents. More recently, InGaP and SiGe versions have become available operating to well over 20 GHz. The Darlington pair has a current gain equal to the product of the individual transistor current gains and gives an increased input impedance and larger S_{21} bandwidth. The circuit diagram of a typical Darlington pair amplifier is shown in Figure 10.11(b). The two emitter resistors provide series feedback while the left-hand transistor's bias resistors double as the parallel feedback network. This provides a well-matched broadband amplifier with good stability. The other bias circuitry is often implemented off-chip so that the full bandwidth potential can be realised. It should be noted that if a broadband amplifier is used for convenience in a low-frequency application, the circuit must be laid out appropriately and the packaged device must be grounded effectively at high frequencies to avoid instability. Also, a very high gain cascade of wideband amplifiers should be avoided unless absolutely necessary because of the noise bandwidth and the potential for unwanted feedback.

10.5.4 Long-Tailed Pair

The circuit shown in Figure 10.11(c) is simplified but forms the basic core of many RFIC circuits. The long-tailed pair share a constant current source, which leads to the all-important differential amplifier operation. As the current in one transistor increases, the other must decrease and so the voltages at the two collectors (as shown) are anti-phase. The constant current source is realised with the active load arrangement described earlier, as indeed are the output bias circuits very often. This is the core circuit, but there are many improvements that can be made by taking advantage of the very low parasitics of transistors on ICs. The current source may be realised with a digitally controlled current source, to give gain control, a cascode arrangement may be used for the amplifying transistors, and negative feedback may be applied by cross-coupling transistors of suitable scaled periphery. The excellent book by T.H. Lee should be consulted for further information on this important topology [2].

10.5.5 Distributed Amplifier

The distributed, or travelling-wave, amplifier was proposed by Percival in the 1930s [10] as a method of absorbing the capacitance of tube devices into a ladder structure for broadband operation. The technique is particularly applicable to FETs, where the gate-source capacitance dominates the input impedance, while the series input resistance is low and the R_{ds}, the parallel output resistance, is high. Figure 10.12 shows a four-section distributed amplifier using FETs. The input signal is input on the left and travels down the gate line, feeding each transistor in turn. The gate-source capacitances form the shunt capacitors in a low-pass ladder network. Any remaining input signal is absorbed in a resistive termination at the end of the gate line. In fact, the input resistance of the FETs causes loss on the gate line, so the input signal is steadily attenuated as it travels along the gate line, and a modest portion is absorbed in the load. Each FET creates a current due to its transconductance and feeds this into the drain line. The current must split to the left and right but, because of the phase delay along the gate line, it is only the currents travelling to the right-hand side that add in-phase, creating the amplified signal output. Some current does travel to the left-hand end of the drain line, especially at low frequencies where the phase delay is small, but it is absorbed by the drain-line termination.

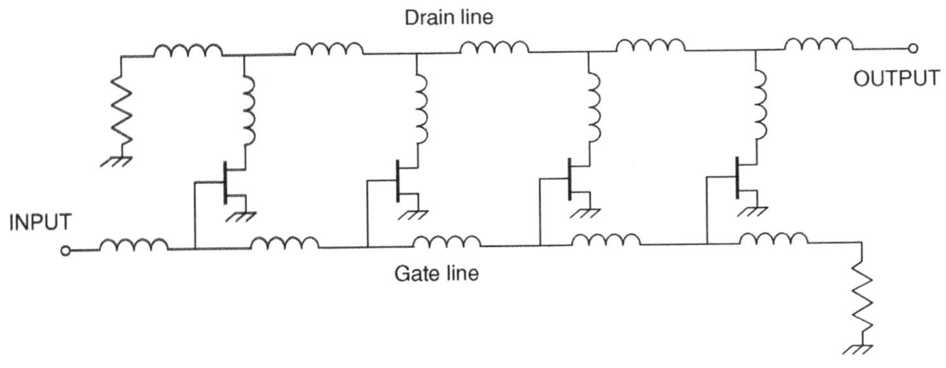

Figure 10.12 A four-section distributed amplifier using FETs.

C_{ds} is typically significantly smaller than C_{gs}, and without some form of equalisation of the drain line will have a much higher cut-off frequency than the gate line and the signals will not add constructively at the output over the full band. In Figure 10.12, small series inductors have been added at the FET outputs to achieve this, although there are various other techniques. In GaAs pHEMT technology a distributed amplifier can comfortably cover 1 to 40 GHz. The gain-bandwidth product is determined by the device technology since the devices can be made smaller for higher cut-off frequency of the gate-line, but then they have lower transconductance. Obtaining significant power from a distributed amplifier is very challenging because the devices are not isolated from one another, as they are in true power combining. So, as the drain line signal increases in voltage, the FET on the right-hand end has the full output signal across its drain-source, which is connected to the output load, and the breakdown voltage therefore restricts the maximum power. The number of sections that can usefully be employed in the design is limited by the gate and drain line loss. Beyond the optimum number, adding more sections does little to increase gain because the last FET on the gate line receives very little input signal due to gate line loss. On the other hand, adding that extra section adds more drain line loss, so it can actually degrade the performance.

10.5.6 Balanced Amplifier

A microwave balanced amplifier consists of two identical amplifiers fed 90° out-of-phase via quadrature 3 dB couplers, as shown in Figure 10.13. One of the attractions of this topology is that the input and output matching are very good by symmetry, which frees the designer to optimise the individual amplifiers for low noise figure, high power or flat gain, all of which normally involve trade-offs with good matching in a standard reactively matched amplifier. The balanced configuration eliminates any mismatch reflections by using the directional nature of the couplers. It is the reflected signals from the two individual amplifiers that are anti-phase, at the balanced amplifier's input and output ports – the amplifiers themselves are operating with signals 90° out-of-phase. The 3 dB quadrature couplers can be realised in a number of different ways, as described in Chapter 6. One way is the branch line coupler, although this method has the disadvantage of its large size and narrow bandwidth. Another method is the use of the Lange coupler. Other advantages of the balanced amplifier include improved stability,

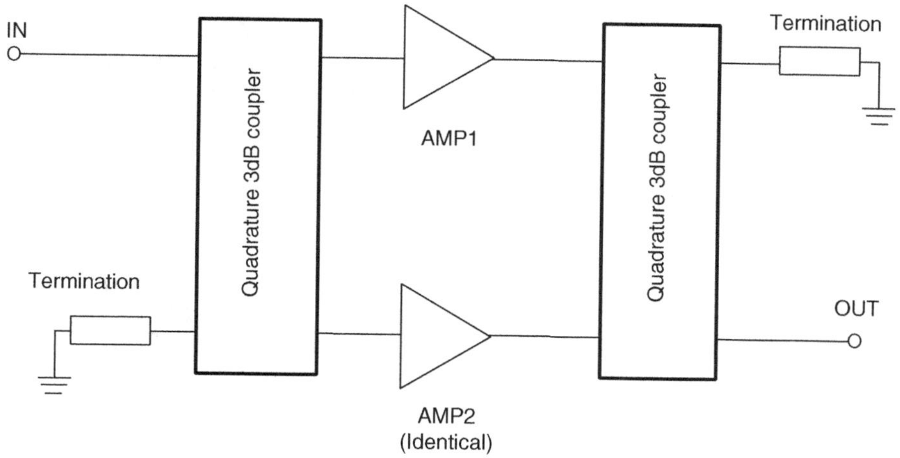

Figure 10.13 Balanced amplifier configuration.

double the output power (+3 dB but minus the output coupler loss) and a degree of redundancy. However, it is more expensive due to the need for two amplifiers and two couplers, and it has twice the DC power consumption.

10.6 Low-Noise Amplifiers

The noise figure of an amplifier is an important figure of merit that determines the weakest signal that can be received. The noise figure and 1 dB gain compression point together determine the dynamic range of the amplifier (i.e., the signal power range over which it works usefully). The noise figure is a measure of how much the signal-to-noise ratio (SNR) is degraded by the amplifier. This degradation is inevitable since the amplifier will not only amplify the input signal, but it will amplify the input noise by the same amount and add some more noise of its own making:

$$N_{OUT} = G \times N_{IN} + N_{ADDED} \tag{10.21}$$

where G is the gain of the amplifier (not in dB).

For a 50 Ω system at room temperature, the thermal noise floor is at –204 dBW Hz^{-1}. On a spectrum analyser, the degradation in SNR can be easily seen, and a display like Figure 10.14 would be observed. The noise floor is composed of white noise, which is random in the time domain and equally distributed over the frequency domain. The actual noise power is determined by the bandwidth over which it is measured. The spectrum analyser's vertical-axis is representing power spectral density, with units dBm Hz^{-1}; that is, there is a certain amount of noise power in each 1 Hz worth of the band. As described extensively in Chapter 2, while the low-noise amplifier (LNA) adds noise, the receiver chain as a whole dramatically improves the SNR prior to demodulating the signal by focusing on the bandwidth. Nevertheless, the noise figure of the LNA is very important in determining the minimum signal that can be detected.

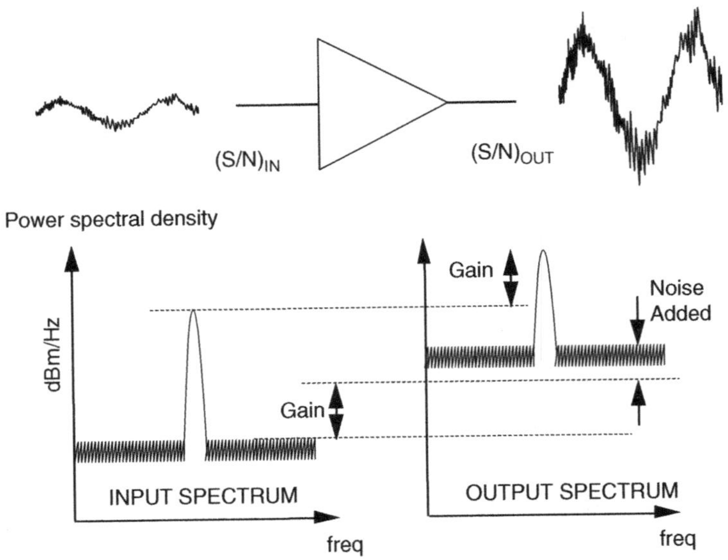

Figure 10.14 Noise figure of an LNA in terms of the display on a spectrum analyser.

The two terms 'noise figure' and 'noise factor' are often used interchangeably, but they should not be confused. Noise factor is used to refer to the ratio of output and input SNR:

$$\text{Noise factor, } F = \frac{(\text{SNR})_{in}}{(\text{SNR})_{out}} \tag{10.22}$$

While the noise figure is the noise factor expressed in decibels:

$$\text{Noise figure, } NF = 10 \log F \tag{10.23}$$

10.6.1 Cascaded Noise Figures

As described in Chapter 2, the noise factor of a cascade of components is easily calculated from the formula of Friis:

$$\text{Noise factor, } F = F_1 + \frac{F_2 - 1}{G_1} + \frac{F_3 - 1}{G_1 G_2} + \frac{F_4 - 1}{G_1 G_2 G_3} \cdots + \frac{(F_n - 1)}{G_1 G_2 \cdots G_{n-1}} \tag{10.24}$$

$$\text{Gain} = G1\ G2\ G3\ G4 \tag{10.25}$$

The first part of an LNA design is, therefore, performed at a system level. Only once the key system parameters have been defined and a transceiver architecture chosen can the receiver target noise figure be decided. Losses (in dB) of all the components that are placed before the LNA, such as the antenna feed, diplexer, isolator, and so forth, will add directly to the receiver noise figure (assuming they are at room temperature). The 'second stage' contribution to noise figure from the components connected after the LNA must then be determined: the mixer, IF or baseband circuits and any intervening filters. Finally, a target specification for the LNA gain and overall noise figure can then be calculated. It is then necessary to choose the right device technology and number of stages, considering economic and practical considerations and intermodulation performance (discussed in the following section). Once a suitable strategy for the LNA design has been found, attention can be turned to the problem of designing the actual amplifiers, with particular attention being paid to the input matching network of the first stage. At extremely high frequencies, if the gain of the transistors is low, then the noise performance of subsequent stages may also need to be carefully considered in the design.

10.6.2 Noise Circles

Conjugate matching at the input of a transistor does not give optimum noise performance because of the added complication of noise from the source resistance and device itself. Instead, a transistor has an optimum source impedance, Z_{opt} (or reflection coefficient Γ_{opt}) that it needs to be presented with in order to achieve the lowest possible noise figure, F_{min}. The measurement of these noise parameters has been transformed by advanced new network analyser capability, for example in the PNA-X from Keysight Technologies, and they can be measured on-wafer using sophisticated automated tuners, such as those from Maury Microwave (described in Chapter 18).

As the source impedance moves away from Z_{opt}, the noise figure increases and it is found that circles of constant noise figure appear on the Smith chart, as illustrated in Figure 10.15. With the automated tuners, a parabola can be constructed in three dimensions with the Smith chart in the x–y plane and noise figure plotted on the z-axis. This method allows the minimum point (Z_{opt} and F_{min}) to be determined with high precision. The additional parameter, R_n, called the equivalent noise resistance, is related to how quickly the noise figure increases as the source impedance moves away from Z_{opt} and these three parameters are sufficient for LNA design. If the amplifier is not conjugately matched, then the input return loss may be poor and

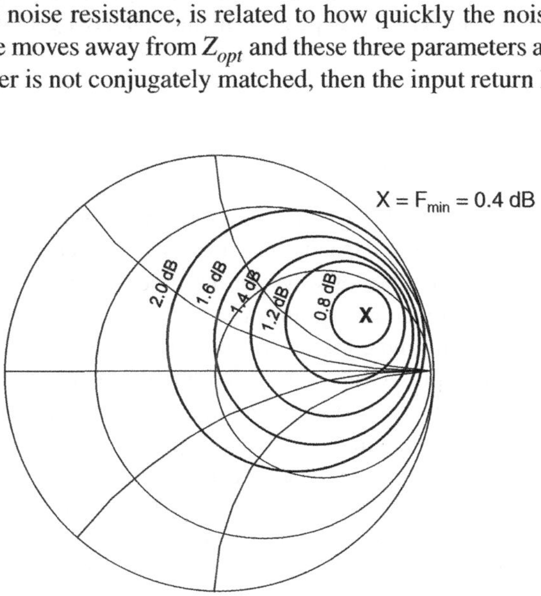

Figure 10.15 Example noise circles on the Smith chart.

this can be solved with a number of techniques. An isolator or balanced amplifier topology is often used, but the loss of the isolator or 3 dB coupler will add to the noise figure, and of course there is additional complexity and cost. In RFIC and MMIC design, it is possible to use series feedback and transformer feedback to bring the Γ_{opt} and conjugate match point (gm_1, see earlier) together, yielding a simultaneous match LNA. In a hybrid MIC this is much more difficult due to the lack of choice of device geometry and package parasitics, but may be possible at microwave frequencies, depending on the package style and other practical layout factors.

10.7 Nonlinearity and Intermodulation

The linearity of amplifiers is of great concern in microwave and millimetre-wave communications systems. In an ideal amplifier there is no distortion or change in waveform, the output signal is a perfect amplified version of the input signal. An amplifier is said to be linear if the graph of the output versus input signal amplitude is a straight line, and its gradient is the gain of the amplifier. Of course, such a component cannot be realised in practice and every amplifier has nonlinear behaviour stemming from the active devices. Nonlinearity in a communications system leads to the presence of unwanted signals as well as distortion of the wanted RF signal. Some of the unwanted signals cannot be eliminated by filtering as they appear side-by-side with the signals being amplified. And yet, with strict controls over what can be transmitted to avoid interference with other users, these signals must be reduced in some way. In modern RF communications, the wanted signal may itself be a sophisticated multicarrier signal with high-level digital modulation, leading to a very significant variation in signal envelope with time, referred to as a high peak-to-average ratio (PAR), and this places even more stringent requirements on the linearity of the system. This is especially challenging for the transmitter in mobile devices, where the battery life depends critically on having a high-efficiency power amplifier.

10.7.1 Nonlinear Characteristics

The nonlinear behaviour of an amplifier is a single phenomenon relating to the transistors' limited voltage and current handling capacity and nonlinear I–V curves. However, it can be represented and modelled in a number of ways. In the time domain, when the amplitude of the sine wave input to a low frequency op-amp is increased, eventually the output voltage will start to reach the supply rails and clipping will be very evident. However, in the RF case, reactive loads mean that an amplifier can produce an output voltage greater than the supply rail, and in any case the RF engineer is more concerned with the frequency domain. The two most important representations of nonlinear behaviour that are generally considered are the signals on a spectrum plot and an input-output characteristic that shows the power levels of key signal components as a function of power level, in dBm. Considering only the fundamental sinusoidal input signal, such a graph is shown in Figure 10.16. At low powers the input-output characteristic is linear, but inevitably the output power is limited and so it starts to level off, eventually reaching the maximum, referred to as the saturated output power. The amplifier will be highly nonlinear at saturation and a more common performance metric is the 1 dB gain

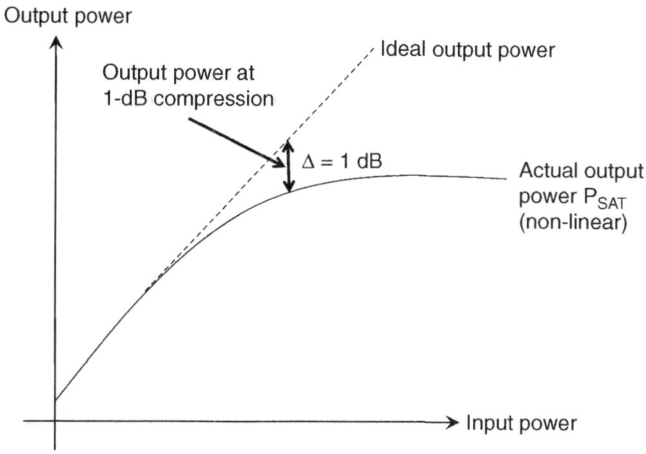

Figure 10.16 Input-output power characteristic illustrating the 1-dB gain compression point.

compression point, which is the output power level at which the gain of the amplifier is 1 dB below its linear gain, as shown in Figure 10.16.

As indicated earlier, distortion of the output waveform can be represented in the frequency domain by a number of frequency components. For a purely sinusoidal input, these are the harmonics, but harmonic distortion is not a sufficiently thorough representation of power amplifier linearity for RF applications. To study linearity, the first step is to represent the output voltage as a power series of the input voltage. The output signal from a nonlinear amplifier can be represented as:

$$V_{out} = a_0 + a_1 V_{in} + a_2 V_{in}^2 + a_3 V_{in}^3 + \ldots\ldots a_n V_{in}^n \tag{10.26}$$

The first term, a_0, is a real number and represents a DC bias term. The second term $a_1 \times V_{in}$ is the linear term and a_1 is the voltage gain. The other terms represent unwanted components at different frequencies and generate various forms of distortion. They are represented in the curved region of the gain curve in Figure 10.16.

10.7.2 Second-Order Intermodulation

If the input signal is a single-frequency sinusoid, the second-order term, $a_2 V_{in}^2$, leads to the production of the second harmonic of the fundamental signal. However, a better understanding of the effect of nonlinearity is obtained by considering a two-tone input and evaluating the intermodulation products at the output. Therefore, consider a signal V_{in} that contains two tones with different amplitude and frequency, as follows:

$$V_{in} = A\cos\omega_1 t + B\cos\omega_2 t \tag{10.27}$$

Applying this for V_{in} in the squared term gives:

$$V_{out} = a_2(A\cos\omega_1 t + B\cos\omega_2 t)^2 \tag{10.28}$$

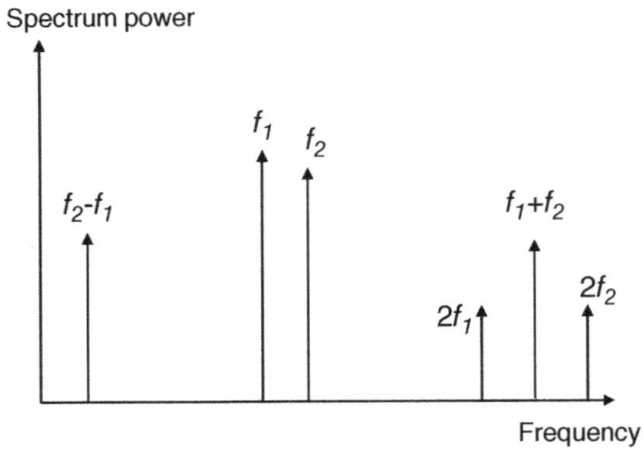

Figure 10.17 Frequency spectrum illustrating second-order intermodulation products.

Hence:

$$V_{out} = a_2\left(\frac{A^2+B^2}{2} + \frac{A^2}{2}\cos 2\omega_1 t + \frac{B^2}{2}\cos 2\omega_2 t + AB[\cos(\omega_1-\omega_2)t + \cos(\omega_1+\omega_2)t]\right) \tag{10.29}$$

Here, the first term is the DC term, while the second and third terms are the second harmonics of each signal. The last terms are the second-order intermodulation products (intermods). The frequencies of these intermods are the sum and difference of the fundamental frequencies. As shown in Figure 10.17, the second harmonics and the second-order intermodulation products can be removed by filtering, using a band-pass or band-stop filter, because their frequencies are sufficiently far from the input signal frequencies.

The amplitude of the harmonic and second-order intermodulation terms at the output will increase at a greater rate than the amplitude of the fundamental frequencies. The reason is that their amplitude increases in proportion to the square of the input signal amplitude, whereas that of the fundamental frequency will increase linearly. The power of the intermods is less than that of the fundamental, as shown in Figure 10.18, which is plotted in dBm, but the slope is 2:1, which means that if the line is extrapolated it will intersect. The point of intersection is called the second-order two-tone intercept point, and may be specified as a certain power level, referred to the input or output.

From Figure 10.18, if P_{IM2} is the second-order output intermodulation power level, OIP_2 is the second-order output intercept point and P_0 is output power, then:

$$OIP_2 = P_0 + P_0 - P_{IM2} \tag{10.30}$$

$$OIP_2 = 2P_0 - P_{IM2} \tag{10.31}$$

Hence,

$$P_{IM2} = 2P_0 - OIP_2 \tag{10.32}$$

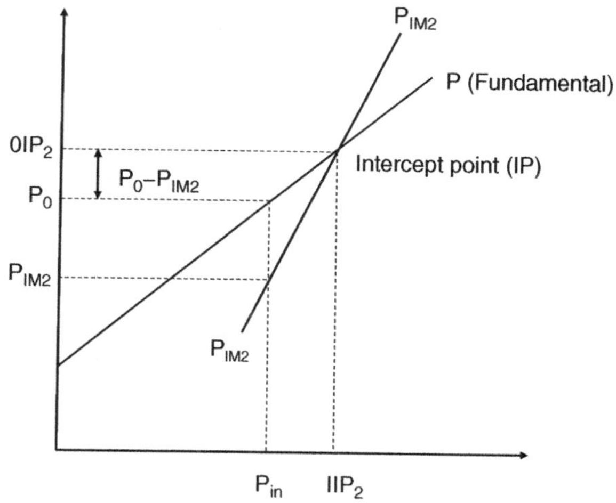

Figure 10.18 Graphical representation of second-order intercept point with fundamental and intermodulation signals.

Representing OIP_2 with IP_2, we get:

$$P_{IM2} = 2P_0 - IP_2 \tag{10.33}$$

This shows that the power of the second-order intermodulation products (P_{IM2}) depends on the second-order intercept point. So, simply specifying the intercept point is sufficient to define the second-order intermodulation behaviour. The output intercept point is related to the input intercept point through the gain.

10.7.3 Third-Order Intermodulation

The third-order intermodulation product comes from the cubic term of Equation (10.26).

$$V_{out(3)} = a_3 V_{in}^3 = a_3 \times V_{in}^2 \times V_{in} \tag{10.34}$$

Therefore, applying the same two input tones as before gives:

$$V_{out} = a_3 a_2 \left\{ \frac{A^2 + B^2}{2} + \frac{A^2}{2}\cos 2\omega_1 t + \frac{B^2}{2}\cos 2\omega_2 t \right.$$
$$\left. + AB[\cos(\omega_1 - \omega_2)t + \cos(\omega_1 + \omega_2)t] \right\}(A\cos\omega_1 t + B\cos\omega_2 t)) \tag{10.35}$$

When expanded, this equation contains two terms that are the fundamental frequencies but with amplitudes that are a nonlinear function of the input amplitudes. There are four terms which are the third-order two-tone intermodulation products, and two terms that are the third

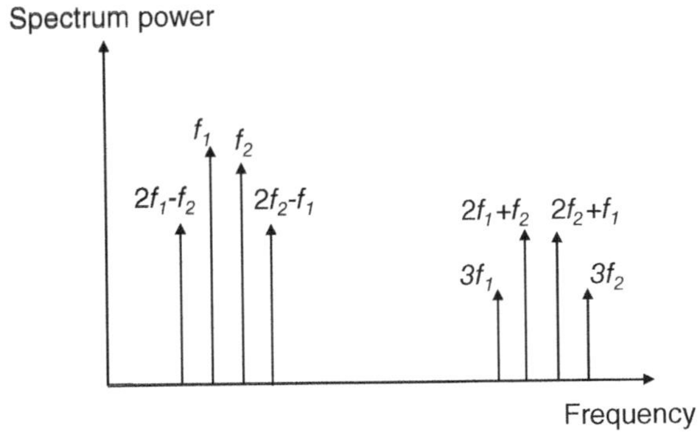

Figure 10.19 Frequency spectrum of illustrating third-order intermodulation products.

harmonics. Extending this approach, the general form of intermodulation products can be shown to be

$$m\omega_1 + n\omega_2 \tag{10.36}$$

where m and n are integers, positive or negative, and $|m| + |n|$ = the order.

For the third-order intermodulation products, this yields frequencies of $2\omega_1+\omega_2$, $2\omega_1-\omega_2$, $2\omega_2+\omega_1$ and $2\omega_2-\omega_1$ and the harmonics $3\omega_1$, $3\omega_2$.

Figure 10.19 shows the frequency spectrum illustrating these third-order intermodulation product terms. Two of the four third-order intermods are particularly troublesome since they are adjacent to the fundamental frequencies – the input signals. These are the terms with frequencies of $(2f_1-f_2)$ and $(2f_2-f_2)$. These are very problematic because they are almost impossible to eliminate through filtering. Furthermore, in a multicarrier system there will be other channels at these frequencies and interference will occur. This is why power amplifiers often have to be linearised so that they can be run close to maximum output power, for high efficiency, but with the intermodulation products minimised in a number of ways, as discussed later. The power level of the third-order intermodulation products rises at three times the rate of the fundamental signal outputs, as shown in Figure 10.20. Again, the line with 3:1 slope is extrapolated to give the third-order intercept point, which can be referred to the input or output and is usually expressed in dBm units. As shown in Figure 10.20, if P_{IM3} is the third-order output intermodulation power level, OIP_3 is the third-order output intercept point, and P_0 is output power, then it follows from geometry that:

$$OIP_3 = P_0 + \frac{P_0 - P_{IM3}}{2} \tag{10.37}$$

$$OIP_3 = \frac{2P_0 + P_0 - P_{IM3}}{2} \tag{10.38}$$

$$2OIP_3 = 3P_0 - P_{IM3} \tag{10.39}$$

Thus,

$$P_{IM3} = 3P_0 - 2OIP_3 \tag{10.40}$$

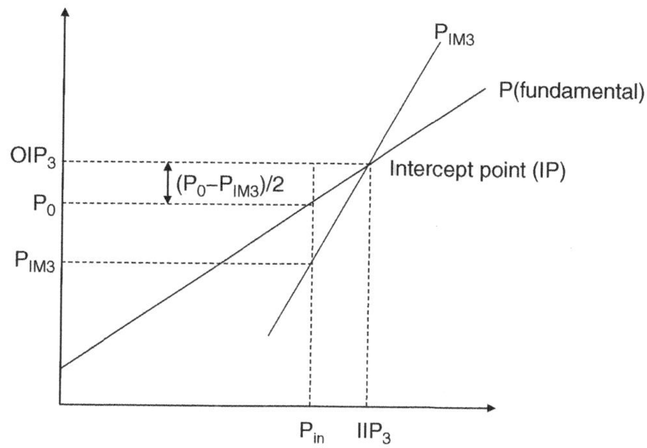

Figure 10.20 Graphical representation of third-order intercept point with fundamental and intermodulation signals.

or

$$P_{IM3} = 3P_0 - 2IP_3 \tag{10.41}$$

Again, the power of the intermodulation terms can be determined from the third-order intercept point, which makes it a standard figure of merit for amplifiers, mixers and other active circuits.

10.7.4 Higher-Order Terms

Higher-order terms such as v_{in}^4 to v_{in}^n will produce additional harmonics and intermodulation products given by the general form of Equation (10.26). However, they are usually small due to the highly steep slope of their output power against fundamental input power characteristic. These higher-order terms do affect the second-order and third-order terms. The higher even-order terms account for the curvature in the second-order characteristic at high levels, while the higher odd-order terms add to the steepness in the third-order characteristic at higher levels. These effects can often be observed when making two-tone intermodulation measurements; on the spectrum analyser, it is often found that the spectrum is asymmetrical and the intermods below the input signals are not the same amplitude as those above. In an intercept diagram, these effects often lead to graphs that are nothing like the ideal straight lines shown here, with effects such as nulls or 'sweet spots' that can be used to advantage in some cases.

10.7.5 Intermodulation in Cascaded Circuits

The intercept point of cascaded modules can be determined by obtaining the intercept of each individual module and combining them, using their respective gains, in an expression to determine the overall intercept point. This expression, relating the cascade intercept point to the individual intercept points, will first be obtained for two cascaded modules and then be extended to the general case. In a similar manner to Friis' formula for cascaded noise figures, this is one of those occasions where dB units are not suitable.

The power at the second circuit's output is the power output of the cascade, given by the first circuit's output power multiplied by the gain of the second module. The expression for the overall output intercept point is given by:

$$\frac{1}{POIP_{n,cas}} = \frac{1}{g_2 POIP_{n,1}} + \frac{1}{POIP_{n,2}} \tag{10.42}$$

where $POIP_n$ is the power of the output intercept point of the n order term and g_2 is the gain of the second module. This expression is for general case.

For coherent addition of the intermods, the expression for the overall intercept point of the cascade is given by:

$$\frac{1}{POIP^{1/2}_{n,cas}} = \frac{1}{g_2 POIP^{1/2}_{n,1}} + \frac{1}{POIP^{1/2}_{n,2}} \tag{10.43}$$

Note that the power of the input intercept point is related to the output intercept point by the gain as shown:

$$POIP_n = PIIP_n + Gain \tag{10.44}$$

The intercept point of the cascade can be determined by assuming the intermods add coherently or randomly as in the case of two-module cascade. The expression of the overall intercept for adding coherently across the cascade is given by:

$$\frac{1}{POIP^{\frac{n-1}{2}}_{n,cas}} = \sum_{i=1}^{q} \frac{1}{(g_{i+1,q} POIP_{n,i})^{\frac{n-1}{2}}} \tag{10.45}$$

where $POIP_n$ represents power of the output intercept point for the nth order term.

For random addition:

$$\frac{1}{POIP^{n-1}_{n,cas}} = \sum_{i=1}^{q} \frac{1}{(g_{i+1,q} POIP_{n,i})^{n-1}} \tag{10.46}$$

A wide range of simple analysis tools and spreadsheets are available for calculating the intermodulation performance of cascaded networks, often combined with Friis' formula for the cascaded noise figure calculation. The two subjects are related because whilst sufficient LNA gain is required in a receiver to reduce second-stage noise contributions (e.g., the mixer plus IF amplifier noise figure combination), if the gain is not spread out carefully in a receiver the intermodulation performance will be poor, as well as noise power becoming excessive if there is too much gain in high-bandwidth parts of a system.

10.7.6 Other Measures of Distortion in RF Amplifiers

The analysis of two-tone intermodulation products is important but gives a rather simplistic view of nonlinearity problems in RF transceivers. Other forms of distortion in radiofrequency systems can be in form of spurs, noise, frequency response distortion, phase distortion, distortion of the IQ constellation and spectral regrowth.

Spurs are unwanted spurious frequencies that are generated by various components of transceiver systems. These can be present at high power levels relative to the wanted signals, making them highly problematic. Therefore, it is usually best to design a frequency plan in order to avoid spurs falling into the RF and IF bands. When a spur with power level greater than the RF and IF signal saturates the transceiver, this phenomenon is called 'de-sensing'.

Frequency response distortion is distortion of a wideband signal due to the non-flat frequency response of the transceiver. Phase distortion refers to the change in a waveform due to a phase response that is not linear with frequency. This means different frequency components of the signal start to spread out in time.

Nonlinear distortion of digitally modulated (IQ) signals leads to distortion of the signal constellation and spectral regrowth. Distortion of the constellation will lead to bit errors, while spectrum regrowth can lead to adjacent channel interference and failure to meet the spectrum regulations, typically defined in great detail in terms of acceptable spectral masks defined by licensing and regulatory agencies such as the FCC. Figure 10.21(a) shows an example of the spectrum of a QAM signal at the input and output of a power amplifier. The sidelobes, which have been carefully suppressed in the modulator by baseband filtering, start to reappear at the amplifier output. As data rates have increased, more and more sophisticated IQ modulation schemes have been deployed, such as 256-QAM, and distortion of the constellation must be very low indeed since symbols are close together. For that reason, the error vector magnitude (EVM) has been widely adopted as a key figure of merit for linearity. The EVM is the distance between the received symbol and the ideal case, as illustrated in Figure 10.21(b). EVM can be expressed in a number of ways. Arguably, it is best expressed as a percentage and root-mean-square averaged over the full constellation, as follows:

$$\mathrm{EVM} = \sqrt{\frac{\sum_1^N |Error_n|^2}{\sum_1^N |s_n|^2}} \times 100\% \tag{10.47}$$

where N is the number of symbols, and $Error_n$ is the distance between received symbol number n and the corresponding ideal constellation point, s_n.

There are many contributors to EVM in a complete system, as discussed in Chapter 15. Of all of these sources of distortion to the constellation, power amplifier nonlinearity is one of the most problematic since high efficiency of the PA is required, especially in battery-powered portable devices. Hence, a huge amount of work has gone into designing more efficient amplifiers, and linearising amplifiers, as discussed in the following sections.

10.7.7 Intermodulation in Other Components

So far, this section has focussed on intermodulation in amplifiers since third-order intermodulation in amplifiers is very troublesome. However, it is important to note that nonlinear effects

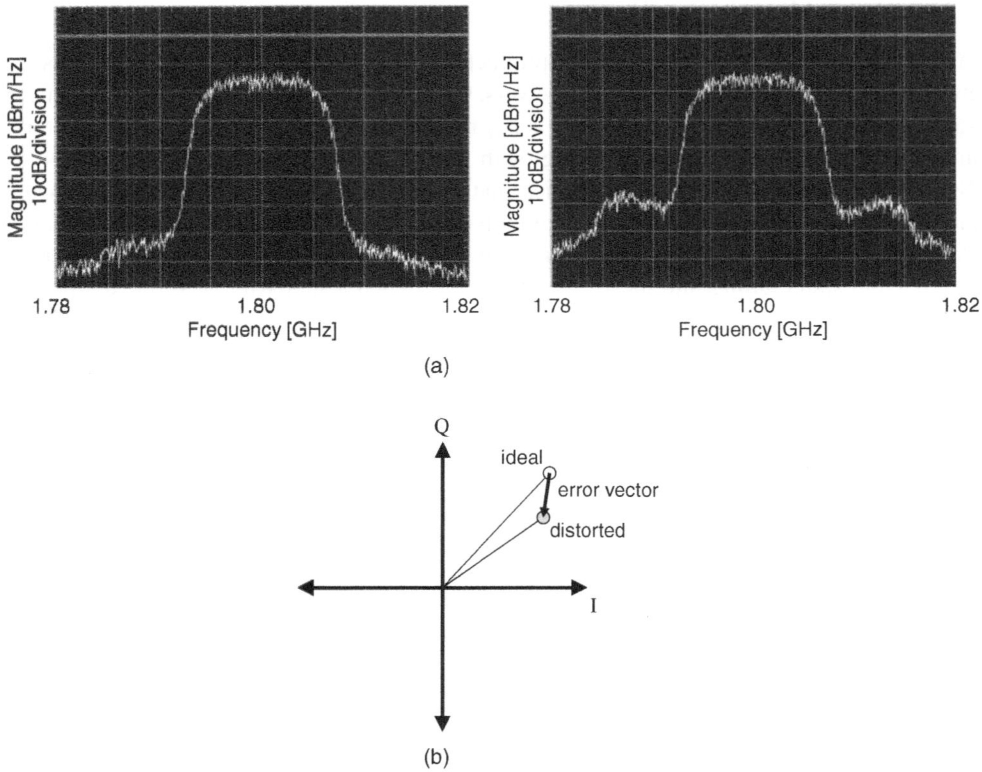

Figure 10.21 Nonlinear distortion of IQ-modulated signals. (a) Spectral regrowth. (b) Error vector magnitude (EVM).

and intermodulation occur in other components in the RF system. The parameters of intercept point apply equally to other components using semiconductor devices, such as mixers, modulators, switches, variable attenuators and phase shifters. For these, it is very important to be sure whether the specified third-order intercept point is referred to the input or the output, since they generally have loss and the higher figure is achieved at the input and may be used for marketing purposes. Furthermore, many variable-control circuits may have much lower IP3 at some settings than at others. Finally, note that at very high power levels even passive components can create intermodulation products. This phenomenon of passive intermodulation is caused by ferromagnetic materials, which are naturally nonlinear, and junctions between two dissimilar metals, or corrosion, that create a diode-like behaviour that causes nonlinear behaviour. This can be encountered in the filters, cables, connectors, antenna and fixings, and even unrelated metal parts near the antenna, such as roofing and drainpipes.

10.8 Power Amplifier Classes of Operation

Amplifiers are classified based on how the output waveform behaves over the operation of a full input signal cycle. This is as a result of differences in circuit configuration, DC bias, linearity and dynamic load impedance. The class of operation of the transistors in the power

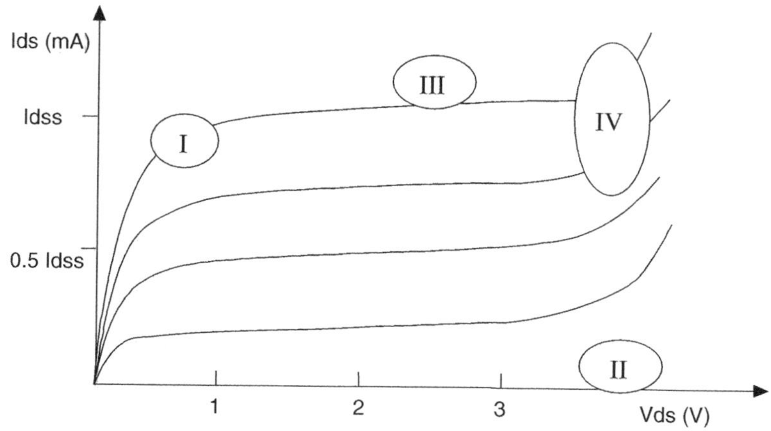

Figure 10.22 Generic I–V curves for a transistor.

amplifiers has very important implications in terms of linearity and efficiency. There is always a trade-off between linearity and efficiency, and many research reports have been made on power amplifiers and efficiency improvement techniques, as well as a number of essential texts [11–13].

In power amplifier design, different types of transistors can be used (FETs, HEMTs, BJTs, HBTs, MOSFETs and LDMOS). However, the BJT transistor amplifier circuit will be used here in discussing the various operations of the amplifiers. The same principles apply in other transistor types. Power amplifiers are classified as class A, B, AB, C, D, E, F, G, H and S. Classes A, B and AB are classified as linear amplifiers, while classes C, D, E, F, G, H and S are classified as nonlinear or switching amplifiers and are specially targeted at achieving high efficiency. Before examining these classes individually in detail, it is important to remember that the transistor in a power amplifier will be driven hard, within the various limits defined by its I–V characteristics. Figure 10.22 shows a set of generic I–V characteristics for a transistor. In region I, the knee voltage limits the output voltage swing. In region II, the transistor is near cut-off and the input voltage swing is restricted. In region III the device is firmly into saturation and the output current is limited. Finally, in region IV any further increase in output voltage will take the device past its breakdown voltage. This breakdown voltage is one of the fundamental limits on output power and the reason for GaN's great success. With a high breakdown voltage, high power can be generated without a large-periphery, high-current device. This provides significant advantages (smaller device, less distributed effects, more convenient impedance levels, high efficiency, etc.) and these reduce the point at which power-combining of multiple amplifiers is needed, giving further advantages of higher efficiency. The objective in high-efficiency classes of operation is to choose the dynamic trajectory on the I–V curves in such a way that the power dissipated in the transistor itself is minimised over the signal cycle.

10.8.1 Class A

This is the most linear class of operation of a power amplifier and is often used in amplitude modulation (AM) and single side-band (SSB) transmitters where the signal amplitude must

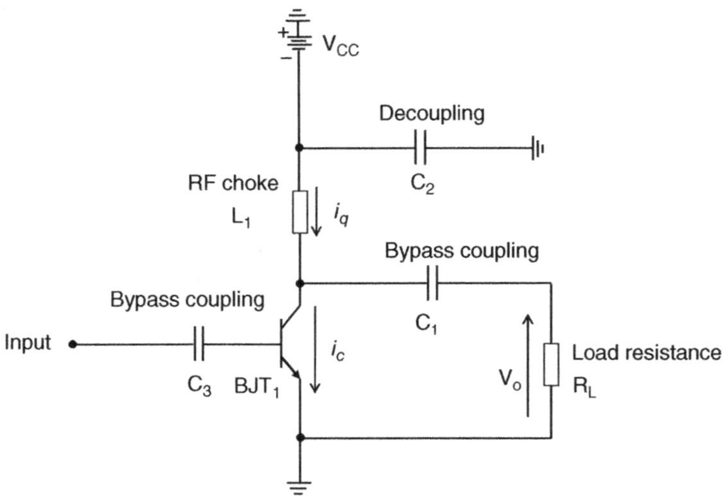

Figure 10.23 Class A common-emitter BJT amplifier.

be preserved with minimum distortion. In class A, the transistor conducts for the full 360° of the signal cycle. One form of class A amplifier circuit is shown in Figure 10.23. The DC bias current, I_q, of the transistor is greater than the amplitude of the signal current. The output current flows in the collector for the complete duration of the input cycle. The input signal and the operating point of a class A amplifier are determined in such a way that the output current flows all the time. Figure 10.24 shows the waveform of the output current of the class A amplifier.

The power efficiency calculation can be determined as follows. First:

$$\text{DC Power Consumption} = P_{dc} = \frac{V_{cc}^2}{R} \tag{10.48}$$

Note that it is independent of the output signal amplitude.

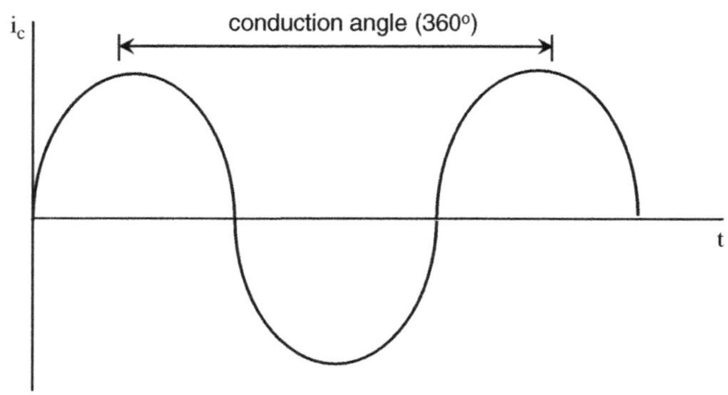

Figure 10.24 Waveform of the collector current of the Class A amplifier.

The signal power is given by:

$$P_L = \frac{V^2}{2R} \tag{10.49}$$

where V is the maximum AC voltage across the load, R. Hence:

$$\text{Efficiency} = \frac{P_L}{P_{dc}} = \frac{v^2}{2V_{cc}} \tag{10.50}$$

This shows that the maximum efficiency is 50%, but in practice it is much less. Although the efficiency of the Class A amplifier is very low, it has good linearity characteristics.

10.8.2 Class B

The key to improving the efficiency of the amplifier is to avoid letting the transistor conduct through the whole RF cycle. In a class B amplifier, the DC operating point is adjusted so that the quiescent current with no input signal is small. As a result, though, amplification takes place for only one-half of the cycle when the excitation is sinusoidal. Figures 10.25 and 10.26 show a Class B amplifier circuit topology and the output waveform, respectively. A single-ended class B amplifier conducts for 50% of its input signal; this results in a sinusoid that is distorted. The quiescent bias current, I_q, is set at zero and the conduction angle is 180°. Class B amplifiers are less linear than the class A amplifier but have higher efficiency, theoretically about 78% maximum. The output waveform is not a pure sinusoid, so a resonant circuit is applied to regenerate the complete waveform and eliminate harmonics.

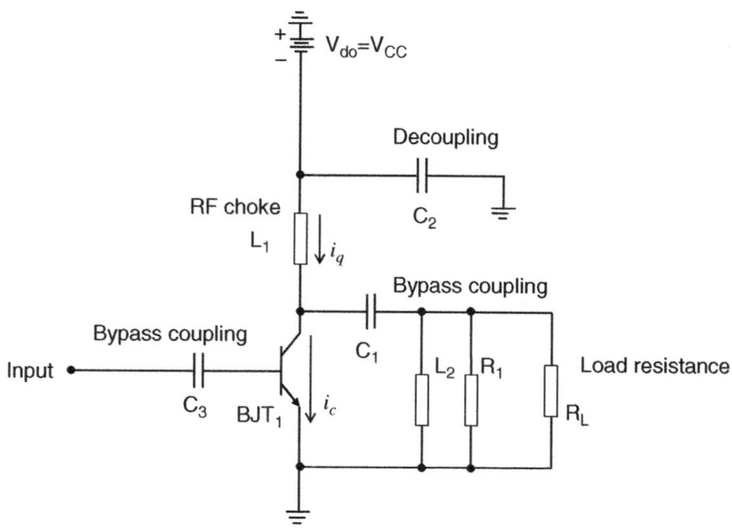

Figure 10.25 Class B narrowband amplifier circuit diagram.

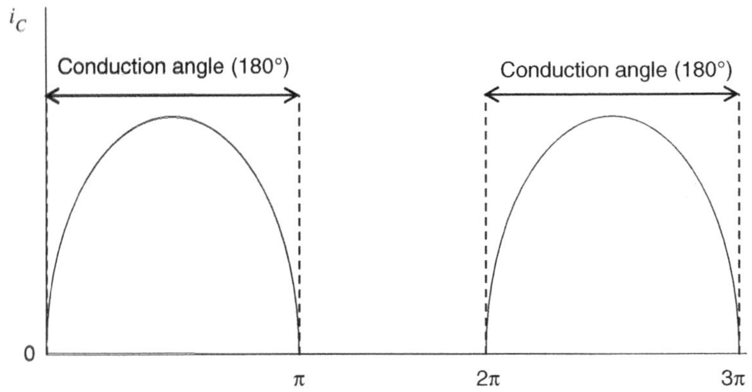

Figure 10.26 Current waveform in the class B amplifier.

The drawback of using the resonant circuit is that it gives a band-pass response that restricts the bandwidth. To solve this problem, a second transistor that conducts for negative half-cycles of the sinusoidal input can be added to the configuration, as shown in Figure 10.27(a). Most power amplifiers built to operate below 400 MHz use a complementary p-n-p/n-p-n transistor pair that operates in the push-pull configuration. In this configuration, one transistor pushes current into the load when the input is positive, while the other transistor sinks the current when the input is negative. The result is that each transistor amplifies the signal for just under half of the input signal cycle. However, the forward diode voltage drop of the transistors means

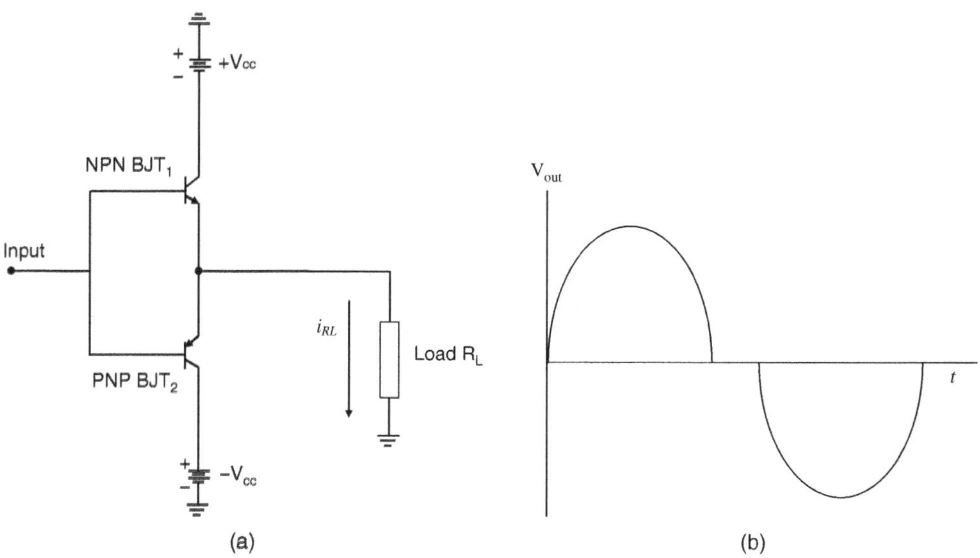

Figure 10.27 Class B push-pull topology. (a) Circuit diagram. (b) Waveform showing cross-over distortion.

that the is a spell where neither transistor conducts, leading to cross-over distortion, as shown in Figure 10.27(b).

At RF frequencies, it is normally impossible to find matching transistor pairs of equivalent performance and in this case the push-pull operation can be realised using baluns. The input balun of the class B push-pull amplifier feeds two identical transistors with 0/180° signals. These are then amplified individually, with one transistor amplifying the positive half-cycles and the other the negative half-cycles. Because of the phase shift, both signal components are positive-going for the transistors and they do not need to be a complementary pair. The outputs are then combined together by an output balun so that the full sine wave is reconstructed at the output.

10.8.3 Class AB

In a class AB amplifier, the cross-over distortion of the class B type is reduced by allowing the transistors to run with a small quiescent current. This mode of operation is therefore a combination of class A and class B. The distortion produced is less than that of a class B stage but the efficiency is significantly greater than that of class A. The conduction angle is greater than 180° but significantly less than 360°. The typical output current waveform of a class AB amplifier is shown in Figure 10.28.

10.8.4 Class C

In this class of amplifier, the operating point is chosen such that the output voltage or current is zero for more than one-half of the input signal cycle. This produces distortion of the signal waveform. As a result, the class C amplifier is not suitable for linear amplification although efficiency is greatly improved. In the class C amplifier, the transistor is biased so that it conducts for less than 180° of the input signal cycle. The tank *L-C* circuit will, however, provide a full cycle of the output signal for the fundamental frequency. Class C amplifiers are not suitable for signals with envelope variation or audio amplifiers due to the heavy distortion present. The output current waveform of class C amplifier is shown in Figure 10.29.

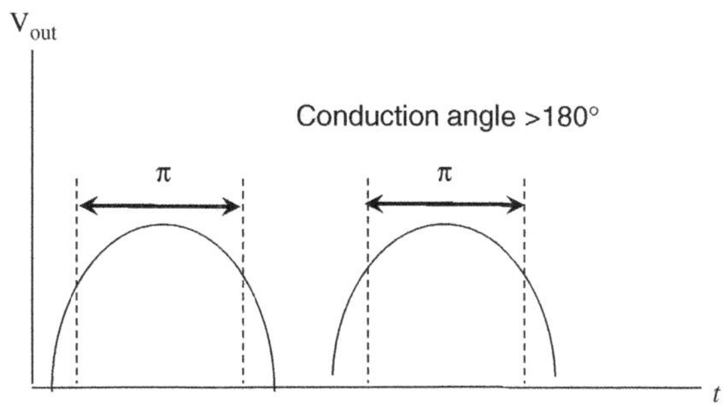

Figure 10.28 Current waveform of a class AB amplifier.

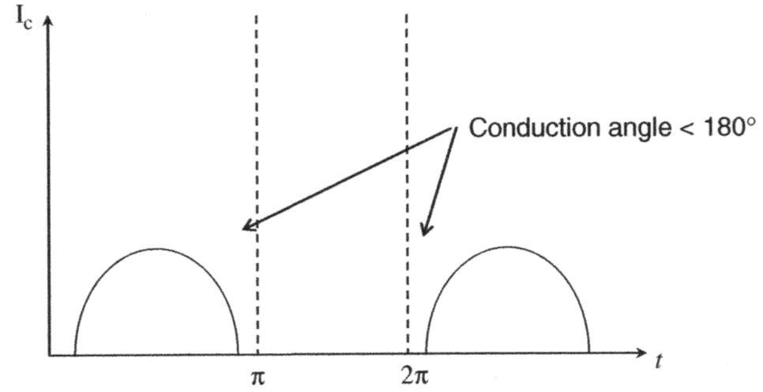

Figure 10.29 Class C amplifier waveform.

10.8.5 Class D

This is a nonlinear, or switching, class of amplifier where the aim is to avoid power dissipation in the transistor. By using the transistor as a switch, rather than in a linear mode, the efficiency of the amplifier can approach 100%. Fundamentally, a perfect switch has either no voltage across it (on) or no current flowing (off) and so there is no power dissipation in the switch and all the power is delivered to the load. However, the operation of the class D amplifier requires the input signal to be converted into a pulse waveform that has only high and low levels. This can be achieved with a pulse-width modulation approach or a pulse-density modulator. The pulse-width waveform is used to then drive the load, but it has to be converted back to its sinusoidal form. Figure 10.30 shows the block diagram of one method that uses a triangle wave generator and comparator to convert the input signal to pulse type and then converts it back to sinusoidal form using a low-pass filter. Figure 10.31 shows the pulse-width modulated waveform. A low-pass filter can be used to convert the pulse-width modulated signal back to a sine wave. High efficiency of up to 90% is possible with class D amplifiers. However, in practice, some factors such as nonlinearity in PWM generation, unwanted characteristics of switching devices, and the parasitic effects of reactive components reduce the efficiency.

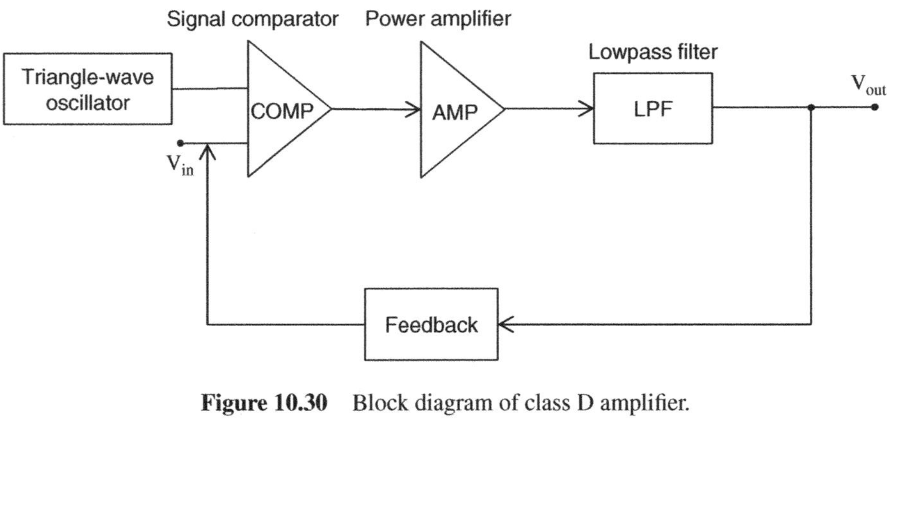

Figure 10.30 Block diagram of class D amplifier.

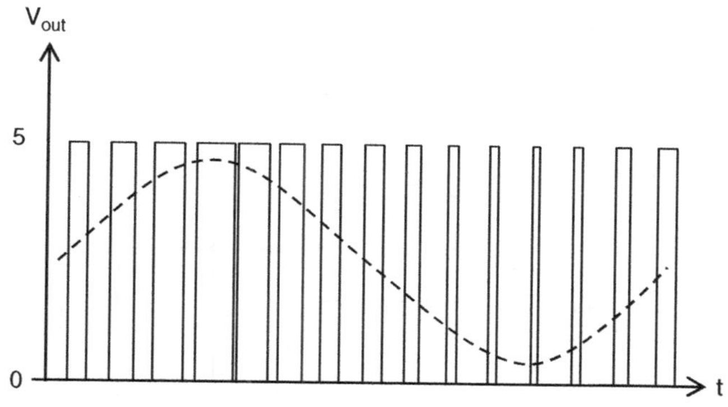

Figure 10.31 Class D amplifier PWM waveform.

10.8.6 Class E

Operating the transistor as a near-ideal switch may not be practical at high frequencies, and the aim of class E and F amplifier design is to manipulate the signal and its harmonics in order to avoid the transistor current and voltage waveforms overlapping. In Class E, the amplifier consists of the transistor acting as a switch, in parallel with a capacitor C_1, and the output is connected via a series *L-C* circuit, as shown in Figure 10.32. The *L-C* circuit resonates at the fundamental frequency, and as a result only sinusoidal current components pass to the load. However, with the transistor operating as a switch, for half the cycle the capacitor C_1 is shorted, meaning that the voltage and current waveforms do not overlap. The Class E amplifier can produces efficiency close to 100%. In practical applications, saturation voltage and switching time can reduce this efficiency.

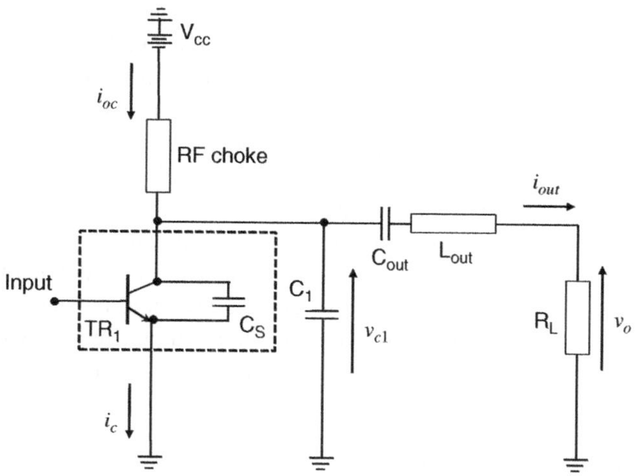

Figure 10.32 Class E amplifier configuration.

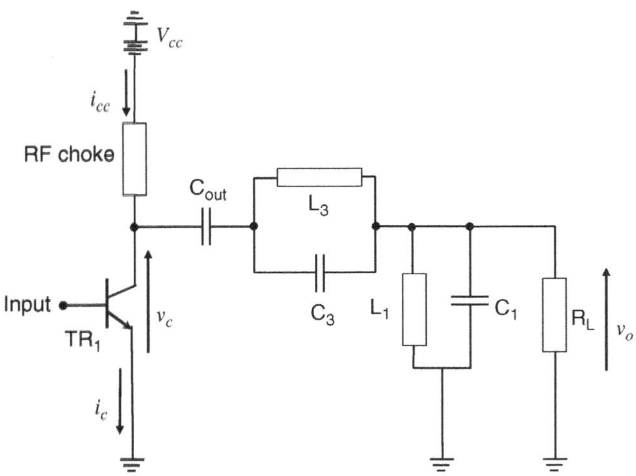

Figure 10.33 Class F amplifier.

10.8.7 Class F

In order to achieve a waveform that approaches an ideal square wave, in class F, as shown in Figure 10.33, the impedance of the load network at the fundamental frequency and a number of harmonics is carefully selected. The output circuit typically aims to short-circuit the even harmonic frequencies while presenting an open circuit to odd harmonics. If more harmonic frequencies are loaded in this way, the output waveform approaches a square wave and the efficiency is improved.

10.8.8 Other Classes

A class G amplifier uses more than one voltage supply, using the low-voltage supply when there is a low-level output and switching to the high-voltage supply automatically when the signal envelope is large. This reduces the average power consumed and increases efficiency. In class H, the power supply circuit is controlled dynamically, either in steps or continuously, according to the signal envelope. In a Doherty amplifier, a pair of RF amplifier stages deals separately with small and large envelope parts of the signal: a high-linearity stage amplifies the signal at low amplitude, whilst as the envelope increases a class C parallel stage kicks in to deliver high output power. Finally, the class S amplifier [14] has been proposed as an extension of class D. It explicitly uses a pair of transistor switches, allowing four-quadrant operation, each fed from a pulse width modulated digital waveform. The RF analogue output waveform is reconstructed with a low-pass filter. Very fast devices are required since the frequency at which transistors must switch is much larger than the final output frequency of the amplifier, so this class of amplifier is currently only suited for microwave frequencies.

10.9 Power-Combining Techniques

The output power available from a single transistor is inevitably limited by thermal effects, distributed effects and low impedances, as discussed in Chapter 8. In many applications, it therefore becomes necessary to combine the output powers of a number of amplifiers. This requires matched power modules operating in parallel, fed with power splitters at the input and with the outputs connected with combiners at the output. Power combining can be performed on a chip, in a package, or in a microstrip module, but the loss of planar transmission lines has a very deleterious effect on the efficiency of the combined amplifier. Many attempts have been made to realise solutions to this, such as micromachined and suspended substrate transmission lines, but it is very difficult to achieve the extremely low loss achievable with coaxial or waveguide combiners. With these, power-combining techniques are widely employed in order to develop solid-state replacements for very high-power vacuum devices such as travelling wave tube amplifiers, magnetrons and klystrons. This is attractive in many applications in order to overcome the need for high-voltage power supplies and achieve higher reliability. In a power-combined replacement, the end-of-life degradation is more graceful as some power is still available even if some modules are damaged.

10.9.1 Planar Combining Techniques

This approach is attractive because power combining can be performed at the substrate level using planar components such as the Wilkinson divider, branch-line coupler, hybrid ring, or Lange couplers, as described in Chapter 6. The most common topology is the corporate structure, shown in Figure 10.34 with Wilkinson power dividers/combiners, which is also known as the binary or tree combining technique. A disadvantage of this approach is that the isolation resistors in the Wilkinson combiners are series components and it is difficult to realise them in a form that can handle significant RF power. One alternative is the Gysel power splitter/combiner which has only shunt resistor terminations on isolated ports [15]. Alternatively, the balanced amplifier approach can be applied, as illustrated in Figure 10.35. In this case, as well as achieving power combining, the balanced approach aids broadband matching and can improve stability. It is important to recognise that isolation between the parallel-connected amplifier modules is a very important aspect of achieving stable operation.

Chip-level combining can use these same techniques but it is also possible to be more creative and make use of the flexibility of device layout that is afforded to MMIC/RFIC designers. A number of circuit design techniques have been reported, including cluster matching, enhanced cluster matching [16], bus-bar power combining and the distributed active transformer [17,18]. In all of these methods, the total device area required to achieve the required output power is

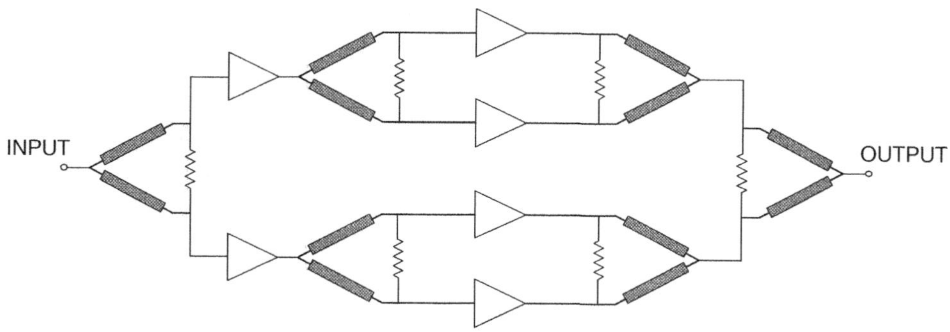

Figure 10.34 Power amplifier using corporate or binary power splitting/combining.

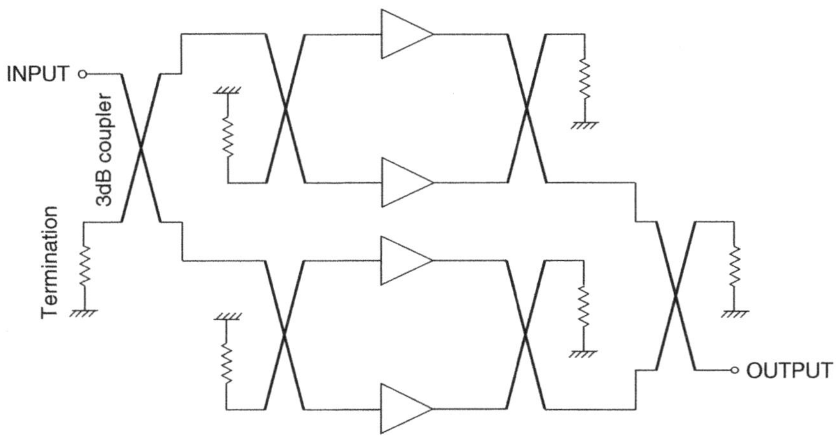

Figure 10.35 Power combining using the balanced amplifier technique.

realised by splitting the device into 'cells' in order to distribute the heat, avoid low impedance levels, and ensure uniform signal distribution across the devices, both in amplitude and phase. In the enhanced cluster matching technique, active devices with a common input and twin outputs are used as active power splitters, as illustrated in Figure 10.36.

With active power splitting, significant passive circuitry is avoided and there is isolation between the two outputs across a very wide frequency range, which greatly aids stability. The stability of parallel-connected devices needs careful investigation at the design stage. It is evident from Figure 10.36 that there are internal loops in the circuit that can lead to unexpected feedback. For example, consider a single transistor pair combined in parallel: in the anticipated mode of operation, the two input and output signals are in-phase and of equal amplitude, and flow in one direction from input to output via parallel paths. This is referred to as the 'even mode', with the two output nodes at the same voltage at all times. However, there is also a feedback loop, not present with a single device, which goes through one transistor in the forward direction, and back through the other transistor in the reverse direction. This creates a second mode of operation in which the two outputs are not in-phase and this is called the 'odd mode' of operation. In a circuit simulator, where everything is ideal, this mode will not be seen from a regular S-parameter simulation, and even a time-domain simulation may fail to see problems that are caused in practice by slight asymmetry in the circuit (e.g., normal process variations between devices). Hence, a number of analysis and simulation techniques can be employed to carefully check the stability. With more than two transistors, even more unwanted modes of operation can be present [19]. Provided that they are investigated thoroughly, a number of methods such as shunt resistors across the devices can minimise this problem. These shunt resistors draw no current in the wanted even-mode but can damp out unwanted oscillations. Figure 10.37 shows the photograph of a GaAs pHEMT 1 Watt power amplifier chip operating at 42 GHz. Careful study of the layout will reveal that this work of art makes use of many of the techniques described in this section, with eight power devices power-combined in the output stage.

Finally, it should be noted that planar transmission lines generally have relatively high loss and the combining loss leads to a major reduction in the power amplifier efficiency. Trying

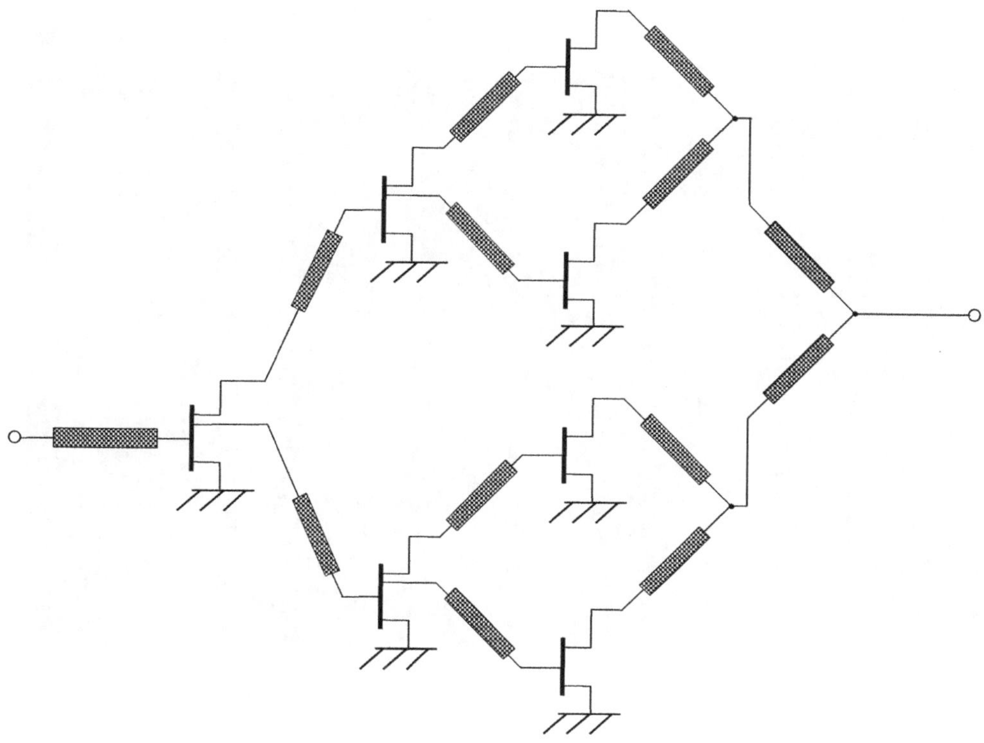

Figure 10.36 Simplified illustration of the enhanced cluster matching technique.

to combine more than eight devices will often be self-defeating due to the additional loss. Whilst 1 dB loss may be tolerable in terms of gain, when it comes to absolute output power and efficiency it is not tolerable as far too much power is dissipated as heat in the combining network. For example, 4 × 5 W RF power with 1 dB combining loss leads to only 15.9 W output power, with 4.1 W dissipated as heat in the combiners. If the individual 5 W amplifiers were 50% efficient, the 15.9 W combination would have 40% efficiency. If two such 4 × 5 W modules were combined together with another 1 dB loss, the output power would be 25.2 W and the efficiency only 31.5%.

10.9.2 Coaxial and Waveguide Combiners

In order to minimise combining losses, TEM coaxial lines or rectangular waveguide operated in TE_{10} mode may be used. The advantages of this power-combining technique are that the number of active devices can be increased without excessive combining loss. Furthermore, without the restrictions of a planar circuit, the binary power-combining structure is not needed and single N-way power combiners can be realised, leading to a very important reduction in combiner loss. Such combiners include the well-known radial combiner [20] as well as conical [21] and various 3-D designs that are part coaxial and part waveguide, where the waveguide output is required for its extremely high power handling and compatibility with antenna feeds.

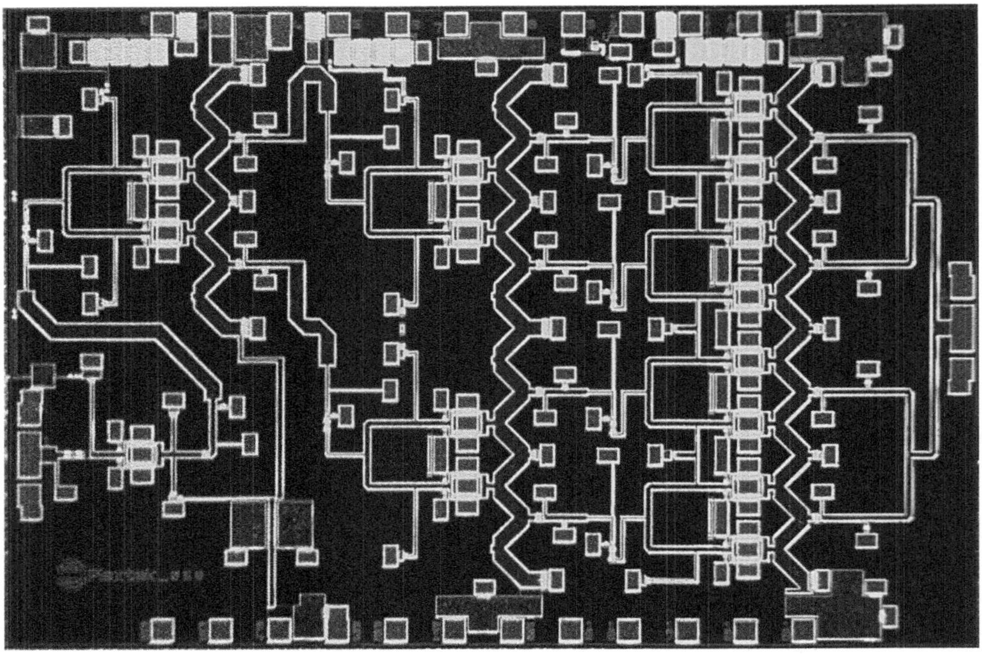

Figure 10.37 GaAs pHEMT 1 Watt power amplifier chip operating at 42 GHz. Illustration courtesy of Plextek Ltd., © 2016. Used with permission; all rights reserved.

In millimetre-wave design it is often an elegant solution to integrate a waveguide power combiner into the metal housing, with the rest of the power amplifier realised in planar form on a substrate. A low-loss microstrip-to-RWG transition is then essential in order to benefit from the low combining loss of the waveguides. The substrate-integrated waveguide (SIW) has an important role to play in this and, indeed, hollow or air-filled SIW has been proposed as a means of realising low-loss power combiners and other high-Q components within a single substrate [22, 23].

10.9.3 Spatial and Quasi-Optical Power Combining

Spatial or quasi-optical power combining involves combining different power sources simultaneously in a manner that avoids the losses of guided-wave structures. The power divider may stay in planar form since its loss is not critical, but the key is to collect the output power from many modules or devices in free-space, as shown in Figure 10.38, or in an over-moded waveguide where the conductor loss is minimal. A major advantage of the spatial power-combining technique is that efficiency is independent of the number of devices being combined and so, in principle, hundreds or thousands of devices could be combined. For example, a wafer scale array of devices could perhaps be power-combined to realise a useful terahertz source operating at room temperature with approximately 1 W output power. Of course, such an approach may not be economically viable in many cases and the individual devices must be efficient. A practical example of quasi-optical power combining is the use of a 2-D array of amplifiers with

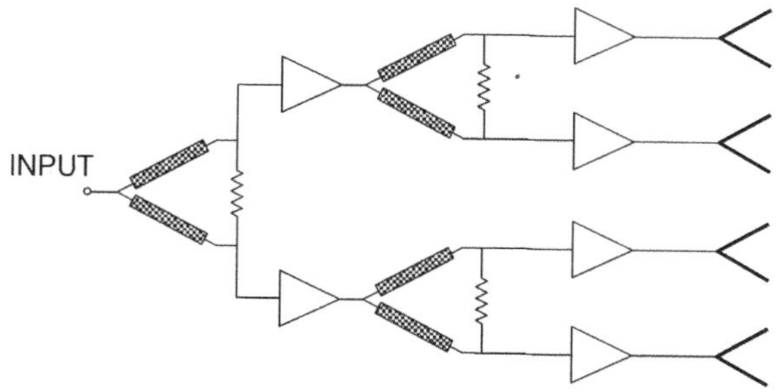

Figure 10.38 Spatial power-combining principle.

finline antennas at both input and output, placed together in a coaxial or waveguide structure [24, 25].

10.10 Power Amplifier Linearisation

Linearisation techniques are used to compensate for a power amplifier's nonlinear behaviour in some way so that it can be operated closer to saturation without the output signal being excessively distorted. Linearisation techniques can be classified as either open loop or closed loop. Closed-loop systems involve various forms of feedback systems that are capable of comparing the output signal with the input signal. This approach is highly effective but is restricted to signals with a certain maximum modulation bandwidth due to the loop delay time. An open-loop system, on the other hand, is capable of handling a broader modulation bandwidth but is generally capable of less precise linearisation, since the condition of the output signal is not known. So, for example, it cannot cope so easily with ageing and drifting of the components, whereas a closed-loop system can adapt dynamically. The main classes of linearisation technique applied to a nonlinear amplifier are feedback, predistortion and feed-forward linearisation.

10.10.1 Feedback Techniques

In traditional analogue design, feedback is used to increase bandwidth and make the gain of an amplifier less sensitive to variations in components of the circuit. Negative feedback can also reduce distortion and noise. Figure 10.39 presents the basic block diagram for a classical feedback amplifier.

The output signal is fed back to the input through a feedback factor β, so:

$$V_{in} = V_s - \beta V_{out} \tag{10.51}$$

and, therefore the closed-loop gain,

$$A_{FB} = \frac{A}{1 + \beta A} \tag{10.52}$$

where A is the open-loop gain of the amplifier.

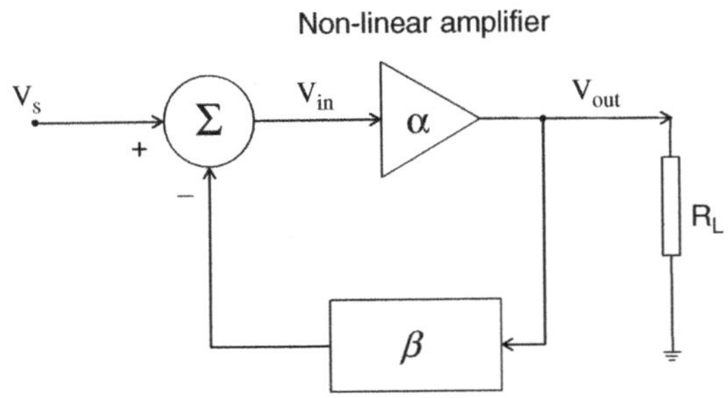

Figure 10.39 Feedback system.

Hence, for large open-loop gain, the feedback amplifier's gain is approximately equal to $1/\beta$. This means that the feedback amplifier has a gain dictated largely by the feedback circuits, not the active devices. The stability of a feedback amplifier is determined by how much the loop gain varies with frequency. In certain situations, the feedback can become positive and cause instability or oscillations in the amplifier.

In a communications application, the signal is modulated and a number of more sophisticated feedback techniques can be employed to linearise the amplifier. In envelope feedback, a sample of the input and output signals are passed to envelope detectors and the amplitude from the detectors is compared. The difference in amplitude is then used to control an amplitude modulator. In Cartesian-loop feedback, a demodulator is placed at the output to sample a portion of the transmitted signal. The amplitude and phase can then be fed back in Cartesian (IQ) form and compared with the input values, and an error signal used for correction. The disadvantage of this Cartesian loop is that it is generally restricted to narrow band application. In polar loop feedback the instantaneous amplitude and phase information is fed back.

10.10.2 Predistortion Linearisation

As the name suggests, in predistortion linearisation the amplifier nonlinearity is characterised or modelled in some way, after which the input signal is deliberately distorted so that when it passes through the amplifier it becomes the desired transmitted signal again. Predistortion can be applied at baseband, at an IF, or directly at the RF frequency. Predistortion is generally not restricted in bandwidth and can correct for both amplitude and phase distortion. Furthermore, since there is no additional circuitry at the output of the amplifier, it does not reduce the efficiency of the amplifier. However, in practice the amplifier nonlinear characteristics will change with temperature and ageing, so some type of adaptive predistortion is necessary. Digital predistortion has become a method of choice for transceivers and is described in detail in Chapter 15.

10.10.3 Feedforward Linearisation

This is a powerful technique that compares the output signal of the amplifier with the input, in order to determine the error that nonlinearity has introduced [26]. This error is then inverted and fed back into the output so that it can be cancelled out. This is achieved with two loops, as

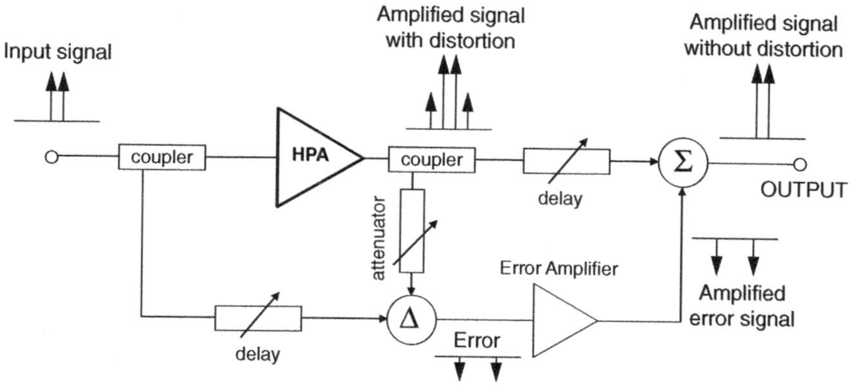

Figure 10.40 Feedforward linearisation block diagram.

illustrated in Figure 10.40. The first loop subtracts an attenuated portion of the output signal from the input signal, in order to create the error signal. The second loop then amplifies this error signal up to the appropriate level and injects it into the output to cancel out the distortion. In practice, the hardware is complex and the amplitude and phase of the two loops need to be carefully controlled in order to achieve an improvement in linearity. Also, the error amplifier may consume considerable DC power and the output coupler introduces loss, so the advantage in efficiency improvement may be marginal compared to the complexity. Cancellation of the distortion signal relies on subtracting two signals from different paths, and this requires equal magnitude and opposite phase within tight constraints, otherwise the benefits are not significant. This really requires an adaptive control system. Also, for wide bandwidth, the time delay of the two loops must be equal. The advantage of the feedforward linearisation technique is that the system can be waveform agnostic; that is, there is no need to know anything about the input signal, and with appropriate digital control of the two loops excellent performance can be achieved.

10.11 Conclusion

This chapter has summarised some of the key principles involved in the design of low-noise, general-purpose and high-power amplifiers. The classical theory of amplifier matching and stability has been described, although the increasing use of more sophisticated design techniques has tended to make this simple approach much less relevant to modern wireless design. The effects of nonlinearity such as distortion, harmonics, and intermodulation have been discussed, and these are applicable to other components in a wireless transceiver, even including passive components if they are operating at sufficiently high power. The nature of second- and third-order intermodulation products and their effects in the RF system have been considered. Several topologies of power amplifier have been described. The linearity and efficiency of power amplifiers has attracted a great deal of attention since the battery life of mobile terminals is strongly affected by their power consumption. This has led to several new classes of amplifier, far beyond the traditional classes A, B and C. Furthermore, techniques are widely employed for linearisation of the power amplifier, including feedback, predistortion and feedforward methods. Since these techniques benefit greatly from digital control methods and, in any case, are being applied to digitally modulated signals, further coverage of this important topic is provided in Chapter 15.

References

1. Lucyszyn, S. (1997) Power-added efficiency errors with RF power amplifiers. *International Journal of Electronics*, **82** (3), 303–312.
2. Lee, T.H. (2004) *The Design of CMOS Radio Frequency Integrated Circuits*, 2nd edition. Cambridge University Press.
3. Casier, H., Steyaert, M., and Van Roermund, A.H.M. (eds) (2008) *Analog Circuit Design: Sensors, Actuators and Power Drivers; Integrated Power Amplifiers from Wireline to RF; Very High Frequency Front Ends*. Springer, Netherlands. ISBN: 978-1-4020-8262-7 (print); 978-1-4020-8263-4 (online).
4. Ellinger, F. (2008) *Radio Frequency Integrated Circuits and Technologies*, 2nd edition. Springer. ISBN-10: 3540693246.
5. Rollett, J.M. (1962) Stability and power-gain invariants of linear two ports. *IRE Transactions on Circuit Theory*, **CT-9** (1), 29–32.
6. Edwards, M.L. and Sinsky, J.H. (1992) A new criterion for linear 2-port stability using a single geometrically derived parameter. *IEEE Transactions on Microwave Theory and Techniques*, **40** (12), 2303–2311.
7. Lombardi,G. and Neri, B. (1999) Criteria for the evaluation of unconditional stability of microwave linear two-ports: a critical review and new proof. *IEEE Transactions on Microwave Theory and Techniques*, **47** (6), 746–751.
8. Norton, D.E. (1975) High dynamic range transistor amplifiers using lossless feedback. Proceedings of the 1975 IEEE Symposium on Circuits and Systems, pp. 438–440.
9. Niclas, K.B. (1982) On the design and performance of lossy match GaAs MESFET amplifiers. *IEEE Transactions on Microwave Theory and Techniques*, **30** (11), 1900–1907. doi: 10.1109/TMTT.1982.1131341.
10. Percival, W.S. (1937) Thermionic Valve Circuits. British Patent Specification no. 460,562, filed 24 July 1936, granted January 1937.
11. Cripps, S.C. (1999) *RF Power Amplifiers for Wireless Communications*. Artech House Publishers.
12. Kenington, P.B. (2000) *High-Linearity RF Amplifier Design*. Artech House Publishers.
13. Boylestad, R.L. and Nashelsky, L. (2006) Power amplifiers, in: *Electronic Devices and Circuit Theory*, 9th edition. Pearson Education.
14. Wentzel, A., Meliani, C., and Heinrich, W. (2010) RF class-S power amplifiers: State-of-the-art results and potential. IEEE MTT-S International Microwave Symposium Digest, 23–28 May, pp. 812–815. doi: 10.1109/MWSYM.2010.5517402.
15. Gysel, U.H. (1975) A new N-way power divider/combiner suitable for high-power applications. IEEE-MTT-S International Microwave Symposium, 12–14 May, pp. 116–118. doi: 10.1109/MWSYM.1975.1123301.
16. Pavlidis, D., Archambault, Y., Efthimerou, N., Kaminsky, D., Bert, A., and Magarshack, J. (1983) A new specifically monolithic approach to microwave power amplifiers. IEEE Microwave and Millimeter-wave Monolithic Circuits Symposium, Boston, USA, pp. 54–58.
17. Aoki, I., Kee, S.D., Rutledge, D.B., and Hajimiri, A. (2002) Distributed active transformer – a new power-combining and impedance-transformation technique. *IEEE Transactions on Microwave Theory and Techniques*, **50** (1), 316–331. doi: 10.1109/22.981284.
18. Ying He, Lianming Li, and Reynaert, P. (2010) 60 GHz power amplifier with distributed active transformer and local feedback. Proceedings of the European Solid-State Circuits International Research Conference (ESSCIRC), 14–16 September, pp. 314—317. doi: 10.1109/ESSCIRC.2010.5619707.
19. Ohtomo, M. (1993) Stability analysis and numerical simulation of multidevice amplifiers. *IEEE Transactions on Microwave Theory and Techniques*, **41** (6), 983–991. doi: 10.1109/22.238513.
20. Schellenberg, J. and Cohn, M. (1978) A wideband radial power combiner for FET amplifiers. Solid-State Circuits Conference. Digest of Technical Papers. XXI IEEE International Symposium, 15–17 February, pp. 164–165. doi: 10.1109/ISSCC.1978.1155840.
21. de Villiers, D.I.L., van der Walt, P.W. and Meyer, P. (2007) Design of a ten-way conical transmission line power combiner. *IEEE Transactions on Microwave Theory and Techniques*, **55** (2), 302–308.

22. Parment, F., Ghiotto, A., Tan-Phu Vuong, Duchamp, J.-M., and Wu, Ke (2015) Air-filled substrate integrated waveguide for low-loss and high power-handling millimeter-wave substrate integrated circuits. *IEEE Transactions on Microwave Theory and Techniques*, **63** (4), 1228–1238. doi: 10.1109/TMTT.2015.2408593.
23. Jin, L., Lee, R.M.A., and Robertson, I.D. (2014) Analysis and design of a novel low-loss hollow substrate integrated waveguide. *IEEE Transactions on Microwave Theory and Techniques*, **62** (8), 1616–1624. doi: 10.1109/TMTT.2014.2328555.
24. Alexanian, A. and York, R.A. (1997) Broadband waveguide-based spatial combiners. IEEE MTT-S International Microwave Symposium Digest, 8–13 June, vol. 3, pp. 1139–1142. doi: 10.1109/MWSYM.1997.596528.
25. Pengcheng Jia, Lee-Yin Chen, Alexanian, A., and York, R.A. (2003) Broad-band high-power amplifier using spatial power-combining technique. *IEEE Transactions on Microwave Theory and Techniques*, **51** (12), 2469–2475. doi: 10.1109/TMTT.2003.819766.
26. Pothecary, N. (1999) *Feedforward Linear Power Amplifiers*. Artech House Publishers.

11

Oscillators

11.1 Introduction

For a wireless communication transceiver, oscillators are important subsystems that are required to supply a clean sinusoidal signal for modulators, demodulators and frequency mixers. 'Clean' means that the output signal should be a pure sinusoidal wave with a stable carrier frequency, with no other unwanted signals or noise appearing at the output. In practice, this is very challenging, and oscillator theory, design and measurement is a vast subject in its own right, especially because communications systems have placed ever more stringent requirements on performance and tuning capability, meaning that synthesisers have become more and more advanced. This one chapter cannot do justice to all these topics and instead focuses on the fundamental theory and design principles [1–3] that are the core knowledge for those working in microwave and millimetre-wave circuit design for wireless communication systems.

11.2 Basic Concepts

11.2.1 One-Port Model: Negative Resistance

This one-port descriptive model is based on a linear system approximation. It is simple to understand and provides a quick approach for the analysis and design of an oscillator. This model is found to be appropriate for high-frequency design. The alternative two-port model is less applicable to high-frequency oscillator design because it is based on the assumption that the feedback network and active circuit do not load each other, which is not valid at high frequency as the network impedance is frequency-dispersive. Furthermore, the two-port model is not suitable for high-frequency oscillator design because of the prominent effect of parasitic components at high frequency that result in extra feedback paths. This makes the task of separating the forward part and the feedback network rather difficult. As a result, the one-port model based on a negative-resistance technique is more popular for high-frequency oscillator analysis and design.

Microwave and Millimetre-Wave Design for Wireless Communications, First Edition.
Ian Robertson, Nutapong Somjit and Mitchai Chongcheawchamnan.
© 2016 John Wiley & Sons, Ltd. Published 2016 by John Wiley & Sons, Ltd.

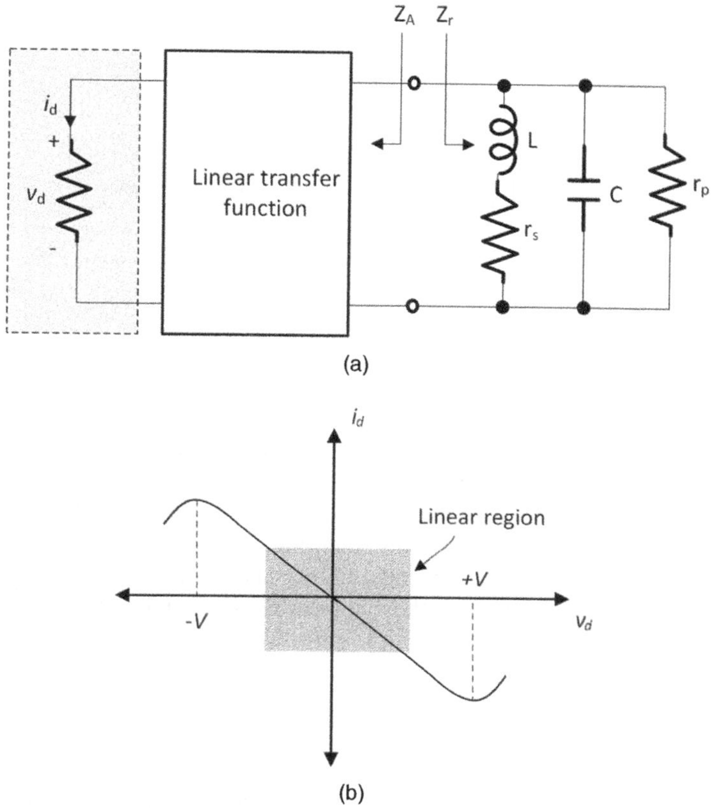

Figure 11.1 (a) A one-port negative-resistance oscillator model. (b) Characteristic function of a non-linear resistor.

The beauty of the one-port model is its simplicity. Figure 11.1(a) shows the one-port model for oscillator analysis based on the negative-resistance approach. The oscillator is a nonlinear circuit that consists of active devices connected with a linear resonant network. The nonlinear mechanism in an oscillator stems from nonlinearity in the active devices. Here, this nonlinearity is simply modelled with a nonlinear resistor. The frequency response of the active network is determined by a matching network and the parasitic elements of the semiconductor device. The loading resonator network is modelled as a parallel *LC* network with the conductor loss in the inductor modelled with a series resistor r_s and dielectric loss in the parallel capacitor modelled with the parallel resistor r_p. The assumed I–V characteristic of the nonlinear resistor is shown in Figure 11.1(b). With a voltage drop v_d, the current that flows through the nonlinear resistor is i_d. At a sufficiently small current i_d, v_d is linearly related with i_d. In this case, the amplitude of voltage swing linearly increases with the current. The negative slope of the I–V curve in the linear region indicates the reversed current. This implies that the active device, when a DC bias is applied, behaves as a signal source. The linear characteristic is not valid when i_d increases. The voltage swing is limited to $\pm V$. From Figure 11.1(a), the signal spectrum from the nonlinear resistor will be shaped with the transfer function of the active network.

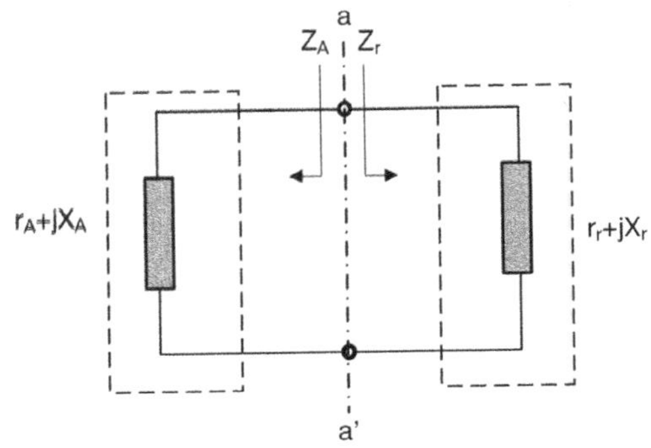

Figure 11.2 A simplified one-port linear circuit model for an oscillator.

The output response of the active network will be applied to the resonant network of resonant frequency f_0. The output signal is a sinusoidal signal at the desired oscillating frequency.

As previously described, a negative linear slope, giving a negative resistance of the device, is valid for a small-signal condition only. This condition only occurs during the initial stage of oscillation behaviour, that is, the startup stage. In this case, a nonlinear oscillator model can be simplified as a one-port circuit where the nonlinear resistor is modelled as a linear resistor with negative value. The value of this resistor is given by:

$$r_d = \frac{\partial v_d}{\partial i_d}. \tag{11.1}$$

With this simplification, linear network theory can be applied. The nonlinear circuit model shown in Figure 11.1(a) becomes a one-port linear network, as shown in Figure 11.2. The cut a-a' shows the reference plane separating the active circuit and the resonator. The driving-point impedance functions of the active and the resonator networks are denoted as $Z_A(s)$ and $Z_r(s)$, where s is the Laplace variable. Based on the linear network theory, $Z_A(s)$ and $Z_r(s)$ can be simply written as:

$$Z_A(s) = r_A(s) + jX_A(s) \tag{11.2}$$

$$Z_r(s) = r_r(s) + jX_r(s) \tag{11.3}$$

where $r_A(s)$ and $r_r(s)$ are the resistive part while $X_A(s)$ and $X_r(s)$ are the reactive part of $Z_A(s)$ and $Z_r(s)$, respectively. The negative resistance r_d defined in Equation (11.1) is embodied in $r_A(s)$. The oscillation condition of the one-port model shown in Figure 11.2 is obtained when the magnitude of current in the circuit at the oscillating frequency, ω_0, increases exponentially. This can occur if – and only if – the total impedance of the network is zero; that is:

$$Z_A(j\omega_0) + Z_r(j\omega_0) = 0 \tag{11.4}$$

Note that the Laplace term s is substituted with $j\omega_0$. This is under the assumption that the signal in the one-port network is sinusoidal with sufficiently small amplitude. The condition is reasonable because the signal in the closed-loop system at the startup condition is a sinusoidal signal with very small amplitude. The complex equation given above leads to two real conditions as follows:

$$r_A(j\omega_0) + r_r(j\omega_0) < 0 \tag{11.5}$$

$$X_A(j\omega_0) + X_r(j\omega_0) = 0 \tag{11.6}$$

It is noted that the resistance condition given in Equation (11.5) is not derived exactly from Equation (11.4). The equation is modified such that the total resistance in the oscillator network is negative. This modification is to ensure that the oscillation can start up in practice. The first condition is the unique condition specifically for an oscillator circuit. If the resonant network is a passive circuit, its resistive part will be positive and hence r_A must be negative. In other words, the active circuit must provide sufficient negative resistance to overcome all resistive losses in the matching network, parasitic elements and the resonator. Hence, to design an oscillator, the negative resistance of the active device should be set such that its magnitude is larger than the resistive loss in all passive networks (r_T); that is:

$$|r_d| > r_T \tag{11.7}$$

The second oscillation condition shown in Equation (11.6) is easily met because the reactance of a capacitor and inductor have opposite signs. For example, if X_A at f_0 is inductive (positive), adding a suitable series capacitor to X_r can cancel it out. On the other hand, a series inductance can add to X_r to cancel out if the X_A at f_0 is negative. In fact, the second condition is just simply a condition for a passive resonant network. In order to understand how to apply these two oscillation conditions defined for a one-port model, let us consider the following example.

Example: The schematic circuit diagram of a Clapp oscillator with a common-collector bipolar junction transistor (BJT) is shown in Figure 11.3. For convenience, the DC bias circuitry is not included. Derive the oscillation conditions using the one-port model.

The Clapp oscillator shown in Figure 11.3 uses a common-collector configuration to generate negative resistance. The reference plane a-a' separates the active circuit and series-L-C

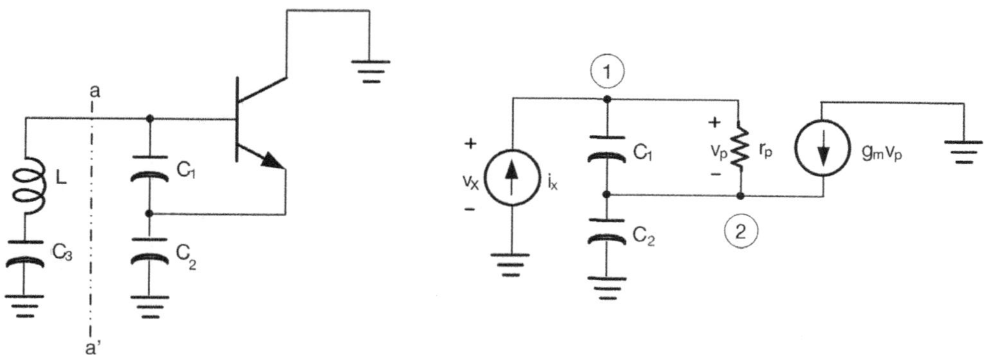

Figure 11.3 Clapp oscillator schematic circuit diagram and small-signal model.

resonator (L, C_3). The active circuit is composed of two tapped capacitors C_1, C_2 and the common-collector BJT. Applying a simple small-signal BJT model, we obtain the small-signal circuit shown on the right. The negative resistance is obtained by computing the driving-point impedance at the right-hand side of the cut plane a-a'. The negative resistance of this oscillator is obtained by calculating the ratio of voltage v_x with the current i_x injected from the current source to the active network:

$$r_A = \frac{v_x}{i_x} \tag{11.8}$$

Applying Kirchhoff's current law at the node 1 gives:

$$\left(\frac{1}{r_\pi} + j\omega C_1\right) v_x - j\omega C_1 v_2 = i_x \tag{11.9}$$

At node 2,

$$-j\omega C_1 v_x + \left(\frac{1}{r_\pi} + j\omega C_2\right) v_2 = g_m v_\pi \tag{11.10}$$

From Figure 11.3(b) we have

$$v_\pi = v_x - v_2 \tag{11.11}$$

Substituting v_π into Equation (11.10) gives:

$$-j\omega C_1 v_x + \left(\frac{1}{r_\pi} + j\omega C_2\right) v_2 = g_m \left(v_x - v_2\right) \tag{11.12}$$

Normally r_π is large, hence Equation (11.12) can be simplified to:

$$v_2 = \left(\frac{g_m + j\omega C_1}{g_m + j\omega C_2}\right) v_x \tag{11.13}$$

Substituting v_2 into Equation (11.9) gives:

$$j\omega C_1 v_x - j\omega C_1 \left(\frac{g_m + j\omega C_1}{g_m + j\omega C_2}\right) v_x = i_x \tag{11.14}$$

Simplifying Equation (11.14) further, yields:

$$\frac{v_x}{i_x} \approx \frac{-1}{\omega C_1}\left(\frac{g_m + j\omega C_2}{\omega C_1 + \omega C_2}\right) \tag{11.15}$$

The negative resistance, r_A, is the real part of the above driving-point impedance:

$$r_A \approx -\frac{g_m}{\omega^2 C_1}\left(\frac{1}{C_1 + C_2}\right) \tag{11.16}$$

and the reactive part of the active network is:

$$X_A = -\frac{1}{\omega C_1}\left(\frac{C_2}{C_1 + C_2}\right) \tag{11.17}$$

The series *L-C* resonant network has a resistor loss in the inductor r_s. This is related to the unloaded Q of the inductor L by:

$$Q = \frac{\omega L}{r_s} \tag{11.18}$$

and the reactance of series *L-C* resonant network is:

$$X_r = \omega L - \frac{1}{\omega C_3} \tag{11.19}$$

Hence, the conditions for the Clapp oscillator to oscillate at resonant angular frequency ω_0 are:

$$\omega_0 L - \frac{1}{\omega_0}\left[\frac{1}{C_3} + \frac{1}{C_1}\left(\frac{C_2}{C_1 + C_2}\right)\right] = 0 \tag{11.20}$$

and

$$-\frac{g_m}{\omega_0^2 C_1}\left(\frac{1}{C_1 + C_2}\right) + \frac{\omega_0 L}{r_s} < 0 \tag{11.21}$$

For a given resonant frequency and a chosen inductor value, the capacitors in a Clapp oscillator can be determined from the first condition. Meanwhile, the second condition is used for designing the bias point of the BJT transistor to obtain the desired value of g_m. With these two necessary conditions, all the initial design parameters for a Clapp oscillator can be determined.

11.2.2 Two-Port Model: Positive Feedback

Based on linear control theory, the oscillator mechanism can be simply described as a positive feedback system, as shown in Figure 11.4. The model consists of a closed-loop system connected with a passive matching network, $C(\omega)$. The closed-loop system consists of an active network, $A_{OL}(\omega)$, and a feedback passive network $B(\omega)$. The transfer function at frequency ω of the active network is given by:

$$A_{OL}(\omega) = \frac{S_o(\omega)}{S_s(\omega)} \tag{11.22}$$

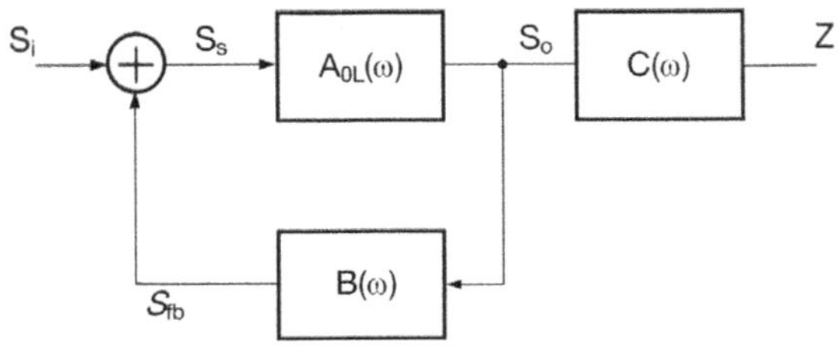

Figure 11.4 Two-port oscillator model: positive feedback.

where $S_o(\omega)$ and $S_S(\omega)$ are the electrical signals at frequency ω which can be either in voltage or current form. Considering the feedback network, the output signal $S_o(\omega)$ is fed back to the feedback network giving the output signal $S_{fb}(\omega)$ given by:

$$S_{fb}(\omega) = B(\omega) S_o(\omega) \tag{11.23}$$

The feedback signal is summed with the external input $S_i(\omega)$ to provide $S_s(\omega)$:

$$S_s(\omega) = S_i(\omega) + S_{fb}(\omega) \tag{11.24}$$

Substituting $S_{fb}(\omega)$ from Equation (11.23) into Equation (11.24) gives:

$$S_s(\omega) = S_i(\omega) + B(\omega) S_o(\omega) \tag{11.25}$$

Hence, from the open loop gain definition shown in Equation (11.22), we obtain:

$$A_{OL}(\omega) = \frac{S_o(\omega)}{S_i(\omega) + B(\omega) S_o(\omega)} \tag{11.26}$$

The closed-loop gain, $A_{CL}(\omega)$, is defined by the ratio between $S_i(\omega)$ and $S_o(\omega)$. Hence:

$$A_{CL}(\omega) = \frac{S_o(\omega)}{S_i(\omega)} = \frac{A_{OL}(\omega)}{1 - A_{OL}(\omega) B(\omega)} \tag{11.27}$$

Equation (11.27) demonstrates the relationship between the input and the output of oscillator. One may wonder how a two-port model can be applied, since at first glance an oscillator has no external input signal. In the oscillation mechanism, an oscillator circuit can be triggered by either internal shot or thermal noise sources. Another possible trigger signal is a transient signal when switching on the DC power supply of the oscillator. In either case, the spectrum shape of $S_i(\omega)$ is frequency independent over a broad spectrum range. With this, $S_o(\omega)$ becomes:

$$S_o(\omega) = \frac{A_{OL}(\omega)}{1 - A_{OL}(\omega) B(\omega)} S_i(\omega) \tag{11.28}$$

The magnitude of the noise spectrum $S_i(\omega)$ is very small. For a single-frequency of oscillation, the transfer function of the feedback network should have a narrow-band band-pass response where the closed-loop gain would be infinite. The first case is achieved using a frequency-selective network such as a high-Q resonator in the feedback network, $B(\omega)$. For the latter case, the closed-loop transfer function is infinite at the oscillation frequency, ω_0:

$$A_{OL}(\omega_0)B(\omega_0) = 1 \tag{11.29}$$

It is known that $A_{OL}(\omega)$and $B(\omega)$ are complex functions; that is, $A_{OL}(\omega) = |A_{OL}(\omega)|e^{j\emptyset_{ALO}(\omega)}$, $B(\omega) = |B(\omega)|\, e^{j\emptyset_B(\omega)}$. Consequently, the equation can be expanded to both conditions which are necessary to satisfy as follows:

$$|A_{OL}(\omega_0)||B(\omega_0)| = 1 \tag{11.30}$$

and

$$\emptyset_{ALO}(\omega_0) + \emptyset_B(\omega_0) = 2n\pi \tag{11.31}$$

where n is any integer number which is greater than 0. The first condition requires that the product of the magnitude of open loop gain and the feedback network response at the oscillation frequency, called the loop gain, is equal to 1. The second condition requires that positive feedback must occur in the loop. These two conditions – known as Barkhausen's criterion – are necessary but not sufficient for stable oscillation because it is only valid for a single frequency and small-signal conditions.

Barkhausen's criterion is for checking the stability condition after the oscillator start-up period. To ensure that oscillation does start up, the magnitude of the loop gain in practice should be greater than 1; that is, $|A_{OL}(\omega_0)||B(\omega_0)| > 1$. Figure 11.5 shows the stages of oscillation, from the start-up to the stable state. Initially, noise or a transient from turning on the power supply is present. The amplitude at the oscillation frequency is increased by the loop gain, a process that is repeated many times round the loop. After a certain time, when the amplified signal in the loop is sufficiently large, the active network will operate nonlinearly at the oscillating frequency. In this nonlinear operation, the gain of the active network is reduced (or the automatically controlled gain function in the oscillator operates) and hence the loop gain is reduced. In this situation, the amplitude of oscillation gradually decreases. The oscillator finally settles in a stable state, where the amplitude reaches an equilibrium level.

Figure 11.6 shows the realisation of an oscillator circuit using the feedback approach. The active network is a transistor amplifier while the feedback network is a resonant network. The resonant network commonly has a π- or T-network topology formed of passive components such as inductor/capacitor, dielectric resonator, transmission line or crystal. The π-network topology is preferred to the T-topology for low-power application due to a single series passive component. The amplifier shown in Figure 11.6 uses a common-source FET, but other configurations can be applied such as common-base or common-drain configurations. A common-base configuration is widely used in wireless transceivers. In Figure 11.6, the oscillator feedback utilises the shunt-shunt topology. This is evident since parallel connections appear between the active and the feedback networks for both input and output ports. However, series-series, series-shunt, and shunt-series feedback topologies can also be applied.

For high-frequency applications, a good SNR is very important in a wireless transceiver. The I–Q constellation points from a modulator fed by an oscillator with high phase noise

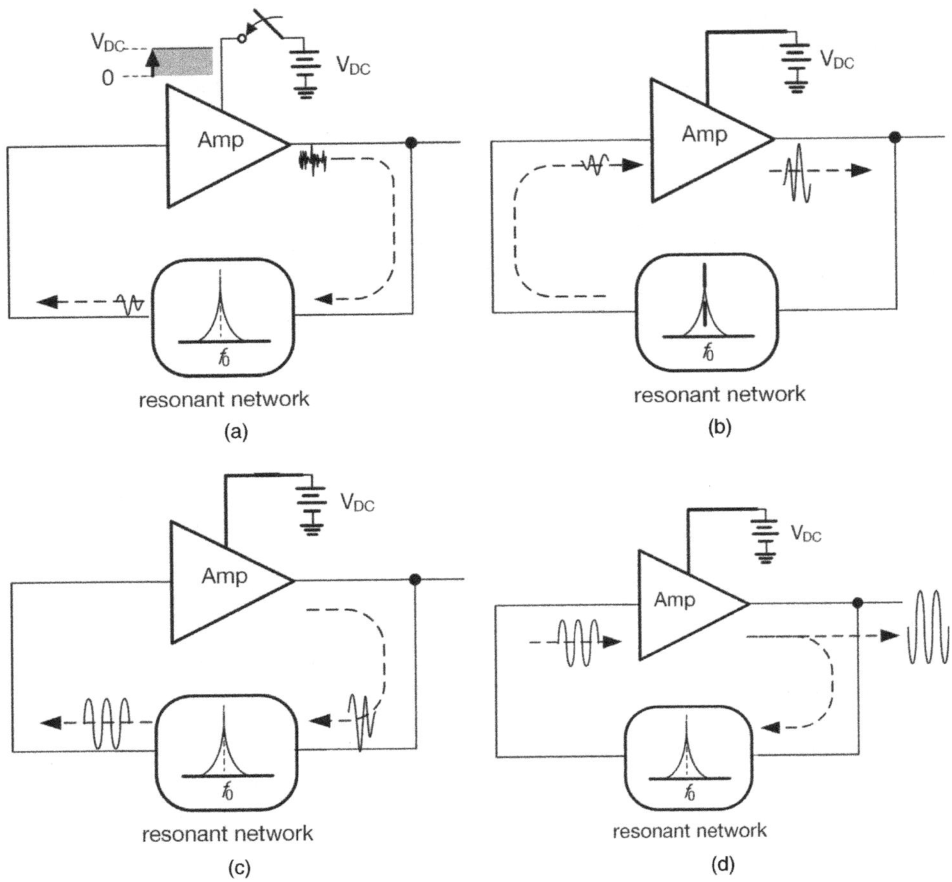

Figure 11.5 Oscillation mechanism periods, from (a) initial, (b) amplified, (c) feedback, and (d) the stable state.

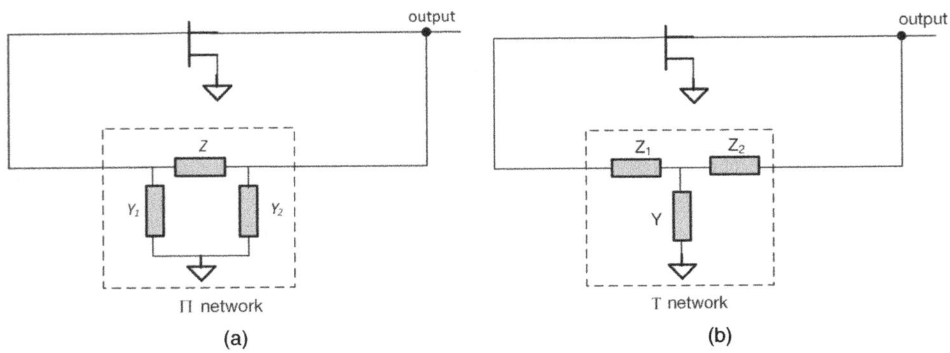

Figure 11.6 Oscillator based on a shunt-shunt feedback topology of (a) π and (b) T.

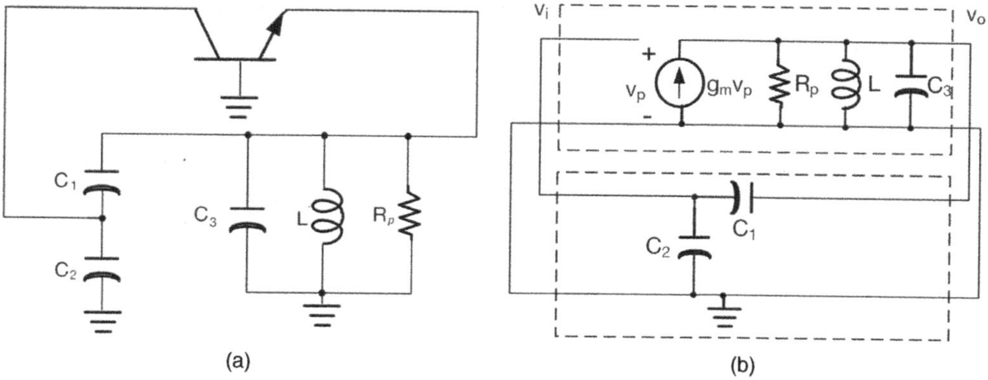

Figure 11.7 Colpitts oscillator. (a) AC circuit. (b) Small-signal feedback model.

are scattered around a circle arc, as described in Chapter 2. In a wireless receiver, a high bit-error-rate is obtained when demodulating data with a high phase-noise carrier. Therefore, a resistive network is not chosen for a feedback-resonant network in a high-frequency oscillator.

Example: A Colpitts oscillator, implemented with a common-base BJT, is shown in Figure 11.7(a). The DC bias circuitry is not shown. The feedback circuit is a capacitive divider formed of C_1 and C_2. A parallel resonant network is tuned to the oscillating frequency, ω_0. Derive the design equation for this Colpitts oscillator from the two-port oscillation condition.

It is assumed that the transistor operates in its mid-band range, and the small-signal model of the BJT is used. The small-signal feedback model is shown in Figure 11.7(b). The active network is composed of the transistor amplifier loaded with the resonant network (R_P, L, and C_3). It is noted that R_p is related to the quality factors of the inductor L and capacitor C_3. The feedback network is a capacitive voltage divider (C_1 and C_2). Obviously, the closed-loop system is naturally positive feedback (emitter-collector feedback). The input and output voltages of the active network are denoted by v_i and v_o, respectively. For small-signal and open-loop conditions, we obtain:

$$v_o(\omega) = g_m Z_L(\omega) v_i(\omega) \tag{11.32}$$

where

$$Z_L(\omega) = \frac{1}{G_p + j\left(\omega C_3 - \frac{1}{\omega L}\right)} \tag{11.33}$$

G_p is the conductance related to R_p, i.e. $G_p = \frac{1}{R_p}$. Hence, the open-loop gain is:

$$A_{OL}(\omega) = \frac{v_o(\omega)}{v_i(\omega)} = \frac{g_m}{G_p + j\left(\omega C_3 - \frac{1}{\omega L}\right)} \tag{11.34}$$

The feedback network is a capacitive voltage divider. Thus, the ratio between the output and input voltages is given by:

$$B(\omega) = \frac{v_0(\omega)}{v_i(\omega)} = \frac{C_1 + C_2}{C_1} \tag{11.35}$$

The resonant frequency, ω_0, is determined by the parallel resonant network L and C_3 by:

$$\omega_0^2 = \frac{1}{LC_3} \tag{11.36}$$

Hence, we will obtain the condition to allow oscillation at ω_0 from the first feedback condition as:

$$g_m R_p = \frac{C_1}{C_1 + C_2} \tag{11.37}$$

The phase condition at ω_0 is already satisfied since the phase responses of both the active network and feedback network are zero.

11.2.3 Nyquist Theory Applied to the Oscillation Condition

It is noted that the conditions developed from the negative resistance or feedback models are derived for start-up condition and also while the oscillation is developing. Unfortunately, these conditions are not sufficient as the oscillation mechanism is frequency-dependent and nonlinear, while the aforementioned conditions are valid for a single frequency and small-signal conditions. An additional frequency-dependent condition is recommended to ensure the oscillator condition in the initial design stage. The condition is based on the Nyquist criterion from linear system theory. The analysis starts with the equation of a closed-loop system with the Laplace term s:

$$A_{CL}(s) = \frac{G(s)}{1 + G(s)H(s)} \tag{11.38}$$

where $G(s)$ is the open-loop transfer function and $H(s)$ is the feedback network transfer function. The loop gain function $LG(s)$ is defined by:

$$LG(s) = G(s)H(s) \tag{11.39}$$

The closed-loop transfer function can be written in a generalised polynomial form as:

$$A_{CL}(s) = K\frac{\prod_{i=1}^{m}(s + z_i)\prod_{j=1}^{n}\left(s^2 + 2a_j s + \omega_{z,j}^2\right)}{\prod_{k=1}^{q}(s + p_k)\prod_{l}^{r}\left(s^2 + 2b_l s + \omega_{p,l}^2\right)} \tag{11.40}$$

All of the constants in this characteristic equation (K, z_i, p_k, a_j, b_l, $\omega_{z,j}$ and $\omega_{p,l}$) are real numbers. As shown in Equation (11.40), a feedback system can be characterised by its

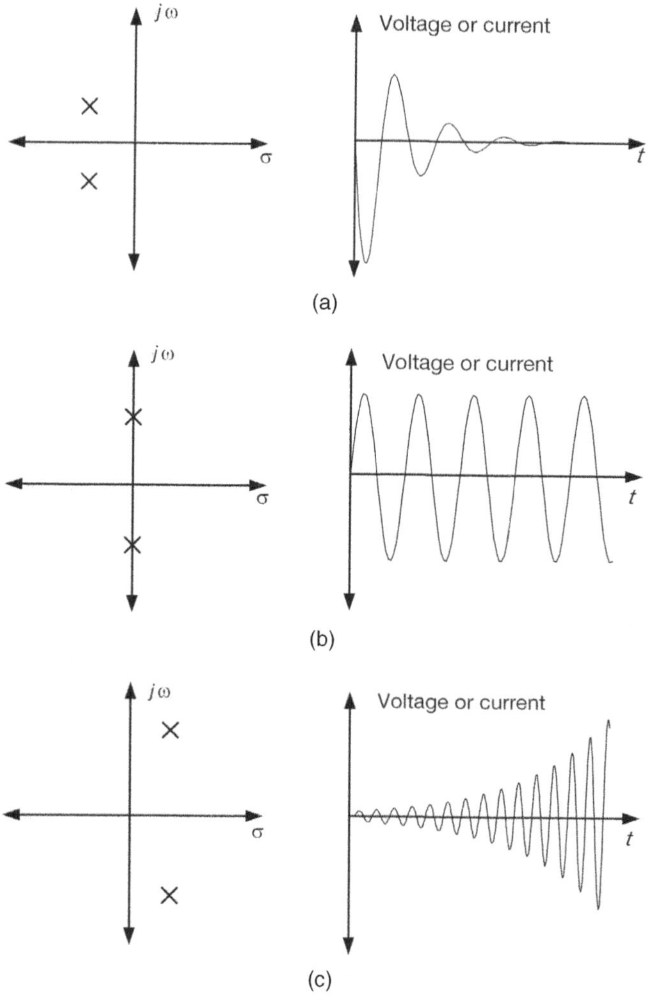

Figure 11.8 Conjugate poles and the unit step responses for (a) $\sigma < 0$, (b) $\sigma = 0$ (critically damped), and c) $\sigma > 0$.

poles and zeros which lie in the complex s-plane. In the general characteristic equation [Equation (11.40)], there are m real zeroes, n complex conjugate pairs of zeroes, q real poles and r complex conjugate pairs of poles. Theoretically, these zeros and poles can be purely real, imaginary, or complex. The positions of the complex conjugate poles in the s-plane can indicate the stability of the system. There are three possible positions of a pair conjugates poles. Figure 11.8 shows these three different cases and the corresponding unit step responses. A system that has a pair of conjugate poles in the right-half plane (RHP) will build up an oscillation. Referring to the equations defined in Equation (11.40), a feedback system will be unstable if the denominator term is zero. Hence, all roots of the term $\prod_{k=1}^{q}(s + p_k)\prod_{l}^{r}(s^2 + 2b_l s + \omega_{p,l}^2)$, which are the poles of the closed-loop system, should lie in the RHP of the complex plane.

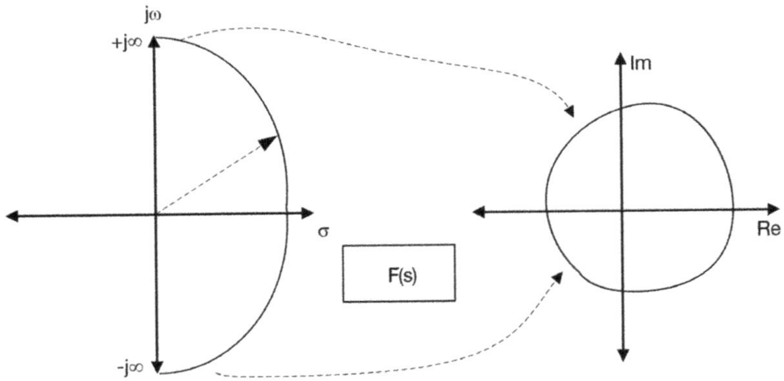

Figure 11.9 Mapping from a Nyquist contour on the *s*-plane through a function *F(s)* to produce a Nyquist diagram.

The denominator term in Equation (11.40) is equivalent to $1 + LG(s)$, the denominator term in Equation (11.38). The zeros of $1 + LG(s)$ are also the zeros of the loop gain $LG(s)$. Consequently, a loop gain of any oscillator system should have at least a pair of conjugate zeroes in the RHP.

Knowing the poles of the closed-loop transfer function can determine the stability condition, but this requires an algorithm to factor Equation (11.40). This is rather difficult to do. A simpler way is to use a Nyquist plot, which is a polar plot of $F(s)$. The function $F(s)$ plotted can be chosen to be either $1 + LG(s)$ or the loop gain $LG(s)$. The Laplace term s is substituted with $j\omega$, where ω is varied from $-\infty$ to $+\infty$. The Nyquist contour is created to enclose the entire RHP of the s-plane. The poles and zeros of $F(s)$ are excluded from the Nyquist contour. Once the Nyquist contour is created, the next step is to map a Nyquist contour in the s-plane through a function $F(s)$ to create a Nyquist diagram in a complex plane. This procedure is shown in Figure 11.9.

The Nyquist diagram of the open-loop gain $F(s)$ can indicate the stability condition of the closed-loop system. This is based on Cauchy's argument principle, referring to the mapping function $1 + LG(s)$ which states that:

$$N = Z - P \tag{11.41}$$

where N is the number of encircles of the origin point (0, 0) by the contour of the Nyquist diagram, Z is the number of zeros of $F(s)$ in the RHP, and P is the number of poles of $F(s)$ in the RHP.

On the other hand, if we plot the Nyquist diagram by mapping through the loop function $LG(s)$, N will be the number of encircles around (–1, 0). Since our goal is to determine the stability condition of a closed-loop system $A_{CL}(s)$, this means that no pole of $A_{CL}(s)$ exists in the RHP. From the definition of $A_{CL}(s)$, relating to $1 + LG(s)$ and $LG(s)$ shown in Equations (11.38) and (11.39), the open-loop function either $1 + LG(s)$ or $LG(s)$ should have no zero in the RHP. From Equation (11.41),Z, the number of zeros of the open-loop function, for a stable system, is computed from:

$$Z \equiv N + P = 0 \tag{11.42}$$

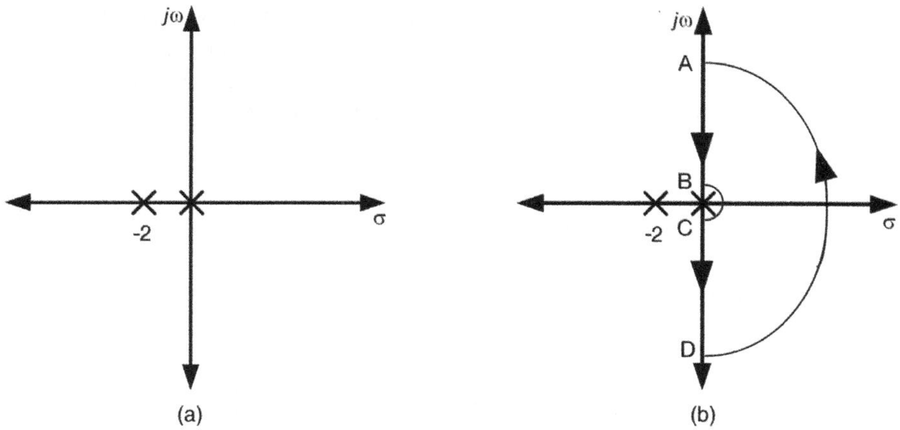

Figure 11.10 From s-plane to Nyquist contour. (a) Pole positions. (b) Nyquist contour.

The number of encircles around the critical point N given in Equation (11.42) is obtained by plotting a complete contour of the Nyquist diagram of $F(s)$. Meanwhile, the number of RHP poles is obtained by plotting poles of function$F(s)$ in the s-plane. If there is no encircling and no RHP pole, the system is stable, or vice versa. A procedure to plot the Nyquist diagram is demonstrated in the following example.

Example: Considering a closed-loop system which has a loop gain $\frac{10}{s(s+2)}$, determine the stability condition of the system.

From the loop gain given, the poles are located at the origin and at (–2,0), and the zero is at infinity. The pole positions of the loop gain are plotted in Figure 11.10(a).

From the plot in Figure 11.10(a), the Nyquist contour is created in the s-plane covering the entire RHP. The contour consists of a Nyquist path starting from $+j\infty$ along the imaginary axis toward $-j\infty$ and closed by a semicircle in the right half s-plane. If poles and zeros of open-loop gain function occur, they will not be included in the Nyquist contour. For the loop gain in this example, the Nyquist contour in the s-plane is shown in Figure 11.10(b). The contour starts from point A$(0, +j\infty)$ to point B via the straight line, semi-circles around the origin with negligible radius to point C towards point D$(0, -j\infty)$, and finally ends at point A with the semicircle of radius ∞. By inspecting the Nyquist contour shown in Figure 11.10(b), the number of poles in the RHP, P, is equal to 0. Next, we substitute $s = j\omega$ into $\frac{10}{s(s+2)}$, giving:

$$F(j\omega) = \frac{10}{-\omega^2 + j2\omega} \tag{11.43}$$

The magnitude and phase functions are:

$$|F(j\omega)| = \frac{10}{\omega\sqrt{\omega^2 + 4}} \tag{11.44}$$

$$\angle F(j\omega) = -\frac{\pi}{2} - \tan^{-1}\frac{\omega}{2} \tag{11.45}$$

The next step is to map the Nyquist contour (A-B-C-D-A) in the s-plane shown in Figure 11.10(b) to the complex plane. The contour point A, B, C, D and paths AB, BC, CD, and DA are used to create the Nyquist diagram via the functions defined in Equations (11.44) and (11.45). The point A from the s-plane is mapped to the point A′ in the complex plane as shown:

$$|F(j\infty)| \angle F(j\infty) = 0\angle - \pi \tag{11.46}$$

Any point that lies on the straight line AB $(0, +jh)$ where $0 < h < \infty$ is mapped and the result is:

$$|F(jh)| \angle F(jh) = \frac{10}{h\sqrt{h^2+4}} \angle \frac{-\pi}{2} - \tan^{-1}\frac{h}{2} \tag{11.47}$$

Equation (11.47) shows that the path AB is mapped to a semicircle A′B′ with infinite radius. The point B from the s-plane, $(0, j\varepsilon)$, is mapped to B′ in the complex plane by substituting into Equations (11.44) and (11.45). This leads to:

$$\lim_{\varepsilon \to 0} |F(j\varepsilon)| \angle F(j\varepsilon) = \infty\angle - \frac{\pi}{2} \tag{11.48}$$

Any point on the path BC is mapped to the complex plane. Using polar coordinates, a point on the path BC can be written as $ae^{j\theta}$, where a is very close to zero and $-\frac{\pi}{2} < \theta < \frac{\pi}{2}$. The path B′C′ is governed by:

$$F(ae^{j\theta}) \approx \infty e^{-j\theta} \tag{11.49}$$

With $a \to 0$ and θ ranging from $-\frac{\pi}{2}$ to $\frac{\pi}{2}$, the mapped path B′C′ is a semi-circle with infinite radius. The path B′C′ starts from $-\frac{\pi}{2}$ to $+\frac{\pi}{2}$. The point C in the s-plane is mapped to C′ in the complex plane, which is:

$$\lim_{\varepsilon \to 0} |F(-j\varepsilon)| \angle F(-j\varepsilon) = \infty\angle\frac{\pi}{2} \tag{11.50}$$

The straight line CD is converted to C′D′ by mapping point $(0, -jh)$ on the path CD, where $0 < h < \infty$. The result is a semicircle with infinite radius given by:

$$|F(-jh)| \angle F(-jh) = \frac{10}{h\sqrt{h^2+4}} \angle \frac{-\pi}{2} + \tan^{-1}\frac{h}{2} \tag{11.51}$$

The point D $(0, -j\infty)$ in the s-plane is converted to point D′ in the complex plane by:

$$|F(-j\infty)| \angle F(-j\infty) = 0\angle\pi \tag{11.52}$$

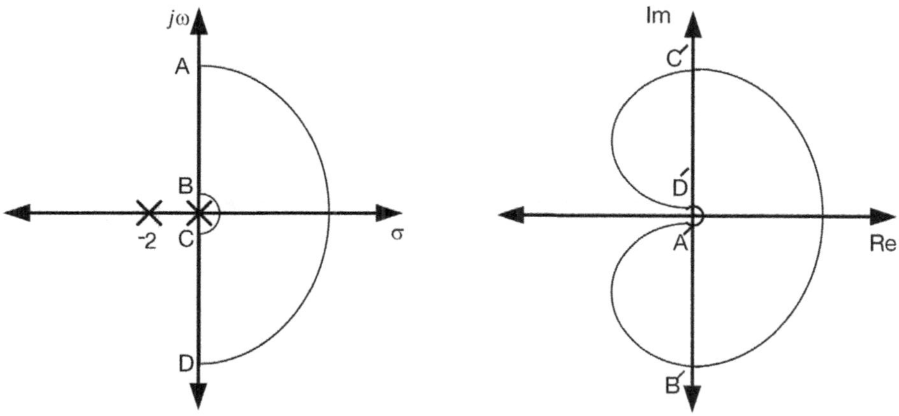

Figure 11.11 Nyquist contour and Nyquist diagram of loop gain $\frac{10}{s(s+2)}$.

Finally, the semi-circle DA is mapped to D′A′ by substituting $be^{j\theta}$ for $b \rightarrow \infty$ and θ ranging from $-\frac{\pi}{2}$ to $\frac{\pi}{2}$; this results in a circle with negligible radius, as shown:

$$F(be^{j\theta}) \approx 0e^{-j2\theta} \tag{11.53}$$

The complete Nyquist diagram in the complex plane obtained by mapping via the loop gain function $\frac{10}{s(s+2)}$ with the Nyquist contour plot is shown in Figure 11.11. The points and paths of the Nyquist contour and Nyquist diagram are summarised in Table 11.1. There is no pole in the RHP from the Nyquist contour and no encircling around (–1,0) in the Nyquist diagram. Hence, the feedback system with this loop gain function is stable.

Table 11.1 Summary of points and paths of the example Nyquist contour and diagram.

s-plane			Complex plane		
Point	Path	Value	Point	Path	Complex-plane
A	–	$(0, +j\infty)$	A′	–	$(0, 0)$
–	AB	Straight line traversing along imaginary axis	–	A′B′	Semi-circle with infinite radius
B	–	$(0, j\varepsilon)\,;\varepsilon \rightarrow 0$	B′	–	$(0, -j\infty)$
–	BC	Semi-circle with negligible radius	–	B′C′	Semi-circle with infinite radius
C	–	$(0, -j\varepsilon)\,;\varepsilon \rightarrow 0$	C′	–	$(0, j\infty)$
–	CD	Straight line traversing along imaginary axis	–	C′D′	Semi-circle with infinite radius
D	–	$(0, -j\infty)$	D′	–	$(0, 0)$
–	DA	Semi-circle with infinity radius	–	D′A′	Circle with negligible radius

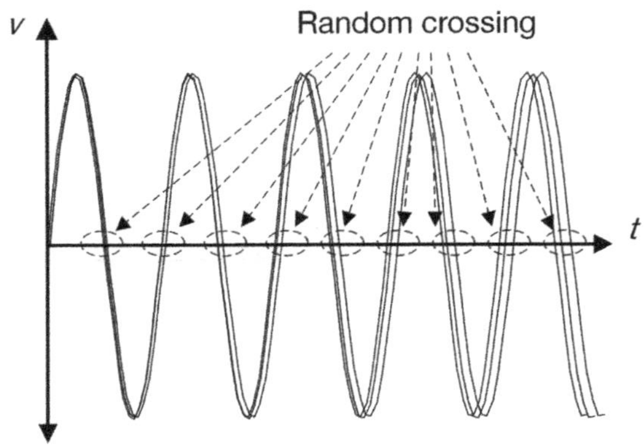

Figure 11.12 Timing jitter.

It is highly recommended that designs are tested with this mathematical tool. For oscillator design, the technique is sufficient to guarantee the oscillation condition during the start-up stage. However, this procedure to prove the stability condition using Nyquist criterion is quite tedious, very complicated, and definitely prone to error. Fortunately, there are design tools available in modern commercial software such as Advanced Design System (ADS^{TM}), Microwave Office^{TM} as well as mathematical tools in MATLAB^{TM} and MathCAD^{TM}.

Note that there is a situation in which the amplitude of oscillation decreases and the oscillation stops. The conditions derived from linear system theory are not sufficient to guarantee stable oscillation. Nonlinear modelling methods may be needed for more accurate evaluation in the stability condition and oscillation parameters such as amplitude and frequency.

11.2.4 Timing Jitter

Timing jitter is the random deviation of the zero-crossings of a signal, caused by phase noise, as illustrated in Figure 11.12. The phase noise variation transfers to the variation of timing jitter by:

$$\sqrt{\sigma_T^2} = \frac{\sqrt{\sigma_\phi^2}}{f} \tag{11.54}$$

where σ_T^2 and σ_ϕ^2 are the jitter and phase noise variances.

Jitter in a system will degrade the communication performance. For example, consider an ADC with a sampling clock that has high jitter, as shown in Figure 11.13. The ADC block consists of a sampling circuit modelled as a switch with a switching speed determined by a clock oscillator. Timing jitter exists in the clock signal due to the phase noise in the oscillator. The amplitude distribution of jitter time is random and can be modelled as a Gaussian function. This jitter causes a random error in sampling time (Δt) and hence creates an error in amplitude signal ΔV. Therefore, the quantised signal from the quantiser is distorted, as shown in Figure 11.13. A high-speed sampling circuit and high-resolution quantiser, as required in many high data rate communications systems, requires a clock signal with very low jitter.

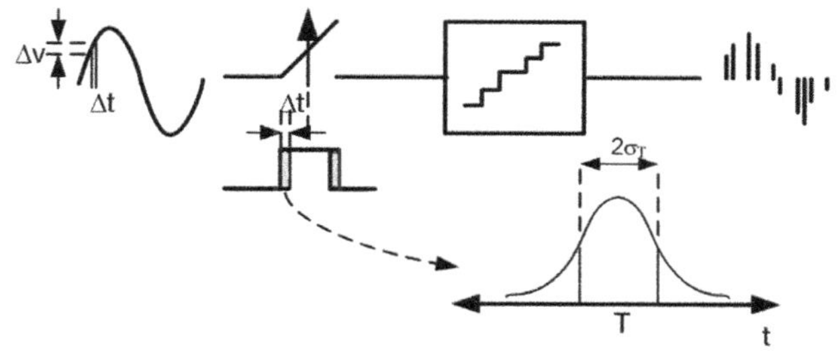

Figure 11.13 Effect of timing jitter in an analogue-to-digital converter.

11.2.5 Phase Noise

Oscillators are required to produce a sinusoidal carrier signal for modulation and demodulation, as well as for the local oscillator (LO) for mixers in heterodyne systems. Theoretically, the oscillator signal is a sinusoidal signal with an ideal line spectrum (a delta function) located at f_0. In Figure 11.14, it is assumed that two signals appear at the mixer RF port; these are the desired signal and some adjacent-channel interference. The adjacent channel interference may come from an undesired user occupying the adjacent frequency channel, for example. In several wireless standards, the channel bandwidth is very small as compared to the carrier frequency. These RF signals are modelled as two sinusoidal signals at the RF input port. The spectrum of the input and output signals when mixed with an ideal LO is shown in Figure 11.14(a). The downconverted outputs from the mixer have two clearly separated spectrum components, with no interference. Unfortunately, an oscillator cannot provide an ideal single-line spectrum like this.

In practice, there are noise sidebands around the discrete spectrum, as shown in Figure 11.14(b). The incoming signals, when mixed with this LO, are downconverted to a low-frequency spectrum with noise sidebands. The sideband noise from the in-band and adjacent channel both leak to the desired communication channel and degrade the SNR of the receiver. This sideband noise from the oscillator is called 'phase noise', and many wireless standards clearly declare the phase noise requirement for an oscillator to meet the communication specifications.

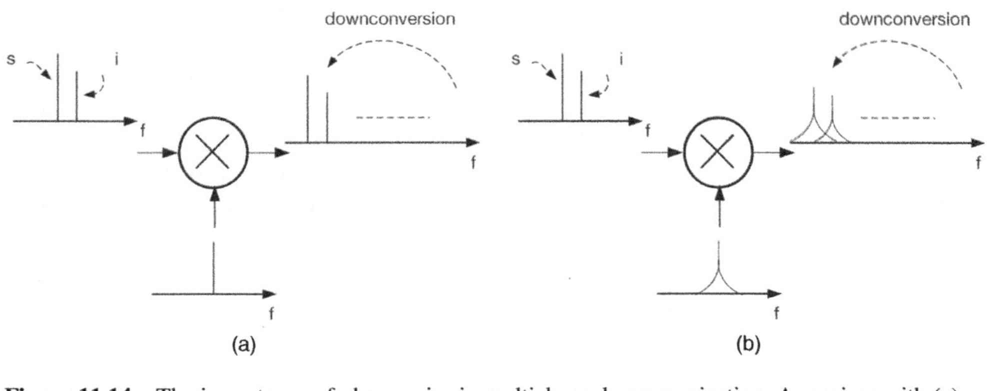

Figure 11.14 The importance of phase noise in multichannel communication. A receiver with (a) an ideal local oscillator and (b) an oscillator with phase noise.

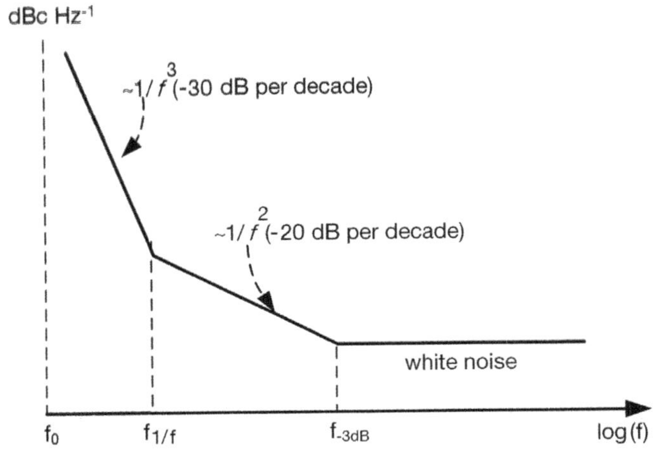

Figure 11.15 Generic noise spectrum for an oscillator.

For the sideband noise in the oscillator spectrum, the Fourier transform shows in the time domain that the oscillator signal is perturbed in both amplitude and phase by noise. The amplitude noise can be eliminated by using a limiter or automatic gain control circuit. The noise sources occurring in an oscillator are the three types which were already described in Chapter 2, namely thermal noise, shot noise and flicker noise. These noise signals are produced from active and passive devices in the oscillator circuit. The double-sideband noise spectrum is a result of nonlinear mechanisms in the circuit that upconvert these noise signals to the oscillation frequency. This double-sideband noise around the carrier frequency is called the phase noise of the oscillator. It is a key parameter that indicates the short-term stability of the oscillator. The upper and lower noise sidebands are nearly symmetrical, and the typical frequency characteristic of one sideband is shown in Figure 11.15.

The power spectrum density (in dBc Hz^{-1}) is plotted relative to the oscillation frequency f_0 on a log-frequency scale. The spectrum density near to f_0 increases to DC with a slope of f^{3} or 30 dB per decade. This comes from the $1/f$ flicker noise effect from active devices in the oscillator. The corner frequency $f_{1/f}$ in the phase noise spectrum relates to the corner frequency of the flicker noise. Thermal noise and shot noise from the devices lead to the other regions: the normally flat spectrum is shaped by the frequency response of the second-order response of the resonator, leading to a $1/f^2$ characteristic. The corner frequency, $f_{-3\mathrm{dB}}$, is in fact determined by the resonator bandwidth. Beyond this frequency, the response is not governed by the resonator and the noise spectrum density is flat, giving the noise floor of the oscillator spectrum.

Based on a linear time-invariant assumption, D.B. Leeson famously proposed the following formula for the single-sideband noise spectral density of phase noise in a feedback oscillator with a second-order resonator [4]:

$$L\left(\Delta f_m\right) = \frac{FKT}{P_S}\left(1 + \frac{\Delta f_{1/f}}{\Delta f_m}\right)\left(\left\{1 + \frac{f_0}{2Q_L \Delta f_m}\right\}^2\right) \tag{11.55}$$

where Δf_m is the offset frequency from the oscillation frequency, f_0, in Hz, Q_L is the loaded Q of the resonator, and P_S is the average power of the oscillation signal. The Leeson model provides a clear picture of which noise mechanisms produce phase noise in an oscillator. In the Leeson model, the factor $\frac{FKT}{P_S}$ is the ratio between a thermal noise source amplified by the noise figure of the oscillator F and the carrier power P_S. The factor $(1 + \frac{\Delta f_{1/f}}{\Delta f_m})$ is the flicker noise term. If the offset frequency Δf_m is less than $\Delta f_{1/f}$, that is, $\frac{\Delta f_{1/f}}{\Delta f_m} > 1$,the noise power is mainly from the flicker noise and the noise slope is –30 dB per decade. For $\Delta f_m > \Delta f_{1/f}$, the noise power is mainly from the spectrum of white noise $\frac{FKT}{P_S}$shaped by the term $\{1 + \frac{f_0}{2Q_L \Delta f_m}\}^2$.It is noted in the factor $\{1 + \frac{f_0}{2Q_L \Delta f_m}\}^2$, the term $\frac{f_0}{2Q_L}$ is the bandwidth of the second-order parallel resonant network,f_{-3dB} . The effect of term $\{1 + \frac{f_0}{2Q_L \Delta f_m}\}^2$ will be negligible when $\Delta f_m > f_{-3dB} \equiv \frac{f_0}{2Q_L}$. This means that the spectrum of phase noise is flat when the offset frequency is greater than the resonant bandwidth. It is noted that the power of 2 in the factor $\{1 + \frac{f_0}{2Q_L \Delta f_m}\}^2$ results from the second-order resonant network. If an nth-order resonant network is used, the factor term $\{1 + \frac{f_0}{2Q_L \Delta f_m}\}^2$ should be changed to $\{1 + \frac{f_0}{2Q_L \Delta f_m}\}^n$. This changes the slopes of the noise power spectrum in the flicker noise and white noise bands to n+1 and n, respectively.

11.3 Resonators

11.3.1 Dielectric Resonator

A dielectric resonator is a cylindrical resonator produced from a number of ceramic materials with low loss characteristics. The important parameters are the quality factor, Q, temperature coefficient of resonant frequency (τ_f) and dielectric constant (ε_r). The quality factor is approximately equal to the reciprocal of the loss tangent of the material. Natural material provides a trade-off between these parameters. Therefore, dielectric materials are mixed to provide new compounds that can offer optimum performance. These compound materials have high dielectric constants from 20 to 1000, and a high unloaded Q can be obtained at microwave frequencies. For example, Al_2O_3 has ε_r of 9.8 and the unloaded Q at 10 GHz is 50 000. The maximum frequency range is more than 18 GHz where the temperature coefficient is from 0 to 6. The resonant frequency of a dielectric resonator depends on its dimensions and surroundings. The range of resonant frequency of commercial dielectric resonators is from 1 GHz up to 100 GHz. The dielectric resonator provides a highly stabilised oscillator solution and has been long established for microwave and millimetre-wave applications.

11.3.2 Transmission Line Resonator

A transmission line can be used as a resonator for microwave up to millimetre-wave applications by open- or short-circuiting the ends. If a transmission line is lossless, the input impedance (Z_{in}) of the transmission line of characteristic impedance Z_0 and length l, terminated with a complex load Z_L, is given by:

$$Z_{in} = Z_0 \frac{Z_L + jZ_0 \tan(\beta l)}{Z_0 + jZ_L \tan(\beta l)} \tag{11.56}$$

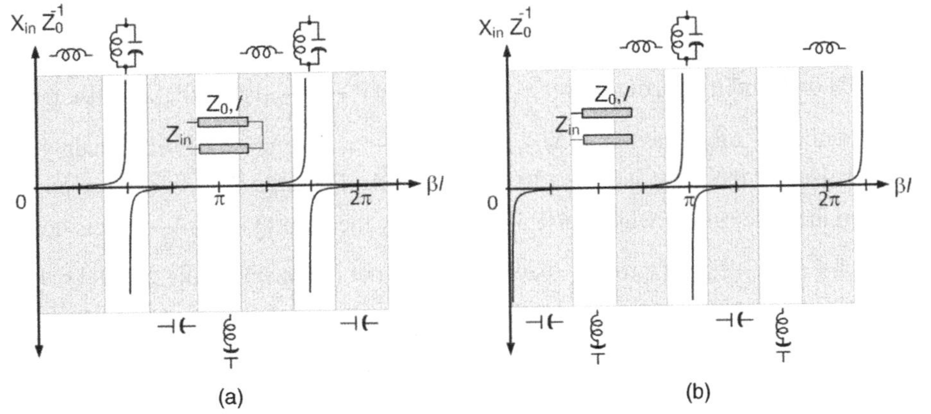

Figure 11.16 Characteristics of (a) short-circuited and (b) open-circuited transmission lines.

Hence, a transmission line resonator can be implemented in two main forms, as shown in Figure 11.16, and the driving-point impedances for the short and open-end transmission line resonator are:

$$Z_{in,short} = jZ_0 \tan(\beta l) \tag{11.57}$$

$$Z_{in,open} = jZ_0 \cot(\beta l) \tag{11.58}$$

The tangent and cotangent characteristics of input impedance of the short- and open-end transmission lines depend on the product between the phase constant, β, and l. The reactance functions are π periodic. Figure 11.16 plots the reactances normalised with Z_0 against the electrical length of the transmission line (βl) from 0 to 2π. A short- or open-circuited transmission line resonator behaves as a resonator when $\beta_0 l = \frac{n\pi}{2}$, where β_0 is the phase constant at the resonant frequency, ω_0. For compactness and low attenuation, we should implement either a parallel or series L-C network resonating at ω_0 with a quarter-wave transmission line. Therefore, a parallel resonator is usually implemented with a short-circuited quarter-wave transmission line while a quarter-wave open-circuited transmission line is chosen to implement a series resonator. The quality factor of a transmission line resonator is computed from the transmission line propagation constant, γ:

$$\gamma = \alpha + j\beta \tag{11.59}$$

where α is the attenuation constant. For a short-circuited transmission line resonator, the input impedance is given by:

$$Z_{in} = Z_0 \frac{1 - j\cot(\beta l)\tanh(\alpha l)}{\tanh(\alpha l) - j\cot(\beta l)} \tag{11.60}$$

For a quarter-wave transmission line, that is, $l = \frac{\lambda_g}{4}$, its phase angle is:

$$\beta l = \frac{\omega}{v_p}\frac{\lambda_g}{4} \tag{11.61}$$

The guided wavelength λ_g is related to v_p and ω_0 by:

$$\lambda_g = 2\pi\frac{v_p}{\omega_0} \tag{11.62}$$

Substituting Equation (11.62) into Equation (11.61) and simplifying, the phase angle is:

$$\beta l = \frac{\omega}{\omega_0}\frac{\pi}{2} \tag{11.63}$$

At an angular frequency near ω_0, that is, $\omega = \omega_0 + \Delta\omega$, we have:

$$\beta l = \frac{\omega_0 + \Delta\omega}{\omega_0}\frac{\pi}{2} = \frac{\pi}{2}\left(1 + \frac{\Delta\omega}{\omega_0}\right) \tag{11.64}$$

Substituting Equation (11.64) into Equation (11.60) and simplifying, this leads to:

$$Z_{in} \approx \frac{Z_0}{\alpha l + j\frac{\pi\Delta\omega}{22\omega_0}} \tag{11.65}$$

Therefore, the input impedance and the quality factor of a parallel resonant network at frequency near ω_0 are given by:

$$Z_{in} = \frac{R}{1 + j2\Delta\omega RC} \tag{11.66}$$

$$Q = \omega RC \tag{11.67}$$

Therefore, the quality factor of a quarter-wave short-circuited transmission line at the resonant frequency is:

$$Q\big|_{\omega_0} = \frac{\beta_0}{2\alpha} \tag{11.68}$$

For a quarter-wave open-circuited transmission line resonator, it can be shown that the quality factor is also given by Equation (11.68).

11.3.3 Quartz Crystal Resonator

In the frequency-control market, the quartz crystal is by far the mostly popular resonator technology due to its accuracy, environmental stability and very low cost. They have not only been used in consumer products such as smart phones and digital cameras in billion-plus

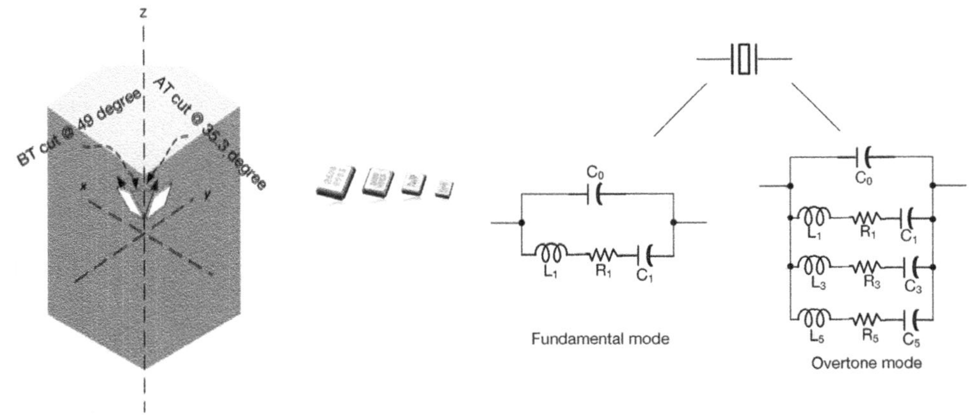

Figure 11.17 Quartz crystal, AT and BT cuts, and equivalent circuits.

large-volume production, but also appear in base stations and network equipment. The crystal device uses the piezoelectric effect to generate a highly stable signal. The external electric field applied to the device causes deformation in the crystal structure. A signal with high frequency precision and accuracy results from this atomic vibration. The signal frequency, specified as the natural frequency of the device, depends on the crystal shape, cut and size. There are several cutting styles which depend on the specific angle relative to the crystallographic axes. Different crystal cuts lead to several vibration modes which influence the frequency stability and thermal characteristics. The AT- and BT-cut quartz crystals, which are 35.3° and 49.0° angle cuts from the z-axis (as shown in Figure 11.17) are mostly used due to their high-temperature stability. The thickness of a finished quartz device determines the resonance frequency. If a quartz crystal with a high resonant frequency is desired, a thin cut will be needed. Other cuts are also available but there are some trade-offs with temperature stability and some properties related to mechanical-electrical characteristics.

The symbols for a quartz crystal and its equivalent circuit for the fundamental and overtone modes are shown in Figure 11.17. The circuit model defined for the fundamental mode is composed of a resonant circuit which can resonate in series or parallel [5]. L_1 and C_1 are called the motional inductance and capacitance. The capacitor C_0 is from the electrodes, package and the leads. The value of C_0 depends on the crystal size and packaging type. The resistor R_1 represents a bulk loss and has to be taken into account in the design to prevent the failure of oscillation start-up. The first frequency resonance is from the series resonance of L_1 and C_1, that is:

$$f_1 = \frac{1}{2\pi\sqrt{L_1 C_1}}. \tag{11.69}$$

The second resonance is a parallel resonance with frequency:

$$f_2 = \frac{1}{2\pi\sqrt{L_1 \frac{C_0 C_1}{C_0 + C_1}}}. \tag{11.70}$$

The parallel mode provides a higher resonant frequency than the series resonance. If the parallel resonance mode is used, a proper loading capacitor should be connected in parallel with the quartz crystal. Apart from the fundamental mode operation, a quartz crystal resonator can resonate at higher frequency in the overtone mode. This mode is actually the harmonic mode which occurs at only odd multiples, that is, third, fifth, and so forth. Such modes are inherent in all quartz crystals. For a high-frequency crystal-stabilised oscillator design that needs to prioritise cost, using an overtone-mode crystal rather than an expensive high-frequency operated fundamental model device would fulfil the need.

11.4 Some Oscillator Circuits

11.4.1 Colpitts

11.4.1.1 Common-Base Configuration

The classic Colpitts oscillator is an *L-C* oscillator that is used in many applications from low frequency up to the millimetre-wave range. The Colpitts oscillator has gained much interest as only one inductor is needed (ignoring bias networks). Figure 11.18 shows a Colpitts oscillator designed with a BJT. A common-base topology provides the necessary positive feedback and the transistor base is AC-grounded; for simplicity's sake, the DC bias circuitry is not shown. The oscillation frequency is controlled by a resonant network formed from the capacitor voltage divider, C_1 and C_2, and the inductor L. The output of the Colpitts oscillator is connected to a resistive load, R_L. Assuming that the oscillation frequency is in the mid-range of the transistor's operating frequency, and that the inductor has the resistive loss R_s and the dielectric losses in C_1 and C_2 are negligible, the AC small-signal model of a Colpitts oscillator is as shown in Figure 11.18(b). The reference plane a-a′ is defined at the interface between the voltage-controlled current source with transconductance g_m, and the resonant network is loaded with R_L.

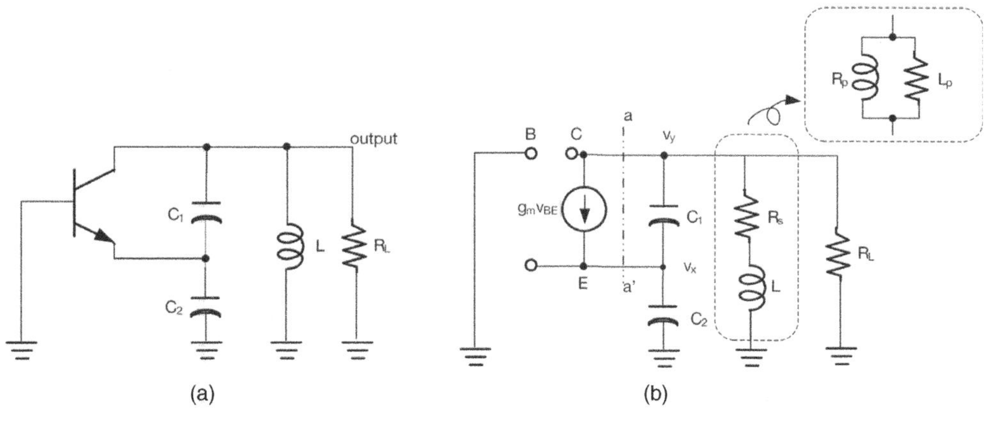

Figure 11.18 Schematic of (a) Colpitts oscillator, and (b) its AC small-signal model.

The series RL network from the resistive loss in the inductor can be converted to a parallel RL network as follows:

$$L_P = L\left(1 + \frac{1}{Q^2}\right) \quad (11.71)$$

$$R_P = R_s(1 + Q^2) \quad (11.72)$$

where

$$Q = \frac{R_P}{\omega L_P} = \frac{\omega L}{R_S} \quad (11.73)$$

If the inductor has a large quality factor, that is, $Q \gg 1$, L_P and R_P can be approximated by:

$$L_P \approx L \quad (11.74)$$

$$R_P \approx R_s Q^2 \quad (11.75)$$

This parallel resistance shunts with R_L. Therefore, the total parallel resistor R_{eff} can be approximated by:

$$R_{eff} \approx \frac{R_L R_s Q^2}{R_L + R_s Q^2} \quad (11.76)$$

The finalised AC model for the Colpitts oscillator is shown in Figure 11.19(a). At node v_y we obtain:

$$\left(R_{eff}^{-1} + [j\omega L]^{-1} + j\omega C_1\right)v_y - j\omega C_1 v_x = g_m v_x \quad (11.77)$$

At node v_x we obtain:

$$j\omega(C_1 + C_2)v_x - j\omega C_1 v_y = -g_m v_x \quad (11.78)$$

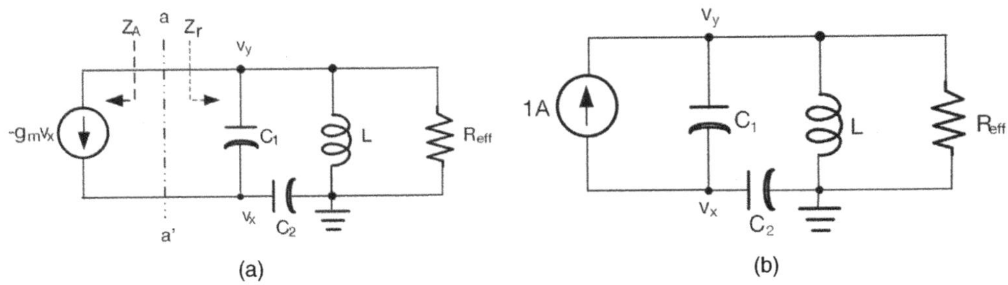

Figure 11.19 Colpitts oscillator. (a) Simplified AC model. (b) Equivalent circuit for calculating $Z_r(\omega)$.

Hence,

$$\frac{v_y}{v_x} = \frac{g_m + j\omega(C_1 + C_2)}{j\omega C_1} = \frac{C_1 + C_2}{C_1} - j\frac{g_m}{\omega C_1} \tag{11.79}$$

The driving-point impedance $Z_A(\omega)$ is:

$$Z_A(\omega) \equiv \frac{v_y - v_x}{-g_m v_x} = -\frac{C_2}{g_m C_1} + j\frac{1}{\omega C_1} \tag{11.80}$$

Injecting a current source of 1 Ampere, as shown in Figure 11.19(b), the impedance Z_A is calculated from the difference voltage $v_y - v_x$. At node v_y,

$$\left(R_{eff}^{-1} + j\left(\omega C_1 - \frac{1}{\omega L}\right)\right) v_y - j\omega C_1 v_X = 1 \tag{11.81}$$

At node v_x,

$$j\omega(C_1 + C_2)v_x - j\omega C_1 v_y = -1 \tag{11.82}$$

By equating Equations (11.79) and (11.80) and rearranging, this leads to:

$$\left(R_{eff}^{-1} + j\left(\omega C_1 - \frac{1}{\omega L}\right)\right) v_y - j\omega C_1 v_X = -j\omega(C_1 + C_2)v_x + j\omega C_1 v_y \tag{11.83}$$

Hence,

$$v_y = \frac{-j\omega C_2}{R_{eff}^{-1} - j\frac{1}{\omega L}} v_x \tag{11.84}$$

Substitute into equation node v_x,

$$v_x = \frac{1}{\frac{\omega^2 C_1 C_2}{R_{eff}^{-1} - j\frac{1}{\omega L}} - j\omega(C_1 + C_2)} \tag{11.85}$$

Then,

$$v_y = \frac{-j\omega C_2}{R_{eff}^{-1} - j\frac{1}{\omega L}} \frac{1}{\frac{\omega^2 C_1 C_2}{R_{eff}^{-1} - j\frac{1}{\omega L}} - j\omega(C_1 + C_2)} \tag{11.86}$$

Thus,

$$Z_r \equiv v_y - v_x = \frac{-R_{eff}^{-1} + j\left(\frac{1}{\omega L} - \omega C_2\right)}{\omega^2 C_1 C_2 - \frac{C_1 + C_2}{L} - j\omega(C_1 + C_2)R_{eff}^{-1}} \tag{11.87}$$

If the inductor has large Q, X_r is:

$$X_r = \frac{\left(\omega^2 C_1 C_2 - \frac{C_1+C_2}{L}\right)\left(\frac{1}{\omega L} - \omega C_2\right)}{\left(\omega^2 C_1 C_2 - \frac{C_1+C_2}{L}\right)^2} \tag{11.88}$$

At the resonance frequency, the total reactance of the oscillator circuit shown in Equation (11.88) should be zero. Hence, the circuit will oscillate at ω_0 which is determined by:

$$\omega_0 = \frac{1}{\sqrt{LC_{eff}}} \tag{11.89}$$

where

$$C_{eff} = \frac{2C_1 C_2}{2C_1 + C_2} \tag{11.90}$$

11.4.1.2 Common-Emitter Configuration

A common-emitter Colpitts oscillator is shown in Figure 11.20(a). The DC bias circuitry is not shown here to simplify the analysis. The capacitive voltage divider is connected between the collector and base. The inductor L is to create a resonance at the desired oscillating frequency. The AC equivalent circuit model is shown in Figure 11.20(b). The magnitude of the current source is controlled by the voltage at capacitor C_2, which is v_x. The biasing point of the transistor defines the transconductance, g_m. Applying the negative resistance technique, the total reactance in the circuit at the resonant frequency is zero. This total reactance is computed from the reactance of *RL* series circuit and the reactance seen from the series *RL* subnetwork. The impedance of the series *RL* network is:

$$Z_{RL}(\omega) = R_s + j\omega L \tag{11.91}$$

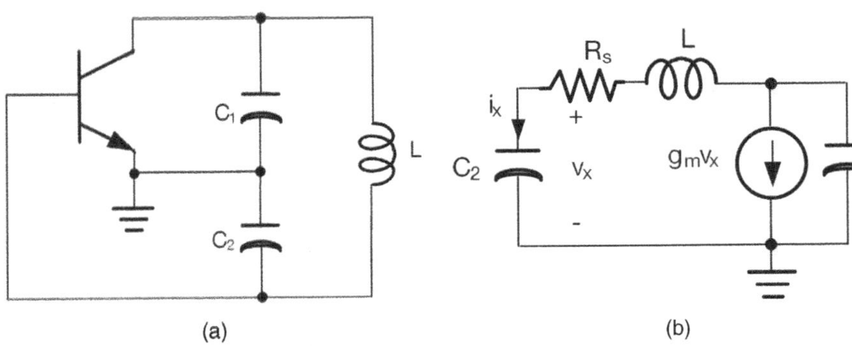

Figure 11.20 Common-emitter Colpitts oscillator. (a) Circuit schematic. (b) AC model.

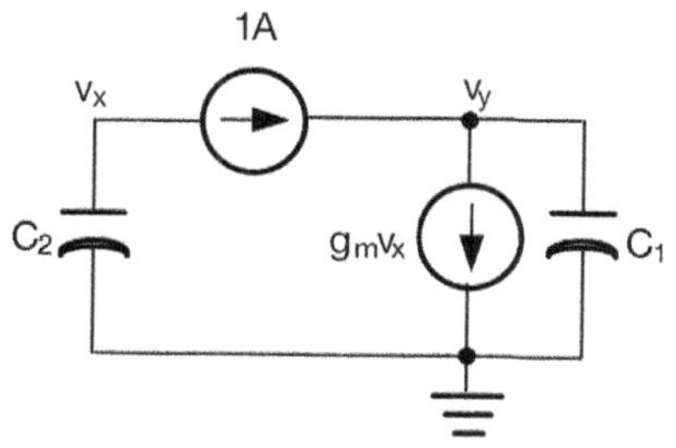

Figure 11.21 Circuit model for computing driving-point impedance seen from the *RL* subnetwork.

The driving-point impedance seen from the series *RL* subnetwork is computed from the model shown in Figure 11.21. Here, a constant current source of 1 A is injected between node v_x and v_y. From Figure 11.21, the voltage v_x is then:

$$v_x = \frac{-1}{j\omega C_2} \tag{11.92}$$

Applying Kirchhoff's current law at node v_y, we obtain:

$$j\omega C_1 v_y = 1 + \frac{g_m}{j\omega C_2} \tag{11.93}$$

Therefore, the driving-point impedance is:

$$Z_D(\omega) \equiv v_y - v_x = \frac{1}{j\omega C_1} + \frac{-g_m}{\omega^2 C_1 C_2} + \frac{1}{j\omega C_2} \tag{11.94}$$

To allow the oscillator to start-up at ω_0, the transconductance of the transistor should satisfy the following condition:

$$g_m > \omega_0^2 R_s C_1 C_2 \tag{11.95}$$

The oscillating frequency of the common-emitter Colpitts oscillator can be computed when the total reactance at ω_0 is zero, that is:

$$X_{RL}(\omega_0) + X_D(\omega_0) \equiv \omega_0 L - \left(\frac{1}{\omega_0 C_1} + \frac{1}{\omega_0 C_2}\right) = 0 \tag{11.96}$$

Hence, the value of ω_0 is calculated from:

$$\omega_0 = \frac{1}{\sqrt{LC_{eff}}} \tag{11.97}$$

where

$$C_{eff} = \frac{C_1 C_2}{C_1 + C_2} \tag{11.98}$$

11.4.2 Cross-Coupled Oscillator

A cross-coupled oscillator is routinely used in CMOS integrated circuit technology. The concept can be easily explained with the positive feedback theory (two-port oscillator model). The oscillator is formed by using an identical tuned amplifier as the active circuit and feedback network. A tuned amplifier, usually found in radio and TV applications, employs a resonant-circuit load to achieve a narrowband amplifier. The schematic of a CMOS-tuned amplifier is shown in Figure 11.22(a). The FET M_1 is common-source biased. A small-signal input is applied at the gate and the output signal is obtained at the drain. The tank circuit at the drain allows narrowband amplification. The resistor R_p represents mainly the loss of the inductor, not a real physical resistor. The frequency-selective amplification is obviously evident when considering the small-signal circuit model shown in Figure 11.22(b). The small-signal voltage gain of the tuned amplifier is:

$$A_v(\omega) \equiv \frac{v_o(\omega)}{v_i(\omega)} = \frac{-g_m}{\frac{1}{R_p} + j\left(\omega C - \frac{1}{\omega L}\right)} \tag{11.99}$$

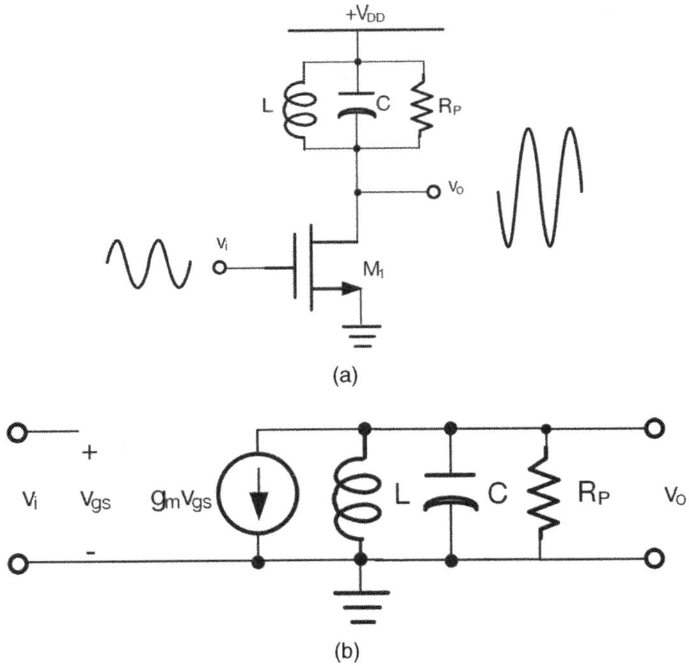

Figure 11.22 Schematic of (a) the tuned amplifier, and (b) its AC small-signal model.

Hence, the magnitude and phase response of the tuned amplifier are:

$$|A_v(\omega)| = \frac{g_m}{\sqrt{\left(\frac{1}{R_p}\right)^2 + \left(\omega C - \frac{1}{\omega L}\right)^2}} \tag{11.100}$$

$$\phi_A(\omega) = \pi - \tan^{-1}\frac{\omega C - \frac{1}{\omega L}}{1/R_p} \tag{11.101}$$

As stated previously, the cross-coupled oscillator is realised using the same tuned amplifier for both the active circuit and the feedback network, as shown in the conceptual diagram in Figure 11.23(a), giving the circuit shown in Figure 11.23(b). The two transistors M_1 and M_2 are assumed to be identical. Therefore,

$$A_{OL}(\omega) = B(\omega) = A_v(\omega) \tag{11.102}$$

The oscillation frequency of a cross-coupled oscillator is equal to the resonant frequency of the tuned amplifier composing the circuit. Applying the oscillation condition based on the two-port oscillator model, the magnitude condition at the oscillation frequency provides the design equation:

$$g_m = \frac{1}{R_p} \tag{11.103}$$

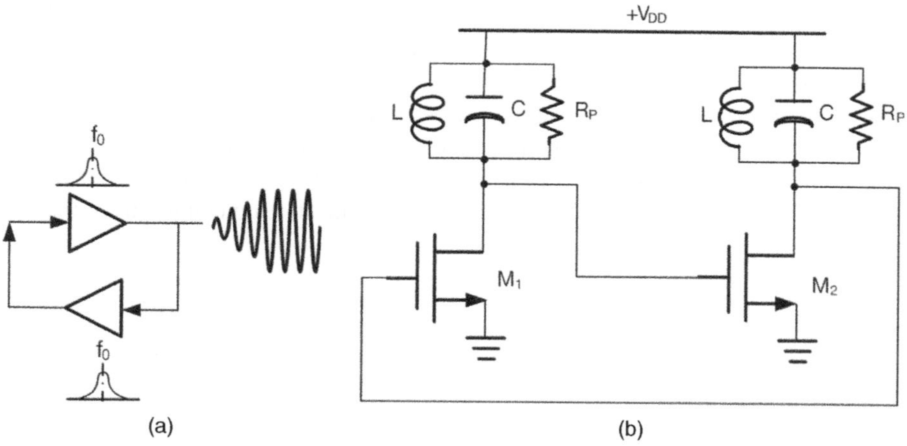

Figure 11.23 Cross-coupled oscillator formed from two tuned amplifiers. (a) Conceptual diagram. (b) Schematic circuit diagram.

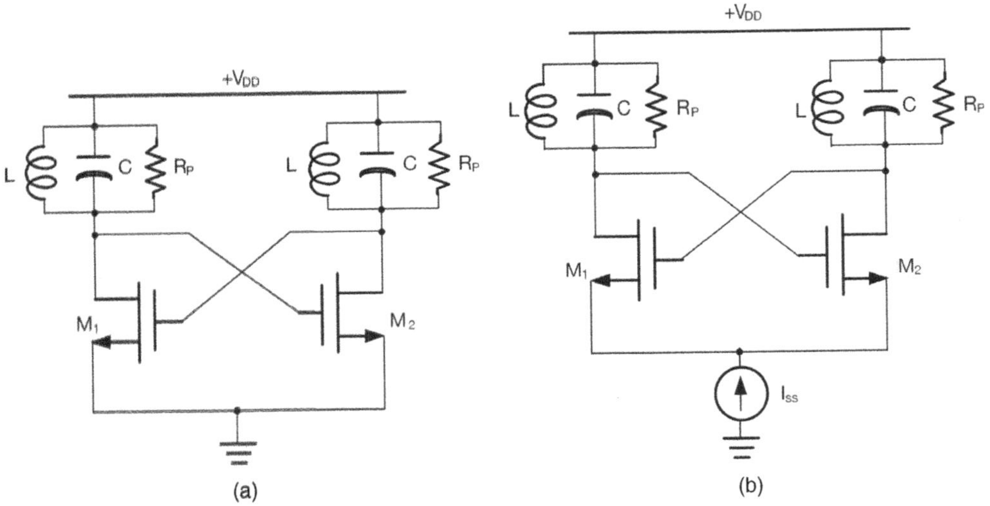

Figure 11.24 Cross-coupled oscillator circuit with (a) emitter-to-ground and (b) current source-to-ground.

where the phase condition confirms the positive feedback in the closed loop and is obtained from:

$$\phi_A(\omega_0) + \phi_F(\omega_0) = 2\pi \tag{11.104}$$

If the cross-coupled oscillator circuit shown in Figure 11.23(b) is rearranged by vertically flipping the transistor M_1, the circuit shown in Figure 11.24(a) is obtained. This topology will be immediately recognisable to RFIC design engineers and the term 'cross-coupled' comes from the arrangement of the two transistors. Better phase-noise performance can be obtained if a constant current source, I_{ss}, replaces the grounding connection at the sources, as shown in Figure 11.24(b).

For this circuit arrangement, the oscillator model based on a one-port network can also be applied to derive the necessary condition. This may be done by breaking the loop at the transistor drains, as shown by the reference plane a-a′ in Figure 11.25. The driving-point impedance at the oscillation frequency is then derived. The start-up condition for the cross-coupled oscillator is:

$$g_m > \frac{1}{R_p} \tag{11.105}$$

Equation (11.105) implies that the circuit can oscillate if the resonator loss R_p is compensated by the transistor transconductance. This condition is met by selecting a high-Q resonator or applying an appropriate transistor biasing point.

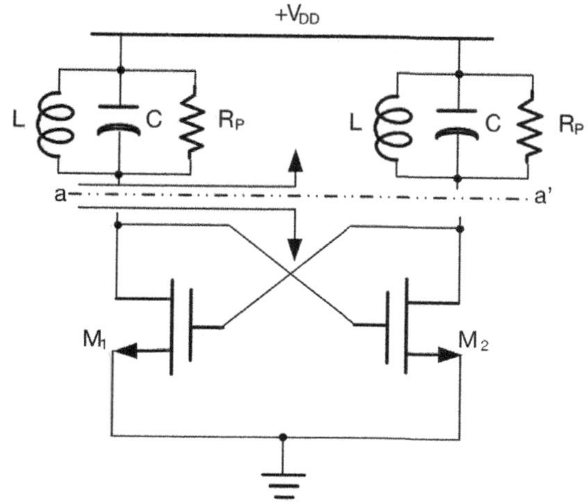

Figure 11.25 Applying the one-port oscillator model (negative resistance technique) to the cross-coupled oscillator.

11.4.3 Voltage-Controlled Oscillator

A voltage-controlled oscillator (VCO) is a free-running oscillator that has tuning capability and is an important building block for a wireless transceiver. The tuning of a VCO can be implemented by adding a tuning mechanism to the resonator. This can be achieved by adding tuning devices that may be electrical or mechanical to a fixed-frequency oscillator circuit. A variable capacitor is commonly used rather than a variable inductor, due to its simpler implementation. In designing a VCO, one of the most important design parameters of the VCO is the VCO gain (K_v) in Hz V^{-1}, or in Hz A^{-1} if a current tuning technique is used. Here, we assume that VCO tuning is performed with a control voltage. The degree of frequency change Δf is related to the tuning voltage by:

$$\Delta f = K_v v_{ctrl} \tag{11.106}$$

A large VCO gain means a large tuning range. The tuning range (TR) of a VCO is calculated from the total frequency range and the average operating frequency:

$$TR = \frac{f_{max} - f_{min}}{\left(\frac{f_{max}+f_{min}}{2}\right)} \tag{11.107}$$

Generally, a typical value of TR for a VCO is around 20–30%, depending on the impedance range of the tuning device. An electrical tuning device like a varactor diode is far more practical than any mechanical device. Using a simpler electrical tuning device provides a simple solution for integration with other circuitry. In some IC technologies, a mechanical tuning device like a MEMS varactor may be available, but this requires a special fabrication process, leading to higher costs. A tuneable capacitor can be easily obtained by applying a reverse DC bias voltage to a PN-junction or MOS diode in a standard IC technology, and this is the most common solution for wireless communications applications.

We will first consider a PN junction diode, made from P- and N-type semiconductor junctions. When applying a reverse bias to the PN junction, a depletion region is created which causes a junction capacitance that acts like a parallel capacitance. The junction capacitance (C_j) is related to the reverse-bias junction voltage (V_j) by:

$$C_j = \frac{C_{j0}}{\left(1 - \frac{V_j}{\varphi_0}\right)^{\gamma}} \tag{11.108}$$

where γ depends on the device doping profile, C_{j0} is a junction capacitance at zero bias, and φ_0 is the barrier potential of the junction.

The Q of a PN junction diode, especially at high frequency, is often low due to the bulk resistance, and its tuning capacitance ratio is limited. Therefore, a VCO design for high-frequency application using the PN junction diode may not meet the target specifications in terms of phase noise and TR.

The MOS varactor is an alternate solution, with which the tuning capacitance can reach 2:1 and the quality factor is better than the PN-junction diode. The MOS varactor and its cross-section are shown in Figure 11.26. The MOS varactor is usually on a p-substrate where the device is grown in an n-well. The drain and source channels are heavily doped with n-type material (n+), and are connected together as bulk (B). When a forward bias is applied, the capacitance that appears is from the oxide layer. On the other hand, when a reverse bias is applied to the varactor terminal the depletion region is developed in the n-well. The junction capacitance formed by the depletion region and the oxide layer depends on the dimensions.

An accurate electrical model for the varactor diode is required to predict VCO performance parameters such as K_V, tuning range, phase noise, oscillation amplitude and power consumption. For IC design, most foundries supply a parameterised circuit model for a diode. If a VCO is designed in hybrid form (e.g., microstrip on softboard), high-performance discrete varactor diodes are available, usually encapsulated in surface mount technology (SMT) form. The varactor diode model including parasitic effects from the diode package is then needed from the manufacturer. In either case, the equivalent circuit model of the varactor diode and any parasitic elements must be used for the design. As described earlier, a VCO design can

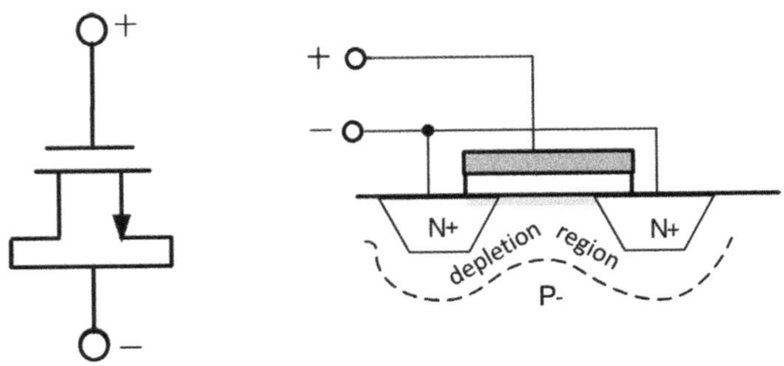

Figure 11.26 MOS varactor symbol and device cross-section.

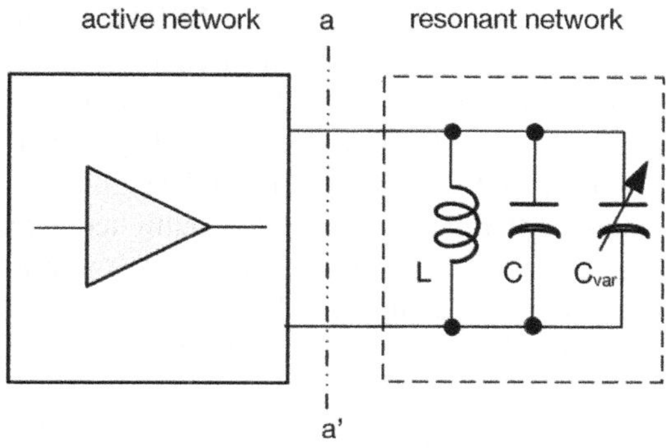

Figure 11.27 Conceptual diagram of a VCO.

be realised by putting the tuning device in the resonant circuit. The conceptual diagram of a VCO design based on the one-port oscillator model is shown in Figure 11.27. The schematic of this VCO is similar to the fixed oscillator, with the fixed capacitor in the resonator network changed to a tuning capacitor.

11.4.4 Phase-Locked Loops

In a wireless communication system that supports multiple users or devices, a frequency-tuneable oscillator is required in the transmitter and receiver. In such a case, a fixed-frequency, highly stable, stand-alone source such as a crystal-controlled or dielectric-resonator stabilised oscillator is not, by itself, suitable. On the other hand, while a VCO provides frequency tuning capability, it is very sensitive to environmental changes. The DC control voltage input of a VCO is susceptible to drift, noise, EMI (both conduction and emission effects), and other interference. All of these effects make the frequency unstable, which in turn degrades the communication performance. A phase-locked loop (PLL) system is a solution that provides both frequency-tuning capability and a high-stability (low phase noise and jitter) signal.

The block diagram of a basic PLL is shown in Figure 11.28. The term 'phase-locked' indicates that the technique employs a phase detector to lock a poorly stabilised high-frequency signal to a very stable fixed low-frequency signal. The system is a feedback control system. The reference oscillator generates a high-stability and high-purity signal at a fixed frequency f_{ref}. At low frequency, it is not difficult to design a high-stability oscillator since a high-Q

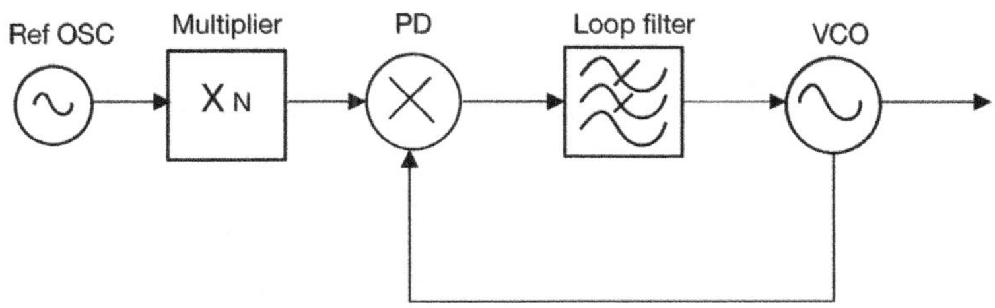

Figure 11.28 Phase-locked loop block diagram.

resonator technology such as quartz, ceramic, or even MEMs, is generally available. The signal at frequency ω_{ref} is multiplied up N-times with a frequency multiplier. In Figure 11.28, the output frequency from the multiplier $v_{Mul}(t)$ will be equal to the output frequency ω_0 of the PLL system, that is, $\omega_o = N\omega_{ref}$. The frequency-multiplied signal is applied to a phase detector (PD), which is implemented with a frequency mixer in microwave and millimetre-wave applications. The phase difference between the multiplied reference signal and the feedback output signal $v_f(t)$ is processed in the PD block. Without any loss of generality, the amplitudes of $v_{Mul}(t)$ and $v_f(t)$ are assumed to be equal to 2 and 1, respectively. Hence, the time-domain output signal of PD, $v_{PD}(t)$, can be written as:

$$v_{PD}(t) = v_{Mul}(t) \cdot v_f(t) = 2\cos(\omega_o t + \varphi_{Mul}(t)) \cdot \sin(\omega_o t + \varphi_f(t)) \tag{11.109}$$

Using the trigonometric function, the result is given by:

$$v_{PD}(t) = \sin(\varphi_{Mul}(t) - \varphi_f(t)) + \sin(2\omega_o t + \varphi_{Mul}(t) + \varphi_f(t)) \tag{11.110}$$

As clearly shown in Equation (11.110), the PD output has two components, a low-frequency and high-frequency signal component. The output from the PD is applied to the loop filter (LF) which is low-pass type. The high-frequency spectrum component in Equation (11.110), which is much higher than the cut-off frequency ($f_{\text{c,LF}}$) of the LF, is thus eliminated. If the phase difference is sufficiently small the output signal of the LF, which has a low-pass function, will be a DC signal:

$$v_{LF}(t) \approx \lim_{T \to 1/f_{c,LF}} \frac{1}{T} \int_{-T/2}^{+T/2} \{\varphi_{Mul}(t) - \varphi_f(t)\}\, dt \tag{11.111}$$

The DC signal is obtained by integrating the phase noise difference over the period $1/T$, or the bandwidth of LF ($f_{\text{c,LF}}$). If the phase noise of the reference oscillator is excellent, the instantaneous DC output of the LF will depend on the average of phase noise of the feedback signal from the VCO over the range $f_{\text{c,LF}}$. This DC signal is applied to the VCO where its centre frequency is tuned at carrier frequency and is linearly increased to the DC input with the VCO gain K_V (in rad s^{-1} V^{-1}) by:

$$\omega_o(t) = \omega_c + K_V \lim_{T \to \infty} \frac{1}{T} \int_{-T/2}^{+T/2} \{\varphi_{Mul}(t) - \varphi_f(t)\}\, dt \tag{11.112}$$

Finally, the output of this PLL system is:

$$v_{PLL}(t) = \sin\left[\omega_c t + K_V \lim_{T \to \infty} \frac{1}{T} \int_{-T/2}^{+T/2} \{\varphi_{Mul}(t) - \varphi_f(t)\}\, dt\right] \tag{11.113}$$

Equation (11.113) shows that the frequency of the signal output is controlled by an instantaneous DC voltage from the LF. This DC signal tracks the average of the phase difference between the phase of the reference signal and the phase of the feedback signal. With the integration function, which is naturally a low-pass filter function, the amplitude fluctuation

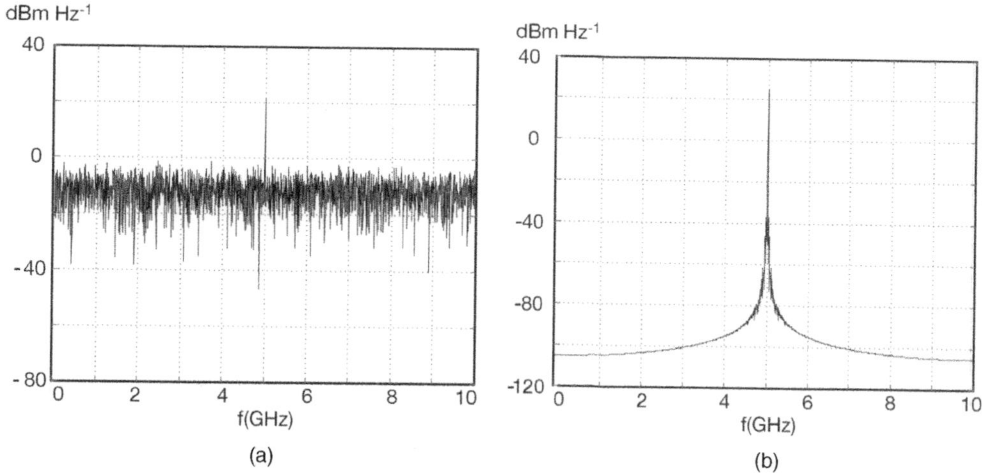

Figure 11.29 Spectrum of (a) free-running VCO with random white phase noise, and (b) PLL output.

and the phase noise occurred in the reference oscillator and the VCO are eventually eliminated when the PLL approaches steady state. If the PLL is stable, the output signal from the PLL will have very low phase noise and the oscillation frequency over the PLL bandwidth is stabilised. The signal from the PLL shown in Equation (11.113) is plotted against a free-running VCO at the same oscillation frequency. We assume that the carrier frequency is 5 GHz and the phase noise of the free-running VCO is white noise with zero mean and variance 1. The output spectrum of the free-running VCO is shown in Figure 11.29, which clearly shows the noise power. Applying a third-order Butterworth LPF at $f_c = 20$ MHz and assuming that the VCO gain is 1, the spectra were calculated using MATLAB™. In Figure 11.29(a), the free-running oscillator has a poor carrier-to-noise ratio, while the carrier-to-noise ratio of the phase-locked carrier is enhanced considerably.

11.4.4.1 PLL System Applied to Coherent Demodulator

In a communications receiver that uses a coherent demodulator, the oscillator required for demodulation has to be phase-synchronised with the received signal, otherwise the SNR in the receiver will be degraded. To illustrate this, assume that a received AM voltage signal denoted by $v_{AM}(t)$ is given by:

$$v_{AM}(t) = A_m(t)\cos(\omega_c t) \tag{11.114}$$

Given that the information is sinusoidal signal with frequency ω_m and a carrier at frequency is ω_c.The oscillator voltage at the receiver is assumed to be:

$$v_{osc}(t) = 2\sin(\omega_c t + \emptyset) \tag{11.115}$$

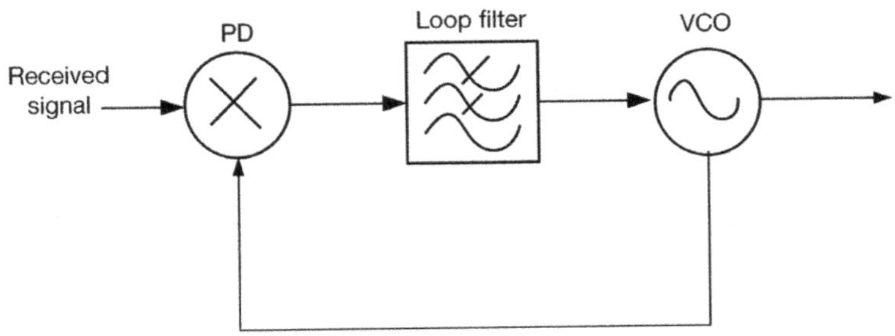

Figure 11.30 Phase recovery PLL.

It should be noted that the amplitude 2 in Equation (11.115) is chosen for the sake of mathematical simplification. The demodulated output is then:

$$v_{demod}(t) = 2\mathrm{A}_m(\mathrm{t})\cos(\omega_c t)\sin(\omega_c t + \emptyset) \tag{11.116}$$

After filtering with a LPF, the output is:

$$v_{LPF}(t) = \mathrm{A}_m(\mathrm{t})\sin(\emptyset) \tag{11.117}$$

The information signal can be perfectly reconstructed if the oscillator phase at the demodulator is coherent with the carrier phase at the receiver input. Otherwise, the signal will be degraded. To clarify this, if Ø in Equation (11.117) is zero or 180°, the voltage at LPF will be zero. The signal can be perfectly reconstructed when $\emptyset = 90$, that is $v_{osc}(t) = 2\cos\left(\omega_c t\right)$. In this case, we need a special circuit that can generate a phase-synchronised oscillator and, certainly, a PLL system can solve the issue.

An application of the PLL for phase recovery system is shown in Figure 11.30. In this system, the received signal in a receiver is applied to the PD. Given that the signal is modulated with phase-shift keying as shown, the input signal is assumed to be:

$$v_{psk}(t) = 2\cos(\omega_c t + \Theta) \tag{11.118}$$

The phase difference between the feedback carrier from the VCO and the received signal is applied to the LF. In this system, the VCO is pretuned at the carrier frequency which is shown in Equation (11.112). Hence, the output from PD is the product between $v_{AM}(t)$ and $v_{osc}(t)$, this leads to:

$$v_{PD}(t) = 2\mathrm{A}_m(\mathrm{t})\cos(\omega_c t + \Theta)\sin(\omega_c t + \emptyset) \tag{11.119}$$

Only the low-pass signal is considered which, in practice, is limited by the PD bandwidth. This gives:

$$v_{PD}(t) = \mathrm{A}_m(\mathrm{t})\sin(\emptyset - \Theta) \tag{11.120}$$

In a mathematical sense, the output of the LF is equivalent to integrating Equation (11.120) over the filter bandwidth (1/*T*):

$$v_{LF}(t) = \frac{1}{T}\int_{-T/2}^{T/2} [\mathrm{A}_m(\mathrm{t})\sin(\emptyset - \Theta)]\,\mathrm{dt} \tag{11.121}$$

If the PLL is designed properly, the system will track the input phase. The equation above will be simplified to:

$$v_{LF}(t) \approx \frac{1}{T}\int_{-T/2}^{T/2} \left\{A_m(t)\left[\emptyset - \Theta\right]\right\}\mathrm{dt} \tag{11.122}$$

The term $\emptyset - \Theta$ is the phase difference between the phases of the received and free-running signals. If the PLL is designed properly, the PLL system will track the received signal phase and this leads to a reduced DC signal until zero is reached. Hence, the locked condition is reached when,

$$\emptyset = \Theta \tag{11.123}$$

This shows the application of PLL as a phase-recovery system.

11.4.4.2 PLL Frequency Synthesiser

To tune a wireless transceiver basically requires a low-phase noise oscillator that is highly stable and yet frequency-tuneable with very small frequency steps. The subsystem that provides this function is called a 'frequency synthesiser', which uses a PLL block. Many modern communication standards use sophisticated techniques such as frequency hopping and OFDM to counteract interference and signal impairments in the hostile wireless environment. Therefore, a minimum set of specifications that today need not only include spectral purity (phase noise and spurious) and frequency shifting but also covers fast frequency switching, low power consumption and wide bandwidth.

One of the most common PLL-based synthesisers is an integer-N PLL. The block diagram of the synthesiser is shown in Figure 11.31. There is a frequency divider with the predetermined integer division ratio, *N*, in the feedback path. The reference signal is a stabilised low-frequency oscillator, for example using a quartz crystal. The loop locks the VCO to the reference signal such that the output frequency is:

$$f_{out} = Nf_{ref} \tag{11.124}$$

With a programmable divider, *N* can be changed arbitrarily. Therefore, the output frequency of the synthesiser can be tuned to be any integer multiple of the reference frequency. The frequency step size of the integer-*N* synthesiser is, therefore, equal to the reference frequency. One of the other main disadvantages of the integer-*N* synthesiser is its low switching speed. Improving this speed requires wider bandwidth of the loop filter which increases phase noise and degrades

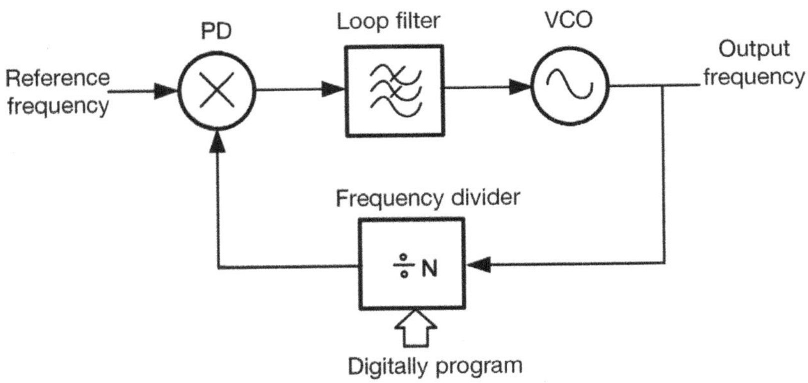

Figure 11.31 An integer-N PLL-based frequency synthesiser.

the loop stability. Many designers now employ a fractional-N PLL-based frequency synthesiser that offers a faster switching speed, a smaller frequency step size, and a wider bandwidth. Readers can explore this in more detail in, for example, references [6–8].

11.5 Oscillator Design Procedure

11.5.1 Design Flow

Conventionally, a successful design flow for oscillator circuit design relies on both linear and nonlinear analysis approaches. The linear analysis and design procedure for an oscillator is summarised in Figure 11.32. Starting with the desired specifications, primarily the oscillating frequency, the next step is to choose a technology and a circuit topology that are appropriate. Choosing between a discrete or IC solution depends on issues such as cost, time to market or expected product volume, and will usually be dictated by the customer or end-user. If mass production is needed, an IC solution is normally selected. Active device technology is another important issue that needs to be carefully considered. A HEMT device is suitable for a low-noise amplifier, but phase noise is a different matter and bipolar devices are usually much better. Meanwhile, a CMOS device is mostly selected for IC technology due to its low cost. In either case, the f_T of a transistor is a transistor parameter to indicate the maximum oscillation frequency. This parameter is a good guide for selecting an appropriate transistor. The power consumption is also considered. After the device has been chosen, the next concern is to decide which oscillator topology suits the required specifications. The pros and cons of each topology need to be considered. For both discrete and IC technologies, Colpitts and Clapp oscillators have gained much attention. On the other hand, cross-coupled and ring oscillators have been commonly selected as the preferred oscillator for a wireless transceiver chip. Moreover, there are some other circuit topologies such as differential-pair or quadrature designs. Once a topology is chosen, the next step is to decide on a suitable technology for the resonator which will lead to a low-noise and low-power oscillator design. For IC implementation, a high-Q on-chip inductor is one of the important keys to achieve the low-noise and low-power goal. If the discrete technology or off-chip solution is chosen, a high-Q resonator such as dielectric resonator, waveguide cavity, coaxial cavity, or MEMS device should definitely be used.

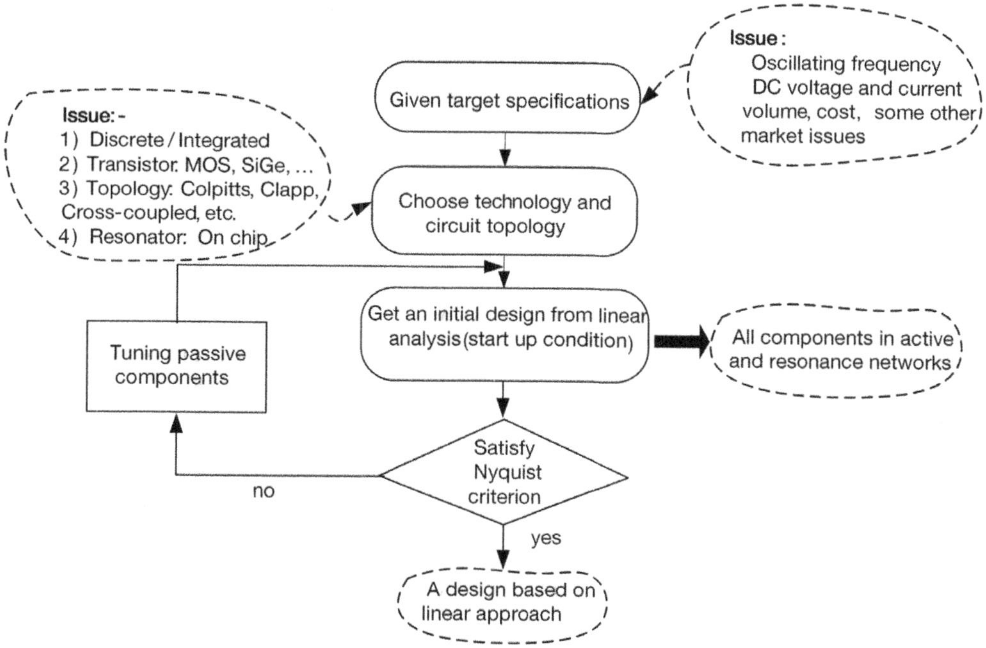

Figure 11.32 Flowchart for oscillator design.

At the desired oscillating frequency, a given set of initial circuit parameters for the circuit topology, which is chosen in the previous design step, can be obtained from a set of design equations derived based on the linear system theory developed from the two-port or one-port oscillator model. A linear CAD software tool, if available, can be applied to facilitate this step. For a transistor amplifier, a biasing point of the transistor which determines the transconductance g_m is selected to match the start-up condition. The parameters of all passive components for a given g_m will be obtained. The next step is to check the oscillation condition to guarantee the oscillation stability. The Nyquist theory described in the earlier section is recommended. If the design does not pass the Nyquist criterion, the parameters in the open-loop system should be adjusted until the design passes both the start-up condition and the stability condition for oscillation.

Phase noise can be analysed using either a nonlinear or linear time-variant (LTV) approach. After this step, a nonlinear analysis approach such as the harmonic balance technique (as described in Chapter 17) is applied to check the oscillation amplitude and the oscillator harmonics. An accurate nonlinear model for the active device and a linear model including parasitic components for the passive devices are required. An optimisation process is suggested to obtain the circuit parameters that meet the target specifications. This completes the oscillator circuit design part. The next step is to layout the design, and the results should be verified by an electromagnetic simulation software tool. Optimisation is also needed to guarantee that the design layout will meet the target specifications. The design flow of the oscillator after the linear design procedure is summarised in Figure 11.33.

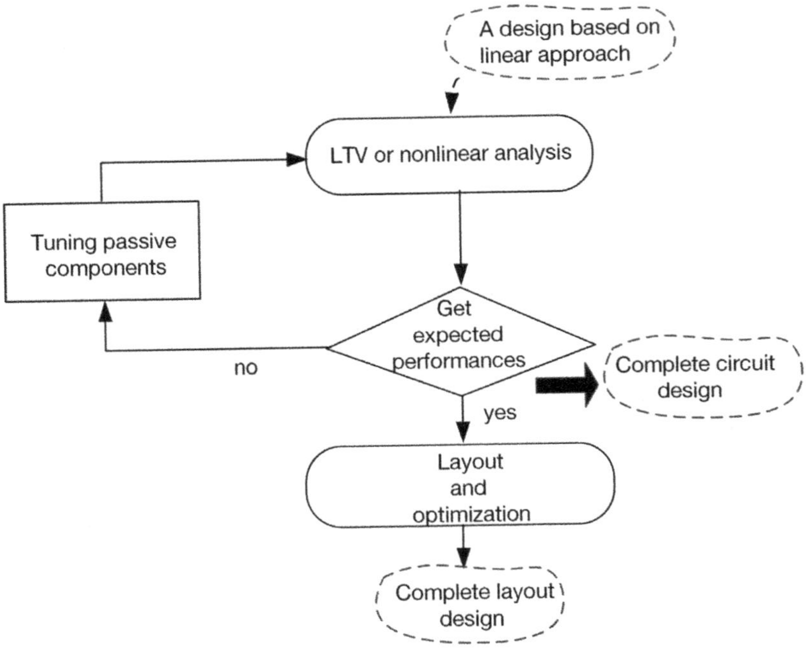

Figure 11.33 Nonlinear oscillator design and layout optimisation.

11.5.2 Negative Resistance Technique Applied to Microwave Oscillator Design

The design flow in Figure 11.32 shows a linear approach which provides an initial design oscillating at the desired frequency. In this section, the negative resistance technique which was described earlier in Section 11.2.1 is extended to clarify the design steps for microwave and millimetre-wave oscillators. The concept of this technique begins with a two-port microwave amplifier which is terminated at the input and output ports, as shown in Figure 11.34.

The input and output ports are terminated with the source impedance Z_s and load Z_L. With the characteristic impedance Z_0, the reflection coefficients at source, load, input and output are

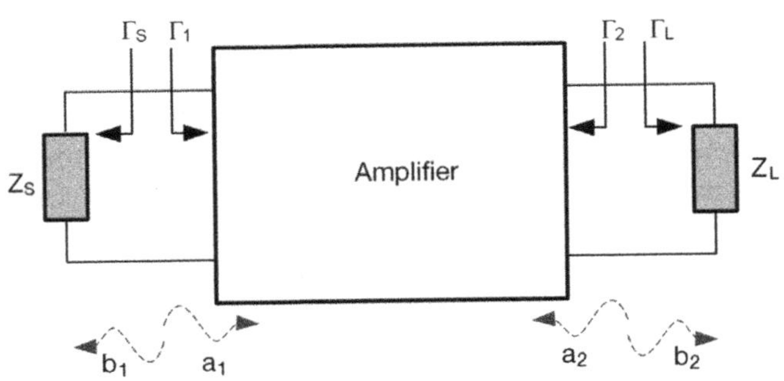

Figure 11.34 Oscillator design description with scattering parameters.

$\Gamma_S, \Gamma_L, \Gamma_1$, and Γ_2, respectively. It is noted that Γ_1, and Γ_2 are related to the input and output impedances of the amplifier when terminating with source and load impedances, (Z_{in}, Z_{out}). These are given by:

$$\Gamma_1 = \frac{s_{11} - \Delta\Gamma_L}{1 - s_{22}\Gamma_L} \quad \text{and} \quad \Gamma_2 = \frac{s_{22} - \Delta\Gamma_S}{1 - s_{11}\Gamma_S} \tag{11.125}$$

where Δ is the determinant of S parameters of the amplifier.

Let a_{i} and b_{i} ($i = 1, 2$) be the incident and reflected waves at port i. If b_s and b_1 are the reflected waves from source and load, a_1 and b_1 are related to b_s at port 1 by:

$$a_1 = b_s\left(1 + \Gamma_1\Gamma_S + [\Gamma_1\Gamma_S]^2 + [\Gamma_1\Gamma_S]^3\right) + \ldots \equiv \frac{b_s}{1 - \Gamma_1\Gamma_S} \tag{11.126}$$

$$b_1 \equiv \Gamma_1 a_1 = \frac{b_s\Gamma_1}{1 - \Gamma_1\Gamma_S} \tag{11.127}$$

Similarly, at port 2,

$$a_2 = b_L\left(1 + \Gamma_2\Gamma_L + [\Gamma_2\Gamma_L]^2 + [\Gamma_2\Gamma_L]^3\right) + \ldots \equiv \frac{b_L}{1 - \Gamma_2\Gamma_L} \tag{11.128}$$

$$b_2 \equiv \Gamma_2 a_2 = \frac{b_L\Gamma_2}{1 - \Gamma_2\Gamma_L} \tag{11.129}$$

Equations (11.127) and (11.129) show the steady-state wave. This implies that, at the steady-state oscillation condition, two following conditions should be satisfied:

$$\Gamma_1\Gamma_S = 1 \tag{11.130}$$

$$\Gamma_2\Gamma_L = 1 \tag{11.131}$$

This is relevant to the condition of the loop gain condition derived from the feedback theory. During the start-up condition of oscillator, the wave magnitudes in the oscillator should continually grow. Therefore we need:

$$\Gamma_1\Gamma_S > 1 \tag{11.132}$$

$$\Gamma_2\Gamma_L > 1 \tag{11.133}$$

The source and load are passive devices, therefore $|\Gamma_S| < 1$ and $|\Gamma_L| < 1$. From Equations (11.132) and (11.133), this means that $|\Gamma_1| > 1$and $|\Gamma_2| > 1$. However, we know that

$$\Gamma_1 = \frac{Z_{in} - Z_0}{Z_{in} + Z_0} \tag{11.134}$$

$$\Gamma_2 = \frac{Z_{out} - Z_0}{Z_{out} + Z_0} \tag{11.135}$$

Applying $|\Gamma_1| > 1$ to the definition of Γ_1, we obtain:

$$R_{in} < 0 \tag{11.136}$$

Similarly, from the condition $|\Gamma_2| > 1$ we obtain

$$R_{out} < 0 \tag{11.137}$$

These two conditions show that the start-up oscillation condition needs the input and output resistance to be negative. In Figure 11.34, the amplifier needed to design an oscillator should be unstable. At high frequency, an amplifier is normally designed with transistors, either with BJT, HEMT, or CMOS technology. If a transistor is used, a technique to destabilise the device may need to be applied. The stability factor of the transistor should be computed. This stability factor (K) is calculated from:

$$K = \frac{1 - |s_{11}|^2 - |s_{22}|^2 + \Delta^2}{2|s_{21}s_{12}|} \tag{11.138}$$

If K is less than 1, the transistor is potentially unstable, and this is suitable to design an oscillator. Otherwise, a feedback technique is applied to destabilise the device. This can be achieved by either using a series or shunt feedback technique, as shown in the examples applied to a common-source transistor. Figure 11.35(a) and (b) show the series and shunt feedback techniques applied to a FET.

From the condition which determines stability or instability, we now develop the equations for the locus of the desired load and source that provide oscillation. Let c_i and c_o be the centres

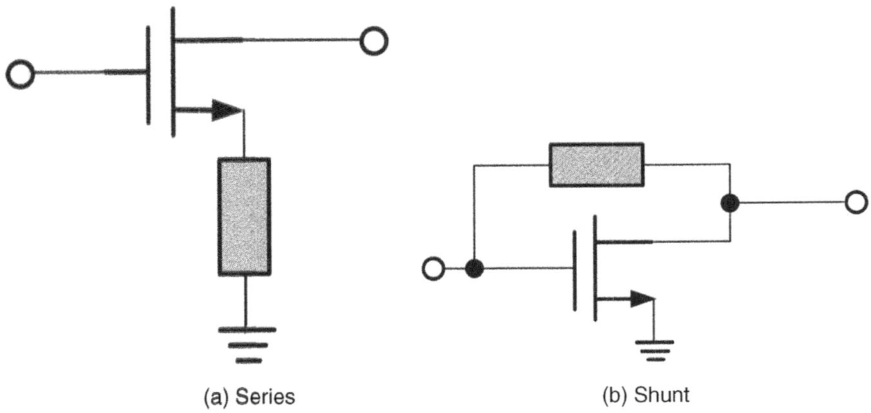

Figure 11.35 Feedback techniques applied to destabilise a transistor.

of the input and output stability circles; r_i and r_o are the radiuses input and output stability circles, the stability circles are defined from:

$$c_i = \frac{\left(s_{11} - \Delta \cdot s_{22}^*\right)^*}{|s_{11}|^2 - |\Delta|^2} \tag{11.139}$$

$$c_o = \frac{\left(s_{22} - \Delta \cdot s_{11}^*\right)^*}{|s_{22}|^2 - |\Delta|^2} \tag{11.140}$$

$$r_i = \left| \frac{s_{12}s_{21}}{|s_{11}|^2 - |\Delta|^2} \right| \tag{11.141}$$

$$r_o = \left| \frac{s_{12}s_{21}}{|s_{22}|^2 - |\Delta|^2} \right| \tag{11.142}$$

The input and output stability circles of the transistor can be drawn on a Smith chart. The instability regions can be defined and Γ_S and Γ_L terminating at the transistor to allow oscillation will be given. Assuming that Γ_L is previously chosen from the unstable region defined from the output stability circle, then Γ_i can be calculated from:

$$\Gamma_i = s_{11} + \frac{s_{12}s_{21}\Gamma_L}{1 - s_{22}\Gamma_L} \tag{11.143}$$

The source resistance and reactance can be calculated from the real and imaginary parts of Γ_i, (R_{in}, X_{in}), using:

$$R_s = -\frac{R_{in}}{3} \tag{11.144}$$

$$X_s = -X_{in} \tag{11.145}$$

It should be noted that the above conditions are in fact derived from the oscillation condition defined in Equation (11.130). The factor 1/3 was suggested by Pozar [9] to ensure the oscillation is able to build up at the initial stage.

Example: A common-gate transistor has the following S-parameters at 10 GHz:

$s_{11} = 1.118\angle 165.6°$; $s_{12} = 0.162\angle 166.9°$; $s_{21} = 2.068\angle - 12.723°$; $s_{22} = 1.154\angle - 3.535°$

First, calculate Δ from the S-parameters of the device.

$$\begin{aligned}
\Delta &= s_{11}s_{22} - s_{12}s_{21} \\
&= (1.118 \times 1.154)\angle(165.6 - 3.535)° - (0.162 \times 2.068) \\
&\quad \angle(166.9 - 12.723)° \\
&= 0.959\angle 164.809°
\end{aligned}$$

Hence, K is calculated as:

$$K = \frac{1-(1.118)^2-(1.154)^2+(0.959)^2}{2\times 0.162\times 2.068} = -0.988$$

Since $K < 1$, the common-gate transistor is potentially unstable, so the S-parameter of this transistor is suitable to design an oscillator at 10 GHz. The centre and radius of the output stability circle are given by:

$$c_O = \frac{\left(s_{22}-\Delta\cdot s_{11}^*\right)^*}{|s_{22}|^2-|\Delta|^2} = 0.237\angle 35.432$$

$$r_o = \left|\frac{s_{12}s_{21}}{|s_{11}|^2-|\Delta|^2}\right| = 0.815$$

The output stability circle is considered. Since $|s_{11}| > 1$, the stable region of the device is in the circle in the shaded area shown in Figure 11.36. The load is chosen to be 150 Ω, which gives Γ_L as 0.5. Therefore

$$\Gamma_i = 1.118\angle 165.6^\circ + \frac{0.162\angle 166.9^\circ \times 2.068\angle -12.723^\circ \times 0.5}{1-1.154\angle -3.535^\circ \times 0.5} = 1.5\angle 161.397^\circ$$

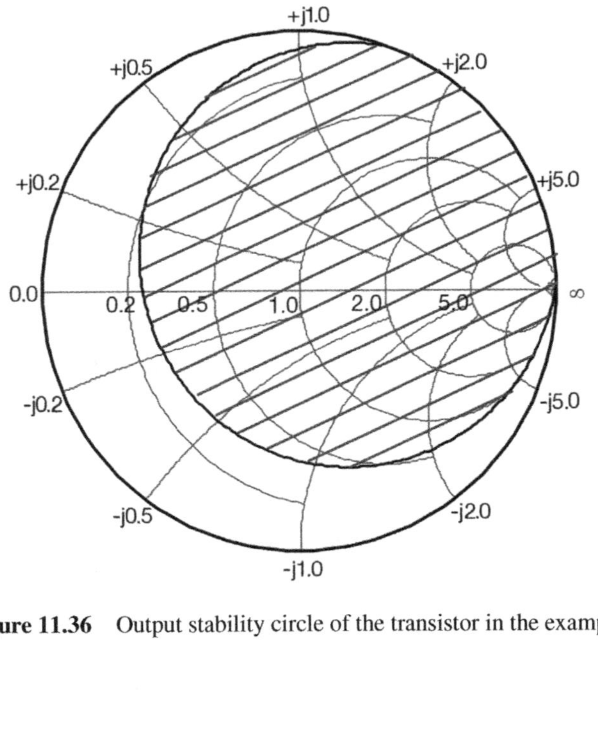

Figure 11.36 Output stability circle of the transistor in the example.

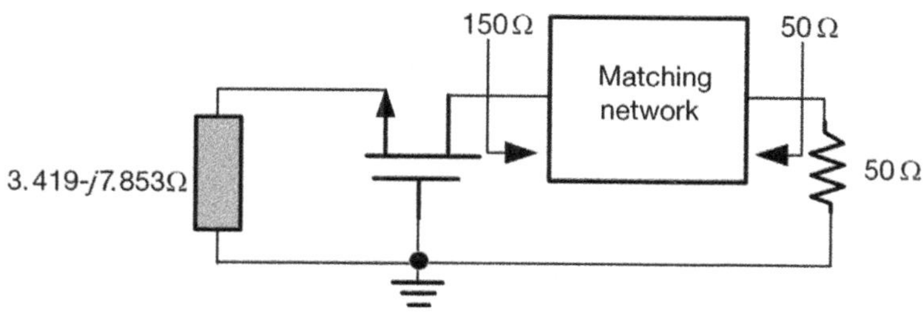

Figure 11.37 Final design from the example.

The input impedance (Z_{in}) of the device is calculated from:

$$Z_{in} = Z_0 \frac{1+\Gamma_i}{1-\Gamma_i} = -10.257 + j7.853$$

From Equations (11.144) and (11.145), the source resistance and reactance terminated at the transistor input are then:

$$R_s = 3.419$$

$$X_S = -7.853$$

The circuit diagram of the final design is shown in Figure 11.37. The DC biasing circuit is not included. The impedance terminating the transistor source is 3.419-j7.853 Ω. The microwave load is 50 Ω, so a matching network is employed to transform the 150 Ω impedance to a 50 Ω load.

11.6 Conclusion

Some important concepts have been described for the analysis and design of oscillators. Two key models, based on negative resistance and positive feedback, have been thoroughly discussed. These two models are very fundamental for describing the mechanisms at work in an oscillator. The Nyquist condition for checking the stability of oscillator has been introduced and a simple numerical example has been given to help show how the Nyquist condition is applied to check the stability of the design. Choosing a suitable resonator is very crucial for obtaining a high-Q factor, and Leeson's model shows how important this is in determining phase noise. Today, piezoelectric and dielectric resonators are commonly used in microwave applications for this reason. This chapter has also described some of the commonly used circuit topologies, including the Colpitts and cross-coupled topologies which are often chosen for high-frequency applications. The concepts of VCO and PLL have been introduced conceptually, but not in great detail. The contents presented in this chapter should provide readers with the background needed to understand modern oscillator design concepts for microwave and millimetre-wave applications. However, modern wireless systems have dramatically increased the need for highly advanced frequency synthesiser design techniques, such as fractional synthesis, which we cannot possibly do justice to here. Hence, readers are strongly advised to consult dedicated modern texts on frequency synthesis for further information [6–8].

References

1. Da Silva, E. (2011) *High Frequency and Microwave Engineering*. Butterworth-Heinemann.
2. Gilmore, R. and Besser, L. (2003) *Practical RF circuit Design for Modern Wireless Systems*, Volume II. Artech House Publishers, Boston.
3. Pookaiyaudom, S. (2003) *Negative Feedback Circuits and Oscillators* (in Thai).
4. Leeson, D.B. (1966) A simple model of feedback oscillator noise spectrum. *Proceedings of the IEEE*, **54** (2), 329–330. doi:10.1109/PROC.1966.4682.
5. Hewlett-Packard. *Fundamentals of Quartz Oscillators*, Application Note 200-2. Available at: www.keysight.com.
6. Lacaita, A., Levantino, S., and Samori, C. (2007) *Integrated Frequency Synthesisers for Wireless Systems*. Cambridge University Press.
7. Razavi, B. (2014) *RF Microelectronics*. Pearson, India.
8. Razavi, B. (2003) *Phase-Locking in High-Performance Systems: From Devices to Architectures*. Wiley-IEEE Press.
9. Pozar, D.M. (1998) *Microwave Engineering*, 2nd edition. John Wiley & Sons Inc.

12

Mixers and Modulators

12.1 Introduction

The mixer is a central part of most transmitters and receivers, playing the key role of frequency-converting the signal between the RF carrier frequency and an IF or baseband. Many designers have been able to avoid too much involvement in mixer design because of the very wide range of modular components that is available, for example from Mini-Circuits®. Up to a few GHz, the ubiquitous double-balanced diode-ring mixer can use baluns realised using wire-wound transformers on a ferrite core. The design of these baluns is sometimes regarded as a black art, but is actually grounded in the well-known principles of transmission line transformers and baluns [1–3] in which the distributed capacitance of the windings is accounted for, giving broadband operation and a maximum operating frequency much higher than a lumped transformer approach could achieve. For frequencies beyond a few GHz, modular mixers will make use of specialist quasi-3-D structures such as suspended-stripline baluns, although more recently the planar Marchand balun has been used to excellent effect and miniature double-balanced mixers are available, for example, as LTCC components.

Conventional couplers and baluns can be difficult to implement monolithically, and so a wide range of new techniques has been developed for the RFIC or MMIC mixer. The transmission line circuits, couplers and filters used in conventional modular mixers are at best wasteful of space and therefore expensive, and at worst they are nonplanar and hence not realisable. Only at frequencies above about 10 GHz can normal microstrip couplers be used readily on an IC, and even at these frequencies they are a major obstacle to keeping integration levels high. One approach is to use folded and spiral couplers instead, but ultimately stray coupling limits how compact the couplers can be made. Another approach is to replace them with active circuits. These active baluns employ the phase-inversion property of transistors and offer useful advantages of gain and port-to-port isolation.

Besides exploring new techniques for baluns and coupling structures for RFIC and MMIC mixers, active devices are often used in place of diodes. FET mixers have been extensively investigated since this allows integration with other circuits such as amplifiers and oscillators in single-chip transceivers. In RFIC design, the majority of mixer designs are based on the Gilbert cell mixer, which uses four mixing devices in a commutating arrangement to form

Microwave and Millimetre-Wave Design for Wireless Communications, First Edition.
Ian Robertson, Nutapong Somjit and Mitchai Chongcheawchamnan.
© 2016 John Wiley & Sons, Ltd. Published 2016 by John Wiley & Sons, Ltd.

a compact double-balanced mixer. For millimetre-wave Si/SiGe transceivers, the Gilbert cell topology is often combined with passive baluns.

Mixer design is a vast topic since mixers come in so many forms. In some applications, such as a Bluetooth transceiver, circuit size is a primary concern, whereas in specialist applications such as receivers for sub-millimetre-wave sensing, the frequency may be so high that there are no low-noise amplifiers (LNAs) available, in which case the mixer conversion loss is the main factor determining the receiver noise figure, which means a very different approach must be used in which every frequency component in the mixer must be terminated with carefully chosen load impedance in order that no signal energy is wasted. This, therefore, leads to a very different design philosophy. This chapter aims to describe some of the key principles of mixer design, spreading across this wide range of design approaches, and focussing only on the most popular design techniques. First, in order to introduce the mixing mechanism and compare approaches, single-ended mixers are described so that the principles underpinning diode and FET mixers are well understood. After that, balanced and image-rejection mixers are covered at a block diagram level, to focus attention on the reduction of unwanted output signals and improvement in port isolation that can be achieved with different topologies. The various techniques for realising the balun and couplers that are needed for mixer design are then introduced. Finally, all of these building blocks are brought together with a detailed look at the most common mixer circuit topologies. Since there are many choices of device, balanced topology, and balun realisation, the total number of combinations is huge. As an estimate, suppose there are four types of device, six balanced/image-rejection topologies, and four types of balun or coupler: simplistically, the total number of possible design approaches is therefore $4 \times 6 \times 4 = 96$! Actually, this is an underestimate, and it is impossible to fully cover the fascinating subject of mixer design in this one chapter; however, a number of excellent texts are available with further information [4,5]. This chapter will focus on the most common mixer design techniques, especially the double-balanced diode mixer (the most common form of modular mixer) and the Si/SiGe Gilbert cell mixer technique, which is now adopted even beyond 100 GHz in RFIC design. A few more specialist microwave and millimetre-wave mixer and modulator circuits will also be briefly described as they often have performance advantages.

12.2 Single-Ended Mixers

Mixers are the core components for the up- and down-conversion of signals, including use in IQ modulators and direct conversion transceivers. A mixer is a circuit that can be used to frequency-translate a signal by multiplying it with a local oscillator (LO) signal. For example, in the traditional superheterodyne receiver, the difference frequency output signal of the mixer shown in Figure 12.1 is called the intermediate frequency (IF), and this frequency-translated IF signal retains the information content of the RF input. In a fixed-IF receiver, tuning from one channel to another is achieved by varying the LO frequency, and the benefit of translating the RF input down to the fixed IF is that a high-order, high-selectivity band-pass filter can be used, for example realised with SAW/BAW technology. As discussed in Chapter 2, for integrated circuit transceivers there has been a shift towards zero IF architectures so that these filters are not required; however, the mixer remains a crucial component at the heart of any transmitter or receiver.

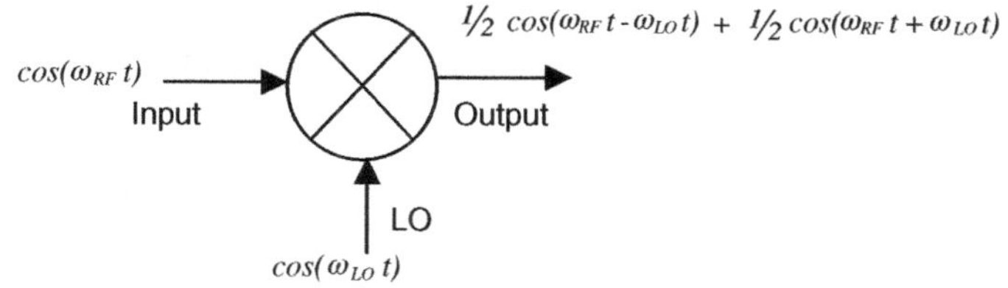

Figure 12.1 Sum and difference frequency outputs from ideal multiplication of RF and LO.

Generally, there are two distinct ways of realising the mixing process, although they overlap. In one method, mixing makes use of the nonlinear behaviour of a diode or transistor, as shown in Figure 12.2, to create a waveform that is a power series containing a significant square-law term that creates the sum and difference frequencies. DC bias is often required in order that

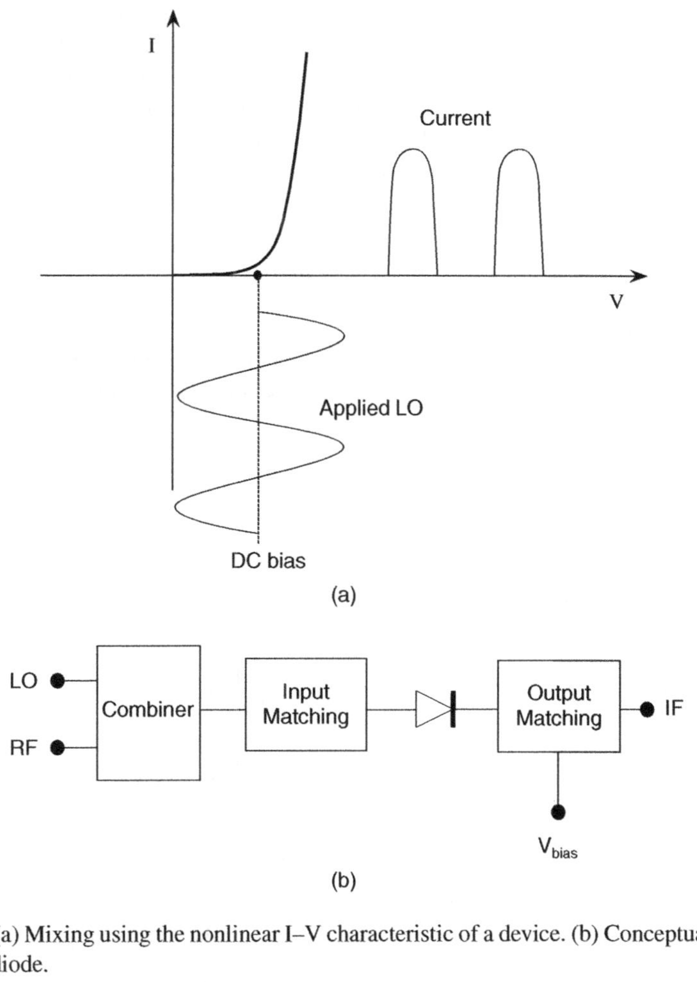

Figure 12.2 (a) Mixing using the nonlinear I–V characteristic of a device. (b) Conceptual single-ended mixer using a diode.

the required degree of nonlinearity is achieved with less LO power. Supposing the LO and RF are sinusoidal signals, combined together and applied to the device, then the square-law term gives an output signal of the form:

$$\left(A\cos\omega_{LO}t + B\cos\omega_{RF}t\right)^2 = A^2\cos^2\omega_{LO}t + AB\cos\omega_{LO}t\cos\omega_{RF}t + B^2\cos^2\omega_{RF}t \quad (12.1)$$

Since $\cos^2\theta = {}^1\!/_2(1+\cos 2\theta)$, the $\cos^2$ terms give a DC output and second harmonics of the LO and RF.

Since

$$cos\,\alpha\,cos\beta = {}^1\!/_2\,[cos\,(\alpha+\beta) + cos\,(\alpha-\beta)] \quad (12.2)$$

the middle term gives the sum and difference frequencies, as follows:

$$AB\cos\omega_{LO}t\cos\omega_{RF}t = {}^1\!/_2AB\cos\left(\omega_{LO}+\omega_{RF}\right)\mathrm{t} + {}^1\!/_2AB\cos\left(\omega_{LO}-\omega_{RF}\right)t \quad (12.3)$$

For a downconverter application, the difference frequency is used to create the desired IF or baseband output, while for an upconverter the sum frequency delivers the required output signal at high frequency. Crucially, any modulation on the RF signal (e.g., if B was amplitude-modulated) is replicated directly on the sum and difference frequency signals, so the information is preserved by the frequency translation. Of course, in practice the device nonlinearity will have higher-order terms, giving a rich spectrum of output signals with frequencies given by:

$$m\omega_{LO} \pm n\omega_{RF} \quad (12.4)$$

where m and n are integers.

Provided that the RF signal is much smaller than the LO, the terms with A in dominate and the output spectrum shown in Figure 12.3 is obtained. In this context, the image

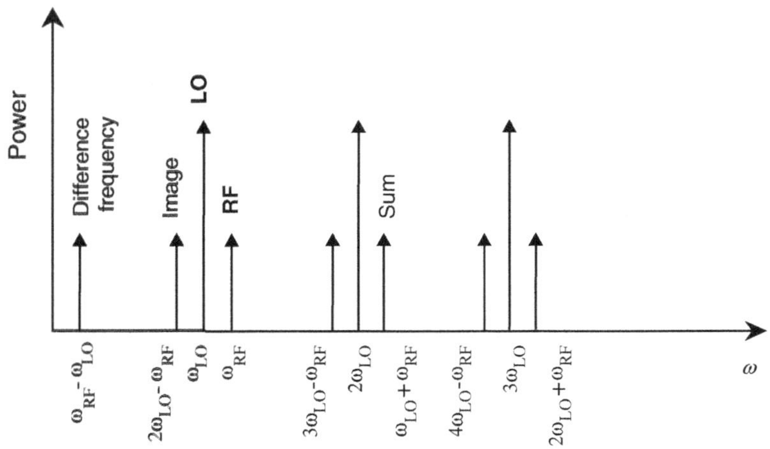

Figure 12.3 Output spectrum of a downconversion mixer with the LO signal power assumed to be significantly larger than the RF.

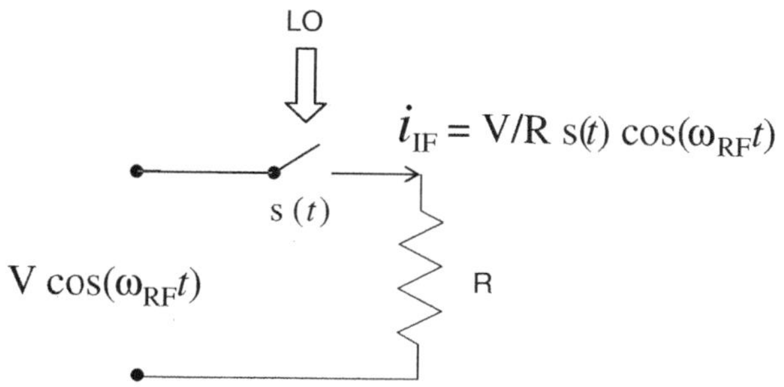

Figure 12.4 Mixer as a switch driven by the LO.

frequency is the frequency $2\omega_{LO} - \omega_{RF}$ and this is generated by the mixer – it is not an input signal as in the context described in Chapter 2. It is evident from Figure 12.3 that there are many unwanted output signals from a mixer, and this is why balanced and double-balanced mixers are so important. Furthermore, remember that this spectrum assumes that the RF amplitude is very small compared with the LO, and the situation is even worse if this is not the case. With so many signals present at the output of a mixer, and remembering that this example only has a single RF input, it can be seen that evaluating the behaviour of the mixer in terms of spurious outputs when at the heart of a transmitter or receiver is very important. This is the purpose of spur analysis, for which a number of spreadsheets and other tools are available, including a very effective graphical version from Marki Microwave [6].

Another method of explaining the mixing of two signals is based on switching the conductance of a device periodically, as illustrated in Figure 12.4. Here, the mixer works by using the LO signal to switch the diode or transistor on and off, modelled as an ideal switch. The switching waveform $s(t)$ is nearly a square wave when the LO power is high, so harmonics are present in practice and a power series analysis will again yield the mixer output signals, as illustrated in Figure 12.3. In fact, considering Figure 12.2(a) it can be seen that the current waveform does approach a square wave, and as in many areas of electronics, the time-domain and frequency-domain methods of understanding a circuit are inextricably linked by the Fourier series. There are situations where a circuit and its behaviour are easier to understand in one or the other domain. For example, in a diode mixer the switching model allows a thorough analysis on the importance of pulse-duty ratio on mixer performance, which is dictated by the LO power and DC bias [7]. The most common form of diode mixer uses four diodes in a commutating action in order to form a double-balanced mixer where the LO and RF do not appear at the IF port. The ubiquitous Gilbert cell mixer uses four transistors switching on and off sequentially in a commuting action.

On the other hand, balanced mixers are generally much simpler to understand in the frequency domain. The balanced (+/–, 0/180°) RF and LO signals are created from the single-ended inputs (e.g., microstrip, CPW or coaxial input ports) by using baluns (as discussed later). As a simple example, by using balanced input signals, the RF and LO can made to cancel out at the IF output, thanks to the symmetry of the circuit simply making use of various trigonometric identities, such as $\cos(\theta + 180°) = \cos\theta$ and $\sin(\theta + 180°) = -\sin\theta$. In other words, if two mixers are fed with anti-phase inputs, the desired signals can be summed at the

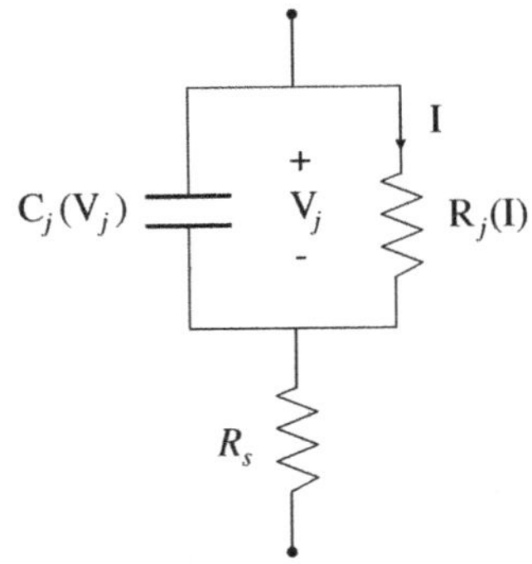

Figure 12.5 Schottky diode model.

common output, while many of the undesired signals are cancelled out. This is discussed in more detail in the next section.

12.2.1 Diode Mixers

Diode mixers are used in many microwave systems, but mostly when using a separate mixer module. When implemented on RFICs or MMICs, diodes are usually created using some derivative of the transistor structure, which results in suboptimal diodes with inferior performance in comparison with the use of discrete devices. Schottky diodes are by far the most common as they have a low barrier height and are majority carrier devices with a high cut-off frequency. In order to have an absolute minimum of package parasitics, microwave diodes are often supplied in a beam-lead package. Figure 12.5 shows the equivalent circuit of a Schottky diode in forward bias conditions, with series resistance R_s, primarily from the Ohmic contacts and material leading up to the junction, a junction capacitance C_j in parallel with the junction resistance R_j. C_j and R_j are strong functions of the applied voltage. The current in the diode is given by:

$$I = I_s\left(e^{q(V-IR_S)/nKT} - 1\right) \tag{12.5}$$

It is the time variation of this current when the LO is applied across the junction that gives the mixing mechanism, whilst variation in C_j will generally lead to unwanted intermodulation products. In order to reduce the required amount of LO power, DC bias is very often used.

12.2.2 Active FET Mixers

For RFIC and MMIC mixers, FETs are used in a variety of mixer topologies. Active FET mixers have the advantage of compatibility with amplifiers, oscillators and so forth, and may offer higher performance than the suboptimal MMIC diodes. There are several mechanisms that can be used to perform mixing. The best performance is usually achieved with a gate mixer mode, in which the transistor is biased in the saturation region and the LO applied to the gate, as illustrated in Figure 12.6(a), so that the LO modulates the transconductance. The

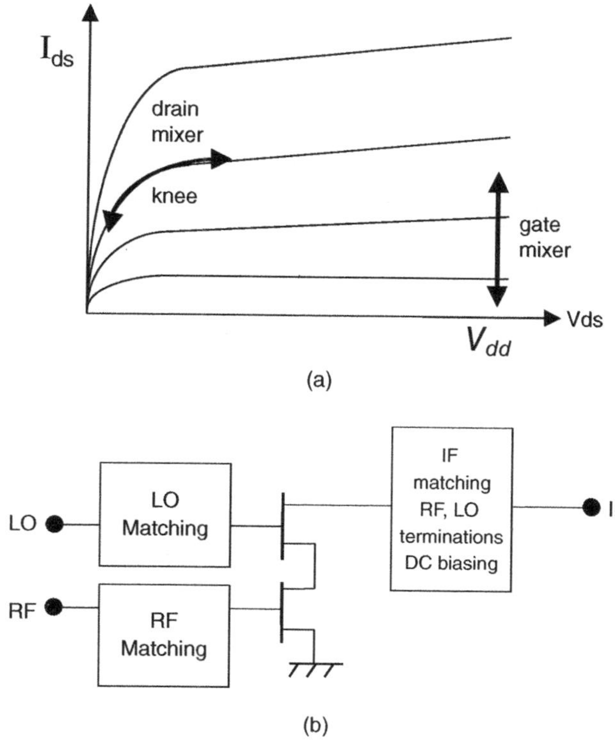

Figure 12.6 (a) FET I–V curves showing mixing modes. (b) A dual-gate FET mixer.

gate mixer has the disadvantage that the LO and RF need to be applied to the same terminal and sometimes, the transistor might be used in the triode region with the LO applied to the source or drain. Dual-gate FETs or a cascade of two FETs provides an attractive solution for IC mixers because the RF and LO can be input to separate gates, as illustrated in Figure 12.6(b), and so the number of passive components can be reduced [8].

12.2.3 Resistive FET Mixers

Whilst the active FET mixer gives good conversion gain, its intermodulation performance can be rather poor; however, this can be addressed by using the FET in a passive mode where the channel resistance of an unbiased FET is modulated by applying the LO to the gate. This is referred to as a 'resistive' FET mixer. It was first reported for GaAs FETs in [9] and the schematic diagram of a generic single-ended resistive mixer is shown in Figure 12.7. Without drain bias, the channel resistance acts as a very linear time-varying resistor when the LO is applied at the gate. This results in a mixer with low intermodulation when the RF is injected at the drain. The IF is extracted from the drain through filtering. The other advantage of this mixer is the low DC power consumption and some natural separation of LO and RF/IF ports. In practice, the most common form of resistive mixer is a double-balanced one using a FET quad (as discussed later).

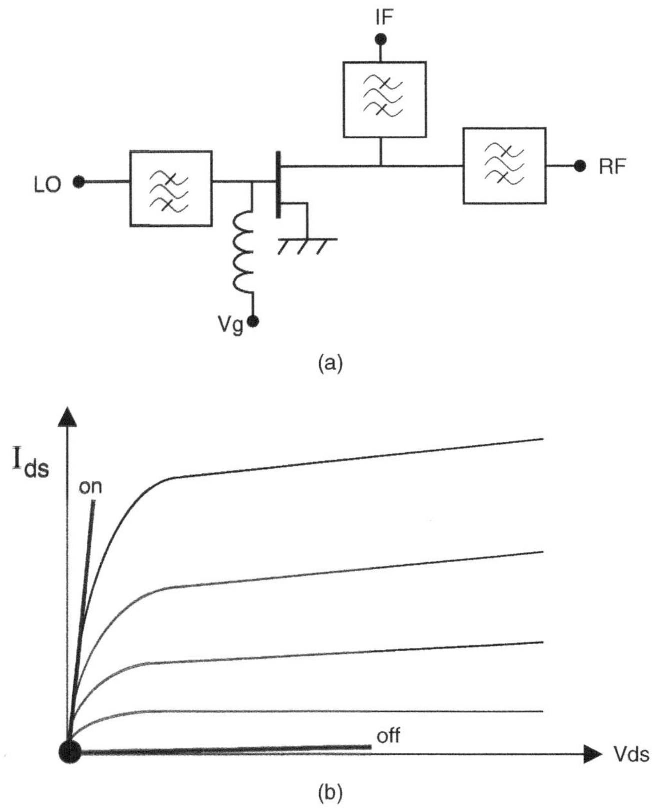

Figure 12.7 (a) Single-ended resistive FET mixer. (b) Operation on I–V curves.

12.3 Balanced and Image-Rejection Mixers

Balanced mixers use two or more mixers fed with at least one anti-phase signal in such a way that cancelation of unwanted signals occurs, providing port-to-port isolation or rejection of unwanted spurious signals. Balanced mixers are extremely important because they can reduce the specifications of filters in a transceiver down to something realisable in a compact, low-cost form. In some specialist applications, such as instrumentation and electronic warfare, the very broad bandwidths used may mean that frequency bands overlap and a balanced mixer is the only means of separating the signals. In a single-balanced mixer, two mixer devices are employed and in a double-balanced mixer there are four mixing elements. The multiple mixing devices may be incorporated into a single network, for example diode quads, or the mixers may be quite separate subcircuits combined together at a higher level. A balanced-to-unbalanced transition is key component to convert single-ended microstrip, coax or CPW inputs into a balanced signal form. Alternatively, an active circuit can be used to create the 0/180° signal outputs, and although they may be independent unbalanced signal outputs, the term 'active balun' is still generally used. In some applications, a 90° coupler is used for the RF/LO and it is the IF signals that are anti-phase in the two mixer devices. The degree of port isolation or unwanted signal cancellation that is achieved is dictated by the amplitude and phase imbalance

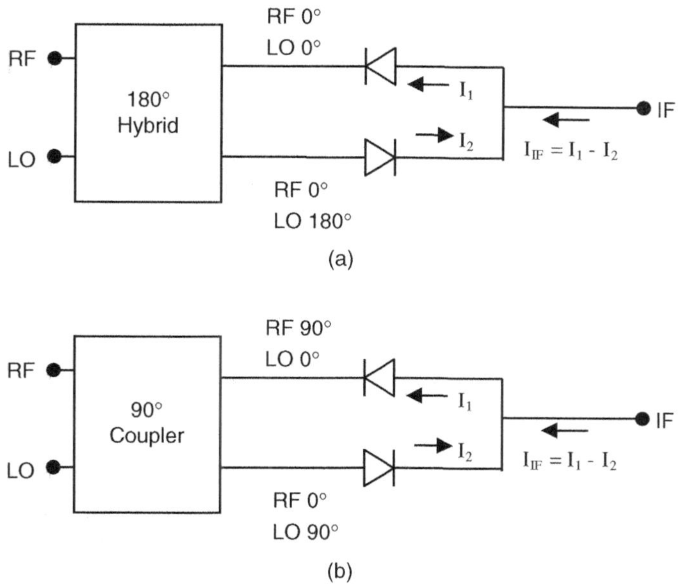

Figure 12.8 Single-balanced diode mixers. (a) Using a 180° hybrid. (b) Using a 90° coupler.

of the baluns, with target levels of around ±0.5 dB and ±5° being required to achieve a useful rejection on the order of 25 dB.

12.3.1 Single-Balanced Mixers

Figure 12.8 shows two forms of single-balanced mixer using diodes. In Figure 12.8(a) the LO signal is fed anti-phase to the two diodes, whilst the RF signal is in-phase. Two anti-phase IF currents are thus created, which are summed at the IF port simply by having the diodes placed noise-to-tail. By operating the diodes with a balanced LO signal, the LO is cancelled out at the IF port, giving LO-IF isolation, and approximately half of the other unwanted output spurs are cancelled out, to a degree determined by the amplitude and phase balance. In Figure 12.8(b), the RF and LO are fed into a quadrature coupler, such as a branch-line or Lange coupler, and fed to the diodes. The IF currents are again anti-phase, but LO/RF-to-IF isolation is not achieved by balancing and has to be achieved via the circuit design. The advantage of this topology is that the coupler gives good input match and LO-RF port isolation.

12.3.2 Double-Balanced Mixers

In a double-balanced mixer there are four mixing devices fed with anti-phase LO and RF inputs, with the four devices connected in such a way that all four combinations of RF/LO signal phase are used in the four mixing devices. In this way, approximately two-thirds of the unwanted output signals are suppressed. Figure 12.9 shows a generic representation of a double-balanced mixer as a commutating switch network. The switches can be realised with

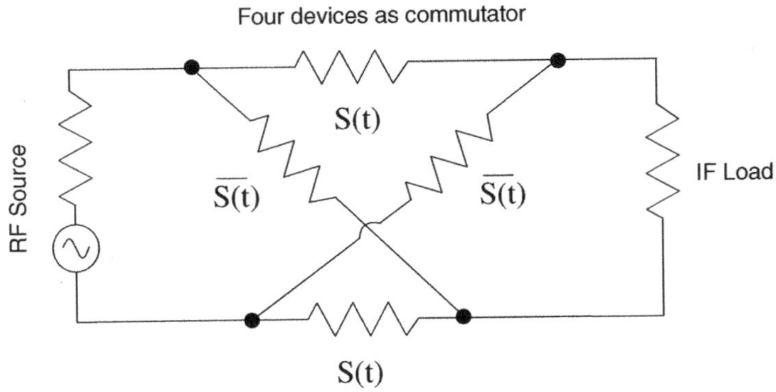

Figure 12.9 Double-balanced mixer representation with four commutating switch elements.

diodes, most often connected in a ring arrangement, or with transistors, most often in the arrangement known as the Gilbert cell. The LO 0/180° signals are switching the four devices, yielding anti-phase switching, whilst the balanced RF signal is fed from the left-hand side. The signal to the IF load, R_{IF}, is balanced and a single-ended signal is created in a diode mixer by placing diodes nose-to-tail, while in FET mixers an IF balun of some form is required. This is illustrated in Figure 12.10 which shows a double-balanced FET mixer that uses dual-gate FETs to give separate LO/RF signal inputs. The four DG-FETs operate as commutating switches with all four combinations of RF/LO 0/180° phase. At the drain outputs, a balanced IF signal results which can either be fed into a differential circuit or converted to a single-ended output with an active or passive balun. Even though this has three baluns, it is not a triple-balanced mixer. A triple-balanced mixer, or double-double-balanced mixer, is most commonly realised as a diode mixer with TWO diode quads and baluns for the RF, LO and IF. This means that it has eight mixer devices and the advantage is that the IF bandwidth is enhanced and the IF band may overlap the RF/LO band since the IF is not tapped off from one of the other baluns but has its own.

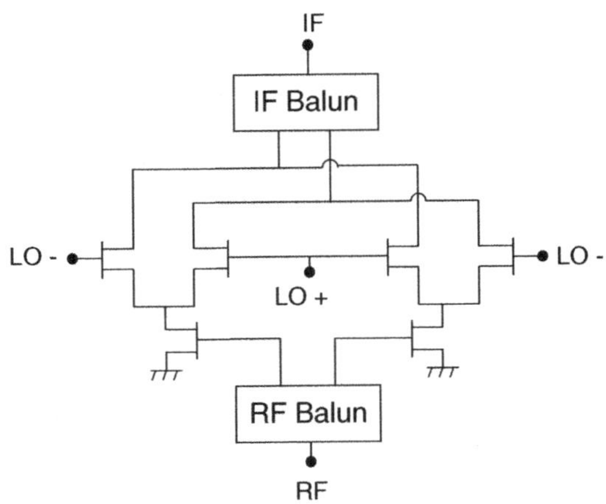

Figure 12.10 Double-balanced dual-gate FET mixer.

12.3.3 Image-Rejection Mixer and SSB Upconverter

As discussed in Chapter 2, the incoming image signal is difficult to deal with, and in an upconverter the output signal will have two sidebands, above and below the LO, which means that quite challenging filter specifications are necessary to meet spectral mask requirements and avoid interference with other users or systems. A phasing-type arrangement can be used to alleviate the problem, in a similar way to other balanced mixers, but a key difference is that in order to separate the lower and upper sidebands a 90° phase difference is required, which represents the Hilbert transform from the signal processing domain. Remembering that all spectra are double-sided around DC, a 90° delay is found to give a phase delay to positive frequency components but a phase advance to negative frequency components. In the frequency translation, this gives the essential means of differentiating between the signals that are above and below the LO. Figure 12.11(a) shows an image-rejection mixer using the topology that is referred to as the Hartley architecture. Figure 12.12 illustrates a 71–86 GHz subharmonic image reject mixer chip using this architecture, with the Lange coupler and Wilkinson divider/combiner instantly recognisable, for the quadrature and in-phase splitter/combiners.

Another approach, which realises the 90° IF phase shift by employing a second pair of mixers, fed with a different LO 0/90° pair, as shown in Figure 12.11(b), is referred to as the 'Weaver architecture' and is sometimes used in wideband receivers. The degree of image or sideband suppression that can be achieved depends primarily on the amplitude and phase

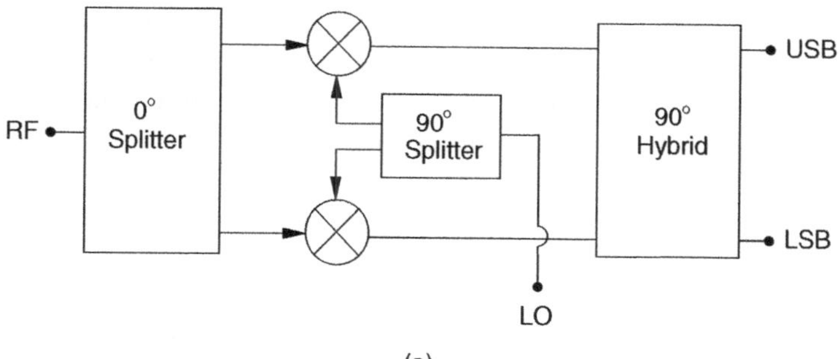

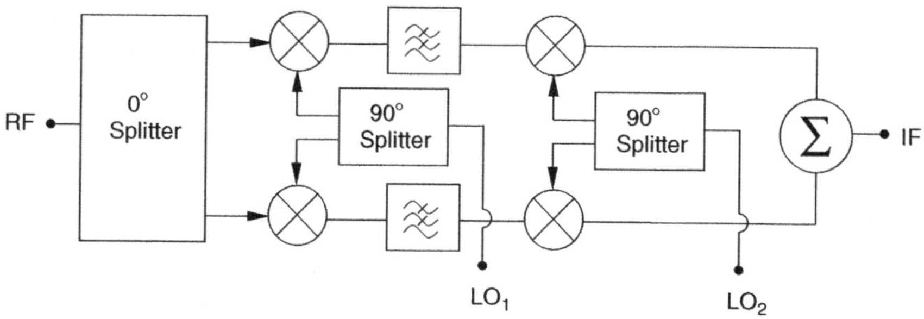

Figure 12.11 Image-rejection architectures. (a) Hartley. (b) Weaver.

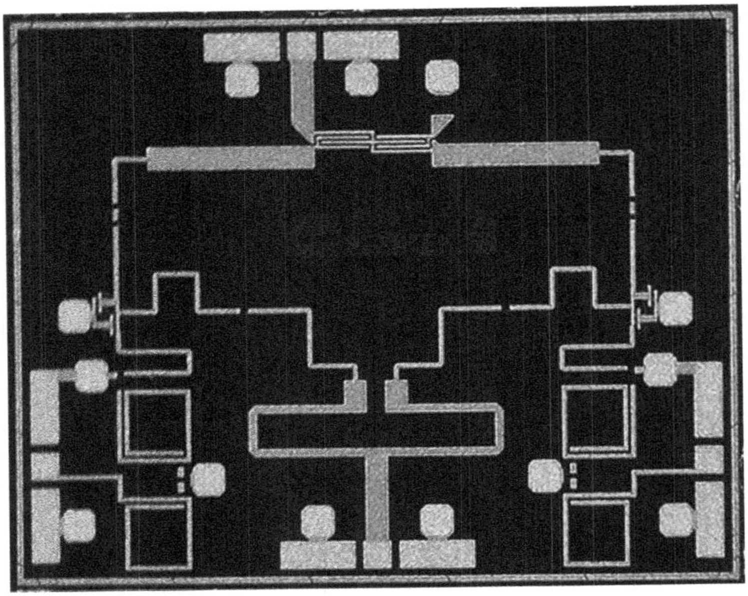

Figure 12.12 71–86 GHz subharmonic image reject mixer chip photograph. Illustration courtesy of Plextek RFI. Used with permission; all rights reserved.

balance of the splitters and combiners. Achieving more than 30 dB over any significant bandwidth is very challenging. Indeed, it can be shown that even the return loss of the various components will limit the cancellation of unwanted signals at the output. Adaptive control loops have received some attention (as discussed later).

12.4 Baluns and Couplers

Traditionally, balanced mixers have used wire-wound components with ferrite cores at low microwave frequencies and suspended-substrate baluns at higher frequencies (microwave to low millimetre-wave). Millimetre-wave mixers often still employ waveguide components such as magic-Tees. All of these techniques are thoroughly described in a number of key sources, and this section focuses on techniques for realising baluns and couplers in microstrip MICs and on RFICs and MMICs.

12.4.1 Transmission Line Baluns

The planar version of the Marchand balun, shown in Figure 12.13, is highly attractive due to its planar structure and wideband performance. It consists of two coupled sections, which may be realised using microstrip coupled lines, Lange couplers, multilayer coupled structures, or spiral coils [10–14], These baluns are usually designed through circuit simulations using 3-D-planar electromagnetic analysis (as described in Chapter 17). There are other methods

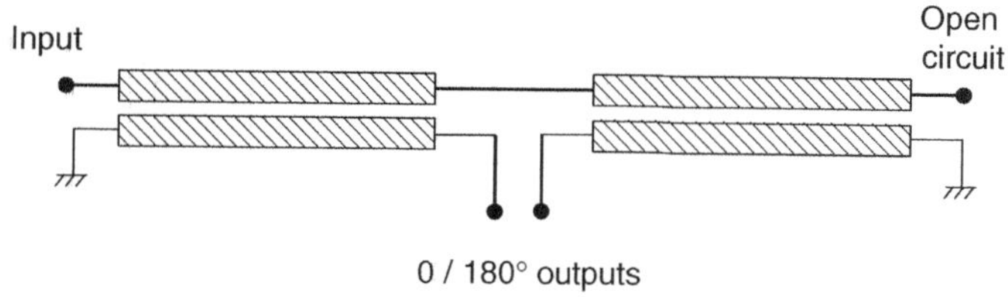

Figure 12.13 Planar Marchand balun.

of realising baluns using transmission line transitions such as CPW-slotline and CPW-CPS transitions that can be used at much higher frequencies.

12.4.2 Lumped-Element Splitters and Combiners

In the microwave frequency range, couplers are too large for integrated circuits. Here, lumped-element splitters and combiners can be used [15, 16]. Figure 12.14 shows the circuit diagram of a lumped-element 90° phase splitter, for example, which consists of a lumped Wilkinson divider followed by a complementary low-pass/high-pass filter pair that give 90° difference in phase over a wide band. This could replace the Lange or branch-line coupler used in image-rejection mixers. Lumped spiral transformers, consisting of interwound spiral inductors, can be used in balanced diode mixers in place of conventional bi-filar wound transformers. The frequencies that this technique can be used at are limited by the size and DC resistance at low frequency, and the interspiral capacitances at higher frequencies.

12.4.3 Active Splitters and Combiners

To save space, instead of using passive baluns the phase inversion of the FET can be used for creating a balanced signal pair. At low frequencies the long-tailed pair is very well known and can form an integral part of a balanced mixer. At higher frequencies, the FET capacitances start to degrade the phase balance, but it has been shown that an active balun design can be realised at microwave frequencies by using a pair of FETs, one in common-source and one in common-gate configuration, as shown in Figure 12.15 [17]. In-phase splitters and combiners

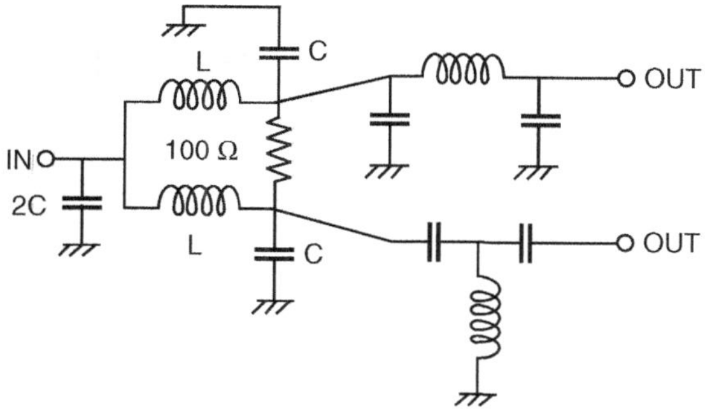

Figure 12.14 Lumped-element 90° splitter.

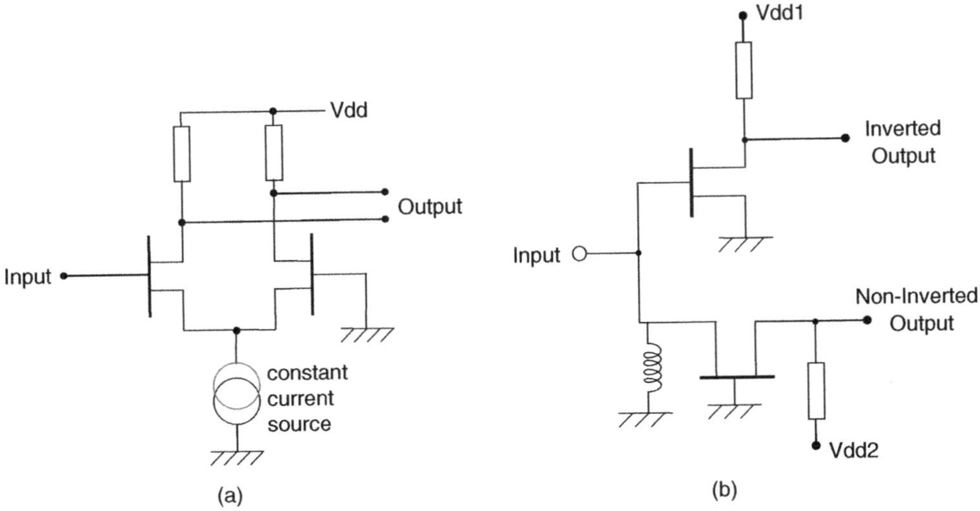

Figure 12.15 Examples of active balun circuits. (a) Differential pair. (b) Common-source/common-gate pair.

simply use pairs of transistors with common inputs or outputs, respectively. Active splitters and combiners have useful advantages such as reverse isolation, gain, and high isolation between inputs or outputs.

12.5 Common Mixer Circuits

12.5.1 Passive Double-Balanced Diode Mixer

This type of mixer, shown in Figure 12.16, has formed the core of many microwave systems for decades. The diode quad is akin to a full-wave bridge rectifier. The diodes are switched on and off sequentially by the LO signals, and the commutating action ensures that the IF current

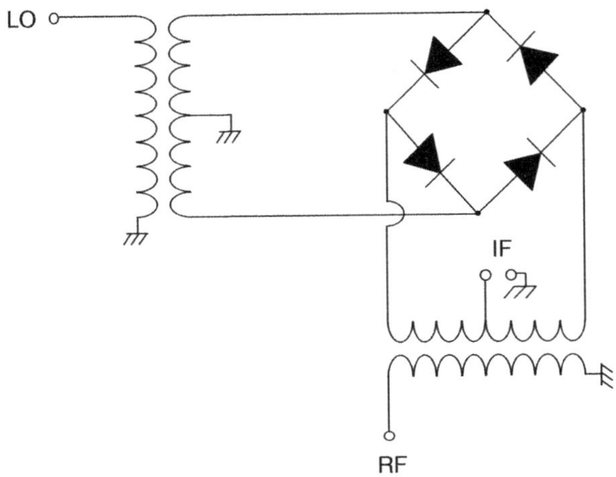

Figure 12.16 Double-balanced diode ring mixer.

flows into the load in the same sense at all times. The ports are isolated, and being double-balanced ensures low intermodulation and rejection of many spurious output signals. With four diodes, however, the LO power requirement is increased compared with single-ended mixers. The IF is typically extracted from the RF balun, at a point where symmetry ensures there is a virtual ground. To ease assembly and ensure that the diodes are well matched, diodes are available from manufacturers in packaged quad form, with a variety of package styles, including beam-lead style for low parasitics. The conversion loss of double-balanced mixers with Schottky barrier diodes is typically 6 to 8 dB and the LO power required is a minimum of 7 dBm. For high-level mixers, series-connected diodes are used, which further increases the LO power. High-level mixers with 30 dBm third-order input intercept point, but 27 dBm LO power, are readily available if desired.

12.5.2 Gilbert Cell Mixer

The Gilbert cell mixer forms the basis of the vast majority of RFIC mixers [18]. Strictly speaking, Gilbert's main contribution was to propose the use of diodes in the four-quadrant analogue multiplier in order to compensate for the nonlinear behaviour of the transistors. Most often, the mixer is operated in a switching mode, with a mixer core that consists of four switching transistors, driven by the LO 0/180° signals and a transconductance cell that gives RF gain at the input. In Figure 12.17, T1, T2, T3 and T4 form the mixer core, and T5 and T6 are the RF gain at the input. With a constant current source serving all four switching transistors, the current flows through T1–T4 in a commutating action, driven by the LO signal.

12.5.3 Double-Balanced Resistive FET Mixer

The resistive FET mixer has been shown to have excellent intermodulation performance, and since it is a simple mixing element it forms a natural alternative to the diode-ring quad in a double-balanced mixer [19]. Figure 12.18(a) shows the circuit diagram of this topology, with

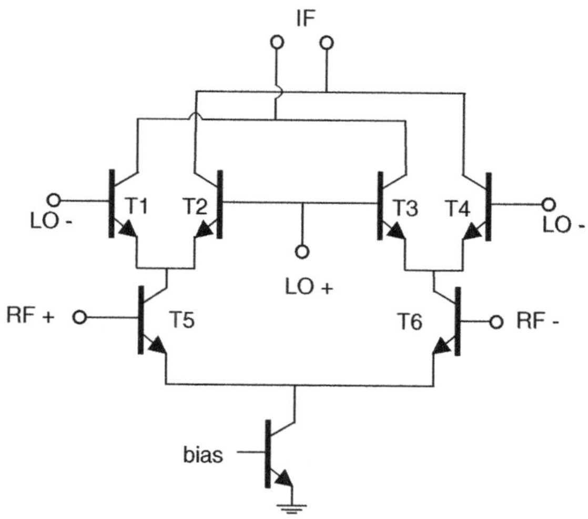

Figure 12.17 Gilbert cell mixer.

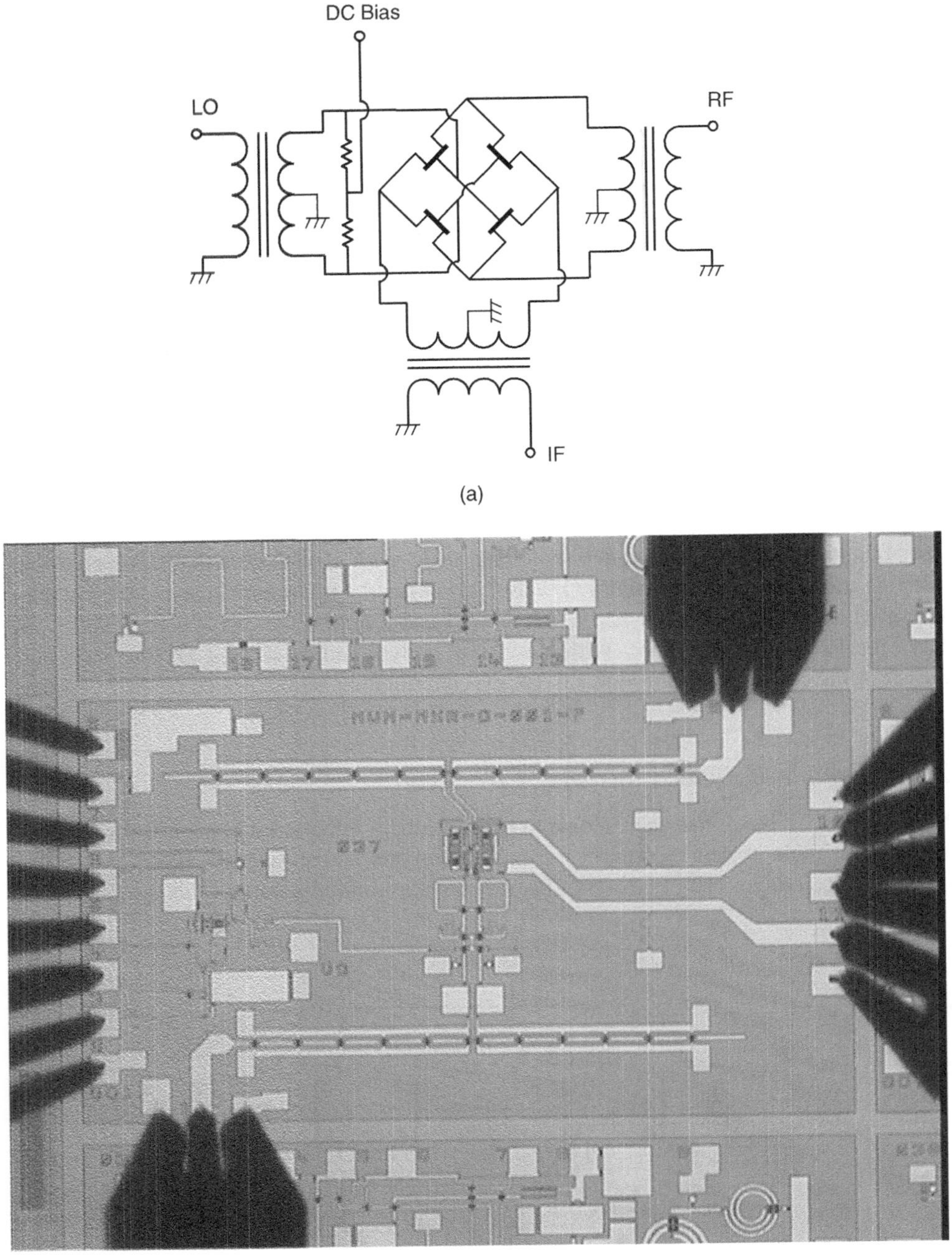

Figure 12.18 (a) Schematic circuit diagram of a double-balanced resistive FET mixer. (b) An example of a GaAs MMIC. Illustration courtesy of Plextek RFI. Used with permission; all rights reserved.

four FETs operating as the commutating switch. The only bias needed is on the gates, so that the LO power required to switch the transistors on and off can be reduced. The drains and sources have DC returns to ground through the baluns. Figure 12.18(b) shows a GaAs MMIC resistive mixer that uses uniplanar Marchand baluns at the RF and LO ports and has on-chip active bias circuitry to set the FET gate bias from a –5 V supply. In this case the IF output is differential.

12.5.4 Distributed FET Mixer

The distributed or travelling-wave mixer was first demonstrated by Tang and Aitchison [20] in hybrid form, using single-gate FETs with a passive coupler to combine the RF and LO signals. The principle is similar to the distributed amplifier, with the gate and drain capacitances incorporated into artificial transmission lines, so that a multi-octave bandwidth is achieved. Dual-gate FETs can also be used so that the coupler is not needed [21]: the mixer then has a gate-line for the RF, a gate-line for the LO and a drain-line for the IF, as shown in Figure 12.19(a).

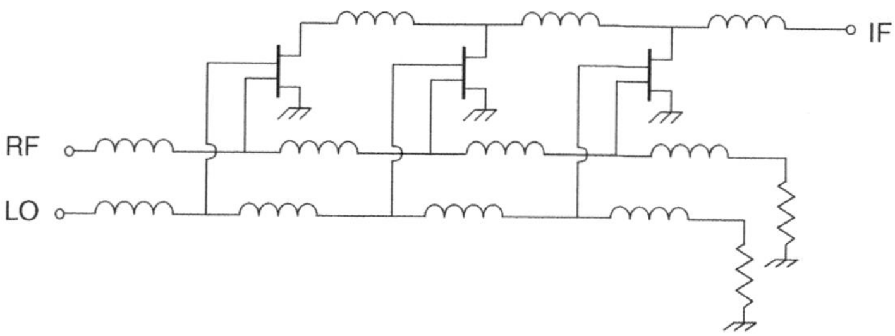

(a)

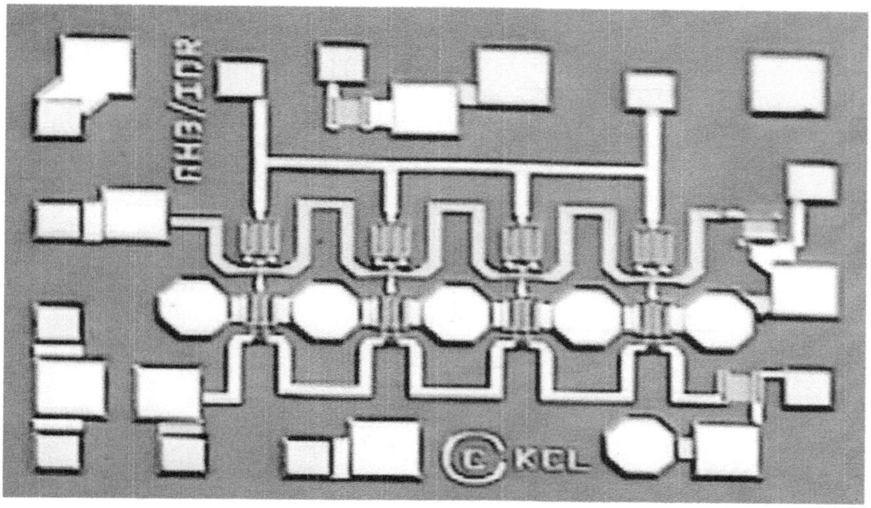

(b)

Figure 12.19 (a) Dual-Gate FET distributed mixer. (b) A GaAs MMIC distributed mixer.

Because this has some in-built LO/RF isolation, it is very suitable for MMIC implementation. A wide variety of circuits can be realised in distributed form, including splitters, combiners, active baluns and a distributed Gilbert cell mixer. Figure 12.19(b) shows a GaAs MMIC distributed mixer with FETs in cascode and narrow microstrip lines as inductors. A CMOS distributed mixer can use other transmission lines, such as a coplanar strip transmission line [22].

12.5.5 Millimetre-Wave Single-Balanced Diode Mixer

With passive diode mixers being able to achieve conversion loss of the order of 6 dB, it is not at all uncommon that active mixers have significantly worse noise figure. There is no direct benefit in an active mixer delivering a conversion gain if the noise figure is high and the comparison of active and passive mixers requires a proper analysis of the noise and intermodulation performance of the LNA, mixer and IF amplifier combination. Furthermore, diode mixers can operate to frequencies well over 200 GHz, where LNAs may not be available, or at least are not economically viable for the application. Hence, diode mixers remain of tremendous interest to microwave and millimetre-wave engineers, especially in applications where performance is a high priority. The single-balanced mixer is a highly practical solution since it provides LO/RF isolation without filters, while a double-balanced mixer will generally have too much loss in the baluns, require too much LO power, and generally be too complex to realise on an integrated circuit.

Figure 12.20 shows the typical layout of a millimetre-wave mixer employing a pair of Schottky diodes in a single-balanced arrangement using a branch-line coupler to combine

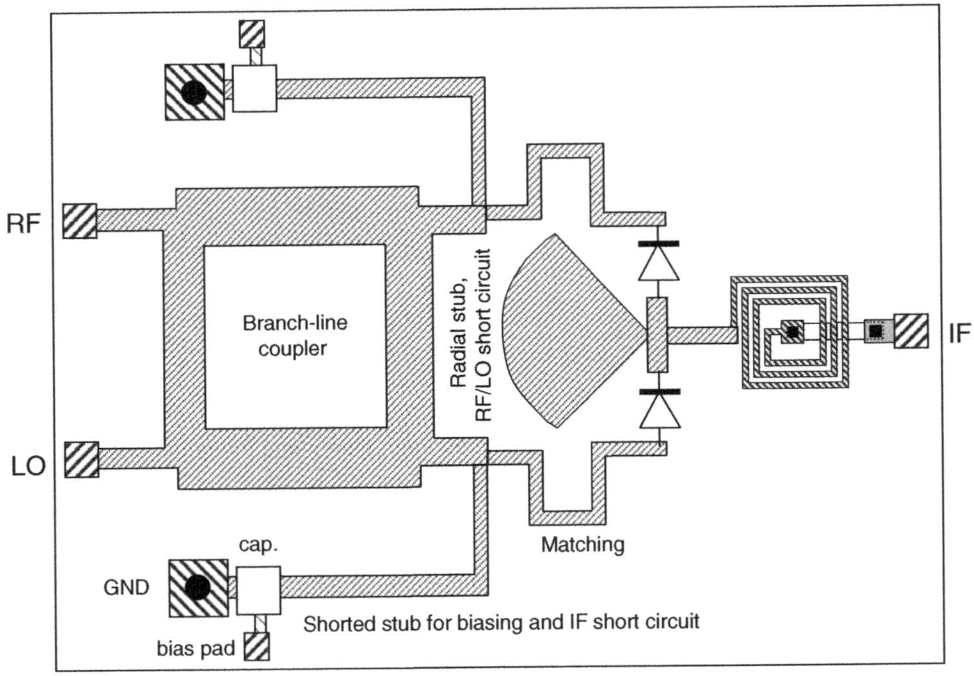

Figure 12.20 A generic millimetre-wave single-balanced diode mixer.

the LO and RF. This brings together a number of the key concepts that are important in mixer design. First, with LO power at a premium and a need to minimise conversion loss, the designer needs to carefully consider the impedance presented to the diode at a number of key frequencies. At the common node between the diodes, which is also the IF output, the LO and its second harmonic are short-circuited. As well as improving LO-IF port isolation, this ensures that the precious LO power is driving the diode junction fully. Next, the diodes are DC-biased to reduce the LO power requirement. By applying this bias at the end of a short-circuited stub that is $\lambda_g/4$ long at the LO frequency, an IF short is applied to the RF/LO end of the diodes without affecting the LO and RF signals. This ensures that the IF current flows to the IF port, improving conversion loss. A matching network is used between the coupler and the diodes.

In some specialist applications, such as millimetre-wave and submillimetre-wave astronomy, even this single-balanced mixer requires too much LO power and has too much loss before the diodes, leading to increased conversion loss. In these applications, it is still common to use hollow rectangular waveguides and to mount the diode inside using the absolute bare minimum of lossy dielectric material by mounting the circuit on a membrane, for example.

12.6 Modulators

The need for quadrature amplitude modulation of an RF carrier is covered extensively throughout this book. By modulating both the amplitude and phase of the carrier, a very high level of spectral efficiency (b/s/Hz) is achieved and the trend towards high-level modulation schemes (256-QAM and beyond) can be expected to continue as the demand for mobile internet data grows. As illustrated in Figure 12.21(a), these signals are most conveniently created by representing the signal in Cartesian form, with an in-phase and quadrature channel. Vector modulators are actually used in a number of other important applications, making them increasingly important in wireless communications. In any form of adaptive control loop (as discussed in other chapters for various applications), some form of amplitude and phase control is required to tune the loop. In an electronically steered antenna, individual elements have to be controlled in amplitude and phase for beam-steering. The direct connection of an I-Q modulator to a DSP/DAC subsystem gives a well-understood means of controlling the amplitude and phase of the RF signal with high precision, shifting much of the effort to the DSP code. Working hand-in-hand with DSP means that the I-Q modulator can be calibrated using a look-up table or feedback loop. Only in a few applications will the DSP approach require too much DC power, chip area, or be too costly, and RF MEMS or other RF control circuits may offer some advantages.

12.6.1 Vector Modulators using Mixers

The most common modulator topology in wireless communications, by a huge margin, is the I-Q modulator illustrated in Figure 12.21(b), in which the input RF carrier is split into a quadrature signal pair using a 90° power divider, such as a Lange coupler, and separate mixers then modulate the orthogonal carrier components with the I and Q baseband waveforms. The modulated signal components are then summed together using an in-phase combiner, such as a Wilkinson combiner, to create the vector-modulated signal.

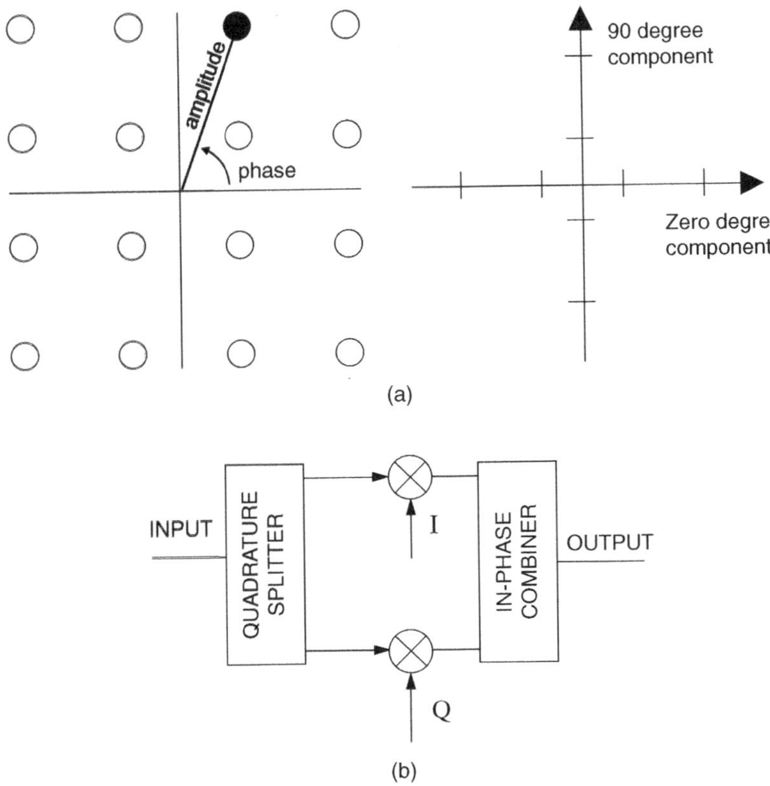

Figure 12.21 (a) I-Q constellation for 16-QAM. (b) I-Q Vector modulator using mixers.

The role of the mixer in this modulator needs careful examination before proceeding to design, or even choose, a suitable mixer. First, the terminology can be confusing; the I and Q signals are at baseband and the 'IF' port of the mixer must operate down to DC. It is the I and Q inputs that are modulating the RF carrier, and the mixers are actually being used as bi-phase amplitude modulators, each one individually covering the horizontal and vertical axes of the constellation of Figure 12.21(a). This means that the conversion loss versus baseband input voltage characteristic must be bipolar and have good linearity, with symmetry about DC. Diode mixers tend to behave poorly in these respects, and most I-Q modulators will be realised with Gilbert cell mixers on an integrated circuit. This has the additional advantage that the 90° LO/RF split can be realised with high precision, which is important for reducing spurious outputs at the modulator output, and the I and Q baseband subsystems (DACs, spectrum shaping filters, etc.) can be co-located [23].

12.6.2 Other Forms of Vector Modulator

In principle, an amplitude control circuit can be cascaded with a 360° phase shifter to realise a vector modulator. In practice, 360° phase shift is rather difficult to achieve without a lot of insertion loss, and significant variation will be observed in signal amplitude with phase

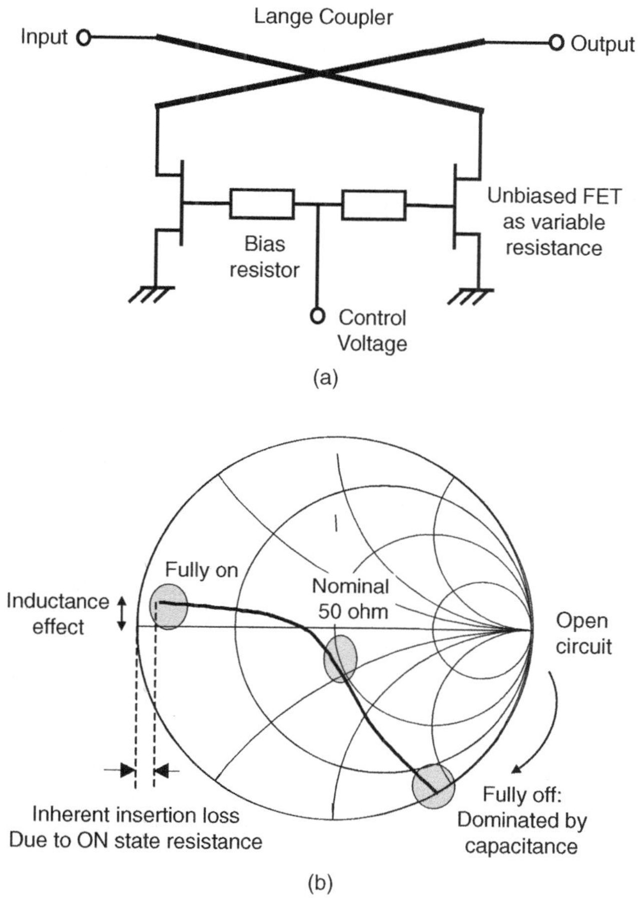

Figure 12.22 (a) Circuit topology of the bi-phase amplitude modulator. (b) Typical GaAs HEMT S_{11} versus bias voltage, at 60 GHz.

state. Likewise, many forms of amplitude-control circuit can be used, but the insertion phase may vary considerably with amplitude state. This amplitude-phase approach will, therefore, generally require something like a look-up table to calibrate these errors out.

Amplitude- and phase-control circuits that have been used successfully include the tried-and-tested analogue reflection-type topology shown in Figure 12.22. In this method, the PIN diodes, FETs, varactor diodes or other device, are used as variable-impedance reflection terminations, connected to the coupled and direct ports of a 3 dB quadrature coupler. The reflected signals cancel at the input port, but superimpose at the isolated port, which is thus the output port. The circuit's S_{21} can be shown to take on the magnitude and phase of the termination device pair's reflection coefficient. Hence, if a varactor diode is used, then in theory only the phase changes, whereas if a passive FET is used as a variable resistance only the magnitude changes ideally. In practice, the device parasitics dominate at high frequencies. For example, Figure 12.22(a) shows a reflection-type amplitude modulator circuit using passive (cold, unbiased) FETs, while Figure 12.22(b) shows the typical FET S_{11} versus gate bias, plotted on a Smith chart. At low values of channel resistance, the S_{11} follows the real axis quite well on the left-hand side, but as the channel resistance increases the capacitance of the FET starts to dominate, as shown on the left-hand side. An important technique to overcome this is the balanced modulator [24].

It is evident from Figure 12.22(b) that the modulator characteristic is rather good on the left-hand side of the Smith chart and so this circuit operates well as an amplitude modulator, but it is not well-behaved as a bi-phase modulator. Furthermore, it is found that the intermodulation performance is very poor in some states. Going back to basics, it can be observed from Figure 12.21(a) that it is entirely possible to realise a vector modulator with four separate *amplitude-only* control circuits, provided that four signal phases are created (0°, 90°, 180°, 270°), amplitude-modulated individually, and then summed. This does entail significant insertion loss (6 dB minimum) but is used when very low-amplitude/phase error is required without calibration/look up tables. For example, in principle four variable-gain amplifiers with significant maximum output power could be used to realise a modulator with very low intermodulation. Finally, in a similar vein, note that just one pair of such amplitude-only control circuits can be used in the 'shifted-quadrant' vector modulator [25], which is a technique with tremendous potential. In summary, it is evident that there are almost limitless possibilities for realising vector modulators without requiring mixers, which is important since better performance, such a low intermodulation, can potentially be achieved.

12.7 Mixer Linearisation and Adaptive Signal Cancellation

Despite the fundamentally nonlinear mode of their operation, the linearity of mixers seldom attracts any attention from engineers dealing with signals with large envelope variations, such as OFDM. As discussed in the Introduction, provided that the RF is much smaller than the LO, the mixer can be considered as a linear frequency-translating device. Only in a few specialist applications will there be any need to improve the intermodulation and distortion performance of the mixer, which is traditionally achieved by using a large LO power and series-connected devices. However, linearisation has been studied using feed-forward techniques [26] and methods of injecting the difference frequency [27, 28]. What has attracted more attention is the use of control loops to adjust the amplitude and phase balance of mixers and dynamically optimise the rejection of unwanted signals. For example, self-calibrating image-rejection mixers have attracted much interest [29, 30], while for direct-conversion receivers a number of techniques have been used for combating effects such as gain and phase imbalance, DC offsets and second-order intermodulation distortion, which can cause troublesome interference.

12.8 Conclusion

It was suggested earlier that there were on the order of 96 distinct mixer topologies since there are various devices, baluns and balancing methodologies that can be employed. Having surveyed the main design techniques used for microwave and millimetre-wave mixers, it can now be seen that 96 is an underestimate. This chapter has only scratched the surface of mixer design, and other techniques that have not been covered include harmonic and subharmonic mixers, specialist waveguide mixers for millimetre-wave and submillimetre-wave applications, self-oscillating mixers and BJT/HBT mixers.

Perhaps because mixer design can be challenging, in the distant past mixers were often treated as a 'black box' in a system and modular components were purchased, with double-balanced diode mixers being the natural choice. However, as levels of integration have increased it has become necessary to integrate mixers into MCMs and onto integrated circuits. This has led to a great deal of research into new mixer topologies. In RFIC applications, the Gilbert cell mixer is routinely used, but as transceiver specifications have become more demanding so the need to improve the performance of mixers has developed. Spectrally efficient modulation schemes have, for example, led to mixer linearity being a focus of attention, although the number of papers published every year is still far below that for power amplifier linearisation and efficiency improvement. Already, direct conversion receiver architectures have been demonstrated with sophisticated control loops to dynamically optimise performance and address mixer imperfections.

References

1. Marchand, N. (1944) Transmission line conversion transformers. *Electron*, **17** (12), 142–145.
2. Ruthroff, C.L. (1959) Some Broadband Transformers. *Proceedings of the IRE*, **47**, 1337–1342.
3. Guanella, G. (1944) New method of impedance matching in radio frequency circuits. *Brown Boveri Review*, **47**, 327–329.
4. Maas, S. (1993) *Microwave Mixers*, 2nd edition. Artech House Publishers. ISBN-10: 0890066051.
5. Henderson, B. and Camargo, E. (2013) *Microwave Mixer Technology and Applications*. Artech House Publishers. ISBN-10: 1608074897.
6. http://www.markimicrowave.com.
7. Ang K.S. and Nightingale, S.J. (2001) Mixers, in: *RFIC and MMIC Design and Technology* (eds S. Lucyszyn and I.D. Robertson), Chapter 7. IET Publishers.
8. Tsironis, C., Stahlmann, R., and Meierer, R. (1984) Dual-gate MESFET Mixers. *IEEE Transactions on Microwave Theory and Techniques*, **32** (3), 248–255.
9. Maas, S.A. (1987) A GaAs MESFET mixer with very low intermodulation. *IEEE Transactions on Microwave Theory and Techniques*, **35** (4), 425–429. doi: 10.1109/TMTT.1987.1133665.
10. Chen, T., *et al.* (1991) Broadband monolithic passive baluns and monolithic double balanced mixer. *IEEE Transactions on Microwave Theory and Techniques*, **MTT-39**, 1980–1986.
11. Nishikawa, K., Toyoda I., and Tokumitsu, T. (1999) Compact and broad-band three-dimensional MMIC balun. *IEEE Transactions on Microwave Theory and Techniques*, **MTT-47**, 96–98.
12. Gokdemir, T., Economides, S.B., Khalid, A., Rezazadeh, A.A, and Robertson, I.D. (1997) Design and performance of GaAs MMIC CPW baluns using over-laid and spiral couplers. IEEE International Microwave Symposium Digest, pp. 401–404.
13. Yoon, Y.J., *et al.* (1999) Design and characterization of multilayer spiral transmission-line baluns. *IEEE Transactions on Microwave Theory and Techniques*, **MTT-47**, 1841–1847.
14. Niknejad, A.M. and Meyer, R.G. (1998) Analysis, design, and optimization of spiral inductors and transformers for Si RF ICs. *IEEE Journal of Solid-State Circuits*, **33** (10), 1470–1481. doi: 10.1109/4.720393.
15. Gupta, R.K. and Getsinger, W.J. (1984) Quasi-lumped-element 3- and 4-port networks for MIC and MMIC applications. IEEE International Microwave Symposium Digest, pp. 409–411.
16. Vogel, R.W. (1992) Analysis and design of lumped- and lumped-distributed-element directional couplers for MIC and MMIC applications. *IEEE Transactions on Microwave Theory and Techniques*, **MTT-40**, 253–262.
17. Pavio, A.M., Halladay, R.H., Bingham, S.D., and Sapashe, C.A. (1988) Double balanced mixers using active and passive techniques. *IEEE Transactions on Microwave Theory and Techniques*, **MTT-36**, 1948–1957.
18. Keng Leong Fong and Meyer, R.G. (1999) Monolithic RF active mixer design. *IEEE Transactions on Circuits and Systems II: Analog and Digital Signal Processing*, **46** (3), 231–239. doi: 10.1109/82.754857.
19. Chen, T.H., *et al.* (1995) Broadband single- and double-balanced resistive HEMT monolithic mixers. *IEEE Transactions on Microwave Theory and Techniques*, **MTT-43**, 477–484.
20. Tang, O.S.A. and Aitchison, C.S. (1985) A practical microwave travelling wave MESFET gate mixer. IEEE MTT International Microwave Symposium Digest, pp. 605–608.

21. Howard, T.S. and Pavio, A.M. (1986) A distributed 1–12 GHz dual-gate FET mixer. IEEE MTT International Microwave Symposium Digest, pp. 329–332.
22. Garg, N., Lok, L.B., Robertson, I.D., Chongcheawchamnan, M., and Worapishet, A. (2006) 1 to 20 GHz CMOS distributed mixer using asymmetric coplanar strip transmission lines. IEEE Radio Frequency Integrated Circuits (RFIC) Symposium, 11–13 June, pp. 220–224. doi: 10.1109/RFIC.2006.1651130.
23. Mayer, U., Wickert, M., Eickhoff, R., and Ellinger, F. (2011) Multiband mixed-signal vector modulator IC. IEEE Radio Frequency Integrated Circuits Symposium (RFIC), 5–7 June, pp. 1–4. doi: 10.1109/RFIC.2011.5940599.
24. Nam, S., Ashtani, A.E., Passiopoulos, G., Lucyszyn, S., and Robertson, I.D. (1998) A 60 GHz 256 QAM balanced vector modulator for short range LOS communication applications. IEEE Radio Frequency Integrated Circuits (RFIC) Symposium, 11–18 June, pp. 271–274. doi: 10.1109/RFIC.1998.682096.
25. Kpogla, D.K.A., Ng, C.Y., and Robertson, I.D. (2003) Shifted-quadrant microwave vector modulator. *Electronics Letters*, **39** (14), 1058–1059. doi: 10.1049/el:20030696.
26. Ellis, T.J. (1998) A modified feed-forward technique for mixer linearization. IEEE MTT-S International Microwave Symposium Digest, 7–12 June, vol. 3, pp.1423–1426. doi: 10.1109/MWSYM.1998.700641.
27. Chongcheawchamnan, M., Ng, C.Y., and Robertson, I.D. (2002) Difference frequency injection linearisation technique for mixer systems. *Electronics Letters*, **38** (23), 1450–1452. doi: 10.1049/el:20021003.
28. Cheng, K.K.M. and Chung-Fai Au-Yeung (2004) Novel difference-frequency dual-signal injection method for CMOS mixer linearization. *IEEE Microwave and Wireless Components Letters*, **14** (7), 358–360. doi: 10.1109/LMWC.2004.828017.
29. Paez-Borrallo, J.M., Quiros, F.J.C., and Zazo, S. (1996) Self calibrating low if digital image rejection receiver for mobile communications. European Signal Processing Conference, EUSIPCO, 10–13 September, pp. 1–4.
30. Cetin, E., Kale, I., and Morling, R.C.S. (2004) Adaptive self-calibrating image rejection receiver. Communications, 2004 IEEE International Conference, 20–24 June, vol. 5, pp. 2731–2735. doi: 10.1109/ICC.2004.1313027.

13

RF MEMS

13.1 Introduction

The field of radiofrequency micro-electromechanical system (RF MEMS) refers to a range of electronic devices based on miniature mechanical structures on the micrometre scale that perform an RF function. RF MEMS can implement a wide variety of RF functions, with switches being by far the most common application. RF MEMS technology [1, 2] has been proposed as a way of realising highly integrated RFIC and MMIC system-on-chip (SoC) products, with new architectures made possible by the realisation of very low-loss switches for tuneable filters, beam-forming networks, and so forth. Being based on established microelectronic technologies, RF MEMS is produced at the wafer level and has many advantages such as low cost, small size, low mass and near-zero power consumption. RF MEMS is available for signal processing in microwave and millimetre-wave circuits and has already had a significant impact on existing radio communications systems. With the advent of the 5G+ era, low-power and ultra-compact new devices with planar structures are urgently needed which are capable of integration with signal processing and transceiver circuits for wireless communications, especially in mobile and satellite communications.

RF MEMS devices can be grouped into two categories: (i) passive MEMS without moving parts; and (ii) active MEMS with moving parts, which are moved or deformed by various mechanisms under electronic control. As wireless communications transceivers are expected to be ever-more adaptable and reconfigurable, RF MEMS is one of the most significant recent developments in RF technology. At present, the number of discrete passive components required for a communication system, such as inductors, variable capacitors, filters, couplers, phase shifters and switch arrays, sets a limitation on the further reduction of the size of the hardware. RF MEMS technology is expected to solve this problem. By using RF MEMS fabrication technology, passive devices and active circuits can be integrated directly within the same chip, which offers high levels of integration and eliminates discrete components. In this way, RF MEMS technology makes tuneable and reconfigurable circuits possible within a system and can significantly enhance the performance of the system.

Microwave and Millimetre-Wave Design for Wireless Communications, First Edition.
Ian Robertson, Nutapong Somjit and Mitchai Chongcheawchamnan.
© 2016 John Wiley & Sons, Ltd. Published 2016 by John Wiley & Sons, Ltd.

13.2 Novel Transceiver Architectures using RF MEMS

In modern wireless communication systems, where size and form factor are of nearly equal importance to performance, RF MEMS have become a most promising alternative to existing off-chip passive components [3,4]. In particular, in wireless transceivers the numerous passive components – inductors, tuneable capacitors, resonators, filters, and so forth – take up a lot of space in the physical circuit. However, conventional IC technologies have proved incapable of meeting the high Q-factor requirements for sharp selectivity and low phase noise that ensure equal or better performance than their discrete counterparts. Low power consumption, higher data rates, low weight, low cost and increased functionality are a few other factors that stimulate continued research for improvements in high-performance wireless systems, and many of the aforementioned problems have been offset by continued integration of the active functions in transceiver systems without significant loss of performance. The size and component count reductions that have been attained by this integration and chip scaling are currently being hindered by the apparent inability of on-chip variants of the passive components to meet existing performance criteria.

The basic MEMS building blocks are switches, inductors and varactors, and they are used to realise other passive functions including filters, resonators, impedance matching networks, phase shifters, and so forth. These match the conventional PIN diode or FET switches, high-Q inductors and high-Q varactor diodes, respectively, in performance and offer the added advantage of further RFIC/MMIC integration of these indispensable devices, and subsequently low cost. These RF MEMS devices make it possible for conventional communication systems to be made reconfigurable with superior performance. RF MEMS have advantages over semiconductor devices in the areas of high Q-factor, linearity and low power consumption, amongst others. Figure 13.1(a) and (b) shows an example of a tuneable RF transceiver module making extensive use of RF MEMS.

13.3 Micromachined Transmission Lines and Passive Elements

The electrical properties of a semiconductor substrate, serving as a mechanical support, affect the figures of merit of the RF devices. The loss properties of an on-chip transmission line fabricated on a semiconductor substrate indicate the suitability of the substrate. There are usually three dominating transmission line losses: radiation loss, dielectric loss and conductor loss. To decrease losses, especially at high frequencies, one attractive technique is to utilise micromachined transmission lines. Recent improvements in multilayer electroplated metal micromachining fabrication have made it conceivable to make low-loss transmission lines such as rectangular-coaxial (recta-coax or rectax) and coplanar waveguide (CPW).

One of the biggest advantages of using RF MEMS passive components over discrete components is that they have very high performance and low fabrication costs. Some examples of such MEMS components are capacitors, inductors and switches. MEMS inductors have wide tuning ranges and good linearity, while MEMS switches can offer high signal isolation and very low insertion loss. RF MEMS passive components can have many applications. MEMS tuneable capacitors and inductors can be simply used to implement high-Q-factor tuneable filters, whilst phase shifters can be used in electronically steered antennas, and switches can be used in antenna systems for diversity switching and tuning or reconfiguring the antenna.

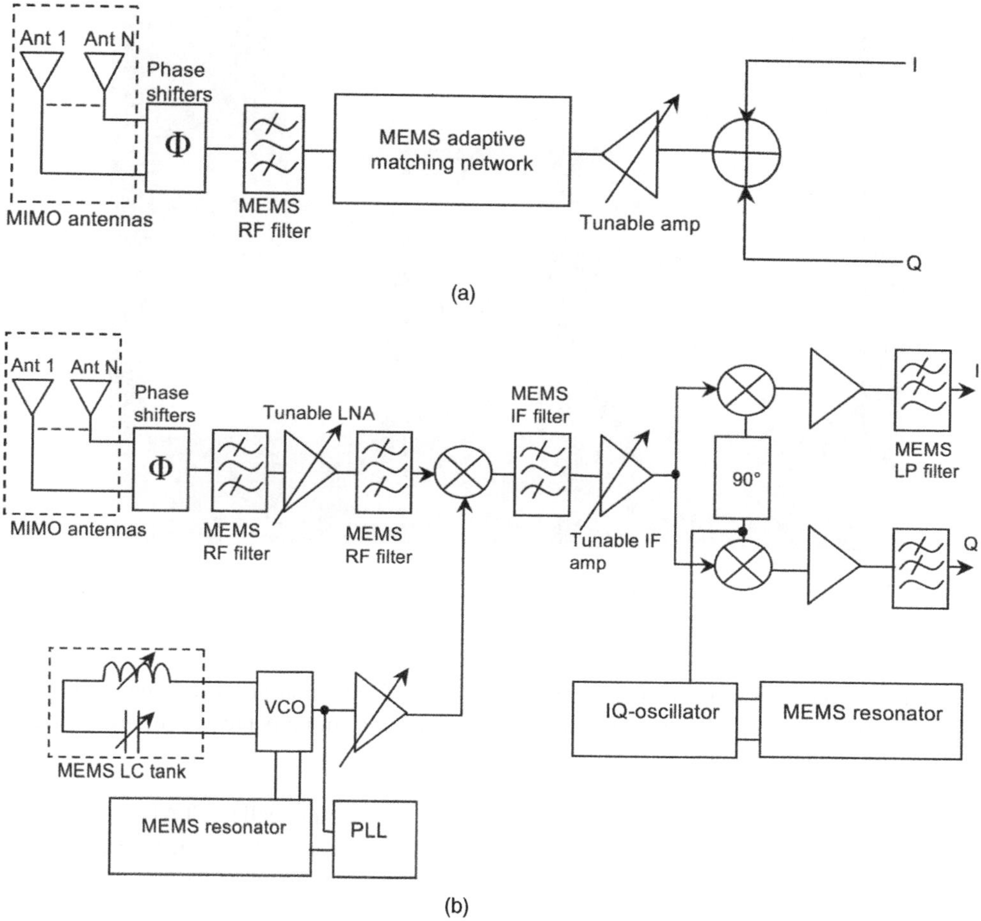

Figure 13.1 A block diagram of an RF MEMS reconfigurable transceiver. (a) Transmitter module. (b) Superheterodyne-based receiver.

13.3.1 Rectangular-Coaxial Transmission Lines

Recta-coax lines, shown in Figure 13.2, have a rectangular centre conductor which is suspended in air and is enclosed by a ground plane. They are manufactured utilising a thin-film fabrication process on top of a substrate. There are three main advantages of recta-coax lines over both CPW and microstrip transmission lines:

1. Recta-coax has a TEM and not quasi-TEM mode of propagation, meaning that it can operate over a broader frequency range (typically 0–200 GHz).
2. The ground plane encloses the centre line, which means adjacent lines can be placed very close to each other and even crossed over one another without causing significant unwanted coupling.

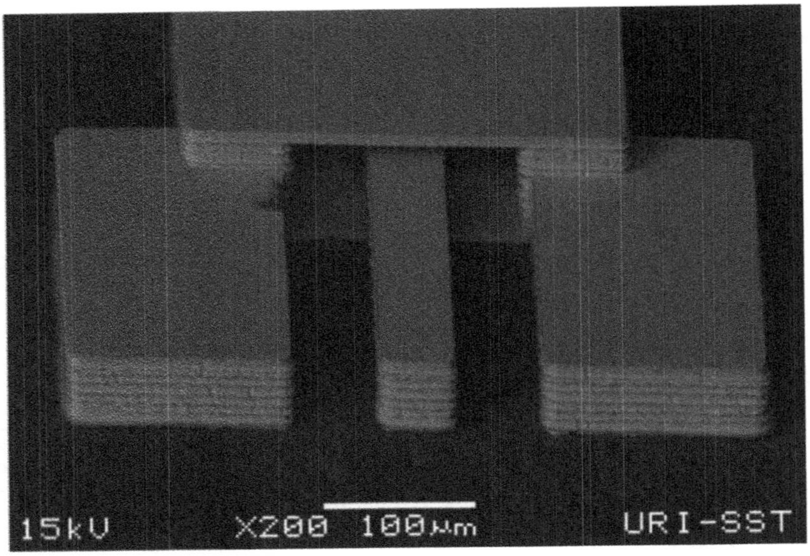

Figure 13.2 A micromachined on-chip recta-coax line [5]. *Source*: Reid, J.R. *et al.* (2006). Reproduced with permission of IEEE.

3. The design and operation is completely independent of the substrate, which means there is no need for substrate thinning and means designs can be implemented on many different substrates without changing anything.

In Ref. [5] the recta-coax lines were fabricated using a technique called EFAB, which uses layering to create the transmission line. It uses three different layers that are placed on a suitable planar surface. The first layer is a patterned sacrificial layer, the second is the structural layer, which is deposited on top of the sacrificial layer, and the third layer is the top surface, which is planarised to create a single layer with a uniform thickness. The sacrificial layer, which serves as temporary mechanical support material, is removed by means of chemical etching to leave only the structural layers. However, the layering technique can lead to scalloping at the layer interfaces [5]. To decrease any effects this may have on the impedance, the lines were designed so that the ratio of line width to gap thickness is larger than the ratio of thickness to gap width which is:

$$\frac{w}{t_g} > \frac{t}{w_g} \tag{13.1}$$

This means that the capacitances of the top and bottom ground planes dominate the capacitance of the centre signal line and the capacitance from the sidewalls is decreased.

13.3.2 Micromachined Coplanar Waveguide

There are various types of micromachined CPW, including micro-shield line and CPW suspended in the air on top of a silicon dioxide membrane above a micromachined channel.

Both of these CPW structures have their own specific advantages. The microshield line has a large bandwidth, low dispersion and near-zero dielectric loss, whereas the suspended CPW is compatible with CMOS fabrication processes currently available in commercial foundries and, thus, is suitable for monolithic integration with commercial CMOS devices and circuits.

The microshield line is constructed from a central metallic strip with two CPW ground planes, which are supported on a dielectric membrane and backed by a metal shielding cavity. It is fabricated by bonding two silicon wafers together. The first wafer holds a thin air-suspended membrane with a CPW on the top surface. The wafer is selectively etched from the bottom, which creates a cavity, and the sidewalls of the etched cavity are metallised. The capacitances from the overlap between the top ground plane and the metallised walls are large, and so at microwave frequencies the overlap region performs as a short circuit at the designed frequency. Another wafer is metallised on the top and is bonded to the first to form a metallised bottom wall to the cavity.

13.3.3 MEMS Inductors

MEMS technology has been used to improve on-chip inductor performance by, for example, removing the substrate underneath the inductor or lifting the inductor structure above the substrate, as shown in Figure 13.3. This technique reduces the substrate loss and parasitic capacitance and, hence, gives a higher Q-factor and self-resonant frequency.

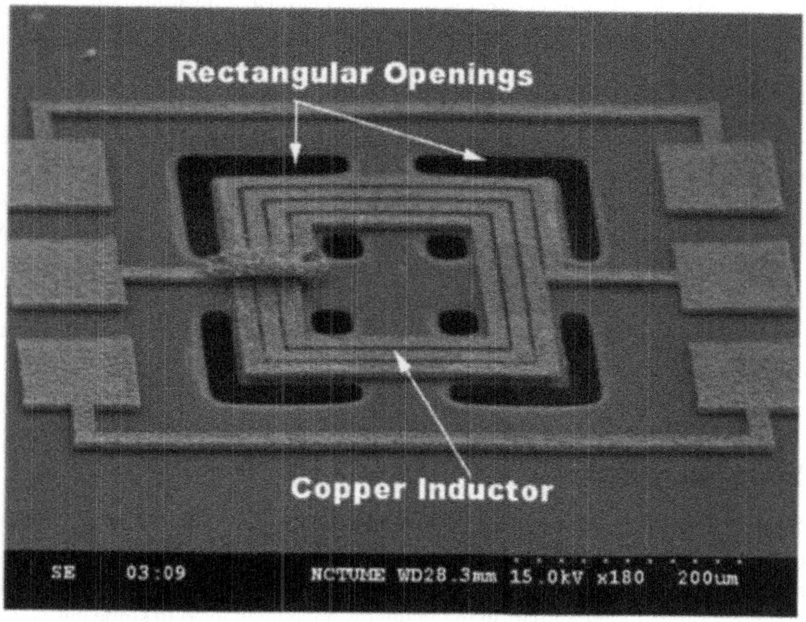

Figure 13.3 Scanning electron microscopy images of robust inductor [7]. *Source*: Jr-Wei Lin, Chen *et al.* (2005). Reproduced with permission of IEEE.

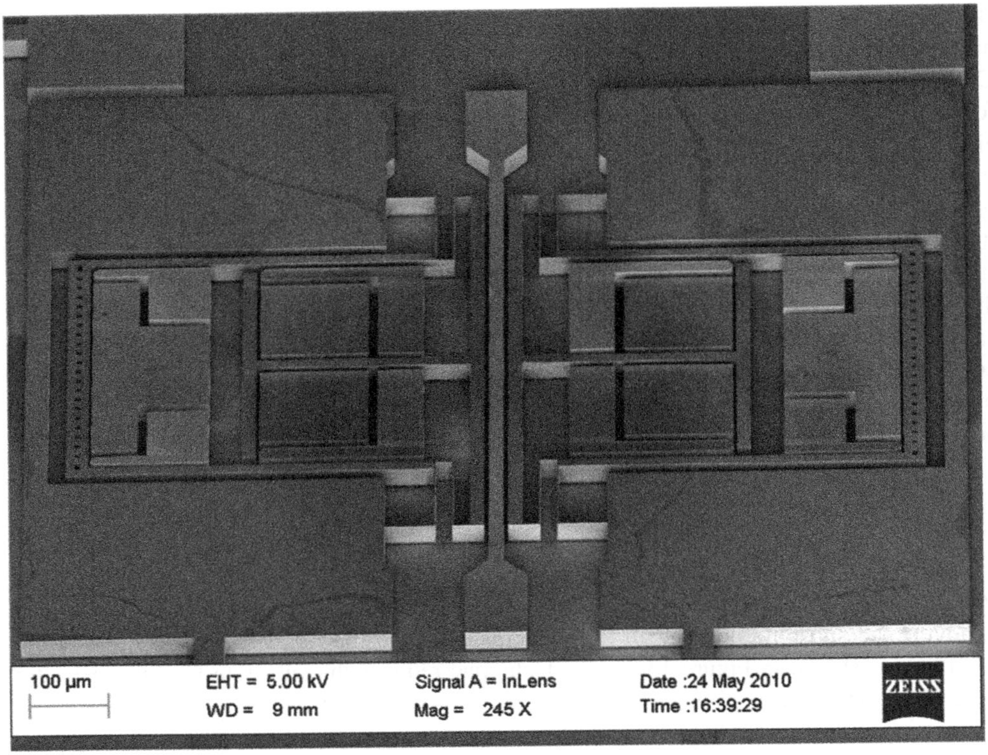

Figure 13.4 MEMS tuneable capacitor based on in-phase moveable sidewalls of the ground layer.

13.3.4 MEMS Tuneable Capacitor

Voltage-tuneable capacitors (varactors) are widely used in RF circuits. However, conventional semiconductor diode varactors have various disadvantages, including low Q-factor, small tuning range, and sometimes significant power consumption. MEMS varactors can overcome these drawbacks and can offer superior RF performance. To date, various types of MEMS varactors have been reported, including interdigitated, movable-dielectric and digitally controlled parallel-plate varactors. Figure 13.4 shows the basic concept of one type of MEMS tuneable capacitor, based on in-phase moveable sidewalls of the ground layer of a 3-D micromachined coplanar transmission line, displaced by integrated MEMS electrostatic actuators [8].

13.4 RF MEMS Switches

RF MEMS switches can be categorised into two groups, shunt and series topology, according to the operating mechanism [9, 10]. RF MEMS switches typically rely on electrostatic actuation by a microactuator that displaces a movable structural element, in a manner similar to an electromechanical relay. Other actuation mechanisms have been used in some cases, including electromagnetic, piezoelectric and electro-thermal. Most often, either a cantilever or a bridge forms the moving part of a switch. A cantilever is anchored at one and is deflected at the other by the actuator, while a bridge is anchored at both ends and actuated in the middle. Sometimes, a diaphragm is used, where a thin membrane is fixed around the edges and suspended over a cavity.

The most important differentiator in MEMS switches is the contact mechanism, which can be capacitive or ohmic fashion: metal-insulator-metal or metal-to-metal. In both types, some or all of the metallic parts are suspended over another metal contact beneath. When the actuation voltage is applied, the electrostatic force causes mechanical deformation and, whatever the polarity of the actuation voltage, the upper metal is pulled down. The bending leads to an opposing tensile force, and when a certain threshold voltage is reached the electrostatic force exceeds the tensile force, which means the top contact bends enough to touch the bottom contact. When the two contacts are touching the switch is in the 'on' state, and when they are not actuated the switch is in the 'off' state. If the applied actuation voltage is reduced, the top contact will eventually rise so that the two contacts are no longer touching. However, the switch has a hysteresis characteristic, and so the voltage at which the top contact rises is less than the threshold voltage needed to force them to touch originally.

There are two factors that affect the actuation of switches. The voltage required to pull down the switch depends on the stiffness of the cantilever or bridge. The release of the switch depends critically on the stiction between the contacts. The stress is generally tensile and increases the spring constant, which increases the electrostatic force/voltage needed to make the two contacts touch. Stiction is exhibited when the top and bottom contact bind together due to the microscopic surface forces. This is most problematic in metal-to-metal contacts. Using a capacitive switch in which a thin dielectric layer is placed between the contacts reduces the likelihood of stiction. A capacitive switch is sufficient for most RF applications. Applying a DC signal between the electrodes creates the pull-in force and the force between top and bottom of the actuation electrodes is calculated by:

$$F = \frac{\varepsilon A V^2}{2\left[g + \frac{t_d}{\varepsilon_r}\right]} \tag{13.2}$$

where V is the voltage between upper and lower electrodes, g is the gap distance between electrodes, A is the area of the electrodes, t_d is the thickness of the bottom electrode dielectric cover and ε_r is the relative dielectric constant of bottom electrode dielectric cover.

Forces on the order of μN are usually sufficient to close the switch contacts, with electrode voltages of 20–80 V being typical. The actuation electrodes of the switch have capacitance, and a small amount of energy is therefore required to actuate the switch, given by $\frac{1}{2}CV^2$. The power dissipation then depends on the switching speed and is given by $\frac{1}{2}CV^2 \times f_s$ where f_s is the switching frequency. The mechanical resonant frequency, f, of a switch is given by:

$$f = \left(\frac{K}{M}\right)^{1/2} 2\pi \tag{13.3}$$

where K and M are the equivalent spring constant and mass, respectively.

The switching time of the MEMS switch and the electrostatic force required increase as the gap decreases. It is much quicker for the bridge to drop than it is for it to rise to the up state, and therefore the longest time is usually given for the limitation of RF MEMS switching

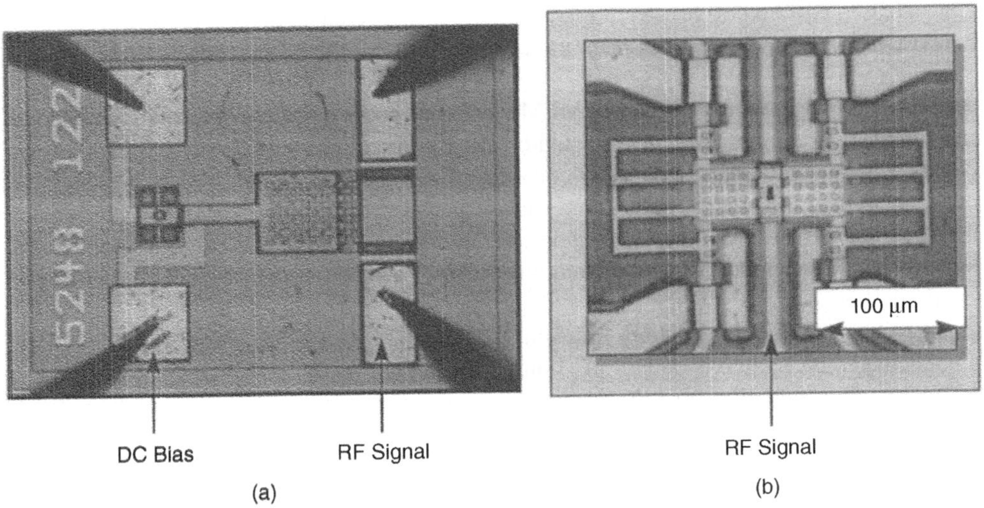

Figure 13.5 Switches developed at (a) Hughes Research Laboratories and (b) Rockwell Science Centre [11]. *Source*: Brown, E.R. (1998). Reproduced with permission of IEEE.

speed. The above has assumed that the switch is operating with a DC voltage. With an AC voltage, the membrane will behave in the same way as it does for DC voltage as long as the signal frequency of the voltage is much lower than the natural mechanical frequency of the switch. This means switching is induced when the amplitude of the signal reaches the threshold voltage. If the frequency of the AC signal is much greater than the mechanical resonant frequency of the switch, then the membrane will respond to the r.m.s voltage between the electrodes. If the r.m.s voltage is large enough to induce switching without the need for a DC bias, then RF-induced switching is often displayed.

Two types of RF MEMS switches are discussed in this section. In the first type, a metal-to-metal contact is used in a series single-pole-single-throw (SPST) configuration, with a cantilever or bridge holding a contact over a gap in the signal conductor. In the second type, a metal-insulator-metal contact is realised using a bridge structure, RF grounded at both ends and help above the signal conductor, to realise a SPST switch connected to ground. Figure 13.5 shows series-configured switches developed at Hughes Laboratories and Rockwell Science Centre. For these switches, the switch is in the *on* state when the metal beam is in the down position, with thin-film dielectric layer on top of the transmission line underneath the beam. By separating the actuation electrode from the RF contacts, they can have a large area which allows a lower actuation voltage for a long physical displacement. This is important because a significant gap between the contacts is required in the *off* state in order to give high isolation.

Figure 13.6 shows an RF shunt capacitive switch developed by Texas Instruments. The parallel plate approximation gives the values for C_{on} and C_{off} as follows:

$$C_{\text{on}} = \left[\frac{h_A}{\varepsilon_0 A} + \frac{h_D}{\varepsilon_0 A}\right]^{-1} \tag{13.4}$$

$$C_{\text{off}} = \frac{\varepsilon_0 A}{h_D} \tag{13.5}$$

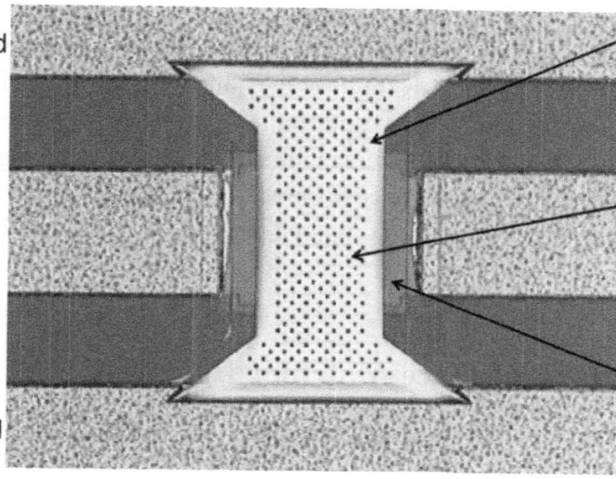

Figure 13.6 Switch developed at Texas Instruments [11]. *Source*: Brown, E.R. (1998). Reproduced with permission of IEEE.

The ratio of C_{off} to C_{on} is important because, as stated above, C_{on} needs to be as low as possible and C_{off} needs to be as high as possible; therefore, the ratio needs to be very high for good isolation and insertion loss. The ratio is given by:

$$\frac{C_{\text{off}}}{C_{\text{on}}} = 1 + \frac{\varepsilon_D h_A}{\varepsilon_0 h_D} \tag{13.6}$$

The isolation for the series configuration degrades as the frequency increases, whereas the isolation improves with frequency for the parallel configuration. The isolation is defined as the power at the load that is divided by the delivered power from the source in the *off* state; which is also equal to $1/|S_{21}|$. For a series configuration, the isolation is given by:

$$I = \frac{(\omega Z_0 C_s)^2}{\left[1 + (\omega Z_0 C_s)^2\right]} \tag{13.7}$$

and for parallel configurations the isolation is given by:

$$I = \frac{1}{\left[1 + (\omega Z_0 C_s)^2\right]} \tag{13.8}$$

It can be seen from these equations that, as the frequency tends to zero, the series configuration has isolation that tends towards 0 (-∞ dB), while for the parallel configuration the isolation tends towards 1 (unity, 0 dB).

The performance of the RF switch is defined based on parameters that include minimal power consumption, insertion loss, low intermodulation, high port isolation between the output switching points and, of course, cost. The use of air gaps in the fabrication of RF MEMS switches results in very high isolation due to minimal off-state capacitance at high frequencies, and the device exceeds the p-i-n and FET alternatives with respect to the aforementioned parameters. However, apart from the stiction issue, several challenges are still being tackled by enhancements in the design, fabrication and use of RF MEMS switches. This includes the inability to handle high-power RF signals, high-voltage drive creating the need for a voltage up-converter, relatively low switching speed and limited reliability in terms of the number of expected switching cycles. The most prominent challenge, however, is the high packaging cost as it requires an inert gas and low-humidity environment, otherwise its lifetime is reduced and this makes RF MEMS less attractive when compared to the p-i-n diode and FET switch costs.

13.5 Reconfigurable Impedance-Matching Networks

Reconfigurable impedance-matching networks are used to tune the input and output impedances of an RF circuit or component in order to tune the nominal frequency or optimise the performance for different initial conditions. There are various types of MEMS reconfigurable impedance-matching networks. One of the most attractive solutions comprises a loaded transmission line having identical switched MEMS capacitors, or more precisely capacitive switches, whereby the impedance of the line is changed by varying the capacitive loading. The capacitors are digitally switched between the up-state and down-state and will have different capacitances in each state, giving a capacitance ratio C_{UP}:C_{DOWN}. The switched capacitors are formed by combining MEMS capacitive switches in series configuration with fixed metal-air-metal capacitors. Figure 13.7 shows a loaded CPW transmission line with eight switched capacitors [12] which, in practice, could be any number of elements, *N*. By digitally switching each of the MEMS capacitive switches up or down, a total number of 256 impedance points (2^N) can be obtained. The impedance (Z_L) and effective dielectric constant ($\varepsilon_{\text{reff-L}}$) of the

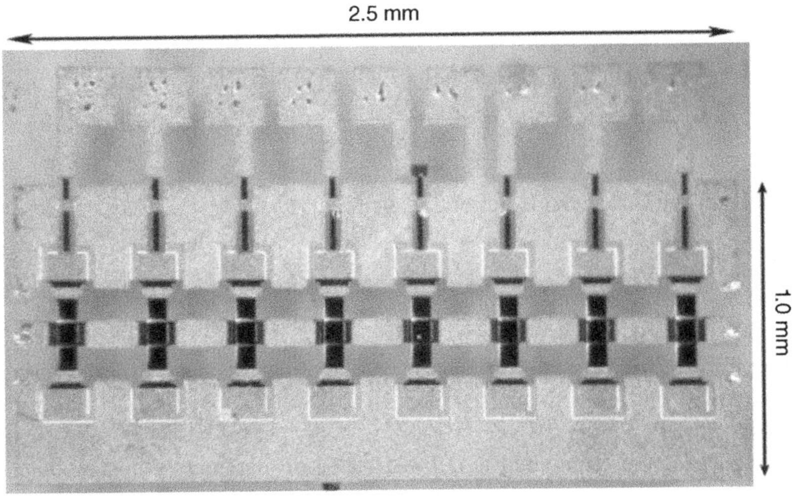

Figure 13.7 Circuit diagram of an eight-element reconfigurable matching network [12]. *Source*: Vähä-Heikkilä, T. and Rebeiz, G.M. (2004). Reproduced with permission of IEEE.

capacitively-loaded transmission line is computed using [12]:

$$K = 1 + \frac{cZ_oC}{s\sqrt{\varepsilon_{\text{reff}}}} \tag{13.9}$$

$$\varepsilon_{\text{reff-L}} = \varepsilon_{\text{reff}}\text{K} \tag{13.10}$$

$$Z_L = Z_U, Z_D = Z_0/\sqrt{\text{K}} \tag{13.11}$$

where c is the speed of light in free space, C is the capacitance of the switched capacitor (C_{UP}, C_{DOWN}) $\varepsilon_{\text{reff}}$ is the effective dielectric constant of the unloaded line, s is the period of loading structure, Z_U is the impedance of the line in up-state of MEMS switches, Z_D is the impedance of the line in down-state of MEMS switches, and Z_0 is the impedance of the unloaded line.

Also, the Bragg frequency, which determines the highest operational frequency, can be calculated by:

$$f_B = \frac{cZ_L}{\pi s Z_o\sqrt{\varepsilon_{\text{reff}}}} \tag{13.12}$$

Although the previously reported MEMS impedance-matching networks achieved a wide range of impedance tuning, their performances are still quite limited in terms of high-power handling because of the MEMS capacitive bridges or switched capacitors on the transmission line, which normally suffer from self-actuation with large RF power levels. These impedance-tuning networks also exhibit a much higher insertion loss when a greater number of tuning elements is used to cover a wider impedance tuning range.

13.6 MEMS Phase Shifters

Loaded-line MEMS phase shifters use MEMS structures [13–16], especially shunt capacitive switches, to change the phase velocity of a transmission line and produce a phase shift. Due to the low insertion loss of MEMS devices, a phase shifter constructed using MEMS has a considerably lower loss than traditional phase shifters constructed on, for example, GaAs using FETs. For a state-of-the-art three-bit MEMS phase shifter, the average insertion loss is 0.9 dB from 10 to 14 GHz, which is approximately 3 dB better than GaAs FET designs. The improvement increases with higher design frequencies. The reduction of loss may mean that amplifiers are not required, which gives a great saving in the required power of a transceiver and makes new architectures possible. The performance of a phase shifter is denoted by its insertion loss, useable bandwidth, power consumption, phase shift range, and so forth.

Based on the control method, phase shifters can be classified as either analogue or digital. In the analogue type, the transmission properties of the transmission line, especially its phase velocity, are continuously varied as a function of the actuation voltage. In a capacitively loaded line, the phase velocity decreases as the shunt load capacitance increases. Compared to the unloaded transmission line, the signal requires a longer propagation time along the loaded transmission line.

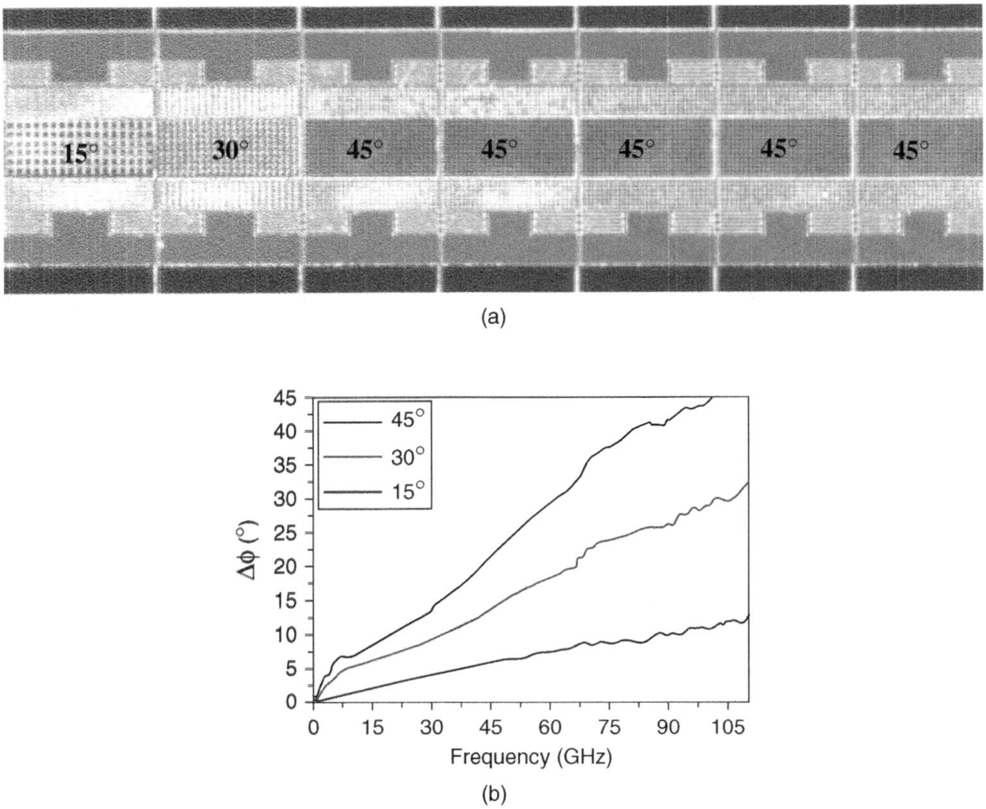

Figure 13.8 An example of a digital MEMS phase shifter and its phase response.

While analogue phase shifters give a phase shift that varies continuously, digital phase shifters give a phase shift that can be changed in discrete steps, as shown in Figure 13.8. For a digital phase shifter design, the challenge is to eliminate or limit the signal attenuation to its very minimum by transmitting the RF signal through the best possible or shortest routes. There are four conventional MEMS phase shifter architectures: the switched-line phase shifter; loaded-line phase shifter; switched low-pass and high-pass filter phase shifter; and the reflection-type phase shifter.

13.7 Tuneable Filters

Filters are typically realised by some form of resonating component, including lumped-element *L-C* type, transmission line resonators, cavity resonators, or acoustic wave resonators. The *L-C* tank circuit usually resonates at a fixed frequency as determined by the values of *L* and *C*, but with MEMS technology the fixed value capacitance can be replaced with a varactor. Therefore, RF MEMS switches and varactors are promising elements for tuneable filters [18]. The resonant frequency of the resonators can be varied as much as the capacitor capacitance tuning range permits. For the most part, the tuning functions of these filters are achieved using

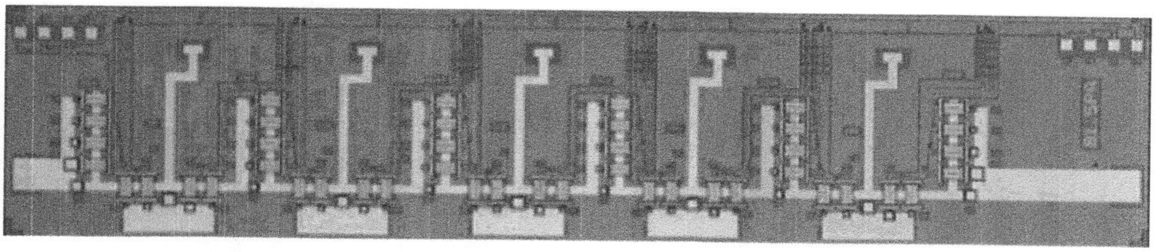

Figure 13.9 Photograph of a UHF five-pole MEMS reconfigurable filter [19]. *Source*: Brank, J. *et al.* (2001). Reproduced with permission of IEEE.

tuneable capacitors or switched capacitor banks. RF MEMS filters have been reported to have low intermodulation levels which make them have less issues of nonlinearity.

Figure 13.9 shows an example of a tuneable UHF five-pole filter using capacitive switches [19]. The filter combines fixed capacitors with MEMS capacitive switches to create a switchable capacitance for the resonators, which had fixed inductances of 2.9 nH. This is a key part of the design, since if a resonator is required to have a tuning range of:

$$\propto = \frac{f_{max}}{f_{min}} \tag{13.13}$$

then the capacitors must have a tuning range of approximately:

$$\frac{C_{max}}{C_{min}} = \propto^2 \tag{13.14}$$

The UHF filter successfully tuned from 885 to 986 MHz with a bandwidth of 180 MHz using four-bit control of the variable capacitors.

13.8 MEMS Antennas

Reconfigurability has become a highly desirable property of communications subsystems and components, no less so for antennas [20, 21]. A number of properties of antennas can be tuned, such as radiation pattern, polarisation, operating frequency, to suit certain operating conditions. For example, the radiation pattern could be manipulated in some applications to deliberately avoid interference or extend the beam-steering capability of large phased array antennas. In the majority of cases, an antenna is required to change the operating frequency whilst maintaining its radiation characteristics. Many mobile handsets already use an antenna tuning element for this purpose.

Antenna reconfigurability is, therefore, another important application of MEMS technology. MEMS devices are preferred to other methods of reconfiguring antennas due to their superior linearity performance and resultant lower signal distortion as compared to semiconductor devices [14]. A few methods of reconfiguring antennas involving MEMS devices will now be highlighted. The first method involves changing the shape of the effective antenna radiating structure by including MEMS switches in the antenna circuitry which, when switched on or off, will change the pattern of radiation or the frequency. It is well known that antenna

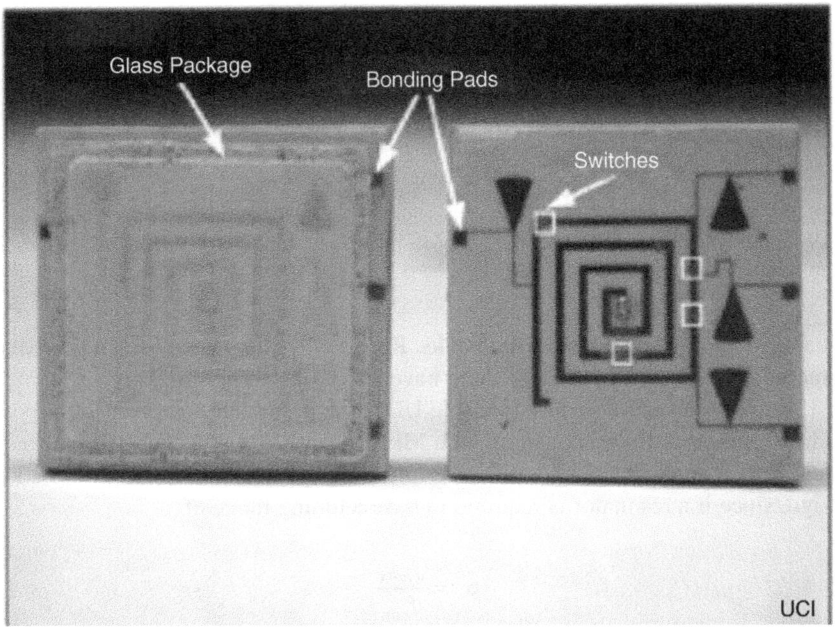

Figure 13.10 Reconfigurable rectangular spiral antenna [22]. *Source*: Haridas, N. *et al.* (2008). Reproduced with permission of IEEE.

operating frequency is related to the electrical length of the antenna, and this relationship is exploited to alter the frequency by changing the effective electrical length of the antenna. Figure 13.10 illustrates one implementation of this concept, in which four switches are included along the length of a spiral antenna to create five sections. By switching on/off each switch, progressive lengths of the antenna can be added to the antenna radiating structure. Another method involves integrating the antenna with MEMS capacitors which can tune the resonant frequency. This involves capacitively loading the antenna structure by varying the capacitances of the integrated varactors, as shown in Figure 13.11.

13.9 RF MEMS Fabrication and Packaging

The fabrication of MEMS devices is based on many established processes from the highly developed semiconductor fabrication industry. Processes such as photolithography, deposition and chemical etching are used for the fabrication of MEMS devices [24–28]. For better integration with other IC devices, the selection of wafer to be used in MEMS fabrication is critical. High-resistivity silicon is normally used as it ensures minimal substrate losses and a better integration with CMOS devices. A key process that is used to fabricate MEMS devices is generally referred to as 'micromachining'. Two variants of this process exist. In the first variant, structural components of the MEMS device are patterned, deposited and etched entirely on the substrate surface by the use of sacrificial layers. This process is termed 'surface micromachining' and allows greater flexibility in the choice of substrate material as

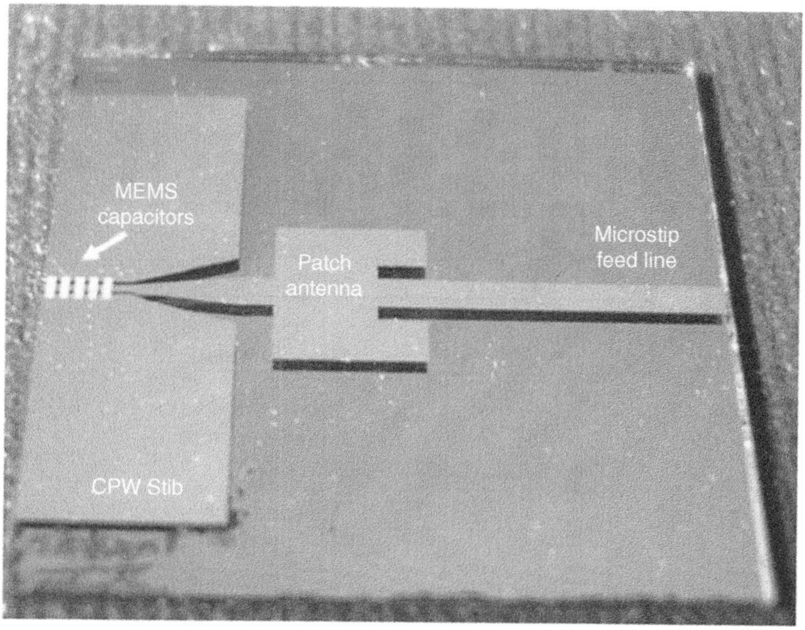

Figure 13.11 Micromachined patch antenna integrated with MEMS varactors [23]. *Source*: Erdil, E. *et al.* (2007). Reproduced with permission of IEEE.

the structural layers are formed externally to the substrate and thus the substrate mostly acts as a rigid mechanical base. Glass and plastic are alternative substrates that have been used to this effect. An example of the surface micromachining technique is illustrated in Figure 13.12.

The selective removal of the sacrificial layer is an important step in surface micromachining, and care must be taken to ensure that this is achieved without damaging the more important structural layer. Gold, aluminium and copper are common metals used for contact metals and electrodes; the dielectric, where required, is most often formed using silicon nitride. The sacrificial layer is usually formed from photoresist, which can be removed conveniently by a specific etching method such as O_2 plasma, while the number of mask layers required is usually defined by the complexity of the structure to be manufactured, with each mask producing a different pattern. The second commonly used micromachining technique is referred to as 'bulk micromachining', and differs from surface micromachining in that the substrate material is patterned and shaped to effectively form the microstructures within the substrate. Single-crystal silicon is typically used as the substrate, as it forms an important functional component of the resulting device and can be anisotropically etched to high precision.

Post fabrication, devices produced on different substrates may need to be bonded together. Silicon, glass and metal can be bonded using eutectic bonding, tin-solder bonding, or another convenient bonding mechanism for mechanical support and protection. MEMS devices are very sensitive to external forces, mechanical shock and thermal expansion especially, as they contain movable fragile parts that may be damaged by otherwise harmless vibrations, or humidity – which can cause contact parts to stick together. This makes it imperative that MEMS devices are packaged in a controlled environment. Standard wafer dicing, for instance,

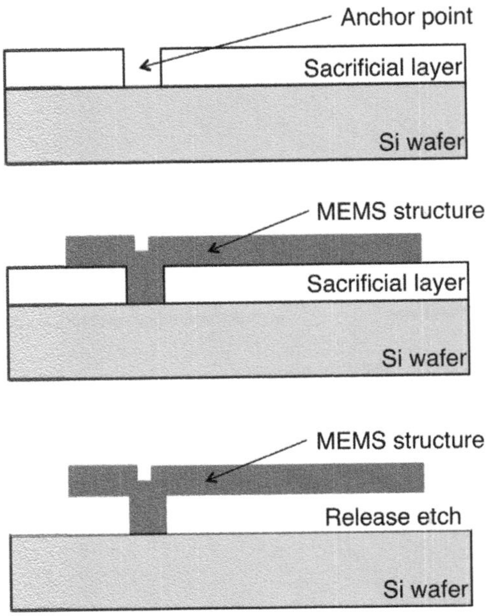

Figure 13.12 Surface micromachining technique.

will destroy the MEMS device and, as a result, the packaging of MEMS devices is often performed on the wafer during wafer processing. This packaging technique is termed 'wafer-level packaging' or 'zero-level packaging', and it allows for hermetic cavity encapsulation of the movable parts of the MEMS device on the wafer prior to dicing. In addition to hermeticity, the sealed cavity must be strong, and equipped with feed-throughs for electrical signals. The cavity is formed usually by a capping wafer added to the MEMS structure, as illustrated in Figure 13.13.

To bond the capping chip to the MEMS wafer, benzocyclobutene (BCB) is utilised which also acts to seal off the formed cavity. Benzocyclobutene is chosen because if possesses a low loss tangent, high resistivity and a low dielectric constant. The process of mounting the capping wafer and bonding it to the MEMS wafer is called 'flip-chip mounting'. A glass-frit sealant can be used instead of BCB. Tin solder bonding and localised silicon fusion and eutectic bonding have also been used to good effect. The dicing of the wafer is carried out after the MEMS has been sealed in a cavity.

13.10 Reliability and Design Consideration of RF MEMS Devices

The most important feature of RF MEMS devices is their ability to make reconfigurable components of small size. However, the small moving parts need delicate handling, thereby raising issues with respect to reliability [29, 30], life span, reproducibility, cost-effective packaging, power control and management, efficient yield for economy of scale, and so forth. So far, the common practice for evaluating reliability has involved a cycling process which may not necessarily be suitable for assessing the operation of devices in the field. Hence, there

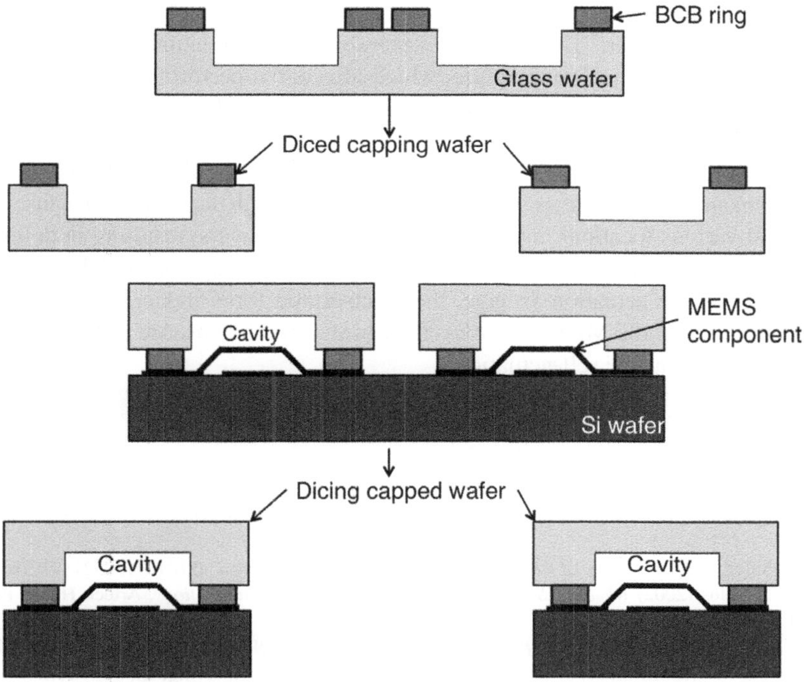

Figure 13.13 Zero-level packaging process flow.

is a high demand for information about degradation factors to help define the standards for high reliability, in contrast to other well-established IC-related technologies.

One of the first known attempts to improve the performance of MEMS devices was the micromachining method introduced to overcome substrate parasitic effects by creating a quasi-free-space behaviour. Other techniques following this featured several designs based on electrostatic actuation, which introduces a number of performance concerns to be considered during the design. Trapped charge is experienced when, as a result of trapped charges in the free interface of the MEMS device, a shift in the actuation voltage occurs, also causing a time-dependent lag in the switching characteristics of the moving part of the device. An attempt to tackle these issues included the first standard measurement techniques which had no ability to improve down-time performance but did acquire valuable performance information through a capacitance–voltage curve. This method was, however, limited by its inability to capture the deformation characteristics resulting from uneven charge distribution which caused actuation voltage ramps, and this led to the modification of the measurement method to reflect the effect of uneven charge distribution. Specialised devices such as electrostatic fringing-field actuators (EFFA) can be used to apply stress on the substrate, thereby narrowing the capacitance–voltage characteristics curve. This shows that both charging of the substrate and the free interface affect the performance of the MEMS device.

In packaging MEMS devices, the need to protect them from damage extends beyond the presence of impurities, to the environmental conditions in which they are being packaged. One core element among these conditions is the thermal factor, which may have a damaging effect on the dielectric of the device and result in faster harmful charging and degradation or outright breakdown during actuation. There is also a possibility of RF detuning due to permanent deformation from an elevated temperature, which affects the performance and reliability of the MEMS device. All of these packaging considerations need to be carefully considered during the design of the device.

RF power cannot be controlled in packaging and it affects the MEMS device through the resulting current and voltage effects. As the current flows through the structure it causes heating and hotspots at various locations, especially in thin metallic layers, resulting in deformation, especially for poorly designed devices. Additionally, if the voltage of the RF signal is high enough, relative to the actuation voltage, the electrostatic force induced by the RF signal can unwantedly actuate the MEMS cantilevers or bridges in the absence of the DC actuation voltage; this is known as RF self-actuation.

References

1. Rebeiz, G.M. (2003) *RF MEMS: Theory, Design, and Technology*. John Wiley & Sons. ISBN: 978-0-471-20169-4.
2. Lucyszyn, S. (ed.) (2010) *Advanced RF MEMS*. Cambridge University Press. ISBN: 9780521897716.
3. Santos, H.J. de los (2002) *RF MEMS circuit design for wireless communications*. Artech House Publishers. ISBN-10: 1580533299.
4. Tilmans, H.A., De Raedt, W., and Beyne, B. (2003) MEMS for wireless communications: from RF-MEMS components to RF-MEMS-SiP. *Journal of Micromechanics and Microengineering*, **13**, S139–S163.
5. Reid, J.R., Marsh, E.D., and Webster, R.T. (2006) Micromachined rectangular-coaxial transmission lines. *IEEE Transactions on Microwave Theory and Techniques*, **54** (8), 3433–3442.
6. Simons, R. (2001) *Coplanar Waveguide Circuits, Components and Systems*. Wiley-IEEE Press, New Jersey.
7. Wei Lin, Jr, Chen, C.C., and Yu-Ting Cheng (2005) A robust high-Q micromachined RF inductor for RFIC applications. *IEEE Transactions on Electron Devices*, **52** (7), 1489–1496.
8. Shah, U., Sterner, M., and Oberhammer, J. (2013) Multi-Position RF MEMS tunable capacitors using laterally moving sidewalls of 3-D micromachined transmission lines. *IEEE Transactions on Microwave Theory and Techniques*, **61** (6), 2340–2352.
9. Rebeiz, G.M. and Muldavin, J.B. (2001) RF MEMS switches and switch circuits. *IEEE Microwave Magazine*, **December**, 59–71.
10. Dimitrios, P., Pacheco, S.P., Sarabandi, K., and Katehi, L.P. (2003) Electromechanical considerations in developing low-voltage RF MEMS switches. *IEEE Transactions on Microwave Theory and Techniques*, **51** (1), 259–270.
11. Brown, E.R. (1998) RF-MEMS switches for reconfigurable integrated circuits. *IEEE Transactions on Microwave Theory and Techniques*, **46** (11), 1868–1880.
12. Vähä-Heikkilä, T. and Rebeiz, G.M. (2004) A 4–18-GHz reconfigurable RF MEMS matching network for power amplifier applications. *International Journal of RF and Microwave Computer-Aided Engineering*, **14** (4), 356–372.
13. McFeetors, G. and Okoniewski, M. (2006) Distributed MEMS analog phase shifter with enhanced tuning. *IEEE Microwave and Wireless Components Letters*, **16** (1), 34–36.
14. Pillans, B., Eshelman, S., Malczewski, A., Ehmke, J., and Goldsmith, C. (1999) Ka-band RF MEMS phase shifters. *IEEE Microwave and Guided Wave Letters*, **9** (12), 520–522.
15. Rebeiz, G.M., Tan, Guan-Leng, and Hayden, J.S. (2002) RF MEMS phase shifters: design and applications. *IEEE Microwave Magazine*, **3** (2), 72–81.
16. Somjit, N., Stemme, G., and Oberhammer, J. (2009) Binary-coded 4.25-bit W-band monocrystalline–silicon MEMS multistage dielectric-block phase shifters. *IEEE Transactions on Microwave Theory and Techniques*, **57** (11), 2834–2840.
17. Somjit, N., Stemme, G., and Oberhammer, J. (2011) Power handling analysis of high-power W-band all-silicon MEMS phase shifters. *IEEE Transactions on Electron Devices*, **58** (5), 1548–1555.

18. Kamran, E. and Rebeiz, G.M. (2005) A 12–18-GHz three-pole RF MEMS tunable filter. *IEEE Transactions on Microwave Theory and Techniques*, **53** (8), 2566–2571.
19. Brank, J., Yao, J., Eberly, M., Malczewski, A., Varian, K., and Goldsmith, C. (2001) RF MEMS-based tunable filters. *International Journal of RF and Microwave Computer Aided Engineering*, **11**, 276–284.
20. Schoebel, J., *et al.* (2005) Design Considerations and Technology Assessment of Phased-Array Antenna Systems with RF MEMS for Automotive Radar Applications. *IEEE Transactions on Microwave Theory and Techniques*, **53** (6), 1968–1975.
21. Greg, H. and Bernhard, J.T. (2006) Integration of packaged RF MEMS switches with radiation pattern reconfigurable square spiral microstrip antennas. *IEEE Transactions on Antennas and Propagation*, **54** (2), 464–469.
22. Haridas, N., Erdogan, A.T., Arslan, T., Walton, A.J., Smith, S., Stevenson, T., Dunare, C., Gundlach, A., Terry, J., Argyrakis, P., Tierney, K., Ross, A., and O'Hara, T. (2008) Reconfigurable MEMS Antennas. NASA/ESA Conference on Adaptive Hardware and Systems (AHS '08), 22–25 June, pp. 147–154.
23. Erdil, E., Topalli, K., Unlu, M., Civi, O.A., and Akin, T. (2007) Frequency tunable microstrip patch antenna using RF MEMS technology. *IEEE Transactions on Antennas and Propagation*, **55** (4), 1193–1196. doi: 10.1109/TAP.2007.893426.
24. Judy, J.W. (2001) Microelectromechanical systems (MEMS): fabrication, design and applications. *Smart Materials and Structures*, **10**, 1115–1134.
25. Yao, Z.J. (1999) Micromachined low-loss microwave switches. *IEEE Journal of Microelectromechanical Systems*, **8** (2), 129–134.
26. Tilmans, H.A., *et al.* (2001) Wafer-level packaged RF-MEMS switches fabricated in a CMOS Fab. IEEE Electron Devices Meeting (IEDM '01), December, pp. 921–924.
27. Cheng, Y.T. (2000) Localized silicon fusion and eutectic bonding for MEMS fabrication and packaging. *IEEE Journal of Microelectromechanical Systems*, **9** (1), 3–8.
28. Goyal, A., *et al.* (2004) Tin-based solder bonding for MEMS fabrication and packaging applications. *Journal of Micromechanics and Microengineering*, **14** (6), 819–925.
29. Melle, S., *et al.* (2003) Reliability overview of RF MEMS devices and circuits. 33rd European Microwave Conference Proceedings (IEEE Cat. No.03EX723C).
30. Rottenberg, X., Czarnecki, P., Tilmans, H., De Raedt, W., and De Wolf, I. (2008) Multi-physics simulation and reliability analysis for RF-MEMS devices. EuroSimE 2008: International Conference on Thermal, Mechanical and Multi-Physics Simulation and Experiments in Microelectronics and Micro-Systems, pp. 1–6.

14

Antennas and Propagation

14.1 Introduction

Although the telegraph, a communication system using electrical signal over wires, was first introduced in 1844, it was not until 1896 that the first public transmission of wireless signals was demonstrated by Marconi, based on the electromagnetic theory developed by Maxwell and the pioneering work of Hertz and others. During World War II, wireless communication began to develop at a particularly fast pace. Nowadays, electronics and communication are essential to modern life and we are surrounded by electromagnetic waves emitted from a wide range of wireless devices [1]. Antennas are one of the main components in any wireless device. They provide the coupling between electromagnetic waves guided by transmission lines and those in free space, or vice versa. In addition, thanks to a significant effort to reduce their size, nowadays there is often more than one antenna in many wireless communication devices [2].

An antenna is a passive device that provides the transition of electromagnetic waves from a guided wave to a free-space wave, or vice versa, carrying power from a transmitter into the free space in three dimensions, or from free space to a receiver. The properties of an antenna can be summarised as follows:

- It is acts as a sensor to electromagnetic waves.
- It can be used to radiate electromagnetic energy in different directions.
- It is used to interface both the transmitter and receiver to the free space channel so as to reduce signal interference and noise.
- It can convert energy from one form to another (it acts as a transducer).

The key properties of an antenna are:

Radiation resistance, *R*: This is the resistance needed to dissipate a given amount of the radiated power when signals are radiated. It is obtained

Microwave and Millimetre-Wave Design for Wireless Communications, First Edition.
Ian Robertson, Nutapong Somjit and Mitchai Chongcheawchamnan.
© 2016 John Wiley & Sons, Ltd. Published 2016 by John Wiley & Sons, Ltd.

using the expression:

$$R = \frac{Radiated\ Power}{\frac{i_{av}}{\sqrt{2}}} \tag{14.1}$$

where i_{av} is the average applied current fed to the antenna.

Antenna effective length, l_{eff}: The effectiveness of an antenna in propagating and receiving a signal can be measured by either using the effective area, for parabolic and horn antennas where physical size is a major contributor to the aperture, or effective length for monopole and dipole antennas where there is no direct link to the physical size of the aperture. The longer the effective length, the higher the receiving power of the antenna since power is proportional to the supply voltage. l_{eff} is given by:

$$l_{eff} = \frac{V_0}{E_s} \tag{14.2}$$

where E_s is the electric field intensity or strength of the antenna and V_0 is the supplied voltage. It is relevant to note that the effective length is lower than the actual length.

Intensity of radiation: This is a measure of the amount of power radiated per unit angle (solid angle in steradian) and it is independent of distance. It is the product of power density (S_r) and the square of the distance from the antenna, r:

$$RI = S_r \times r^2 \tag{14.3}$$

The S.I. unit is Watts per steradian.

Gain, G: This is the ratio of the power radiated in a particular direction to the power radiated by an ideal isotropic antenna. An isotropic antenna is one that radiates equally all directions. The gain is a measure of both the antenna's efficiency and its ability to focus power in a certain direction.

Antenna efficiency ($e_{antenna}$): This is the ratio of output power (radiated power) to the input power:

$$e_{antenna} = \frac{power_{radiated}}{power_{input}} = \frac{w_r}{w_i} \tag{14.4}$$

Directivity, D: This is the ratio of the power radiated in a particular direction to the power radiated by an ideal isotropic antenna that is transmitting the same amount of radiated power. If the antenna has loss, the theoretical isotropic antenna has to radiate more power in this definition. Directivity does not take into account the efficiency of the antenna. In contrast, the gain, G, takes both the directivity and efficiency of the antenna into account and is given by:

$$G = D \times e_{antenna} \tag{14.5}$$

Radiation pattern: This is a plot of the gain or directivity versus angle, either in three dimensions, or plotted in 2-D in certain planes that are defined by convention (e.g., E- and H-plane for certain antenna types).

Beamwidth: This is the width of the radiation pattern in degrees, normally defined 3 dB down on each side from the peak.

Antenna bandwidth: This is the frequency range over which the antenna parameters such as standing wave ratio, impedance, radiation pattern and so forth meet pre-defined targets. It is normally measured in Hz and measured at the –3 dB peak radiated power points or 2:1 input VSWR points.

Front-to-back ratio: This is the ratio of the power radiated in the desired forwards direction to that radiated in the reverse (opposite) direction. It is usually the ratio of the front lobe to back lobe gain of the radiation antenna.

Polarisation: This is a measure of the direction/orientation of the electric and magnetic field of the radiating wave. The radiated wave can have horizontal, vertical, circular or elliptical polarisation depending in the type of antenna and how it is fed.

Near-field and far-field region: Three fields surround the antenna: reactive near-field, radiating near-field and the far-field region. In the near-field, the radiation field changes rapidly, include radiating and reactive energy and the near-field boundary can be calculated from:

$$R(nearfield\ radius) = \frac{2L^2}{\lambda} \tag{14.6}$$

where L is the diameter of the antenna.

The radiation pattern of an antenna is characterised by the far-field region. Thus, the far-field region is normally the major concern when designing and measuring an antenna, and it is where the radiation pattern shape does not change with distance. However, the fields decay at $\frac{1}{R}$ and the power density at $\frac{1}{R^2}$ (where R is distance). In the far-field region, three conditions must be satisfied:

$$R > \frac{2L^2}{\lambda}, \quad R \gg L, \quad \text{and} \quad R \gg \lambda \tag{14.7}$$

Radio-wave propagation is a crucial contributor to the overall performance of a communications system. Propagation studies allow the radio channel to be measured and modelled so that optimum performance is achieved. Propagation issues may be considered to be simple, but in practice propagation is quite complicated. This chapter will describe a number of propagation mechanisms that determine the propagation of electromagnetic waves indoors and outdoors. In order to operate a reliable channel, an efficient estimation of channel parameters is required. Some key models are illustrated in order to study the propagation loss in particular environments.

14.2 Antenna Systems

Antennas can be divided into seven categories: wire antennas; reflector antennas; travelling-wave antennas; microstrip antennas; log-periodic antennas; aperture antennas; and other antennas. The antenna types discussed in this section are wire antennas, travelling-wave antennas, reflector antennas, microstrip antennas and the special case of near-field communication (NFC) antennas.

14.2.1 Wire Antennas

Wire antennas are the most familiar type of antenna for most people as they are used widely in everyday applications. This type of antenna can be subdivided into many variants; those that will be discussed in this section are dipole antennas, monopole antennas and loop antennas.

14.2.1.1 Dipole Antenna

The dipole antenna consists of two conducting wires or rods, which are fed with the signal through the feeder, which is positioned at the middle of the conducting wires. Figure 14.1 shows a basic diagram of a dipole antenna. There are many derivatives of dipole antennas, which are: short dipole antenna; half-wave dipole antenna; and folded dipole antenna. The adjective before the word 'dipole' represents how the conducting wires or rods are laid out. In a dipole, the size of the conducting wires or rods combined is equal to half the wavelength being transmitted. Having the electrical length half of the wavelength forms a resonant circuit, therefore cancelling imaginary impedances.

14.2.1.2 Short Dipole Antenna

Short dipole antennas, also known as Hertzian antennas, are the simplest dipole antennas. A dipole antenna is basically an open-circuited wire of length l with a signal fed into the middle

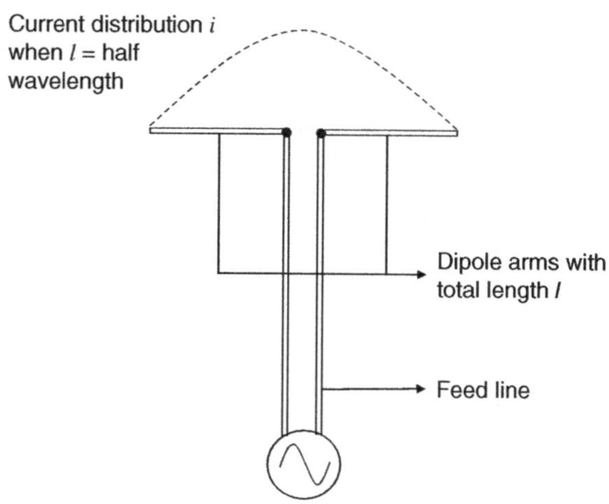

Figure 14.1 Dipole antenna.

of the wire, dividing the wire into two sections of length $l/2$. A 'short' dipole is one with length between $\lambda/50$ and $\lambda/10$, where λ is the wavelength of the signal transmitted. If the middle of the wire is placed on the origin of the z-axis, the current in the wire is equal to:

$$I(z) = I_0\left(1 - \frac{2|z|}{l}\right) \tag{14.8}$$

and the fields radiated in the far field are given by:

$$E_\theta = \frac{j\eta k I_0 l e^{-jkr}}{8\pi r}\sin\theta \tag{14.9}$$

$$H_\phi = \frac{E_\theta}{\eta} \tag{14.10}$$

$$E_r = H_r = E_\phi = H_\theta = 0 \tag{14.11}$$

where E_θ is the electric field, H_ϕ is the magnetic field, η is the free-space impedance, k is the wavenumber, and r is the distance from the antenna. Figure 14.2 shows the radiation pattern of the short dipole antenna.

Figure 14.2 shows that the radiation pattern mainly depends on the *sine* in the equation; when theta is 0° then minimum electrical field is emitted, whereas when theta tends to 90° the electrical field increases to maximum and then decreases down to 0, as theta tends to completing a half cycle.

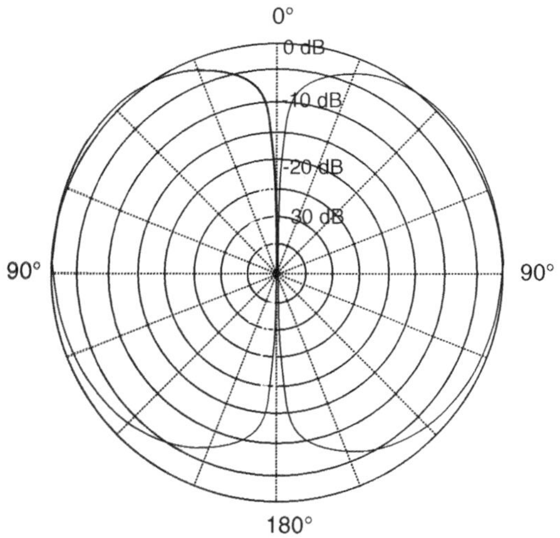

Figure 14.2 Radiation pattern of a short dipole antenna.

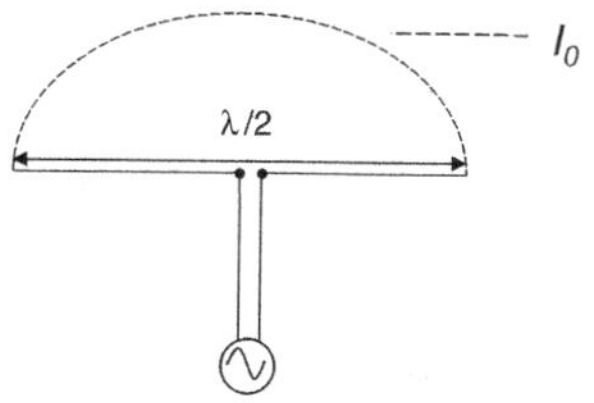

Figure 14.3 Current distribution of a half-wave dipole antenna.

14.2.1.3 Half-Wave Dipole Antenna

Half-wave dipole antennas are similar to the short dipole antenna in geometry, but the length is half the wavelength of the transmitted signal. So, for example, if the frequency of the transmitted signal is 1 GHz, then the wavelength would be 0.3 m and the half wavelength dipole antenna length is 0.15 m. Figure 14.3 shows the current distribution along the wires of the antenna. The E and H-field from the half-wave dipole antenna are given by:

$$E_\theta = \frac{j\eta I_0 e^{-jkr} \cos\left(\frac{\pi \cos\theta}{2}\right)}{8\pi r \sin\theta} \tag{14.12}$$

$$H_\phi = \frac{E_\theta}{\eta} \tag{14.13}$$

14.2.1.4 Folded Dipole Antenna

Folded dipole antennas are similar to the simple dipole antenna but a conductor is used to connect the two ends to form a complete loop of wire. Since the antenna looks like it is folded back on itself, it is called folded dipole antenna. Figure 14.4 shows a diagram of the folded dipole antenna structure.

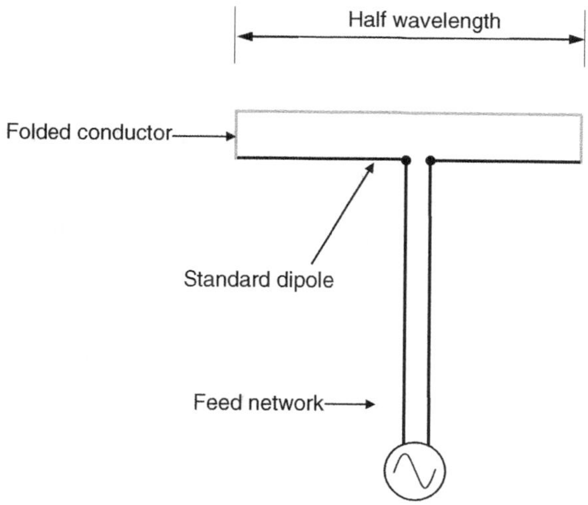

Figure 14.4 Folded dipole antenna.

One of the primary reasons for utilising a folded dipole antenna instead of the basic dipole antenna is the increase in the input impedance it provides at the feed point. If the conductors in the primary dipole and the second one, the folded conductor, have the same diameter, then it is found that there is four times the input impedance. In a normal dipole antenna, the current flowing along the dipole arms are in phase and therefore the radiation fields do not cancel each other out. In the folded dipole antenna, the folded dipole could be considered as an extension to the basic dipole with the ends of the dipole folded and the ends connected to each other. Therefore, the current in the folded dipole would be moving in the same direction as in the normal dipole. As a result it could be seen that the current flowing along both half-waves are in phase and do not cancel each other out, and therefore the antenna would radiate just like a half-wave dipole antenna.

The impedance is increased because the power supplied to the antenna is evenly shared by the two sections which create the antenna. Therefore, if the currents were measured, the values read in the folded dipole would be half the values in the basic dipole antenna, and since the same power is supplied, the resistance is increased by a factor of 4 to balance the following equation:

$$P = I^2 R \tag{14.14}$$

The advantages of the folded dipole antennas are as follows:

Higher feed impedance: Folded dipole antennas provide a significant increase in the impedance levels, enabling the antenna to be matched more easily.

Wide bandwidth: The folded dipole antenna offers a flatter frequency characteristic, and therefore it is capable of functioning in a wider bandwidth.

Due to these advantages the folded dipole antenna is used in domestic TV applications and for Very High Frequency (VHF) FM radio antennas.

14.2.1.5 Monopole Antenna

A monopole antenna is a half-plane version of a dipole antenna, consisting of a single length of wire mounted above a ground plane, as shown in Figure 14.5. By image theory, an identical virtual wire exists mirrored in the ground plane, giving the equivalent of a dipole. The radiation pattern of a monopole antenna above the ground plane is similar to the dipole antenna pattern. The difference is the impedance; the impedance of a monopole antenna is half the impedance of the dipole antenna. This could be understood since only half of the voltage is required to feed the monopole antenna at the same current as the dipole antenna. If the directivity of the dipole antenna is D_1 dB, then the directivity of a monopole antenna is (D_1+3) dB. This is due to the ground plane below the monopole antenna; since the radiation is in one direction, unlike dipole antennas, the monopole antennas are twice as directive. Monopole antennas are also half the size of dipole antennas and so they are attractive when a small antenna is needed.

14.2.1.6 Loop Antennas

Loop antennas are made of wires and are connected in a circular shape, as shown in Figure 14.6. These antennas have low radiation resistance and high reactance, so it is difficult to match

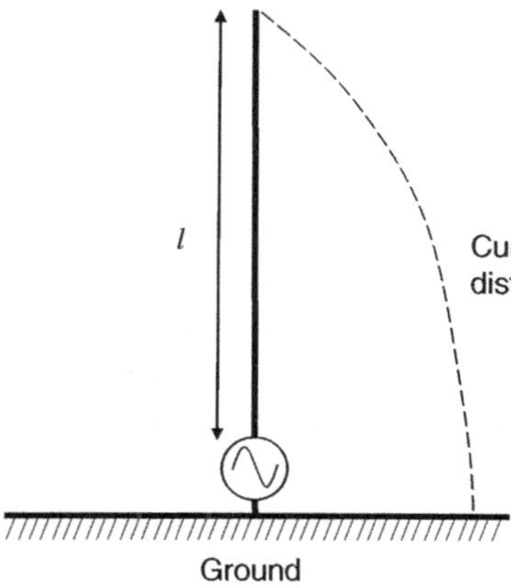

Figure 14.5 Monopole antenna.

their impedance. Therefore, this type of antenna is mainly used in the receiver side of the communication link where mismatch loss can be tolerated. The loop antenna is considered as electrically small; therefore, the current flow along the loop can be calculated approximately as a constant. Hence, the fields from a small circular loop are calculated from:

$$E_{\emptyset} = \frac{\eta k^2 a^2 I_0 e^{-jkr} \sin\theta}{4r} \tag{14.15}$$

$$H_{\theta} = \frac{-E_{\emptyset}}{\eta} \tag{14.16}$$

Generally, the larger the loop, the better the antenna performance becomes. A loop antenna will become resonant as the perimeter of the loop approaches one wavelength. The one-wavelength

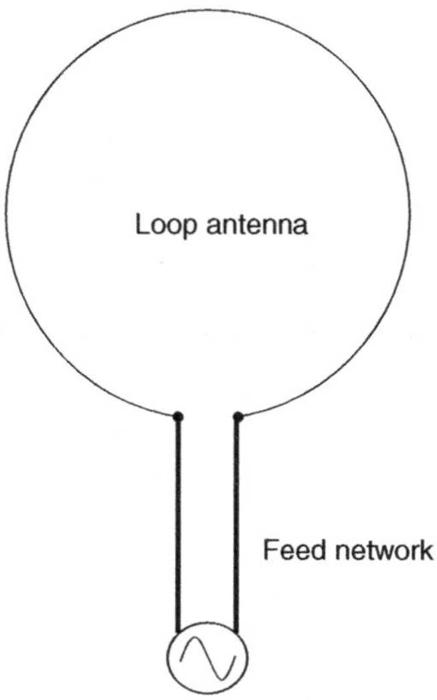

Figure 14.6 Loop antenna.

perimeter loop antenna has similar characteristics to a folded dipole antenna with impedance higher than that of a half-wavelength dipole antenna.

14.2.2 Travelling-Wave Antennas

Travelling-wave antennas use travelling waves on a guiding structure as the main radiating method. There are two main types of travelling-wave antenna: slow-wave antennas and fast-wave antennas. The latter are also known as leaky-wave antennas. Slow-wave antennas deal with guided waves with a propagation speed lower than the speed of light in free space. Such waves do not radiate naturally, and the radiation only occurs at discontinuities of the antenna. On the other hand, the propagating wave on a leaky-wave antenna may have a phase velocity larger than the speed of light. Travelling-wave antennas radiate constantly along their length and therefore the propagation constant is complex, consisting of both a phase and attenuation constant.

There are different variants of this antenna type, such as helical antennas, Yagi–Uda antennas and spiral antennas. Yagi–Uda antennas will be discussed further as this well-known structure is used for more applications than the rest. One of the most familiar is the Yagi–Uda antenna used in popular consumer RF receivers, including TV reception, where it is designed for high directivity. The Yagi–Uda antenna enables better levels of signal-to-noise ratio (SNR) to be achieved and the high directivity also reduces interference levels, remembering from Chapter 1 that there are typically many TV transmitters spread across the country. Figure 14.7 shows the Yagi–Uda antenna structure.

There are three main elements in the Yagi–Uda antenna:

1. **The driven element**: this element of the antenna is fed with the RF power. It is normally a half-wave dipole or a folded dipole.

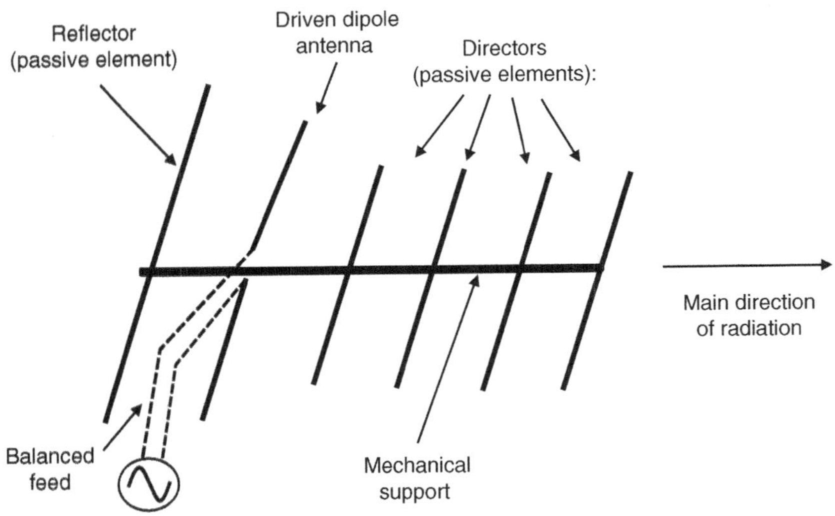

Figure 14.7 Yagi–Uda antenna.

2. **The reflector**: The Yagi–Uda antenna will generally have one reflector, which is behind the driven element. The reflector is used to add a few dBs of gain in the forward direction (4 dB to 5 dB).
3. **The director**: The director is placed in front of the driven element. Usually there would be more than one director as each director is used to add approximately 1 dB to the gain in the forward direction. Adding more directors would not increase the gain by the same value (the benefit of adding more elements keeps decreasing).

Yagi–Uda antennas have four main advantages:

1. High gain, allowing for low power signals to be received.
2. Low interference due to the high directivity.
3. Simple construction, using only conducting rods.
4. The construction makes it easy to mount the antenna vertically on any pole with standard mechanical fixings.

Even though these advantages make the Yagi–Uda antenna attractive, it also has some disadvantages. A high-gain antenna is physically long and the gain achieved is typically limited to 20 dB for a single antenna.

14.2.3 Reflector Antennas

Reflector antennas [3] have been used far more widely during recent years due to the increased adoption of satellite TV. One of the best-known types of reflector antennas is the parabolic reflector antenna, also known as a 'dish' antenna. A reflector antenna, as shown in Figure 14.8, is an antenna with electromagnetic field in the aperture that is formed by the reflection of propagating electromagnetic waves from a metal surface or a special mirror or reflector. Therefore, this type of antenna is not easy to construct due to complications of the feed mechanism, but at the same time the reflector antenna has a high gain and, consequently, a narrow beam, which is more focused on the source or target of radiation. Therefore, they are

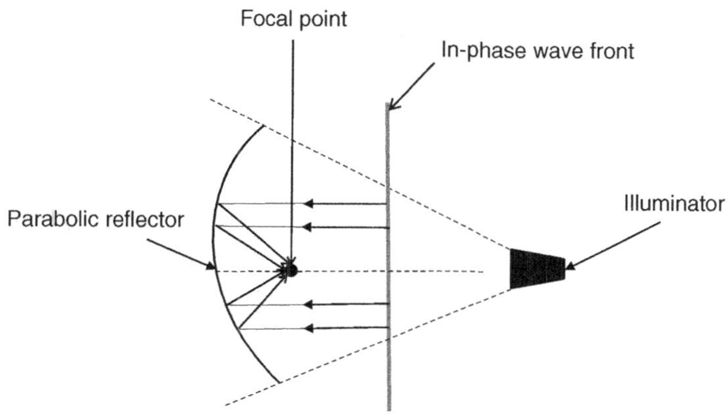

Figure 14.8 Parabolic antenna.

widely used in radio and wireless applications at frequencies above 1 GHz. There are various reflector antenna modifications, for example making use of a greater number of reflectors.

The source of the electromagnetic wave is normally a smaller-sized antenna, called the illuminator, as illustrated in Figure 14.8. The electromagnetic wave emanates from the illuminator, and when it reaches the conductive surface of the reflector it excites currents that create the secondary field, commonly called the 'reflected wave field'. The illuminator should emit only in one hemisphere in the direction of the mirror and not radiate to the other direction. In the aperture, the reflected wave typically has a flat wave front to give a narrow radiation pattern.

Like any antenna, parabolic antennas have advantages and disadvantages. These need to be considered against other types of antenna in order to make an informed choice. The advantages are:

High gain: Dish antennas are capable of providing very high levels of gain. The larger the dish in terms of wavelengths, the higher the gain achieved. The gain, G, of the reflector antenna can be calculated by:

$$G = 10\log_{10} k\left(\frac{\pi D}{\lambda}\right)^2 \tag{14.17}$$

where k is the efficiency coefficient of the reflector antenna, typically ranging from 50–60%, D is the diameter of the reflector, and λ is the free-space wavelength.

High directivity: As with the gain achieved by dish antennas, the directivity obtained from this type of antenna is very high. The higher the gain, the narrower the beamwidth. This is a major advantage in applications where the power is required to be directed over a small area. This would prevent it from interfering with other users, and this is an important feature when communicating with satellites because it enables satellites using the same frequency bands to be separated by distance and resolved by the pointing angle of the antenna.

The disadvantages are:

Requires reflector and drive element: The parabolic reflector is part of the antenna, but it requires a feed to be placed at the focus of the parabola.

Cost: This type of antenna needs to be manufactured with care, as the reflector needs to have a smooth and undistorted shape to reflect the radio signals properly. On top of this, a feed is needed which adds to the total cost.

Size: The size of this antenna is large at low frequencies.

14.2.4 Horn Antennas

A horn antenna has a wide bandwidth and high gain, so it is widely used in a range of applications, including the feed for a satellite dish antenna. An example of a pyramidal rectangular waveguide horn antenna is shown in Figure 14.9. There are various other forms

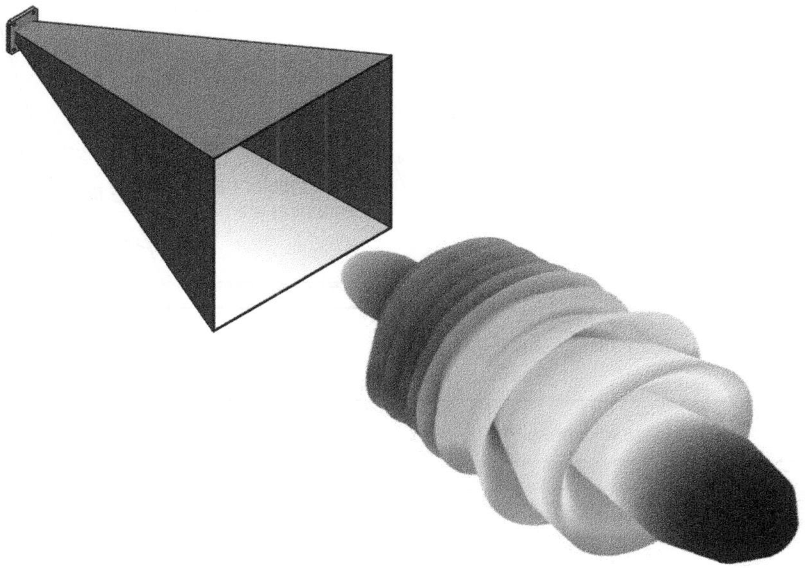

Figure 14.9 Pyramidal waveguide horn antenna and its radiation pattern.

of this type of antenna, including sectorial, conical, corrugated and ridged, but the basic idea is always the same; to have a waveguide that tapers to an open-ended aperture that radiates. Many modifications to the basic structure have been used to improve particular electrical characteristics. Horn antennas are very broadband, consistent with the feeding waveguide. In fact, the bandwidth of the antenna is usually determined by the properties of the excitation of the waveguide. Low levels of the rear lobe (up to –40 dB) are typical for a horn antenna due to the fact that there is almost no RF current on the outer side of the horn. Provided that the horn is electrically smooth and made from (or plated with) a high-conductivity metal that does not tarnish or oxidise, the horn has a low loss so its directivity is roughly equal to its gain.

14.2.5 Microstrip Patch Antennas

Patch antennas are a popular form of microstrip antenna and are becoming increasingly useful because they can be printed on a circuit board directly, or even on the surface of a package if the frequency is high enough. They are becoming very common within the mobile phone field as well, often fabricated as a 3-D structure on a plastic component within the handset. As shown in Figure 14.10, a microstrip patch antenna consists of a metal patch on top of a substrate, with a ground plane underneath. In microstrip, the signal conductor and ground plane are made of a high-conductivity material such as copper. Actually, the radiating structure may have one of many geometric shapes, perhaps rectangular, circular, triangular, elliptical or circular. Moreover, these structures can use either resonant or nonresonant modes and can be excited by other feed structures, such as stripline, coaxial cable, and so forth [4].

The rectangular patch has two important parameters, namely the length (L) and width (W); it sits on top of the substrate, which also has two important parameters, the height (h) and relative

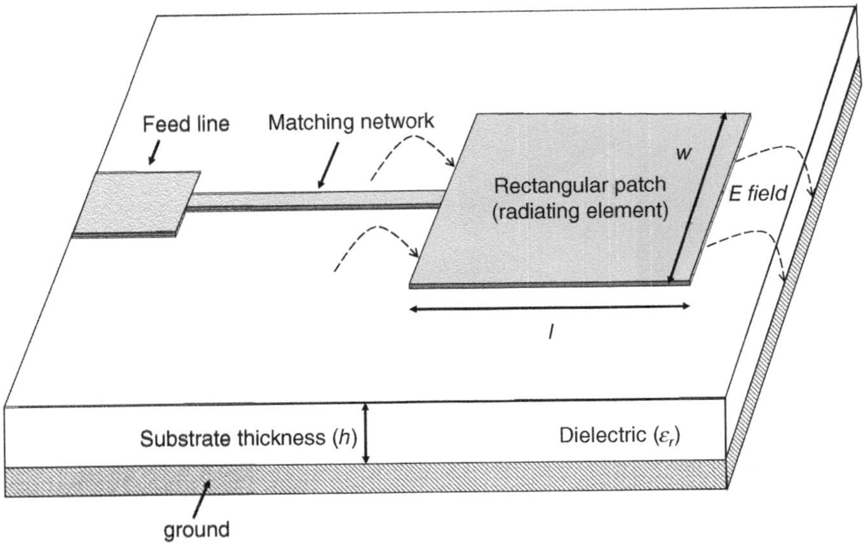

Figure 14.10 Microstrip patch antenna.

permittivity (ε_r). Usually, h is much smaller than the wavelength of operation. The length L of the patch determines the resonant frequency of the patch antenna. The centre frequency is given by (approximately):

$$f_c = \frac{c}{2L\sqrt{\varepsilon_r}} \tag{14.18}$$

The width of the microstrip patch controls the input impedance and the bandwidth possible from the antenna. To reduce the impedance, the width of the patch antenna is increased. The width also controls the radiation pattern. Usually, the bandwidth of patch antennas is very small and is sometimes affected by fringing fields. Various techniques can be employed for improving the matching and bandwidth, such as using a feed that is slightly inset into the patch.

Figure 14.10 shows a rectangular patch antenna, which is the most typical patch antenna because it is easy to construct and there is a direct correlation between the practical and ideal gain. However, there is wide range of other forms, namely triangular, square, ellipsoid, circular, the choice of which affects several properties of the antenna such as the bandwidth, radiation pattern, return loss, and gain [5]. Additionally, the shape has a significant effect on the polarisation of the antenna. By itself, a square or rectangular antenna is linearly polarised, although it can be fed from quadrature inputs along orthogonal edges to give circular polarisation.

Patch antennas generally have the following advantages and disadvantages:

Advantages:

- Low fabrication cost
- Easy to form a large array of antennas spaced at half-wavelength or less
- Light weight

Disadvantages:

- Limited bandwidth
- Low power handling
- Low efficiency
- Sidelobes, for example due to substrate modes

The size of a patch antenna is not considered a major advantage because the size is inversely proportional to its frequency. At low frequencies, a microstrip patch becomes a pointless option due to the size required, and the fact that the ground plane needs to be much larger than the patch.

14.2.6 NFC Antennas

Near-field communication (NFC) is a set of standards for smartphones and similar devices to communicate with each other by bringing them close to each other. The applications of NFC include swipe proximity payments and information exchange over short distances. NFC most often operates at a frequency of 13.56 MHz. This indicates that the wavelength of the signal transmitted is 22 m; if a half-wave dipole antenna were to transmit this signal then the antenna size would be 11 m! NFC is based on the mutual inductance of a pair of inductors; when current is flowing through an inductor and it is brought near to a second inductor, a current will be induced in the second inductor. Therefore, an NFC antenna could be thought of as an inductor (loops of wire, usually a planar spiral for ease of fabrication and integration), which covers as much of the device as possible. It is important to note that an inductor has a magnetic field around it, but unlike an antenna it does not radiate in the far field. Instead, when the two inductors are sufficiently close the second antenna is in the near-field of the first and the magnetic field of the first one induces a current in the second inductor. This is, therefore the action of an air-cored transformer, with weak coupling (transformers themselves use a magnetic core to strongly couple the two coils). Working in the near-field means that the received power drops off very quickly. This is the essence of NFC communications and provides the advantages of low interference between users and improved security.

14.3 Transmission Equations, Free-Space Path Loss and Link Budget Calculation

Choosing the type of antennas for a communications link is an essential part of the system design. Different types of antenna have different gain values, and the transmitter and antenna gain are both extremely important in determining the received power of the signal. This section will describe the transmission equation, free-space path loss and link budget calculations which are used to predict the required transmitter power, the path loss, the gain required from the antennas, and the received signal strength in a communications system.

14.3.1 Transmission Equation

As well as spreading out according to the antenna directivity, the transmitted RF signal power is affected by the channel and by many other factors as it propagates in the environment. As a

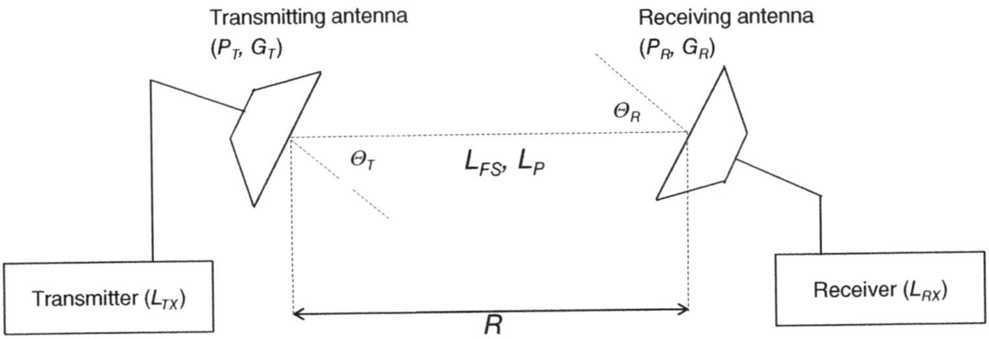

Figure 14.11 Transmission system parameters for Friis' transmission equation.

starting point, it is essential to calculate the power received by the receiver's antenna, with gain G_R for a signal transmitted from another antenna, and with gain G_T when they are separated by distance R and operate at frequency f, as shown in Figure 14.11. To perform this calculation, Friis' transmission equation is used.

To derive the Friis transmission equation, let us assume that there are two antennas: the transmitter, with power P_T, has an antenna that is assumed to be omnidirectional and lossless, while the receiver antenna is assumed to be in the far-field of the transmit antenna. Then, the power density p (in Watts per m^2) of the plane wave incident on the receiver antenna, which is a distance R away from the transmitter, is given by:

$$p = \frac{P_T}{4\pi R^2} \tag{14.19}$$

If the transmit antenna has gain G_T in the direction of the receiver antenna, then the power density becomes:

$$p = \frac{P_T}{4\pi R^2} G_T \tag{14.20}$$

The gain and loss factors are accounted for when calculating the power density. Following that, it is assumed that the receiver has an effective aperture given by A_{ER}. The power received by this antenna is then given by:

$$P_R = \frac{P_T}{4\pi R^2} G_T A_{ER} \tag{14.21}$$

where A_E for any antenna is given by:

$$A_E = \frac{\lambda^2}{4\pi} G \tag{14.22}$$

Therefore, the resulting received power is given by:

$$P_R = \frac{P_T G_T G_R \lambda^2}{(4\pi R)^2} \tag{14.23}$$

This formula is known as the Friis transmission equation. It gives the received power, taking into account the free-space path loss, antenna gains, signal wavelength and transmitter power.

14.3.2 Free-Space Path Loss

Free-space path loss (FSPL) is used in many communications links to predict the strength of the radio signal in the receiver. FSPL does not include all the real characteristics of most terrestrial situations because of the effects of obstacles and the ground between the transmitter and receiver, but it is still quite useful for the basic understanding of actual propagation environments. FSPL is the loss in signal strength that happens when an electromagnetic wave propagates along a line-of-sight path in free space. Under these conditions, there are no obstacles or surfaces that could reflect or refract the transmitted signal. FSPL calculations only take into account the loss caused by the path itself, without considering any other factors such as antenna gains, transmitter power or the receiver sensitivity level. To understand the nature of the loss in free space, consider a signal that is spreading from the transmitter; the signal will move away in a spherical manner (in all directions). Since the signal is made of energy, it is logical to think that as the sphere size increases, the energy per-unit-area of the signal decreases, and this simply explains why there is free space path loss in radio systems, and it is often referred to as the 'spreading loss'. As a result, it is evident that the signal power decreases in a way that is inversely proportional to the square of the distance between the transmitter and receiver. This is the case for free space, and to consider obstacles between the transmitter and receiver the relationship between the signal and the distance should be altered. To consider the obstacles, the exponent must be increased to 4 or even 6, depending on the radio system environment.

The term in Equation (14.22) that is independent of all the hardware is the parameter defined as the FSPL, given by:

$$\text{Free Space Path Loss} = \left(\frac{4\pi d}{\lambda}\right)^2 \tag{14.24}$$

where d is the distance, and λ is the signal wavelength.

It should be noted at this point that the $1/\lambda^2$ term in Equation (14.23) gives a rather false impression that high frequencies (e.g., millimetre-waves) will automatically suffer very high path loss. To clarify this issue, it is useful to rewrite Equation (14.23) in decibel form using the FSPL parameter, which gives:

$$P_{R(dB)} = P_{T(dBm)} + G_{T(dB)} - \text{FPSL}_{(dB)} + G_{R(dB)} \tag{14.25}$$

In this form, it is much clearer that any apparent increase in FSPL with frequency is actually countered by the receiver antenna gain. In other words, the conventional FSPL is defined in terms of power density at the receiver antenna and has effectively assumed the receiver antenna is isotropic (0 dB gain). Spreading loss is, therefore, probably a better terminology since Equation (14.24) is referring only to the geometric spreading of the transmitter signal over the surface of a sphere. The receiver antenna aperture is, of course, a key factor in determining how much of that power is collected and fed into the receiver. For millimetre-wave propagation there can, however, be a very high path loss due to atmospheric absorption, particular at the oxygen and water absorption peaks and rain fading.

14.3.3 Link Budget

As the name indicates, a link budget is a summation of all the signal gains and losses associated in a wireless communication system, yielding the receiver signal power or SNR. The link budget takes into account all communication elements between the transmitter and receiver to characterise the signal strength when it arrives at the receiver. The link budget calculation may include the following parameters:

- Transmitter power
- Antenna gains (transmitter and receiver)
- Antenna feeder losses (transmitter and receiver)
- Path losses
- Receiver sensitivity

The link budget equation takes the following form:

$$\text{Received Power (dBm)} = \text{Transmitter power (dBm)} + \text{gains (dB)} - \text{losses (dB)} \quad (14.26)$$

Building on this, the link budget needs to include all types of gains and losses possible in the communications system. Generally, to perform a link budget calculation it is necessary to explore all areas where signal gains and losses may be introduced between the transmitter and receiver. From this, it could be deduced that each communication system would have a link budget equation of its own (depending on the gains and losses between the transmitter and receiver), but the equation below shows the generic link budget equation:

$$P_{RX} = P_{TX} + G_{TX} + G_{RX} - L_{TX} - L_{FS} - L_P - L_{RX} \quad (14.27)$$

where:

P_{RX} = Received power (dBm)
P_{TX} = Transmitted power (dBm)
G_{TX} = Transmitter antenna gain (dBi)
G_{RX} = Receiver antenna gain (dBi)
L_{TX} = Transmit feeder and associated loses (dB)
L_{RX} = Receiver feeder and associated loses (dB)
L_{FS} = Free space path loss (dB)
L_P = various signal propagation loses (dB).

14.4 Other Propagation Effects

When an RF signal is transmitted in an indoor or urban area it propagates via many different paths, due to the obstacles between the transmitter and receiver that reflect and/or diffract the signal. Signals can also be scattered in different directions instead of going straight to the receiver antenna. These propagation mechanisms have a major effect on the received signal power, and will be discussed in this section. Typical propagation effects are free space propagation, reflection, diffraction, and scattering [6].

As discussed above, free space is a region where there is no obstacle between the transmitter and receiver. Only outer space is a good approximation to free-space, where there are no obstacles to get in the way of the signal to scatter it or absorb the signal's energy. As explained previously, the signal would lose energy as it spreads equally in all directions, as the distance travelled increases the loss increases, as given by the FSPL equation (14.24).

14.4.1 Reflection and Refraction

In many situations, some of the transmitted signal power will be reflected off obstacles instead of going directly to its supposed destination. A radio signal could be reflected from buildings, trees, the Earth and many other obstacles. When a radio signal in a dielectric medium (e.g., air) strikes another dielectric medium that has different electrical characteristics (given by the permittivity or refractive index), then the signal is partially reflected and partially refracted (the refracted signal goes through the second medium). The Fresnel reflection coefficient (Γ) is used to give the relationship between the electric field intensity of the reflected signal in relation to the incident signal in the original medium. This reflection coefficient depends on the properties of the original material, the angle of incidence and the frequency of the incident signal.

Refraction is the deflection of signals when they pass through different mediums, as the air/atmosphere causes a change in the velocity of propagated signals. This phenomenon causes the waves to bend back to the Earth when they travel through the Earth's atmosphere at constant speed. When the waves enter dense, electrically charged layers, the part which enters first tends to travel faster than other parts of the waves, causing them to bend back towards the Earth at an angle. The amount of refraction and the angle of bending will depend on the density of the charged layers (which changes through the day), the frequency of the transmitted waves and the wave's entry angle to the layers. The Fresnel equations are based on Snell's law, which relates the incident and refracted (transmitted) angles. Snell's law is given by:

$$\frac{n_1}{n_2} = \frac{\sin\theta_2}{\sin\theta_1} \tag{14.28}$$

where n_1 and n_2 are the refractive indices of medium 1 and medium 2 respectively. θ_1 and θ_2 are the incident and transmitted signal angles.

14.4.2 Diffraction

Transmitted signals can propagate around curved surfaces and behind obstructions by the process of *diffraction*. This means that a signal sent from a transmitter could be received at the receiver even though it may be blocked by a large obstacle between them (shadow fading). This phenomenon is more significant for signals with a large wavelength (low frequency), and therefore long-wavelength signals can be received at longer distances. To understand this phenomenon, Huygens' principle must be explained. This states that every point on a spherical wave front can be considered as a source of a secondary wave front. Even though the area behind an obstacle is a shadow zone, the diffracted signal will start to fill the void. The effect of diffraction is found to be greater when the obstacle becomes sharper and more like a 'knife edge' [7]. For example, a mountain ridge would have a greater diffraction effect on the signal than would a rounded hill. As noted above, signals with a low frequency diffract more than signals with a high frequency, and this is why signals with a low frequency could cover a larger range when transmitted in a hilly terrain.

14.4.3 Scattering

When an obstacle has dimensions that are comparable to the signal wavelength, scattering can occur. This may be due to reflection off a structure that has surface features of significant size compared to the wavelength, or to diffraction from small features in an inhomogeneous medium. The scattered wave propagates in many different directions with different phase shift. The effect of scattering depends on the angle of the incident wave and the wavelength as compared to the obstacle's size and orientation. Predicting this phenomenon is considered to be the most difficult task compared to other propagation mechanisms.

14.5 Millimetre-Wave and THz Propagation

Use of the millimetre-wave part of the electromagnetic spectrum has significantly increased during recent years. The first applications of millimetre waves were in radio astronomy and military systems, but as the technology matured engineers turned their attention to this range of frequencies for a wide range of new applications. The millimetre-wave spectrum occupies the range from 30 GHz to 300 GHz, and laying as it does between the microwave and infrared ranges the behaviour of the signals shares many common features with both the microwave range and the optical domain [8]. As the signals can penetrate through different materials they can be used in certain medical imaging applications such as diagnostic techniques [9]. In addition, the antennas for millimetre-wave systems are very small due to their short wavelength. The millimetre-wave range provides a high bandwidth for telecommunications systems, such as the 57–64 GHz unlicensed band. However, millimetre-wave propagation characteristics must be considered carefully as the signals suffer from weak diffraction and high atmospheric absorption loss. Unlike traditional mobile communications signals of around 1–3 GHz, it cannot be taken for granted that millimetre-waves can propagate even through thin materials.

The terahertz part of spectrum lies between the optical and millimetre-wave regions. It extends above 300 GHz up to 10 THz, although 300 GHz to 3 THz is also often referred to as the 'submillimetre-wave region'. These frequencies have not been exploited as much as the optical and microwave ranges, mainly because of the lack of suitable signal sources. Electronic devices generally have maximum output power that falls as $1/f^2$ (or worse), while lasers require a semiconductor material with a band gap that equates to photons of the desired signal wavelength. Hence, a great deal of effort has been put into realising quantum cascade lasers that fill the 'terahertz gap' between electronics and photonics [10].

Many scientists believe that combining optics and electronics will generate a new era of terahertz applications such as transmitting terabit-per-second data rates, along with medical, industrial and environmental sensing. The main issue for this spectral region is the high absorption in the atmosphere, primarily due to moisture content. Above a frequency of 2.2 THz the absorption increases significantly, reaching 2×10^5 dB km^{-1} at 8 THz [11]. Of course, for very short-range wireless communications the high loss is potentially advantageous, allowing frequency re-use over a short distance and a low probability of interference for improved security. Currently, however, this technology is most often used in the fields of astronomy and atmospheric sensing, where specific species of molecules are of interest and have resonances in the terahertz range.

14.6 Indoor Propagation

Recently, there has been great interest in modelling indoor propagation, especially for personal communication systems. Users demand a service that provides suitable coverage inside buildings [12]. It has been shown that indoor propagation is influenced by features such as materials used in the building construction, the building layout, the construction methods and the number and size of the windows. Indeed, even the type of glass is found to make a difference at very high frequencies.

It is possible that the wave travels from the transmitter to the receiver in many different ray paths [13]. Thus, the same mechanisms occur as outdoors: diffraction, reflection, transmission and scattering, but the obstacles are closer together and the signal strength varies rapidly with distance because of constructive and destructive interference. The signal level will even depend on whether doors are open or closed in the building. An antenna located at desk level will receive a different signal level to one mounted on the ceiling. Indoor channels may be divided as the line of sight or obstructed with different degrees of clutter [14]. Some indoor propagation models, which are applied to predict indoor propagation loss, will be presented here.

14.6.1 Long-Distance Models

This model is influenced by parameters such as frequency and the distance from the transmitter to the receiver. Two long-distance path loss models have been developed: the first model assumes a linear dependence of the loss on distance, while the second model suggests a logarithmic dependence. Both can be used in case of non-line-of-sight propagation.

14.6.2 Partition Losses on the Same Floor

The internal and external structure of different buildings has a variety of obstacles and partitions. There are two types of partition, namely hard partitions that are considered as part of the building structure, and soft partitions that are movable. It is difficult to form a general model for all indoor installations because the physical and electrical characteristics of the partitions vary widely. Nevertheless, researchers have created an extensive database of loss for different partition types.

14.6.3 Partition Losses Between Floors

The materials and external dimensions of the building affect the loss between floors. Indeed, even the number of windows can impact the loss value. It has been observed that attenuation in the first floor is higher than the incremental attenuation, which is caused by each additional floor.

14.6.4 Motley–Keenan Model

This model is widely used for non-line-of-sight propagation in dense buildings such as offices. The loss can be expressed as the summation of the free space loss, L_0, losses due to the obstacles in the direct line of sight path, and a constant, L_c:

$$L = L_0 + L_c + \sum_{j=1}^{N} N_j L_j + N_f L_f \tag{14.29}$$

where N_j represents the number of walls of type j, L_j is the loss of wall type j, N_f represents the number of tiles which occur along the line-of-sight path and L_f is the tiles loss. All of these parameters are recorded as either tables or databases.

14.6.5 Ericsson Multiple Breaking Model

This model was obtained from different measurements in a multifloor building. It has been estimated that the attenuation at distance 1 m is 30 dB. Instead of using a log-normal shadowing component, this model provides deterministic limits of path loss at a specific distance. Figure 14.12 shows a plot of in-building attenuation from the Ericsson model as a function of distance.

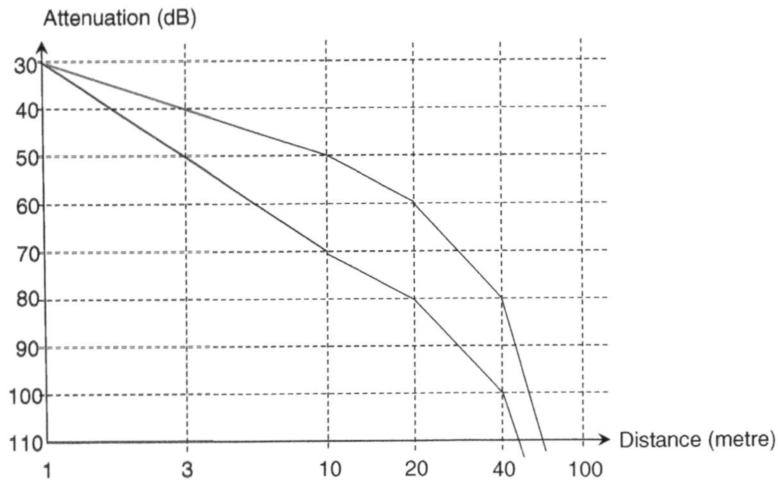

Figure 14.12 Ericsson in-building path loss model.

All of the above models are empirical models for estimating the indoor loss, while some research groups prefer to use deterministic models such as the ray-tracing method. This method depends on the image theory to take into account the propagation mechanisms (reflection and diffraction). The ray-launching method, which is based on 3-D models, can also be used but is not popular due to the long and complex calculations required.

14.7 Outdoor Propagation

The quality of wireless systems depends on having accurate propagation prediction models for path loss to optimise the system implementation. Designing without propagation models would not give path loss parameters unless by using field measurements, which is considered too expensive and time-consuming [16]. In general, propagation prediction models divide into three main types: theoretical, empirical and semi-empirical.

14.7.1 Theoretical Propagation Prediction Models

These models are based on a combination of physical laws and adequate approximations. They require different methods such as finite difference and finite elements, parabolic equation and geometrical optics to solve Maxwell's equations. The main drawback of theoretical models is the required computation time, and such models can be considered as references of some cases. The variable used in these models are deterministic, and therefore such models are referred to as 'deterministic models'. In few situations, they may take random variables. As an example of theoretical models, diffraction over-rooftops is derived from physical optics by assuming a spacing of buildings and uniform heights.

14.7.2 Empirical or Statistical Models

Empirical models are based on statistical analysis and sets of equations, which are derived from experimental measurements. These models are efficient, simple and accurate for environment conditions similar to the environment when the measurements were taken. As a main drawback, empirical models cannot be used when the environment is changed unless the model is modified [17].

14.7.3 Semi-Empirical Models

Semi-empirical models combine physical phenomena such as reflection, diffraction and scattering with statistical experimental measurements. These models do not account for the improbability of independent variables; therefore, they are more complex than the purely empirical models.

14.7.4 Okumura Model

This model is an empirical model based on measurements made in and around Tokyo by Okumura and his collaborators, who measured signal strength over a wide range of frequencies extending from 150 to 1920 MHz. In addition, they used antenna heights of between 30 and 1000 m with cell diameters ranging from 1 to 100 km. They plotted the field strength as a set of curves versus the distance for specific heights at different frequencies.

To calculate path loss with this model, free space loss of the test points is first determined, extracting the value of median attenuation, A_{mu}, and adding it to the free space loss, L_f, with a correction factor according to the type of terrain. The model can be expressed as:

$$L_{dB} = L_f + A_{mu} - G(h_t) - G(h_r) - G_{area} \tag{14.30}$$

where $G(h_t)$ is the gain of transmitter antenna with height h_t, $G(h_r)$ is the gain of receiver antenna with height h_r, and G_{area} is the gain of the type of environment.

Further correction graphics are provided to enhance the model in rural areas, urban areas and irregular terrain. The completeness of the model and the ability to use it in different terrains lead this system to be the most common and, indeed, it has become a standard by which others are judged. However, the data are available only as curves, which makes it inconvenient to use.

14.7.5 Hata Model

The Hata model is an empirical equation of the graphical curves provided by Okumura. It is valid for a frequency from 150 to 1500 MHz. Hata presents the propagation loss of an urban area as a formula and supplies correction formulas for other situations. The loss in this model for urban area is calculated as:

$$L_{50}(dB) = 69.55 + 26.16logf_c - 13.82logh_t - a(h_r) + (44.9 - 6.55logh_t)log(d) \tag{14.31}$$

where f_c is the frequency, typically from 150 to 1500 MHz, d is the distance, normally from 1 to 20 Km, and $a(h_r)$ is the correction factor for receiver height.

For medium and small cities:

$$a(h_r) = (1.1logf_c - 0.7)h_r - (1.56logf_c - 0.8) \tag{14.32}$$

For large cities:

$$a(h_r) = 8.29(log1.54h_r)^2 - 1.1 \text{ for } f_c \leq 300MHz \tag{14.33}$$

$$a(h_r) = 3.2(log11.75h_r)^2 - 4.97 \text{ for } f_c \geq 300MHz \tag{14.34}$$

14.7.6 Other Models

Rustako *et al*. presented a six-ray model for a line-of-sight urban area that predicts the results accurately. The six rays are the direct line-of-sight ray, one reflected ray from the ground, two rays with double-wall reflection and two rays from one-wall reflection. The propagation model presented by Lee is an area-to-area model, which is used for a flat terrain region. Many parameters are considered and, if the situation changes, a correction factor is calculated and added to the original formula. Lee proposed another propagation model, which is a point-to-point model, and in which a small account of the terrain is obtained. Research groups pay attention to the ray-tracing model for its highly accurate calculations. This model obtains the uniform theory of diffraction (UTD), which modelled the obstacles along the propagation path by simple geometrical objects. There are many techniques for this model, the most significant one being the imaging technique, which takes into account all of the propagation paths besides the direct path.

14.8 Multipath Propagation

When a signal is transmitted, the most preferred propagation path is the line-of-sight path, but in an indoor or urban area this is almost impossible. Even outdoors, hills and trees can block the signal. In reality, what happens is that the transmitted signal reflects, refracts and scatters off the obstacles between the transmitter and receiver, which causes the signals to propagate in multiple paths. The receiver therefore receives multiple versions of the transmitted signal, with each version propagating along a different path and therefore arriving at a slightly different time instant. The presence of multipath in a communications system results primarily in inter-symbol interference, along with signal distortion and data loss. This propagation effect leads to what is known as 'multipath fading'. In the frequency domain, multipath fading can be considered as constructive and destructive interference of radio waves that have travelled different distances and have a varying phase relationship. This leads to peaks and nulls in the channel's frequency response. Digital communications engineers need to combat the inter-symbol interference (ISI) in order to increase the data rate. The time-domain impulse response of the channel will show the received signal having multiple time delays. In real communications channels, multipath propagation is time-varying and must be modelled with statistical techniques. The key parameter that characterises the quality of the channel is the r.m.s delay spread, which is a measure of the extent to which the energy of the ideal impulse is spread in time at the receiver.

In some scenarios, multipath propagation can be used to advantage. Schemes such as multiple-input and multiple-output (MIMO) communications use multipath propagation to increase the capacity of the system. As the spectrum is becoming more scarce, it is important that the spectrum is used more efficiently and MIMO provides significant improvements in channel capacity. Multipath propagation will be discussed in this section, including multipath basics, fading and interference. Multipath propagation occurs in all terrestrial radio links. The radio signals do not propagate only in a line-of-sight path from the transmitter to the receiver; rather, the signal can propagate in different directions and reach the receiver's antenna by reflection and refraction due to objects such as buildings, trees, hills and reflective surfaces such as water, the ground, and so forth. The signal would therefore reflect off many objects and the receiver would receive multiple versions of the transmitted signal.

14.8.1 Multipath Fading

As mentioned above, a transmitted signal will reflect off objects between the transmitter and receiver, and therefore the receiver will receive different versions of the transmitted signal. When all signals arrive at the receiver they superimpose to form the receiver input signal. The signals can superimpose either constructively or destructively, depending on the different relative path lengths that they have taken. This can vary due to motion of the transmitter, receiver and/or objects in between. As the signal phases change, the resulting received signal strength will vary. Rayleigh fading is a statistical model for this time-varying property of the signal, caused by multipath fading. Fading affects radio communication links very badly, and can be categorised as either 'flat fading' or 'selective fading'. Flat fading has an effect on all frequencies in a wireless channel in the same manner, which is why it is called 'flat' fading. When flat multipath fading occurs, the signal will have amplitude changes, either rising or falling, over a period of time or with movement from one position to another.

Selective fading is multipath fading that has a different effect on different frequencies in the wireless channel. This indicates that the amplitude and phase through the channel can change significantly in-band. Sometimes, deep nulls might occur, which would cause severe distortion in a wideband-modulated signal. Simply sustaining the overall power of the received signal cannot overcome the effect of selective fading and equalisation is required. Some digital signal formats, such as orthogonal frequency division multiplexing (OFDM), are able to spread the data over the entire channel, therefore mitigating the effect of nulls on the received signal.

Cellular communications systems are susceptible to multipath fading for a number of reasons. First, the mobile station of a user could be moving and therefore the path lengths of all the signals transmitted are changing. Second, any objects between the transmitter and receiver could be moving, which would have a significant effect on the received signal. Multipath fading that dominates wireless and mobile communication networks is normally fast fading, as this type of fading occurs over a short distance. Slow fading occurs when an obstruction between a handset terminal and the base station is present and, thus, the signal slowly fades out.

Multipath propagation could increase interference, which could lead to a reduced SNR and higher bit error rate (BER) for digital signals. One reason for signal degradation at the receiver is the multipath fading (as explained above), though there are many other ways that multipath propagation could degrade the signal and affect system reliability. When multipath propagation occurs, as the receiver receives the different signal versions, it might contain a line-of-sight signal and other reflected signals. As these signals are received at different times, due to the different propagation path lengths, the property of each received signal can be different because the signals are transmitted at slightly different times. Hence, when the two signals are received at the receiver together, distortion would occur if they have similar signal strength levels. ISI is caused by multipath propagation, and if the delay between two received signal components is equal to a significant proportion of the symbol duration, then one symbol will start to overlap with the next and data can be corrupted.

14.9 Antenna Arrays

An antenna array is a set of two or more antennas working together to give better performance compared to that produced by a single antenna. An antenna array can be used to:

- Increase the overall gain.
- Provide reception diversity.
- Cancel out interference from a specific direction.
- Steer the antenna array electronically.

The number of antenna elements in an antenna array can be anything from two to thousands; for example, the Square Kilometre Array (SKA) telescope is one of the most ambitious antenna arrays ever attempted [18]. Antenna arrays are already used extensively on cellular mobile base stations [19], while in 60 GHz wireless systems small patch antenna arrays have been used to create a switchable beam to avoid blocking. In the future, 5G+ systems will make increasing use of antenna arrays since, with sophisticated DSP, it is possible to create multiple beam antennas and achieve very high area spectral efficiency. In general, the performance of an antenna array improves as the number of elements increases, the only drawback being that it increases the cost, size and complexity. The output of an antenna array [20] is given as follows:

$$Y = \sum_{n=1}^{N} w_n X_n \tag{14.35}$$

where N is the number of antennas in the array, X_n is the output of each antenna element, w_n is the weight of each output and Y is the total output from the array.

To understand the basic principle of antenna arrays, consider a set of three antennas placed along the z-axis. These antennas are at the receiver, which receives a signal at an angle relative to the z-axis of θ. The first antenna is placed at coordinate (0, 0, $\lambda/2$), the second one at (0, 0, 0), and the third one at (0, 0, $-\lambda/2$). The E-field of the plane wave, which is assumed to have constant amplitude, can be expressed as:

$$E(x, y, z) = e^{-j(k_x x + k_y y + k_z z)} \tag{14.36}$$

$$E(x, y, z) = e^{-j|k|(\sin\theta\cos\emptyset\cdot x + \sin\theta\sin\emptyset\cdot y + \cos\theta z)} \tag{14.37}$$

$$E(x, y, z) = e^{-jk\cdot r} \tag{14.38}$$

where k is the wave factor that characterises the variation of the signal phase as a function of position.

Furthermore, assume that the antennas, either at the receiver or transmitter, radiate isotropically, and that the signal received at each receiver is then equal to the E-field at the antenna location. Therefore, for antenna n, the received signal is:

$$X_n = e^{-j\frac{2\pi}{\lambda}\cos\theta z_n} \tag{14.39}$$

Only the z-axis value is used in the equation because x and y are kept at 0. The received signals are distinctive by a complex phase factor, depending on the separation of each antenna and the angle of arrival of the electromagnetic wave. If the signals received are summed together, the result would be:

$$Y = e^{-j\frac{2\pi}{\lambda}\cos\theta z_1} + e^{-j\frac{2\pi}{\lambda}\cos\theta z_2} + e^{-j\frac{2\pi}{\lambda}\cos\theta z_3} \tag{14.40}$$

$$Y = \sum_{i=1}^{3} e^{-j\frac{2\pi}{\lambda}\cos\theta z_i} \tag{14.41}$$

$$Y = \sum_{m=-1}^{1} e^{-jm\pi\cos\theta} \tag{14.42}$$

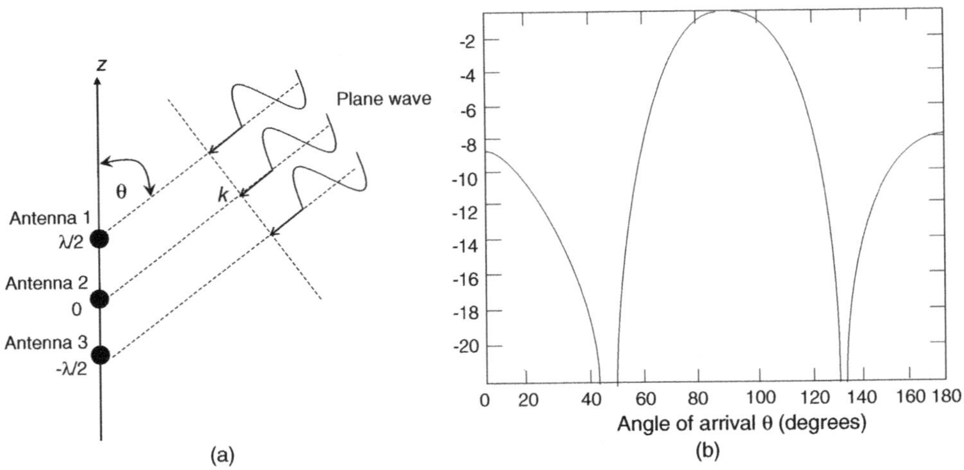

Figure 14.13 (a) Three-element array. (b) Calculated radiation pattern of the array.

If the magnitude of Y is plotted against the phase angle (θ), the result is shown in Figure 14.13.

As the graph shows, the antenna array processes the signal received better at some angles than others. For this example, the antenna array receives the signal with best signal quality when the angle is 90°, whereas when the reception angle is 45° or 135° the antenna array produces no output power. In this way, a directional radiation characteristic can be achieved, even though the antenna elements are themselves isotropic [21–23].

14.10 Multiple-Input and Multiple-Output Systems

Multiple-input and multiple-output (MIMO) is an extremely important development for wireless communications systems. The transmitter and receiver have multiple antennas, as shown in Figure 14.14. MIMO is an enhanced wireless communication technique that has been used

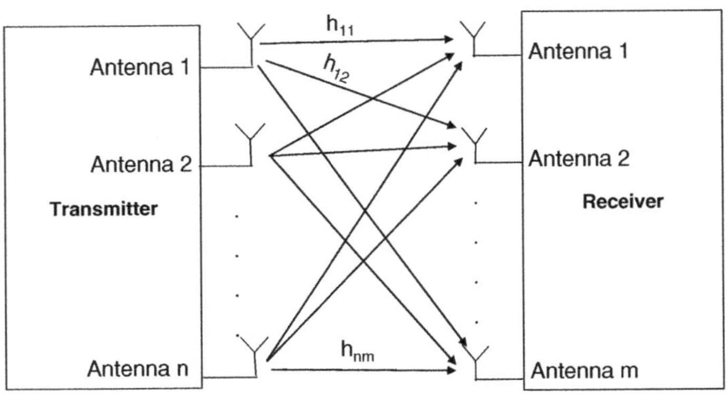

Figure 14.14 Basic MIMO system.

in many commercial systems, such as Wi-Fi (802.11n and beyond) and Long-Term Evolution (LTE). As explained before, multipath propagation causes Rayleigh fading, which will affect a signal transmitted and this will have an impact on the SNR and therefore degrade the BER. The problem faced by a single-input and single-output (SISO) system is that the entire OFDM signal transmitted is affected by multipath, and the receiver will have only one version of the signal transmitted. This makes it more difficult for the demodulator at the receiver side to reconstruct the original signal. For example, with frequency-selective fading, some parts of it are lost completely.

MIMO offers diversity, and this principle provides many versions of the signal transmitted to the receiver. If each transmitted signal is affected in a different way by multipath, then the probability of the signals being affected in the same way is reduced. Therefore, diversity allows a reduction in BER. For example, if the signal is transmitted from four transmitters, each signal is affected by multipath in a different way and therefore the receiver is able to reproduce the original signal.

There are different types of diversity that could be employed [24]. The first type is 'time diversity', where the signal may be transmitted at different times, while the second type is 'frequency diversity', where the signal transmitted may be transmitted at different frequencies, which is achieved by transmitting on different subchannels for the case of OFDM. The third type of diversity is 'space diversity', which is the basis of MIMO, in which the signal is transmitted in different directions and actually takes advantage of multipath.

MIMO systems have more than one antenna at the transmitter and receiver side, and this allows different paths for the signals, so that each signal path is affected in a different way. One of the main advantages of MIMO systems is that the system can increase the channel capacity significantly while still obeying Shannon's law. By increasing the number of antennas on both sides, it is possible to increase the information throughput of the wireless channel linearly. This makes MIMO technology essential in many communications systems, to improve their capacity.

Shannon's law states that there is a limit to the rate at which data can be transmitted in a channel in the presence of noise. MIMO offers a method of increasing data rates beyond what is achievable in a regular wireless channel; therefore, exploring this law is important. Shannon's law defines the maximum rate of which error-free data can be transmitted over a certain bandwidth with the presence of various noise signals. Shannon's law is expressed as follows:

$$C = W \cdot \log_2 \left[1 + \frac{S}{N}\right] \tag{14.43}$$

where C is the channel capacity in bits-per-second, W is the bandwidth in Hertz, and S/N is the SNR.

From this formula it can be seen that the limit of the data rate that can be transmitted depends on the bandwidth available and the SNR of the received signal. From these limits, it can be seen that a decision should be taken with regards to the manner of transmission. The modulation scheme used is a major factor determining the capacity; using higher-level modulation schemes, increases the capacity but this would require a better SNR. Thus, a balance is required between the data rate and the acceptable error rate, SNR and transmitter power. Even though this option of changing the modulation scheme is possible, it is not easy to achieve and could cost more in practice. Therefore, other methods should be utilised to increase the capacity of the channel, which brings us to MIMO technology that offers a capacity increase. To allow the enhanced throughput capability, MIMO employs a set of multiple antennas.

MIMO wireless systems use a matrix mathematical approach to achieve the advantage of the additional throughput. Data streams t_1, t_2, ... t_n are transmitted from antennas 1, 2,...n. There are various propagating paths that can be used, and each path has its own unique channel characteristics. To enable the receiver to discriminate between the different signal paths it is necessary to use a variable to represent the different channel characteristics. This variable can be represented by h_{ij}, which is the channel characteristic from transmitter antenna i to receiver antenna j. To explain further, assume we have three antennas in the transmitter and receiver side, then we are able to set up a matrix as follows:

$$r_1 = h_{11}t_1 + h_{21}t_2 + h_{31}t_3 \tag{14.44}$$

$$r_2 = h_{12}t_1 + h_{22}t_2 + h_{32}t_3 \tag{14.45}$$

$$r_3 = h_{13}t_1 + h_{23}t_2 + h_{33}t_3 \tag{14.46}$$

where r_1 is the signal received at antenna 1, r_2 is the signal received at antenna 2, and so forth.

This can be represented in a matrix form as:

$$\boldsymbol{R} = \boldsymbol{H} \cdot \boldsymbol{T} \tag{14.47}$$

To restore the transmitted information at the receiving terminal, it is important to include many steps of digital signal processing. The first step is that the system decoder has to predict the specific channel characteristic h_{ij} and create the channel transfer matrix. Once the channel estimations are completed, the matrix [$\boldsymbol{H}$] could be produced and that can be used to reconstruct the transmitted information streams by multiplying the received vector by the inverse of the estimated matrix:

$$\boldsymbol{T} = \boldsymbol{H}^{-1} \cdot \boldsymbol{R} \tag{14.48}$$

This represents the basis of MIMO wireless systems, but in practice the situation is more complicated as the signal propagation is not simple. In addition, each variable in the equation consists of a continuous information stream, which means that it is not constant for a long time.

References

1. Carr, J. (2001) *Practical Antenna Handbook*. McGraw-Hill, New York.
2. Chen, Z. (2007) *Antennas for Portable Devices*. John Wiley Sons Ltd, Chichester.
3. Rahmat-Samii, Y. (1988) Reflector antennas, in *Antenna Handbook*. Springer, USA, pp. 949–1072.
4. Stutzman, W.L. and Thiele, G.A. (1998) Resonant antennas, in *Antenna Analysis and Design*, 2nd edition. John Wiley & Sons, USA, pp. 164–225.

5. Sharma, S., Bhushan, B., Gupta, S., and Kaur, P. (2013) Performance comparison of microstrip antennas with different shape of the patch. *International Journal of Science and Technology*, **6** (3), 13–22.
6. Sizun, H. and de Fornel, P. (2005) *Radio Wave Propagation for Telecommunication Applications*. Springer, Berlin, Heidelberg.
7. Parsons, J.D. and Parsons, P.J.D. (2000) *The Mobile Radio Propagation Channel*, volume 2. John Wiley & Sons Ltd, New York.
8. Lioubtchenko, D., Tretyakov, S., and Dudorov, S. (2003) *Millimeter-Wave Waveguides*. Kluwer Academic.
9. Seybold, J.S. (2005) *Introduction to RF propagation*. John Wiley & Sons Ltd.
10. Chakraborty, S., Davies, A.G., and Linfield, E.H. (2006) *THz Quantum Cascade Laser*. United Kingdom Patent Application No. 0614833.2.
11. Bründermann, E., Hübers, H.W., and Kimmitt, M.F. (2012) *Terahertz Techniques*. Springer.
12. Cheung, K.W., Sau, J.M., and Murch, R.D. (1998) A new empirical model for indoor propagation prediction. *Vehicular Technology, IEEE Transactions*, **47** (3), 996–1001.
13. Rappaport, T.S. (1996) *Wireless Communications: Principles and Practice*, volume 2. Prentice-Hall PTR, New Jersey.
14. Stavrou, S. and Saunders, S.R. (2003) Factors influencing outdoor to indoor radio wave propagation. Twelfth International Conference on Antennas and Propagation, pp. 581–585.
15. Sarwate, V.V. (2006) *Electromagnetic Fields and Waves*. New Age International Ltd, Noida.
16. Iskander, M.F. and Yun, Z. (2002) Propagation prediction models for wireless communication systems. *IEEE Transactions on Microwave Theory and Techniques*, **50** (3), 662–673.
17. Mittra, R., Chan, C.H., and Cwik, T. (1988) Techniques for analyzing frequency selective surfaces – a review. *Proceedings of the IEEE*, **76** (12), 1593–1615.
18. Dewdney, P.E., Hall, P.J., Schilizzi, R.T., and Lazio, T.J.L.W. (2009) The Square Kilometre Array. *Proceedings of the IEEE*, **97** (8), 1482–1496.
19. Barba, M., Page, J.E., Encinar, J.A., and Montejo-Garai, J.R. (2006) A Switchable Multiple Beam Antenna for GSM-UMTS Base Stations in Planar Technology. *IEEE Transactions on Antennas and Propagation*, **54** (11), 3087–3094.
20. Blake, L.V. and Long, M.W. (2009) *Antennas: Fundamentals, Design, Measurement*, 3rd edition. SciTech Publishing Inc., Raleigh.
21. Patel, P.D. (2007) *Fundamentals of antenna arrays*. Astron, The Netherlands.
22. Wolff, C. (1998) Phased antenna arrays. *Radar tutorial*, GNU free documentation.
23. Reljic, B.M. and Jokanovic, B. (2002) A new class of antenna arrays with passive elements. *Microwave Review (Serbia and Montenegro IEEE MTT-S Chapter)*, **8** (1), 8–13.
24. Siwiak, K. (1998) *Radiowave Propagation and Antennas for Personal Communications*. Artech House, Inc.

15

Digital Signal Processing for Transceivers

15.1 Introduction

This chapter presents some key digital signal processing (DSP) techniques that are used in modern wireless communication transceivers. With the ever-increasing demand for high-speed communication, many new wireless communication standards have been introduced. These standards define specifications that are challenging to meet for engineers who are in charge of wireless transceiver design. A communication system that can support high-speed data must meet the bandwidth and SNR dictated by stringent specifications yet, especially in consumer electronics, a low-cost and low-power communication transceiver is essential. Silicon integrated circuit technology is the technology of choice because it is low cost and the digital–analogue integration capability is key to addressing many of the performance limitations of the analogue circuitry. Advanced signal formats and high data rates place exceedingly high requirements on linearity, phase noise, interference rejection and I–Q imbalance. However, the RF circuits are constrained by many technical difficulties, such as low-Q passive components, substrate noise, nonlinearity and so forth, which can never be completely overcome even with the very best design techniques. Using technology such as Si/SiGe BiCMOS, digital techniques for compensating these imperfections in the RF circuits can be realised on the same chip. A great deal of research and development has been conducted into the feasibility of such techniques, which include predistortion, feedback and feedforward control loops, look-up tables, and so forth. The explicit recognition that less-complex RF circuitry can be employed, with DSP techniques used to correct for the imperfections, has been recognised as a new design paradigm [1].

15.2 RF Performance Challenges

With the emergence of a number of key digital compensation techniques, various even more sophisticated digital techniques (either coding, modulation or diversity), that were not practical

Microwave and Millimetre-Wave Design for Wireless Communications, First Edition.
Ian Robertson, Nutapong Somjit and Mitchai Chongcheawchamnan.
© 2016 John Wiley & Sons, Ltd. Published 2016 by John Wiley & Sons, Ltd.

in the past, have received widespread attention. Hence, we have witnessed, during the past decade, communication standards employing extremely sophisticated analogue and digital techniques, including LTE-advanced, the 802.11 standards, and 5G. The result is that new and emerging communication standards continually push the limits of the performance of the RF front end, leading to a greater need to address RF performance challenges using DSP techniques.

15.2.1 Imperfections in the RF Front End

Designing RF front-end circuits operating at high frequency is naturally challenging, since the wavelength is very small and comparable to the dimension of components. Typically, an analogue front end consists of a quadrature modulator, preamplifier, power amplifier, oscillator, mixer and filters. In a zero-IF digital transmitter, a quadrature modulator translates the baseband signal directly to RF. The modulator has in-phase and quadrature signal paths, which should ideally have 90° phase difference and equal gains. In practice, amplitude and phase imbalance causes direct distortion of the signal constellation, as shown in Figure 15.1(a), which ultimately deteriorates the SNR and causes bit errors. In a low-IF wireless transmitter, the quadrature modulator upconverts the I and Q baseband signals to a complex signal at a low IF frequency. In this transmitter architecture, the amplitude and phase imbalance does not distort the signal constellation but instead creates an unwanted odd-order spectrum component, as shown in Figure 15.1(b). The magnitude of this unwanted signal depends on the degree of imbalance. Removing the image through filtering is not an attractive solution. In modern IC transceivers, the baseband signals are produced from a DSP processor and analogue baseband chip. Therefore, a simple digital compensation technique can be applied to correct for the imperfections of the quadrature modulator in both architectures.

Another common problem is encountered in image-rejection receivers employing a quadrature demodulator. The imbalances in amplitude and quadrature phase are very critical to the

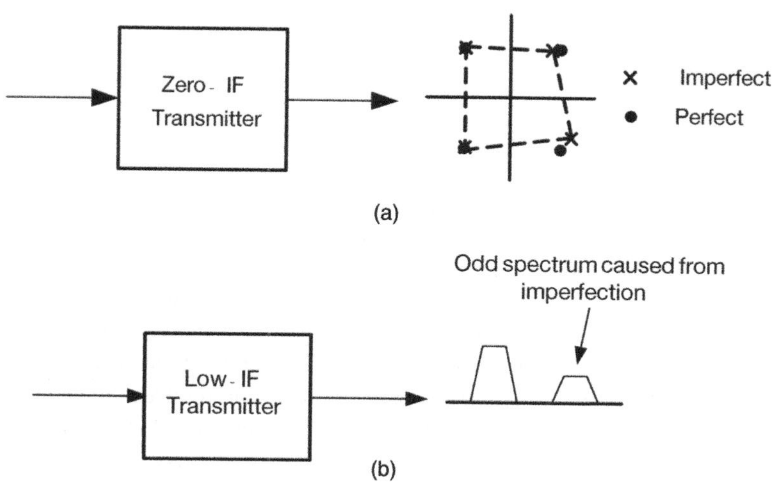

Figure 15.1 Effect of imbalance in the quadrature modulator for (a) zero-IF, and (b) low-IF transmitters.

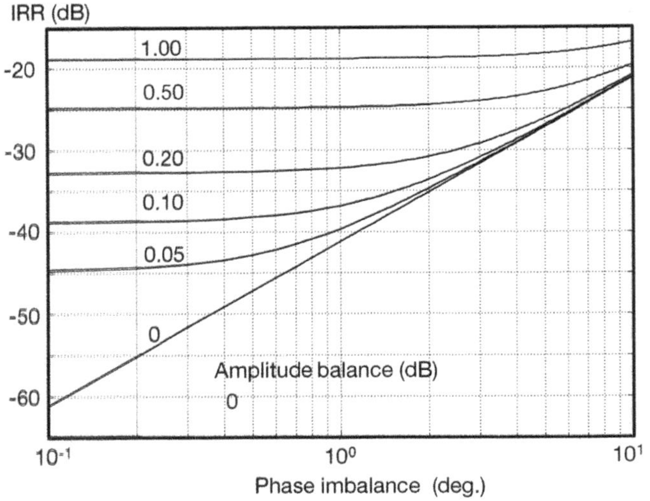

Figure 15.2 Effect of *I* and *Q* mismatch to the IRR performance of image rejection receiver.

level of image signal cancellation. The image rejection ratio (IRR), which is defined by the ratio of the image and desired signal powers, is given by:

$$\text{IRR} = \frac{1 - 2g\cos\emptyset + g^2}{1 + 2g\cos\emptyset + g^2} \tag{15.1}$$

where g is the gain imbalance ratio and $\emptyset$ is the phase imbalance from 90°.

The plot of the IRR in Figure 15.2 shows how stringent the requirements are for amplitude and phase balance in a quadrature modulator. With amplitude and phase imbalances of 0.05 dB and 0.5°, the receiver can achieve an image rejection ratio of only 44 dB. If 60 dB IRR is needed, the *I* and *Q* waveform amplitudes should be exactly equal with a phase error less than 0.1°. This demanding specification is virtually impossible to achieve unless a compensation technique using digital signal processing is implemented.

15.2.2 Cost and Low Power Consumption

A portable communication device has become the most ubiquitous consumer electronics product in the whole world. The transceiver must be power-efficient so that it can work with a small battery. A single wireless communication IC chip is a common solution and can be very cheap if its chip area is small and few off-chip components are used. As discussed throughout this book, many circuit design techniques contribute to achieving this. In particular, the zero-IF transceiver architecture is one of the most important developments in wireless communication ICs. The transceiver topology removes the IF circuitry and reduces the need for off-chip filtering. The transceiver topology reduces not only the cost but also power consumption, due to the elimination of IF circuits. Unfortunately, a zero-IF transceiver has the problem of a DC offset that can be static or dynamic. This problem occurs from LO leakage, even-order harmonic distortion and interference in the analogue front end. Using a high-pass filter is a straightforward solution but it unavoidably consumes the die area (due to large capacitor area) and hence increases the cost. The DC offset can be calibrated out using a compensation technique via the digital domain. The digital solution is attractive because the digital compensation algorithm can be implemented in a digital hardware resource already, which is available in the baseband processor.

15.2.3 Linearity

Modulation techniques can be classified into two types, linear and nonlinear. Linear modulation, such as 16-QAM or 64-QAM, is often specified in communication standards due to its high bandwidth utilisation efficiency (BUE). Unfortunately, the modulated RF signal has a nonconstant envelope and therefore is sensitive to the linearity of a system. Generally, the power amplifier (PA) is the key circuit that determines the linearity of the transmitter. The modulated spectrum, when fed through a nonlinear power amplifier, will exhibit spectral regrowth which will lead to adjacent channel interference that degrades the SNR of neighbouring channels. The simple solution is to operate the PA backed-off, with output power that is somewhat less than the full power of the PA. This trade-off between the PA linearity and its power efficiency is not acceptable, however, since the battery of the device will go flat quickly. Several techniques have been proposed to improve linearity as well as efficiency. These are the pre-distortion technique, adaptive feedforward, Cartesian feedback, and the 'linear using nonlinear components' (LINC) technique, to name a few. All of these techniques implemented with DSP techniques can successfully improve the linearity of the PA while achieving good power efficiency.

15.3 DSP in Modern Wireless Communications Systems

With the continual demand for higher data rates, many new and emerging wireless communication standards employ advanced communication schemes that rely heavily on sophisticated digital signal-processing techniques. Not so long ago, the arrival of digital communication transformed the telecommunications industry. The transmission of high bit-rate digital data with low bit-error-rate (BER) requires synchronous communication, with both carrier and bit synchronisation, and a source encoding technique. The role of digital techniques in the early days was to serve these basic communication functions. More recently developed schemes include OFDM (Orthogonal Frequency Division Modulation), OFDMA (Orthogonal Frequency Division Multiple Access), MIMO (Multiple Input-Multiple Output), adaptive modulation techniques and beam-forming techniques for electronically steered antennas. All of these schemes require a good understanding of digital communications and knowledge of signal-processing techniques. These systems require a close integration of RF analogue circuits with digital signal-processing hardware and software. For example, consider the block diagram of a MIMO transceiver, a technology that is used in the LTE-advanced standard. The basic diagram of a 2×2 MIMO system is shown in Figure 15.3. The system consists of two transmitters and two receivers, while any number of transmitters and any number of receivers can be used, depending on the desired application.

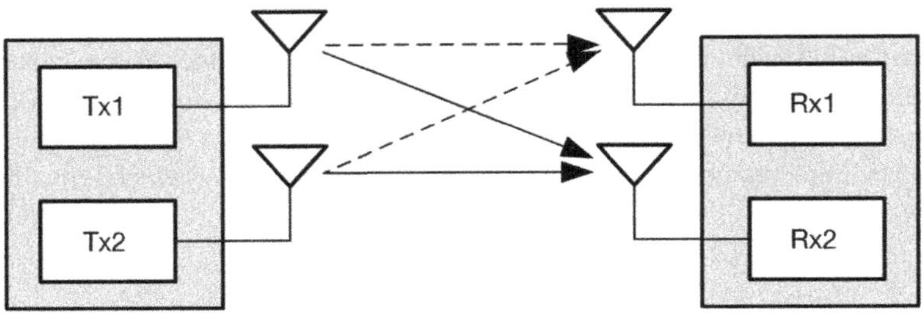

Figure 15.3 2 × 2 MIMO transceiver.

The concept behind MIMO is simple, being basically a diversity technique, but with channel estimation techniques used to improve the signal quality. Increasing the number of transmitters and receivers increases the number of signal paths. The source information is delivered to R × 1 and R × 2 by both transmitters, T × 1 and T × 2. If the received signals from R × 1 and R × 2 are combined, the received power increases and hence the SNR improves. On the other hand, the MIMO receiver can select a good quality signal, either from R × 1 or R × 2. This mode is equivalent to a selective diversity technique. The key to MIMO technology is employing communication and signal-processing techniques to mitigate the hostility of the wireless propagation channel. Channel estimation algorithms and techniques to combat channel impairments are embedded in a digital part in the communication transceiver chip.

As discussed previously, there are challenging issues in wireless transceiver RF front-end design. In microwave and millimetre-wave wireless communication applications, solutions based on digital signal processing are very important. Hence, engineers who are working in this area should not only have a background in circuit design and electromagnetics but should also be familiar with the basics of communication theory and DSP, as discussed in Chapter 1. Although the DSP field is much newer than the analogue counterpart, its distinct advantages have brought much attention from research groups around the world who continuously develop new theories and algorithms that further extend the applications of DSP in wireless communication systems, from source encoding through to sophisticated smart antennas. Rather than attempt to survey the ocean of knowledge of DSP techniques, this chapter will cover a few key principles of DSP which include sampling theory, quantisation noise, along with A/D and D/A conversion. The later sections in the chapter will introduce some key DSP compensation techniques that can support the design of wireless communication devices. These are the predistortion technique for linearising a power amplifier, the OFDM technique, MIMO, digital correction from *I–Q* imbalance imperfection, and MIMO transceivers.

15.4 Signal Conversion and Processing

15.4.1 Analogue-to-Digital Conversion

The process of converting an analogue signal to a digital signal is called analogue-to-digital conversion. The process consists of three main steps which are sample and hold (S/H), quantisation, and coding (as shown in Figure 15.4). An analogue input is applied in the analogue-to-digital converter (ADC) and digital data appears at the ADC output. The operation of each part of the ADC is now briefly described.

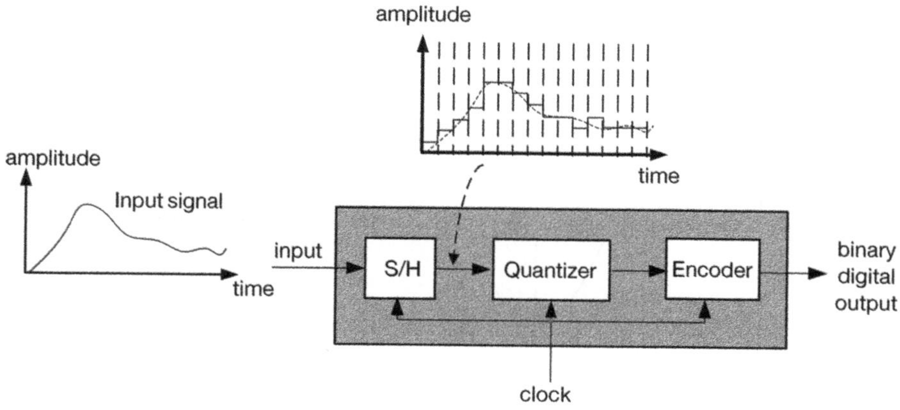

Figure 15.4 Analogue-to-digital conversion.

15.4.1.1 Sample and Hold (S/H)

Based on sampling theory, the S/H circuit converts the analogue waveform to a number of samples. The foundations of sampling theory began when Augustin-Louis Cauchy, a famous mathematician, proposed the idea of reconstructing an analogue waveform using periodic samples of the waveform. But it was more than 84 years until Harry Nyquist and Claude Shannon showed how to reconstruct an original waveform from a minimum number of samples, through contributions now recognised as the Nyquist–Shannon Sampling Theorem. This theory states: 'If a function $x(t)$ contains no frequencies higher than W rad/s, it is completely determined by giving its ordinates at a series of points spaced 1/2 W seconds apart'. This can be written in mathematical form as:

$$f_s \geq 2f_c \tag{15.2}$$

where f_s is the sampling frequency, the rate of sampling, and f_c is the highest frequency contained in the signal.

To illustrate this, consider a baseband sinusoidal signal with a frequency of 1 MHz. The waveform is sampled with two different sampling frequencies of 4 and 1.5 MHz. The obtained sample sequence according to the two chosen values of sampling frequency, f_s, are plotted together along with the original waveform at 1 MHz in Figure 15.5(a) and (b), respectively. For the sampling frequency of 4 MHz, a sample occurs every 0.25 μs, but if the signal is sampled at 1.5 MHz the samples occur every 0.67 μs, as shown. Reconstruction of the waveform from the sample sequence in Figure 15.5(a) is possible, but this is not possible for the under-sampled case shown in Figure 15.5(b).

Let $x(t)$ and $x_s(t)$ be the band-limited signal to be sampled and the sampled signal, respectively, and T_s be the sampling time which is the reciprocal of sampling frequency. The sampling process is equivalent to the signal to be sampled multiplied by a train of delta function signals, $\delta(t)$, occurring at every T_s, as follows:

$$x_s(t) = x(t) \sum_{n=-\infty}^{\infty} \delta(t - nT_s) \tag{15.3}$$

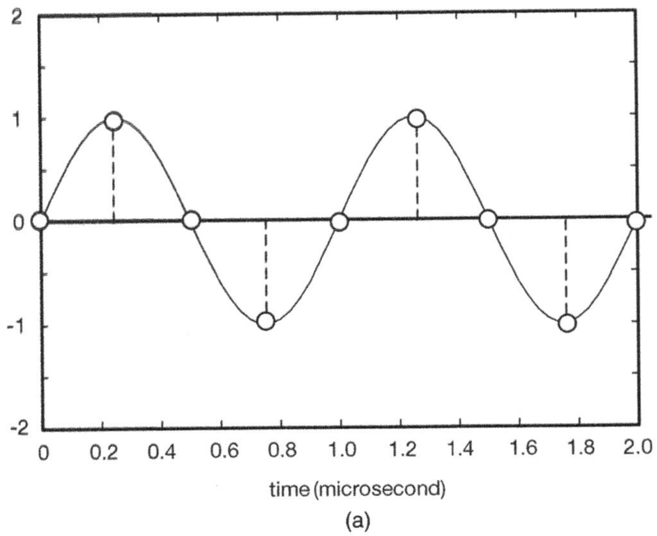

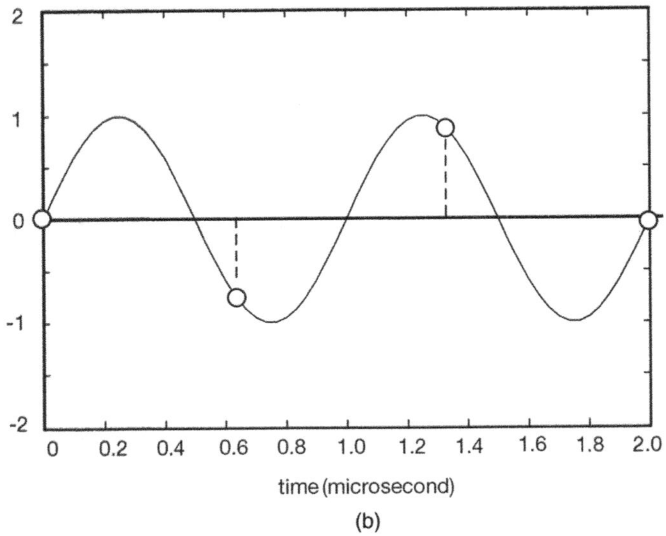

Figure 15.5 A 1 MHz sinusoidal waveform which is sampled at (a) 4 MHz and (b) 1.5 MHz.

Using the Fourier transform, the spectrum of a sampled signal is found to be:

$$X_s(f) = f_s \sum_{n=-\infty}^{\infty} X(f - nf_s) \tag{15.4}$$

The spectrum of the sampled waveform $X_s(f)$ has the shape of the original spectrum $X(f)$ but is repeated every f_s, as shown in Figure 15.6. According to Equations (15.3) and (15.4), Figure 15.6(a) and (d) shows a periodic signal (which is undersampled) and the sampled signal in the time- and frequency domains. The Nyquist–Shannon theorem indicates that any

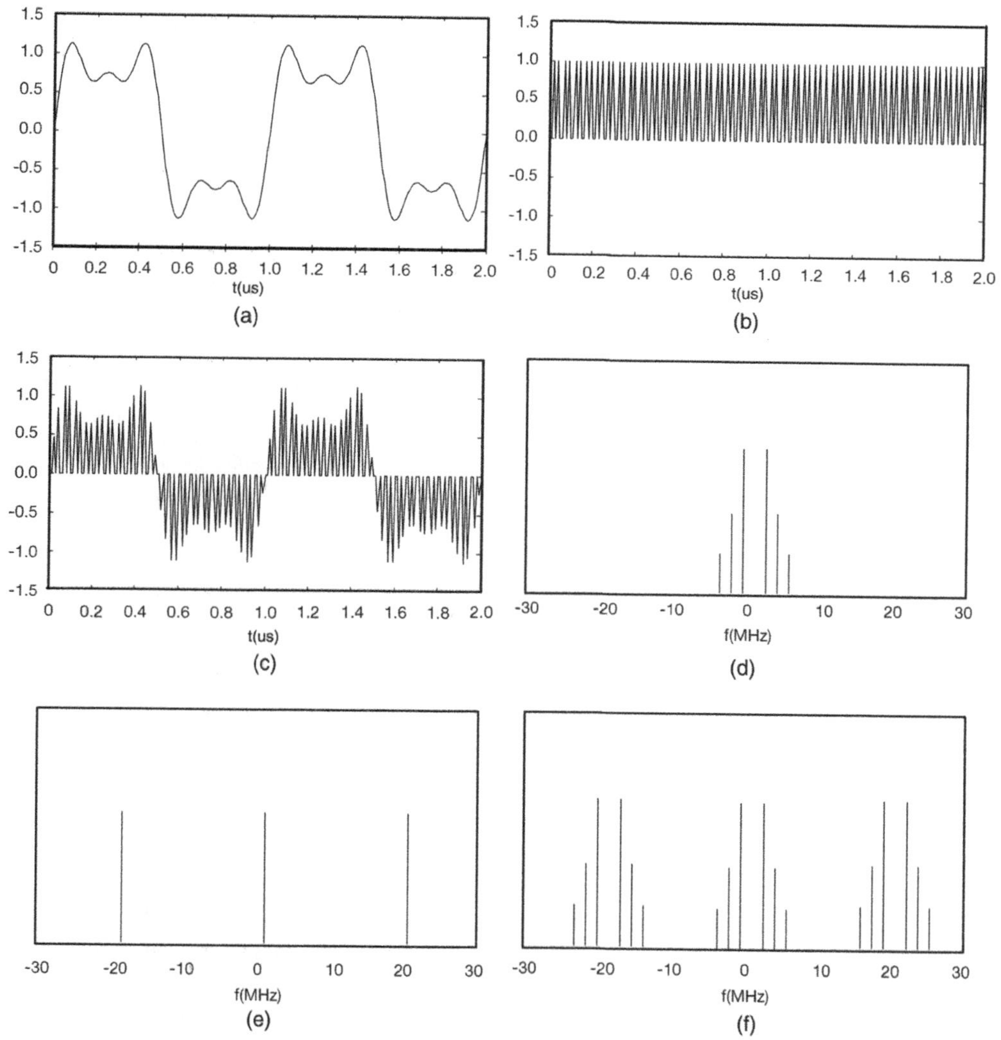

Figure 15.6 Time- and frequency-domains of waveforms which is sampled and the sampled signal: (a) $x(t)$; (b) $\delta(t)$; (c) $x_s(t)$; (d) $X(f)$; (e) $\delta(f)$; (f) $X_s(f)$.

time-varying band-limited signal, which is periodically sampled at a rate of at least twice the highest-frequency of any sinusoidal component contained in the signal, can be perfectly reconstructed from the periodic samples. Therefore, if either the sampled signal is not band-limited or the frequency sampling does not follow the Nyquist–Shannon theorem (as is the case in Figure 15.2(b)), a reconstruction error called 'aliasing' occurs. To avoid aliasing, the signal must be band-limited and the sampling frequency must be chosen according to the theorem. The first condition can be met by applying a filter which limits the signal bandwidth. This filter, called an anti-aliasing filter, will remove all signal components whose frequencies are above $f_s/2$. An anti-aliasing filter is implemented with a low-pass filter with a cut-off frequency of $f_s/2$, so that no frequency components beyond a half of the sampling frequency

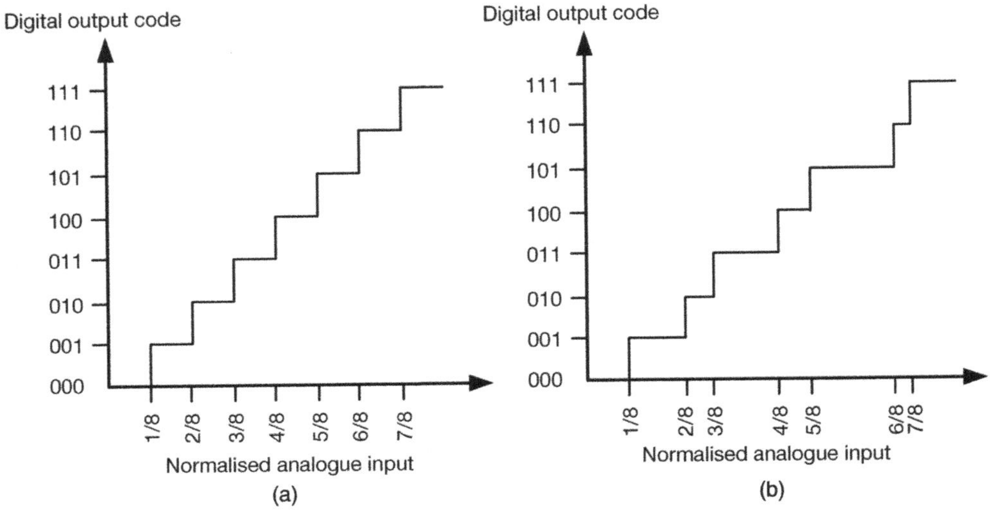

Figure 15.7 Uniform and nonuniform quantisation.

are input into the sampling circuit. Therefore, the conditions defined by the Nyquist–Shannon theorem are satisfied.

15.4.1.2 Quantiser

The next part of the ADC is the quantiser, which takes the sampled data and converts the amplitude of each sample to one of a finite range of discrete values. The number of discrete values in the finite set, called the number of quantisation levels, is 2^N, where N is the number of bits used to represent each sample. The range of a quantisation level is called the quantisation step, Δ. According to the characteristic of Δ, a quantiser can be classified into two types, namely a uniform and nonuniform quantiser. The input and output relations of these two quantisers are shown in Figure 15.7. The x- and y-axes of the graphs represent the input and output, respectively, of the quantiser. All quantisation steps are equal for the uniform quantiser, while in the nonuniform quantiser the steps are smaller in the centre of the characteristic, so that small signals can be accurately represented. In practice, choosing uniform or nonuniform quantisation depends on the application. It can be shown that, either using uniform or nonuniform quantisation, the quantised value is in fact the approximate value of the sampled amplitude; that is, the value closest to the amplitude of the sampled point. The difference between the output $x_q(t)$ and the input data $x_s(t)$ is referred to as quantisation noise, $n_q(t)$, which is defined by:

$$n_q(t) = x_s(t) - x_q(t) \tag{15.5}$$

Here, we only consider uniform quantisation. If the maximum output range of a uniform quantiser is $\pm A$, the step size for an N-bit uniform quantiser is given by:

$$\Delta = \frac{A}{2^N} \tag{15.6}$$

The statistical characteristic of quantisation noise has a uniform distribution within the range of Δ. The quantisation noise power, $\sigma^2_{n_q}$, can be calculated from the noise variance:

$$\sigma^2_{n_q} = \int_{-\Delta/2}^{+\Delta/2} \frac{n_q^2}{\Delta} dn_q = \frac{\Delta^2}{12} \tag{15.7}$$

If the applied signal is sinusoidal, the root-mean-square value for a full-scale input is:

$$x_{rms-q} = \frac{2^{N-1}}{\sqrt{2}}\Delta \tag{15.8}$$

The maximum possible SNR for an N-bit uniform quantiser operating over the range DC to $f_s/2$ is then given by:

$$SNR_{maxQ} = 20\log_{10}\left(\frac{\frac{2^{N-1}}{\sqrt{2}}}{\sqrt{\frac{\Delta^2}{12}}}\right) = 1.76 + 6.02N \tag{15.9}$$

Equation (15.9) can be used to determine the number of quantisation bits required. This is obtained by rearranging Equation (15.9), giving:

$$N = \frac{SNR_{maxQ} - 1.76}{6.02} \tag{15.10}$$

15.4.1.3 Bit Encoder

The function of the bit encoder is to convert the quantised sequence of samples obtained from the quantiser into a sequence of N-bit binary digital numbers. The binary representation is resilient to noise and the sampled signal can be processed by a digital signal processor. Currently, there are various ADC architectures that can be used. These are sigma-delta, dual-slope, successive approximation (SAR), pipeline, interleaved and flash architectures. The pros and cons of these ADCs are related to trade-offs in maximum conversion speed, accuracy, chip area, power consumption and cost.

15.4.2 Digital-to-Analogue Conversion

A digital-to-analogue converter (DAC) performs the reverse process to the ADC. That is, binary digital data is converted to an analogue signal. The gap between two adjacent digital sample points can be filled by interpolation, which can be done using a low-pass filter. A DAC can be implemented in several ways, such as binary-weighted, R2R ladder, successive approximation, or delta-sigma, to name a few. The criteria for choosing the suitable architecture in a certain application depend on resolution, maximum sampling rate, noise, distortion and monotonicity. During recent years there has been a strong tendency to move towards techniques that are much less reliant on precision analogue design, and therefore more suitable for IC implementation, such as oversampling techniques using a pulse density modulator.

15.4.3 Discrete Fourier Transform

Many advanced DSP techniques used in wireless communication require the signal to be analysed or processed in the frequency domain. In theory, the Fourier transform is the essential mathematical tool for determining the spectrum of a time-domain signal. However, it is not practical because the Fourier transform is defined rigorously with an integral operator. This means that a closed-form solution is expected from the integral Fourier equation. This unavoidably restricts its use in practice because the applied time-domain signal often cannot be written in a closed form, or there is no exact solution for the integral equation. This is especially the case with digital samples obtained by digitising analogue signals. Hence, the Discrete Fourier Transform (DFT) was proposed as a solution for computing the spectrum of a sequence of digital samples, more generally referred to as a discrete-time signal. Given the discrete-time signal defined in Equation (15.3), the DFT algorithm limits the energy of the discrete-time signal $x_s(t)$ by applying the window function $w(t)$ to give:

$$y_1(t) = w(t) \cdot x_s(t) \tag{15.11}$$

The role of the window function is to ensure that energy of the analysed discrete-time signal will be limited. The simplest window function is a rectangular function of length NT_s which is defined by:

$$w(t) = \begin{cases} 1; & 0 \leq t < (N-1)T_s \\ 0; & \text{elsewhere} \end{cases} \tag{15.12}$$

Hence, the discrete-time signal obtained after applying the rectangular window is given by:

$$y_1(t) = \sum_{n=0}^{N-1} x(nT_s) \tag{15.13}$$

The next step of the DFT is to convolve a windowing signal with function $c(t)$. The function $c(t)$ is given by:

$$c(t) = \sum_{m=-\infty}^{\infty} \delta(t - mNT_s) \tag{15.14}$$

This leads to $y_2(t)$ which is given by:

$$y_2(t) = y_1(t) \otimes c(t) = \sum_{m=-\infty}^{\infty} \sum_{n=0}^{N-1} x(nT_s)\delta(t - nT_s - mNT_s) \tag{15.15}$$

where ⊗ represents the convolution operator. This step is unique for DFT calculation. Convolving with $c(t)$ is equivalent to multiplying the windowing signal spectrum with a comb function. This eliminates the integral operator in the calculation. The last step of the DFT is to compute the spectrum, which is the DFT output of discrete-time signal, $x_s(t)$, that is, $X_s(f)$:

$$x_s(t) \overset{DFT}{\Rightarrow} X_s(f) = \sum_{m=-\infty}^{\infty} \sum_{n=0}^{N-1} x(nT_s) e^{-j2\pi k\left(m+\frac{n}{N}\right)} \tag{15.16}$$

It is evident that there is an infinite number of spectrum components. With k and n being integers, the exponential term $e^{-j2\pi k}$ is periodic. The output spectrum repeats with the spectrum data from $k = 0$ to $N - 1$. The integers n and k are time- and spectrum-indices of $x_s(t)$ and $X_s(f)$. We therefore use the notation $x_s(n)$ and $X_s(k)$ to represent the discrete signals for the time and frequency domains, respectively. If we consider only the first period, $m = 0$, then $X_s(k)$ from the DFT can be computed from $x_s(n)$ by:

$$X_s(k) = \sum_{n=0}^{N-1} x_s(n) e^{-\frac{j2\pi nk}{N}} \tag{15.17}$$

The inverse DFT can be written as:

$$x_s(n) = \frac{1}{N} \sum_{k=0}^{N-1} X_s(n) e^{\frac{j2\pi nk}{N}} \tag{15.18}$$

As shown in Equation (15.17), the computation complexity involving complex numbers grows increasingly with approximately N^2. This results from the requirement of $(N - 1)^2$ complex multiplication and $N(N - 1)$ complex addition operators to compute the DFT. The N^2 computation complexity gives large computation time and power consumption. This is not acceptable for high-speed and portable wireless communication.

15.4.4 *Fast Fourier Transform*

The Fast Fourier Transform (FFT) algorithm was proposed to reduce the computation complexity required in the DFT. By considering the DFT defined in Equation (15.17), the kernel function $e^{-j2\pi k}$ is symmetrical. If the data point number is chosen to be 2^n (i.e., $N = 2^n$), the computation can be significantly simplified with the concept of *divide and conquer*. The computation algorithm that developed with this principle is the FFT algorithm. The algorithm concept of *divide* stems from decomposing an N point time-domain signal into N time-domain signals of one point. Then, the concept of *conquer* begins by calculating each spectrum of N single-point time-domain signals. The calculated N spectra are collected to form a spectrum of N data points. In summary, there are three main calculation steps in the FFT algorithm: (i) decomposition; (ii) spectrum calculation; and (iii) forming the spectrum.

The decomposition step in the FFT algorithm is illustrated here with the example of $N = 8$. In this example, a time-domain signal of eight points is decomposed in three stages, as shown in Figure 15.8. The first stage decomposes the eight-point signal to a pair of four-point signals.

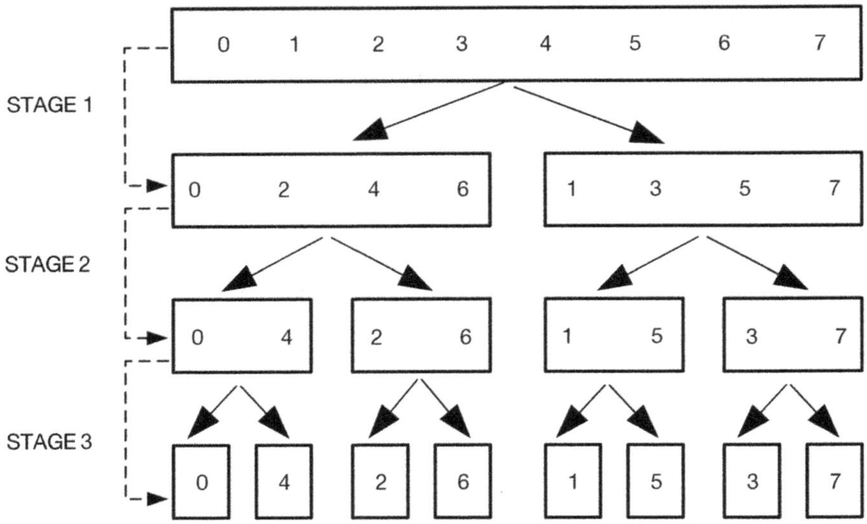

Figure 15.8 Decomposing a time-domain signal of eight data points.

The first group is the signal with the even number samples where the other comprises the odd sample numbers. The second decomposition stage splits further into four groups of two-point signal and the final stage separates further to eight groups of single-point signal.

In digital implementation, the decomposition step can be performed by re-ordering the bit sequence. The sequence number of digital sequence input (stage 1) and the final sequence output (stage 3) in 3-bit binary number is shown Figure 15.9. The input sequence in 3-bit format is applied to the decomposition block while the output bit sequence in 3-bit format appears at the block output. The results in Figure 15.9 clearly show that the decomposition can be implemented simply by reversing data bits. For example, the sequence '100', a sample number '4', is applied to decomposition. By bit reversing, one obtains '001', the sample number '1', at the output of decomposition, as shown in Figure 15.9.

The second step of the FFT is to compute the spectrum of the single-bit time-domain signal. Applying the DFT to a single-point time-domain signal, the DFT output is equal to the input

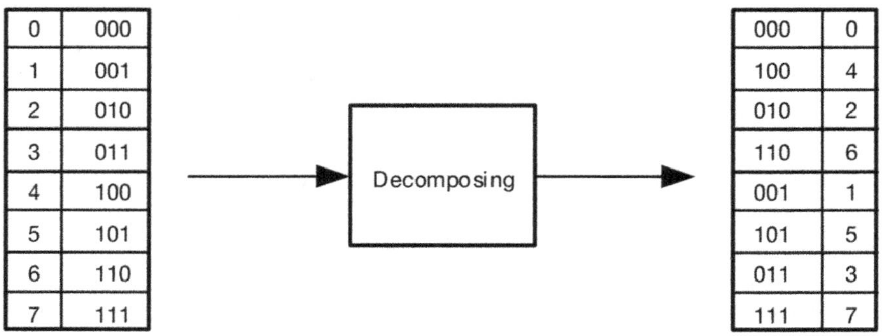

Figure 15.9 Decomposition step which is equivalent to a bit reversion process.

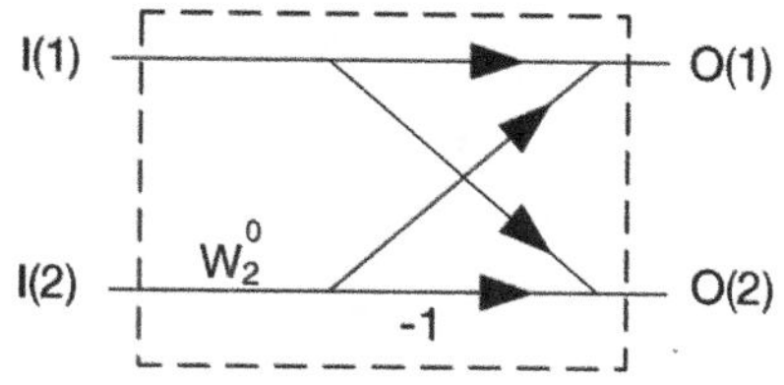

Figure 15.10 Basic FFT cell; a butterfly diagram.

itself. The final FFT step is to synthesise an N-point signal spectrum from N spectra of the single-point signals. The step is to combine the N spectra from the reverse order that resulted from the decomposition in time domain. To do this, a basic calculation element for the FFT, called the FFT butterfly diagram and shown in Figure 15.10, is required. The term W_N^{nk} is defined by:

$$W_N^{nk} = e^{-j\frac{2\pi nk}{N}} \tag{15.19}$$

Given that $I(1)$, $I(2)$ are the inputs to the butterfly cell, the outputs $O(1)$, $O(2)$ can be written as:

$$O(1) = I(1) + I(2)\, W_2^0 \tag{15.20}$$

$$O(2) = I(1) - I(2)\, W_2^0 \tag{15.21}$$

To compute the eight-point spectrum from the FFT, there are three stages, as shown in Figure 15.11. The first stage starts with applying the reshuffled data to the butterfly diagram to form four sets of two-point spectrum. The next step is performed by forming two sets of four-point spectrum data from four sets of two-point spectrum using the butterfly diagram in Figure 15.10. The final stage is to compute an eight-point spectrum from two sets of four-point spectrum data. The output of this stage is the output of the FFT, the eight-point spectrum output. These are denoted by $X[0], X[1], \ldots, X[7]$.

From Figure 15.11, it is shown that the computation complexity is reduced significantly. The number of complex multiplications is only 12, while the number of complex additions (including subtraction) is only 24. This compares very favourably with the mathematical operations required in DFT computation, for which the number of complex multiplication and additions are 49 and 56, respectively. The FFT algorithm offers a considerable reduction in computation complexity. Consequently, the FFT is widely used in practice.

15.4.5 CORDIC Algorithm

When implementing DSP algorithms in wireless communications systems it is important to consider the computational complexity carefully. The main constraints for a design usually concern the size, cost and power consumption, while the specification targets must be strictly met. The COordinate Rotation DIgital Computer (CORDIC) algorithm offers an efficient way to implement mathematical operations concerning trigonometric functions. It has attracted

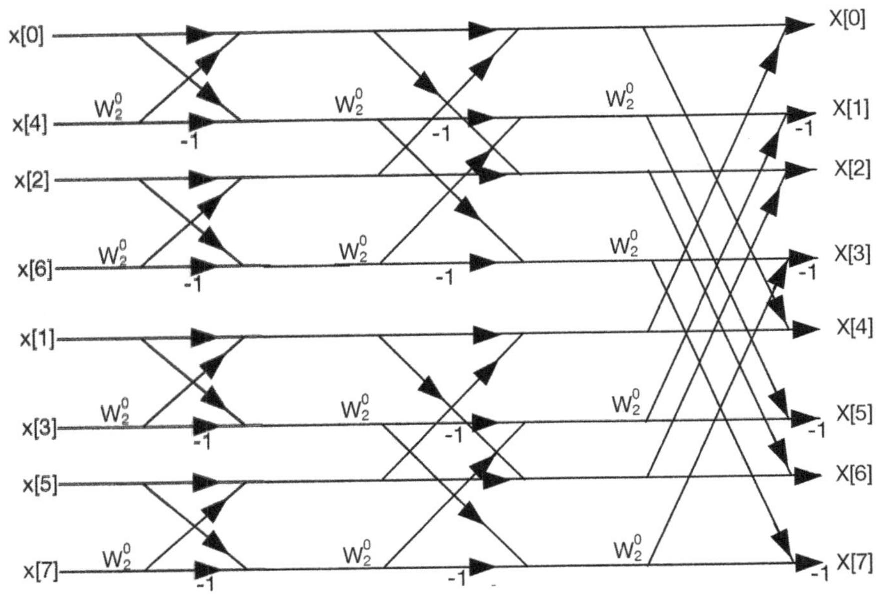

Figure 15.11 FFT applied to compute an eight-point spectrum.

much attention because the CORDIC algorithm can reduce multiplication operations enormously by using an iterative technique. Examples of trigonometric functions that can be computed using CORDICs are sine, cosine, inverse tangent, inverse sine, the magnitude of vector, the transformation from polar-to-rectangular coordinates or vice versa. The reduction in multiplication operations in the CORDIC algorithm is achieved by performing vector rotation with specific angles using 'shift' and 'add' operations.

The CORDIC concept starts with a simple rule for two-dimensional vector rotation. As shown in Figure 15.12, given a point P_0 located at Cartesian coordinate (x_0, y_0), the vector P is to be obtained by rotating vector P_0 through an angle θ. The coordinate of P, (x, y), can be determined from:

$$x = x_0 \cos\theta - y_0 \sin\theta \tag{15.22}$$

$$y = x_0 \sin\theta + y_0 \cos\theta \tag{15.23}$$

The rotation equation above can be written in the matrix form as:

$$\mathbf{p} = \mathbf{R} \cdot \mathbf{p}_0 \tag{15.24}$$

where $\mathbf{p}_0$ is the original vector and $\mathbf{p}$ is the rotated vector.
The rotation matrix $\mathbf{R}$ is:

$$\mathbf{R} = \cos\theta \begin{bmatrix} 1 & -\tan\theta \\ \tan\theta & 1 \end{bmatrix} \tag{15.25}$$

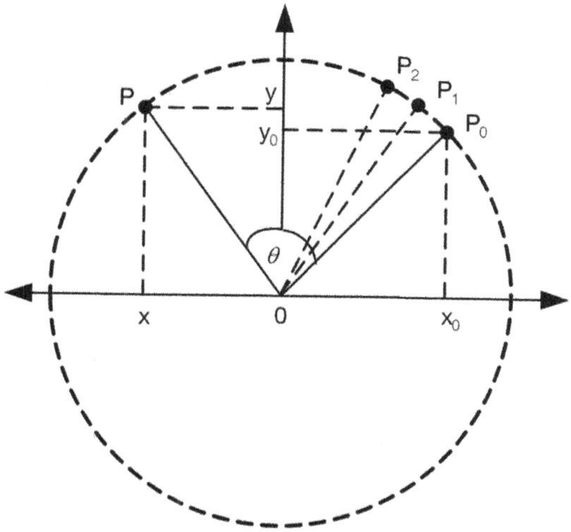

Figure 15.12 Vector P_0 rotates to P with θ.

To perform the rotation directly simply involves calculating the trigonometric functions (cosine and tangent functions). Unfortunately, in a DSP processor this requires a numerical algorithm such as a Taylor's series which uses many multiplication and addition terms. Instead of computing P from P_0 using the rotation vector equation given in Equation (15.25), the CORDIC algorithm avoids the trigonometric computation task with a less direct, but smarter, approach. The CORDIC algorithm first divides the angle θ into smaller angles. Then, there are many intermediate points between P_0 and P that lie on an arc of the unit circle, say P_1, P_2, P_3, and so forth. The computation is iteratively applied to shift from P_i from P_{i-1}. The rotation angles chosen in the CORDIC algorithm are chosen such that the tangent of these angles are in the form of 2^{-n}, where $n = 0, 1, 2,..$ These small angles for rotation are computed from:

$$\theta_n = \tan^{-1}(2^{-n}) \tag{15.26}$$

If this rule is used to set the angle, we obtain $\theta_0 = 45°$, $\theta_1 = 45°/2 = 22.5°$, $\theta_3 = 45°/4 = 11.25°$, and so forth. The cosine function of θ_n is determined from:

$$\cos\theta_n = \frac{1}{\sqrt{1 + (2^{-n})^2}} \tag{15.27}$$

With the rotation equation, we will write the rotation matrix relating $0P_i$ and $0P_{i\text{-}1}$ the angle of rotation θ_n by:

$$\mathbf{R}_\mathrm{n} = \cos\theta_n \begin{bmatrix} 1 & -\tan\theta_n \\ \tan\theta_n & 1 \end{bmatrix} \tag{15.28}$$

Substituting $\tan\theta_n$ and $\cos\theta_n$ into $\mathbf{R}_\mathrm{n}$, we obtain:

$$\mathbf{R}_\mathrm{n} = \frac{1}{\sqrt{1+(2^{-n})^2}} \begin{bmatrix} 1 & -\frac{1}{2^n} \\ \frac{1}{2^n} & 1 \end{bmatrix} \tag{15.29}$$

Therefore, by applying the CORDIC algorithm to divide the rotation angle θ into N small angles, i.e. $\theta_0, \theta_1, \ldots, \theta_N$ the final vector $\mathbf{p}$ rotated with θ from the original vector $\mathbf{p}_0$ can be computed from:

$$\mathbf{p} = \prod_{n=0}^{N} \frac{1}{\sqrt{1+(2^{-n})^2}} \left(\mathbf{R}_0 \cdot \mathbf{R}_1 \cdot \mathbf{R}_2 \cdots \mathbf{R}_{\mathrm{N}-1} \cdot \mathbf{R}_\mathrm{N}\right) \cdot \mathbf{p}_0 \tag{15.30}$$

It may appear at first that the rotation equation provided from the CORDIC approach looks more complicated than the original, due to the product series and the matrix multiplication. Looking first at the product series which, from now on, will be denoted with K, given by:

$$K = \prod_{n=0}^{N-1} \frac{1}{\sqrt{1+(2^{-n})^2}} \tag{15.31}$$

The product series shown above converges to a certain value when N is sufficiently large. From Figure 15.13, when the number of small angles N, implying the number of iterations, is more than 6, K will converge to 0.60725294. Hence, the complexity occurred in K is considerably reduced to a constant value which needs just one store instruction and two multiplication

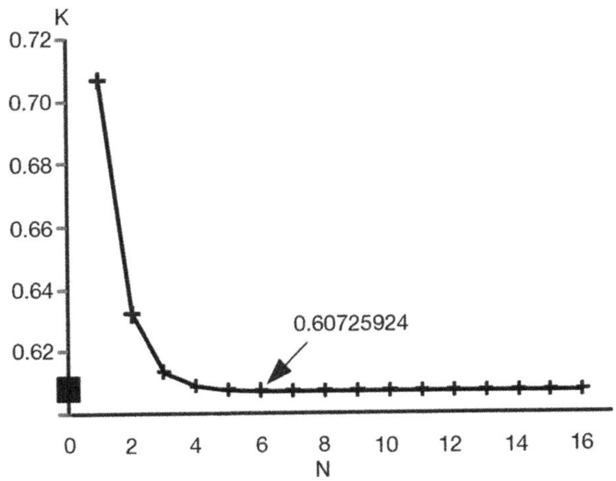

Figure 15.13 Convergence of K versus N.

operations. Next, consider the complexity of the matrix multiplications in Equation (15.30). This can be considered from the case of a single multiplication, for example:

$$\begin{bmatrix} 1 & -2^{-n} \\ 2^{-n} & 1 \end{bmatrix} \begin{bmatrix} a \\ b \end{bmatrix} = \begin{bmatrix} a - 2^{-n}b \\ 2^{-n}a + b \end{bmatrix} \tag{15.32}$$

The term 2^{-n} is calculated without using a multiplication operator but with a bit-shift operation. Hence, from each matrix rotation, we only require two additions (including subtraction) and two bit-shift operations. We also need two more store instructions. In summary, a small rotation of a vector requires six digital operators which are very low in computational complexity: two for addition; two for bit-shifting; and two for data storage operations. Therefore, the total complexity of the CORDIC algorithm, with N iterations, will reduce to:

$$N \times (2\text{ Adding} + 2\text{ Shifting} + 2\text{ Storing}) + 2\text{ Multiplying} + 1\text{ Storing} \tag{15.33}$$

The following example will show how to calculate a trigonometric function using the CORDIC algorithm. This example will explain the CORDIC algorithm and how it is applied.

Example: Compute cos 86.5° using the CORDIC algorithm.

To simplify the computation, we choose $\theta_{-1} = 0$ as the initial angle. Hence, $x_{-1} = 1$ and $y_{-1} = 0$. Choosing $N = 10$, we can therefore use $K = 0.60725294$. The magnitude of angle for rotation and the rotated angle obtained from iteration is shown in Table 15.1. The final rotated angle after 10 iterations is 86.6°, which is 0.1° different from the desired rotated angle.

When applying the CORDIC algorithm for $n = 0$ to 9, it is equivalent to calculating the matrix equation as shown by:

$$\begin{bmatrix} \mathrm{x} \\ \mathrm{y} \end{bmatrix} = 0.60725294 \cdot \begin{bmatrix} 1 & -1 \\ 1 & 1 \end{bmatrix} \begin{bmatrix} 1 & -0.5 \\ 0.5 & 1 \end{bmatrix} \begin{bmatrix} 1 & -0.25 \\ 0.25 & 1 \end{bmatrix} \cdots \begin{bmatrix} 1 & 1/256 \\ -1/256 & 1 \end{bmatrix} \begin{bmatrix} 1 & 1/512 \\ -1/512 & 1 \end{bmatrix} \begin{bmatrix} 1 \\ 0 \end{bmatrix} \tag{15.34}$$

Hence from the CORDIC algorithm, we obtain:

$$x = \cos 86.5° = 0.06 \tag{15.35}$$

which is very closed to the exact value 0.061.

Table 15.1 Parameter details obtained from applying CORIC calculation in the example.

N	0	1	2	3	4	5	6	7	8	9
θ_n	45	26.6	14	7.1	−3.6	−1.8	−0.9	0.45	−0.23	−0.11
2^{-n}	1	1/2	1/4	1/8	1/16	1/32	1/64	1/128	1/256	1/512
$\theta_{n-1} + \theta_n$	45	71.6	85.6	92. 7	89.1	87.3	86.4	86.9	86.7	86.6

15.5 Digital Calibration for *I–Q* Imbalance

Digital calibration techniques can be used to counteract the imperfections in RF front-end ICs. A complex low-pass signal in I and Q channel form, $x_{LP}(t)$, is defined as:

$$x_{LP}(t) = x_I(t) + jx_Q(t) \tag{15.36}$$

As shown in Equation (15.36), the complex signal is formed of in-phase and quadrature signal components. In an ideal RF transmitter, a complex low-pass signal is modulated to high frequency at a carrier frequency f_c. This transmitted band-pass signal at RF frequency can be written in the following form:

$$x_{RF}(t) = \mathrm{Re}\left\{x_{LP}(t)e^{j2\pi f_c t}\right\} \tag{15.37}$$

The formulation is obtained from the real part of the frequency-translated complex signal $x_{LP}(t)$. Expanding the exponential term, this leads to the equation form which is commonly used for implementing an I–Q modulator as follows:

$$x_{RF}(t) = x_I(t)\cos(2\pi f_c t) - x_Q(t)\sin(2\pi f_c t) \tag{15.38}$$

Equation (15.38) is illustrated schematically in Figure 15.14(a). If imbalances occur in both amplitude and phase in the I and Q channels, the effect can be modelled as shown in Figure 15.14(b). The amplitude and phase imbalances in an I–Q modulator are described with ΔA_t and $\Delta\emptyset_t$. The output signal from an I–Q modulator with these imbalances will be:

$$x_{RF,Im} = (1 + \Delta A_t)\cos(2\pi f_c t + \Delta\emptyset_t)x_I(t) - (1 - \Delta A_t)\sin(2\pi f_c t - \Delta\emptyset_t)x_Q(t) \tag{15.39}$$

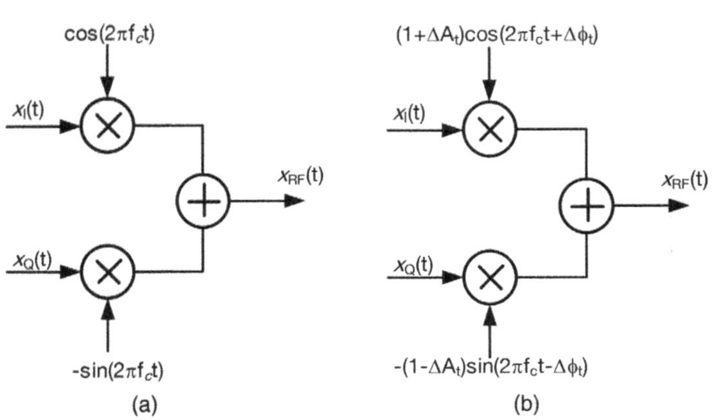

Figure 15.14 (a) Ideal *I–Q* modulator transmitter. (b) *I–Q* imbalance model.

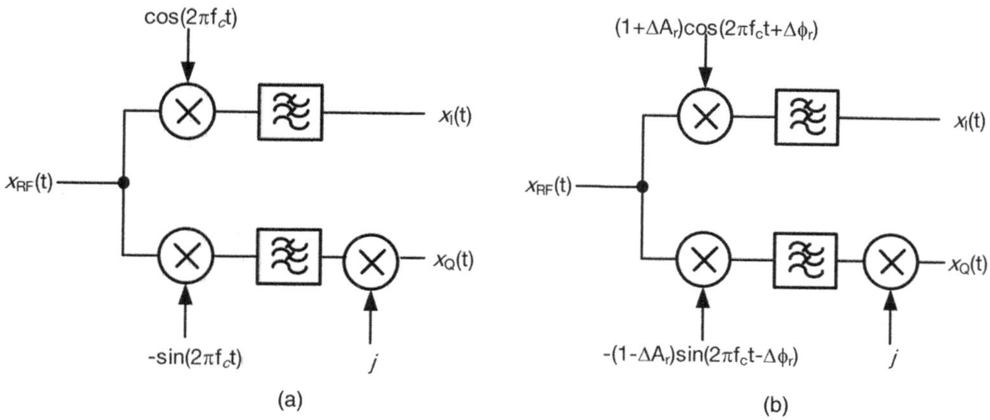

Figure 15.15 Practical implementation forms of (a) an *I–Q* demodulator, and (b) with imbalance effect.

At the RF receiver, the RF band-pass signal is assumed to be demodulated at the same carrier frequency f_c. In mathematical terms, this is equivalent to multiplying $x_{RF}(t)$ with a local oscillator, $e^{-j2\pi f_c t}$. The output signal from this multiplication can be written as:

$$x_{RF}(t)e^{-j2\pi f_c t} = \frac{1}{2}\left[\{x_I(t) + jx_Q(t)\}e^{j2\pi f_c t} + \{x_I(t) - jx_Q(t)\}e^{-j2\pi f_c t}\right]e^{-j2\pi f_c t} \qquad (15.40)$$

The term in the square bracket is another form of $x_{RF}(t)$. Manipulating the right-hand-side of Equation (15.40) leads to:

$$x_{RF}(t)e^{-j2\pi f_c t} = \frac{1}{2}\left[\{x_I(t) + jx_Q(t)\} + \{x_I(t) - jx_Q(t)\}e^{-j2\pi(2f_c)t}\right] \qquad (15.41)$$

There is a high-frequency term at $2f_c$ in Equation (15.41) which, in practice, can be eliminated by applying a low-pass filter. In a practical demodulator, the original *I* and *Q* signals can be applied by multiplying with cosine and sinusoidal terms, as shown in Figure 15.15(a). The amplitude and phase imbalances in an *I* and *Q* demodulator are given by ΔA_r and $\Delta\emptyset_r$. The low-pass signal when demodulating from the *I–Q* demodulator that suffers from imbalance can be written as:

$$x_{LP,Im} = x_{LP}(t)\{\cos\Delta\emptyset_r - j\Delta A_r \sin\Delta\emptyset_r\} + x^*_{LP}(t)\{\Delta A_r \cos\Delta\emptyset_r + j\sin\Delta\emptyset_r\} \qquad (15.42)$$

Equation (15.42) shows that the low-pass signal from an imbalanced *I–Q* demodulator has two main components. One component is the low-pass signal from an ideal receiver, while the other is its complex conjugate. This complex conjugate part distorts the demodulated information.

If imbalance exists in both the *I–Q* modulator and demodulator, the low-pass signal is obtained from:

$$x_{LP,ImTR}(t) = \alpha_{TR}x_{LP}(t) + \beta_{TR}x^*_{LP}(t) \qquad (15.43)$$

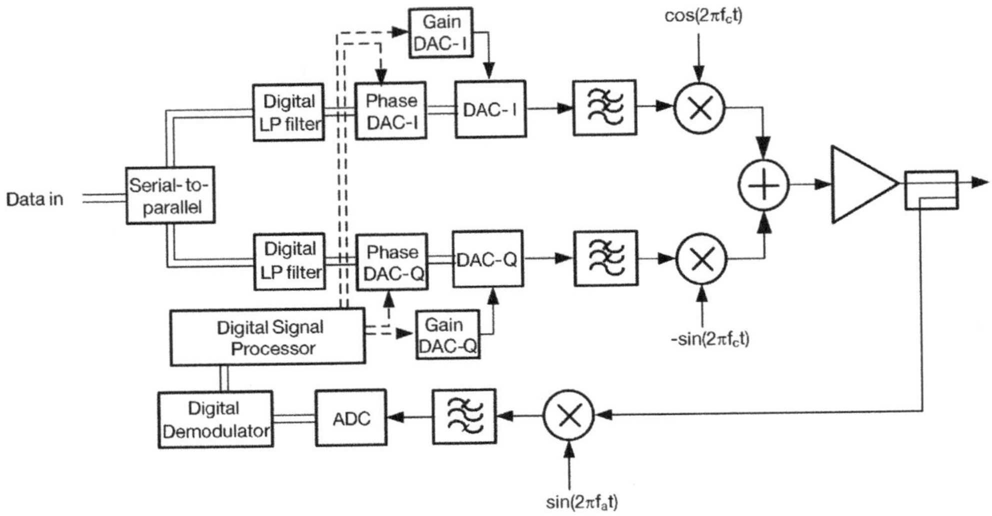

Figure 15.16 Loop back receiver applied to direct conversion transmitter.

where

$$\alpha_{TR} = (1 + \Delta A_t \Delta A_r)\cos(\Delta\emptyset_t - \Delta\emptyset_r) + j(\Delta A_t - \Delta A_r)\sin(\Delta\emptyset_t + \Delta\emptyset_r) \qquad (15.44)$$

$$\beta_{TR} = (\Delta A_t + \Delta A_r)\cos(\Delta\emptyset_t - \Delta\emptyset_r) + j(1 - \Delta A_t \Delta A_r)\sin(\Delta\emptyset_t + \Delta\emptyset_r) \qquad (15.45)$$

A low-pass signal obtained from an imbalanced *I–Q* modulator and demodulator is distorted with both amplitude and phase, as shown in Equation (15.43). If no imperfection occurs, then $x_{LP,ImTR}(t) = x_{LP}(t)$. Obviously, an algorithm for calibration is required. The correction for *I–Q* imbalance at the modulator can be made by applying a demodulator and feedback loop to monitor the constellation diagram of the transmitted signal. To illustrate this, a direct-conversion transmitter with a feedback loop is depicted in Figure 15.16. The single input bit stream is split into two bit streams as usual, using a serial-to-parallel converter. Each bit stream channel is then grouped into a stream of digital words, depending on a chosen digital modulation scheme. For each channel, a digital low-pass filter is applied to limit the spectral bandwidth. Two streams of digital words are applied to two DACs (DAC-I and DAC-Q) whose gains and phases can be adjusted separately as shown with blocks 'Gain DAC-I', 'Gain DAC-Q', 'Phase DAC-I' and 'Phase DAC-Q'. The analogue outputs of the DACs are low-pass filtered to eliminate the out-of-band spectrum before upconverting to RF with the mixer/quadrature LO combination. At the output, a receiver in the feedback path samples the transmitted signal via a coupler. The sampled signal is downconverted back to a low frequency, with a band-pass filter used to eliminate any unwanted noise and limit the signal bandwidth. The filtered analogue signal is applied to an ADC and the digital data output is applied to a digital demodulator to obtain the *I* and *Q* data streams. The digital signal processor reads the *I* and *Q* data streams from the digital demodulator. An algorithm programmed in the DSP part can analyse the *I–Q* imbalance and the amplitude and phase for compensation for *I* and *Q* channel will be calculated and fed to the gain and phase adjustment blocks, as shown in

Figure 15.16. To show how this equation can be used, the following example will explain the method for compensating imbalance in an *I–Q* modulator.

Example: In a transmitter employing an *I–Q* modulator, the modulator has static amplitude and phase errors at a carrier frequency f_c of $\Delta A_t = \alpha$ and $\Delta\emptyset_t = \varepsilon$. Assuming that a feedback loop with ideal receiver is employed, calculate the required gain and phase to be programmed for digital compensation.

The problem states that $\Delta A_t = \alpha$, $\Delta\emptyset_t = \varepsilon$ and $\Delta A_r = \Delta\emptyset_r = 0$. This leads to simplification of Equation (15.43), giving:

$$x_{LP,ImTR}(t) = [\cos\varepsilon + j\alpha\sin\varepsilon]x_{LP}(t) + [\alpha\cos\varepsilon + j\sin\varepsilon]x^*_{LP}(t) \tag{15.46}$$

The original $x_{LP}(t)$ applied to the *I–Q* modulator and the output signal from the modulator $x_{LP,ImTR}(t)$ can be written in the following forms:

$$x_{LP}(t) = I_o(t) + jQ_o(t) \tag{15.47}$$

$$x_{LP,ImTR}(t) = I_m(t) + jQ_m(t) \tag{15.48}$$

Substituting Equation (15.47) into Equation (15.46), we obtain the relationships between the modulated *I* and *Q* waveforms with the original *I* and *Q* waveforms:

$$I_m(t) + jQ_m(t) = [\cos\varepsilon + j\alpha\sin\varepsilon][I_o(t) + jQ_o(t)] + [\alpha\cos\varepsilon + j\sin\varepsilon][I_o(t) - jQ_o(t)] \tag{15.49}$$

Manipulating the complex equation given above, we obtain two real equations:

$$I_m(t) = [(1+\alpha)\cos\varepsilon]I_o(t) + [(1-\alpha)\sin\varepsilon]Q_o(t) \tag{15.50}$$

$$Q_m(t) = [(1+\alpha)\sin\varepsilon]I_o(t) + [(1-\alpha)\cos\varepsilon]Q_o(t) \tag{15.51}$$

We assume further that the phase imbalance occurs in this case is sufficiently small, hence $\varepsilon \approx 0$. This leads to simplify the above equations further with the condition $\sin\varepsilon \approx 0$. Hence,

$$I_m(t) = [(1+\alpha)\cos\varepsilon]I_o(t) \tag{15.52}$$

$$Q_m(t) = [(1-\alpha)\cos\varepsilon]Q_o(t) \tag{15.53}$$

We can write term $\cos\varepsilon$ in terms of $I_o(t)$ and $I_m(t)$ as:

$$\cos\varepsilon = \left(\frac{1}{(1+\alpha)}\right)\frac{I_m(t)}{I_o(t)} \tag{15.54}$$

Substituting Equation (15.54) into Equation (15.53), we obtain:

$$Q_m(t) = \left(\frac{1-\alpha}{1+\alpha}\right)\left(\frac{I_m(t)Q_o(t)}{I_o(t)}\right) \quad (15.55)$$

With some further manipulation, the amplitude imbalance from this case can be calculated from:

$$\alpha(t) = \frac{I_m(t)Q_o(t) - I_o(t)Q_m(t)}{I_m(t)Q_o(t) + I_o(t)Q_m(t)} \quad (15.56)$$

Hence,

$$\varepsilon(t) = \cos^{-1}\left\{\frac{1}{2}\left(\frac{I_m(t)}{I_o(t)} + \frac{Q_m(t)}{Q_o(t)}\right)\right\} \quad (15.57)$$

In practice, the process of digital compensation should be performed prior to establishing the communication link. It begins by feeding two known sets of I and Q waveforms to the I–Q modulator; the output waveforms from the uncompensated modulator are collected and the phase and gain imbalances are calculated from Equations (15.56) and (15.57). These two calibration parameters are then programmed in to the digital part. From the derived gain and phase imbalances, the parameters are time-varying parameters in practice. This is highlighted by denoting the phase and gain parameters as $\alpha(t)$ and $\varepsilon(t)$. In such a case, a robust yet simple statistical signal-processing technique such as least mean squares technique (LMS) should be implemented to obtain the parameters to be programmed. It should be noted that the block diagram presented in Figure 15.16 not only allows compensation for the imperfections from the I–Q modulator, but other imperfections, such as LO leakage and nonlinearity in the power amplifier, can also be calibrated out.

15.6 Digital Predistortion Techniques

With the scarcity of spectrum and the ever-increasing demand for wireless data, many wireless communication standards have been launched that support multiple users by using frequency-division multiplexing. In this case, the allocated spectrum is divided into a number of spectrum slots, depending on the number of concurrent users. Any wireless transceiver must adhere strictly to the spectrum mask of the relevant communication standard and frequency allocation. This is to ensure that there will be no interference appearing in other frequency bands, either being adjacent or nonadjacent channels. Any spurious output signals occurring at frequencies far from the occupied spectrum can be eliminated with an RF band-pass filter in the front end. However, interference appearing close to the occupied spectrum cannot be removed by filtering. The cause of this adjacent interference is primarily the nonlinearity of the power amplifier, and one engineering approach to obtaining a high-linearity power amplifier while maintaining high power efficiency is to apply a linearisation technique.

In a wireless transmitter, the nonlinearity of the power amplifier is more problematic than the I–Q modulator. Several linearisation techniques concentrate on modelling the nonlinearity in the power amplifier. The model for the power amplifier can be memory-less or include memory effects. In the memory-less model, the output amplitude of the power amplifier depends on the input amplitude in a nonlinear fashion; this is the so-called AM/AM conversion. The mathematical model for memory-less PA can be described with a power series:

$$y(t) = \sum_{k=1}^{K} a_k x^k(t) \tag{15.58}$$

The output and input of the power amplifier in time-domain PA are $y(t)$ and $x(t)$, respectively. The coefficient a_k which is real number, defines the nonlinearity characteristic of the power amplifier. The model defined at baseband is given by:

$$y(t) = \sum_{k=1,3,5\ldots}^{K} b_k x(t)\,|x(t)|^{k-1} \tag{15.59}$$

where,

$$b_k = 2^{1-k}\begin{pmatrix} k \\ \frac{k-1}{2} \end{pmatrix} a_k \tag{15.60}$$

On the other hand, the nonlinear amplifier model with memory effects leads to phase distortion in the amplifier output; this is called AM-PM conversion. The model can be described with a Volterra series as follows:

$$y(t) = \sum_{k} \int \cdots \int h_k(\boldsymbol{\tau}_k) \prod_{i=1}^{k} x(t-\tau_i)\, d\tau_1 d\tau_2 \cdots d\tau_k \tag{15.61}$$

where $\boldsymbol{\tau}_k = [\tau_1, \ldots, \tau_k]^{\mathrm{T}}$ and $h_k(\cdot)$ is the Volterra kernel of kth order.

The AM-PM effect can be better observed from the signal constellation diagram than in the frequency domain since it causes in-band distortion. Hence, the model that takes memory effects into account is more appropriate than the memory-less one because it is more generalised and governs all operating conditions of any power amplifier.

There are several proposed DSP-based linearisation techniques, among which the digital predistortion technique (DPD) is perhaps the most effective. The technique has been widely used in ICs and PA modules. The distinct advantage of the DPD technique over others is its simplicity, both in concept and implementation. This is because many of the complicated tasks are implemented in the software part of the system. The keeps the production cost very low compared to other methods, such as feedforward or Cartesian feedback techniques. A DPD technique can be classified as two types, the DPD at IF and at baseband. To apply DPD at IF requires a high-speed DAC, which is not suitable for low-cost solutions. The baseband DPD technique, applied to a direct-conversion transmitter, is shown in Figure 15.17. A baseband functional block called the digital predistorter (PD) is cascaded to apply a nonlinear function to the baseband signal. The predistorted baseband signal is converted to an analogue signal by the DAC, and the output of the DAC is then fed to an *I–Q* modulator and a power amplifier. The nonlinearities of the power amplifier, as well as the *I–Q* modulator, are compensated for

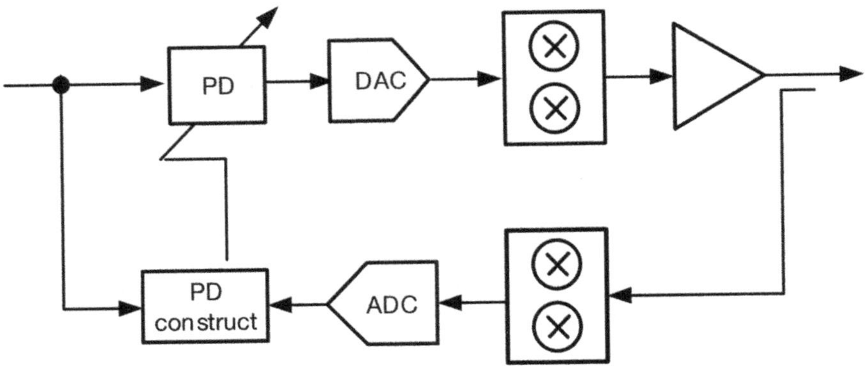

Figure 15.17 Baseband digital predistortion technique.

by the predistorted waveform. The PA can operate at high efficiency while its nonlinearity is precomputed and cancelled out with the PD, and hence the waveform output of the PA will be free from distortion. The feedback path is applied to provide an adaptive function to the DPD. This is necessary to characterise the nonlinearity characteristic and train the PD. With this feedback path, the PD function will operate automatically and any drift with time and temperature in the PA and *I–Q* modulator will be accommodated.

A straightforward way to implement the PD function is to determine the characteristic function of the amplifier and determine its inverse. This can be achieved with two methods: (i) a look-up table (LUT); or (ii) a parametric approach. The LUT approach [2] is implemented in such a way that the characteristic function of a PA is formed with a piecewise linear function. If the amplifier is, instead, directly modelled with nonlinear functions such as a spline or Volterra series, it is termed a parametric approach. The choice between the LUT or a parametric approach depends on the complexity. LUT is more appropriate for a memoryless PA model but it can be extended to apply to a PA with memory effects by trading off with complexity. An example of applying the LUT technique to implement the predistortion function is shown in Figure 15.18. The complex baseband signal $x_{in}(t)$ is applied to the functional block. The envelope of the complex baseband signal is computed to provide $|x_{in}(t)|$.

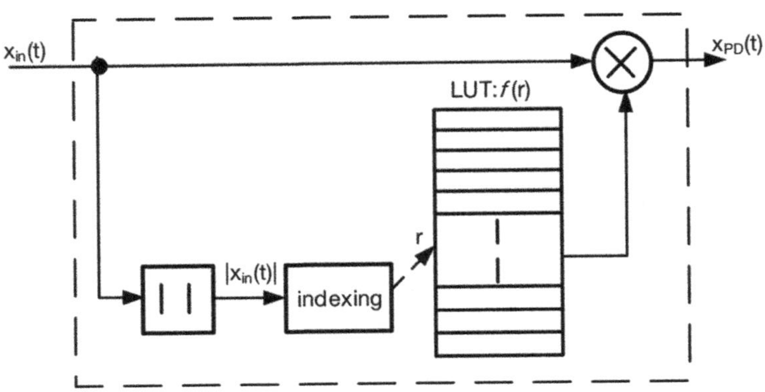

Figure 15.18 LUT implementation for PD function.

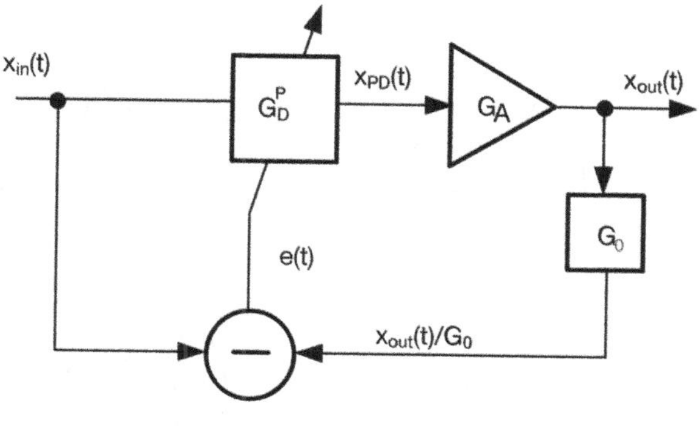

Figure 15.19 Direct adaptation topology.

This signal envelope will index to the LUT block, in which the values of the inverse function defined by $f(r) = \frac{G_0(r)}{G_A(r)}$ are programmed. The output of the LUT is complex-multiplied with $x_{in}(t)$ to produce the predistorted waveform $x_{PD}(t)$. The memory requirement of the LUT is dictated by the stability, accuracy and bandwidth requirements, with simple implementation in a Field-Programmable Gate Array (FPGA) or DSP chip.

The feedback path in the DPD technique is important for counteracting drift due to temperature, ageing and so forth in the PA. A simple technique called the direct adaptation topology is shown in Figure 15.19. The transfer function of the PA is denoted with a complex gain G_A whose linear gain is G_0. The transfer function of the predistorter is denoted as G_{pd}. The original complex input signal, x_{in}, is applied to the predistorter to produce the output, x_{pd}. The predistorted signal is applied to the nonlinear PA, giving the output signal x_{out}. The error signal, e, is the difference between the original signal and the output signal attenuated by G_0, that is, $\frac{x_{out}}{G_0}$. This error signal is used to adjust the predistorter function G_{PD} until there is no difference between the original signal and the attenuated output. With the LUT approach, this topology can be realised by reprogramming the data in the memory of the LUT after each training process of the adaptive algorithm is completed.

15.7 DSP Techniques for OFDM

A combination of MIMO and the OFDM technique has emerged as an important communication solution to address the bottleneck of SNR and bandwidth utilisation efficiency caused by the limited spectrum resource. However, this advanced solution entails significant hardware complexity in both the analogue and digital parts of a radio system. The origin of OFDM dates back to 1962, when it was first used practically in wire-based digital subscriber line technology (xDSL). Since then, its uses have been extended to several similar technologies such as HDSL, ADSL and V(H)DSL. OFDM was mentioned in the introduction as it is used in the terrestrial digital video broadcasting standard (DVB-T) and in the IEEE 802.11 wireless LAN standards.

The technique is based on the fundamental principle of orthogonality between the carrier signals. A message is separated into several small parts, individually modulated onto a number of orthogonal subcarriers and transmitted together as one RF signal. At the receiver, the original message can be recovered perfectly. OFDM is a form of Frequency-Division Multiplexing

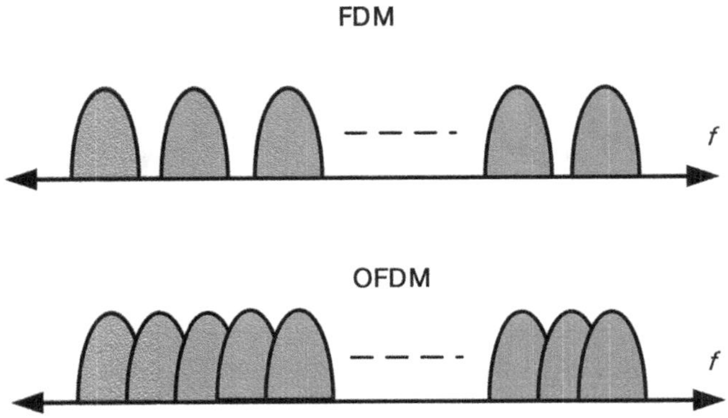

Figure 15.20 Comparison of the number of occupied channels for FDM and OFDM.

(FDM), where the familiar concept is to employ different frequency bands for different communication channels. However, as shown in Figure 15.20, regular FDM requires that there is a clear frequency spacing between each channel, while it is not necessary in OFDM by making use of orthogonal carriers. With the same bandwidth, OFDM therefore offers significantly more channels than FDM.

The key in OFDM to achieving orthogonality is to choose carriers with a *sinc*-shaped spectrum. The *sinc* subcarrier spectrum is zero at equally spaced frequencies, allowing other channels to overlap without interference. The frequency spacing between two channels (Δf_{sc}), which is equal to a half of two zero frequencies of a *sinc* subcarrier, is related to the symbol length (T_s) by $1/\Delta f_{sc}$. Figure 15.21 shows five subcarriers of an example OFDM spectrum. It is evident that the main lobe of each subcarrier has its peak at a zero amplitude point of all the others, so the spectrum of each subcarrier does not interfere with the others, allowing more densely packed carriers and a higher date rate.

OFDM is not practical without DSP techniques. If implemented with analogue techniques, the hardware would be far too complex. Synthesising the orthogonal channels with the *sinc* function requires a discrete Fourier and inverse Fourier transform (DFT, DIFFT). In this case, a FFT algorithm is used. As high-speed transmission is very sensitive to the multipath effects resulting from indoor propagation, a process called a 'cyclic prefix' was proposed to reduce inter-symbol interference (ISI). This prefix, which is inserted at the transmitter and removed at

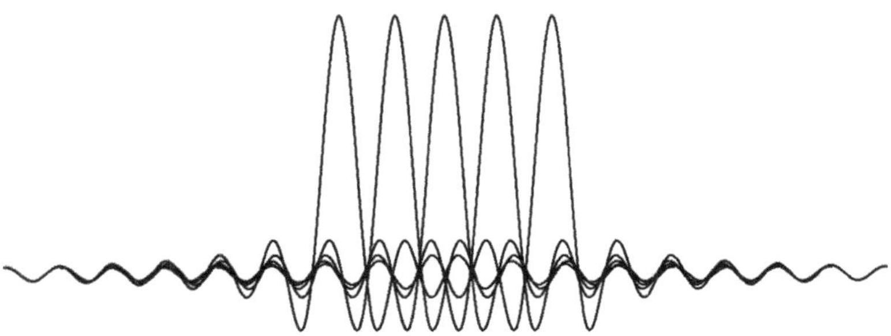

Figure 15.21 Multiple *sinc* spectrum components in the OFDM technique.

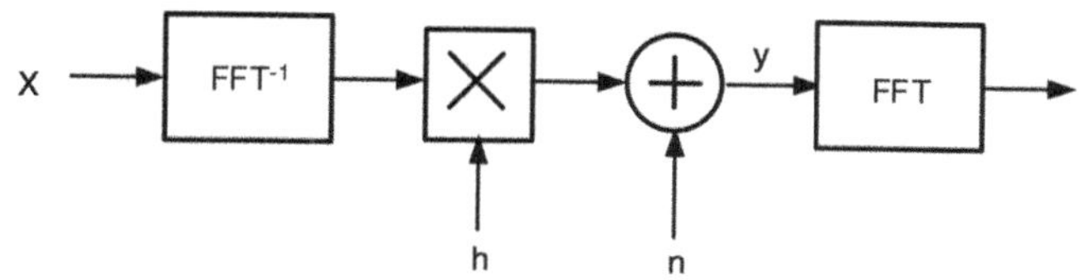

Figure 15.22 Mathematical model for OFDM.

the receiver, performs as a guard time to alleviate the ISI between two adjacent symbols. The term cyclic stems from a duplication of the prefix at the beginning and the end of each OFDM symbol. This will prevent interference between two adjacent symbols. The OFDM technique, combined with multicarrier modulation, has opened various possibilities to support very high bit rate communication in a hostile wireless channel. From a mathematical viewpoint, an OFDM system can be modelled as shown in Figure 15.22. In an OFDM transmitter, a vector signal $\boldsymbol{x}$ is applied to the inverse FFT (IFFT) block. The length of vector $\boldsymbol{x}$ is equal to the number of orthogonal subcarriers. To counteract the multipath propagation, the vector $\boldsymbol{x}$ will be cyclically convolved with the predetermined multipath channel impulse response vector ($\mathbf{h}$), producing a transmitted OFDM signal. The noise vector $\mathbf{n}$ models the noise in a hostile communication channel. Each row element in vector $\mathbf{n}$ represents noise in each subcarrier channel. Assuming that the noise vector is the additive type, then the received signal vector $\mathbf{y}$ is the addition between the output vector from the convolution block and the noise vector $\mathbf{n}$. To obtain the original message vector, the FFT is applied to $\mathbf{y}$ for demodulation.

15.8 MIMO

MIMO technology was introduced to the IEEE 802.11n wireless LAN standard to increase communication reliability and bandwidth utilisation efficiency (BUE). Previous studies [3] first presented the implementation in ASIC (Application Specific Integrated Circuits) technology for MIMO combined with OFDM. The technology could achieve very high BUE of up to 9.6 bps Hz^{-1}. Other studies on ASIC implementation have demonstrated very high throughput up to 170 Mbps [4]. The capability of the MIMO technique to boost the BUE can derived from the channel capacity of MIMO. It has been shown that the BUE at least linearly increases with the number of transmitting and receiving antennas. If N_t and N_r are the number of transmitting and receiving antennas, respectively, the number of communication channels from this configuration is N_tN_r. The channel matrix ($\mathbf{H}$) of dimension $N_t \times N_r$ is defined by [5]:

$$\mathbf{H} = \mathbf{R}_{\mathrm{Tx}}^{1/2} \cdot \mathbf{N} \cdot \left(\mathbf{R}_{\mathrm{Rx}}^{1/2}\right)^{\mathbf{T}} \tag{15.62}$$

The operator $(\cdot)^{1/2}$ means the matrix square root and $(\cdot)^{\mathrm{T}}$ means the transpose.

By assuming a white noise environment, $\mathbf{N}$ is the $N_t \times N_r$channel noise matrix and its elements are independent and Gaussian-distributed. $\mathbf{R}_{\mathrm{Tx}}$ and $\mathbf{R}_{\mathrm{Rx}}$ are the spatial correlation matrix at the transmitter and the receiver, respectively. These two matrixes are square with dimensions equalling the number of antennas. The elements of the matrix are the correlation coefficients between the antennas, depending on the antenna array structure. The correlation element, denoted by ρ_{ij} is a complex number where its magnitude and phase relate to the amount of coupling power and phase delay transferring from antenna i to j. Since an antenna is a passive device, we can conclude that $\rho_{ij} = (\rho_{ji})^*$. Consequently, an $N \times N$ correlation matrix is defined as:

$$\mathbf{R} = \begin{bmatrix} \rho_{11} & \rho_{12} & \cdots & \rho_{1N} \\ \rho_{12}^* & \rho_{22} & \cdots & \rho_{2N} \\ \vdots & \vdots & \ddots & \vdots \\ \rho_{1N}^* & \rho_{N2}^* & \cdots & \rho_{NN} \end{bmatrix} \tag{15.63}$$

For simplicity, we will assume that no information of channel state is given. At the MIMO transmitter with N_t antennas, the transmitter transmits a modulated signal of row vector $\mathbf{s}$, where $\mathbf{s} = [\, s_1 \; s_2 \; \cdots \; s_{N_t} \,]$ in the AWGN channel of vector $\boldsymbol{n}$. Therefore, the transmitted signal, $\boldsymbol{y}$, will be:

$$\mathbf{y} = \mathbf{H} \cdot \mathbf{s} + \mathbf{n} \tag{15.64}$$

The upper bound of the channel capacity (C) is given by:

$$C \leq \min\{N_t, N_r\} \log_2 \left(1 + \eta \frac{N_r}{\min\{N_t, N_r\}} \mathrm{E}\langle |h|^2 \rangle \right) \tag{15.65}$$

where $\min\{N_t, N_r\}$ gives the minimum value between N_t and N_r, η is the average SNR which is the ratio of total transmitted power and the average noise power, and $\mathrm{E}\langle |h|^2 \rangle$ is the expectation value over the channels.

Meanwhile, the channel capacity of a conventional Single-In-Single-Out (SISO) architecture has a channel capacity given by:

$$C_{\mathrm{SISO}} = \log_2(1 + \eta \mathrm{E}\langle |h|^2 \rangle) \tag{15.66}$$

Comparing these two capacities, MIMO provides a capacity gain over the SISO with the minimum between N_r and N_t. If a 2×2 MIMO transceiver is used, the upper bound will be:

$$C \leq 2 \log_2(1 + \eta \mathrm{E}\langle |h|^2 \rangle) \equiv 2C_{\mathrm{SISO}} \tag{15.67}$$

which is double the SISO capacity. This is a prominent advantage over the SISO system. An enhancement of channel capacity, or BUE, will be obtained if the channel state information (CSI) is known. This requires DSP algorithms for channel estimation and signal detection.

The implementation of MIMO is complicated in the receiver process. Considering the MIMO signal model shown in Equation (15.64), a MIMO receiver should be able to detect the signal $\mathbf{s}$ from $\boldsymbol{y}$. Figure 15.23 shows a diagram of a MIMO receiver. All the components are implemented in the digital domain. The MIMO detector unit is to detect the signal $\mathbf{s}$ from $\boldsymbol{y}$ buried in noise. There are many techniques proposed to recover the signal $\mathbf{s}'$. One of the

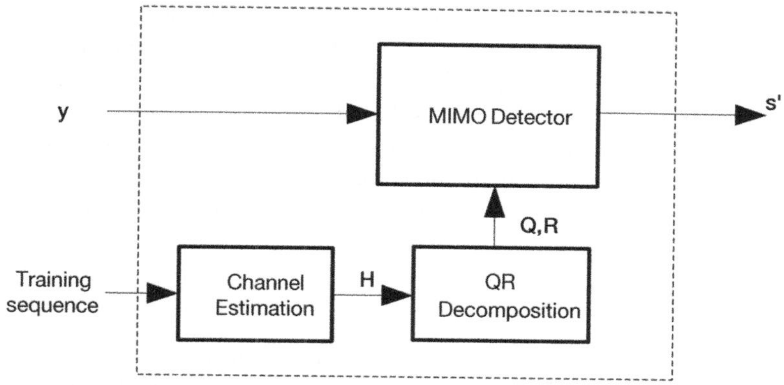

Figure 15.23 Simplified diagram of a MIMO receiver.

techniques to apply is the Maximum-Likelihood (ML) technique which provides an estimation result with minimum variance. It is estimated by:

$$\mathbf{s}' = \arg\min \|y - \mathbf{H} \cdot \mathbf{s}\|^2 \tag{15.68}$$

Solving the equation of the ML estimator is rather complicated for a high-level signal constellation and large number of antennas. Detecting the signal, **s**, requires a knowledge of the channel matrix **H**. The computation can be simplified if the QR composition technique is applied to **H**. Hence, if it is, we can write:

$$\mathbf{H} = \mathbf{Q} \cdot \mathbf{R} \tag{15.69}$$

The ML estimator turns out to be the following equation:

$$\mathbf{s}' = \arg\min \|y' - \mathbf{R} \cdot s\|^2 \tag{15.70}$$

where

$$y' = \mathbf{Q}^{\mathrm{T}} \cdot y \tag{15.71}$$

In Figure 15.23, the known training sequences are fed in to the 'Channel Estimation' block to estimate the channel information **H**. This channel information matrix is applied to perform QR decomposition to provide matrix **Q** and **R**. The complex algorithm for QR decomposition is rather sophisticated and can cause communication latency. The Modified QR decomposition algorithm is suggested due to its better numerical stability. In addition, the algorithm complexity can be reduced, for example avoiding a large number of multipliers by implementing the CORDIC algorithm.

15.9 Conclusion

In this chapter, the importance of DSP techniques in modern wireless communication has been described. The topics that have been covered are only a tiny part of the DSP field, but a brief introduction has been given to analogue-to-digital and digital-to-analogue conversions, including some of the theory behind sampling and quantisation. The importance of the FFT in modern signal processing has been introduced and some applications of DSP to combat nonlinearity and *I–Q* imbalance in the transceiver. DSP is no longer confined to the modulation, demodulation and baseband operations in a transceiver. Rather, DSP and RF design have become ever more closely integrated, as digital techniques are used to compensate for the imperfections in the high-frequency analogue circuitry. With the use of massive MIMO and electronically steered antennas as solutions for Gb s^{-1} mobile wireless communications systems, the trend for RF and DSP integration will continue for many years to come.

References

1. Fettweis, G., Lohning, M., Petrovic, D., Windisch, M., Zillmann, P., and Rave, W. (2005) Dirty RF: a new paradigm. Personal, Indoor and Mobile Radio Communications (PIMRC). IEEE 16th International Symposium, 11–14 September, vol. **4**, pp. 2347–2355. doi: 10.1109/PIMRC.2005.1651863.
2. Cavers, J.K. (1999) Optimum table spacing in predistorting amplifier linearizers. *IEEE Transactions on Vehicular Technology*, **48** (5), 1699–1705.
3. Perels, D., *et al.* (2005) ASIC implementation of a MIMO-OFDM transceiver for 192 Mbps WLANs. Proceedings, 31st European Solid-State Circuits Conference (ESSCIRC), France, pp. 215–218.
4. Burg, A., Borgmann, M., Wenk, M., Zellweger, M., Fichtner, W., and Bolcskei, H. (2005) VLSI implementation of MIMO detection using the sphere decoding algorithm. IEEE Journal of Solid-state Circuits, **4** (7), 1566–1577.
5. Hottinen, A., Tirkkonen, O., and Wichman, R. (2003) *Multi-antenna transceiver techniques for 3G and beyond.* John Wiley & Sons.

16

Packaging and Assembly

16.1 Introduction

This chapter describes the main techniques by which integrated circuits (ICs), discrete semiconductor devices and passive components are integrated to make microwave and millimetre-wave modules. In many markets, such as aerospace and defence, modules often still take the form of machined metal housings with coaxial connectors or waveguide flanges, with components of various types assembled onto a microwave substrate inside. Often, there are significant advantages in terms of thermal performance and EMC shielding that make this approach necessary. In consumer electronic products, there is an overriding need for low cost, high-volume manufacture and small size, so there has been a significant shift towards sophisticated multichip module or system-in-package approaches making use of 3-D chip- and wafer-scale packaging techniques. The microwave and millimetre-wave design engineer is required, therefore, to be familiar with an exceptionally wide range of packaging and assembly techniques, from surface mount assembly on regular PCBs through to esoteric techniques integrating thin-film circuits into rectangular waveguides.

The chapter starts with a discussion of how an optimum transmitter-receiver requires careful integration of passive components with active devices and ICs that may be drawn from a wide variety of semiconductor technologies. This is followed by a summary of the main substrate types, including PCB and laminate boards, thin-film, thick-film, and low-temperature co-fired ceramic (LTCC) technologies. Laminate technology is increasingly popular for its ability to achieve high performance and low cost, but ceramic technologies are useful for operation at high temperatures and in harsh environments. A summary of the key device packaging techniques is then given, followed by an introduction to mainstream electronics manufacture using surface mount components on PCBs. As wireless products have moved into the consumer electronics domain, it is essential that RF engineers understand these principles of mass production. Finally, the concept of a system-in-package or system-on-substrate is introduced. This approach is invaluable for millimetre-wave design since it means that engineers can take advantage of high-technology multilayer substrate technologies to create compact designs with all the required active and passive components in a single module, often with integrated

Microwave and Millimetre-Wave Design for Wireless Communications, First Edition.
Ian Robertson, Nutapong Somjit and Mitchai Chongcheawchamnan.
© 2016 John Wiley & Sons, Ltd. Published 2016 by John Wiley & Sons, Ltd.

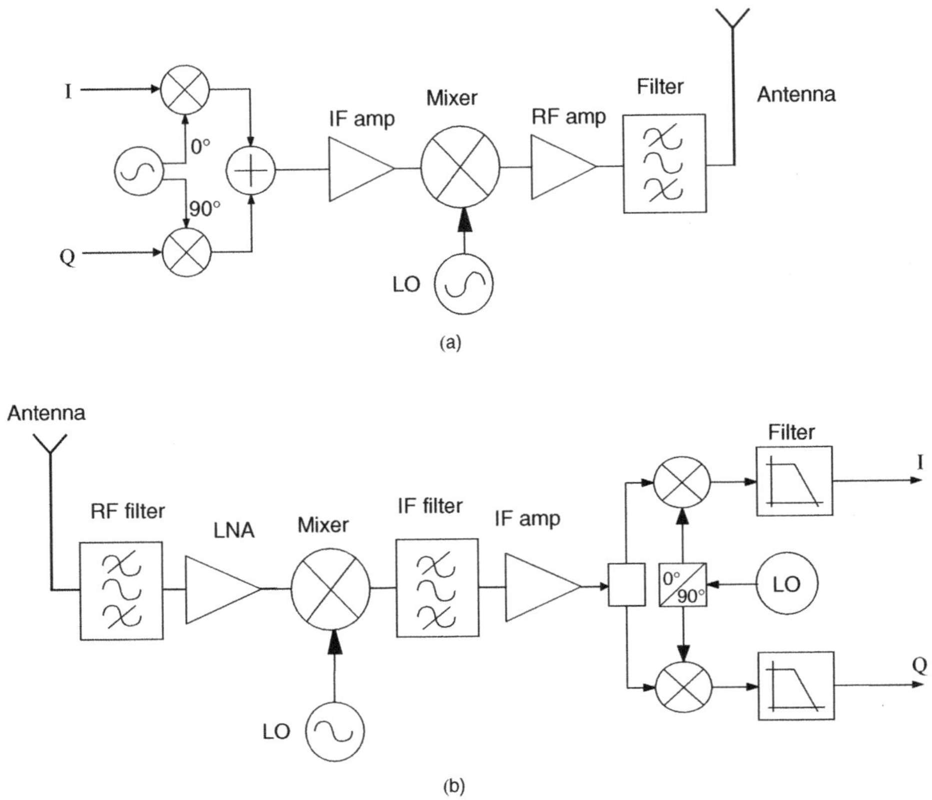

Figure 16.1 Simplified (a) transmitter and (b) receiver block diagrams.

antennas. This, in turn, makes it easy for customers to integrate the module into a regular PCB production line and get price-competitive products to market quickly.

16.2 Technology Options and System Partitioning

Figure 16.1 shows a generic simplified block diagram of a transmitter-receiver. At the core of almost all products is a sophisticated silicon system-on-chip that integrates, at the very least, an embedded processor and analogue baseband circuits. This immediately means that the 'stratification' of the industry, as introduced in Chapter 1, starts to become very important. In standard applications, where volume manufacture and cost are not at the 'smartphone' level of intensity, the system architect and RF designers can survey available 'chipsets' from fabless design houses, IC manufacturers and distributors in order to choose the IC(s) forming the core of the product, and then design the system around that chipset. However, with this approach, there is some loss of competitive advantage since other companies can use the same chipset.

In contrast, at the 'smartphone' level of market volume and competitiveness, an application-specific IC (ASIC) can be designed, so that as much of the RF circuitry as possible is integrated onto the same chip as the digital and baseband electronics, minimising the overall parts count and reducing the cost. However, the 'cost of entry' to the world of silicon RFIC design is very high due the cost of the EDA tools, the cost of fabrication, mask costs and the number of experienced engineers that are required, covering a range of areas from embedded systems and DSP to advanced millimetre-wave design. This means that industry partnerships start to become important for mutual benefit, since the time-to-market is extremely short and products have a brief life-span in the market. For a system integrator, there is a significant advantage in working closely with chipset design and manufacture companies in order that the chipset, package and allied circuitry are all developed in parallel at an accelerated rate. The IC manufacturer may still be able to sell the chipset to other customers, but the partner that has worked hand-in-hand with them to develop it will have a significant advantage in getting a product with the state-of-the-art chipset into the marketplace.

Nevertheless, it is very rare that an integrated RFIC transceiver can work without any off-chip components. Some of the cases where it is possible would include RFID and NFC applications and some short-range wireless standards. However, in the majority of applications, even if an RFIC forms the core of the transceiver system, it will need to be supplemented with a range of off-chip components in order that high-Q resonators and filters, efficient high-power amplifiers (e.g., GaAs and GaN), low-noise amplifiers and other such high-performance components can be used. The challenge, then, is to find the correct level of integration for the particular application. For moderate volumes, where low NRE (development cost) is a primary requirement, then the off-chip components can be comfortably integrated onto a microstrip substrate with, where necessary, transitions to coax or waveguide when the very highest performance is needed for filters and so forth. If volume manufacture and production costs are the primary drivers, then such an approach will not be competitive and it is necessary to develop a sophisticated system-in-package module where the external active and passive components are integrated into a multilayer 3-D substrate. In this case, partitioning of the system must be examined at the very start to identify which parts are realised by a silicon RFIC, which parts require GaAs/GaN/InP and so forth components, which passive components can be embedded into the substrate, and which must be sourced as separate surface-mounted device (SMD) components. This partitioning of the module is illustrated conceptually in Figure 16.2. Only once this partitioning and selection of components has been studied can an optimum choice of substrate technology be made.

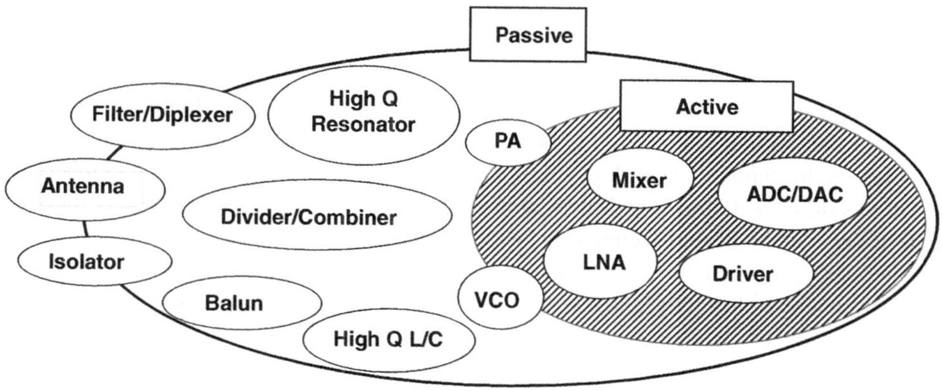

Figure 16.2 Conceptual partitioning of active and passive components for a transceiver module.

16.3 PCB/Laminate Technology

FR-4, which is the industry-standard composite material for PCBs, consists of epoxy-resin strengthened with an embedded woven fabric of glass fibres. FR-4 has a number of limitations for high-frequency applications, including a high loss tangent, inhomogeneity, a high coefficient of thermal expansion, significant variation of dielectric constant with temperature, and susceptibility to moisture absorption. In many microwave and millimetre-wave products, it is common to use other substrate materials such as Duroid^{TM}, liquid crystal polymer (LCP) or flexible polyimide sheets, but in most respects the basic fabrication process is the same. Copper is deposited on large fibre-glass reinforced, or other woven or PTFE-based boards, using electrodeposition or rolling. Due to the large area, the photoresist is usually applied by laminating a preprepared dry film onto the substrate. For small circuits, a liquid form of photoresist might be applied by dipping in a tank, or by spinning, and then drying in an oven. The photoresist is then exposed to ultraviolet (UV) light via a photomask, or with a digital imaging tool, and then developed. In an 'etch back' process the copper is then etched away where it is not protected by the patterned photoresist. The etch-back process is relatively coarse, and fine lines cannot easily be fabricated. In mass production, a more sophisticated process is employed for better yield and resolution. The steps to produce high-quality double-sided boards with plated through-holes on large panels (e.g., 18 × 24″) for consumer product applications are typically as follows:

1. Drill the board with a computer numerically controlled (CNC) drill.
2. Apply electroless copper to the holes.
3. Apply a layer of negative photosensitive dry film.
4. Expose the photosensitive material and develop it to form the trace pattern.
5. Electrochemically plate up the exposed copper areas and holes.
6. Finish the surface – typically by plating with tin.
7. Remove the dry film and strip off (etch) the unwanted copper (traces and holes are protected by the tin plating).
8. Apply the soldermask layer (also known as the solder resist); this is a special protective coating and only areas that require soldering are left exposed.
9. Solder coat the pads; exposed component pad areas are plated with solder, typically by dipping into a solder tank and then levelling the surface using a hot-air knife.
10. Print the text (legends or ident) by silk-screening, liquid photoimaging (LPI) or direct legend printing (DLP).

Modern circuits are usually far too complicated to use just single- or double-sided PCBs, and it is not uncommon to employ PCBs with eight layers; indeed, some PCB technologies offer 24 layers. In standard technology, these layers are all fabricated one-by-one, with copper patterning as described above, and then laminated together to form a single board. This has become a very sophisticated technology, with very large board sizes and yet with 50 μm tracks and 100 μm-diameter laser-drilled via holes. In the lamination process, a core of FR-4 is laminated with multiple layers of conductor foil and 'pre-preg' (which means 'pre-impregnated'), an epoxy resin that is impregnated with glass fibres. The layers are laminated together using heat and high pressure to form the multilayer board, as illustrated in Figure 16.3. This advanced technology is referred to as MCM-L (laminate) and Figure 16.4 shows an example of a set of circuits sharing a single panel.

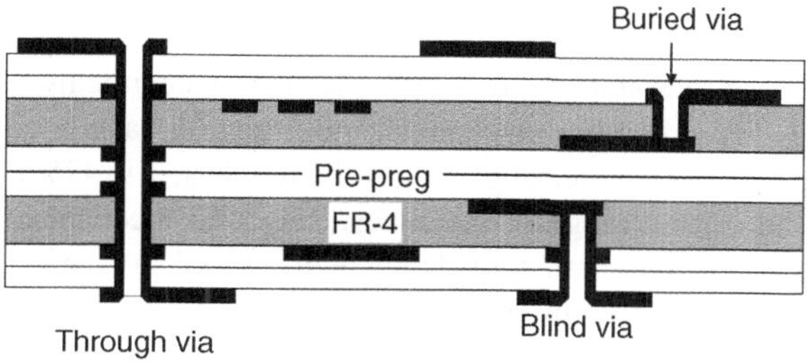

Figure 16.3 Cross-section of a typical multilayer PCB.

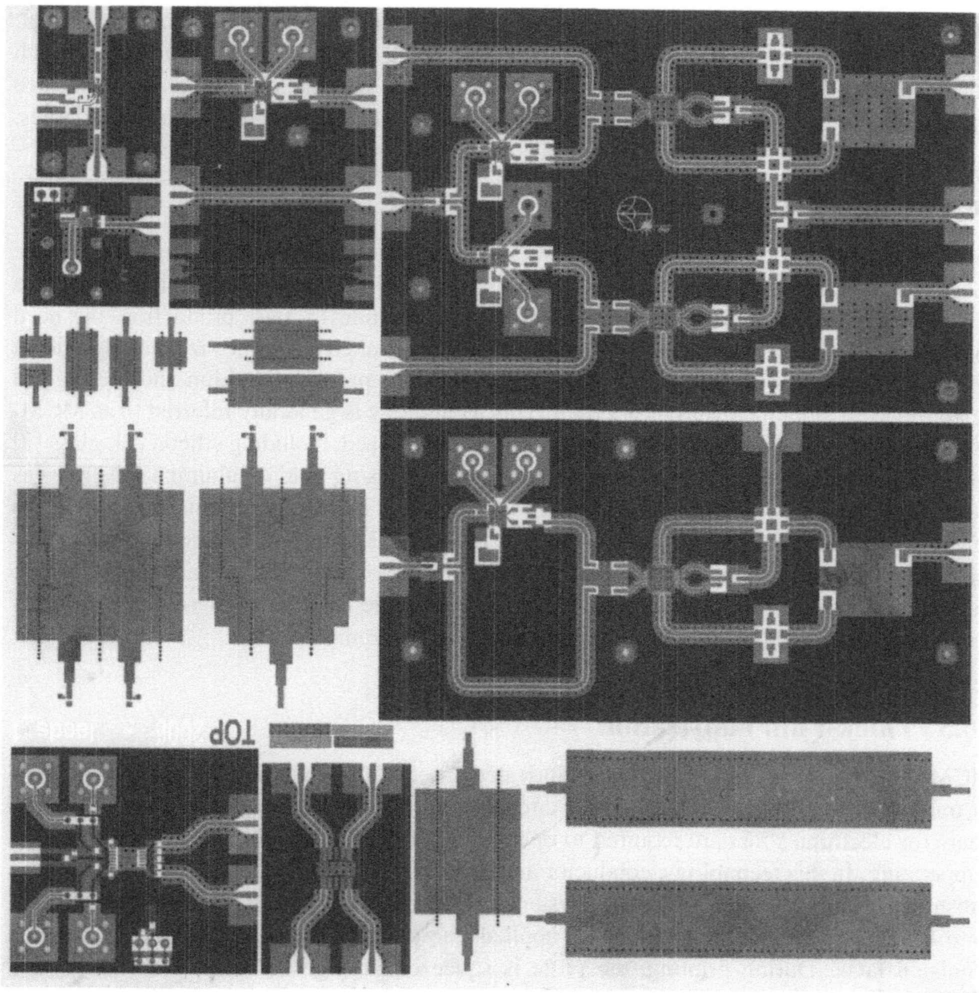

Figure 16.4 Multilayer laminate PCB (8 × 8″) containing a number of microwave circuits, prior to mounting of components.

16.3.1 Rapid Prototyping of PCBs

For the fastest possible turnaround, or to avoid any use of chemicals, a PCB prototyping machine may be used. These can use mechanical milling, with basic models costing around £5000, or a sophisticated laser-patterning system. LPKF, based in Germany, are the world leaders in this field and offer a range of machines. The S-63 mechanical machining model features a 60,000 rpm motor, automatic tool change with 15 tool slots, precision Z-axis control and 0.25 μm resolution. It is well-suited for precision circuits, including RF and microwave applications. Being a mechanical milling machine, it can also be programmed to machine circuit housings and engrave front panels.

The Protolaser U3 uses a UV laser and patented copper removal process for even faster prototyping of PCBs and a smoother board surface: for example, a typical 10 × 10 cm board can be processed in less than 5 minutes. An additional advantage of the laser machine is that, by careful selection of laser pulse parameters, a wide range of materials can be processed, such as ceramics. LPKF also provide a range of chemicals and equipment that enables the user to take the machined boards and combine them into multilayer boards with plated through-holes.

16.4 Thin-Film Fabrication

In thin-film technology, metal deposition techniques such as sputtering and evaporation are used, possibly also with electroplating for increased metal thickness. The equipment used is relatively expensive, and the substrate must be in a vacuum. In mass production, the need to wait for the chamber pressure to drop down, and the limited substrate size, are significant drawbacks. However, thin-film technology gives the best pattern definition and the highest performance if suitable materials are used. This technology is generally referred to as MCM-D (deposited) and a number of substrates have been employed, including silicon [1], glass [2] and polished alumina. For many years, thin-film circuits using gold on alumina substrates as shown in Figure 16.5 were the mainstay of the microwave industry since this approach gives low microstrip line losses and permits the chip-and-wire assembly of semiconductor devices and specialist high-frequency components. Thin-film fabrication still has an important role to play in high-performance applications but is not suitable for most high-volume consumer applications, where laminate (or softboard) technologies dominate.

16.5 Thick-Film Fabrication

MCM-C refers to multichip modules fabricated in ceramic technology. This has its roots in traditional thick-film ceramic hybrid circuit technology, which has been used for many years for electronics that are required to operate in harsh environments, for example at high temperature. In this technology, conductor and dielectric pastes are applied to a ceramic base substrate (usually alumina) using screen/stencil printing. A screen is a fine metal wire mesh, onto which a photographic emulsion is applied; the circuit pattern is reproduced onto this emulsion layer. During printing, the paste is squeezed through the mesh where there are emulsion openings onto the substrate, as shown in Figure 16.6. The paste is then dried and fired at around 850 °C. Successive layers can be printed to form multilayer circuits, although the layers often need to be fired individually to permit the escape of organic binders and this means that the processing time will, in practice, limit the number of layers than be employed.

Figure 16.5 A thin-film circuit using a gold conductor on an alumina substrate, mounted in a test fixture with SMA launchers.

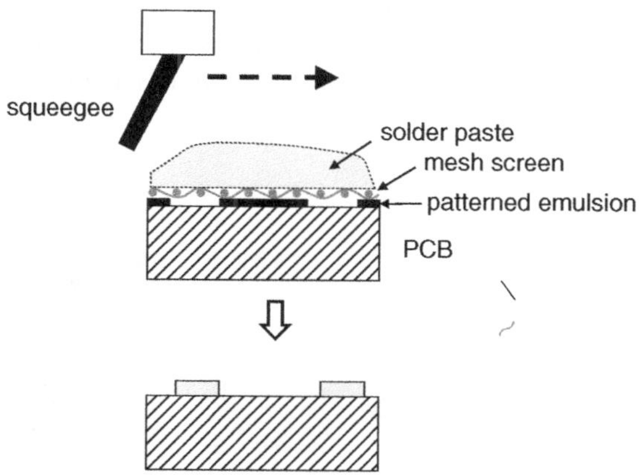

Figure 16.6 Screen printing of solder paste using a mesh (or screen).

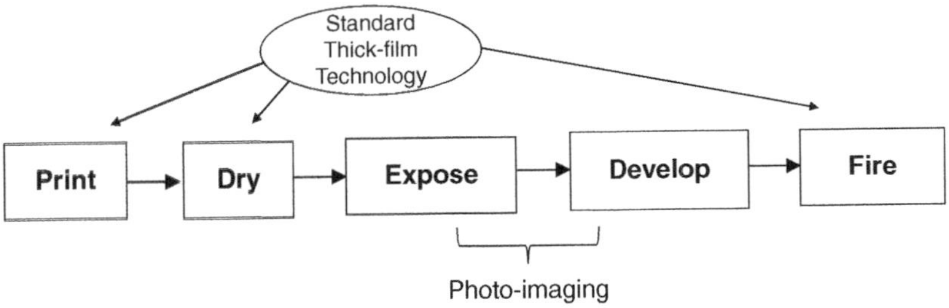

Figure 16.7 The photoimageable thick-film process.

16.5.1 Photoimageable Thick-Film Process

One drawback of conventional thick-film technology is that the minimum conductor dimensions are limited to about 100 μm because of the screen's jagged edges. Furthermore, the rheology of the paste (or ink) has to be carefully optimised to avoid 'slump' of the printed patterned structure, and this makes it challenging to produce high-conductivity inks that give the low surface roughness needed for high-frequency circuits. The photoimageable process is an extension of conventional thick-film technology, as shown in Figure 16.7, in which standard thick-film paste is replaced by a photosensitive material. Photosensitive thick-film pastes have been developed by combining a photosensitive vehicle and metal-glass powders, both of which affect the electrical properties and resolution characteristics. The thick-film paste is first printed everywhere over the total area of the substrate. Since blanket coverage is required, and not a patterned layer, the levelling properties of the photosensitive pastes can be optimised to provide a uniform thick film with a very smooth and dense surface, free of pinholes and other printing defects. After drying, the paste is exposed to UV light through a mask and then developed, leaving the conductor or dielectric pattern. Figure 16.8 shows examples of microwave structures that were fabricated using the Hibridas^{TM} conductor and dielectric pastes [3, 4]. The ability to fabricate arbitrarily shaped chip-cavities using photoimageable dielectric is of considerable interest for applications that require bare-chip ICs and single-layer capacitors to be wire-bonded into a circuit.

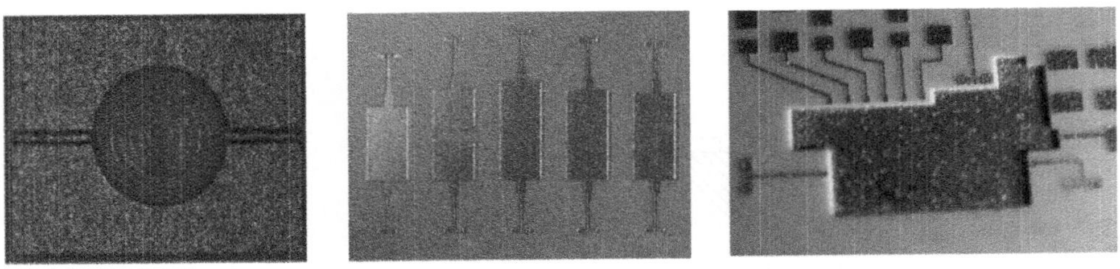

Figure 16.8 Examples of microwave components fabricated with photoimageable thick-film technology. From left to right: spiral inductor; SIW; chip cavity using photoimageable dielectric.

16.6 LTCC Technology

Co-fired multilayer ceramic technology is of immense importance to the electronics industry with most SMD passive components and ceramic IC and MEMS packages being manufactured in this way. Currently, more than 1000 billion SMD passive components, including multilayer ceramic capacitors (MLCC), are manufactured every year. Ceramic packaging provides hermeticity and a mechanically stable base for ICs, with a low-temperature coefficient of expansion, making it suitable for operating at high temperatures and in harsh environments. By using a firing temperature below 1000 °C, low-temperature co-fired ceramic (LTCC) technology permits the use of high-conductivity metals such as gold and silver that are needed for microwave circuits, whereas high-temperature co-fired ceramic (HTCC) uses firing temperatures as high as 1600 °C and is generally restricted to using metals such as tungsten that have high melting points but low conductivity. Copper is not widely used in LTCC, however, due to its oxidation at high temperature.

In LTCC technology, the ceramic is processed in its unfired 'green' state, and is a flexible sheet. Figure 16.9 shows a simplified summary of the LTCC process. First, the green tape is cut into the correct size and then preconditioned, which means it is dried by baking or is left in a nitrogen atmosphere for approximately 24 h. Via-holes are then made by mechanical punching, or laser machining, and filled with a special conductor paste which is then dried. Subsequently, the conductor patterns are screen-printed onto each sheet individually. The via-fill and conductor pastes are typically different formulations because of the different requirements for conductivity and shrinkage behaviour. Once a set of metallised layers has been produced, they are stacked together using a dedicated stacking machine or alignment jig. The multilayer substrate is then laminated under high pressure using a specialist isostatic laminator or basic hot uniaxial press. Isostatic lamination typically entails placing the substrate in a sealed bag and placing it in hot, high-pressure water so that an improved uniformity and density of the layers is achieved that reduces delamination, for example. The final 'co-firing' step is used to complete the process, most often with a belt furnace that has 'zones' to provide a carefully controlled temperature profile as the substrate moves along. LTCC technology has the advantage over MCM-L that SMD resistors and capacitors are already manufactured using

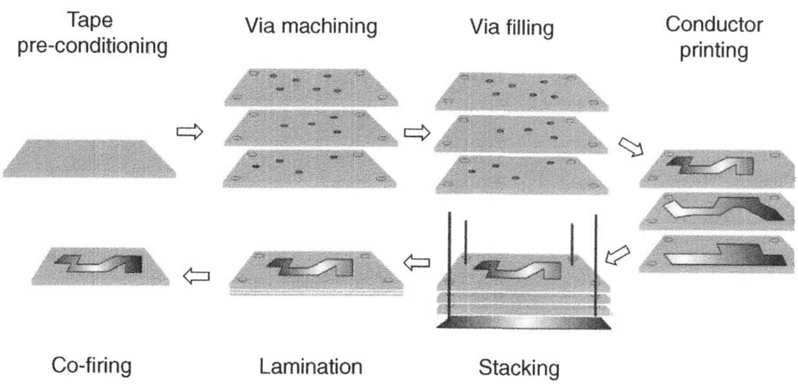

Figure 16.9 Diagrammatic representation of the main LTCC processing steps.

the same sort of process. Therefore, embedded passive components can be fabricated in the module, which reduces component count and costs. With this capability of producing compact multilayer boards with embedded passive components, the cost of an overall product might be lower in some cases due to the reduced component count, but LTCC is much more expensive per mm^2 than FR-4 and similar laminates. One of the greatest challenges in LTCC technology is controlling shrinkage during the sintering process, or at least accounting for it at the design stage. This has led to a number of techniques being developed to avoid shrinkage, including pressure-assisted sintering (PAS) and self-constraining tape [5,6].

16.7 Chip Packaging

For global compatibility, packaging standards are laid down by JEDEC (Joint Electron Device Engineering Council) – now known as the JEDEC Solid State Technology Association. They work alongside other industry and professional associations, including the Electronic Industries Alliance (EIA), Electronic Industries Association of Japan (EIAJ), International Electrotechnical Commission (IEC), IPC (Association Connecting Electronics Industries, previously known as the Institute of Printed Circuits), IEEE and the International Microelectronics and Packaging Society (IMAPS) on industry-wide issues such as standards in assembly, components, connectors, environmental initiatives such as lead-free, Restriction of Hazardous Substances (RoHS), Waste Electrical and Electronic Equipment (WEEE), as well as research, education and training. Today, many varieties of package styles and technologies are available, and those which are relevant to RF engineers are described briefly in the following sections.

16.7.1 Chip-on-Board

In chip-on-board or chip-and-wire assembly, the chip is epoxied or soldered down onto the substrate and the bondpads on the chip are wire-bonded onto the printed tracks. Wire-bonding involves attaching tiny wires (thinner than a human hair) with a machine that uses some combination of heat, pressure and ultrasonic energy to fuse the metal wire onto the bondpads and tracks. A drawing of these tiny wires connected to a chip is shown in Figure 16.10. For production assembly, automated wire bonders are used which can attach 12 or more wires every second. Once the wires are attached, the chip is buried with a blob of epoxy; this is often referred to as 'glob top packaging'. In many microwave applications the chip will be left bare so that there are no unintended device parasitics from the epoxy, and then the substrate must itself be housed appropriately.

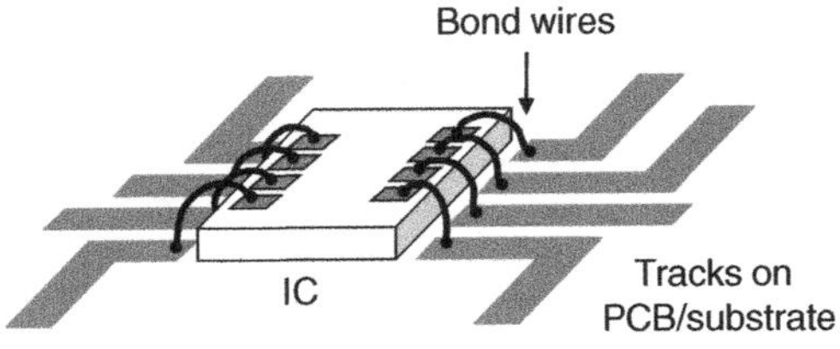

Figure 16.10 Chip-on-board assembly.

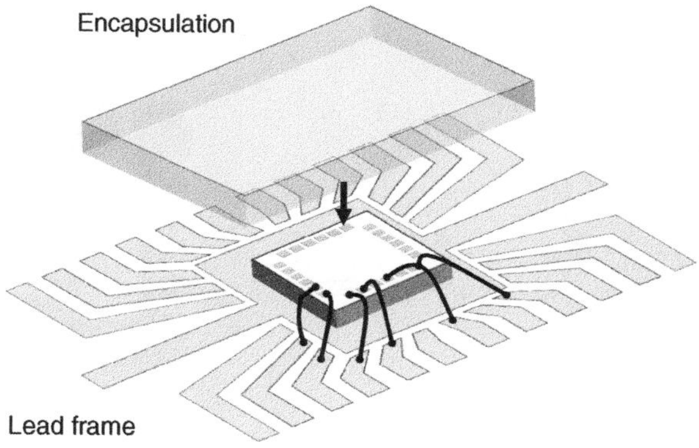

Figure 16.11 Lead frame package.

16.7.2 Lead Frame

In this package style, the chip is attached to a metal part that incorporates the legs or pins of the package. The chip is wire-bonded to the inside part of the metal lead frame, and the whole assembly is then encapsulated in plastic or ceramic, as illustrated in Figure 16.11. Figure 16.12 shows a microwave MMIC in a plastic over-moulded QFN (Quad Flat No-leads) package, soldered onto a laminate PCB. The leads of the QFN package do not extend outside, giving a small footprint.

16.7.3 Pin-Grid Array (PGA) and Ball-Grid Array (BGA)

This type of package is used for complex chips such as microprocessors, FPGAs and RFICs that require many input and output connections. The package essentially contains a miniature PCB, allowing key passive components to be placed very close to the chip. As illustrated in Figure 16.13, the chip is mounted on a multilayer substrate that routes the chip bondpads to an array of package inputs and outputs. The bottom of the package has a square, or rectangular, grid of pins or solder balls on the underside of the substrate. For the BGA, during reflow the surface tension of the molten solder helps keep the package aligned to the PCB solder pads, but BGA packages require specialist expertise in manufacture to ensure long-term reliability. For example, the solder joints are not visible, requiring X-ray inspection for quality assurance purposes.

16.7.4 Flip-Chip

In this method, which includes Chip-on-Film or flex (COF) and Chip-on-Glass (COG) technologies, the chip's bondpads have solder bumps so that they can be directly soldered onto an underlying printed circuit, as shown in Figure 16.14. Using a special tool, the chip is held upside-down and placed down onto the PCB's pads and heat applied to melt the solder. In some variants, a special conducting adhesive is used – this is deliberately anisotropic so that it only makes electrical connections in a certain direction. To create a stronger mechanical joint and improve thermal dissipation, a nonconducting underfill material may be applied between the chip and the board.

(a)

(b)

Figure 16.12 (a) Microwave MMIC in a plastic over-moulded QFN package. (b) Microwave MMIC mounted on a laminate PCB. Illustration courtesy of Plextek RFI, used with permission. All rights reserved.

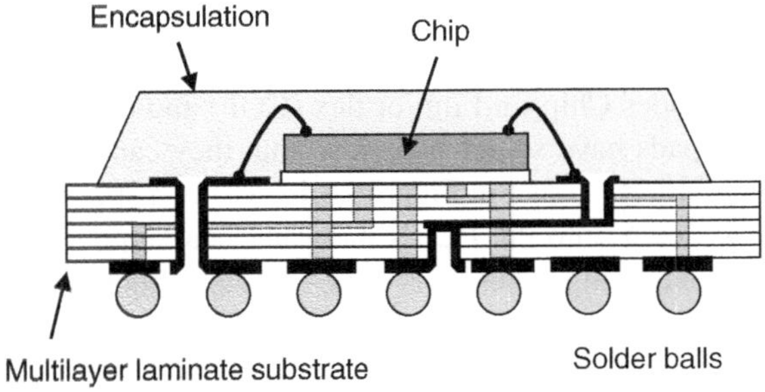

Figure 16.13 Cross-section through a typical ball-grid array (BGA) package.

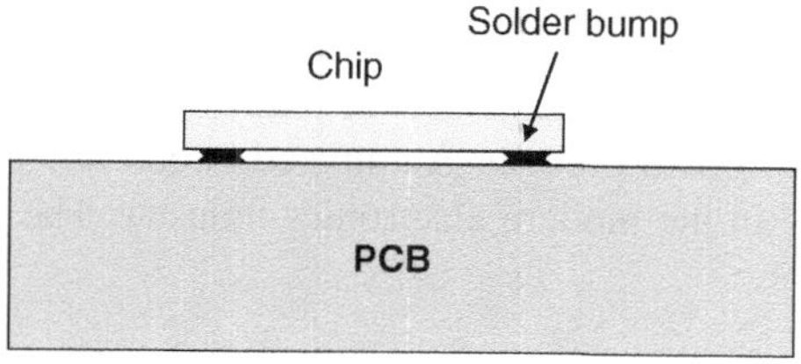

Figure 16.14 Flip-chip assembly.

16.7.5 3-D Packaging

Chip packaging can be applied at several levels. For example, a MEMS device may be encapsulated at wafer level, for hermeticity, before singulation (dicing of the wafer). This is referred to as zero-level packaging or wafer-level packaging. The previous examples, where a chip is mounted into its own ceramic or plastic package, are referred to as first-level packaging. However, a packaged device could, in turn, be integrated into a MCM-L multichip module, which may itself be assembled onto a regular PCB. A more efficient and space-saving approach is to take a holistic view of the 3-D packaging opportunities. A stack of integrated circuits can be interconnected by adding a redistribution layer (RDL) to the top of each chip and using through-silicon vias (TSVs) to make connections between chips. If necessary, an interposer layer can be inserted between chips to provide further flexibility in the routing of signals. A conceptual 3-D millimetre-wave active antenna array realised using 3-D stacking of ICs with TSVs is illustrated in Figure 16.15. The International Technology Roadmap for Silicon (ITRS) includes an extensive report of the opportunities for 3-D integration of ICs, package-on-package (PoP) and system-in-package techniques [7].

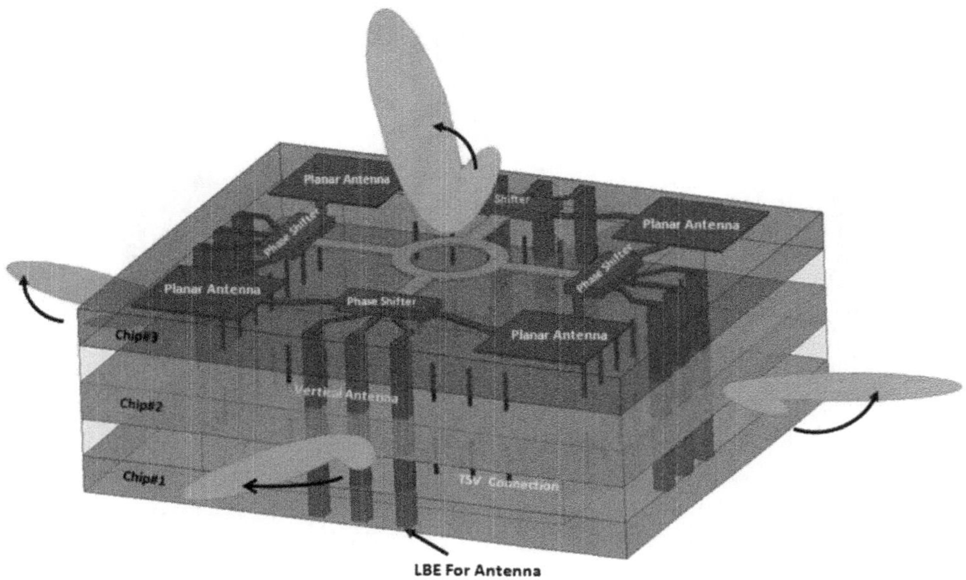

Figure 16.15 3-D integration of a millimetre-wave module using TSVs and stacked RFICs.

16.8 Manufacturing using Surface Mount Technology

Once the PCB, ICs and other components are all ready, the next step is to integrate them together and assemble the finished product. Automated circuit assembly is essential to meet the cost pressures involved in the modern electronics industry. The main steps with surface mount technology (SMT) are:

1. Stencil print solder paste onto the PCB pads.
2. Populate the PCB with ICs, passive components, and other SMDs.
3. Heat the solder paste in a reflow oven.
4. Assembly the printed circuit assembly (PCA) with other electronic and mechanical components into the product's metal or plastic case or body.

Populating the board is performed by pick-and-place machines. To facilitate this, components are supplied by the manufacturers in tape-and-reel form, as illustrated in Figure 16.16. The pick-and-place machine can then use vacuum tools to pick components off the reels and place them down onto the correct pads of the PCB. This is all achieved under computer control, with vision systems used to ensure precise alignment. A pick-and-place machine is capable of placing thousands of components on boards every hour. In a factory the boards will be fed down the production line on a conveyor belt, from stencil printing, to population, to soldering.

16.8.1 Solder Deposition

Screen or stencil printing is an important technique for PCBs. In circuit fabrication, it is used for printing the soldermask layer, the component labels (legends) and also for depositing solder

Figure 16.16 Components supplied in tape-and-reel form for automated assembly.

paste in the desired areas. A screen comprises a wire or fabric mesh stretched over a rigid metal frame, with a pattern of open areas that can allow the ink or paste to be squeezed through the fine mesh onto the PCB. The pattern is normally created on a photographic emulsion for mesh screens. The technique is widely used for printing coloured inks onto T-shirts, baseball caps, mugs, and so forth. A stencil is a thin stainless steel sheet with a pattern of holes made by etching, electroforming or laser cutting.

16.8.2 Solder Reflow

Once the PCB has all the components in place, the solder paste must be melted with a carefully controlled temperature profile that ramps up steadily to avoid thermal shock and damage to the components. As for thick-film pastes, a belt furnace is often used with individual zones along its length to set a temperature profile with preheat, thermal soak, reflow and cooling zones. The correct profiles and temperature ramps are necessary to ensure that the flux does it job while the paste does not spatter, oxidise, or clump into balls. Programmable, infrared reflow ovens may also be used to good effect. The final construction of the product (soldering of connector, cables, etc.) is in most cases too awkward to be automated, and the final stages of the product assembly are carried out by hand. However, industrial robotics companies are being increasingly applied to the manual aspects of electronics assembly; ABB's YuMi® is just one example.

16.9 System-in-Package and System-on-Substrate Technology

System-in-package (SiP) or system-on-substrate technology [8, 9] is vital for the realisation of miniature microwave, millimetre-wave and terahertz systems. Whilst the 'GaAs versus Silicon' debate that persisted during the 1990s [10] is now long forgotten, it is widely recognised that a multitechnology platform is still required by which ICs, devices and components from any technology (Si/SiGe, GaAs, GaN, InP, graphene, ferroelectrics, optoelectronic, microfluidic, bionanoelectronic, etc.) can be integrated to form a system, as illustrated conceptually in Figure 16.17. Whilst many pioneering advances have been made recently in this area [11], there remain significant challenges to be addressed. On the one hand, there is an urgent need to push down the cost of millimetre-wave systems, since this frequency range is exploited for 5G+ systems where sophisticated steerable antenna arrays or massive MIMO are required. On the other hand, there is a significant technical challenge to be overcome in order to push the performance envelope so that SiP solutions can be employed for systems beyond 100 GHz, including terahertz systems, as and when the necessary devices become commercially available. Only with access to this SiP/SoS capability can companies fully benefit from the explosion in applications of the millimetre-wave and terahertz bands for applications such as 5G+, IoT, m2m communications, biosensing, imaging and security screening.

Another trend in systems integration is to realise multifunction and multifrequency systems by creating switching and tuning functions that are embedded into the passive components, so as to obviate the need, for example, to employ PAs to cover different bands. Figure 16.18 illustrates a conceptual example of a multichip module integrated in a 'smart substrate' that incorporates tuneable filters, antenna beamforming and programmable 'pixelated' patches. The

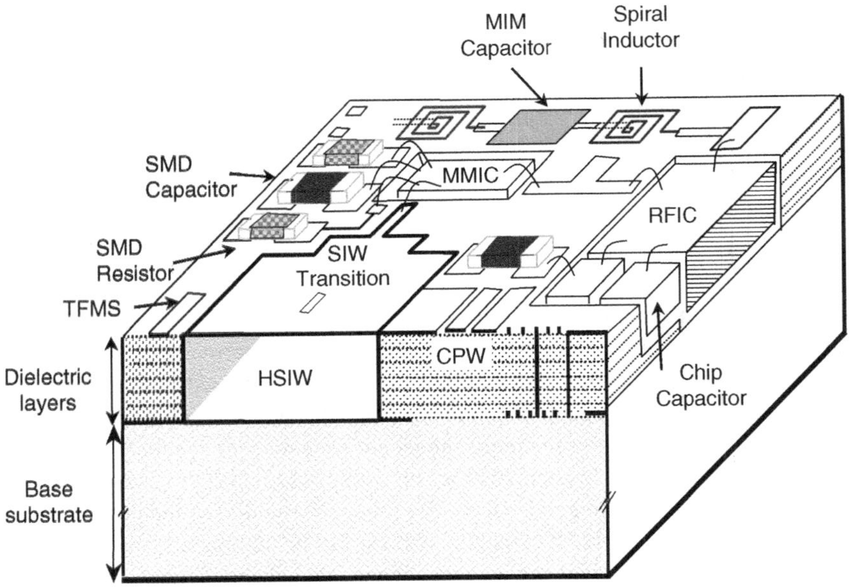

Figure 16.17 A conceptual system-in-package using a multilayer substrate.

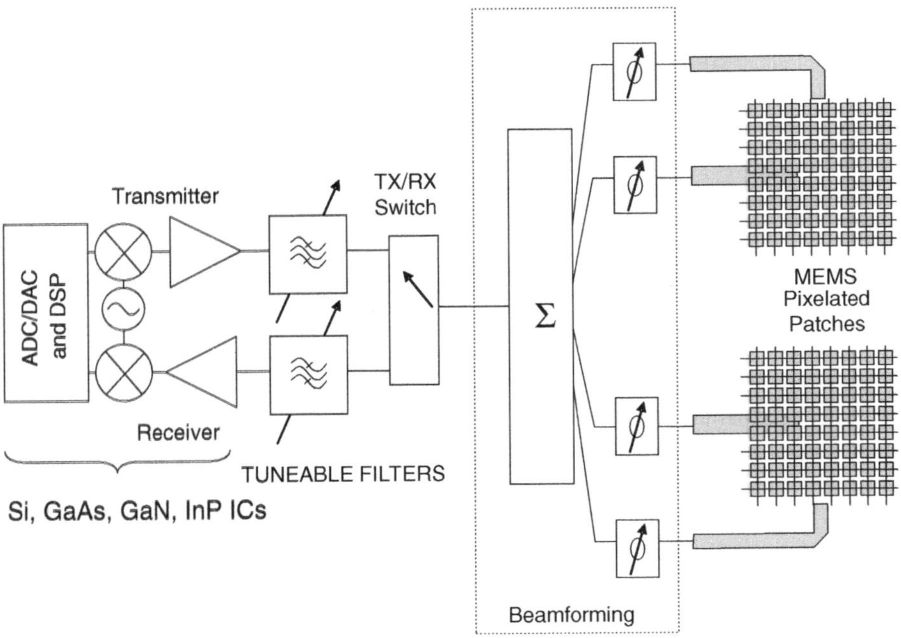

Figure 16.18 Conceptual system-on-substrate multi-band transceiver.

technologies that can make such an approach feasible include ferroelectric materials [12] and large-area MEMS [13]. In the case of extremely low-cost manufacture of IoT applications and wearables, considerable interest has been expressed in manufacturing microwave components on paper [14], textiles [15] and plastic [16]. Indeed, plastics-based electronics will surely emerge as a key technology for these applications, since electronic devices can be integrated onto the substrate via the low-temperature fabrication of organic transistors [17]. A 6 × 6″ test substrate of microwave components fabricated on an acetate sheet with carbon-based conductor ink is shown in Figure 16.19.

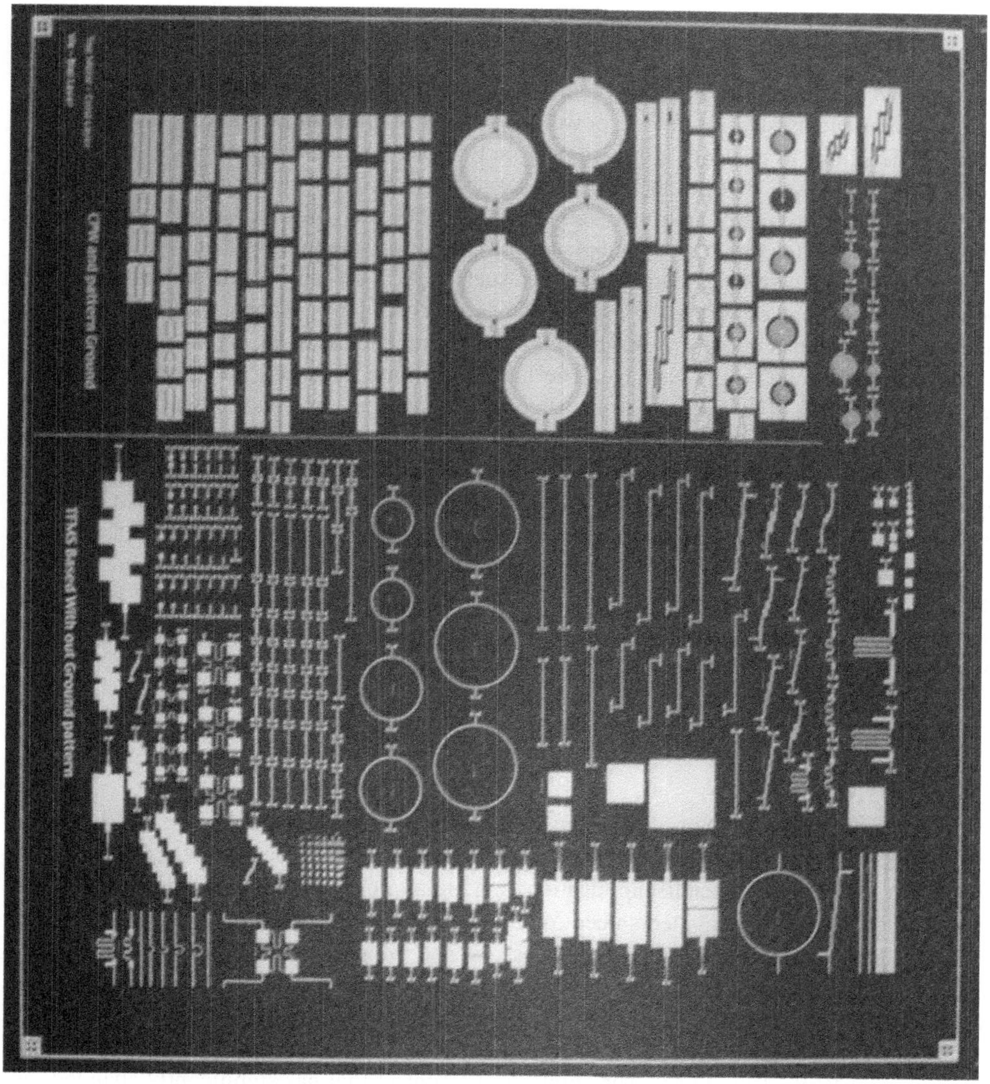

Figure 16.19 Microwave components printed onto acetate sheet with carbon-based ink.

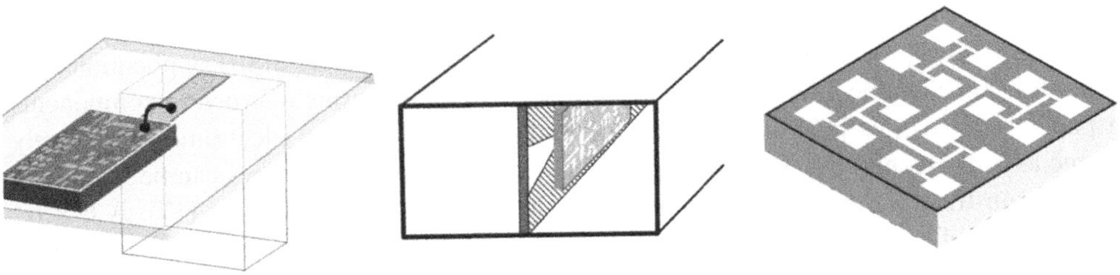

Figure 16.20 Examples of transitions and antenna integration techniques.

16.10 Transitions and Antenna-in-Package Techniques

Since active circuitry often needs to be integrated economically with very high-performance passive components such as filters, power combiners and antennas, there is an increasing need for designers to consider a transition from microstrip to waveguide, or similar, in their design, in such a way that production line assembly is straightforward and does not require any tuning. Some examples are illustrated in Figure 16.20. RWG-to-microstrip transitions can employ a slot in the microstrip ground plane, with the signal conductor above forming a probe into the RWG beneath, or E-plane techniques such as the microstrip-to-finline transition can be used. At frequencies where the antennas are small enough, it is increasingly common to integrate the antennas into the package itself – for example, by realising a patch antenna array on the chip or the lid of the package. The latter technique is gaining widespread adoption, leading to the idea of an antenna-in-package. This is entirely practical for millimetre-wave applications where the wavelength is very short, and 32-element antenna arrays have been integrated into a package to form a 60 GHz wireless HDMI transceiver module for consumer applications [18]. Established technology beyond 100 GHz is esoteric and costly, but to match it on performance requires a step-change in the performance that can be achieved with the SiP approach. Figure 16.21 shows a conceptual terahertz transceiver, based on a core Si/SiGe RFIC as the 'motherboard' but with specialist devices such as InP MMICs, RTDs, HBVs and Schottky diodes used to increase the operating frequency to the terahertz range, making use of flip-chip assembly [19] and back-face patch antennas on the III–V chips [20].

16.11 Conclusion

This chapter has provided a brief overview of the very wide range of packaging and assembly techniques used in modern microwave and millimetre-wave circuit design. Whether the product is for a low-volume specialist application, or a high-volume consumer application, the end result is that modern RF engineers need to be familiar with a wide range of high-technology products. There is a continual push for lower cost and improved performance in all application sectors. In consumer wireless applications, the prices for individual components can be shockingly low, and a close integration of DSP and RF circuitry is essential in order to realise a competitive product. Silicon technology has become dominant for increasing numbers of functions, but RFIC design has a very high 'cost of entry' that has significantly changed the way in which the industry works. Whilst chipsets are widely available for many applications,

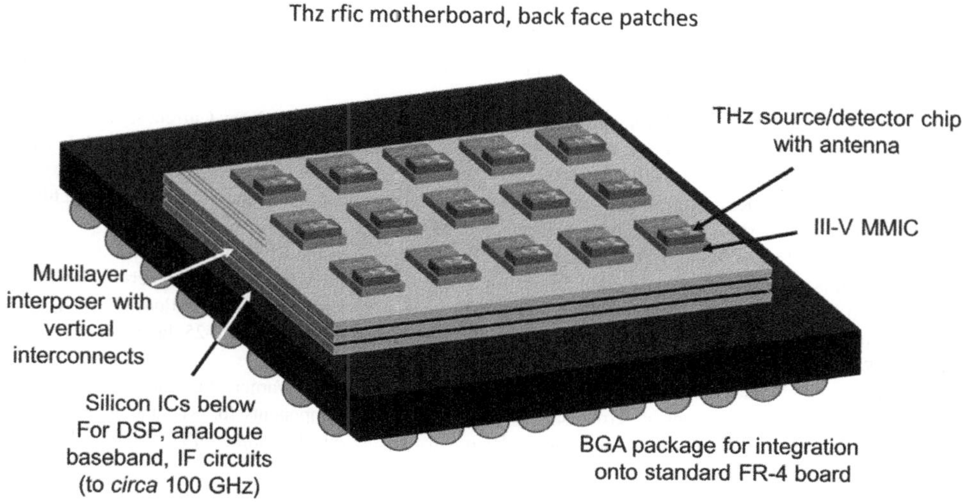

Figure 16.21 Conceptual terahertz system-on-substrate using III–V ICs on a Si/SiGe RFIC motherboard.

there are still many design choices that need to be made. For example, a millimetre-wave silicon RFIC transceiver may require additional high-performance components such as GaAs, GaN or InP devices in order to meet the link budget requirements and systems specifications. Furthermore, multiband transceivers and multiple and/or switched antennas are increasingly important, and there remains a need for extensive research and development into innovative system-in-package, system-on-substrate and antenna-in-package solutions. This is especially true in millimetre-wave design, where today these sophisticated technologies are rapidly making huge new parts of the electromagnetic spectrum available for mass market applications. This is in contrast to the past, when this frequency range was largely the preserve of the aerospace and defence sectors.

References

1. Arnold, R.G. and Pedder, D.J. (1992) Microwave characterization of microstrip lines and spiral inductors in MCM-D technology. *IEEE Transactions on Components, Hybrids, and Manufacturing Technology*, **15** (6), 1038–1045. doi: 10.1109/33.206928.
2. Iezekiel, S., Soshea, E.A., O'Keefe, M.F., and Snowden, C.M. (1995) Microwave photonic multichip modules packaged on a glass-silicon substrate. *Microwave Theory and Techniques, IEEE Transactions*, **43** (9), 2421–2427. doi: 10.1109/22.414598.
3. Muckett, S. and Minalgene, J. (1998) Hibridas photoimageable thick film process and materials for microwave and sensor component applications. *IEMT/IMC Symposium*, **15**–17 April, pp. 154–160. doi: 10.1109/IEMTIM.1998.704546.
4. Samanta, K.K. and Robertson, I.D. (2011) Advanced multilayer thick-film system-on-package technology for miniaturized and high performance CPW microwave passive components. *IEEE Transactions on Components, Packaging and Manufacturing Technology*, 1 (11), 1695–1705. doi: 10.1109/TCPMT.2011.2167231.
5. Bartsch, H., Grieseler, R., Muller, J., Barth, S., and Pawlowski, B. (2010) Properties of high-k materials embedded in low-temperature cofired ceramics. Electronic System-Integration Technology Conference (ESTC), 13–16 September, pp. 1–5. doi: 10.1109/ESTC.2010.5642935.

6. Stekovic, M. and Sandera, J. (2013) 3D structure with opened cavities in low shrinkage LTCC. 36th International Spring Seminar on Electronics Technology (ISSE), 8–12 May, pp. 74–78. doi: 10.1109/ISSE.2013.6648218.
7. www.itrs.net: SiP White Paper. The next Step in Assembly and Packaging: System Level Integration in the package (SiP).
8. Feger, R. and Stelzer, A. (2015) Millimeter-wave radar systems on-chip and in package: Current status and future challenges. IEEE Topical Conference on Wireless Sensors and Sensor Networks (WiSNet), 25–28 January, pp. 32–34. doi: 10.1109/WISNET.2015.7127407.
9. Deslandes, D. and Wu, Ke (2003) Single-substrate integration technique of planar circuits and waveguide filters. *IEEE Transactions on Microwave Theory and Techniques*, **MTT-51** (2), 593–596.
10. GaAs vs. Silicon: The Technology Battleground, Workshop, Microwaves and RF Conference, 1996.
11. Wang, H. (2008) Current Status and Future Trends for Si and Compound MMICs in Millimeter-Wave Regime and Related Issues for System on Chip (SOC) and/or System in Package (SIP) Applications. IEEE Topical Meeting on Silicon Monolithic Integrated Circuits in RF Systems (SiRF 2008), 23–25 January, p. 21. doi: 10.1109/SMIC.2008.12.
12. Miranda, F.A., Subramanyam, G., Van Keuls, F.W., Romanofsky, R.R., Warner, J.D., and Mueller, C.H. (2000) Design and development of ferroelectric tunable microwave components for Ku- and K-band satellite communication systems. *IEEE Transactions on Microwave Theory and Techniques*, **48** (7), 1181–1189. doi: 10.1109/22.853458.
13. Robben, D., Peik, S.F., Henning, T., Becker, M., and Froehner, K. (2012) Laser machined microsystems for active frequency selective surfaces. IEEE MTT-S International Microwave Symposium Digest, 17–22 June, pp. 1–3. doi: 10.1109/MWSYM.2012.6259596.
14. Li Yang, Rida, A., Vyas, R., and Tentzeris, M.M. (2007) RFID Tag and RF structures on a paper substrate using inkjet-printing technology. *IEEE Transactions on Microwave Theory and Techniques*, **55** (12), 2894–2901. doi: 10.1109/TMTT.2007.909886.
15. Lanlin Zhang, Zheyu Wang, and Volakis, J.L. (2012) Embroidered textile circuits for microwave devices. IEEE Antennas and Propagation Society International Symposium (APSURSI), 8–14 July, pp. 1–2. doi: 10.1109/APS.2012.6348932.
16. Hettak, K., Ross, T., James, R., and Momciu, A. (2015) Low-cost mm-wave coplanar waveguide bandpass filter using inkjet printing of silver nano-particles on flexible plastic substrate. IEEE MTT-S International Microwave Symposium (IMS), 17–22 May, pp. 1–3. doi: 10.1109/MWSYM.2015.7167092.
17. Reed, M.A. (1999) Molecular-scale electronics. *Proceedings of the IEEE*, **87** (4), 652–658. doi: 10.1109/5.752520.
18. Gilbert, J.M., Doan, C.H., Emami, S., and Shung, C.B. (2008) A 4-Gbps uncompressed wireless HD A/V transceiver chipset. *IEEE Micro*, **28** (2), 56–64. doi: 10.1109/MM.2008.20.
19. Green, D.S., Dohrman, C.L., Demmin, J., Zheng, Y., and Chang T.-H. (2015) Heterogeneous integration for revolutionary microwave circuits at DARPA. *Microwave Journal*, **June**, 22–24.
20. Robertson, I.D. (1993) Millimetre-wave back-face patch antenna for multilayer MMICs. *Electronics Letters*, **29** (9), 816–818. doi: 10.1049/el:19930545.

17

Electronic Design Automation

17.1 Introduction

Today, there is a very wide range of electronic design automation (EDA) software available for microwave and millimetre-wave design. At the basic end, it is possible to design a simple circuit using a basic open-source simulator with simple microstrip elements and other models. At the professional level, there are fully integrated design environments, such as ADS^{TM} as shown in Figure 17.1, which can be used to simulate a transceiver chip operating in its package environment with direct integration of electromagnetic simulation for 3-D structures. Some of the key vendors in the microwave EDA business are listed in Table 17.1. There is a wide variety of simulation techniques; linear, nonlinear, mixed signal, system level and electromagnetic. The fundamentals of the main types of RF-oriented simulation techniques are described in this chapter. It should not be forgotten, of course, that a modern transceiver may form a part of a sophisticated system-on-chip (SoC), but the wide range of digital design tools used is outside the scope of this chapter. Furthermore, there is also a growing need for multiphysics simulation: for example, to achieve optimum power efficiency and linearity from a power transistor it is necessary to include thermal, electromagnetic, nonlinear and thermal modelling [1].

The design flow for a microwave or millimetre-wave circuit starts with the definition of the specifications from the systems requirement perspective, followed by selecting a circuit topology, modelling the active and passive devices, and drawing the circuit layout. Each step needs a different analysis technique, from system-level simulation to electromagnetic modelling. Furthermore, the process will be highly iterative, since in distributed circuits the layout has an absolutely critical effect on the performance of the circuit. Commercially available high-frequency software provides all the simulation 'engines' mentioned above, for completing the design task and shortening the time-to-market. It is necessary for students and engineers who work in this area to become familiar with the analysis techniques used in the simulators in order to understand their limitations. Understanding the functionality and limitations of each of these techniques is an essential part of choosing a simulation strategy to fulfil the design task. Hence, the main aim of this chapter is to provide the background theory behind the most common simulation engines used in commercial EDA software.

Microwave and Millimetre-Wave Design for Wireless Communications, First Edition.
Ian Robertson, Nutapong Somjit and Mitchai Chongcheawchamnan.
© 2016 John Wiley & Sons, Ltd. Published 2016 by John Wiley & Sons, Ltd.

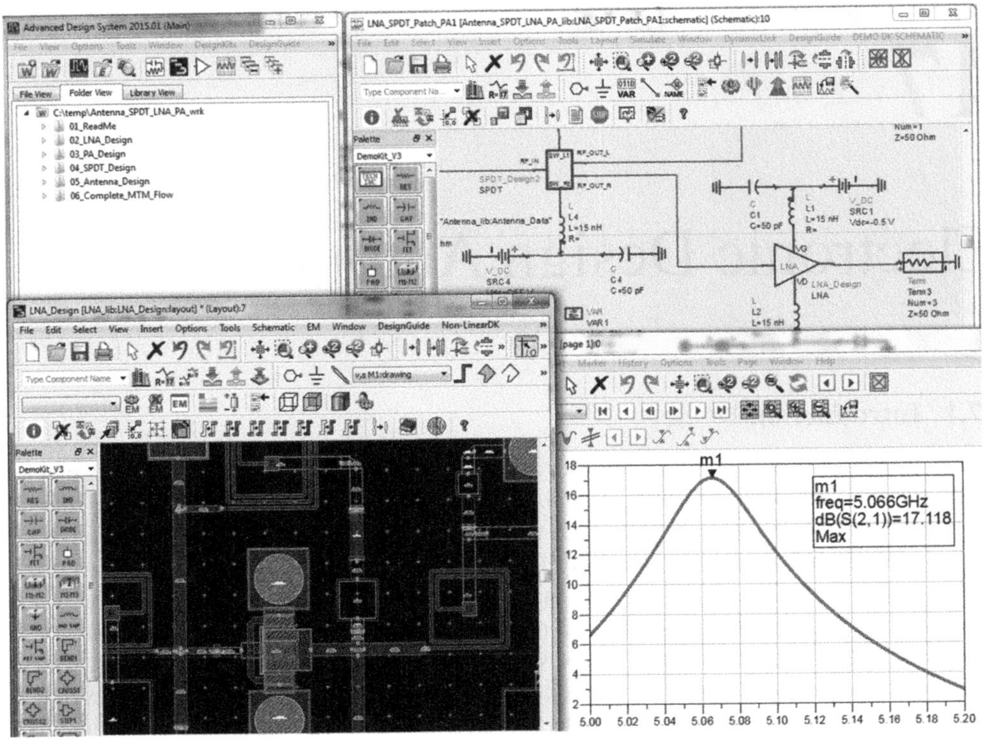

Figure 17.1 MMIC simulation in ADS™.

The linear frequency-domain analysis technique solves for the frequency response of a circuit, usually presenting results in the form of *S*-parameters. From a mathematical aspect, the linear simulation technique is relatively straightforward, for example making use of *Y*-parameter matrix techniques for nodal analysis, with matrix operations and linear algebra used to solve for the desired results. Most students and engineers are already familiar with the underlying principles of the necessary circuit theory. The time-domain analysis technique is very crucial for analysing nonlinear circuits, modelling the transient response and nonlinear behaviour of active devices. However, pure time-domain techniques have limitations for distributed circuits, and so the harmonic balance technique is introduced. This technique has been developed for analysing a circuit operating under large signal conditions that has a significant number of distributed (transmission line) elements. One circuit that is often analysed with the harmonic balance method is a power amplifier. With the transistor operating with a large signal, the circuit must be regarded as a nonlinear system, and yet the transmission line matching networks are not amenable to analysis with transient time-domain simulation (the reasons for this ought to be obvious from the reflection examples in Chapter 5, Section 5.2). For analysing a circuit in the condition that large and small signals are excited, a technique called 'large-signal/small-signal analysis' is described. Mostly, the technique is applied for analysing mixer conversion loss and noise in microwave/millimetre-wave circuits.

The direct analysis of structures based on Maxwell's equations has become a steadily more important technique. This has been driven by the demand for miniaturisation through

Table 17.1 A selection of some commercial simulators (all trademarks acknowledged).

Company	Product	Type
Keysight Technologies	ADS	Full linear, nonlinear and system suite
	Momentum	3-D planar
	EMPro	3-D
	SystemVue	System level
	GoldenGate	RFIC simulation, integrated into Cadence
ANSYS	DesignerRF	Integrated environment
	HFSS	Full 3-D
NI AWR	Microwave Office	Linear/nonlinear microwave suite
	Analog Office	Analog/RFIC design
	VSS	System level
	AXIEM	3-D planar electromagnetic
	Analyst	3-D FEM
Sonnet	Sonnet Suites *em*	3-D Planar
	Blink	Planar em tool for use with Cadence
Cadence	Virtuoso	RFIC design suite
	SpectreRF	Nonlinear transistor-level simulation
Mentor Graphics	DXD/Expedition	Board-level design linking to ADS and Microwave Office
	Questa ADMS RF	Mixed RF/DSP simulation
	EldoRF	Transistor level RFIC design
Silvaco	Atlas	Physical-level device simulation
	SmartSpiceRF	Spice-based design and layout
	Utmost	Device model parameter extraction
IMST	MultiLib	Multilayer model library for ADS
	Empire	3-D electromagnetic
CST	Studio Suite	3-D electromagnetic
COMSOL	Multiphysics	MEMS, 3-D EM and others

3-D-integration, the pressure on getting products to market quickly without design iterations, and the need to manufacture products in large volume without any production-line tuning. Electromagnetic modelling is especially important for millimetre-wave design where many basic assumptions, such as the quasi-TEM behaviour of microstrip, are often inadequate and full-wave solutions are needed for accurate simulation results. Personal computers with sufficient computational power are so affordable that most modern commercial EDA software now has a planar electromagnetic (EM) solver available to analyse critical parts of the circuit at the layout stage. The term planar is employed to specify that the technique is for use with a planar circuit only. The chapter will cover only 2-D EM simulation in detail since that provides sufficient background to the kinds of technique used, which is normally the method of moments (MoM) for planar solvers. However, it is recognised that full 3-D simulation using techniques such as the finite element method (FEM) and finite-difference time-domain (FDTD) are extremely important in modern RF engineering, and some of these commercially available tools are listed in Table 17.1. However, efficient computational modelling of electromagnetic fields is a fast-moving area of research and development and there is a wide range of other specialist software available, which is not covered here.

17.2 Linear Frequency-Domain Analysis

This type of simulator generally uses Y-parameter matrix techniques to solve for the steady-state frequency response of linear circuits. Since individual components are treated essentially as frequency-dependent admittances within a nodal matrix, solution for the overall circuit response is a case of using matrix reduction techniques to end up with an overall set of Y-parameters. These can then be converted to S-parameters, and so forth, for display. The simulator does not need to calculate the currents or voltages in the circuit, and the analysis is very fast as a result. In particular, working in the frequency domain means that only the steady-state response of distributed elements (transmission lines) needs to be considered. This is very important, because the multiple reflections encountered in a microwave circuit may take a long time to solve in the time domain. Furthermore, frequency-dependence of transmission line parameters (e.g., dispersion, resulting in a change of Z_0 with frequency) can easily be handled. However, nonlinear elements cannot be modelled since the currents and voltages in the circuit are not calculated. The linear simulator 'engines' within ADS, Microwave Office®, and so forth, are based on this method, although it is not the only analysis method available in these cases. Schematic capture is used for entering designs directly into a circuit diagram using symbols. It is easy to visualise the circuit and prevent mistakes, provided that a subcircuit hierarchy is used extensively in order to keep the schematic tidy and easy to follow. For microstrip circuit design, it is necessary to break the circuit down into its constituent transmission lines and discontinuities. This is illustrated in Figure 17.2, which shows the ADS schematic of a microstrip Wilkinson power divider. The T-junction and curves are added manually and an MSub data block defines the substrate parameters.

17.3 Time-Domain Simulation

In order to calculate the true response of nonlinear components such as transistors, it is necessary to work in the time-domain. Direct time-domain methods are generally based on SPICE, with more convenient display options added. Time-domain analysis is also an indispensable tool for engineers when the transient response of the circuit is important, for example the start-up in an oscillator circuit. Theoretically, a circuit can be analysed with a nonlinear differential equation system. To solve the problem with a computer, the nonlinear differential equations are converted to a nonlinear difference equation system. All continuous-time signals are put into sampled form with a certain time step interval. Hence, the original continuous-time problem turns to a discrete-time problem. The system can be universally solved with numerical methods to determine the unknown voltages or currents. There are several numerical methods normally embedded in the time-domain simulators integrated in commercial EDA packages. These numerical algorithms are aimed at solving the integral equation.

Consider a nonlinear circuit consisting of a nonlinear conductance and a linear capacitor, as shown in Figure 17.3. The current flowing through the nonlinear conductance is denoted by

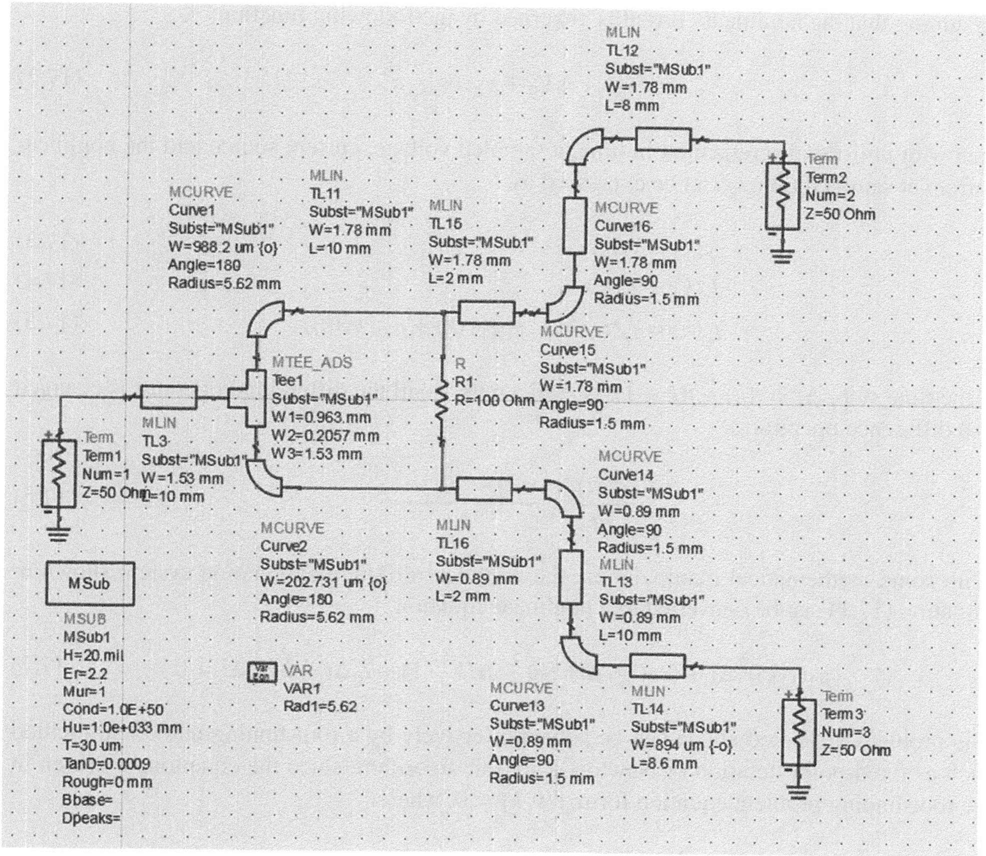

Figure 17.2 Microstrip Wilkinson power divider schematic in ADS™.

$i_g(v)$, where v is the output voltage. The source current and the capacitor current are $I_s(t)$ and $i_c(t)$, respectively. Applying Kirchhoff's current law to Figure 17.3, one obtains:

$$c\frac{dv(t)}{dt} + i_g(v) = I_s(t) \tag{17.1}$$

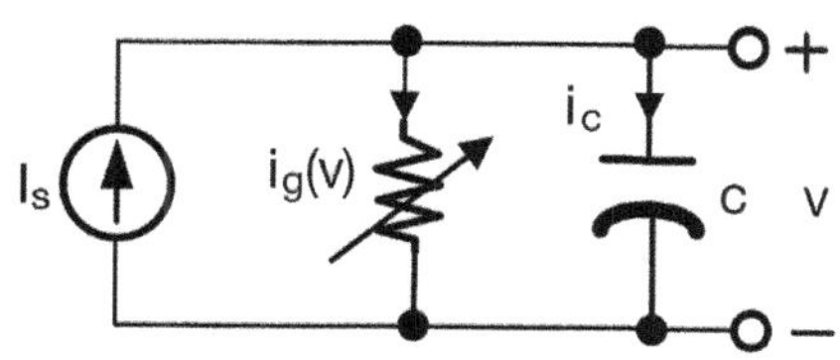

Figure 17.3 First-order nonlinear circuit.

Assuming that the nonlinear current is governed by the following function:

$$i_g(v) = V_0 + g_1 v + g_3 v^3 \tag{17.2}$$

then with uniform discretisation in time, a sampled voltage, current source and the nonlinear current at sampled time t_k can be expressed as:

$$v(k) = v(t_k) \tag{17.3}$$

$$I_s(k) = I_s(t_k) \tag{17.4}$$

$$i_g(k) = i_g(t_k) = I_0 + g_1 v(k) + g_3 \cdot (v(k))^3 \tag{17.5}$$

If the time step, $\Delta t = t(k) - t(k-1)$, is sufficiently small the differential operator is changed to a difference operator:

$$\frac{dv(t)}{dt} = \frac{v(k) - v(k-1)}{\Delta t} \tag{17.6}$$

With some mathematical manipulation, the nonlinear differential equation system shown in Equation (17.1) can be converted to a nonlinear equation:

$$g_3(v(k))^3 \Delta t + (c + g_1 \Delta t) v(k) - c v(k-1) + I_0 \Delta t - I_s(k) \Delta t = 0 \tag{17.7}$$

The problem presented above can be solved effectively by a root-finding numerical method such as fixed-point iteration or Newton–Raphson algorithm, since the equation is written in the root-finding problem equation form, $f(v, k) = 0$, where:

$$f(v) \equiv g_3(v(k))^3 \Delta t + (c + g_1 \Delta t) v(k) - c v(k-1) + I_0 \Delta t - I_s(k) \Delta t \tag{17.8}$$

From Equation (17.8), the evaluation can be performed if the previous time voltage sample is known. The Newton–Raphson algorithm is applied in this case. With an initial guess $v^0(k)$, the voltage at time sample k is calculated from:

$$v^i(k) = v^{i-1}(k) - \frac{f(v^{i-1}, k)}{f'(v^{i-1}, k)} \tag{17.9}$$

where i is the iteration number and $f'(v^{i-1}, k)$ is the derivative of $f'(v^{i-1}, k)$.

The iteration of each k calculation terminates when the magnitude of the updated rate is less than the threshold ε:

$$\mathrm{abs}(v^i(k) - v^{i-1}(k)) < \varepsilon \tag{17.10}$$

The convergence time and accuracy depend on this parameter ε. The technique is quite time-consuming as the iteration is performed for each sample k. For example, if the number of samples is 5000 and each sample requires 20 iterations, the total number of computational steps becomes 100 000.

Another technique to determine the time-domain waveform $v(k)$ is to apply a numerical method for the ordinary differential equation (ODE) method. In this case, the initial voltage value at time t_0 should be given. Simple methods such as the Euler or Runge–Kutta method can be applied, depending on the convergence rate that needs to be achieved. The Euler method is the simplest method to apply. The derivation is from the first-order approximation. The Runge–Kutta method is basically similar to the Euler concept but with a higher-order approximation to converge the solution faster. For simplicity's sake, Euler's technique is applied here to solve the time-domain waveform from the ODE problem. With the initial value, y_0, the ODE equation can be solved numerically. The ODE problem can be written as:

$$y' = f(x, y); y(x_0) = y_0 \tag{17.11}$$

Applying discretisation to the ODE problem, the solution obtained by Euler's method is:

$$y^{i+1} = y^i + \Delta x \cdot f^i \tag{17.12}$$

where

$$y^{i+1} = y(x_i + \Delta x) \tag{17.13}$$

$$y^i = y(x_i) \tag{17.14}$$

$$f^i = f(x_i, y_i) \tag{17.15}$$

The solution is accurate if Δx is sufficiently small. Substituting Equation (17.2) into Equation (17.1), one obtains:

$$c\frac{dv(t)}{dt} + I_0 + g_1 v(t) + g_3(v(t))^3 = I_s(t) \tag{17.16}$$

Rearranging the equation such that only the derivative term, $\frac{dv(t)}{dt}$, appears on the left-hand side, gives:

$$\frac{dv(t)}{dt} = \frac{I_s(t)}{c} - \frac{I_0}{c} - \frac{g_1}{c}v(t) - \frac{g_3}{c} \cdot (v(t))^3 \tag{17.17}$$

In this case, the voltage at initial time t_0, $v(t_0) = v_0$, is assumed to be known. The solution obtained from Euler's method can be written as:

$$v(k+1) = v(k) + \frac{\Delta t}{c} \cdot [I_s(k) - I_0 - g_1 v(k) - g_3 \cdot (v(k))^3] \tag{17.18}$$

where $k = 0, 1, 2, \ldots$

Example: Determine the voltage waveform in Equation (17.17) from $t = 0$ to 10 ns, given that the excitation current is sinusoidal with amplitude 0.1 mA and frequency 10 GHz. The capacitor is 10 fF and the initial voltage across the capacitor is 0 mV. The nonlinear parameters (I_0, g_1, g_3) are $(-0.1 \times 10^{-3}, 0.05 \times 10^{-3}, 0.05 \times 10^{-3})$.

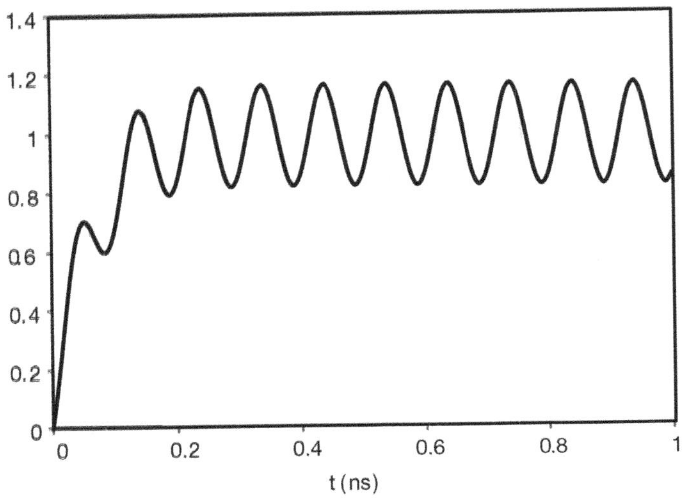

Figure 17.4 Voltage waveform derived from the Euler method.

Since the excitation frequency is 10 GHz, the criterion to choose the step time Δt is obtained by applying the Nyquist sampling theorem; that is:

$$\Delta t < \frac{1}{2 \times 10 \times 10^9} = 0.05 \text{ ns} \tag{17.19}$$

Here, we select $\Delta t = 0.01$ ns. Applying all parameters to Equation (17.18), one obtains:

$$v(k+1) = v(k) + \frac{10^{-11} \times 10^{-3}}{10 \times 10^{-15}} \cdot [0.1 \sin(2\pi \times 10 \times 10^9 t) + 0.1 - 0.05 v(k) - 0.05(v(k))^3] \tag{17.20}$$

Plotting the equation in MATLAB™, the voltage waveform across the capacitor shown in Figure 17.4 is obtained. The voltage abruptly increases from the initial value of 0 V. The steady-state waveform is a sinusoidal wave at 10 GHz, the frequency of the external current source.

It is shown that the time-domain waveform is determined by solving the ODE equation. The waveform at a certain time $t(k)$ is computed from the waveform in the previous time sample, $t(k-1)$. Though the method based on solving with the ODE is less complex than the root-finding technique, the complexity in time-domain analysis technique causes a large computational burden and is very time consuming.

17.4 Harmonic Balance

The handling of transmission line components is difficult in the time-domain because of the long time taken to reach a steady state and the need to consider dispersion. Some of the newer time-domain simulators can deal with dispersion by using a convolutional technique.

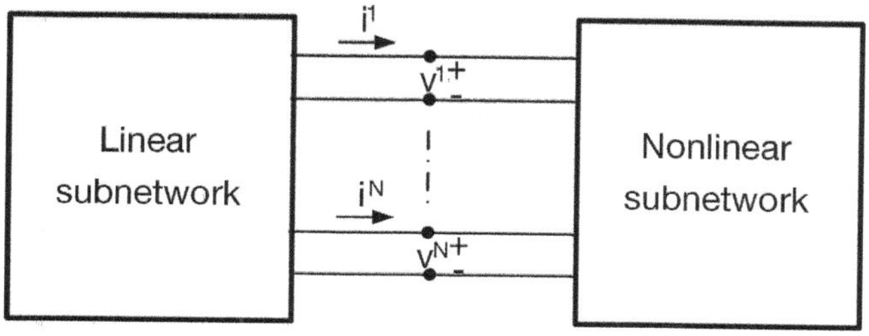

Figure 17.5 The concept of partitioning a network into linear and nonlinear subnetworks.

For steady-state large-signal analysis, such as the gain compression of a power amplifier, the harmonic balance technique is faster than time-domain methods and is now an industry-standard tool. This uses a combination of time- and frequency-domain analysis [2]. The voltages and currents in the nonlinear components are calculated in the time domain, while the linear parts are treated in the frequency domain, as illustrated in Figure 17.5. Fourier transformation is then used to compare the currents and voltages of the nonlinear components with the terminating impedances presented by the linear components at all the harmonic frequencies used in the analysis. Any error is reduced by successive iteration. The signals must be periodic in order to use the Fourier transform. Due to the enormous number of calculations required, and the need for iterations to ensure convergence, this type of analysis requires a fast computer and the simulation time increases significantly as the number of harmonics increases.

17.4.1 Concept

The harmonic balance technique assumes that the sinusoidal solution of a nonlinear differential equation is a linear combination of the excitation frequency and its harmonics. This causes the solution of the system to be periodic. With the sinusoidal excitation, the solution of the system, $x(t) = \mathrm{Re}\{Xe^{-j\omega_0 t}\}$, can be written in the form of a phasor variable X. Hence, the differential equation is converted to an algebraic equation in terms of the phasor variable X and subsequently solved. The phasor variable is put into the form of a combination of multiple frequencies, as stated above. By superposition, and due to the orthogonality of harmonic waves, the algebraic equation is reduced to multiple simple equations for each harmonic. The equation systems will be solved in the balancing of the algebraic equations for each harmonic, which is the origin of the name harmonic balance.

Given a differential equation,

$$f(x,u) = 0 \tag{17.21}$$

where $f(\cdot)$ is continuous and real function, $u(t)$ is the excitation sinusoidal waveform at ω_0 and $x(t)$ is a solution, $u(t)$ can be written using Fourier series as:

$$u(t) = \sum_{k=-\infty}^{\infty} \mathrm{U}(k)e^{jk\omega_0 t} \tag{17.22}$$

Hence, (t), which is real, can be written with Fourier series in the following form,

$$x(t) = \sum_{k=-\infty}^{\infty} X(k)e^{jk\omega_0 t} \tag{17.23}$$

The procedure to perform harmonic balance is described in a step-by-step fashion as follows:

Step 1: Substituting $x(t)$ and $u(t)$ from Equation (17.22) and Equation (17.23) to Equation (17.21), we obtain:

$$f(x,u) = \sum_{k=-\infty}^{\infty} \mathbf{F}(\mathbf{X},\mathbf{U},k)e^{jk\omega_0 t} \tag{17.24}$$

where

$$\mathbf{X} = [\cdots \quad X(-1) \quad X(0) \quad X(1) \quad \cdots]^{\mathrm{T}} \tag{17.25}$$

$$\mathbf{U} = [\cdots \quad \mathrm{U}(-1) \quad \mathrm{U}(0) \quad \mathrm{U}(1) \quad \cdots]^{\mathrm{T}} \tag{17.26}$$

Step 2: Solve the algebraic equation system as shown:

$$\mathbf{F}(\hat{\mathbf{X}},\mathbf{U},k) = 0 \tag{17.27}$$

Step 3: Solve the equation using a numerical method.

A numerical technique such as the Newton–Raphson method is used to determine the solution vector $\hat{\mathbf{X}}$. The Newton–Raphson method, which is based on an iterative technique, is usually chosen due to its fast convergence and robustness. The solution at the $(i+1)^{th}$ iteration, $\mathbf{X}^{i+1}$, can be calculated from the previous solution, $\mathbf{X}^i$, by:

$$\mathbf{X}^{i+1} = \mathbf{X}^i - [\mathbf{J}(\mathbf{X}^i)]^{-1}\mathbf{F}(\mathbf{X}^i,\mathbf{U},k) \tag{17.28}$$

where $\mathbf{J}(\mathbf{X}^i)$ is the Jacobian matrix of $\mathbf{X}^i$ and $[\cdot]^{-1}$ is the inverse matrix operation. The condition to check whether the algorithm converges is computed from the norm of the updated vector. The algorithm is deemed to have reached a solution when the updated rate is less than the threshold ε:

$$||\mathbf{X}^{i+1} - \mathbf{X}^i|| < \varepsilon \tag{17.29}$$

17.4.2 Application to Microwave and Millimetre-Wave Circuits

In microwave and millimetre-wave circuit design, the harmonic balance technique has been widely applied for analysing the steady-state solution of nonlinear circuits such as power amplifiers and mixers. The nonlinear circuits to be analysed are assumed to be stable, and the technique targets the steady-state response. The computational complexity depends on the number of harmonics and circuit nodes. The issue of the nonlinear characteristic of the devices is another concern that affects the convergence. Most modern microwave and millimetre-wave

EDA software offers a harmonic balance engine as one of the circuit design tools. Yet, with the assumption of a steady-state solution, the harmonic balance technique alone is not always enough on its own for the successful design of certain microwave circuits, such as oscillators or mixers. For example, the output signal from an oscillator that is oscillating at f_0 during the start-up period may contain several frequency components that are not only the harmonics of f_0. When analysing a mixer, which is a circuit involving both large and small signals, the harmonic balance technique cannot always precisely predict the conversion gain. Nevertheless, due to its fast computation, the harmonic balance technique is an indispensable tool in commercial EDA software for the modelling of circuits that contain weakly or strongly nonlinear devices.

17.5 Large-Signal/Small-Signal Simulation

The large-signal/small-signal analysis technique is suitable for a nonlinear circuit involving a combination of large and small amplitude signals simultaneously. The technique assumes that large signal drives the nonlinear devices at a certain nonlinear operating point where the small signal perturbs the nonlinear operating points. The circuit response is obtained by linearising the nonlinear devices around the operating point with some perturbations from the small signal. The problem is very similar to the AC small-signal analysis for a transistor amplifier biased at a certain DC value. The quiescent point of the transistors at DC bias voltage is calculated by a nonlinear analysis. Normally, a load line technique is used and small perturbation analysis is applied by considering the AC steady-state condition. The large-signal/small-signal analysis is somewhat similar, but the operating point of the nonlinear devices is time-variant and periodic. If the circuit problem under consideration falls into this category, the large signal/small-signal analysis technique is recommended as a solution technique. For example, in a diode mixer the device is pumped with a large signal from a local oscillator. For a certain LO power, the time-varying capacitance and conductance can be calculated and used in the conversion loss analysis when a small-signal input is applied to the diode. Another problem which is commonly encountered is the analysis of noise in nonlinear circuits. In this case, the desired AC signal is considered to be large while the noise is regarded as a small signal. The large-signal/small-signal analysis can be applied to these two problems and other, similar, applications.

In this section, the large-signal/small-signal analysis technique is applied to a mixer problem [3]. A single-ended diode mixer is used for demonstrating the concept. A simple equivalent model for this diode mixer is shown in Figure 17.6. Let the junction voltage and the diode

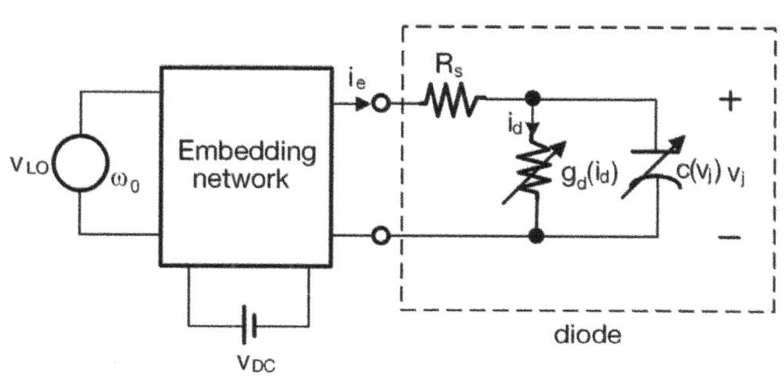

Figure 17.6 Equivalent circuit of single-ended diode mixer.

current be denoted by v_j and i_e, respectively. The nonlinearity of the diode is modelled with a nonlinear capacitor, $c(v_j)$ and diode conductance, $g_d(i_d)$. The linear devices are the series resistance, R_s, and the embedding network. The diode is DC biased and the external LO source at frequency ω_0 is pumped to the diode.

From Figure 17.6, the voltage drops at the diode junction and the diode current can be written in terms of Fourier coefficients as follows:

$$v_j(t) = \sum_{k=-\infty}^{+\infty} V_j(k)e^{jk\omega_0 t};\ V_j(k) = V_j^*(-k) \tag{17.30}$$

$$i_e(t) = i_d(t) + i_c(t) = \sum_{k=-\infty}^{+\infty} I_e(k)e^{jk\omega_0 t};\ I_e(k) = I_e^*(-k) \tag{17.31}$$

The currents $i_d(t)$ and $i_c(t)$ are related to the junction voltage $v_j(t)$ by:

$$i_d(t) = I_0\left(e^{\frac{qv_j}{\eta KT}} - 1\right) \tag{17.32}$$

$$i_c(t) = c_j(v_j)\frac{dv_j}{dt} \tag{17.33}$$

where the junction capacitance $c_j(v_j)$ is given by,

$$c_j(v_j) = c_{j0}\left(1 - \frac{v_j}{\phi}\right)^{-\gamma} \tag{17.34}$$

γ depends on the doping profile of the diode. The small-signal conductance of the diode resulting from a small perturbation is given by:

$$g_d(t) = \frac{di_d}{dv_j} \approx \frac{q}{\eta KT} i_d(t) \tag{17.35}$$

With the LO pump driving the diode, the junction voltage is determined and the spectra of $i_d(t)$, $g_d(t)$ and $c_j(t)$ are obtained by applying the Fourier series expansion. Their Fourier coefficients are denoted by $I_d(k)$, $G_d(k)$, and $C_j(k)$ which are given by:

$$i_d(t) = \sum_{k=-\infty}^{\infty} I_d(k)e^{jk\omega_{LO} t};\ I_d(k) = I_d^*(-k) \tag{17.36}$$

$$g_d(t) = \sum_{k=-\infty}^{\infty} G_d(k)e^{jk\omega_{LO} t};\ G_d(k) = G_d^*(-k) \tag{17.37}$$

$$c_j(t) = \sum_{k=-\infty}^{\infty} C_j(k)e^{jk\omega_{LO} t};\ C_j(k) = C_j^*(-k) \tag{17.38}$$

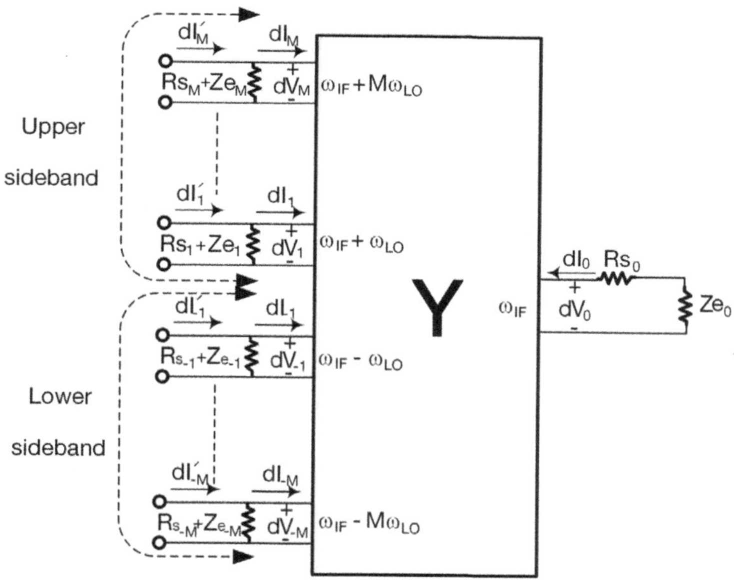

Figure 17.7 Small-signal model of single-ended diode mixer for conversion matrix analysis.

With these results, the small-signal conversion matrix for the diode is formed. Figure 17.7 shows a small-signal model of the mixer as a multiport network. The admittance matrix of the intrinsic diode describing the time-varying junction capacitance and conductance is defined by $\mathbf{Y}$. The extrinsic impedance at port m is denoted with Ze_m. This extrinsic impedance covers any matching networks at input and output ports of a singly diode mixer as well as parasitic elements in the diode package. The series resistance of the diode is modelled with Rs_m which depends on the diode material, the depletion width and the thermal time-constants. The small-signal voltage and current at port m of $\mathbf{Y}$ are denoted by $(\delta \mathrm{I}_m, \delta \mathrm{V}_m)$ where m relates to the mixing frequency $\omega_m = \omega_{IF} + m\omega_{LO}$. Note that ω_{IF} is the intermediate frequency and ω_{LO} is the LO frequency. The conversion matrix denoted by $\mathbf{Y}$, is related to the admittance matrix of the intrinsic diode. It relates the small-signal current and voltage vectors $(\delta\mathbf{I}, \delta\mathbf{V})$ as follows:

$$\delta\mathbf{I} = \mathbf{Y} \cdot \delta\mathbf{V} \tag{17.39}$$

where

$$\delta\mathbf{I} = [\cdots, \delta \mathrm{I}_1, \delta \mathrm{I}_0, \delta \mathrm{I}_{-1}, \cdots]^{\mathbf{T}} \tag{17.40}$$

$$\delta\mathbf{V} = [\cdots, \delta \mathrm{V}_1, \delta \mathrm{V}_0, \delta \mathrm{V}_{-1}, \cdots]^{\mathbf{T}} \tag{17.41}$$

The elements of current and voltage vectors at port m, i.e. $(\delta \mathrm{I}_m, \delta \mathrm{V}_m)$, are defined as shown in Figure 17.7. Each port represents one sideband frequency. The elements of $\mathbf{Y}$ correspond to the sideband components. **Y** can be written as:

$$\mathbf{Y} = \begin{bmatrix} Y_{M,M} & \cdots & Y_{M,1} & Y_{M,0} & Y_{M,-1} & \cdots & Y_{M,-M} \\ \vdots & \ddots & \vdots & \vdots & \vdots & ⋰ & \vdots \\ Y_{1,M} & \cdots & Y_{1,1} & Y_{1,0} & Y_{1,-1} & \cdots & Y_{1,-M} \\ Y_{0,M} & \cdots & Y_{0,1} & Y_{0,0} & Y_{0,-1} & \cdots & Y_{0,-M} \\ Y_{-1,M} & \cdots & Y_{-1,1} & Y_{-1,0} & Y_{-1,-1} & \cdots & Y_{-1,-M} \\ \vdots & ⋰ & \vdots & \vdots & \vdots & \ddots & \vdots \\ Y_{-M,M} & \cdots & Y_{-M,1} & Y_{-M,0} & Y_{-M,-1} & \cdots & Y_{-M,-M} \end{bmatrix} \tag{17.42}$$

where

$$Y_{m,n} = G_{m-n} + j(\omega_{LO} + m\omega_{IF}) \cdot C_{m-n} \tag{17.43}$$

However, $\delta\mathbf{I}$ is the intrinsic diode current vector not the observed current vector, $\delta\mathbf{I}'$. The relationship between $\delta\mathbf{I}'$ and $\delta\mathbf{V}$ is:

$$\delta\mathbf{I}' = \mathbf{Y}' \cdot \delta\mathbf{V} \tag{17.44}$$

As shown in Figure 17.7, $\mathbf{Y}'$ is the admittance matrix including the intrinsic diode and all the passive elements in the single-diode mixer circuit. The matrix $\mathbf{Y}'$ is defined as:

$$\mathbf{Y}' = \mathbf{Y} + \mathbf{Y}_{\mathrm{A}} \tag{17.45}$$

where

$$\mathbf{Y}_{\mathrm{A}} = \begin{bmatrix} (R_{sM} + Z_{eM})^{-1} & \cdots & 0 & 0 & 0 & \cdots & 0 \\ \vdots & \ddots & \vdots & \vdots & \vdots & ⋰ & \vdots \\ 0 & \cdots & (R_{s1} + Z_{e1})^{-1} & 0 & 0 & \cdots & 0 \\ 0 & \cdots & 0 & (R_{s0} + Z_{e0})^{-1} & 0 & \cdots & 0 \\ 0 & \cdots & 0 & 0 & (R_{s-1} + Z_{e-1})^{-1} & \cdots & 0 \\ \vdots & ⋰ & \vdots & \vdots & \vdots & \ddots & \vdots \\ 0 & \cdots & 0 & 0 & 0 & 0 & (R_{s-M} + Z_{e-M})^{-1} \end{bmatrix} \tag{17.46}$$

From this, one can calculate the voltage at all ports from the excitation current source. The impedance matrix is then computed from the inverse function of $\mathbf{Y}'$ by:

$$\mathbf{Z}' = (\mathbf{Y}')^{-1} \tag{17.47}$$

Most often, one of the most desired parameters of a mixer design, which in this case is a single-ended diode topology, is low conversion loss (L) which is defined as the conversion loss of the intrinsic mixer (L') and all losses in the series resistance of the diode at the IF output (K_0) and input signal (K_1) frequencies by:

$$L = K_0 L' K_1 \tag{17.48}$$

where

$$K_0 = \frac{\mathcal{R}e\{Z_{e0} + R_{s0}\}}{\mathcal{R}e\{Z_{e0}\}} \tag{17.49}$$

$$K_1 = \frac{\mathcal{R}e\{Z_{e1}\}}{\mathcal{R}e\{Z_{e1} + R_{s1}\}} \tag{17.50}$$

$$L' = \frac{1}{\mathcal{R}e\left\{\delta V_0 \delta I_0^*\right\}} \frac{\left|\delta I_1'\right|^2 |Z_{e1} + R_{s1}|^2}{4\mathcal{R}e\{Z_{e1} + R_{s1}\}} \tag{17.51}$$

Substituting L', K_0 and K_1 into L and simplifying, one obtains the mixer conversion loss as follows:

$$L = \frac{1}{4\left|Z_{01}'\right|^2} \frac{\left|Z_{e0} + R_{s0}\right|^2}{\mathcal{R}e\left\{Z_{e0}\right\}} \frac{\left|Z_{e1} + R_{s1}\right|^2}{\mathcal{R}e\left\{Z_{e1}\right\}} \tag{17.52}$$

Similarly, the conversion loss for each other mixing frequency output i to the input frequency j is given by:

$$L_{i,j} = \frac{1}{4\left|Z_{ij}'\right|^2} \frac{\left|Z_{e,i} + R_{si}\right|^2}{\mathcal{R}e\left\{Z_{ei}\right\}} \frac{\left|Z_{e,j} + R_{sj}\right|^2}{\mathcal{R}e\left\{Z_{ej}\right\}} \tag{17.53}$$

z_{ij}' is the matrix element in the ith row and j^{th} column, which should be computed by applying the large-signal technique first and subsequently forming a conversion matrix using small-signal analysis.

17.6 Planar Electromagnetic Simulation

These software tools are intended to solve for the S-parameters of arbitrarily shaped microstrip, CPW and other planar structures [4]. Usually, the circuit conductors are divided into subsections and the method of moments (MoM) is used for the electromagnetic analysis [5]. Examples of this type of simulator are *em*™ (Sonnet), AXIEM™, and Momentum™. Often, many metal and dielectric layers can be handled, but these are assumed to be planar, and so the term 3-D planar has been coined. These simulators cannot normally analyse true 3-D structures, such as microstrip-to-waveguide transitions. 3-D planar simulators are used extensively in RFIC and MMIC design to analyse lumped inductors and nonstandard microstrip structures such as coupled bends and meandered lines.

A circuit component is analysed in terms of S-parameters by subdividing the layout geometry into subsections that are much smaller than the guided wavelength. Compared with regular circuit theory, which also provides linear parameters, the planar electromagnetic technique offers a microscopic view by analysing the field distribution that is really present in the design layout. Figure 17.8 shows the layout of a via hole which is subdivided by the simulator into

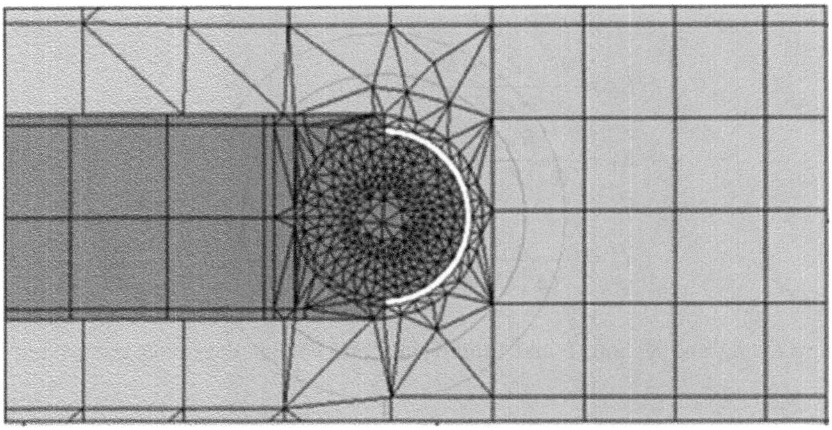

Figure 17.8 Via hole meshing.

many mesh elements. The mesh can have any shape, but is most often rectangular or triangular. The dimensions of the mesh element should be much smaller than the guided wavelength, which in practice ranges from 1/30 to 1/10 of the guided wavelength, depending on the required accuracy, simulation time and available computer memory. A smaller mesh size gives very accurate field distribution but requires a long computing time to cover the area occupied by the mesh. On the other hand, a large mesh may be applied but would give limited accuracy. A good solution is to use a mesh that is nonuniform. Figure 17.8 shows an example in which the software uses a very fine mesh around the via hole, while the signal conductor and the ground plane are meshed with much larger square elements.

The method of moments (MoM) technique is usually employed for planar electromagnetic simulation. The problem is formulated by determining the approximate solution of Maxwell's equations with boundary and initial conditions. The basic concept of the technique is to transform the continuous integral or differential form of Maxwell's equations into matrix form. Assuming that the unknown solution is given by a continuous function, f, this function is approximated by:

$$f \approx \sum_n \alpha_n f_n \tag{17.54}$$

where α_n is the coefficient of the expansion or sometimes called the basis function $f_n(x)$. With the chosen expansion function, the solution is determined by solving α_n. It is shown that the summation is truncated to N terms for algorithm implementation. The general problem formulation is in the form:

$$\mathcal{L}(f) = g \tag{17.55}$$

where $\mathcal{L}$ is a linear operator, g is a known function and f is the unknown function that needs to be determined. Next, we define a set of weighting functions, $w_1, w_2, \ldots, w_N$ in the range of $\mathcal{L}$.

Taking the inner product to the right-hand-side term in Equation (17.54) with the weighting functions, one obtains:

$$\sum_n \alpha_n \langle w_m, \mathcal{L}f_n \rangle = \langle w, g \rangle \tag{17.56}$$

In terms of mathematics, the output from the inner product shown in the right-hand-side of (17.56) is a projection vector of two vectors under the operator. The equation system can be written in the matrix form:

$$\mathbf{A} \cdot \boldsymbol{\alpha} = \mathbf{g} \tag{17.57}$$

where

$$\mathbf{A} = \begin{bmatrix} \langle w_1, \mathcal{L}f_1 \rangle & \langle w_1, \mathcal{L}f_2 \rangle & \cdots \\ \langle w_2, \mathcal{L}f_1 \rangle & \langle w_2, \mathcal{L}f_2 \rangle & \cdots \\ \vdots & \ddots & \vdots \end{bmatrix} \tag{17.58}$$

$$\boldsymbol{\alpha} = \begin{bmatrix} \alpha_1 & \alpha_2 & \cdots \end{bmatrix}^{\mathrm{T}} \tag{17.59}$$

$$\mathbf{g} = \begin{bmatrix} \langle w_1, g_1 \rangle & \langle w_1, g_2 \rangle & \cdots \end{bmatrix}^{\mathrm{T}} \tag{17.60}$$

The expansion function f_{n} for the MoM is a pulse function defined by:

$$f_{\mathrm{n}} = \begin{cases} 1 \text{ in the region of interest} \\ 0 \text{ otherwise} \end{cases} \tag{17.61}$$

The testing or weighting function for the MoM is the Dirac delta function. The advantage of using the Dirac function as a testing function is that it is simple to compute since the inner product of the Dirac function and the vector element is the sample point of vector **g**. We can find the unknown coefficients by inverting the matrix, if **A** in non-singular matrix:

$$\boldsymbol{\alpha} = \mathbf{A}^{-1} \cdot \mathbf{g} \tag{17.62}$$

The coefficient vector $\boldsymbol{\alpha}$ is the approximate solution of the unknown function f. This solution is applied to compute the scattering parameters.

Example: Determine the surface density $\sigma(x, y)$ of an infinitesimally thin metal plate lying in the x-y plane. The plate is centred at the origin as shown in Figure 17.9. It is a square plate with side (a) of 10 cm. The potential of the plate is 10 V.

Generally, the electrostatic potential function of a plate at $z = 0$ is:

$$\Phi(x, y) = \frac{1}{4\pi\varepsilon} \int_{-a/2}^{a/2} \int_{-a/2}^{a/2} \frac{\sigma(x', y')}{\sqrt{(x - x')^2 + (y - y')^2}} \mathrm{d}x' \mathrm{d}y' \tag{17.63}$$

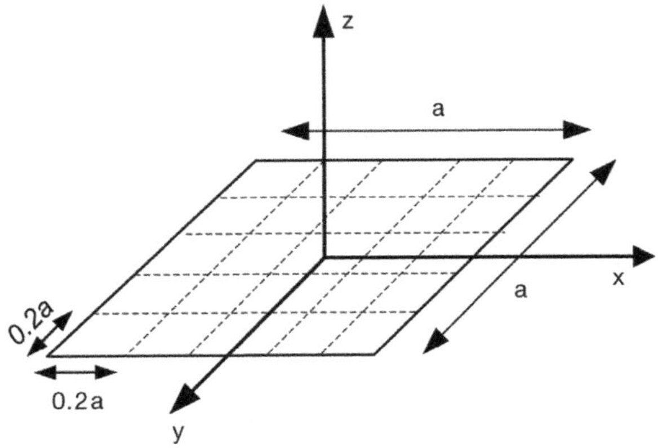

Figure 17.9 Square plate with 25 mesh elements.

The metal plate is assumed to be a perfect conductor and hence is an equipotential plane, so:

$$\Phi(x, y) = V_0 \tag{17.64}$$

The surface density function is the unknown function, that is:

$$f(x, y) \equiv \sigma(x, y) \tag{17.65}$$

The plate is divided into 25 square mesh elements, that is, $\Delta S_1, \Delta S_2, \ldots, \Delta S_{25}$. Each element has a dimension of $0.2a$. The unknown surface density $\sigma(x, y)$ can be approximated with the 2-D basis function $f_n(x, y)$ by:

$$\sigma(x, y) \approx \sum_{n=1}^{N} \alpha_n f_n(x', y') \tag{17.66}$$

The basis function is a 2-D pulse function which is defined by:

$$f_n(x', y') = \begin{cases} 1 \text{ at } \Delta S_n \\ 0 \text{ at } \Delta S_m, m \neq n \end{cases} \tag{17.67}$$

Hence, if we rewrite the integral equation in terms of the general formulation defined in Equation (17.55), one obtains:

$$\mathcal{L}(f) = \frac{1}{4\pi\varepsilon} \int_{-a/2}^{a/2} \int_{-a/2}^{a/2} \frac{f(x', y')}{\sqrt{(x - x')^2 + (y - y')^2}} \mathrm{d}x' \mathrm{d}y' \tag{17.68}$$

and

$$g(x, y) = V_0 \tag{17.69}$$

The weighting function for each mesh element (ΔS_m) is a 2-D Dirac function:

$$w_m = \delta(x - x_m)\delta(y - y_m) \tag{17.70}$$

Each element of the matrix **A** is computed from:

$$A_{mn} = \frac{1}{4\pi\varepsilon} \int_{-b/2}^{b/2} \int_{-b/2}^{b/2} \frac{1}{\sqrt{(x_m - x')^2 + (y_n - y')^2}} \mathrm{d}x' \mathrm{d}y' \tag{17.71}$$

Thus,

$$A_{nn} = \frac{4}{\pi\varepsilon} \ln(1 + \sqrt{2}) \tag{17.72}$$

$$A_{mn} = \frac{4}{\pi\varepsilon\sqrt{(x_m - x')^2 + (y_n - y')^2}}, m \neq n \tag{17.73}$$

The boundary condition provides:

$$\mathbf{g} = [10,\ 10,\ 10,\ 10,\ 10]^T \tag{17.74}$$

Hence, we can find the unknown coefficient which is the desired source density function by substituting into Equation (17.57). From this example, we have seen that the matrix size depends on the number of mesh elements. Most electromagnetic modelling software requires a computer with a high-speed processor and a large memory to handle the matrix calculation problem. More importantly, it is essential to be aware that results obtained from the simulator are approximated and careful interpretation on the results should be noted, based on experience of the limitations of EM methods.

17.7 3-D Electromagnetic Simulation

True 3-D structures such as waveguide-to-microstrip transitions and chip packages require software that uses techniques such as the finite element method (FEM) or finite-difference time-domain (FDTD) method to analyse completely arbitrary metal and dielectric structures. This type of analysis is extremely demanding of CPU time and memory, and since most RFIC and MMIC components are basically planar in nature, 3-D planar simulation is used more extensively. Notable 3-D simulators include the High-Frequency Structure Simulator (HFSS™), EmPro™, Analyst™ and CST Microwave Studio™. Whilst one might imagine that 3-D electromagnetic modelling is a case of draw-and-simulate, in practice the designer needs a great deal of experience in order to get the very best accuracy from a simulation.

The two main techniques, FDTD and FEM, have different advantages and disadvantages. Commercial software which is based on the FDTD method includes CST Microwave Studio™ and Remcom XFDTD. In the FDTD method, the problem space to be solved is divided into small boxes or segments called Yee cells [6], which have an electric field along the edges and a magnetic field on the faces. The subdivided region is called a mesh or grid, and the degree of accuracy depends on the mesh size, although accuracy is traded-off against computation time and stability. The problem is also subdivided in the time-domain, using finite difference techniques along with the boundary or initial conditions to find a solution iteratively. The disadvantage of the FDTD method is the numerical dispersion and reduced accuracy when applied to nongeometrical shapes. On the other hand, the FEM is more general since it allows for arbitrarily shaped structures. The method provides a solution of the partial differential equations (PDEs). The physical model is divided into small and finite subsections called 'elements'. With various shapes of elements, the FEM is more versatile than alternative methods, and many engineers use FEM solvers as their design tools to achieve high-accuracy solutions. Commercial software based on the FEM engine includes EMPro™ from Keysight Technologies, the RF module in COMSOL Multiphysics™, and HFSS™ from Ansoft.

17.8 Integrated Circuit Simulation and Layout

The modelling and design of RFICs and MMICs makes use of all of the aforementioned simulation techniques, but a distinctive feature of IC design is that all of the active and passive components are fabricated by the same fabrication process. This means that the layout of every component is strictly controlled by the manufacturer (e.g., a foundry), and the design and layout must make use of the appropriate models and layout structures. The design environment needs to be customised using a process design kit (PDK) in order that all these models and layout elements are available from a library or palette. RFIC and MMIC layouts use the CALMA™ GDSII or OASIS industry standard format for the 'tape out' stage of sending a set of designs for fabrication. Layout packages are usually available within the simulation tools such as ADS™. However, to create a complete mask set reticle often requires a tool such as Cadence™ or Mentor Graphics™. For serious integrated transceiver designs, RF EDA packages often provide their simulator 'engines' for use with these industry-standard interfaces. Whilst IC manufacturers provide a PDK, it is important to be aware of how the models are typically created in order to understand their limitations compared to the wide range of other component modelling approaches that can be used.

17.8.1 Process Design Kits for RFIC and MMIC Design

The various types of microwave EDA package have been briefly described already. In a fully integrated EDA package for RFIC and MMIC design, there is seamless integration of the schematic and layout. In order to achieve this level of design integration, the simulator must be customised to the IC manufacturer or foundry. This means that the models and layout library components for the specific process that is going to be used must form a core part of the simulator's element library.

The creation of a new process design kit starts with the layout of a test, or characterisation, mask-set. This will include a large number of active and passive components and a number of test structures for process monitoring, as shown in Figure 17.10. The layout of the active devices will rely on experience, engineers' intuition and perhaps physics-based device simulations; the mask-set will certainly undergo a few iterations during the development of the process. Wafer testing and process characterisation starts with DC and low-frequency tests on the PCM (process control monitor) components. These tests yield vital process information such as carrier concentration, doping profile, sheet resistivity, contact resistance, metal conductivity,

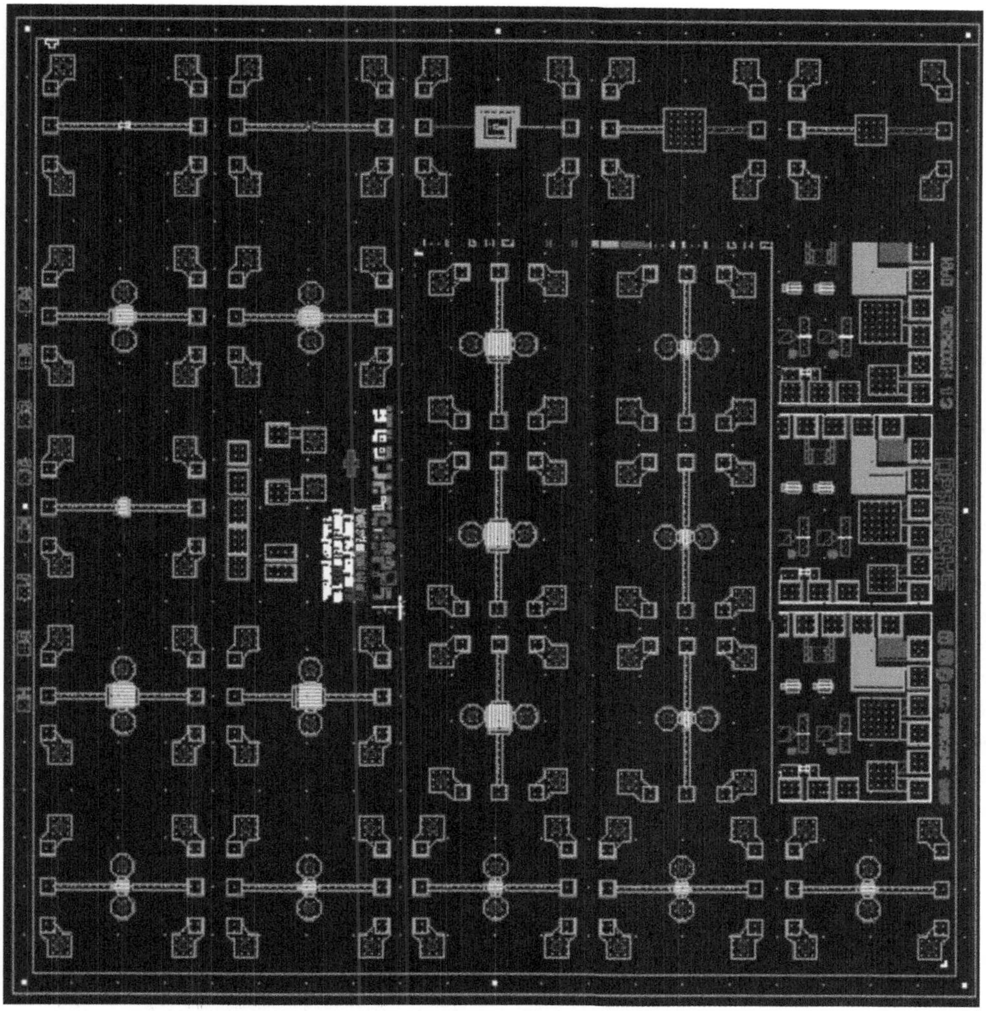

Figure 17.10 Part of a typical test mask for PDK development.

and dielectric permittivity. From these results, key enhancements and changes can be made to various stages of the process to centre the process for optimum yield. RF-on-wafer (RFOW) testing is conducted to investigate the performance of the active and passive components, as described in Chapter 18.

17.8.1.1 Passive Component Equivalent Circuit Models

Measured data is not ideal as it cannot be easily used for optimisation, and designers will need values between those that have been measured. Equivalent circuit models provide a way of producing a variable component value; the parasitic elements are derived from the measurements and curve-fitting used to interpolate between the measured data. A parameterised library component (often referred to as a PCell) is one in which the designer chooses a key parameter, such as the number of turns of an inductor, after which the model is programmed into the EDA tool to create the layout of that component and calculate the values for all the elements in the equivalent circuit model. There may well be underlying assumptions and approximations in these models. Once again, lumped spiral inductor models stand out as being potentially inaccurate above the self-resonant frequency, which is why tools such as Blink™ from Sonnet Software® are important.

17.8.1.2 Active Device Models

From measurements of a wide range of device geometries, a scalable bias-dependant model is reasonably easy to generate. The geometry refers to the number of gate/emitter fingers, the length and width of each, the spacing, and so forth, as described in Chapter 8. The basic device within – the intrinsic device – is not different, and hence the parameters of the small-signal model can easily be scaled with the area (gate width). For a HEMT, the intrinsic parameters scale closely with size: g_m is proportional to the gate width; the capacitances C_{gs}, C_{ds}, C_{dg} are proportional, and the resistance R_{ds} is inversely proportional. The other model parameters – the extrinsic parameters – do not scale so simply and are extracted from device measurements on a range of geometries, or by using distributed or electromagnetic models. For large-signal models, every parameter will vary with input/output signal level and a model typically has more than 30 parameters to find. Assumptions in the model (say, tanh dependence of I_{ds} on V_{gs}) may make it nearly impossible to get a good fit at all bias points. Fortunately, parameter extraction software exists which can help automate the process (e.g., IC-CAP, IVCAD, UTMOST and others). Behavioural models have gained much attention and X-parameters, which are the large-signal extension of S-parameters, are described briefly in Chapter 18.

17.8.1.3 Integrating the Model Library into the EDA Software

Once suitable models are found for the components, it is necessary to programme the model, link it into the simulator, create a layout representation, and set up a customised user interface with customised symbols, menus, palettes and icons. The procedure is entirely dependent on the specific EDA package being used. For example, it may be necessary to programme the model in the 'C' language, compile the model, and then link it into the simulator. Files describing the icons, library palette, help files and so forth must also be created in the correct format. As the amount of work involved is extremely significant, if a design house wishes to use a certain foundry (for reasons of cost, performance, etc.) then they would have no choice but to adopt the EDA tools supported by that foundry. A design house using different Foundries may then have a large EDA budget, and various flexible licensing payment schemes have been introduced to address this. Figure 17.11 shows the completed layout of a GaN power amplifier MMIC.

17.8.1.4 Model Limitations

Foundry PDK models are the most reliable means of simulating a circuit since they are based on measurements of components made with the relevant RFIC or MMIC process. However,

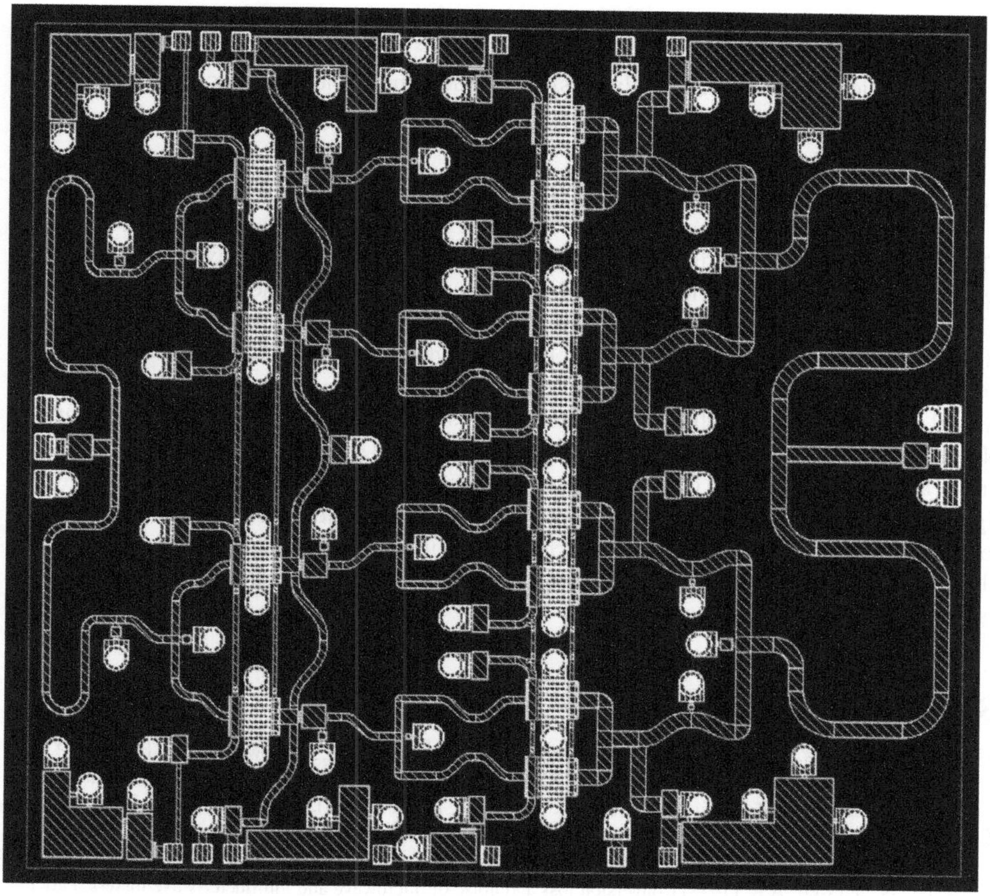

Figure 17.11 Layout of a 25 W X-band GaN PA ready for 'tape out'. Illustration courtesy of Plextek RFI. Used with permission; all rights reserved.

the models cannot be expected to cater for every structure that is encountered in a circuit, and model uncertainties become increasingly important for designs above 20 GHz. There are some potential limitations that the designer must be aware of:

- Differences from the test mask layout: for example, a capacitor might be used with different feed line widths, or with multiple feeds.
- Distributed effects not incorporated into the model: for example, most spiral inductor models are lumped models and are not accurate at frequencies above self-resonance.
- Measurement uncertainty: as well as the vexed questions surrounding calibration techniques, a small probe placement inconsistency during measurement can result in significant phase discrepancy at >30 GHz.
- Grounding problems: even mutual inductance between neighbouring via-hole grounds has been shown to be potentially significant.

- Coupling between components: all components are considered isolated from one another unless electromagnetic simulations are used.
- Active device model limitations: large-signal models should not be expected to be accurate under all bias and signal conditions.

17.9 Conclusion

This chapter has described the various EDA techniques used in microwave and millimetre-wave circuit design, including an examination in some detail at the theory behind them. It has been shown that RFIC and MMIC library elements are derived from an extensive characterisation programme requiring a coherent programme of measurements and equivalent circuit parameter extraction. The result is a set of models that are easy to use and represent the typical components whilst allowing statistical analysis to be performed for yield optimisation. Nevertheless, the models are based on measurements on a set of known standard structures, and great care must be exercised if using nonstandard layout features, such as multiport capacitors, tightly meandered lines, and so forth. Electromagnetic simulation packages provide a useful tool for analysing such nonstandard elements and for investigating coupling effects. There are also some newer techniques such as envelope simulation methods that have not been included, as these topics are beyond the scope of this book.

References

1. Aaen, P.H., Bridges, D., Zhang, L., Johnson, E., Pla, J.A., Barbieri, T., Wood, J., Snowden, C.M., Everett, J.P., and Kearney, M.J. (2012) Multiphysics modeling of RF and microwave high-power transistors. *IEEE Transactions on Microwave Theory and Techniques*, **60** (12), 4013–4023. doi: 10.1109/TMTT.2012.2224366.
2. Kundert, K.S. and Sangiovanni-Vincentelli, A. (1986) Simulation of non-linear circuits in the frequency domain. *IEEE Transactions on Computer-Aided Design*, **5** (4), 521–535.
3. Held, D.N. and Kerr, A.R. (1978) Conversion loss and noise of microwave and millimeter-wave mixers: Part 1 – Theory. *IEEE Transactions on Microwave Theory and Techniques*, **MTT-26** (2), 49–55.
4. Swanson, D.G., Jr and Hoefer, W.J.R. (2003) *Microwave Circuit Modeling Using Electromagnetic Field Simulation*. Artech House Publishers.
5. Tsang, L., Kong, J.A., Ding, K.H., and Ao, C.O. (2001) *Scattering of Electromagnetic Waves*. John Wiley & Sons, New York.
6. Yee, K. (1966) Numerical solution of initial boundary value problems involving Maxwell's equations in isotropic media. *IEEE Transactions on Antennas and Propagation*, **14**, 302–307.

18

Measurement Techniques

18.1 Introduction

This chapter contains a brief description of the main types of test equipment that are encountered in microwave and millimetre-wave measurements. A vast range of test equipment is available because there are many different forms of measurement that need to be made – from complete systems to individual devices on wafers, and from simple DC testing through to 100 Gb s^{-1} testing of millimetre-wave and terahertz components and subsystems. An additional issue is that, in the manufacturing domain, it may take significant time to make the full set of measurements required to check that a product meets quality assurance standards. So, there are many dedicated measurement systems available to suit rapid testing of certain products; for example, a dedicated system for measuring wireless handsets quickly in the factory, often using a modular architecture based on the PXI or similar bus for rapid data transfer. In a wireless communications R&D laboratory, the following equipment is likely to be encountered:

- Oscilloscopes
- Arbitrary waveform generators
- LCR meters and component analysers
- Power meters
- Spectrum analysers and vector signal analysers
- Signal generators: CW, analogue and digital modulation
- Vector network analysers (VNAs)
- Noise figure analysers
- RF-on-wafer (RFOW) probe stations
- Automatic load-pull systems

The following sections provide an outline of what each instrument is for, its basic controls, and some of the pitfalls that must be avoided when using them. Equipment from Keysight Technologies is featured as they are the world's leading test and measurement equipment company with more than 75 years of history, previously being part of Hewlett Packard and

Microwave and Millimetre-Wave Design for Wireless Communications, First Edition.
Ian Robertson, Nutapong Somjit and Mitchai Chongcheawchamnan.
© 2016 John Wiley & Sons, Ltd. Published 2016 by John Wiley & Sons, Ltd.

then Agilent Technologies. The controls on models from different manufacturers may vary enormously in layout but the main functions are generally similar. After some practice on different equipment types, it should be possible to use the basic functions of most well-designed equipment without referring to the manuals. This is increasingly the case as modern equipment usually includes a touch-screen or soft-key menu system, so that the number of buttons is reduced considerably and the more complex functions are out of sight in the menu system. The following is a general guide only, and if in doubt the manual should be consulted. The biggest difference is found with modular instruments, which are generally controlled entirely from within software such as Keysight VEE, LabVIEW™ from National Instruments™ or The Mathworks' MATLAB™ with optional Instrument Control Toolbox™ or other add-ons.

18.2 The Oscilloscope

All electronic circuits have voltages that vary with time – either with analogue waveforms or digital data. The oscilloscope ('scope') is excellent for analysing these signals because it measures instantaneous voltage and displays a graph of it as a function of time. In other words, a scope displays waveforms, enabling many measurements to be made, including the period or frequency of the signal. For many years scopes used a cathode ray tube (CRT), and much of the terminology related to how CRTs create an image by scanning an electron beam remains in use today, even though modern instruments use newer types of display such as LCDs. In some measurement applications the performance limitations of a lower specification instrument may become very evident. In particular, an economy scope has a limited maximum number of samples it can store, due to a modest 'memory depth'. This means that the number of measured waveform samples it can process is rather limited. The low effective sampling rate will become problematical if attempts are made to measure modulated waveforms, since the combination of a high-frequency carrier, but low-frequency envelope, requires many samples. Most oscilloscopes, including the Keysight DSOX3000 series shown in Figure 18.1, have an FFT function in order to display the spectrum of the input signal, but the quality of the results depends very much on the bandwidth of the scope and the memory depth. A feature of most Keysight scopes is that the data can be transferred to the Vector Signal Analyser (VSA) software, running on the instrument or a controller. Figure 18.2 shows an example screenshot with constellation, spectrum, and so forth. The software has powerful analysis function for a vast range of modern wireless standards such as ZigBee, GSM, 3G, LTE, Wi-Fi, and RFID.

The input resistance of oscilloscopes is usually high (1 MΩ is standard) but with a large capacitance in parallel (10–20 pF is typical). Furthermore, if a coaxial cable is used to connect the circuit under test to the scope, the capacitance of the cable will load the circuit as well (the order of 20 pF per foot (30 cm) is typical). To avoid these loading effects, passive probes are commonly used. A 'X10' passive probe has a 9 MΩ series resistor, giving a voltage division factor of 10 due to the potential divider effect with the scope's 1 MΩ input resistance. In addition, so that the 9 MΩ does not form a low-pass filter with the scope's input capacitance, it is bypassed with a capacitance. A small tuneable capacitor in incorporated within the probe or its connector for fine-tuning the response. The process of compensating the probe means tuning this capacitor, usually by measuring a good square-wave reference signal (e.g., the CAL output) and optimising the display to match. The 'X10' voltage division factor of the probe is

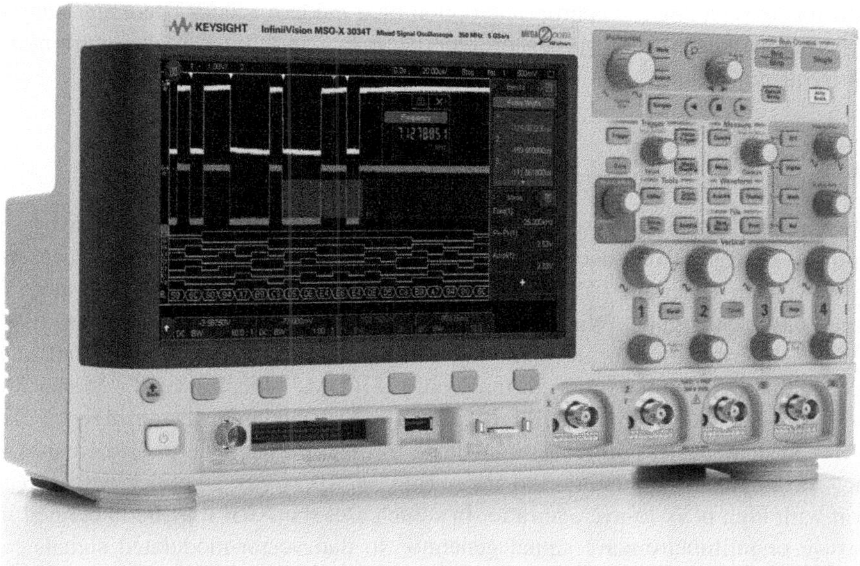

Figure 18.1 Keysight DSOX3000 series digital oscilloscope. © 2016 Keysight Technologies, Inc. Reproduced with permission.

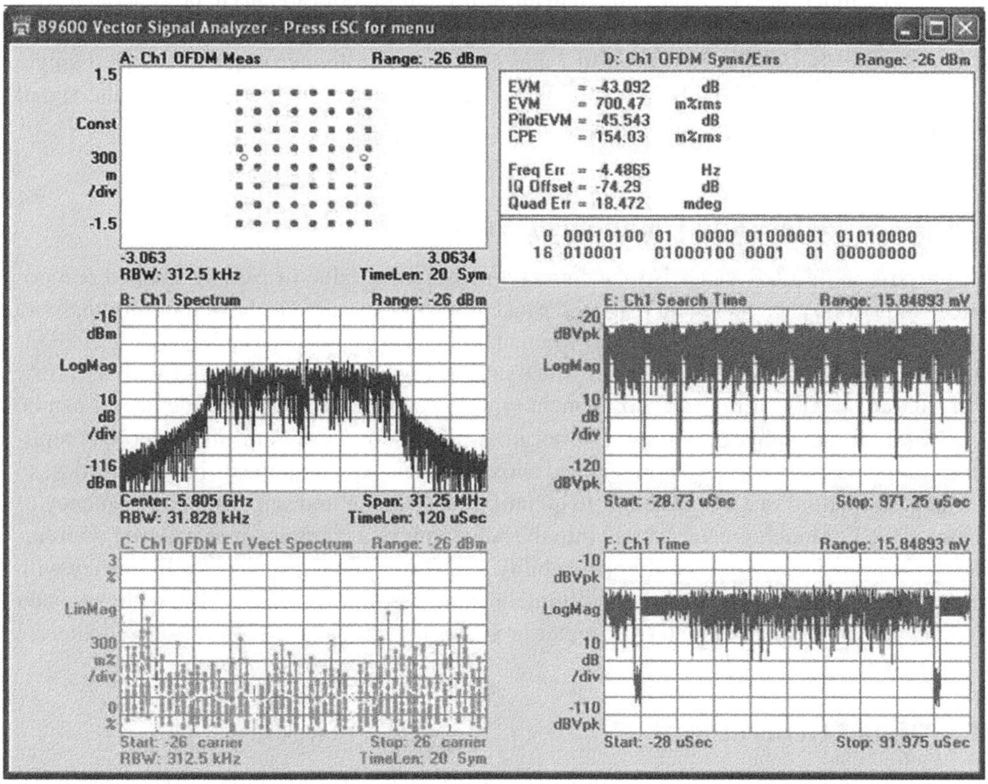

Figure 18.2 Example screenshot of the Keysight VSA software. © 2016 Keysight Technologies, Inc. Reproduced with permission.

accounted for by selecting the appropriate option on a modern scope. There are many other types of probe, including active probes, differential probes, multipoint probes, current probes and high-frequency probes.

18.3 Function Generator and Arbitrary Waveform Generator

Function generators are widely used for low-frequency test signals such as sine waves, square waves, pulses, triangle waves and ramps. For wireless communications test, a dual-channel arbitrary waveform generator (AWG or 'ARB') is an essential means of generating *I* and *Q* analogue baseband signals. With appropriate software such as MATLAB, LabVIEW and SystemVue, *I–Q* waveforms for the full range of wireless communications standards can be generated, including multitone waveforms such as OFDM. With Keysight's Signal Studio, real-world signals can be emulated for IEEE and other standards. Increasingly, there is a need to link system and circuit-level simulations to measurements on physical hardware in real-time; for example, simulating and measuring an RF power amplifier with a realistic waveform with high peak-to-average ratio. In some cases the ARB may be integrated inside a microwave or millimetre-wave signal generator so that vector-modulated signals can be generated conveniently and with high precision. The key parameters that define an ARB's capabilities are the sample rate (bandwidth), the number of bits-per-sample (resolution), and the maximum sequence length (waveform memory). When emulating real-world signals the waveform memory must be large enough to allow pseudo-random signals to be generated with a long sequence. Many ARBs, like the one shown in Figure 18.3, can be linked and triggered off one another in order to synchronise *I* and *Q* or generate longer sequences than a single instrument is capable of. Without a sufficiently long sequence, the periodicity of the signal becomes evident in the spectrum of the signal.

18.4 LCR Meters and Component Analysers

The purpose of an LCR meter is to accurately measure the value of passive lumped components – inductors, capacitors and resistors. Most precision instruments have four terminals, so that the leads can be fully calibrated out when making measurements. The Keysight E4980AL 20 Hz to 1 MHz precision LCR meter illustrated in Figure 18.4 can be used with a variety of test fixtures so that a range of component types – such as surface-mount devices – can be measured accurately. Specialist test fixtures are available for measurements on more complicated components such as transformers and active devices. All components have parasitics, so an important feature of LCR meters is to quantify the *Q*-factor and self-resonant frequency of components. For much greater insight into the behaviour of a component or material with frequency, an impedance analyser offers the ability to measure impedance versus frequency with great precision, and has the ability to accurately characterise from mΩ to MΩ, with precision beyond what is possible with an LCR meter or vector network analyser (as discussed later).

18.5 Signal Generators

The RF signal generator is used to provide a test signal, which may be a pure sine wave (continuous wave, CW) or may be modulated. Increasingly, signal generators are required

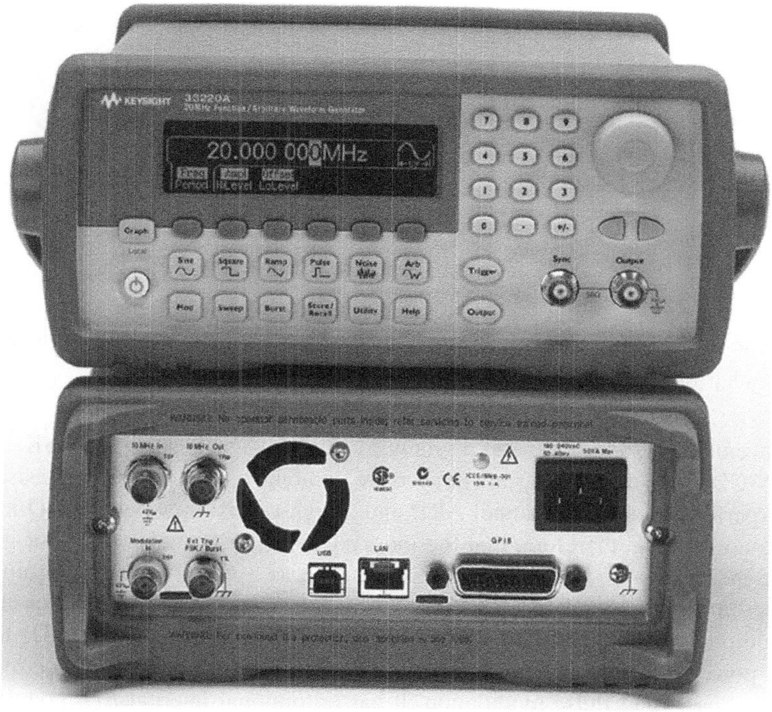

Figure 18.3 Keysight 33220A function/arbitrary waveform generator. © 2016 Keysight Technologies, Inc. Reproduced with permission.

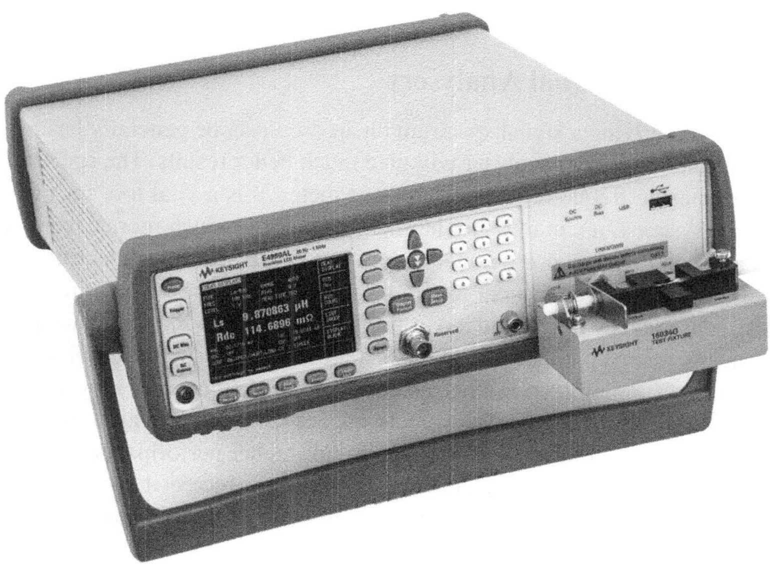

Figure 18.4 Keysight precision LCR meter with test fixture for surface-mount devices. © 2016 Keysight Technologies, Inc. Reproduced with permission.

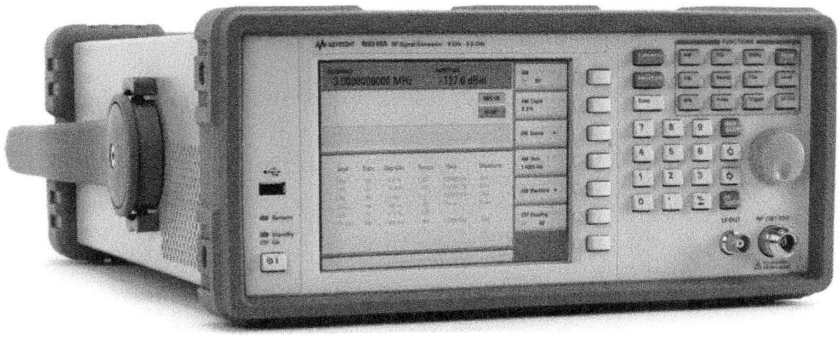

Figure 18.5 Keysight N9310A 9 kHz to 3 GHz signal generator. © 2016 Keysight Technologies, Inc. Reproduced with permission.

that can conveniently produce general digitally modulated signals such as QPSK, 16-QAM, GSM and W-CDMA. Such an instrument needs to combine a high-purity synthesised source with a high-performance, dual-channel ARB and a high-precision wideband *I-Q* modulator. It should be no surprise, therefore, that the most advanced instruments (e.g., for generating 5G signals) are expensive. At lower frequencies, such as 3 GHz, basic test equipment has become much more economical as RFICs and ASICs have become available for critical RF front-end components. The N9310A, shown in Figure 18.5, operates from 9 kHz to 3 GHz, with output power controllable from −127 to +13 dBm. It has built-in analogue modulation for AM, FM, Phase, and Pulse modulation. It has an optional internal *I-Q* modulator, with 40 MHz bandwidth. This gives a powerful capability for experiments and project work on wireless systems. Note that the *I-Q* inputs have 1 V maximum DC levels, and so extreme caution must be exercised or the *I-Q* modulator will be damaged.

18.6 Spectrum and Signal Analysers

The FFT method of deriving a signal spectrum on an oscilloscope generally has many limitations, and a dedicated spectrum analyser will give much better results. The spectrum analyser is essentially a tuneable frequency-selective power meter. It is a vital tool for measuring the output signals of systems; it can measure harmonic distortion much more accurately than an oscilloscope, and can measure noise down to very low levels. A swept spectrum analyser, as illustrated in Figure 18.6, is equivalent to a radio receiver with the LO continually sweeping across the desired frequency range, and the instrument displays on the screen what it measures in a certain narrow received band around each frequency. The simplified block diagram in Figure 18.6 shows a single mixer, but because it is required to cover a very wide frequency range it is possible to have as many as four frequency conversion stages in a swept spectrum analyser in order to avoid spurious responses. For higher performance and faster screen updates, especially at low-resolution bandwidth settings, it is increasingly common to have a digital IF in a spectrum analyser and a combination of sweeping the LO and using a FFT may be used.

The received band of the instrument is determined by the IF or baseband filtering inside the spectrum analyser's receiver, and this parameter is referred to as the *resolution bandwidth.* The shape of the displayed signal is a very strong function of resolution bandwidth, and when

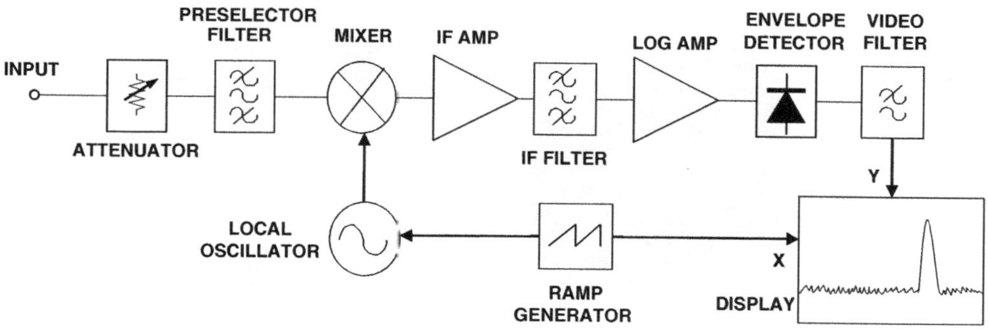

Figure 18.6 Swept frequency spectrum analyser block diagram.

measuring noise, random signals and modulated signals it is important to ensure that the resolution bandwidth is set appropriately to display the signal – both its fine features and its power level. As shown in Figure 18.7, a sine wave in the time domain does not, therefore, lead to just a single 'spike' (δ function) at the signal frequency on the spectrum analyser display: when the receiver is tuned to a frequency far away from the signal, to the left or right, the receiver will pick up noise in the received band. As the instrument sweep reaches the frequency of the signal itself, the trace first rises smoothly as the signal partially enters the received band. The display peaks when the signal is in the centre of the received band, and then falls smoothly as the measurement band sweeps past the signal.

Figure 18.8 shows the Keysight N9320B spectrum analyser, which operates from 9 kHz to 3 GHz. The main control keys are to select the FREQUENCY, SPAN and AMPLITUDE. These allow the user to set the desired measured frequency range (the horizontal axis), and the amplitude scale of the display (the vertical axis). The amplitude scale will normally use decibel units (as described in the next section). The 'BW/AVG' key allows the user to set the resolution bandwidth parameter (RBW) and apply averaging to the measurement. The

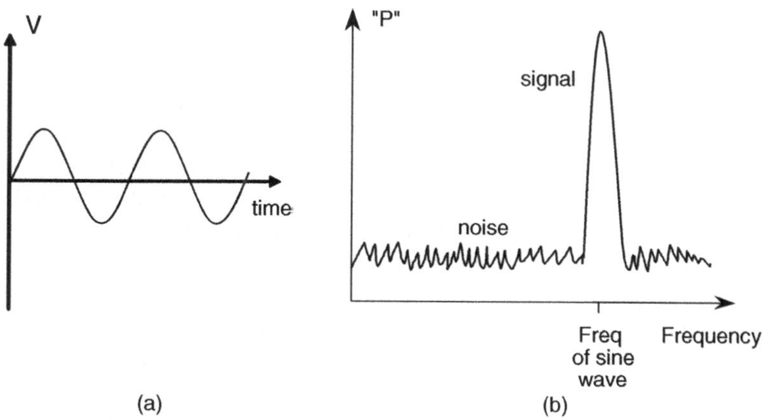

Figure 18.7 (a) Oscilloscope (time-domain) measurement. (b) Spectrum analyser (frequency domain) measurement of a sine wave signal.

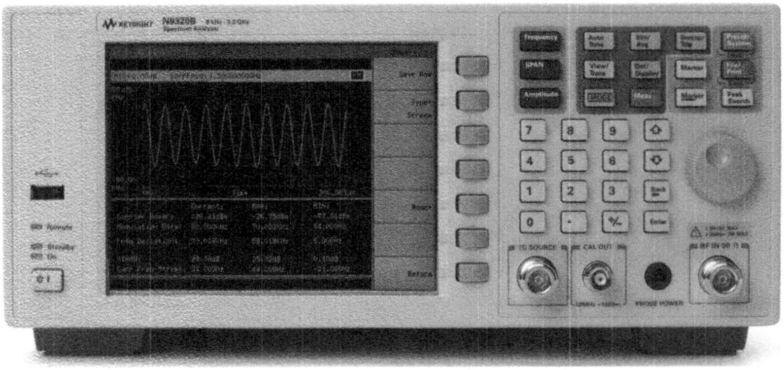

Figure 18.8 Keysight N9320B 3 GHz spectrum analyser. © 2016 Keysight Technologies, Inc. Reproduced with permission.

speed at which the instrument sweeps across the measurement frequency range is related to the resolution bandwidth; the parameters are normally linked by setting the RBW/sweep rate to 'AUTO'. The averaging function is quite different; whatever the setting of sweep rate and RBW, the averaging function takes multiple measurements at that setting and averages the chosen number of trace points to give a result.

The signal input connector is at the bottom right. Next to that is a socket for powering active probes, then a CALIBRATION test signal output (BNC connector), then an optional TRACKING GENERATOR output. The tracking generator option means that the frequency response of components such as amplifiers and filters can be conveniently measured without a separate signal generator and without manually stepping through the frequencies. The instrument has softkeys to the right of the colour LCD, and the menu system provides access to a range of powerful functions that include channel power, occupied bandwidth, adjacent channel power and intermodulation. In spectrum analyser measurements, it is common to encounter the unit 'dBm/Hz', which expresses the fact that noise and random data signals have continuous spectra and the power is spread out over a range of frequencies.

18.6.1 Vector Signal Analysers

Modern communications transceivers have moved steadily away from analogue techniques towards the extensive use of DSP. It is no surprise that test equipment architectures have also followed this trend and, indeed, often push the limits of performance further as R&D engineers require tools to measure advanced new wideband signal formats for emerging standards. In particular, the traditional swept spectrum analyser is rapidly being superseded by the vector signal analyser, which has essentially the same RF front end but quickly converts the received signal to digital form and then performs the remainder of the signal processing in the digital domain, as illustrated in Figure 18.9. The flexibility that working in the digital domain offers means that the signal can be represented as a spectrum, as a constellation, as a baseband waveform pair or eye diagram, and the signal can even be demodulated and decoded so that the packets of data can be inspected. Keysight's VSA software can be running within the signal analyser itself, or the data can be transferred to an external controller. For MIMO and

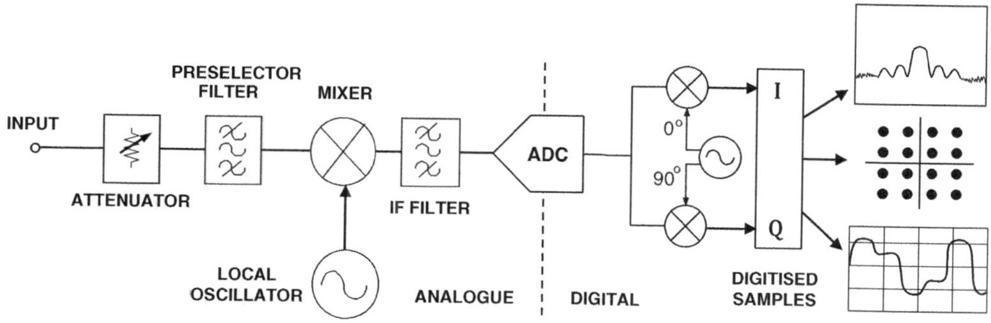

Figure 18.9 Simplified block diagram of a vector signal analyser with digital IF.

other highly sophisticated applications, a modular approach is often required, using multiple downconverter modules in a rack system with a high-speed data bus such as a PXI interface.

18.6.2 Harmonic Mixers

For measurements on components for millimetre-wave and submillimetre-wave systems, such as 5G backhaul links, it is often necessary to employ external waveguide mixers. Harmonic mixers are available to over 1 THz from various suppliers, including Virginia Diodes, Inc. These require a spectrum or signal analyser that has specially configured LO output and IF input ports (which may only be available as optional upgrades at the time of purchase) so that the external mixer module can be used as the front-end mixer of the instrument. There are several challenges with this approach: first, since the LO frequency of the instrument is relatively low, a harmonic mixer is required, and so the conversion loss will be high, leading to reduced dynamic range. Second, the signal level calibration relies on a precise knowledge of the conversion loss of the external mixer (which is managed by the instrument in the case of Keysight's smart mixers). Third, preselectors are rarely available at the very highest frequencies, and so care needs to be exercised to identify spurious signals on the display.

18.7 Vector Network Analysers

To fully understand the operation of a vector network analyser (VNA), it is necessary to have a good background in transmission lines and *S*-parameters. When making high-frequency measurements:

- Proper terminations are required to prevent reflections.
- Impedance is a function of position along a line.

These are the major reasons why microwave engineers have shunned conventional measurements of voltage and current. For many years, measurements of voltage standing wave ratio (VSWR) and impedance were made using slotted lines, as illustrated in Figure 18.10. In this case, a straight section of coaxial line, slabline or rectangular waveguide has a slot machined

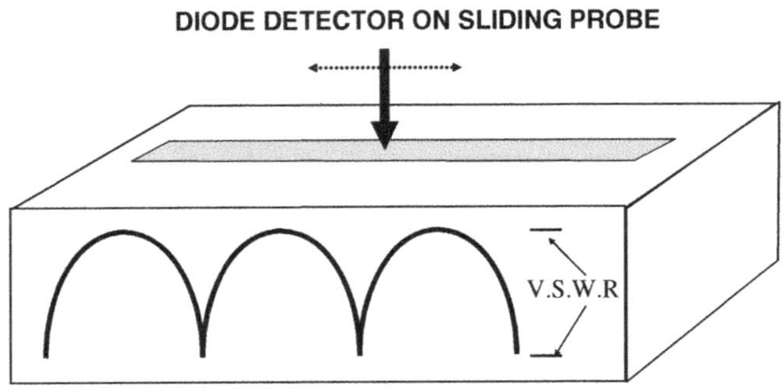

Figure 18.10 Voltage standing wave ratio (VSWR) and impedance measurement with a slotted line.

down one side so that a diode detector can be slid up and down on a precision carriage. This means that the standing wave pattern can be measured directly. With suitable consideration of reference planes, the impedance of an unknown load can be determined. However, making these measurements over a wide range of frequencies is very laborious. A VNA, such as the Keysight PNA-X shown in Figure 18.11, measures the *S*-parameters of a device-under-test (DUT) over a wide frequency range. In order to make accurate measurements it is necessary to be familiar with calibration techniques. Calibration refers to the procedure whereby the unknown reflections and losses from cables, connectors, couplers, bias tees, probes and so forth are determined via measurements on a series of known 'calibration standards'.

Figure 18.12 shows the typical block diagram of a VNA that has full *S*-parameter measurement capability. In order to measure the *S*-parameters of a two-port network without turning the DUT around, the VNA requires hardware on both ports that separates and samples the waves coming out of the DUT (b_1, b_2) from the waves going in (a_1, a_2), as well as a switch to direct the stimulus signal alternately to port 1 and port 2. This can make a VNA rather

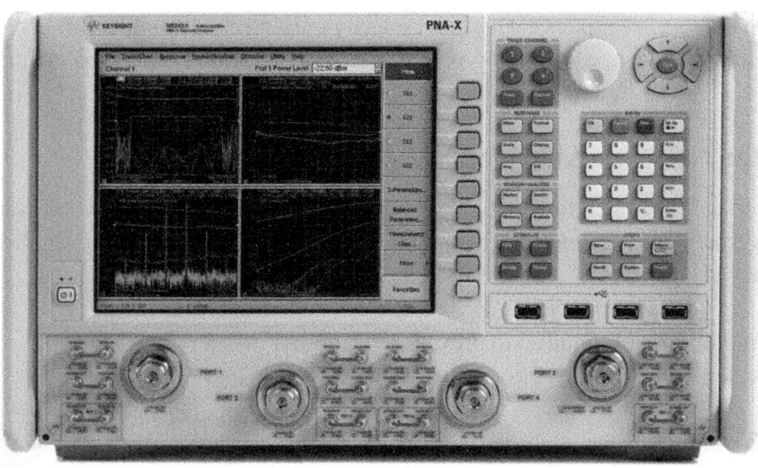

Figure 18.11 Keysight PNA-X vector network analyser. © 2016 Keysight Technologies, Inc. Reproduced with permission.

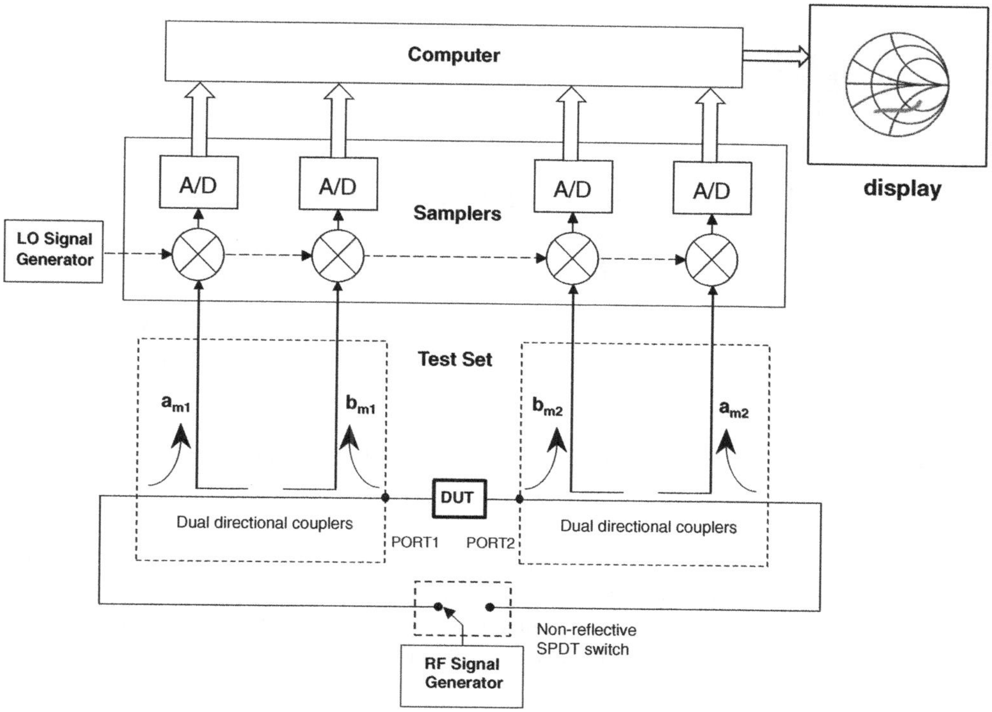

Figure 18.12 Simplified block diagram of a VNA with *S*-parameter test set.

costly, and single-path (transmission/reflection) options or even scalar network analysers are sometimes offered at lower cost.

18.7.1 VNA Calibration

'Calibration' normally refers to the process where the accuracy of an instrument is checked annually by the manufacturer by measuring reference standards such as voltages, power levels and so forth with traceability to national standards held institutions such as NIST, NPL, PTB and EURAMET. In network analyser measurements, however, calibration refers to the day-to-day process of correcting for the unwanted reflections and losses in the measurement system. For example, it may be necessary to make a measurement using a long pair of cables with adaptors on the ends, as shown in Figure 18.13(a). In this case, the instrument itself is not affected, but the cables and adapters certainly need to be calibrated out.

The reflections from adapters, cables, the analyser components themselves and the DUT add constructively and destructively at different frequencies, giving rise to measurement ripple, as shown in Figure 18.13(b). The overall measured reflection coefficient is the vector sum of all the individual components, and any errors will vary with frequency. It is only possible to remove the resulting errors in a measurement by using vector measurements (magnitude and phase) and performing a calibration routine in order to determine the error vectors. All the unwanted reflections and losses must be determined and then calibrated out in order to measure

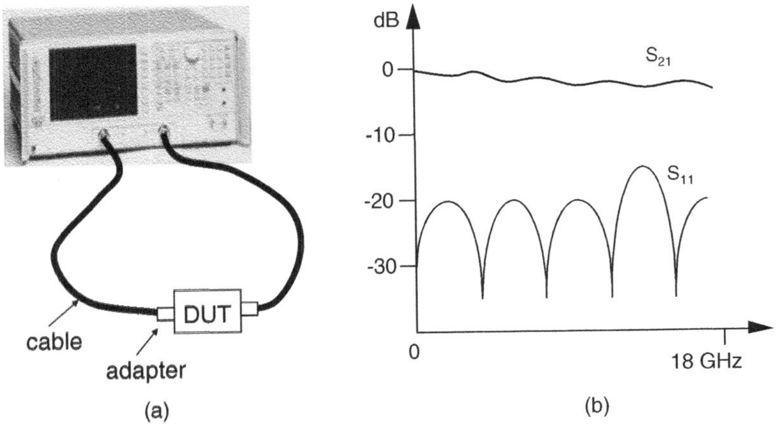

Figure 18.13 (a) Typical VNA measurement of a two-port DUT, using cables. (b) Example of measurement ripple (through line measurement).

the DUT accurately. There are many methods for calibrating a vector network analyser. The simplest to understand is the method using a short circuit, open circuit, 50 Ω load and a through line. These must be precision calibration standards with known characteristics. In the ideal case, these calibration standards would have the following S-parameters:

SHORT	$S_{11} = -1$	
OPEN	$S_{11} = +1$	
LOAD	$S_{11} = 0$	
THRU	$S_{11} = 0$	$\lvert S_{21} \rvert = 1$, angle $= 0°$

By measuring these known standards, the errors in the measurements can be extracted. Once the errors have been calculated (as vectors) they can be extracted from measurements of the DUT. This calibration method is referred to as the 'SOLT' method (short-open-load-thru). In order to analyse how the error correction is achieved, the sources of error must be identified and represented as a mathematical model. Figure 18.14 shows the sources of error for a one-port measurement, and these are numbered as follows:

1. The **directivity** term represents leakage through the non-ideal directional coupler or bridge because it does not have perfect isolation.
2. The **reflection tracking** term represents the loss and electrical length of all the components between the DUT and the sampler.
3. The **source match** term is necessary because the system itself does not have a perfect 50 Ω impedance and there are also reflections before the DUT from cables, adaptors, and so forth.

All these various unwanted signal components are lumped together into what is known as an 'error model'. The one-port error model is shown in Figure 18.15; each signal is a branch in a

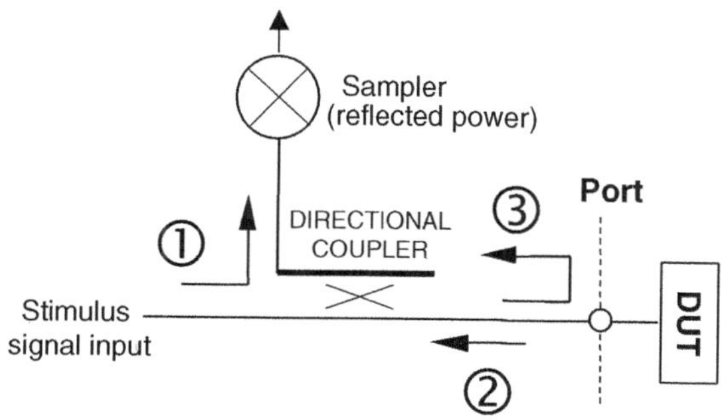

Figure 18.14 Errors in the one-port measurement system.

signal flow graph. The three error terms are:

E_D, the directivity term ①

E_{RT}, the reflection tracking term ②

E_S, the source match term ③

These are all vectors and vary with frequency. They can be found by first connecting the calibration standards one-by-one and recording the measurements in memory, yielding three sets of measured S_{11} for three known values of $S_{11\ DUT}$. When they are all stored, a calibration algorithm can then be used to calculate the three unknown error vectors by solving simultaneous equations (at each measurement frequency). Once E_D, E_{RT} and E_S have been calculated, the data displayed on the screen can be corrected in real time, with the effects of the errors removed.

The calibration algorithm can be determined by referring to the signal flowgraph in Figure 18.15, which shows the error model for a one-port measurement, with the true S_{11DUT} on the right and the measured S_{11} (with errors) on the left. Using Mason's rule gives:

$$S_{11meas} = E_D + \frac{S_{11DUT}E_{RT}}{1 - E_S S_{11DUT}} \tag{18.1}$$

For the LOAD measurement, $S_{11DUT} = 0$, leading to:

$$S_{11meas} = E_D \tag{18.2}$$

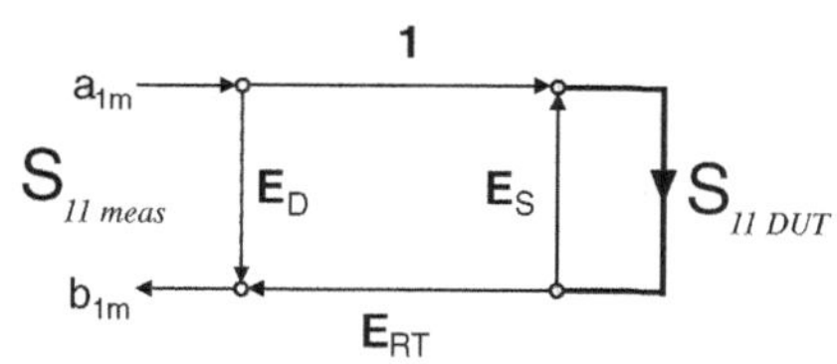

Figure 18.15 One-port error model flowgraph.

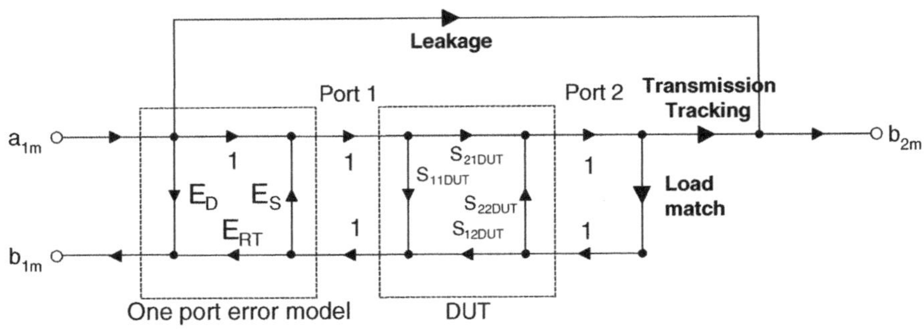

Figure 18.16 12-term error model (forward path) for two-port measurements.

For the OPEN measurement, $S_{11\text{DUT}} = +1$, yielding:

$$S_{11meas} = E_D + \frac{E_{RT}}{1 - E_S} \tag{18.3}$$

For the SHORT measurement, $S_{11\text{DUT}} = -1$, and then:

$$S_{11meas} = E_D - \frac{E_{RT}}{1 + E_S} \tag{18.4}$$

Once the values (magnitude and phase) of the error terms are known at each frequency, the computer within a VNA can almost instantly correct them out and display the true value of $S_{11\text{DUT}}$. This one-port example shows the principles, but for two-port devices the error model has to be extended and have eight, 12 or 16 error terms, depending on the degree of accuracy that is required. For precision measurements, a VNA with an *S*-parameter test set is necessary: this includes a switch to swap the stimulus from port 1 to port 2, giving a forward and return path measurement without disturbing the DUT. This means that all four parameters can be measured nearly simultaneously and the calibration applied to the displayed data. Figure 18.16 shows the 12-term error model in the forward path, showing that forward path loss and phase (transmission tracking), imperfect isolation of the switch (leakage) and imperfect port 2 load (load match) are now included as error terms, yielding six error terms for each path. The full details of the two-port calibration algorithms can be found in Keysight application notes, such as *Applying Error Correction to Network Analyzer Measurements* [1].

18.7.2 Calibration Standards

In practice, there are no ideal calibration standards. For example, the open circuit has some fringing capacitance, which is frequency-dependent, and the short circuit has some inductance. In addition, there is an important issue of *reference planes*. Because phase is now being measured, and the signal is at a very high frequency (short wavelength), even a very simple adapter has significant electrical length. So, the correct position of the S, O, L and T must be maintained throughout the calibration. Different connectors and adapters cannot be mixed when calibrating, and the calibration reference planes must be absolutely consistent.

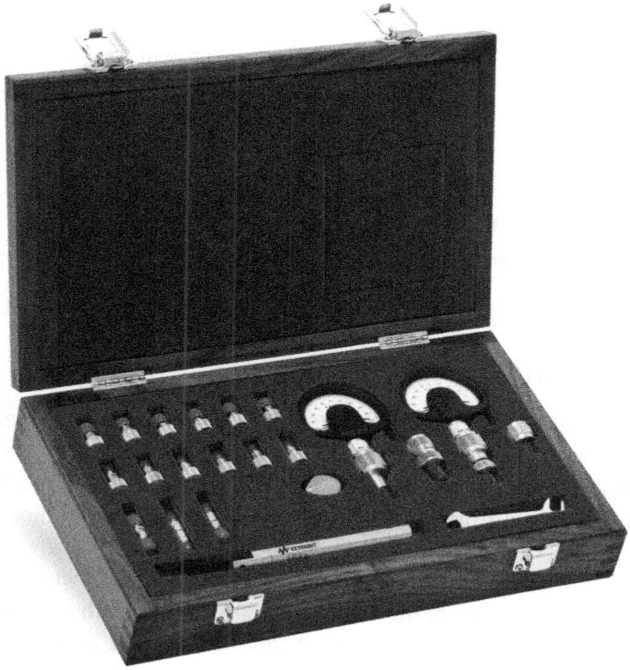

Figure 18.17 A mechanical calibration kit in a box. The kit contains calibration standards, adapters, spanners and gauges. © 2016 Keysight Technologies, Inc. Reproduced with permission.

A mechanical calibration kit consists of all the calibration standards, plus a file or files describing their electrical characteristics. As shown in Figure 18.17, the kit is supplied in a fine wooden box, and there may be other components such as connector gauges and male/female adaptors. Connector gauges are used to check there is no unacceptable protrusion or recession of the centre pin of any coax connectors. A 'cal kit' data file has to be loaded at some point into the VNA. This file contains the electrical characteristics of each calibration standard: coax open and short circuits are usually offset from the connector reference plane. A waveguide calibration set may contain set lengths of guide and offset shorts. Also, in real life, the open, short and load are not perfect. For the open and short circuit, a common solution is to have an equivalent circuit for them: for example, the open could be treated as having a certain value of fringing capacitance, which is frequency-dependent and often modelled as a polynomial function. The short can be modelled with an inductance.

For the 50 Ω load, any residual reflection will cause an error in the calculated value of the directivity term. The traditional solution to this is to use a *sliding load*; the load element is moved inside the coax and S_{11} is measured at a number of different physical positions. It is then found that the measured values of S_{11} lie on a circle on the Smith Chart, with its centre is exactly at E_D, and the radius is the magnitude of the residual S_{11} of the 50 Ω load. By measuring at seven or more positions, as illustrated in Figure 18.18, a wide frequency range is covered. Sliding loads are expensive precision mechanical devices, and have mostly been replaced with more modern calibration techniques, as described in following sections.

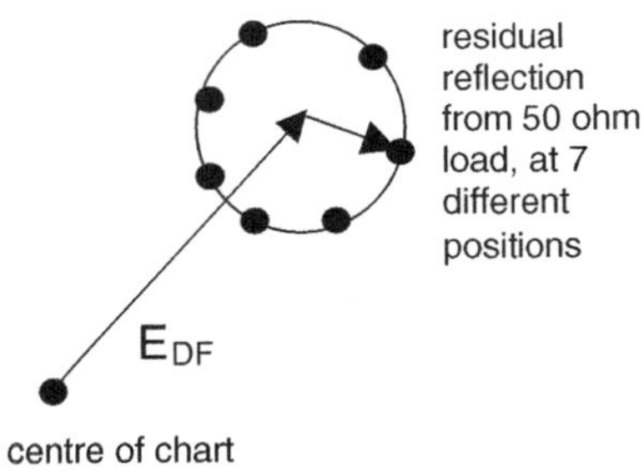

Figure 18.18 Determining the directivity term using a sliding load.

18.7.3 Electronic Calibration

For most coaxial connector DUT measurements, mechanical calibration standards have been replaced by electronic calibration modules. These use an electronically switched circuit, usually a MMIC, that can switch between one of a number of load impedances, a thru line, and a switched attenuator formed from a transmission line with shunt PIN diodes, as illustrated in Figure 18.19. The module can therefore produce a number of known impedance and transmission response states, which are accurately measured at the factory against traceable standards and stored in the memory of the module or VNA. These are then referred to as transfer standards: the *S*-parameters of the module in all states need to be known precisely to solve the simultaneous equations and find the error vectors, but their specific value is not crucial, provided that they cover a suitably wide area of the Smith chart, across the full frequency range, and are traceable. Typically, as many as seven states are measured, which gives more simultaneous equations than necessary, and a least-squares fitting technique is employed in order to give the best possible results. Whilst electronic calibration modules are extremely convenient to use, with VNA software controlling the process, the need for good connector care for repeatable connections and the need to know the exact location of the measurement reference planes, especially when using adapters, is not diminished. Electronic calibration gives fast and accurate results, largely removing human error and reducing wear on connectors and standards. It is especially convenient for making three- or four-port measurements, for which using a mechanical calibration kit is extremely tedious and prone to errors.

18.7.4 Calibration Example

The screen shots in Figures 18.20 and 18.21 are taken from a 6 GHz 8753 VNA, before and after performing a calibration. The results powerfully demonstrate the importance of calibrating

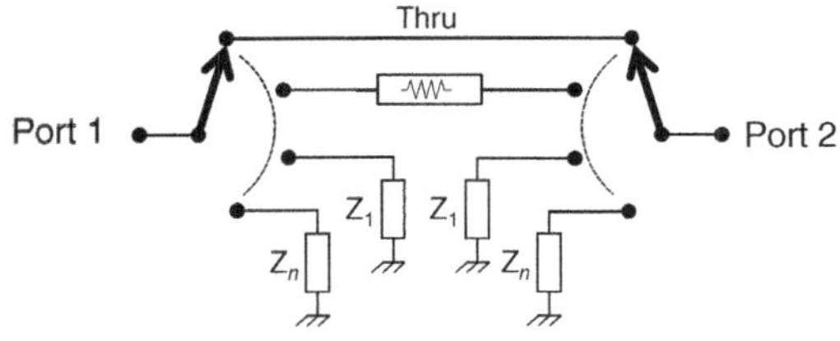

Figure 18.19 Simplified block diagram of a typical electronic calibration module.

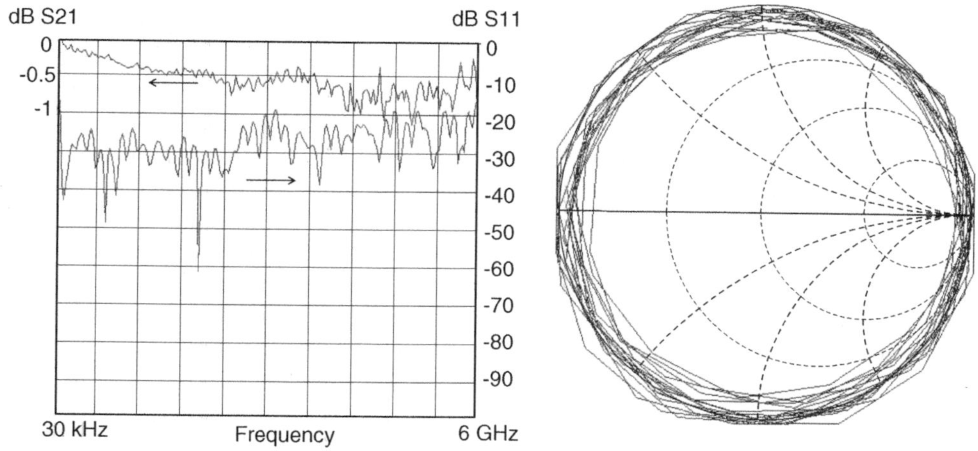

(a) (b)

Figure 18.20 VNA screenshots before calibration. (a) S_{21} (0.5 dB per division) and S_{11} (10 dB per division) for a thru line (the reference line is 0 dB). (b) S_{11} of the offset short circuit standard on the Smith chart display.

the instrument. In Figure 18.20(a), because the VNA is not calibrated these measurements are almost completely useless: in Figure 18.20(b), although the magnitude of S_{11} is approximately unity, the phase is going round and round the Smith chart and it is impossible to tell if this is an open or a short. In Figure 18.21, the calibration routine has been performed and error correction is turned on. For the through line, the S_{21} trace is now almost exactly at 0 dB and the S_{11} is below 60 dB, which is at the connector repeatability limit. The S_{11} of the short-circuit

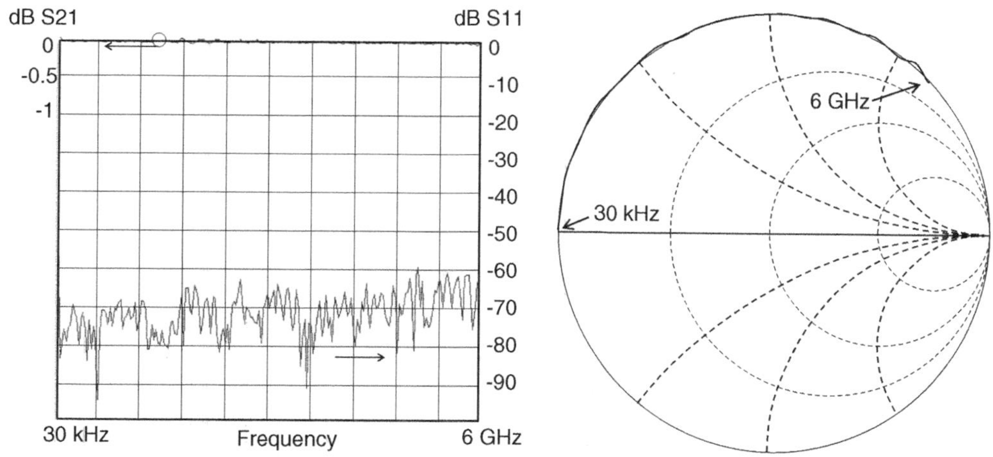

(a) (b)

Figure 18.21 VNA screenshots after calibration. (a) S_{21} and S_{11} for the thru line. (b) S_{11} of the offset short circuit standard on the Smith chart display.

standard is now in the right location on the Smith chart. It may appear that there is a problem because the S_{11} of the short-circuit standard moves around the Smith chart quite significantly. However, this result is correct because the APC3.5 mm short-circuit standard that was used actually has the short circuit placed a small distance away from the connector reference plane (the offset is defined in the calibration kit as a certain delay in picoseconds). Indeed, one should always be suspicious of measurements that are 'too good'. It is all too easy to compare a calibration standard with itself: in this case, if the S_{11} of the offset short-circuit standard had been a perfect 'dot' at the short-circuit point on the Smith chart, it would show that the calibration was not valid because the delay of the offset short-circuit calibration standard had not been accounted for in the calibration. This example clearly demonstrates that a proper understanding of calibration techniques is important. Even here, where the data only goes to 6 GHz, there is the possibility of very large errors and misinterpretation of results.

18.7.5 Connector Care

Some apparently simple connectors and adaptors can cost thousands of pounds (e.g., 1 mm coaxial connectors, which operate to 110 GHz), so they must be handled with great care. Connector care is an important aspect of measurements, and an RF/microwave engineer must develop excellent connector skills to get good results. Dirty or damaged connectors will lead to results which are not repeatable and can damage other connectors – including the calibration standards and VNA ports. It can cost several thousands of pounds to replace the connector on the front of a network analyser. Once the connector is damaged, the analyser's raw performance is degraded and it may become unusable as calibrations are unsuccessful or drift quickly.

Most microwave connectors, such as the APC3.5 connector shown in Figure 18.22, have a threaded-type outer conductor collar to ensure a tight connection. The connector 'sex' is determined by the centre conductor; the one shown, with a pin, is the male end. It is also common to refer to a jack (female) and a plug (male) – think of a mains plug and you should be able to remember which is which. The following points must be noted when using high-frequency connectors:

1. It takes just one mistake to damage a connector permanently.
2. Dirty or damaged connectors will lead to results which are not repeatable.

Figure 18.22 APC3.5 mm connector: This is the male end (or plug). © 2016 Keysight Technologies, Inc. Reproduced with permission.

3. Clean the connectors before use with a lint-free bud and cleaning agent.
4. When screwing connectors together, the outer collar must be rotated whilst keeping the inner body fixed.
5. If the centre pin rotates, it will get damaged or worn very quickly.
6. The outer collar (or nut) must always be done up gently by hand *before* a torque spanner is applied. Starting with a spanner instead of delicate use of fingers will often lead to crossed threads.
7. A torque spanner is used, only for the last quarter of a turn or so, to ensure that the connector is always done up with the same tightness, giving a repeatable connection.
8. Cables must be used gently and not bent significantly. They must be allowed to flex gently by themselves. Do not bend the cables at all near the connectors.
9. There are many types of connector: RF engineers must recognise all of them and know which ones are compatible. Trying to connect incompatible connectors or adapters can cause thousands of pounds of damage.
10. When done correctly, mating of coaxial connectors is an awkward process requiring considerable skill and dexterity.

The Keysight Technologies web site has more information on connector care [2]. There is a lot more to learn, especially when making measurements above 10 GHz, such as the use of connector gauges, pin recession limits and eccentricity.

18.7.6 Time-Domain Measurements

VNAs normally measure in the frequency domain using a swept or stepped signal source. The reason for this is that the receiver(s) can then use heterodyne detection, giving a low noise floor and a better dynamic range. However, many VNAs can perform an inverse Fourier transform of the frequency domain data to present the measurement in the time domain. This is mostly useful for troubleshooting; for example, if you have a long cable which has an unacceptable return loss measurement, in the time domain you could clearly identify where along the cable the problem lay. S_{11}, for example, in the time domain will show individual reflections separated in time, as shown in Figure 18.23.

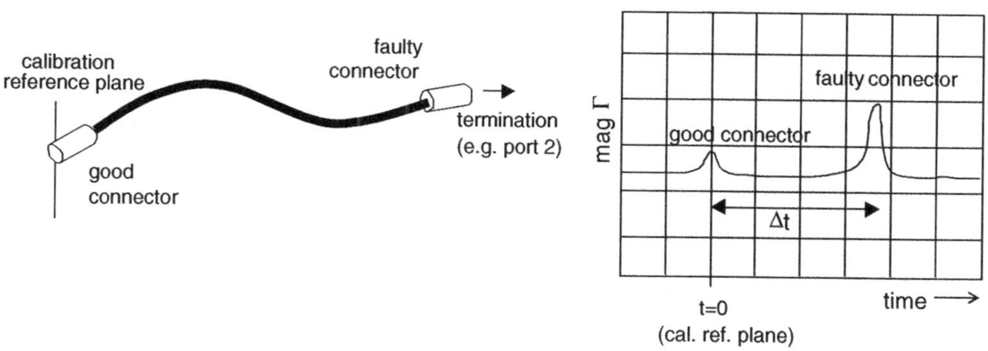

Figure 18.23 Time-domain measurement example.

In this cable example, the two reflections are separated by a significant distance compared with the wavelength at the highest frequency of the measurement. In fact, when looking at reflection data, the time between pulses, Δt, is equal to twice the delay of the cable. The delay can be calculated from the dielectric constant of the material in the cable (or effective dielectric constant for microstrip/coplanar waveguide (CPW)/etc.) as follows:

$$delay = \frac{l \times \sqrt{\varepsilon_r \text{ or } \varepsilon_{eff}}}{c} \tag{18.5}$$

where l is the physical length and c is the speed of light *in vacuo*.

Note that the two reflections each have a Gaussian shape on the display; this is a function of the limited bandwidth of the frequency domain measurement and the windowing function used in the Fourier transform. As the two reflections get closer and closer together, they become impossible to resolve. The only way to get a major improvement in resolution is to increase the measurement bandwidth. With a 45 MHz to 110 GHz measurement, for example, it is just possible to identify the different air-bridges on a GaAs MMIC CPW line if they have a separation of approximately 200 μm. Many VNAs offer a choice of low-pass or band-pass transform; the band-pass is most often used, and gives the type of graph already described. The low-pass type has the advantage that it yields a sign of Γ, so that inductive and capacitive discontinuities can be distinguished.

Direct time-domain measurements are sometimes used: modern systems employ the ultra-short pulses from a laser to create impulses on a chip with approximately THz equivalent bandwidth. The impulses are often created by illuminating photoconductive switches, and the pulses detected using opto-electronic sampling. These research-laboratory systems have been used to measure devices up to tens of THz. However, the frequency resolution is not as good as a frequency-domain measurement.

18.8 Microstrip Test Fixture Measurements

For test fixture measurements on a microstrip-based DUT, there are several ways of making measurements with reference planes located in the microstrip line [3, 4]. These methods can be classified into two groups:

1. A coaxial calibration is performed, with the network analyser's reference planes located at the end of its coaxial cables; de-embedding (or 'stripping') is then used to shift the measurement reference planes to the DUT.
2. Calibration standards are placed directly within the test fixture, using techniques such as TRL (through-reflect-line) or similar calibration.

De-embedding can be performed using an equivalent circuit model of the test fixture, or by employing a secondary calibration routine (such as a TRL calibration) on the test fixture itself. Test fixtures are still widely used for device measurements because they offer a great deal of flexibility, such as allowing DC bias circuitry to be located next to the device. Often, the DUT must be assembled in this manner to operate properly, for reasons of stability or thermal dissipation. An illustration of a basic two-port test fixture for measuring an MMIC

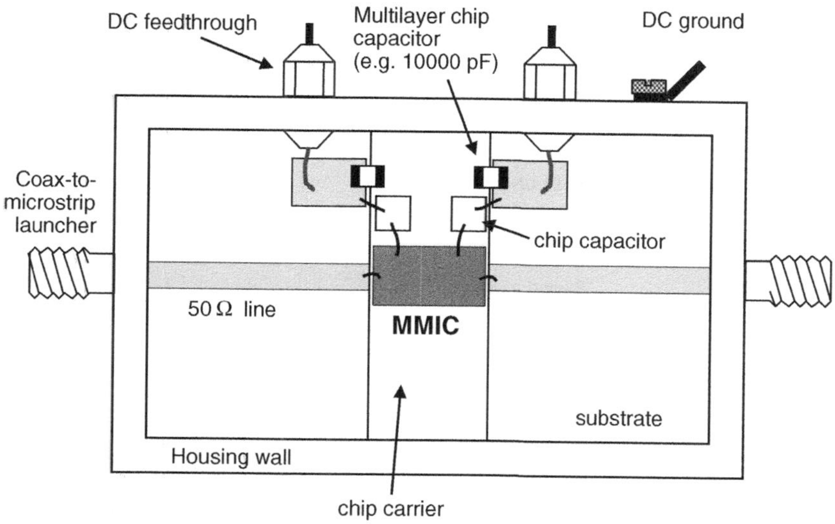

Figure 18.24 A typical two-port test fixture.

is shown in Figure 18.24. The MMIC is permanently attached to the chip carrier with either conductive epoxy glue or solder. The metal housing is employed to hold the chip carrier and the connector/launchers in place. Because of the fixed nature of this set-up, it would be difficult to use in-fixture calibration standards, and de-embedding would be easier. The parameters for de-embedding could be obtained by measuring a microstrip thru line and an MMIC thru line, so that all the connector, fixture and microstrip-MMIC transition effects can be modelled (using tuning and optimisation on a CAD package). In principle, time-domain gating could also be used for de-embedding, but in practice the limited network analyser bandwidth usually means that the spatial resolution is too coarse to make this reliable.

18.8.1 In-Fixture TRL Calibration

The SOLT method has serious limitations for microstrip measurements, and many investigations have been conducted to develop other techniques. When calibrating in microstrip, or on a semiconductor wafer, sliding loads are clearly not convenient. The TRL calibration technique is an important solution, from which a number of others are derived. The standards used are a Thru, a Reflection standard, and a known length of Line. Note that not all VNAs can implement the TRL technique – usually a full *S*-parameter test set is required, with four samplers so that a_1, b_1, a_2, b_2 can be measured simultaneously. In-fixture TRL-type calibration gives much better performance than de-embedding using equivalent circuits. Typically, a special purpose test fixture is used which allows the speedy but repeatable insertion of the required microstrip calibration standards. For TRL calibrations, the calibration standards consist of a number of thru lines of different lengths and a noncritical reflection standard (e.g., a short or open circuit), as shown in Figure 18.25. In order to cover a useful frequency range it is necessary to employ a number of different delay line lengths to overcome phase ambiguity at all of the measurement frequencies. In order to accommodate these different lengths of calibration standard, the

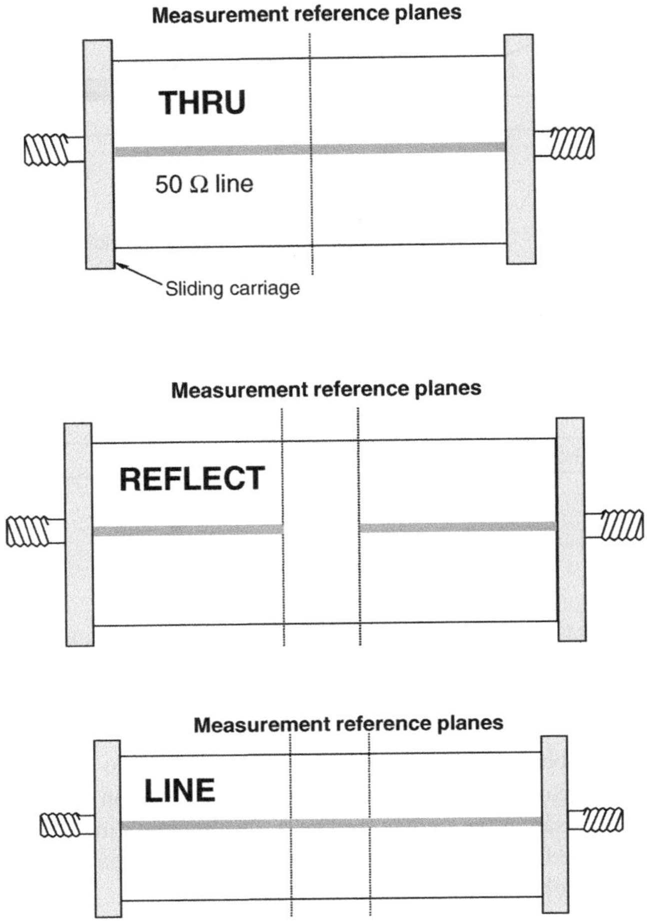

Figure 18.25 TRL in-fixture calibration example.

test fixture must either have coaxial-to-microstrip launchers fixed onto sliding carriages, or it must use a split-block design with an interchangeable centre section. For example, the Anritsu universal test fixture uses the sliding-carriage approach and allows accurate measurements to be made from 0.01 to 60 GHz.

18.9 Probe Station Measurements

Accurate RFIC/MMIC measurements can easily be made using a microwave probe station. The key to the high accuracy is the use of CPW probes, as shown in Figure 18.26, which allow a smooth transition to be made from the network analyser's coaxial connectors (or waveguide flanges) down to the tiny probe pads of the chip, with 50 Ω characteristic impedance maintained throughout by tapering the CPW centre conductor width and ground-plane spacing. Coaxial connectors are currently restricted to 145 GHz, and rectangular waveguide still

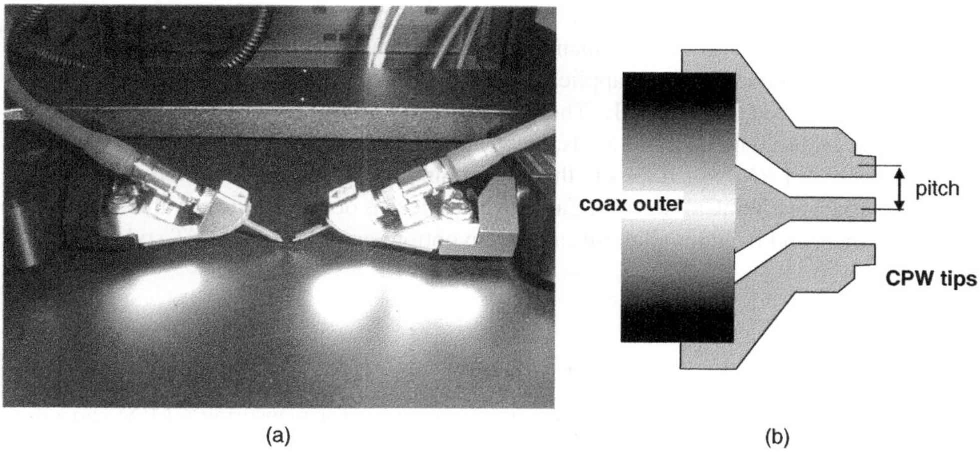

Figure 18.26 (a) Cascade Microtech® 110 GHz probes. (b) Coplanar waveguide (CPW) tips.

plays an important part in millimetre-wave and sub-millimetre-wave measurement systems. Currently, Rohde & Schwarz, Virginia Diodes and OML produce frequency extender modules covering standard submillimetre-wave waveguide bands. The mm-wave module comprises predominantly directional couplers, frequency multipliers to multiply the ~50 GHz source signal up, and harmonic mixers. 'Test port combining' refers to the arrangement whereby a coaxial S-parameter test set is operated in tandem with a waveguide test set, in order to achieve a higher frequency range with a single connection to the DUT. Test port combining can be external, using a diplexer to separate the coax and waveguide test set signals; this has the advantage that the coax and waveguide test sets can be operated independently when required, making the system more flexible. However, the diplexer after the test sets (and the adaptors, etc.) significantly degrade the raw performance of the VNA (especially the directivity) and the user may find that the calibration drifts off more rapidly due to temperature changes. To improve directivity, internal combining can be used, as in the (now obsolete) Agilent 8510XF system.

18.9.1 Probe Calibration Techniques

Probe stations can use SOLT, TRL/LRL (line-reflect-line), LRM (line-reflect-match) or LRRM (line-reflect-reflect-match) calibration procedures. The calibration standards can be on a wafer or on a precision impedance standard substrate (ISS). On-wafer standards can characterise the probe-to-MMIC transition more accurately, whereas ISS standards are less subject to manufacturing variations and can be more traceable. Generally, the SOLT method has poor accuracy above 50 GHz because of the difficulty in realising good-quality standards (especially the open circuit). The LRL and TRL techniques require a minimum of two delay lines and a noncritical reflection standard (usually an open or short circuit), such as the microstrip in-fixture calibration already described. The reference impedance is taken from the characteristic impedance of the delay lines. In order to cover a wide frequency range, a number of different delay line lengths are required because each delay's electrical length is limited to a range between approximately 20° and 160°. This means that the probe separation has to be adjusted during the calibration procedure. For many applications, such as automated test systems, this is a major limitation and for these applications the LRM calibration, developed by Cascade Microtech, is preferred to LRL/TRL. The multiple delay lines required with the LRL/TRL calibration are replaced by matched loads, representing an infinitely long delay line. LRRM is an enhancement to LRM which uses both the short and open standards to improve accuracy. The software package WinCal XE from Cascade Microtech is dedicated to on-wafer calibration and measurement and includes all these calibration options as well as the NIST-style multiline method [5].

18.9.2 On-Wafer Measurements of Microstrip DUTs

Most MMICs use microstrip, not CPW, and an arrangement for measuring a MIM capacitor for modelling purposes is shown in Figure 18.27. The CPW-to-microstrip transitions must be accounted for to give accurate measurements for the capacitor in a microstrip environment. One method of removing their effects is to perform a calibration with on-wafer microstrip standards. Microstrip standards are problematical, however, because the characteristic impedance is affected by dispersion and is critically dependent on the wafer thickness. Some points to be aware of are:

- The line Z_0 depends on substrate thickness; the wafer thickness is surprisingly poorly controlled (e.g., ± 10%).
- The measured data may be referenced to the line Z_0, not 50 Ω (depending on the software used).

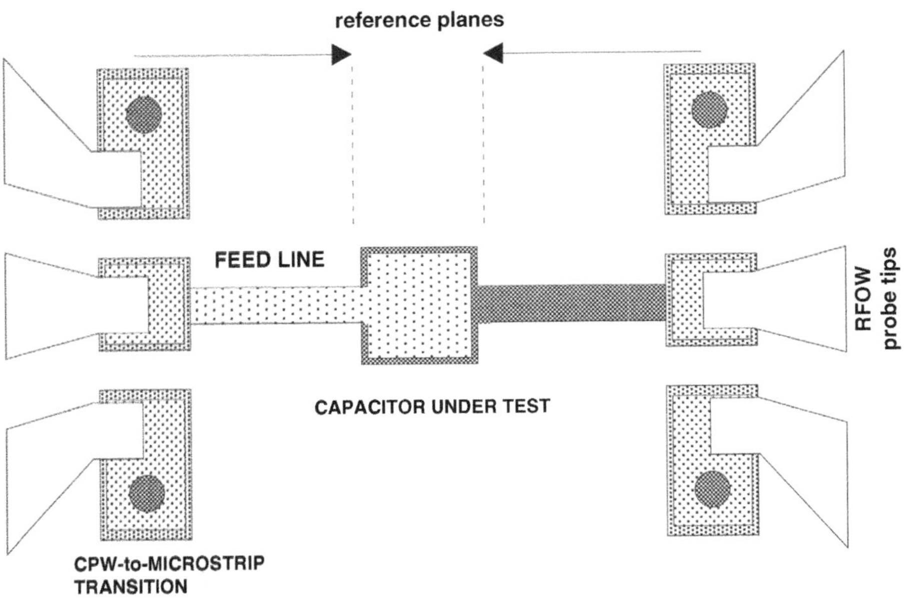

Figure 18.27 RF-on-wafer (RFOW) measurement of a capacitor.

- The 'Thru' must have sufficient length to ensure evanescent modes at the probe-microstrip transition have decayed away.
- The opens must be far enough apart that no coupling occurs.
- Multiple LINE lengths will be required for wideband measurements, and their lengths must be carefully chosen. The NIST MultiCal™ method carefully smoothes over the band edges between the useful range of each line standard.
- At ≫100 GHz the LINE standard length is comparable to the over-travel skate distance. Probe placement errors become a severe limiting factor.
- Substrate modes can be excited above ~100 GHz, depending on the substrate height.

18.10 Mixed-Mode *S*-Parameters

Coaxial connectors and cables dominate measurement applications, and microstrip has been preferred in MIC and MMIC design for many years. Like coax, microstrip is a single-ended transmission medium, with a conductor strip over a ground plane, as illustrated in Figure 18.28(a). Microstrip is popular because it is well understood, behaves well in practice and can be used to integrate a wide range of active and passive components into a subsystem. GaAs MMIC technologists had to specifically develop through-substrate via-hole fabrication techniques for low-inductance grounding of devices in microstrip. Significant effort has been expended on CPW modelling and design to avoid the need for via holes and to realise other benefits such as low dispersion and low-inductance grounding of devices for operation well beyond 100 GHz. However, despite these advantages, CPW is much harder to design with than microstrip: at the very least, regular connections across the ground planes are needed to prevent unwanted coupled-slotline modes from propagating [6]. In contrast, in microstrip the ground plane 'looks after itself', and whilst microstrip can suffer from unwanted modes of propagation they are relatively easy to avoid by following well-known guidelines for maximum substrate thickness and track width.

Differential, or balanced, circuits are of immense importance for mixers, modulators, frequency multipliers and power amplifiers where balanced operation is a convenient way of achieving cancellation of unwanted output signals – LO leakage at the IF port of a mixer, for example. In microstrip design, it is sometimes necessary, therefore, to use a planar Marchand balun [7] to convert between the normal single-ended signal and a differential one, but microstrip remains the medium of choice elsewhere. For RFICs, the story is rather different because the design philosophy originated from the silicon analogue circuits community, where differential active circuits were already the norm. Silicon analogue designers routinely use differential circuits as this gives valuable cancellation of interference from digital circuitry on the same chip – the so-called 'substrate bounce' encountered in CMOS. As RFICs have reached

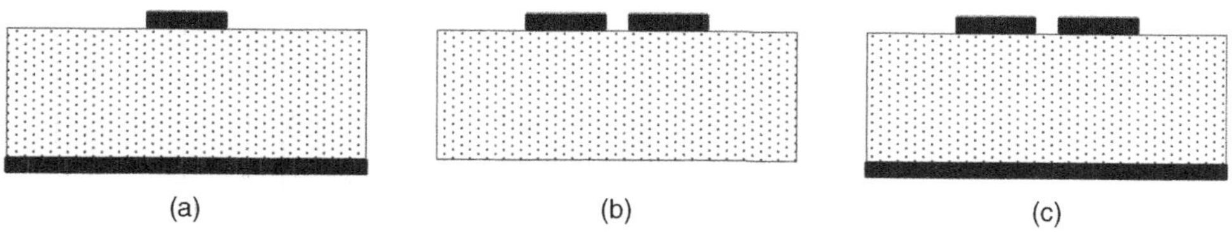

Figure 18.28 (a) Microstrip, single-ended. (b) Differential CPS line. (c) CPS with ground plane.

higher and higher operating frequencies, we find the interesting intersection of two worlds: 'old RF' [8] which is the world of Smith charts, transmission lines and *S*-parameters; and 'new RF', which is the world of differential analogue circuits, DAC/ADCs and DSP. This has led to the juxtaposition of differential active circuits with transmission line components and interconnects. The coplanar strip (CPS) transmission line, as illustrated in Figure 18.28(b), is the obvious candidate for a planar balanced transmission line. CPS is very similar to the flat twin-lead transmission line used in some RF aerial applications, which normally has 300 Ω characteristic impedance, higher than the 50 and 75 Ω encountered for coax. The characteristic impedance of CPS can also be high because the side-by-side conductors have little capacitance between them unless the substrate has a high ε_r. When CPS is above a ground plane, which is almost inevitable as part of the package design, it becomes clear that the differential mode is only one of the modes that can propagate: grounded CPS, as shown in Figure 18.28(c), is the same structure as parallel-coupled microstrip and must, therefore, also have odd and even modes of propagation. The odd mode is the differential mode, with a virtual ground between the strips. The even mode is now a parasitic unwanted mode, and differential input signals can easily excite this mode, leading to unwanted common-mode signal propagation. Even if the circuit does not have a ground plane, it must be mounted into some kind of box or package and any nearby ground will perturb the differential mode. As a result, experience shows that, as a transmission medium, CPS is extremely sensitive to its environment and should generally be avoided if possible.

Suppose, now, that it is necessary to make accurate *S*-parameter measurements on a differential RFIC amplifier. The first problem is that to realise a truly differential VNA is fraught with difficulty. It is probably only feasible with some form of active probe arrangement where the samplers are close to the DUT, because of the challenge of maintaining the purity of the differential mode of propagation. So, one approach is to use a single-ended (coaxial) VNA and use baluns to connect to the differential DUT. However, the second problem is that there are no precise differential traceable standards at high frequency and so it is difficult to calibrate out the balun. Furthermore, the balun will not have perfect amplitude and phase balance and so will perturb the differential behaviour. In any case, even if it was feasible to construct and calibrate a differential VNA, it is also important to be able to measure the extent to which the DUT itself creates or rejects common-mode signals, since cancellation of unwanted signals, noise and interference is desirable. For example, the common-mode rejection ratio (CMRR) is an immediately recognisable parameter of the differential amplifier DUT. The solution to this measurement challenge is to use mixed-mode *S*-parameters [9, 10]. Figure 18.29 shows the labelled mixed-mode *S*-parameter matrix for a two-port network. There are four quadrants: one

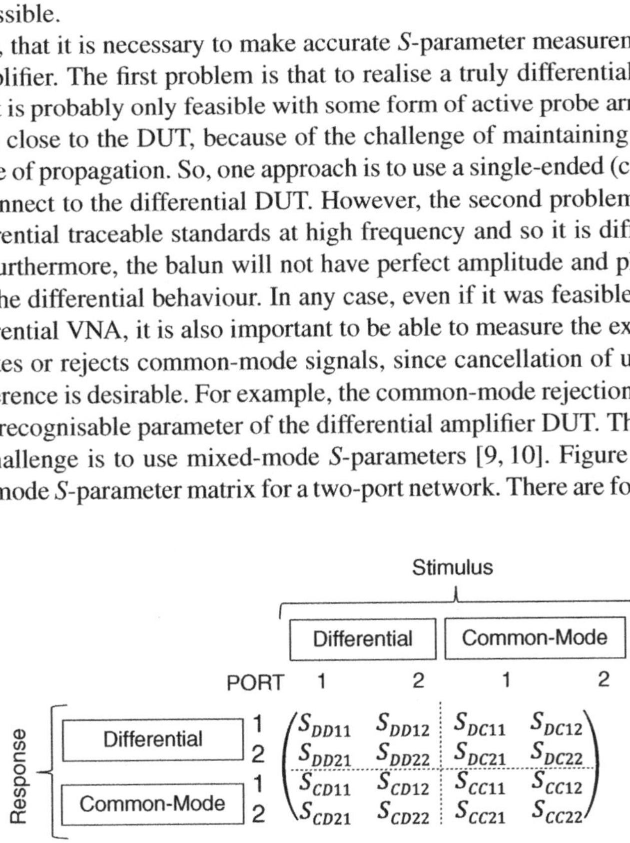

Figure 18.29 Mixed-mode *S*-parameter matrix.

representing the pure differential mode behaviour in reflection and transmission (S_{DDnm}); one representing the single-ended (common-mode) behaviour (S_{CCnm}), and two for the conversion from one mode to the other – which, depending on the nature of the DUT, may be desirable (e.g., in a balun) or undesirable (e.g., poor common-mode rejection in a differential amplifier which would lead to noise pickup).

The technique most widely adopted for characterising DUTs with one or more differential ports is based on deriving the mixed-mode S-parameter matrix from a set of single-ended multiport measurements. So, for example, to measure a balun – which is a two-port network with a single-ended input and a balanced output – the DUT is measured as a device with three single-ended ports (ideally using a multiport VNA with electronic calibration module); the two-port mixed-mode S-parameters are then derived mathematically. This has the major advantage of being a technique in which the signals to, and from, the DUT are well behaved and the calibration standards are traceable. A true balanced measurement is not necessary for most practical applications, but it is important to note that the mathematical balanced/unbalanced conversion assumes a linear DUT. The Keysight PNA-X has the option to perform true differential measurements, even extendible to millimetre-waves, making use of twin signal sources with adjustable phase offset. Other capabilities of this pinnacle of VNA development include pulsed S-parameters, error-corrected noise figure measurements [11], X-parameters (this is a registered trademark of Keysight Technologies), frequency-translation device measurements, intermodulation measurements and vector signal analysis capability. Cascade Microtech offers a range of dual channel/differential probes and calibration substrates, so that differential circuits can be measured on-wafer.

18.11 Source- and Load-Pull Measurements

For active device characterisation, a source- or load-pull system allows the device parameters to be measured as a function of the source and/or load impedance. This is especially important for finding the Γ_{opt} and F_{min} for low-noise amplifier design and for characterising high-power devices. The necessary parameters cannot be determined from S-parameter measurements alone [12, 13]. A passive automated tuner comprises a coaxial/slabline (Figure 18.30) or waveguide in which a number of motorised sliding reflective probes can be moved in and out and along the line to create a computer-controlled variable reflection coefficient that covers a large area of the Smith chart across a wide frequency range. By integrating tuners with a noise figure measurement system, noise circles can be conveniently measured on-wafer. Since the tuner has some loss, and may be some distance from the DUT due to the use of RFOW probes, adapters, cables, bias Tees and so forth, even if the probes could provide perfect short circuits the tuner cannot always provide a reflection coefficient sufficiently close to unity for measuring some high-power devices, which have very low impedances. This has led to the development of active and hybrid active-passive load pull systems where a signal is actively injected. Another consideration is that modern power amplifiers employ very deliberate waveform engineering in order to maximise efficiency/linearity and so harmonic and multiharmonic tuners are available, for which the reflection coefficient at the signal frequency and harmonics can be controlled.

18.12 *X*-Parameter Measurements

X-parameters are an exciting extension to the principles of scattering used for S-parameters, with the added dimension of treating the input and output waves as a series of harmonic

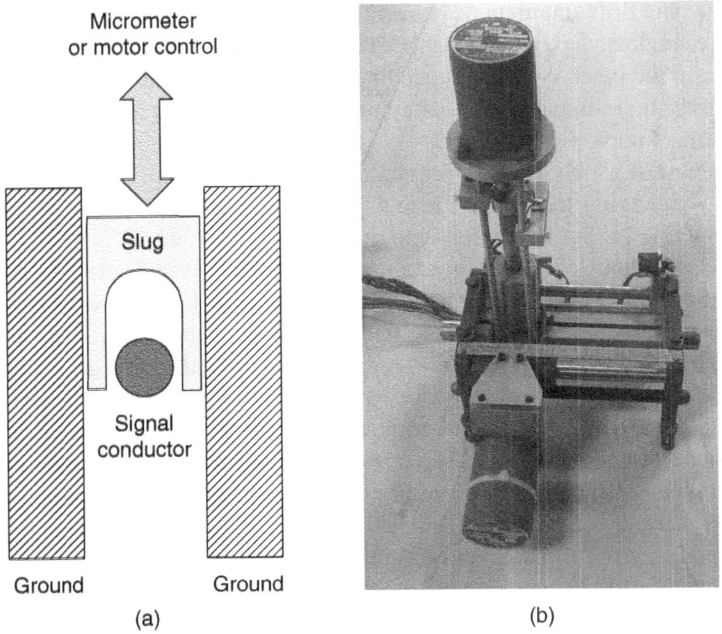

Figure 18.30 (a) Cross-section of a typical slabline tuner. (b) An early prototype of an automated tuner developed at the University of Leeds [12].

components in the frequency domain. Mathematically, X-parameters are based on the poly-harmonic distortion (PHD) framework introduced by Verspecht and Root [14, 15]. The reason that the extension of this work to the representation of devices with X-parameters is so important is that it means, for the first time, that nonlinear devices can be fully represented with a 'black box' of measured or modelled parameters in the same way that linear devices, components and circuits have been represented with S-parameters for more than 50 years.

The principle behind X-parameters is the treatment of each of the scattered waves as a harmonic series, as illustrated simplistically in Figure 18.31. Obviously, if a single tone is input into a nonlinear device then harmonic components will be transmitted and reflected. However, in order to represent a device located anywhere within a nonlinear circuit or subsystem, it is important to create a more general representation and this requires that the input signal is also treated as a harmonic series. These key steps in developing the fully flexible, multipurpose X-parameter representation are illustrated in Figure 18.31. For a two-port nonlinear network, the incoming and scattered waves must all be represented as a Fourier series and the parameters a_1, b_1, a_2 and b_2 from the linear case are developed into matrices containing the magnitude and phase of the Fourier coefficients. The X-parameters characterise the conversion from harmonic-to-harmonic for the scattered wave components. Since the degree to which harmonics are generated is highly dependent on signal level, X-parameters are multidimensional parameters that vary with frequency, signal level and harmonic index. With exceptional elegance, X-parameters are actually a superset of S-parameters, and the linear S_{11}, S_{21}, S_{12} and S_{22} can all be found within the X-parameter set, being those with frequency index [1,1] at low power levels.

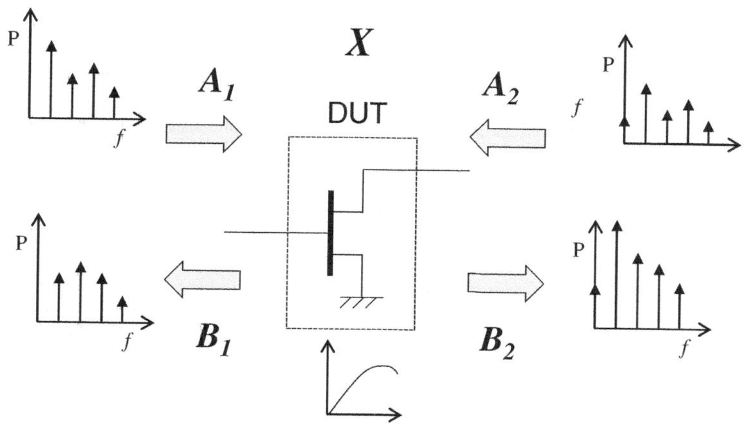

Figure 18.31 *X*-parameter representation of a nonlinear device.

X-parameters have some similarities to the conversion matrices often encountered in the harmonic balance modelling of nonlinear circuits, such as mixers. They assume a steady-state solution and cannot be used for transient simulation, but are able to represent a very wide range of signal waveforms, including modulated signals. In order to derive waveforms from a harmonic series, the phase of each harmonic needs to be known. This provides a new challenge for the network analyser measurement because absolute phase is not determined by conventional VNAs. So, as well as adding absolute power level calibration, an important key step in equipping a VNA to make *X*-parameter measurements is to introduce a phase reference, which Keysight do by injecting a comb generator into the LO. With this, and suitable firmware, the PNA-X can be used as a complete nonlinear VNA (NLVNA) system. Furthermore, the Keysight NLVNA system can be combined with Maury® automated tuners to provide the capability to characterise the *X*-parameters of the DUT over a very wide range of load impedances, providing exceptional capability for power amplifier modelling and design.

References

1. Applying Error Correction to Network Analyzer Measurements, AN 5965-7709E. Available at: www.keysight.com.
2. Millimeter Wave Connector Care, 15th September 2014, Keysight Technologies, Inc. Available at: http://na.support.keysight.com/pna/connectorcare/whnjs.htm.
3. Lucyszyn, S. (2001) Measurement techniques, in: *RFIC & MMIC Design and Technology* (eds I.D. Robertson and S. Lucyszyn), Chapter 12. IEE Publishers.
4. Lucyszyn, S. (2007) RFIC and MMIC measurement techniques, in: *Microwave Measurements* (eds. R. Collier and D. Skinner), Chapter 11, IET Publishers, London, pp. 217–262.
5. Marks, R.B. (1991) A multiline method of network analyser calibration. *IEEE Transactions*, **MTT-39**, 1205–1215.
6. Deal, W.R. (2008) Coplanar waveguide basics for MMIC and PCB design. *Microwave Magazine*, **9** (4), 120–133.
7. Ang, K.S. and Robertson, I.D. (2001) Analysis and design of impedance-transforming planar Marchandbaluns. *IEEE Transactions on Microwave Theory and Techniques*, **49** (2), 402–406.
8. Underhill, M.J. (2001) University of Surrey, Private communication.

9. Bocklemann, D.E. and Eisenstadt, W.R. (1995) Combined differential and common-mode scattering parameters: theory and simulation. *IEEE Transactions on Microwave Theory and Techniques*, **MTT-43**, 1530–1539.
10. Dunsmore, J. and Xin Chen (2104) Measurements for microwave differential and IQ devices. 83rd ARFTG Microwave Measurement Conference, Tampa, FL, 6 June, pp. 1–4.
11. Weatherspoon, M.H. and Dunleavy, L.P. (2005) Vector corrected on-wafer measurements of noise temperature. *IEEE Transactions on Instrumentation and Measurement*, **54** (3), 1327–1332.
12. Pierpoint, M., Pollard, R.D., and Richardson, J.R. (1986) An automated measurement technique for measuring amplifier load-pull and verifying large-signal device models. IEEE MTT-S International Microwave Symposium Digest, Baltimore, MD, USA, 2–4 June, pp. 625–628.
13. McIntosh, C.E., Pollard, R.D., and Miles, R.E. (1999) On the feasibility of using an active tuner for measuring microwave noise parameters. *IEEE Transactions on Instrumentation and Measurement*, **48** (6), 1294–1300.
14. Verspecht, J. and Root, D.E. (2006) Polyharmonic distortion modelling. *Microwave Magazine*, **7** (3), 44–57.
15. Root, D.E., Verspecht, J., Horn, J., and Marcu, M. (2013) *X-Parameters: Characterization, Modeling, and Design of Nonlinear RF and Microwave Components*. The Cambridge RF and Microwave Engineering Series. Cambridge University Press. ISBN: 9780521193238.

Glossary

3GPP 3rd Generation Partnership Project
ACI Adjacent Channel Interference
ADC Analogue-to-Digital Converter
AiP Antenna-in-Package
AM Amplitude Modulation
AMPS Advanced Mobile Phone System
ASIC Application-Specific Integrated Circuit
ASK Amplitude-Shift Keying
ATSC Advanced Television Standards Committee
AWG or ARB Arbitrary Waveform Generator
BAW Bulk Acoustic Wave
BCB Benzocyclobutene
BER Bit-Error Rate
BGA Ball Grid Array
BiCMOS Bipolar Complementary Metal-Oxide-Semiconductor
BNC Bayonet Neill–Concelman connector
BPF Band-Pass Filter
BUE Bandwidth Utilization Efficiency
CCI Co-Channel Interference
CDMA Code-Division Multiple Access
CMOS Complementary Metal-Oxide-Semiconductor
CNR Carrier-to-Noise Ratio
COFDM Coded Orthogonal Frequency Division Multiplexing
CORDIC COordinate Rotation DIgital Computer
CPS Coplanar Strips
CPW Coplanar Waveguide

Microwave and Millimetre-Wave Design for Wireless Communications, First Edition.
Ian Robertson, Nutapong Somjit and Mitchai Chongcheawchamnan.
© 2016 John Wiley & Sons, Ltd. Published 2016 by John Wiley & Sons, Ltd.

DAC Digital-to-Analogue Converter
DBM Double-Balanced Mixer
DC Direct Current
DEMOD Demodulator
DFE Design for the Environment
DFM Design for Manufacture
DFT Discrete Fourier Transform
DIY Do-It-Yourself
DSB Double-Sideband
DSP Digital Signal Processing
DTMB Digital Terrestrial Multimedia Broadcast
DUT Device-Under-Test
DVB-T Digital Video Broadcasting Terrestrial
E-cal Electronic Calibration
EDA Electronic Design Automation
EIA Electronics Industry Association
EIRP Effective Isotropic Radiated Power
EM Electromagnetic
EMC Electromagnetic Compatibility
EMI Electromagnetic Interference
EMS Electronics Manufacturing Services
EPC Electronic Product Code
EURAMET European Association of National Metrology Institutes
EVM Error Vector Magnitude
F Noise Factor
FCC Federal Communications Commission (USA)
FDMA Frequency Division Multiple Access
FDTD Finite-Difference Time-Domain
FEM Finite Element Method
FER Frame-Error-Rate
FET Field-Effect Transistor
FFT Fast Fourier Transform
FM Frequency Modulation
FPGA Field-Programmable Gate Array
FSK Frequency-Shift Keying
FSPL Free-Space Path Loss
GaAs Gallium Arsenide
GaN Gallium Nitride
GPRS General Packet Radio System

GPS Global Positioning System
GSM Global System For Mobile Communications (Previously Groupe Spéciale Mobile)
HBT Heterojunction Bipolar Transistor
HEMT High Electron Mobility Transistor
HSIW Hollow Substrate Integrated Waveguide
IC Integrated Circuit
ICT Inductively-Coupled Telemetry
IEEE The Institute of Electrical and Electronics Engineers
IF Intermediate Frequency
InP Indium Phosphide
IoT Internet of Things
IP Intellectual Property or Internet Protocol
IRR Image Rejection Ratio
ISDB Integrated Services Digital Broadcasting
ISI Intersymbol Interference
ITU International Telecommunications Union
JFET Junction Field-Effect Transistor
LAN Local Area Network
LCP Liquid Crystal Polymer
LDMOS Laterally Diffused Metal-Oxide-Semiconductor
LNA Low-Noise Amplifier
LNB Low-Noise Block
LO Local Oscillator
LOS Line-of-Sight
LRL Line-Reflect-Line
LRM Line-Reflect-Match
LRRM Line-Reflect-Reflect-Match
LTCC Low Temperature Co-Fired Ceramic
LTE Long Term Evolution
LUT Look-Up Table
m2m Machine-to-Machine
MAG Maximum Available Gain
MBE Molecular Beam Epitaxy
MEMS Microelectromechanical Systems
MESFET Metal–Semiconductor Field-Effect Transistor
mHEMT Metamorphic High Electron Mobility Transistor
MIM capacitor Metal-Insulator-Metal Capacitor
MIMO Multiple-Input and Multiple-Output
MMIC Monolithic Microwave Integrated Circuit

MOCVD Metal Organic Chemical Vapour Deposition
MoM Method of Moments
MP3 MPEG Layer 3; audio format
MPEG-2 Motion Picture Experts Group II; video format
MSG Maximum Stable Gain
NF Noise Figure
NFC Near-Field Communications
NIST National Institute of Standards and Technology (USA)
NLOS Non-Line-of-Sight
NMOS N-Type Metal-Oxide-Semiconductor
NMT Nordic Mobile Telephony
NPL National Physical Laboratory (UK)
NTSC National Television System Committee
NTT Nippon Telegraph and Telephone Corporation
ODE Ordinary Differential Equation
ODM Original Design Manufacturer
OEM Original Equipment Manufacturer
OFCOM Office of Communications (UK)
OFDM Orthogonal Frequency Division Multiplexing
OOK On-Off Keying
PA Power Amplifier
PAL Phase Alternating Line
PAPR Peak-to-Average Power Ratio
PAR Peak-to-Average Ratio
PCB Printed Circuit Board
PCM Process Control Monitor
PD Phase Detector (oscillator) or Predistorter (power amplifier)
PDE Partial Differential Equation
PDK Process Design Kit
PGA Pin-Grid Array
pHEMT Pseudomorphic High Electron Mobility Transistor
PLL Phase-Locked Loop
PSK Phase-Shift Keying
PWM Pulse-Width Modulation
QAM Quadrature Amplitude Modulation
QFN Quad Flat No-Leads
QPSK Quadrature Phase-Shift Keying
RF MEMS Radio Frequency Micro-Electromechanical Systems
RF Radio Frequency
RFIC Radio Frequency Integrated Circuit

RFID Radio Frequency Identification
RFOW RF-on-Wafer
RoHS Restriction of Hazardous Substances
RTD Resonant Tunnelling Diode
RWG Rectangular Waveguide
RX Receiver
SAW Surface Acoustic Wave
SECAM Séquentiel Couleur a Mémoire
SFG Signal Flow Graph
SiGe Silicon Germanium
SiP System-in-Package
SIW Substrate Integrated Waveguide
SMD Surface-Mounted Device
SoC System-on-Chip
SOLT Short-Open-Load-Thru
SRD Step-Recovery Diode
SRF Self-Resonant Frequency
SSB Single Side-Band
STB Set-Top Box
TACS Total Access Communication System
TDMA Time-Division Multiple Access
TE Transverse Electric
TEM Transverse Electromagnetic
TFMS Thin-Film Microstrip
TM Transverse Magnetic; or Trademark when superscript
TRL Thru-Reflect-Line
TV Television
TX Transmitter
UHF Ultrahigh Frequency
USB Universal Serial Bus
UV Ultraviolet
VCO Voltage-Controlled Oscillator
VHF Very High Frequency
VNA Vector Network Analyser
VSA Vector Signal Analyser
VSWR Voltage Standing Wave Ratio
W-CDMA Wideband Code-Division Multiple Access
WEEE Waste Electrical and Electronic Equipment
WSN Wireless Sensor Network

Index

Note: Page numbers followed by ff indicate and following pages.

3-D packaging 182, 497, 507
3GPP (3rd Generation Partnership Project) 8

ABCD-parameters 84
active load 310
adjacent channel interference (ACI) 44, 321
admittance parameters 82
Advanced Mobile Phone System (AMPS) 3
Advanced Television Standards Committee (ATSC) 14
amplitude modulation (AM) 36, 49
amplitude shift keying (ASK) 52
analogue-to-digital converter (ADC) 50, 77, 468
antenna 5, 17, 434ff, 512
antenna arrays 458, 512
antenna gain 5, 17, 435, 450
antenna-in-package (AiP) 512
application-specific integrated circuit (ASIC) 496
arbitrary waveform generator (AWG or "ARB") 542
atmospheric attenuation 6, 447

balanced amplifier 314, 318
balanced mixer 398, 405
ball-grid array (BGA) 505, 513
balun 175, 402
band-pass filter (BPF) 35, 111, 194, 211ff
bandwidth utilization efficiency (BUE) 57, 60
Barkhausen criterion 351
Bayonet Neill–Concelman (BNC) connector 157
benzocyclobutene (BCB) 274
bias choke 169, 308
bipolar complementary metal-oxide-semiconductor (BiCMOS) 248, 464
bipolar transistor 263, 301
bit-error-rate (BER) 57, 59, 467
Bode plot 102
bottom-up fabrication 256
branch-line coupler 173, 408
bulk acoustic wave (BAW) 392
Butterworth filter 103, 107, 379

calibration standards 548, 553
carrier-to-noise ratio (CNR) 61, 379
cascaded networks 65, 84, 95
cascaded noise figures 65, 316
Cauer topology 106
cellular mobile communications 3, 20, 249, 458
channel capacity 51, 491
channel estimation 492
characteristic impedance 92, 105, 126, 134, 163, 279
Chebyshev filter 107, 210
chip-on-board 267, 504
Clapp oscillator 347, 382
class A 269, 327
cleanroom 249
cloud robotics 30
coaxial cable 105, 125, 155
coaxial connector types 156
coax-microstrip launcher 164, 559
co-channel interference (CCI) 44

Microwave and Millimetre-Wave Design for Wireless Communications, First Edition.
Ian Robertson, Nutapong Somjit and Mitchai Chongcheawchamnan.

© 2016 John Wiley & Sons, Ltd. Published 2016 by John Wiley & Sons, Ltd.

code division multiple access (CDMA) 22, 544
coded orthogonal frequency division multiplexing (COFDM) 15, 58
combline filter 223ff
common-gate stage 301, 387, 403
common-mode rejection 564
complementary metal-oxide-semiconductor (CMOS) 7, 248, 262, 372
complex *s* operator 102, 355
connector care 556
constant reactance circle 144
constellation diagram 53, 56, 325, 410, 546
contract manufacturer 10
COordinate Rotation DIgital Computer (CORDIC) 477
coplanar strips (CPS) 158, 178
coplanar waveguide (CPW) 177, 418, 560
coupled-slotline mode 178, 563
cross-coupled filters 237
cross-coupled oscillator 372

Darlington pair 312
DC biasing 169, 309
decibels 5, 64, 90, 299
demodulator (DEMOD) 33, 36, 75, 379, 483
design for the environment (DFE) 8
design-for-manufacture (DFM) 8
device-under-test (DUT) 87, 548ff
dielectric image guide 183
dielectric insular image guide 183
dielectric resonator 176, 188, 363
dielectric waveguide 183
digital modulation 50
digital predistortion 340, 486ff
digital signal processing (DSP) 9, 464ff
Digital Terrestrial Multimedia Broadcast (DTMB) 14
digital TV 14
Digital Video Broadcasting-Terrestrial (DVB-T) 14
digital-to-analogue converter (DAC) 473
diode mixer 396, 525
dipole antenna 437
direct conversion receiver 74
directivity 179, 435, 444, 550
discrete Fourier transform (DFT) 474
distributed amplifier 313
distributed mixer 407
Doherty amplifier 334
double-balanced mixer (DBM) 175, 395, 399
double-sideband (DSB) 50, 65
dual-gate FET 262, 397, 400, 407

effective isotropic radiated power (EIRP) 12, 47
electromagnetic interference (EMI) 44
electromagnetic spectrum 3, 452
electronic calibration (E-cal) 554
electronic design automation (EDA) 163, 497, 515ff
Electronic Product Code (EPC) 25
elliptic filter 107, 110, 237
energy per bit (Eb) 51
error correction 60, 550
error vector magnitude (EVM) 325
even and odd modes 172, 207, 336, 564

fabless design 11, 496
far-field region 27, 436, 448
fast Fourier transform (FFT) 15, 475
Federal Communications Commission (FCC, USA) 6, 325
feedforward linearisation 340, 464
FET feedback amplifier 312, 340
FET mixer 396
fibre-radio systems 24
field-effect transistor (FET) 256
Field-Programmable Gate Array (FPGA) 489
finite element method (FEM) 517, 533
finite-difference time-domain (FDTD) 517, 533
finline 182, 339
fixed links 18, 46
flicker noise 41, 77, 362
flip-chip assembly 430, 505
flowgraph 89, 150, 303, 551
folded dipole antenna 439
free-space path loss 442, 449
frequency division multiple access (FDMA) 21
frequency modulation (FM) 50
frequency shift keying (FSK) 54
frequency synthesiser 389
Friis noise factor formula 65, 316
Friis transmission equation 46, 448
front-to-back ratio 436

gain definitions 298, 300
gallium arsenide (GaAs) 248ff
gallium nitride (GaN) 259, 497
gate periphery 261, 300
general packet radio system (GPRS) 22
Gilbert cell mixer 392, 405

Global positioning system (GPS) 17
Global System for Mobile Communications (GSM) 2, 21, 540
group delay 118, 160, 238

hairpin filter 233
harmonic balance technique 522
harmonic mixers 547
Hartley architecture 401
heterodyne receiver 33
heterojunction bipolar transistor (HBT) 248, 265, 302
high electron mobility transistor (HEMT) 248, 258, 411
high power transistors 270
hollow substrate integrated waveguide (HSIW) 182, 272, 510
homodyne receiver 32
horn antenna 444
H-parameters (hybrid parameters) 83, 86

image rejection ratio (IRR) 466
image signal 35, 42, 65, 72, 394, 398, 401
IMPATT diode 276
impedance and admittance inverters 123, 191ff
impulse response 101, 457
indium phosphide (InP) 7, 248ff
indoor propagation 47, 453, 490
inductively coupled telemetry (ICT) 25
integrated circuit 1, 74, 253
Integrated Services Digital Broadcasting (ISDB) 14
intellectual property 8, 11
interdigital filter 217
intermediate frequency (IF) 18, 33, 71, 392
International Telecommunications Union (ITU) 7, 48
Internet of Things (IoT) 1, 22, 25, 509, 511
intersymbol interference (ISI) 14, 44, 61, 457, 490
I–Q Imbalance 76, 464, 482
IQ modulation 57, 484
isotropic antenna 5, 37, 435, 450
I–V curves 258, 270, 327

J-inverter 191ff
jitter 360

K-inverter 122ff, 191ff
Kuroda's identity 203

Lange coupler 173, 314, 335, 399, 411
laterally diffused metal-oxide-semiconductor (LDMOS) 248, 263
lead frame package 505ff
LineCalc 163
line-of-sight (LOS) 18, 46, 69
link budget 61, 447, 450
liquid crystal polymer (LCP) 162, 498
L-networks 284
local area network (LAN) 7, 489
long term evolution (LTE) 7, 22, 461, 540
long-tailed pair 310, 313, 403
look-up table (LUT) 488
loop antenna 441
lossy match amplifier 308
low noise amplifier (LNA) 17, 33, 63, 315
low noise block (LNB) 18
low temperature co-fired ceramic (LTCC) 163, 175, 182, 391, 503ff
low-IF Architecture 77, 465
LRL (line-reflect-line) calibration 561
LRM (line-reflect-match) calibration 561
LRRM (line-reflect-reflect-match) calibration 561
lumped element filter 99ff
lumped-distributed components 180, 199

machine-to-machine communications (m2m) 29, 509
Mason's non-touching loop rule 150, 303, 551
matching network 279ff
maximum available gain (MAG) 306
maximum stable gain (MSG) 306
Maxwell's equations 455, 516, 530
MEMS inductor 419
MEMS packaging 429
MEMS phase shifter 425
MEMS varactor 420
metal organic chemical vapour deposition (MOCVD) 259, 266, 301
metal-insulator-metal (MIM) capacitor 274, 562
metal-semiconductor field effect transistor (MESFET) 257, 300
metamorphic High Electron Mobility Transistor (mHEMT) 259
method of moments (MoM) 517, 529
microelectromechanical systems (MEMS) 415ff
micromachined transmission line 416
microstrip 161ff, 210, 268, 335, 445
microstrip frequency limitations 163

microstrip patch antenna 445
microstrip resonators 176
microstrip test fixture 165, 501, 558
MIMO (multiple-input and multiple-output) 22, 460, 467
mixed-mode S-parameters 563
monolithic microwave integrated circuit (MMIC) 248, 272, 516, 534
molecular beam epitaxy (MBE) 252
Moore's law 1, 249
More than Moore 255
multi-chip module (MCM) 498ff
multipath propagation 14, 21, 45, 457ff, 490

National Institute of Standards and Technology (NIST, USA) 549, 562
National Physical Laboratory (NPL, UK) 549
near-field communications (NFC) 25, 447
NMOS 249, 263
noise circles 317
noise equivalent bandwidth 61
noise factor (F) 62ff, 316
noise figure (NF) 62ff, 316
noise figure of mixers 65
noise performance of cascaded networks 65, 316
noise temperature (T_e) 66
nonradiative dielectric guide (NRD) 183
N-port S-parameters 91
Nyquist criterion 354
Nyquist diagram 356

OFCOM (Office of Communications, UK) 6
On-Off-Keying (OOK) 53
ordinary differential equation (ODE) 521
original design manufacturer (ODM) 11
original equipment manufacturer (OEM) 11
orthogonal frequency division multiplexing (OFDM) 14, 58, 489ff

package styles 267, 504ff
PAL (Phase Alternating Line) 12
parallel stub 168, 290
parallel-coupled line filter 211ff
parallel-coupled microstrip lines 171, 207, 233
parasitics 120
partial differential equations (PDEs) 534
process control monitor (PCM) 534
peak-to-average ratio 59, 298, 318
phase detector 377
phase noise 34, 56, 361ff
phase-shift keying (PSK) 55
phase velocity 134, 425
phase-locked loop (PLL) 377ff
pHEMT (pseudomorphic high electron mobility transistor) 248, 257
photoimageable thick-film process 502
photolithography 249
photoresist 250
PIN diode 276
pin-grid array (PGA) 505
planar electromagnetic simulation 529
planar Marchand balun 175, 402
Plessey Research Caswell 257
power amplifier 269ff, 326ff
power amplifier linearisation 339, 486
power combining techniques 334
power gain 33, 91, 298
predistortion linearisation 340, 486
process design kit (PDK) 534
propagation constant 126, 132, 160
propagation loss 6
pulse-width modulation (PWM) 332

Q-factor 177, 272, 282, 362, 416
Quad Flat No-leads (QFN) 505
quadrature amplitude modulation (QAM) 57, 325, 409
quadrature-phase shift keying (QPSK) 56, 544
quantisation error 51, 468
quarter-wave transformer 169, 293, 296
quartz crystal resonator 365
quasi-optical power combining 338
QUCS (Quite Universal Circuit Simulator) 128

radial and butterfly stub 169
radiation pattern 5, 427, 436
radiation resistance 434
radio frequency identification (RFID) 25
radio frequency micro-electromechanical systems (RF MEMS) 415ff
radio horizon 12
radiofrequency integrated circuit (RFIC) 9, 74, 271ff
rapid prototyping of PCBs 500
rat-race hybrid 174, 181
reactively matched amplifier 311
receiver 32ff

recta-coax 416
rectangular waveguide (RWG) 158, 181, 337, 444
reference planes 88, 552, 560
reflections on transmission lines 126
reflector antenna 443
resistive FET mixer 397
resolution bandwidth 544
resonant tunnelling diode (RTD) 277
reticle 534
return loss 91, 137
RF-on-wafer (RFOW) measurements 535, 560
rib image guide 183
Richards transformation 202
ridge waveguide 182
robotic communications 29

sample and hold 468
satellite communications 15
satellite navigation systems 17
Scattering parameters 81ff, 548
Schottky diode 275
SECAM (Séquentiel couleur à mémoire) 12
second-order intermodulation 319
self-resonant frequency (SRF) 121, 122, 274
semiconductor wafer 249, 254
set-top box (STB) 18
Shannon's Law 59, 461
Shot noise 40, 362
SiGe HBT 265
signal flow graph 89, 150
single-ended mixers 392
single-sideband (SSB) 50, 65, 362, 401
sliding load 553
slotline 178
small-signal equivalent circuit 259, 300
the Smith chart 142ff
Smith chart computer tools 149, 293
SOLT (short-open-load-thru) calibration 550
source- and load-pull 564
S-parameters 81ff, 548
spectral efficiency 9, 20, 58
spectrum licensing 6
spiral inductor 120, 167, 273, 502, 536
stability circles 304
stability factor 305, 386
standing wave ratio circles 147
standing waves 135ff
stepped-impedance low-pass filter 208
step-recovery diode 277
stiction 421
stripline 179
stub matching 290
substrate integrated waveguide (SIW) 181, 338, 502
superheterodyne receiver 37, 71
surface acoustic wave (SAW) 27, 74
surface mount device (SMD) 121, 282, 497, 503
suspended substrate microstrip 179
symbol 53
synchrodyne receiver 33
system impedance 87, 105, 279
system-in-package (SiP) 495, 509
system-on-chip (SoC) 11, 18, 256, 496

TACS (Total Access Communication System) 20
TE (transverse electric) 159, 178
telegraph 3, 434
telegrapher's equation 129
television 3, 12
TEM (transverse electromagnetic) 156, 172, 281, 417
test fixture 165, 501, 558
thermal noise 38, 57, 315
thin-film microstrip (TFMS) 180
third-order intermodulation 321
time division multiple access (TDMA) 21
TM (Transverse Magnetic) 159
top-down fabrication 256
transceiver architectures 71ff
transconductance 258
transfer function 100
transient behaviour of transmission lines 126, 137
transmission-line resonator 176, 188, 363
transmission-line transformer 170
transmitter 32ff
transverse resonances 163
travelling-wave antenna 442
Tri-plate 179
TRL calibration 558
T (transmission)-parameters 95ff
twin-lead 157
twisted pair 157
two-port network 81ff
two-port oscillator model 349
two-port parameter conversion 87

unilateral transducer gain 307
uniplanar transmission lines 177
unit gate width 261

vector modulator 410
vector network analyser (VNA) 94, 547
vector signal analyser (VSA) 540
VNA calibration 549
voltage standing wave ratio (VSWR) 139, 547
voltage-controlled oscillator (VCO) 375

wafer-scale packaging 495
waveguide bands 160
waveguide-to-microstrip transition 512
wavelength 3
W-CDMA (wideband code-division multiple access) 544
wearables and body area networks 29
weaver architecture 401
Wilkinson splitter/combiner 170, 335
wireless sensor network (WSN) 28

Yagi–Uda antenna 442
Y-parameters 82

zero-level packaging 430, 507
ZigBee® 7, 10, 23, 540
Z-parameters 81

Printed and bound by CPI Group (UK) Ltd, Croydon, CR0 4YY
17/09/2026
14964271-0002